D1202121

The McGraw·Hill Companies

Copyright © 2007, 2000 by The McGraw-Hill Companies, Inc. All rights reserved. Printed in the United States of America. Except as permitted under the United States Copyright Act of 1976, no part of this publication may be reproduced or distributed in any form or by any means, or stored in a data base or retrieval system, without the prior written permission of the publisher.

1 2 3 4 5 6 7 8 9 0 DOC/DOC 0 1 2 1 0 9 8 7 6

ISBN-13: 978-0-07-146760-5
ISBN-10: 0-07-146760-2

The sponsoring editor for this book was Larry S. Hager and the production supervisor was Richard C. Ruzycka. It was set in Times by International Typesetting and Composition. The art director for the cover was Handel Low.

Printed and bound by RR Donnelley.

This book is printed on acid-free paper.

McGraw-Hill books are available at special quantity discounts to use as premiums and sales promotions, or for use in corporate training programs. For more information, please write to the Director of Special Sales, McGraw-Hill Professional, Two Penn Plaza, New York, NY 10121-2298. Or contact your local bookstore.

Information contained in this work has been obtained by The McGraw-Hill Companies, Inc. ("McGraw-Hill") from sources believed to be reliable. However, neither McGraw-Hill nor its authors guarantee the accuracy or completeness of any information published herein, and neither McGraw-Hill nor its authors shall be responsible for any errors, omissions, or damages arising out of use of this information. This work is published with the understanding that McGraw-Hill and its authors are supplying information but are not attempting to render engineering or other professional services. If such services are required, the assistance of an appropriate professional should be sought.

CONTENTS

$/20.⁰⁰ B47

Part II   Water Quality Management

Chapter 5. Environmental Management

Part III   Water Pollution Regulations

Chapter 6. Regulatory Standards

Part IV   Water Pollution Control

Chapter 7. Wastewater Treatment Plant Design

Chapter 8. Physical Treatment

Chapter 9. Chemical Treatment 9.1

Chapter 10. Biological Treatment 10.1

Part V Residuals

Chapter 11. Residual Treatment 11.3

Part VI Treatment Summary

Chapter 12. Wastewater Treatment Solutions 12.3

Chapter 13. Pollutant Information 13.1

Part VII   Appendices

PREFACE

Environmental protection has, for at least the last 50 years, been a major concern in the civilized nations of the world. The concern initially was limited to public health but has, since the 1970s, been expanded to include the environment. As stewards of our environment, we are responsible for the protection of the environment, for our own sakes and for the benefit of generations to follow. It is accepted that humankind cannot continue to advance technologically while ignoring the environmental deterioration that occurs when we irresponsibly discharge the waste from our technology. Indeed, the sustainable growth of our civilization requires, as the only reasonable and feasible way to coexist on our planet with nature, that we protect our fragile environment. Much of the legislation promulgated to provide environmental protection has the purpose of not only non-deterioration of present environmental conditions, but the improvement of these conditions so that past abuse is corrected.

For many years we have discharged waste products into the air, the water and the land as if they were limitless reservoirs of storage. We have found of course, that this is not the case, and they have deteriorated our air, water and land quality to the crisis point. Nature fortunately is forgiving, and like the human body, has a remarkable capacity for recovery when abuse ceases. But continuing abuse can literally destroy nature as well as it can our bodies.

Science has advanced to the point where we can fairly accurately predict the risk of pollutant discharge and can develop systems to eliminate the discharges which tend to deteriorate the environment.

Since the environment consists literally of everything that exists on this planet, environmental protection is obviously critical and at the same time difficult to implement. The purpose of the first edition of the *Water Quality Control Handbook* was to recommend certain tools from a wide variety of disciplines to the environmental practitioner so that systems can be developed to protect the environment on a case-by-case basis.

One purpose of the second edition of the *Water Quality Control Handbook* is to introduce and explain new technologies, several of which are unpublished and some of which are published only in technical papers. A second is to update and, in some cases, elaborate on the information provided in the first edition. A third purpose is to greatly expand Chapter 12, "Wastewater Treatment Solutions," as a tool for wastewater professionals to screen and select the optimal treatment systems for various pollutant removals.

In order to allow this new information to be presented in a reasonable-sized handbook, some of the details in Appendix F for organic chemical analysis have been consolidated and the Carbon Absorption Isotherum in Appendix C has been summarized in tabular form. The reader is referred to the first edition for more details in these two areas.

ORGANIZATION

This Handbook is intended to provide industrial, governmental and consulting environmental practitioners a quick reference guide to all major areas of water pollution control. For this purpose, the Handbook is divided into:

Part I. The Theory and Quantification of Water Pollution
Part II. Water Quality Management
Part III. Water Pollution Regulations
Part IV. Water Pollution Control
Part V. Residuals
Part VI. Treatment Summary

In addition, the Appendices are also intended to be a working document, and they are divided into:

Hydraulic Information
Water Characteristics
Chemical Characteristics
Design Information

HANDBOOK USE

The following paragraphs suggest ways to use this Handbook in solving water pollution control problems.

Pollutant Characterization: One of the most important steps in the analysis of a wastewater or the strategy of a design is the quantitative and qualitative characterization. Chapter 3 describes the physical, chemical and biological characteristics of wastewater, and the monitoring, sampling and testing methods involved in their determination. Appendix F gives EPA testing methods in complete detail.

Water Quality: Chapter 4 describes the effects of various wastewater pollutants on the environment, the modeling of some of these pollutants, the classification of receiving streams and the in-situ treatment and management of these bodies of water.

Water Pollution Control Management: Wastewater characterization, water quality and water pollution control information is only academic and is not developed to manage the discharge of pollutants. Chapter 4 under "Watershed Management" and Chapter 5, "Environmental Management," suggest methods of involving top management, establishing policies and requiring documentation to assure the effective management of water pollution control.

Environmental Regulations: Chapter 6 summarizes the applicable water pollution regulatory standards including the WQCA, NPDES, UST, SPCC, SWPPP, RCRA, CERCLA, and SARA and discusses the philosophy and future of regulatory standards.

Water Pollution Control: The following chapters give the theory, design recommendations and practical suggestions for various water pollution control systems.

General Design	Chapter 7
Physical Treatment	Chapter 8
Chemical Treatment	Chapter 9
Biological Treatment	Chapter 10
Residual Treatment	Chapter 11
Wastewater Treatment Solutions	Chapter 12
Pollutant Information	Chapter 13

The purpose of Chapters 12 and 13 is to make use of the design information presented in the previous chapters and present suggestions for optimal treatment methods for various industrial and municipal pollutants and flows.

ACKNOWLEDGMENTS

The genesis of this Handbook is a series of courses in wastewater pollution control taught by me at Vanderbilt University in Nashville, Tennessee; George Washington University, Continuing Engineering Education Program, in Washington, D.C., San Diego, Calif., Indian Head, Md., and London, England; and the Centre for Management Technology in Singapore, Jakarta, Indonesia, and Kuala Lumpur, Malaysia. I would like to thank these institutions for the opportunities to develop the courses and notes which have been expanded and organized into this Handbook.

I would like to thank the employees of E. Roberts Alley & Associates, Inc., who have contributed to the research, writing, and production of information presented in this Handbook.

I would especially like to thank my wife, Marion S. Alley, for the love and encouragement she has shown over the years of engineering, teaching, and authoring.

E. Roberts Alley, P.E.

ABOUT THE AUTHOR

E. Roberts Alley is a professional engineer and founder and chairman of the board of E. Roberts Alley & Associates, Inc., one of the country's leading environmental engineering firms. A diplomate of the American Academy of Environmental Engineers, he is also the author of *Stormwater Management,* co-author of McGraw-Hill's *Air Quality Control Handbook* and author of the first edition of *Water Quality Control Handbook.*

CHAPTER 1
INTRODUCTION

1.1   THE ENVIRONMENT

Our environment consists of physical, chemical, and biological substances, which interact so that the physical and chemical substances support the biological substances and allow them to experience sustainable growth. At the present level of human advancement, mankind is able to negatively and positively influence the balance of these substances, thereby affecting the health of the environment.

The purpose of this Handbook is to provide the environmental professional with a reference which can be used to understand water pollution control, and to make management, design and operational decisions which allow the use of the environment without a negative effect. Existing with the preservation of the environment rather than at the expense of the environment allows humankind to meet the goal of sustainable growth.

Discharges from human activity must be released to the air, the water or the soil. Each of these potential reservoirs can accept a limited amount of physical, chemical and biological substances without significant deterioration. Beyond this point of assimilation, the environment can be deteriorated to the point that sustainable biological growth cannot occur.

This deterioration can be caused by the weather in the form of wind (i.e. dust blown into the air), rain (i.e. stormwater eroding soil into the water and floods depositing solids), lightning (i.e. fires discharging smoke and particulates into the air and water and ash onto the soil), or volcanoes (smoke and particulates discharged into the air and water into the soil). The deterioration can also be caused by vegetation (hydrocarbon vapors discharged into the air and dissolved hydrocarbons running into the water), animals (feces polluting the water, flatus passing into the air), and activities of humans.

The air, as a reservoir, provides no beneficial treatment to pollutants; it only disperses the pollutants. This dispersion either dilutes the pollutant to the concentration at which it is innocuous, or it transfers the pollutant to a downwind location. A factor that makes the modeling of air pollution difficult is the non-predictability of wind direction and velocity. The chemistry which does occur in the air is typically non-beneficial (i.e. NO_x and VOCs are converted to ozone in the presence of ultraviolet radiation). Because of this dispersion proclivity, air pollution can be international in scope.

Air pollution is significant in the study of water pollution since wastewater can volatilize either deliberately or through evaporation and become air pollution.

Water as a reservoir, can provide minimal treatment to certain organic pollutants because of the oxygen and the biota in the water. Water also acts as a disperser of pollutants. Water pollutant dispersion is easier to model since it flows in a defined channel with a predictable velocity. Water pollution can be of regional interest because of this dispersion.

The discharge of pollutants onto or into the soil is normally of only local concern since liquid migrations is soil is slow. Soil pollution is normally only of concern when the pollutant is liquid or is a soluble solid. An insoluble solid will not migrate except through underground channels, nor will it dissolve into groundwater.

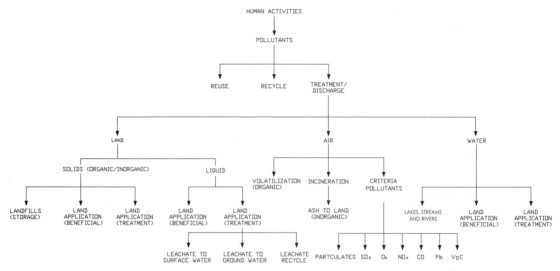

FIGURE 1.1   The flow of pollutants from human activities.

The major concern with soil pollution is the subsequent pollution of the groundwater when the groundwater is used as a source of water for drinking, irrigation or industrial use. The flow of the pollutants caused by human activities is demonstrated in Fig. 1.1.

1.2   WATER QUALITY

In the United States, the Environmental Protection Agency (EPA) has allocated all applicable waterways to a specific use or uses and requires that no degradation of present quality be allowed.

The surface water classifications are:

• Public water supplies
• Propagation of fish and wildlife
• Irrigation
• Recreational purposes
• Navigation

The EPA has assigned concentration and/or mass standards for pollutant discharge into each of the surface water classifications.

Another practical consideration of water quality is the flushing and backflow characteristics of the water. A continually downstream moving waterway will transfer or remove pollution to a lower and usually larger waterway, perhaps better able to assimilate the pollution. A moving waterway is also replenished with dissolved oxygen from the surface to replace oxygen used by organisms in the water to biologically treat organic pollutants. Conversely, a lake or a waterway containing a dam may have zero flow during periods of dry weather where the pollutants will remain at the point of discharge, increasing concentration as stagnation occurs. The residual dissolved oxygen may be depleted during this time and suspended and settleable solids deposited on the bottom of the waterway.

An estuary is a body of water within the transition between a freshwater river and a saltwater body such as an ocean. An estuary offers special challenges for modeling water quality since the flow direction varies with the tides.

1.3 GENERAL CLASSIFICATION OF POLLUTANTS

There are several ways of classifying pollutants in order to predict their effect on water quality and the means of their removal. First they can be classified as a solid, a liquid or a gas, or as one of these mixed with, dissolved in or absorbed onto another. Further each of these can be organic or inorganic. An organic waste can additionally be classified as volatile or nonvolatile, biodegradable or refractory, and of animal, mineral or vegetable origin. Inorganic wastes should be classified as dissolved, suspended or settleable and by pH. Further information needed on a waste includes temperature, volume or quantity. Figure 1.2 shows the various types of waste.

BASIC STATE	MIXED STATE	SUB STATE	EXAMPLE
Solid	Solid	Organic	Sugar
		Inorganic	Salt
	Solid in Liquid	Organic	Food Processing Waste
		Inorganic	Electroplating Waste Treatment
	Solid in Gas	Organic	Coal Dust
		Inorganic	Incinerator Particulates
Liquid	Liquid	Organic	Gasoline
		Inorganic	HCl
	Liquid in Solid	Organic	Garbage
		Inorganic	Uncured Concrete
	Liquid in Gas	Organic	Air Stripper Mist
		Inorganic	Mist
Gas	Gas	Organic	Toluene
		Inorganic	Air
	Gas in Solid	Organic	Methane in Floats
		Inorganic	Hydrogen Sulfide in Floats
	Gas in Liquid	Organic	Anaerobic Decomposition
		Inorganic	Dissolved Air Flotation

FIGURE 1.2   Types of waste.

Before determining the sampling, analytical or treatment method and before understanding the effect of the pollutant on the environment, the above described general characteristics must be determined.

In order to select a treatment system to reduce or remove pollutants, the classification of the pollutants as described above is of more interest than the source or legal category. Whether the waste is from a municipal, industrial or landfill source or is toxic or hazardous is of less interest than its characteristics. Hazardous, toxic, flammable, explosive, corrosive, poisonous or oxidation state are descriptions of the effects or actions of the wastes rather than its makeup or consistency. In waste treatment, there is more interest in the physical, chemical and organic characteristics of the waste since presumably, during transportation and treatment, the waste will be contained so that it cannot exhibit its hazardous, toxic, flammable explosive corrosive, poisonous or oxidizing characteristics.

The reason solid waste may be considered hazardous is that pollutants can be leached out of the waste by groundwater and stormwater percolating through the soil and pollute the underlying groundwater. If the solid waste were fixed as a solid, there would be no leaching, and therefore no pollution. The critical nature of solid waste is therefore its liquid leaching, which means that aside from its handling and storage on the land, solid waste treatment is identical to wastewater treatment in concept and theory.

Before a treatment system can be selected to remove the characterized pollutants, certain additional parameters must be established including geographical, financial, and political factors. Geographical concerns need to be addressed such as the area available for the treatment processes and the topography of the area.

Financial factors include the budget for the project, both in terms of cost and time. Political considerations include zoning, permit availability, community support, and future regulatory legislation.

P · A · R · T · I

THE THEORY AND QUANTIFICATION OF WATER POLLUTION

CHAPTER 2
SOURCES OF WATER POLLUTION

2.1 INTRODUCTION

The life and activities of plants and animals, including humans, contribute to the pollution of the earth, assuming that pollution is defined as the deterioration of the existing state. The purpose of this chapter is to review the various sources of water pollution in order to recognize the opportunities for eliminating, minimizing, reusing or treating these sources so that their negative effect on the environment will be minimized. When pollution control is considered, these questions should be asked and answered.

1. Can the pollution source be eliminated?
   - Is it absolutely necessary?
   - Can it be substituted by another source that accomplishes the same purpose but is less polluting to the environment?
2. Can the pollution source be minimized?
   - Can the source be operated more efficiently to lower pollution?
   - Can the pollutants be converted to another state (gaseous, liquid or solid) which is less polluting to the environment
3. Can the pollutants be reused?
   - Can the pollutants be purified and reused as raw materials?
   - Can relatively pure water be separated from the pollutants and reused?
   - Can the pollutant be recycled to a different source?
4. Can the pollutant be treated?
   - Is the effect on the environment minimized by altering, destroying or concentrating the pollutant?
   - Can the treated pollutant be reused or recycled?

The following are common sources of pollution.

2.2 INDUSTRIAL SOURCES OF WATER POLLUTION

Any industry, in which water obtained from a water treatment system or a well comes in contact with a process or product can add pollutants to the water. The resulting water is then classified as a wastewater. In the United States, the EPA has classified industries into Standard Industrial Classifications (SIC). Industries in any of these classifications can contribute to water pollution as their water supply is used in a process.

The following are examples of industrial water pollution sources:

Non-Contact Water

• Boiler feed water
• Cooling water
• Heating water
• Cooling condensate

Contact Water

• Water used to transport products, materials or chemicals
• Washing and rinsing water (product, equipment, floors)
• Solubilizing water
• Diluting water
• Direct contact cooling or heating water
• Sewage
• Shower and sink water

The wastewater can contain physical, chemical and/or biological pollutants in any form or quantity and cannot adequately be quantified without actual measuring and testing. The wastewater will typically either be discharged directly into a receiving body of water or into the sewerage system of a municipality, or it will be reused or recycled. Normally, a municipality will restrict industrial water pollutants to those listed below for municipalities, using a Pre-Treatment Ordinance.

A municipality is required by the USEPA to limit by Pre-treatment Ordinance, industrial water pollutants to levels which will not 1) harm the municipal wastewater treatment system, 2) pass through the municipal system at levels not meeting the municipal discharge permit, or 3) be deposited in municipal sludge at an illegal concentration.

2.3   MUNICIPAL SOURCES OF WATER POLLUTION

The non-industrial municipal sources of water are typically as follows:

• Dwellings
• Commercial establishments
• Institutions (schools, hospitals, prisons, etc)
• Governmental operations

Table 2.1 lists municipal sources of water in terms of average flows per day and biological strength in BOD_5, the total amount of oxygen used by microorganisms during the first five days of biodegradation.

It is assumed that a non-industrial municipal wastewater source will contain no pollutants except for the following:

• Feces
• Urine
• Paper

TABLE 2.1 Municipal Sources of Wastewater

Classification	Remarks	Average flow/ person/day	BOD_5 person per day
Municipality	Residential	100 gallons	0.20 lb.
Subdivision	Residential	100 gallons	0.20 lb.
Colleges		100 gallons	0.20 lb.
Hospitals	Per bed	200 gallons	0.40 lb.
Nursing homes		100 gallons	0.20 lb.
Schools, high	With cafeteria & showers	25 gallons	0.06 lb.*
Schools, elementary	With cafeteria & showers	20 gallons	0.06 lb.*
Factory or office bldg.	With showers/shift	35 gallons	0.06 lb.
Factory or office bldg.	Without showers/shift	25 gallons	0.06 lb.
Motels	Per unit	100 gallons	0.12 lb.
Restaurants			
Ordinary rest. (not 24 hours)	Per seat	35 gallons	0.20 lb.
24 hour rest.	Per seat	50 gallons	0.28 lb.
24 hour rest. on interstate	Per seat	70 gallons	0.40 lb.
Tavern	Per seat	20 gallons	0.12 lb.
Curb service	Per car space	50 gallons	0.28 lb.
Trailer park	2–1/2 Persons per trailer	50 gallons	0.20 lb.
Country clubs	Per member	50 gallons	0.20 lb.
Shopping center	Without food service or laundry	0.1 gal/sq. ft. of floor space based on flow	200 ppm

*When garbage grinders are used, the BOD_5 loading shall be increased to 0.07 lb. BOD_5 / person.

- Food waste
- Laundry wastewater
- Sink, shower, and bath water

These pollutants are all biological and as such can be readily biodegraded. Any extraneous non-industrial pollutants other than those listed above can be physical or chemical in nature, and ideally should be prevented from entering a municipal system with a Pre-treatment Ordinance, or removed from the municipal wastewater using some method of pre-treatment.

2.4 AGRICULTURAL SOURCES OF WATER POLLUTION

Normally, agricultural water pollutants are transported to an aboveground or underground receiving stream by periodic stormwater. Agricultural wastewater can be of animal or vegetable origin or be from a nutrient, fertilizer, pesticide or herbicide source. Animal or vegetable sources will be limited to biodegradable feces, urine or vegetable constituents. Nutrients or fertilizers will be typically some formulation of carbon, phosphorous, nitrogen and/or trace metals.

Pesticides and herbicides will consist of formulated organic chemicals, many with complex molecular structures, designed to be very persistent in the environment. Pesticides such as Chlorodane and Heptachlor, which consist of a multitude of different organic chemicals, can still exist in the soil around World War II barracks.

Agricultural activities can also allow the runoff of soil into receiving streams. In such cases, pollutants can be any organic or inorganic constituent of the soil.

2.5 NATURAL SOURCES OF WATER POLLUTION

Areas unaffected by human activity can still pollute receiving steams due to stormwater runoff, which can be classified into animal, vegetable and soil sources. Again, animal and vegetable water pollution sources should be readily biodegradable. Soil sources will consist of any organic and inorganic material in the soil.

2.6 STORMWATER SOURCES OF WATER POLLUTION

Stormwater has been mentioned above under agricultural and natural sources of water pollution, but will also transport industrial and municipal water pollutants to a receiving stream or underground water supply.

2.6.1 Industrial Stormwater Sources

Any solid or liquid material or chemical stored, leaked or spilled on the ground from an industrial operation can be transported by stormwater to a recovery stream and become a pollutant. These sources can be from any or all of the following:

- Outside process areas
- Inside process areas which discharge to the outside
- Roof drains
- Parking lots
- Roadways
- Loading/unloading areas
- Storage areas
- Wastewater treatment areas
- Soil runoff
- Spills
- Leaks
- Tank farms

Chapter 6, "Regulatory Standards," explains how the NPDES indirect discharge permitting requirements of the USEPA regulate industrial stormwater runoff.

2.6.2 Municipal Stormwater Sources

Any material or chemical deposited on the ground in a municipality can likewise be transported to a receiving stream as a pollutant. These include:

- Petroleum product spills and leaks
- Garbage and trash
- Soil runoff
- Surfacing underground sewage disposal systems
- Spills and leaks from material or chemical transport

Chapter 6, "Regulatory Standards" explains how the NPDES indirect discharge permitting requirements of the USEPA regulate municipal stormwater runoff.

2.7 LANDFILL WATER POLLUTION SOURCES

Public, private, and industrial landfills can be a source of stormwater pollution because of runoff from the surface and underground leachate. Landfill regulations require daily cover, but during the day, rainfall can cause pollution from surface runoff.

When stormwater leaches through the surface cap and downward through the landfill, the horizontally or vertically migrating discharge from below the landfill is known as leachate and can pollute surface or underground water. Because of the bacteria present in the dirt and in landfill material, there will always be aerobic and anaerobic biological activity occurring in a landfill. As explained in Chapter 10, "Biological Treatment," of this Handbook, aerobic and anaerobic biological activity will emit carbon dioxide. Carbon dioxide in the presence of water form the weak acid, carbonic acid, which tests have shown, will lower the pH in a landfill to around 4.8. This is the reason for the 4.8 pH requirement in the TCLP test as explained in Chapter 6, "Regulatory Standards." This low pH tends to dissolve certain organics and inorganics, which can leach out of the landfill as pollution.

Landfills are normally required to provide leachate, and in some cases, runoff collection and treatment or disposal to prevent contamination of the environment.

2.8 LEAKING UNDERGROUND STORAGE TANK WATER POLLUTION SOURCES

Chapter 6, "Regulatory Standards," explains how all underground petroleum and hazardous waste containing storage tanks are regulated to prevent leaking. If leaking should occur, it is a source of underground and possibly surface pollution.

CHAPTER 3
POLLUTANT CLASSIFICATION

3.1 INTRODUCTION

It is recommended that wastewater pollutants be classified into physical, chemical, and biological constituents. This chapter will describe these classifications and how they are monitored, sampled and tested and how they pass into the environment.

3.2 PHYSICAL POLLUTANTS

3.2.1 Types of Physical Pollutants

3.2.1.1 Introduction. For the purpose of this Handbook, physical pollutants will be categorized as follows:

- Solids content
- Solids type
- Color
- Odor
- Taste
- Conductivity
- Temperature

Each of these categories can have chemical or biological sources and they can possibly be removed by chemical or biological treatment methods.

3.2.1.2 Solids Content

Total Solids. For regulatory and treatment purposes, total solids (TS) can first be classified as suspended or dissolved. Total solids, as defined by Standard Methods[1] and EPA,[2] is the material residue left in a vessel after evaporation of a sample and its subsequent drying in an oven at 103 to 105°C for one hour. See Page F.6 for EPA Method for Total Residue.

Total Suspended Solids. Total suspended solids (TSS) is that portion of the Total Solids that are retained on a no-ash glass fiber filter disc of approximately 0.45 μm pore size.

The wetted and weighed filter disc is placed in a filtering apparatus and a suction is applied (see Fig. 3.1). A measured volume of wastewater is passed through the filter. The filter containing the residue is then dried in an oven for one hour at 103 to 105°C. The sample is then cooled and weighed (see Fig. 3.2). The difference in weight of the dry filter before and after solids are passed through is the TSS milligrams (mg) of suspended solids per liter (l) of wastewater filtered.

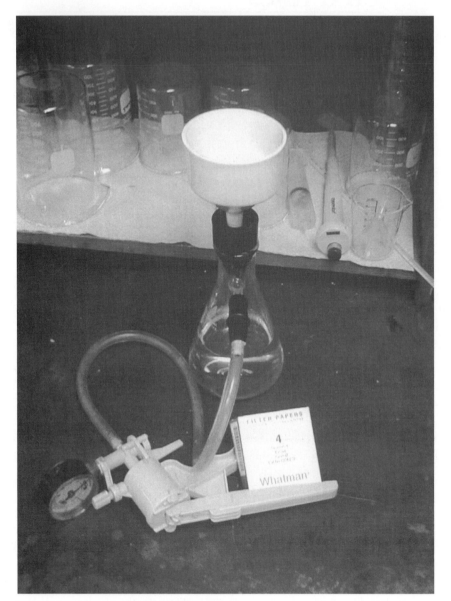

FIGURE 3.1   Total suspended solids filtering apparatus.

This inexpensive TSS test tells much about the wastewater character and can be run in less than two hours with fairly inexpensive equipment.

The TSS test indicates whether it is likely that solids suspended in a wastewater can be removed by settling, floating or filtering. In order to remove all solids that were removed in the TSS Test, a filter similar to the TSS Filter paper (0.45 μm) must be used in full scale. See page F.9 for EPA method for Residue, Non-Filterable.

FIGURE 3.2   Total suspended solids scale.

Total Dissolved Solids.   The total dissolved solids (TDS) are the solids in the filtrate from the TSS test. The liquid which passes through the TSS filter is collected in a weighed dish and evaporated for an hour at 180°C ± 2°C. The dish is then re-weighed with the TDS equaling the difference between the dish weight before and after filling with filtrate and drying, in mg per liter of filtrate.

Again, this inexpensive test can be run in less than two hours and will indicate the chemical or biological solids in a wastewater which cannot be removed through settling, floating or filtration. See page F.7 for EPA method for Residue, Filterable.

Relationship Between TS, TSS and TDS.   These three solids tests can be related with Eq. (3.1):

$$TS = TSS + TD \tag{3.1}$$

3.2.1.3   Solids Type

Introduction.   The most basic solids differentiation is between settleable and floatable solids. An additional important differentiation is between volatile and non-volatile solids. A chemical or biological solid can potentially be removed from a wastewater by conversion through biological or chemical activity or incineration if the solid is organic and volatile (burnable at around 550°C). A volatile solid can also potentially be air or steam stripped or adsorbed. Much effort has been expended in developing an inexpensive test that better indicates biodegradability than volatility, but none has been as successful. It is cautioned that a chemical waste can have varying degrees of biodegradability even though it is completely volatile.

Settleable Solids. A liter of wastewater is settled in an Imhoff cone for one hour and the milliliters of solids accumulating in the bottom is reported as settleable solids (S/S) in ml/l. See Figure 3.3. See page F.12 for EPA method for Settleable Matter.

Flotable Solids. The quickest and simplest test for floatable solids is to test TSS as explained above in a well mixed sample of the wastewater and then let the wastewater float quiescently for one hour. Then take a sample of the wastewater with a pipette from under the float and above any sediment. Run a TSS test on the second pipette sample and report as non-flotable, suspended solids (NFSS). The relation with TSS will be as follows:

$$TSS = FSS + NFSS \tag{3.2}$$

where FSS = Flotable Suspended Solids.

If settable solids are present along with floatable solids, a S/S test must be run so that the relationship will be:

$$TSS = FSS + NFSS + {}^s\!/_s \tag{3.3}$$

Total Volatile Solids. The total volatile solids (TVS) test is used to determine whether a solid is organic, as approximated by being volatile, or inorganic (non-volatile). As explained above, the test is also used to approximate biodegradability. To determine total volatile solids (TVS), the residue from the TS test described above is ignited in a weighed dish in a furnace at $550°C \pm 50°C$ for 15 to 20 minutes and then weighed. The loss of weight is reported as TVS. The remaining weight is total non-volatile solids (TNVS).

The relationship between Total Solids and Volatile Solids is expressed in accordance with Eq. (3.4):

$$TS = TVS + TNVS \tag{3.4}$$

All volatile solids measurements are expressed in terms of mg per liter of wastewater originally filtered.

The total volatile solids test indicates the amount of the total solids which can potentially be destroyed chemically or biologically, volatilized through stripping, or adsorbed. The non-volatile solids are typically inorganic and cannot be destroyed. These solids must be converted or removed by some physical or chemical method explained below.

Total Volatile Suspended Solids. The suspended solids collected on the no-ash filter in the TSS test described above can be ignited in a furnace at $550°C \pm 50°C$ for 15 to 20 minutes to determine the total volatile suspended solids (TVSS). The TVSS will be the loss of weight and the total non-volatile suspended solids (TNVSS) will equal the remaining weight. See page F.13 for EPA method for Residue Volatile.

The relationship between Total Suspended Solids and Volatile Suspended Solids can be expressed in accordance with Eq. (3.5):

$$TSS = TVSS + TNVSS \tag{3.5}$$

Total Volatile Dissolved Solids. If the dried filtrate from the TDS test is ignited in a furnace at $550°C + 50°C$ for 15 to 20 minutes, the residue can be reported as total non-volatile dissolved solids

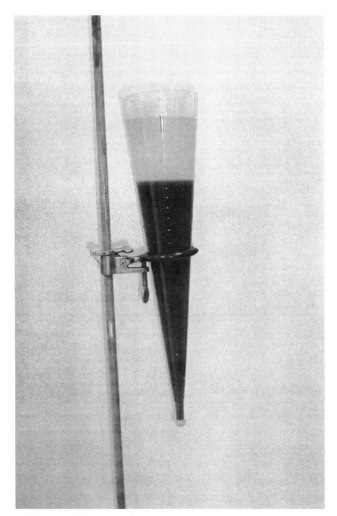

FIGURE 3.3 Settleable solids apparatus.

(TNVDS) and the loss of weight as total volatile dissolved solids (TVDS). Again, formula 3.6 compares these values to TDS: See page F.13 for EPA method for Residue, Volatile.

$$TDS = TVDS + TNVDS \tag{3.6}$$

Turbidity. Another measure of the solids content of water is turbidity, or the lack or clarity. Turbidity in water can be caused by suspended matter such as silt, clay, organic matter, organic compounds, or dissolved inorganics. Turbidity is determined by the optical property that causes light to be scattered, adsorbed or reflected rather than transmitted in a straight line through or into a liquid.

The first method of turbidity measurement was the Jackson Turbidity Unit (JTU) measurement which was the light lost through a wastewater sample from candlelight. These instruments were inaccurate at low turbidities, and have largely been replaced by the nephelometric method or instruments which measure the residual light scattered or reflected from a water. In the nephelometric method, the intensity of scattered light in a sample is compared with the intensity of light scattered by a standard reference solution under the same conditions. The higher the intensity of scattered light, the higher the turbidity. Light dispersing or scattering instruments are shown in Figures 3.4 and 3.5. Light dispersing units are used for turbidity in waters such as potable water, and light scattering units are used for waters containing more turbidity. Se page F.28 for EPA method for Turbidity.

3.2.1.4   Color.   Wastewater is "colored" if it is not completely clear. Color can be suspended color (apparent color) or dissolved color (true color), and may be prohibited in a pre-treatment ordinance.

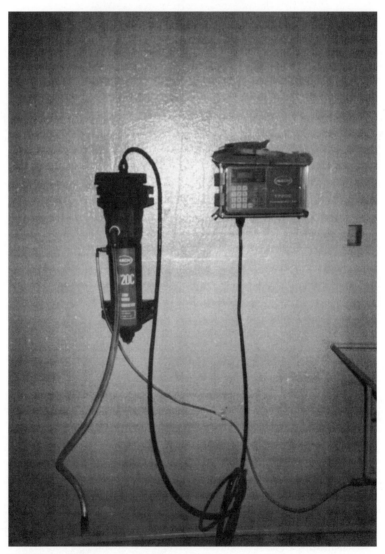

FIGURE 3.4   Light dispersed turbidimeter.

FIGURE 3.5   Surface scatter turbidimeter.

Suspended particles can be removed by settling or filtration, and can be considered as suspended solids regardless of whether they are colored. Dissolved color is of more concern and may be caused by vegetable or mineral dyes, other inorganic industrial wastes, or by organic material from stormwater runoff. Dissolved color should be considered as inorganic or organic chemical water quality and treated as described below as a chemical pollutant.

Color is quantitatively determined by visual comparison with a known concentration of colored solutions or by spectrophotometric methods. See page F.14 EPA method for Color.

3.2.1.5   Odor.   Odor is rare as a permit pollutant, but may be prohibited in a pre-treatment ordinance. Odor, like taste, is a measure of the effect of a stimulating substance on human receptory membranes. Pure unpolluted water is a measure of zero odor. An odor in a wastewater is recognized as a potential environmental hazard.

The accepted odor test is the Threshold Odor Test in which a sample is diluted with pure water until the least perceptible odor is noticed. The Minimum Detectable Threshold Odor Concentration (MDTOC) is reported as units or dilutions required to reduce an odor to its detectable limit. In the following example, a sample is diluted to various concentrations and tested olfactorily for odor:

ml Sample	ml Pure water	Odor
100 ml	0 ml	Present
50 ml	50 ml	Present
25 ml	75 ml	Barely detectable
10 ml	90 ml	Absent

In this example, the MDTOC equals the initial sample volume divided by the barely detectable sample volume:

$$\frac{100 \text{ ml}}{25 \text{ ml}} = 4$$

See page F.17 for EPA method for Odor.

3.2.1.6 Taste. There are only four true tastes that can be recognized by tongue and palate sensory nerves:

- Bitter
- Salty
- Sour
- Sweet

Other so-called tastes are actually odors. Taste is seldom listed as a permit requirement, but if it is, there are three methods of determination:

- Flavor threshold test
- Flavor rating assessment
- Flavor profile analysis

The Flavor Threshold Test is similar to the odor test and is the greatest dilution of a sample using pure water which yields a perceptible taste.

The Flavor Rating Assessment is a scale for rating a drinking water as acceptable or not.

The Flavor Profile Analysis is a comparison between a wastewater taste and that of a documented sensory quality judged acceptable by trained testers.

3.2.1.7 Temperature. The temperature of a water is a physical water quality parameter since temperature can have a negative effect on aquatic life, especially the propagation of fish. Various types of fish require certain temperatures for existence and a lower temperature for propagation.

In general, a wastewater which is too warm to allow appropriate life or propagation, must be cooled to permit temperatures by association with a cooler gas or fluid. See page F.27 for EPA Method for Temperature.

3.2.2 Physical Pollutant Limitations

In the Federal Water Pollution Control Administration's (the predecessor to the EPA) Report of the Committee on Water Quality Criteria,[3] the only physical pollutant recommendations are as shown in Table 3.1.

TABLE 3.1 Physical Pollutant Limitations

	Public water supplies	
Pollutant	Permissible	Desirable
Color (units)	75	<10
Odor	Narrative	Virtually absent
Temperature	Narrative	Narrative
Turbidity	Narrative	Virtually absent

TABLE 3.2 Typical Domestic Wastewater Compositions

Pollutant	Concentration, mg/l		
	Weak	Average	Strong
Total solids	350	800	1200
Total suspended solids	100	240	350
Total dissolved solids	250	500	850
Settleable solids (ml/l)	5	10	20
Volatile suspended solids	80	180	280
Volatile dissolved solids	100	260	300
Ammonia nitrogen	10	20	35
Total nitrogen	20	35	80
Phosphorus	5	10	15
Alkalinity as $CaCO_3$	50	100	250
Oil & grease	50	100	150
5-Day biochemical oxygen demand	120	225	400
Chemical oxygen demand	175	325	575
Total organic carbon	65	125	220

Table 3.2 lists typical compositions of a domestic wastewater with no industrial wastes. Many Publicly Owned Treatment Works (POTW) will use these or similar values to limit or surcharge for industrial waste discharges.

3.2.3 Monitoring Physical Pollutants

Suspended Solids

None of the total suspended solids values can be monitored since they contain inorganics, and instead must be sampled and tested. An estimate of volatile suspended solids which is a more accurate estimate of organic content can be determined with the use of a total carbon (TC) or total organic carbon (TOC) analyzer. This testing apparatus, in the case of TC, incinerates a small sample of wastewater every few seconds and automatically determines the total carbon content. The TOC analyzer subtracts the inorganic carbon (CO and CO_2) and reports the total organic carbon only, which is a more accurate measure of organic content and a close approximation of bacterial mass. A TC or TOC analyzer can be installed on line on a wastewater stream, but must be preceded by a solids filter. This filter will remove much of the suspended solids, therefore indicating only dissolved organic solids. To eliminate this effect, a suspended solids containing sample can be dissolved with acid to solubilize the solids before injection into an analyzer.

Total dissolved solids can be approximated on line using a conductivity meter. A conductivity meter measures electrical conductivity which is proportional to the dissolved solids in the water. Many cooling tower and boiler feed systems use an on-line conductivity meter to blow down a portion of the recirculated water to prevent scaling caused by high calcium and magnesium content.

Color can be monitored continuously using a spectrophotometer which differentiates color using light wavelengths.

Neither *odor* nor *taste* can be continuously monitored.

Temperature can be continuously monitored using a thermometer.

Settleable and *flotable solids* cannot be monitored and must be sampled and tested.

Turbidity can be measured on line with a light dispersed turbidimeter for low solids or a surface scatter turbidmeter for high solids content. See Figs. 3.4 and 3.5 for examples of these types of turbidimeters.

3.2.4   Sampling Physical Pollutants

3.2.4.1   General.   The accuracy of monitoring or testing can be no more accurate than the quality of the sample. The objective of sampling is to collect a sample small enough to handle and transport efficiently and large enough to be representative of the wastewater sampled.

The most common sampling mistake is to include solids which are not part of the normal wastewater stream such as catching sediment from the walls or bottom of the channel.

3.2.4.2   Samples Containing Settleable Solids.   A channel or pipe containing settleable solids at a velocity of less than approximately 2 fps will have a greater concentration of suspended solids closer to the bottom. Therefore, neither the high bottom concentration nor the low top concentration will be representative. A sample should be taken in an area away from surfaces that is typical of the flow and is well mixed. Since settleable solids affect the tests for suspended solids and perhaps turbidity, these cautions are also applicable for those parameters.

3.2.4.3   Samples Containing Flotables.   A wastewater containing flotables will not be typical if it is stratified. This is true with any liquid containing immiscible chemicals of different specific gravities not completely mixed or flowing with a completely turbulent flow.

3.2.4.4   Samples Containing Volatile Solids.   Biological activity can decrease or increase organic and therefore volatile content with time if any cellular material is present. In order to stop biological activity, a volatile-containing sample should be preserved at approximately 4°C between sampling and testing. In a biological treatment plant, since volatile solids are decreased with time, a sample must be taken at a point that is representative of the information desired, i.e. influent, effluent, or a portion of the distance through a basin.

3.2.4.5   Temperature Samples.   Since water is at its densest at 4°C, the depth of a wastewater temperature sample is extremely important. A wastewater can also lose heat throughout a process. Unless multiple depth samples are taken, it is suggested that a representative sample be taken at a vessel influent or effluent depending on the use of the results.

3.3   CHEMICAL POLLUTANTS

3.3.1   Introduction

Chemical pollutants can be organic or inorganic. There are thousands of organic pollutants consisting of various combinations of carbon, hydrogen, and perhaps oxygen and/or many other inorganic or organic molecules. In general, organic pollutants are more biodegradable with fewer carbon and/or other molecules attached. Standard EPA methods for these pollutants are given on pages F.32 through F.129 with chemical pollutants listed first, followed by biological indicators.

Inorganic chemical pollutants can be categorized into pure chemical pollutants and chemical indicators, some of which are described below in context with their environmental effects.

3.3.2   Chemical Pollutants

3.3.2.1   General.   Table 3.3, the Periodic Table of the Elements, lists elements with their respective symbols, atomic number and atomic masses and Table 3.4 describes the various groups in the Periodic Table.

3.3.2.2   Aluminum, Al.   Aluminum is a naturally occurring metal and is commonly used in water and wastewater treatment as alum or aluminum sulfate [$Al_2(SO_4)_3$], or with water [$Al_2(SO_4)_3 \cdot 18H_2O$]. Alum added to water with hydroxide ions forms an aluminum hydroxide [$Al(OH)_2$] precipitate.

Aluminum is suspected to be a contributor to Alzheimer's disease.

TABLE 3.3 Periodic Table of the Elements

Period	IA (1)	IIA (2)	IIIB (3)	IVB (4)	VB (5)	VIB (6)	VIIB (7)	VIIIB (8)	VIIIB (9)	VIIIB (10)	IB (11)	IIB (12)	IIIA (13)	IVA (14)	VA (15)	VIA (16)	VIIA (17)	VIIIA (18)
1	1 H 1.00794																	2 He 4.00260
2	3 Li 6.941	4 Be 9.01218											5 B 10.811	6 C 12.011	7 N 14.00674	8 O 15.9994	9 F 18.99840	10 Ne 20.1797
3	11 Na 22.98977	12 Mg 24.3050											13 Al 26.98154	14 Si 28.0855	15 P 30.97376	16 S 32.066	17 Cl 35.4527	18 Ar 39.948
4	19 K 39.0983	20 Ca 40.078	21 Sc 44.95591	22 Ti 47.88	23 V 50.9415	24 Cr 51.9961	25 Mn 54.9380	26 Fe 55.847	27 Co 58.9320	28 Ni 58.69	29 Cu 63.546	30 Zn 65.39	31 Ga 69.723	32 Ge 72.61	33 As 74.92159	34 Se 78.96	35 Br 79.904	36 Kr 83.80
5	37 Rb 85.4678	38 Sr 87.62	39 Y 88.90585	40 Zr 91.224	41 Nb 92.90638	42 Mo 95.94	43 Tc 98.9072	44 Ru 101.07	45 Rh 102.90550	46 Pd 106.42	47 Ag 107.8682	48 Cd 112.411	49 In 114.82	50 Sn 118.710	51 Sb 121.75	52 Te 127.60	53 I 126.90447	54 Xe 131.29
6	55 Cs 132.90543	56 Ba 137.327	57 *La 138.9055	72 Hf 178.49	73 Ta 180.9479	74 W 183.85	75 Re 186.207	76 Os 190.2	77 Ir 192.22	78 Pt 195.08	79 Au 196.96654	80 Hg 200.59	81 Ti 204.3833	82 Pb 207.2	83 Bi 208.98037	84 Po 208.9824	85 At 209.9871	86 Rn 222.0176
7	87 Fr 223.0197	88 Ra 226.0254	89 †Ac 227.0278	104 Rf 261.11	105 Ha 262.114	106 (Sg) 263.118	107 Ns 262.12	108 Hs (265)	109 Mt (266)									

IA — Alkali metals
IIA — Alkaline earth metals
VIIA — Halogens
VIIIA — Noble or inert gases

Atomic Number
Atomic Mass

* Lanthanide Series

58 Ce 140.115	59 Pr 140.90765	60 Nd 144.24	61 Pm 144.9127	62 Sm 150.36	63 Eu 151.965	64 Gd 157.25	65 Tb 158.92534	66 Dy 152.50	67 Ho 164.93032	68 Er 167.26	69 Tm 168.93421	70 Yb 173.04	71 Lu 174.967

† Actinide Series

90 Th 232.0381	91 Pa 231.0359	92 U 238.0289	93 Np 237.0482	94 Pu 244.0642	95 Am 243.0614	96 Cm 247.0703	97 Bk 247.0703	98 Cf 252.0587	99 Es 252.083	100 Fm 257.0951	101 Md 258.10	102 No 259.1009	103 Lr 260.105

TABLE 3.4   Groups of Elements

Group	Name	Description
I	Alkali metals	Form 1:1 binary compounds with chlorine (does not include H)
II	Alkaline earth metals	Form 1:2 compounds with chlorine and 1:1 compounds with oxygen
Transition elements		Have complex relationships
III		Form 1:3 chlorides and 2:3 oxides
IV		Form 1:4 chlorides and 1:4 hydrides and 1:2 oxides
V		Form binary compounds with hydrogen and oxygen
VI	Chalogens	Form 1:1 compounds with alkaline earth metals and 2:1 compounds with alkali metals (does not include Po)
VII	Halogens	Form 1:1 alkali halides with any alkali metal (does not include At)
VIII	Noble gases	Inert gases
Lanthanide series		Rare earths
Actinide series		Radioactive

3.3.2.3   *Ammonia, NH$_3$, NH$_4^+$.*   Ammonia is present in nature as part of the Nitrogen Cycle as shown in Fig. 3.6. Ammonia, at certain concentrations, is toxic to fish and is therefore considered a pollutant to waters classified for fish and wildlife. Ammonia is primarily in the NH$_3$ or gaseous form below pH 7.0 and primarily in the NH$_4^+$ or ammonium salt form above a pH of 7.0. Ammonia is soluble in water depending on the temperature in accordance with Table 3.5.

3.3.2.4   *Arsenic, As.*   Arsenic is found in nature and in manufactured pesticides. It possesses both metallic and non-metallic properties and is insoluble in water. It is of interest environmentally since it is highly toxic to humans and is chronic, or cumulative in human and animal organs. A chronic chemical accumulates in certain organs and is not discharged with urine or feces.

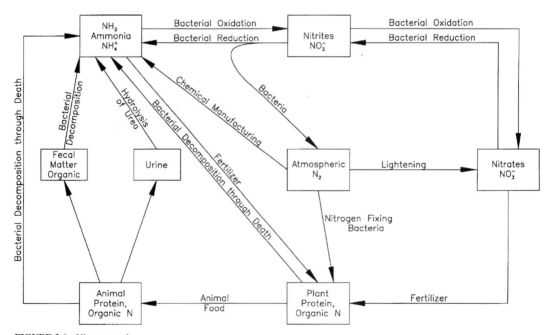

FIGURE 3.6   Nitrogen cycle.

TABLE 3.5 Ammonia Solubility

Weight NH$_3$ per 100 weights H$_2$O	Partial pressure of NH$_3$, mm Hg							
	0°	10°C	20°	25°	30°C	40°C	50°C	60°C
100	947							
90	785							
80	636	987	1450			3300		
70	500	780	1170			2760		
60	380	600	945			2130		
50	275	439	686			1520		
40	190	301	470		719	1065		
30	119	190	298		454	692		
25	89.5	144	227		352	534	825	
20	64	103.5	166		260	395	596	834
15	42.7	70.1	114		179	273	405	583
10	25.1	41.8	69.6		110	167	247	361
7.5	17.7	19.1	50.0		79.7	120	179	26.1
5	11.2	16.1	31.7		51.0	76.5	115	165
4		11.3	24.9		40.1	60.8	91.1	129.2
3			18.2	23.5	29.6	45	67.1	94.3
2.5			15.0	19.4	24.4	37.6	55.7	77.0
2			12.0	15.3	19.3	30.0	44.5	61.0
1.6				12.0	15.3	24.1	35.5	48.7
1.2				9.1	11.5	18.3	26.7	36.3
1.0				7.4		15.4	22.2	30.2
0.5				3.4				

3.3.2.5 Barium, Ba. Barium is known to stimulate heart muscles and affect the gastrointestinal tract and the central nervous system. A fatal dose of barium to humans is reported to be 550 to 600 mg. It is found in mine failing operations.

3.3.2.6 Beryllium, Be. Beryllium is toxic to fish and aquatic life and various plants. Additionally, it inhibits photosynthesis in terrestrial plants.

3.3.2.7 Bicarbonate, HCO$_3$. Bicarbonate, as a source of alkalinity, can be a contributor of iron chlorosis in plants. Bicarbonate reacts with calcium to precipitate as $CaCO_3$.

3.3.2.8 Boron, B. Boron, at approximately 1 to 4 mg/l, can be toxic to plants and at approximately 30 mg/l, can have physiological effects on animals and humans. It can be found in groundwater.

3.3.2.9 Cadmium, Cd. Cadmium affects metabolism and is quite toxic to animals and humans. Cadmium, common in plating wastewaters, is cumulative in the kidney and liver organs and can cause death.

3.3.2.10 Calcium, Ca. Calcium is the most common cause of hardness in water. It can be toxic to fish, but has no known physiological effects to humans.

3.3.2.11 Chloride, Cl. Chloride can be recognized as a taste at 500 to 1000 mg/l. In seawater, chloride is present at approximately 25,000 mg/l and has physiological effects on animals and humans. Chlorides are present in urine and are toxic to plants. High chlorides can act as a laxative.

3.3.2.12 Chlorine, Cl$_2$. Chlorine is highly toxic to fish and other organisms in mg/l concentrations, but as such is highly effective as a disinfecting agent for bacteria. The disinfection efficiency of chlorine is highly pH dependent as explained in Chapter 9. Chlorine can combine with certain organics in wastewater to form carcinogens. Chlorine is not found naturally, but is typically manu-factured from the electrolytic generation of NaCl.

3.3.2.13   Chromium, Cr.   Chromium is found in soil as a heavy metal and in industrial waste. The reduced state trivalent chromium Cr^{+3} is less toxic than oxidized hexavalent chromium Cr^{+6}. Chromium can have its valence changed between these two values using reduction reactions.

3.3.2.14   Copper, Cu.   Copper at levels above 100 mg/L is highly toxic to animals and humans and can cause vomiting and liver damage. High concentrations of copper turns water blue and is used as a paint pigment. Copper sulfate can be used for algae control.

3.3.2.15   Cyanide, CN.   Cyanide is a compound consisting of carbon and nitrogen molecules and can be fatal at 8 mg/L. At low pH, CN forms hydrogen cyanide (HCN) which is a highly toxic, almond smelling mustard gas. Extreme care must be taken in laboratory studies involving CN. Cyanide is present in industrial wastes, especially in electroplating and insecticide production.

3.3.2.16   Fluorine, F.   Fluorine as Fluoride occurs naturally and is non-cumulative. Four grams can be fatal and greater than 15 mg/l can cause tooth mottling. Low levels are known to build tooth hardness and resistance to cavities.

3.3.2.17   Iodine, I.   Iodine at approximately 1 mg/l is known to be desirable for goiter prevention and is the reason for iodized salt.

3.3.2.18   Iron, Fe.   Ferrous iron, Fe^{+2} is soluble in water and ferric iron, Fe^{+3} is insoluble. Therefore ferrous iron must be oxidized to become settleable. In natural groundwater, iron is present as ferrous bicarbonate, $Fe(HCO_3)_2$. The presence of iron can cause a red stain from wastes at 5 to 10 mg/l.

3.3.2.19   Lead, Pb.   Lead is naturally found and is present in paint, solder and industrial waste. Lead is cumulative in animal and human organs with greater than 0.5 mg/l causing lead poisoning. Plumbism is a disease caused by lead which affects the central nervous system of animals and humans.

3.3.2.20   Magnesium, Mg.   Magnesium is common in natural ground and surface waters, causes hardness and has an unpleasant taste. It has a laxative effect as magnesium sulfate.

3.3.2.21   Manganese, Mn.   Manganese behaves somewhat like iron but is more difficult to remove. It can cause a black stain.

3.3.2.22   Nickel, Ni.   Nickel seldom occurs in nature in its elemental form. It is relatively non-toxic to humans, but can be toxic to fish and aquatic life.

3.3.2.23   Nitrate, NO_3.   Nitrate has been used as an indicator for the presence of organics. Nitrates can cause methemoglobinemia at greater than 100 mg/l where a baby cannot breath enough oxygen.

3.3.2.24   Oxygen, O_2.   Oxygen as dissolved oxygen can be analyzed using either the iodometric method (Winkler) or the electrometric method. The iodometric methods is a filtration method and the electrometric method is a membrane method. Dissolved oxygen is important environmentally since it is a necessary source of energy in a biological treatment process and must be present for the survival of fish and aquatic life. Oxygen solubility can be determined as follows:

$$C_s = \frac{468}{(31.6 + T)} \tag{3.7}$$

where C_s = solubility of O_2 in mg/l at one atmosphere
       T = temperature in °C

Appendices C-5 and C-6 gives oxygen saturation values for various temperatures, elevations and salinities.

3.3.2.25 Phenol. Phenol can be found in industrial wastes and will cause taste and odor. If chlorine is present, phenol should be limited to less than 0.001 mg/l.

3.3.2.26 Phosphorous, P. Phosphorous can be a nutrient that encourages the growth of algae, contributing to the eutrofication or rapid aging of a body of water.

3.3.2.27 Potassium, K. High concentrations of potassium can cause heart problems in humans. Potassium can also have an effect on irrigated plants.

3.3.2.28 Selenium, Se. Selenium can be present in pesticides and certain industrial wastewaters. It can cause dental caries and can be toxic to fish and plants.

3.3.2.29 Silica, Si. Silica can cause boiler scale and is a major constituent of sand and glass.

3.3.2.30 Silver, Ag. Silver is usually recovered because of its value but in water is highly toxic to humans.

3.3.2.31 Sodium, Na. Sodium can bind soil if 10 to 20 percent of the cation exchange sites (Ca^{+2} and Mg^{+2}) are taken up by Na. This physical soil deterioration causes low soil permeability.

3.3.2.32 Sulfate, SO_3, SO_4. Sulfates are found in industrial waste and are the oxidized form of sulfides. Sulfates can have a laxative effect over 1000 mg/l.

3.3.2.33 Zinc, Zn. Zinc imparts a metallic taste to water and is found in many industrial wastewaters. Zinc will cause a milky appearance to water over 30 mg/l.

3.3.3 Chemical Indicator Tests

3.3.3.1 General. Certain tests have been developed to indicate water quality based on chemical characteristics which can be simpler, less expensive, or more indicative of water quality than a chemical compound test. These indicator tests are described below: $CaCO_3$ is used as a standard for many of the indicator tests since its molecular weight is 100 and calculations are simplified.

3.3.3.2 Acidity. The acidity of water is an indicator of its capacity to react with a strong base to a designated pH. Titration with a standard alkali solution to an end point of 8.3 pH is used for most wastewaters. For very low pH wastewaters, an end point of 3.7 is occasionally used. Acidity is reported in mg/l of $CaCO_3$.

3.3.3.3 Alkalinity. Alkalinity is primarily a function of the carbonate (CO_3), bicarbonate (HCO_3) and hydroxide (OH) content of a wastewater. Titration with a standard acid to an end point of 8.3 pH is reported as *phenolphthaline alkalinity* and titration to an end point of approximately 4.5 is reported as *total alkalinity*. Alkalinity is measured in mg/l as $CaCO_3$.

3.3.3.4 Conductivity. Conductivity is a quantification of the ability of water to carry an electric current. Most conductivity tests are accomplished with an instrument.

3.3.3.5 Hardness. Total hardness is the sum of calcium and magnesium concentrations expressed in mg/l of $CaCO_3$. It is generally a measure of the capacity of water to precipitate soap.

Hardness is either calculated from the results of separate calcium and magnesium tests or is determined from a color change when titrating a sample with ethylenediaminetetracetic acid (EDTA).

3.3.3.6 Hydrocarbons. In the test described below, hydrocarbons will be indicated as oil and grease. A mixture of all oils and grease in trichlorofluoroethane can be added to silica gel to selectively remove the fatty acids and leave only hydrocarbons for indication.

3.3.3.7 Oil and Grease. Most oil and grease tests quantify substances which are soluble in trichlorotrifluoroethane. These tests will include the presence of certain sulfur compounds, organic dyes and chlorophyll that are not volatilized.

3.3.3.8 Organic Pollutants. There are three tests commonly used to measure the total organic pollutants present in a wastewater; biochemical oxygen demand (BOD), chemical oxygen demand (COD), and total organic carbon (TOC). Since these tests are most often used in conjunction with biological treatment, they will be discussed in section 3.4, "Biological Pollutants."

3.3.3.9 pH. pH is the negative logarithm of the hydrogen ion concentration in a wastewater. pH or hydrogen ion activity is used to indicate the intensity of the acidic or basic character of a solution. pH is normally determined electrometrically, but can be estimated using titration or litmus paper.

3.3.3.10 Salinity. Salinity is generally a unitless measure of the dissolved salts in solution. The most accurate measure of salinity is conductivity as described above.

3.3.3.11 Surfactants. Since surfactants can affect wastewater treatment by emulsifying oils or metals and can cause foaming, they are of interest. About $2/3$ of total surfactants used are anionic. Anionic surfactants can be determined by using the Methylene Blue Active Substances (MBAS) procedure. Nonionic surfactants can likewise be determined with the Cobalt Thiocyanate Active Substance (CTAS) procedure.

3.3.4 Chemical Pollutant Limitations

The Environmental Protection Agency in its "Quality Criteria for Water,"[4] has published recommendations for various pollutant limitations as shown in Table 3.6.

3.3.5 Monitoring Chemical Pollutants

3.3.5.1 General. The only continuous monitoring devices normally used for chemical pollutants are for the indicator tests described above. No continuous chemical compound tests are currently available, nor are there continuous tests for acidity, alkalinity, hardness, hydrocarbons, oil and grease or surfactants.

3.3.5.2 Conductivity. A conductivity meter can be installed to continuously monitor and record conductivity.

3.3.5.3 pH. pH can be continuously monitored and recorded with a meter.

3.3.6 Sampling Chemical Pollutants

3.3.6.1 General. Chemical pollutants can be found in settable, floatable, suspended or dissolved solids and the recommendations for these types of samples in paragraph 3.2.4 should be followed. Sampling preservation and storage procedures listed in Table 3.7 are taken from "Standard Methods for the Examination of Water and Wastewater."[1]

TABLE 3.6   Chemical Pollutant Limitations

Chemical	Drinking water protection	Fish & Wildlife protection	Irrigation of crops
Ammonia	–	0.02 mg/l	–
Arsenic	50 μm	–	100 μg/l
Barium	1 μg/l	–	–
Beryllium	–	11 μg/l in soft water 1100 μg/ in hard water	100 μg/l*
Boron	–	–	750 mg/1*
Cadmium	10 μg/l	4 μg/l in soft water* 12 μg/l in hard water*	–
Chlorine	–	10 mg/l*	
Chromium,	50 μg/l	100 μg/l	
Copper	1 mg/l	0.1 × 96 hr LC_{50}	–
Cyanide	–	5 μg/l	–
Iron	0.3 mg/l	1.0 mg/l	–
Lead	50 μg/l	0.01 × 96 hr LC_{50}*	–
Manganese	50 μg/l	100 μg/l*	–
Mercury	2 μg/l	0.05 μg/l	–
Nickel	–	0.01 × 96 hr LC_{50}	–
Nitrates	10 mg/l	–	–
Oxygen	–	5 mg/l (min)	–
Phenol	1 μg/l	1 μg/l	–
Phosphorus	–	0.10 μg/l	–
Selenium	10 μg/12	0.1 × 96 hr LC_{50}	–
Silver	50 μg/l	0.1 × 96 hr LC_{50}	–
Sulfides	–	2 μg/l	–
Zinc	5 mg/l	0.1 × 96 hr LC_{50}	–
Alkalinity	–	20 mg/l as $CaCO_3$	–
PH	5–9	6.5–9.5	–

•Reader is referred to reference for more information.
* Note: 96 hr LC_{50} is the lethal concentration at which 50 percent of the test organisms die after 96 hours.

TABLE 3.7   Chemical Sample Preservation and Storage Recommendations

Pollutant	Preservation
Ammonia	0.8 ml concentrated H_2SO_4 /l and store at 4°C
Beryllium	1.5 ml concentrated HNO_3/l
Boron	Store in polyethylene or alkali resistant boron free glassware
Chromium	Concentrated HNO_3 to pH < 2.0
Copper	0.5 ml 1 + 1 HCl/100 ml
Manganese	Concentrated HNO_3 to pH < 2.0
Phenol	2 ml concentrated H_2SO_4 /l and store at 4°C
Phosphorous	1 ml concentrated H_2SO_4/l or freeze at < −10°C
Potassium	Concentrated HNO_3 to pH < 2.0, store in polyethylene bottles
Silica	Store in polyethylene bottles
Silver	Concentrated HNO_3 to pH < 2.0
Sodium	Store in polyethylene bottles
Zinc	Analyze within 6 hours or preserve with HCl
Oil and Grease	1 ml concentrated HCl/80 g and store in glass bottle

3.4 BIOLOGICAL POLLUTANTS

3.4.1 Introduction

Biological pollutants have a combination of inorganic and organic constituents and are characterized by being cellular in nature. For biological treatment purposes, only the organic constituents can be destroyed, and since all cellular material uses oxygen for energy, tests involve the measurement of either carbon content or oxygen demand, or an actual bacterial count.

3.4.2 Biochemical Oxygen Demand

The Biochemical Oxygen Demand (BOD) test is an empirical test which uses standard laboratory procedures to determine the oxygen requirements of wastewaters. The BOD, unless treated with an ammonia inhibiting chemical, will indicate the total of carbonaceous and nitrogenous oxygen demand.

A 250 to 300 ml airtight bottle is filled with sample to overflowing. Dissolved oxygen is measured initially after dilution and after incubation at 20°C for the time period of the test. It is necessary to have a population of microorganisms present that is capable of degrading the organic material in the sample. Many unchlorinated domestic or industrial wastewaters will afford sufficient bacteria for this purpose. This population is called "the seed."

One BOD test must be run on a bottle containing seed and dilution water. The dilutions should be made so that the least dilution results in at least a 50 percent dissolved oxygen depletion.

The seed is also added to the actual samples and diluted so that ideally the residual dissolved oxygen is at least 1 mg/l and the uptake at least 2 mg/l after 5 days. Several dilutions of the prepared sample should be made to ascertain that these results will be obtained. Suggestions of initial dilutions are as shown in Table 3.8. Oxygen depletion as a percent of total depletion will typically occur as shown in Table 3.9. Since the majority of the oxygen is depleted in 5 days, this time has been accepted as a standard to minimize the total time for running a BOD test. This test is symbolized by BOD_5.

When seed is used, the BOD_5 is computed as follows:

$$BOD_5 = \frac{(D_1 - D_2) - (B_1 - B_2)}{P} f \qquad (3.8)$$

where BOD_5 = the Biochemical Oxygen Demand after 5 days in mg/l
$D1$ = D.O. of diluted sample immediately after preparation, mg/l
D_2 = D.O. of diluted sample after 5 days of incubation at 20°C, mg/l
P = decimal volumetric fraction of sample used
B_1 = D.O. of seed control before incubation, mg/l
B_2 = D.O. of seed control after 5 days of incubation at 20°C, mg/l
f = ratio of seed in diluted sample to seed in seed control

TABLE 3.8   BOD Dilution Recommendations

Sample source	Percent sample
Strong industrial wastes	0.1 to 1.0
Raw or settled wastewater	1.0 to 5.0
Biologically treated wastewater	5.0 to 25.0
Polluted river waters	25.0 to 100.0

TABLE 3.9 Oxygen Depletion Estimation

Days	Percent oxygen depletion
0	0
5	60 to 80
20	95 to 99
Ultimate	100

3.4.3 Chemical Oxygen Demand

The Chemical Oxygen Demand (COD) test is a measure of the oxygen required from a strong chemical oxidant for the destruction of an organic material. The chemical oxidant selected for the COD test is potassium dichromate ($K_2Cr_2O_7$).

The sample is oxidized by a boiling mixture of $K_2Cr_2O_7$ and sulfuric acid using an excess of $K_2Cr_2O_7$. After digestion, the remaining unreduced $K_2Cr_2O_7$ is measured to determine the amount consumed by titrating with ferrous ammonia sulfate (FAS) until the end point is indicated by a sharp color change. The COD is calculated as:

$$COD = \frac{8000M(A - B)}{\text{ml sample}} \tag{3.9}$$

where COD = Chemical Oxygen Demand, mg/l
  A = ml FAS used for blank
  B = ml FAS used for sample
  M = molarity of FAS

The blank is a test with all reagents without the sample.

The COD test will not indicate ammonia like the BOD test, but it will show an oxygen demand from nitrite (1.1 mg COD/mg NO_2), ferrous iron, sulfide and manganous manganese.

Volatile straight chain aliphatic compounds are not oxidized appreciably in the COD test.

The COD takes about 2 to 3 hours to run compared to 5 days for a BOD test, and can be run with a relatively simple and inexpensive kit.

3.4.4 Total Organic Carbon and Total Carbon

The Total Organic Carbon (TOC) test is a more direct indication of organic content than the BOD or COD test. If a repeatable relationship between the three parameters is determined, TOC can be used to estimate the other parameters.

The TOC test does not provide all of the information that the BOD and COD tests provide. The BOD test indicates the actual oxygen needed for biologically destroying the organic. The COD test indicates the chemical oxidation requirements for destruction, and the TOC test indicates the total organic matter present and is independent of the oxidation state of the pollutant.

This test converts and removes inorganic carbon ($CO + CO_2$) and measures total organic carbon by converting it into CO_2 in a heated reaction chamber. The CO_2 is measured using infrared methods.

An option to this test is for Total Carbon (TC) which eliminates the removal of inorganic carbon from the procedure. The problem with using a TC test as a biological treatment parameter is the presence of CO_2 in the results, which normally is of no interest in this context.

This system can measure a small sample in a matter of seconds, and can therefore practically be used for continuous TOC measurement.

3.4.5   Theoretical Oxygen Demand

The Theoretical Oxygen Demand (ThOD) is the stoichiometrically determined oxygen needed to convert all carbon molecules in pollutants to CO_2, and all NH_3 and NO_2 to NO_3 by balancing equations. This calculation is possible on certain rather pure industrial wastes, but is impractical for most wastewaters, especially those containing domestic sewage or containing vegetable or animal wastes.

3.4.6   Relationship between BOD, COD, TOC and ThOD

For a completely biodegradable wastewater such as glucose, approximately ten percent of the original organics remain as non-biodegradable cellular residues after biological oxidation. Hence, the cellular residues are not measured by the BOD test. Therefore:

$$BODu = 0.9ThOD \tag{3.10}$$

where BODu = ultimate BOD
      ThOD = theoretical oxygen demand
For domestic sewage and some biodegradable industrial wastes, the relationship between BOD_5 and BODu is:

$$BOD_5 = 0.77BOD_u \tag{3.11}$$

where BOD_5 = 5 day BOD
      BODu = ultimate BOD
   For most wastewaters:

$$ThOD = COD \tag{3.12}$$

where ThOD = theoretical oxygen demand
      COD = chemical oxygen demand
since the COD test oxidizes all organics except for those which are totally resistant to dichromate oxidation.

Stoichiometrically, the COD/TOC ratio should be approximately the molecular ratio of oxygen to carbon:

$$\frac{COD}{TOC} = \frac{32}{12} = 2.66 \tag{3.13}$$

The ratio will actually range from zero, when organic material is resistant to dichromate oxidation, to as much as 6.0 when inorganic reducing agents are present.

For raw domestic sewage and some biodegradable industrial wastes, the following ratio of BOD_5/TOC occurs:

$$\frac{BOD_5}{TOC} = \frac{32}{12}(0.90)(0.77) = 1.85 \tag{3.14}$$

where BOD_5 = 5 day BOD
      TOC = Total Organic Carbon
      $0.90 = \dfrac{BODu}{ThOD}$ as per (3.10)
      $0.77 = \dfrac{BOD_5}{BODu}$ as per (3.11)

The author has successfully used this theoretical ratio to initially calibrate a TOC analyzer in an organics chemical plant.

As a wastewater is oxidized through a wastewater treatment plant, the BOD_5/TOC ratio will drop. A treatment plant effluent may have a BOD_5/TOC ratio of as low as 0.5 since the effluent wastewater is so much less biodegradable. (It has already been largely degraded).

The BOD_5 to COD ratio for domestic waste and certain biodegradable industrial wastes can be computed as follows:

$$BOD_5 = 0.7 \ COD \qquad\qquad (3.15)$$

where BOD_5 = 5 day BOD
  COD = chemical oxygen demand
  $0.7 = \dfrac{1.85}{2.66}$
  1.85 = BOD_5/TOC as per (3.14)
  2.66 = COD/TOC as per (3.13)

This ratio can also vary widely depending on the state of biodegradation of the wastewater. The author has found this ratio as low as 0.1 after several days of oxidation. If the BOD of a biodegradable wastewater equals zero, the wastewater will be completely biodegraded. There is some controversy about whether this ever occurs. Many authors will say that the Non-biodegradable Residue (NBDR) is as high as 0.10 as explained in the first paragraph of this section. The author has found that in activated sludge systems with hydraulic detention times in the range of 14 days, there is no accumulation of volatile suspended solids, which indicates that all organics are ultimately degraded under certain anoxic conditions.

3.4.7 Coliform Bacteria Tests

3.4.7.1 Introduction.
Coliform bacteria are some of the most common bacterial species, ranging from pathogenic to innocuous, and are found extensively in soil and in the feces of warm blooded animals. Elevated temperature tests can be used to separate these organisms into those of fecal origin and those from non-fecal sources.

The standard test for coliform bacteria from the intestines of warm blooded animals is the *fecal coliform* test. In a completely treated wastewater, no fecal or non-fecal coliform bacteria should be present after disinfection. All samples for coliform tests must be preserved at 10°C before testing, and should be tested within 24 hours. The samples should be contained in sterile bottles. Ample air space should be left in the sample bottle to facilitate mixing by shaking.

3.4.7.2 Coliform Tests

Multiple Tube Coliform Tests. The multiple tube coliform test is recommended for wastes of high turbidity, salty or brackish waters, as well as muds, sediments and sludges. This test uses multiple capped sterilized lauryl tryptose or lactose broth placed in fermentation tubes with small inverted vials. These tubes are inoculated with sample and incubated at 35°C for 48 hours. The tubes are examined for gas formation or acidic (yellow) growth at 24 and 48 hours. The presence of gas or acidic growth constitutes a positive presumptive reaction.

The positive presumptive cultures are inoculated into fermentation tubes containing inverted brilliant green vials and lactose bile broth and incubated for an additional 48 hours at 35°C. Formation of gas in the fermentation tube vials at any time within 48 hours confirms the tests. Calculation of the MPN is made using Tables 3.10, 3.11 and 3.12.

Membrane Filter Coliform Test. For the membrane filter test, a sufficient sample is filtered to yield between 50 and 200 coliform colonies. The sample is filtered through a membrane under partial vacuum. The filter is then placed in an absorbent enrichment pad in a culture dish and soaked with lauryl tryptose broth. The filter is placed on this pad and incubated for 1.5 to 2 hours at 35°C. The filter is then removed from the enrichment pad and rolled onto the Endo medium (agar) surface.

TABLE 3.10   MPN Index and 95% Confidence Limits for Various Combinations of Positive and Negative Results When 5–10 ml Portions are Used

No. of tubes giving positive reaction out of 5 of 10 ml each	MPN index/100 ml	95% confidence limits (approximate)	
		Lower	Upper
0	<2.2	0	6.0
1	2.2	0.1	12.6
2	5.1	0.5	19.2
3	19.2	1.6	29.4
4	16.0	3.3	29.4
5	>16.0	8.0	Infinite

The dish is then incubated for 20 to 22 hours at 35°C. The typical pink to dark red coliform colonies with metallic surface sheens are counted with the aid of a 10 to 15× microscope. The coliform density is reported as colonies per 100 ml of sample.

3.4.7.3   Fecal Coliform Test.   The fecal coliform test inoculates all positive presumptive cultures from the coliform test described above into fermentation tubes containing inverted vials and E.C. medium broth. These tubes are incubated at 44.5°C for 24 hours and placed into a water bath. Gas production within 24 hours is considered a positive fecal coliform reaction. MPN is calculated as described above for the coliform test. The membrane filter test described above can also be used for the fecal coliform test.

3.4.8   Salmonella Tests

The author has developed a standardized test for Salmonella in water. Since all Salmonella are pathogens and since Salmonella tests are required in the EPA 40CFR Chapter I, Subchapter O, Part 503 Regulations[5,6] for sludge land disposal, this type of test can be used as a direct indicator of pathogens. The test as developed uses a membrane filter to concentrate Salmonella bacteria from a

TABLE 3.11   MPN Index and 95% Confidence Limits for Various Combinations of Positive and Negative Results When 10-ml Portions are Used

No. of tubes giving positive reaction out of 10 of 10 ml each	MPN index/100 ml	95% confidence limits (approximate)	
		Lower	Upper
0	<1.1	0	3.0
1	1.2	0.03	5.9
2	2.2	0.26	8.1
3	3.6	0.69	10.6
4	5.1	1.3	13.4
5	6.9	2.1	16.8
6	9.2	3.1	21.1
7	12.0	4.3	27.1
8	16.1	5.9	36.8
9	23.0	8.1	59.5
10	>23.0	13.5	Infinite

TABLE 3.12   MPN Index and 95% Confidence Limits for Various Combinations of Positive Results When 5 Tubes are Used per Dilution (10 ml, 1.0 ml and 0.1 ml)

Combination of positives	MPN index/100 ml	95% confidence limits Lower	95% confidence limits Upper	Combination of positives	MPN index/100 ml	95% confidence limits Lower	95% confidence limits Upper
0-0-0	<2	—	—	4-2-0	22	9.0	56
0-0-1	2	1.0	10	4-2-1	26	12	65
0-1-0	2	1.0	10	4-3-0	27	12	67
0-2-0	4	1.0	13	4-3-1	33	15	77
				4-4-0	34	16	80
1-0-0	2	1.0	11	5-0-0	23	9.0	86
1-0-1	4	1.0	15	5-0-1	30	10	110
1-1-0	4	1.0	15	5-0-2	40	20	140
1-1-1	6	2.0	18	5-1-0	30	10	120
1-2-0	6	2.0	18	5-1-1	50	20	150
				5-1-2	60	30	180
2-0-0	4	1.0	17	5-2-0	50	20	170
2-0-1	7	2.0	20	5-2-1	70	30	210
2-1-0`	7	2.0	21	5-2-2	90	40	250
2-1-1	9	3.0	24	5-3-0	80	30	250
2-2-0	9	3.0	25	5-3-1	110	40	300
2-3-0	12	5.0	29	5-3-2	140	60	360
3-0-0	8	3.0	24	5-3-3	170	80	410
3-0-1	11	4.0	29	5-4-0	130	50	390
3-1-0	11	4.0	29	5-4-1	170	70	480
3-1-1	14	6.0	35	5-4-2	220	100	580
3-2-0	14	6.0	35	5-4-3	280	120	690
3-2-1	17	7.0	40	5-4-4	350	160	820
4-0-0	13	5.0	38	5-5-0	240	100	940
4-0-1	17	7.0	45	5-5-1	300	100	1000
4-1-0	17	7.0	46	5-5-2	500	200	2000
4-1-1	21	9.0	55	5-5-3	900	9300	2900
4-1-2	26	12	63	5-5-4	1600	600	5300
				5-5-5	≥1600	—	—

wastewater. The filter is enriched with tetrathionate broth for 24 hours and then incubated on bismuth sulfite broth for 48 hours. A count is made of shiny black colonies without halo and of black colonies with halo. The presumptive count is then calculated.

3.4.9   Monitoring Biological Parameters

As explained above, the TOC test is the only test for biological pollutants which can be used for continuous monitoring. These monitoring units yield results in a matter of minutes and have been used by the author to control biological treatment processes.

TOC can be used as an indicator of influent food strength, activated sludge cellular mass (mixed liquor volatile suspended solids) and effluent strength. Figure 3.7 shows a schematic of a TOC controlled system. TOC can also be used to determine the efficiency of different units including break through from a granular activated carbon column. In Figure 3.7 TOC tests are taken from the wastewater sources 1, 2, 3, 4 and 5. The test results are transmitted to a PLC controller which selects the sources to total a pre-set TOC range to enter the Equalization Basin. The food (F) entering the Activated Sludge Basin is sampled at point 6 and the cellular mass in the Activated Sludge basin (M) is sampled at point 7. These values are compared in the PLC controller to preset F/M. The amount

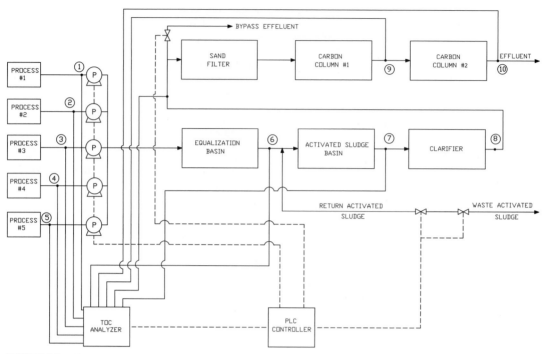

FIGURE 3.7   TOC controlled activated sludge system.

of Return Activated Sludge is selected to maintain this F/M. The clarifier effluent is tested at point 8, the first carbon column effluent, to determine break through at point 9, and the final effluent at point 10. If the final TOC at point 8 meets permit, the filter/carbon column by-pass can be activated by the PLC. If carbon column #1 reaches breakthrough, it can be replaced through valving by the PLC with column #2 and a spare column activated as column #2. In Figures 3.8 and 3.9, the photographs show a TOC Distribution System and Analyzer in an organics chemical plant.

3.4.10   Sampling Biological Parameters

3.4.10.1   General.   Discharge permits will typically require grab and/or continuous sampling. Each of these methods is described below.

3.4.10.2   Grab Sample.   A grab sample is a sample taken at one time in one quantity which is deemed representative of the flow and pollutant concentration. The cautions for sampling procedures described under section 3.3, "Chemical Pollutants" are also applicable for biological pollutants. The grab sample must be taken from a well mixed area to allow all settleable and floatable solids to be included in the sample. Caution must especially be taken in sampling a representative portion of the float if the sample is not well mixed. Care should also be taken to prevent sampling sediment which may contain residue from previous wastewater flow.

Preservation of samples is described above.

3.4.10.3   Composite Samples.   A composite sample is one which is taken proportional to either time or flow. A time composite sample takes an equal quantity of sample each period of time. For instance, a 24-hour time composite sample may take a 100 ml sample each hour for 24 hours. To be

FIGURE 3.8   TOC Distribution System.

truly proportional, this type of sample should be taken only when pollutant concentrations are uniform over the time period in which the samples are taken.

The automatic sampler shown in Fig. 3.10 can be adjusted to either take a discrete sample with time or to add a certain sized sample to a large container at each time period.

This sampler can be packed with ice to preserve the sample for biological sampling.

A flow composite sampler takes an equal quantity of sample with each pre-set number of gallons (liters) passing a flow measuring device, or takes a variable quantity of sample which is proportional to flow passed each uniform time period. The automatic sampler shown in Fig. 3.10 can be programmed as a flow composite supplier if it is connected to a compatible flow measuring device.

3.4.11   Toxicity Tests

3.4.11.1   General.   The purpose of a *toxicity test* is to determine the concentration of a pollutant which can exist in a body of water that will neither kill a test organism (acute toxicity) nor prevent reproduction (chronic toxicity) of the organism. After this determination, an *application factor* is usually multiplied by the test results to obtain a permit limit. Commonly used application factors are

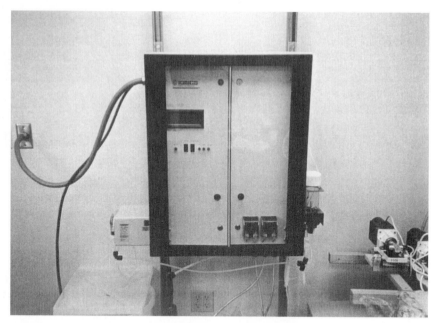

FIGURE 3.9 TOC Analyzer.

FIGURE 3.10 Automatic composite sampler.

0.01 if the pollutant is cumulative in the internal organs, and 0.10 if the pollutant is not cumulative. The use of Standard Methods[1] is recommended for details of toxicity tests.

The most common toxicity tests are for daphnia and fish. Other standard tests are toxicity to algae, phytoplanton, protozoa, coral, annelids, mollusks, acartia tonsa, nacrocrustaceans and aquatic insects. The daphnia and fish tests are described below.

3.4.11.2   Daphnia Toxicity Test.   Daphnia are generally less tolerant to toxic substances than fish and have been used for toxicity studies for over a hundred years. The two most common daphnia tests are for *daphnia magna* and *sera daphnia. Daphnia magna* are the largest of the magna, approximately 5 mm long. Even the young daphnia are about 1.0 mm long and can be observed without optical aids. This is the reason for their popularity in toxicity tests. Since the *daphnia magna* are hardier than smaller daphnia, some regulations require the use of smaller daphnia such as *sera daphnia.*

For the acute toxicity test, dilution water and toxicant solutions are prepared and placed in a series of 125 ml wide mouth glass bottles. Each bottle contains a different concentration of the pollutant with one control containing no pollutant. Then newborn daphnia are introduced into each bottle. The daphnia are observed after 1, 2, 4, 8 and 16 hours and daily thereafter. The number of motile (independently moving after a bottle is rotated) daphnia is recorded at each observation.

The test is continued for 5 days, or as long as most of the daphnia remain motile. The tests should be run in triplicate. The daphnia are not fed during tests as they will live for several days without food in the prepared solution.

The data from the acute test is analyzed for percent survival and reported as an LC50 or TLM value at 24, 48 or 96 hours. The $LC50_{96}$ or TLM_{96} is the "lethal concentration" of the toxicant which will kill 50 percent (or "toxic lethal median") of the daphnia within 96 hours under the test conditions and is typically plotted as follows:

For the chronic or reproductive impairment tests, the test solutions are prepared in the same way as the acute test, except that the water used should be the receiving water (the water into which the tested wastewater will be discharged). The $LC50_{96}$ for the acute test value should be the highest concentration, and subsequent smaller concentrations should be reduced by a factor of three or more for each additional bottle. A control with no toxicant is also tested.

Daily observations of dead or immobilized daphnia are made. The young are removed and the number counted. Observation is continued until six broods of young are produced (around 21 days at 25°C and 30 days at 20°C).

If the number of young produced in the lowest toxicant concentration differs significantly from that in the controls, the test should be repeated with even lower toxicant concentrations until no significant difference is observed. After reaching a toxicant concentration where no significant difference with controls is noted, this concentration is recorded as non-toxic. The next highest concentration is the lowest for which a significant difference in production of young was found.

3.4.11.3   Toxicity Test on Fish.   Many freshwater and saltwater fish have been used for toxicity tests. The main consideration in the selection of fish species is its sensitivity to the toxicant and to the environmental conditions in the receiving water.

The acute test is similar to that described above for daphnia resulting in an LC_{50} (lethal concentration at which 50 percent die) or TLM (Toxic lethal median).

The acute test is normally made with a small species of fish (less than 5 g body weight) such as the fathead minnow for fresh water. Typically 20 fish are exposed to each toxicant concentration and tested at times varying from 24 to 96 hours. The LC_{50} is computed as demonstrated in Fig. 3.11 and described above for the daphnia test.

A life cycle test can be run on the fish as a chronic test. In these tests, embryos or larvae are fed for a period of several weeks and injured or crippled fish are removed. Embryos are removed and counted and after a couple of generations, data is collected on the minimum toxicity which affected the long term growth and spawning of the fish under test conditions.

3.4.11.4   Cautions for Toxicity Tests.   The following cautions for the use of toxicity tests are suggested:

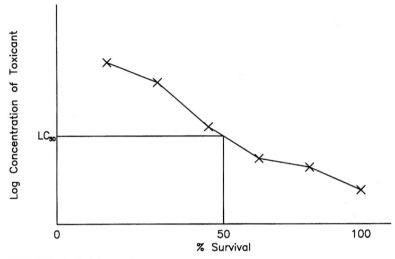

FIGURE 3.11   Toxicity test plot.

- In a receiving stream, sub lethal effects can occur which will drive fish from the area without killing them. Therefore, toxicity tests could show death and there could be no death in a receiving stream.

- Water temperature affects short and long term toxicity for some pollutants. Typically, toxicity increases for some metals such as zinc and copper as temperature increases.

- Receiving water pH, alkalinity, hardness and dissolved oxygen affects toxicity. Most heavy metals are more toxic in waters with less hardness and alkalinity.

- The osmotic pressure of distilled water can kill test species.

- Test water should not typically be aerated since aeration affects the pH and therefore the toxicity.

- If too many organisms are used for a test, all available dissolved oxygen could be depleted, causing toxicity.

3.5   REFERENCES

1. "Standard Methods for the Examination of Water and Wastewater," latest edition, APHA-AWWA-WPCF, American Public Health Association, 1015 Fifteenth St., N.W., Washington, D.C. 20005.
2. "Methods for Chemical Analysis for Water and Wastes," latest revision, EPA Report no. 600/4-79-020.
3. "Report of the Committee on Water Quality Criteria," FWPCA, April 1, 1968.
4. "Quality Criteria for Water," EPA, Report no. EPA-440/9-76-023.
5. E. Roberts Alley & George W. Maloney, "Simple Method Gives Salmonella Density," *Water & Sewage Works,* September, 1974.
6. E. Roberts Alley & George W. Maloney, "How to Identify Salmonellae Colonies," *Water and Sewage Works,* October, 1974.

CHAPTER 4
WATER QUALITY

4.1 GENERAL

The non-degradation principle of the Clean Water Act of 1974 states that the waters of the United States shall not be degraded below their level at the time of the Act. Each type of receiving water has its own specific characteristics which must be considered in the effort to achieve non-degradation while allowing growth in population and industry. This chapter will consider various types of receiving waters, their classified uses, their investigation, their treatment and their management.

4.2 TYPES OF RECEIVING WATERS

4.2.1 Lakes

4.2.1.1 Types of Lakes. A lake is classified as a pool of water, which is trapped behind a natural or artificial dam that normally causes the pool to be contained. If the top of the dam is constant elevation, such as in a natural lake, the water level will equal the dam elevation plus the depth of water spilling over the dam and therefore be fairly consistent.

An artificial lake level can be manipulated as can downstream water flow by adjusting the gates and valves in the dam to allow more or less water passage. These lakes can therefore have large water level variations depending on water flow management. In a very long lake, such as a dammed river, the water level will rise with distance from the dam. The rise of the lake is a factor of the friction head, the shape of the lake and the elevation of the bottom of the lake.

Because of these natural and artificial variations, the water flow through a lake can vary from zero to large flows. In the unusual condition where local rainfalls cause one tributary of a lake to flow at a greater rate than another tributary, the water flow in the second tributary could be reversed.

The 20-year, 3-day flow is a common measurement of an extreme dry weather flow for computation of the assimilative capacity of a body of water. This flow is the historical record of the lowest daily flow for 3 consecutive days for any 20-year period. For many lakes, this flow will be zero, which may cause very stringent permit requirements.

4.2.1.2 Lake Stratification. The density of water is greatest at 4°C; therefore, colder water tends to sink in a lake. This phenomenon occurs until the surface of the lake reaches a temperature close to 4°C in the fall, at which time, a strong wind will overturn the lake stratification and the lake becomes unstratified. During the spring, when the water temperatures begin to rise, thermal stratification begins again. This overturning of a lake can cause the bottom sediment to rise to the water surface. If staining metals such as iron or manganese are present in these bottom sediments, when the lake turns over they will likely stain a dam or waterfall. In periods in which the water temperature in a lake is stable, the water will stratify into the layers shown in Fig. 4.1.

The epilimnion is kept warm by solar radiation and will only be a few feet thick. The thermocline is the transition between the warmer epilimnion and the cooler hypolimnion. The epilimnion will tend to grow algae in the summer, especially in southern climates and in lakes with little or no water

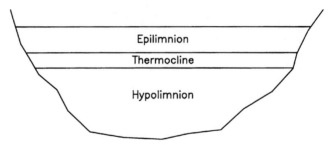

FIGURE 4.1 Lake stratification.

flow. Algal growth will tend to add dissolved oxygen to the epilimnion during the day, but use dissolved oxygen at night. Because of the varying temperatures, dissolved oxygen, and macrophyte presence, different fish and aquatic life will also be found in these different strata.

Due to the stratification of lakes, the water in a lake may be more assimilative to a specific pollutant at a certain temperature. Therefore, thought must be given to the depth of the wastewater discharge.

4.2.1.3 Lake Eutrophication.   Lakes can become **eutrophic** naturally with time or through misuse by humans. A eutrophic lake is one in which the nutrient concentration and biomass production are very high. Natural eutrophication can take several hundreds of years depending on the location and characteristics of the lake. Cultural eutrophication can take place over the course of a few seasons or years, which also depends on its location and characteristics.

The growth of a eutrophying organism, such as algae, is enhanced by the presence of nutrients (carbon, nitrogen, and phosphorous). If excess amounts of these nutrients are present, algae can proliferate and cause eutrophication. Water can be tested to determine which nutrient is the critical or limiting nutrient that causes the algae to proliferate. This nutrient is then typically limited by permit for discharges into the lake. This limitation can be especially complicated when the watershed contains currently unregulated agricultural areas, where high levels of nitrogen and phosphorus are typically found.

4.2.1.4 Water Flow in Lakes.   Besides the overall lake water flow described above which is caused by groundwater, stormwater and artificial controls, a portion of the lake may move at a greater flow. This will normally occur in the original river or streambed of an artificial lake. This higher flow is due to bottom friction and temperature and can be estimated using Manning's Equation. This phenomenon can be readily seen by observing a flooding river which moves rapidly over the riverbed and slowly over the flood plain. Because of this flow variation, many regulators will require a wastewater discharge line (usually called an "outfall") to be carried to the original river bed in an artificial lake, rather than being discharged into more stagnant shallow areas.

If a lake has a tributary which, because of localized storms, has a higher flow than the main river feeding the lake, the flow in the lake upstream from this high flowing tributary may actually flow upstream, transferring pollutant discharges into the lake upstream rather than downstream.

4.2.2 Rivers and Streams

Rivers and streams flow unrestricted by downstream dams and can be modeled fairly accurately with Manning's Equation. The U.S. Corps of Engineers has developed computer programs known as HEC1 and HEC2 for river modeling.

Since rivers are typicallly uncontrolled by dams, the water level will be directly proportional to stormwater and the rivers can have a tremendous variation in flow. A very slowly moving river can stratify similarly to a lake as expained in section 4.2.1.2.

The 3-day, 20-year low flow described above for lakes is also used to determine assimilative capacities of rivers and streams. During dry weather, a river or stream could have a zero low flow. Rivers and streams will tend to re-aerate themselves because of surface turbulence, typically maintaining or recovering higher dissolved oxygen content than lakes. In order to aid in pollutant distribution from an outfall, some permits may require an outfall to discharge at several locations across a river.

4.2.3 Intermittent Streams and Ditches

A stream that only flows during wet weather is called an intermittent stream. These streams will always have a zero low flow and may have correspondingly very stringent permit levels. These stringent permit levels are only justified if the intermittent streams contain aquatic life which can exist in intermittent water flow conditions and would be damaged by pollutant discharge. Otherwise, the permit levels should be controlled by the downstream conditions in a lake, river or stream.

Ditches are artificially created channels, but in many cases are used as a receptor for wastewater. Like intermittent natural streams, ditches have a zero low flow and stringent permit limits. Many ditches, especially those that are lined, will not typically contain aquatic life and should not be subject to such a classification. But, as in the case of intermittent streams, the permit limits may be set by critical downstream conditions.

4.2.4 Estuaries

Estuaries are the transition areas between fresh water and salt water and contain much unique aquatic life such as mollusks and crustaceans. These sensitive areas are protected with stringent discharge permit requirements including outfall restrictions.

4.2.5 Saltwater

Oceans, bays, gulfs, etc. are highly saline and contain unique marine aquatic life. Regulations are gradually being made more strict for saltwater environments as more is discovered about their sensitive environments.

4.2.6 Groundwater

The Environmental Protection Agency has classified all groundwater as drinking water unless it is specifically exempted and, as such, must meet the applicable discharge and clean up standards.

Groundwater may flow very slowly through an impermeable soil such as clay, fairly rapidly through a permeable soil such as sand, or very rapidly through crevices and caves in karst terrain. Groundwater can range from deep, confined aquifers to shallow water table, unconfined aquifers. All water below the ground surface is classified as groundwater and regulated as such.

Due to heterogeneity in subsurface conditions, groundwater flow is not necessarily consistent with overland surface flow. If groundwater migration direction is of interest, well water depths may need to be measured over the area of migration to determine the water table gradient or tracer or dye tests run.

4.3 SIGNIFICANT DETERIORATION OF WATER QUALITY

4.3.1 General

The deterioration of water quality for its classified use can be quantified by physical, chemical or biological monitoring. Possible deterioration is described in the following paragraphs by classification.

4.3.2 Navigation

Since navigation is normally only affected directly by physical pollutants, those such as floatables and settleables are usually included in permits for water of these classifications. These parameters could be in the form of settleable solids or oil and grease. If either of these are evident in a body of water, it could be classified as deteriorated.

4.3.3 Recreation

Since the recreation classification includes bathing, swimming, etc., any pollutant which is transmitted through the skin, nose or mouth of a human can be regulated. Consequently, many waters classified for recreation can be permitted as stringently as drinking water classifications. Recreation, as navigation, may also include settleable solids and oil and grease limitations. Recreation classified water would be considered deteriorated if any of these types of pollutants increase.

4.3.4 Irrigation

Waters classified for irrigation will be considered deteriorated if any pollutant affecting the growth of plants is present. These include boron, chlorides, potassium, selenium and sodium.

4.3.5 Fish and Aquatic Life

Many pollutants are more sensitive to fish and aquatic life than to humans, and regulations will reflect this sensitivity. The deterioration of fish and wildlife classified water can be readily measured by testing for dissolved oxygen or using toxicity tests.

4.3.6 Drinking Water

Chapter 3 lists many pollutants considered toxic to humans and therefore indicative of the deterioration of a water classified as drinking water.

4.4 MODELING OF POLLUTANT DISCHARGE

4.4.1 Dissolved Oxygen Modeling

Most dissolved oxygen models are based on the Streeter-Phelps equation as follows:

$$\frac{dD}{dt} = KL = K_2 D \tag{4.1}$$

where D = dissolved oxygen deficit
L = concentration of organic matter
K = coefficient of deoxygenation
K_2 = coefficient of reaeration

K and K_2 must be determined by measuring dissolved oxygen in the receiving water.

Oxygen sag models are as follows:

$$t_c = \frac{1}{(K_2 - K)} ln\left[\left(\frac{K_2}{K}\right)\left(1 - D_0\left\{\frac{(K_2 - K)}{KL_0}\right\}\right)\right] \tag{4.2}$$

where t_c = time of maximum dissolved oxygen
D_0 = initial dissolved oxygen at the point of waste discharge, mg/L
L_0 = initial BOD$_5$, mg/L

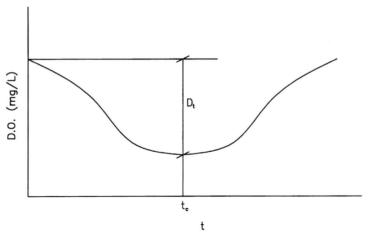

FIGURE 4.2   Dissolved oxygen sag curve.

$$D_t = \left(\frac{KL_0}{K_2 - K} \right)(e^{-Kt} - e^{K_2 t}) + D_0 e^{-K_2 t} \tag{4.3}$$

where D_t = oxygen deficit at time t, mg/L

Figure 4.2 shows a typical dissolved oxygen sag curve.

4.5   NATURAL TREATMENT OF POLLUTANTS IN WATER

Pollutants that can be oxidized by bacteria in the presence of free dissolved oxygen can be aerobically converted to CO_2 and H_2O. These pollutants, such as domestic sewage and many readily biodegradable industrial wastes, can be destroyed in accordance with the oxygen sag equations presented above, if oxygen is naturally reinjected into the water through surface turbulence.

4.6   ARTIFICIAL TREATMENT OF POLLUTANTS IN WATER

Dissolved oxygen can be added artificially to a body of water by aerating the surface with floating aerators. This can provide the oxygen necessary for natural bacterial decomposition of pollutants and for the propagation of aquatic life, can also prevent lake or river stratification minimizing the return of settled pollutants to the water surface. Air diffusers and blowers can also be used for artificially aerating water.

4.7   WATERSHED MANAGEMENT

Since most indirectly discharged pollutants are transmitted to a body of water across the surface, management of the watershed can minimize these pollutants. Also direct discharges can be eliminated or controlled through watershed management. Vegetation, such as trees and grasses, can be used to filter and biologically treat pollutants. In addition, the vegetation serves to stabilize the soil to prevent erosion. Human and even animal activity can be limited in a watershed to minimize associated pollution.

4.8   RISK MANAGEMENT

4.8.1   Introduction

Conditions in the natural environment are not necessarily conducive to man's survival. We are not guaranteed that life will be free from harm; we are faced with risk. In response to adverse conditions, man must adapt the natural environment to meet his needs and wants. He must make decisions to ensure both his own health, and, since he is dependent upon it, the integrity of his environment. Within the context of water quality, risk refers to adverse impacts to organisms and systems of organisms through their exposure to water: drinking water, water bodies, food produced using water, etc.

Risk management is a broad term that applies to any decision-making and implementation process, which has as its end the reduction of damage. When applied to water quality, risks to human health, occupational health, the environment, and even economic stability and growth (both societal and corporate (business)) are key applications. Both regulatory agencies and private institutions address these types of problems.

Risk management often requires risk assessment, which is a process of identifying and characterizing risks and identifying solutions for reducing risk. Risk assessment facilitates the risk management process by providing information about existing risks and potential risks associated with environmental conditions. Risk assessments identify hazards, who is at risk, how they are at risk, and qualify, and if possible, quantify the risk. They also identify solutions for reducing and eliminating risk and identify the effectiveness at reducing risk—how much they can reduce risk—or how much benefit for the effort. Risk assessments tend to fall into two types: health risk assessments for human health, and ecological risk assessments for environmental integrity.

Because risk management is an activity that fits within the context of other activities, risk management decisions require other considerations: the decisions and their implementation must be completed with limited budgets, or limited time, and must fit in with other priorities.

4.8.2   Applications to Water Quality

Because of water's fundamental role in life processes, coupled with its high mobility through the air and land, water quality is of high concern in society. Water quality concerns include the availability of drinkable water and safe foods, and the contamination of rivers, lakes, streams, aquifers, estuaries and saltwater. Addressing these sometimes overwhelming problems requires cooperative work by the public, government, and industry.

The public gives state and federal government authority over large-scale and extensive environmental problems, including water quality. Governments respond by developing, enacting, and enforcing regulations. For effective response, assessing problems in terms of risk provides a common language between the concerned public and those responsible for problems and solutions. As a result, government agencies and academics have developed methodologies to relate risk to decision-making. As this knowledge has been disseminated into the public domain, the value of environmental-risk-based decision-making has been recognized and is continuing to be applied to more specific problems, particularly those related to business and industrial production.

Industrial production requires the use and disposal of water. In addition, stormwater and groundwater may come into contact with industrial substances and transfer them to the surrounding environment. As a result, industries must take action to responsibly dispose of wastewater and to prevent and remedy surface and groundwater contamination. Their effort must be focused on meeting their own needs and goals while answering to the demands of the public. To respond to problems expeditiously, government agencies have historically set requirements in terms of contaminant levels, and this sometimes resulted in remedies that were either too much or too little. The trend in recent years has been to think about site-specific problems in terms of consequence—in terms of actual risk—and risk-based standards are continuing to be developed and applied.

4.8.3 Risk Assessment

Risk assessment is a way of identifying, measuring, and prioritizing risks for decision-making. By providing information about risks, and by estimating the sensitivity of risk to changing conditions, remedies can be developed and implemented effectively. Risk assessments can both initiate and facilitate the risk management process, and they should be tailored to provide information that meets risk management objectives. Risk assessments can be used to address problems related to public health, occupational health and safety, ecological impacts, public welfare, and financial decisions.

The assessment process requires following cause-effect relationships in the environment, which begins with identifying hazards, where they are and might move to, and whether organisms can come into contact with them. An organism exposed to a hazard is referred to as a *receptor*. A hazard without a receptor—actual or potential—is not a risk. Measuring risk requires understanding the effect—toxicity or threat—the hazard poses, measuring or estimating exposure levels, and measuring or calculating the effects or probability of effects on the organism. Risk measures for organisms include safety indicators, such as fatalities and injuries; acute toxicity indicators, such as lethal doses; and health indicators such as incremental cancer cases. For ecological systems, measures include habitat damage, species abundance and diversity, and ecosystem functioning.

Characterizing risks involves summarizing effects, comparing risks, and giving consideration to uncertainty. Risks are prioritized through initial screening of hazards, comparison of risk levels, and considerations of extent and urgency. Given the complexities of life processes and the current state of knowledge, uncertainty is practically unavoidable in the measurement process. These uncertainties should be communicated and if possible, quantified.

Risk assessments are commonly divided into human health assessments and ecological assessments. Their difference in focus requires different methodologies. Human health is addressed by considering effects on the human body. Ecological effects must be addressed by looking not only at organisms, but at the relationships between organisms. Ecological risk assessment methodologies are less established than those for human health, and improvements are being made.

Assessment frameworks help by delineating steps in the process, so that work can be conducted systematically and progress can be monitored. Generally, assessments are divided into the following steps: hazard identification, toxicity or response assessment, exposure assessment, and risk characterization. Often new information will become available, however, that will require returning to completed steps for revision. In fact, some frameworks use a tiered approach, in which steps are repeated with progressively greater levels of detail and accuracy. For example, the ASTM framework for corrective action at petroleum sites has three tiers; the second and third tiers, which are more detailed, only need to be completed if Tier 1 is inconclusive.

4.8.3.1 *Health Risk Assessment.*   The term health risk assessment typically refers to assessment of public health, as contrasted with worker health, which is typically addressed in occupational health and safety assessments. The cause of adverse health effects is contact with or exposure to hazardous agents and in the case of water, this would include drinking, ingesting food, and dermal exposure. Health effects are typically classified as either cancer or non-cancer. Cancer risks often drive the regulatory process because of public perception. Cancer risk is typically expressed as an incremental probability of either contraction of the disease or death, and limits are typically probabilities of 10^{-4} to 10^{-6}. Non-cancer risks are not probabilistic, but are considered definite above a threshold dose. Limits are set as some fraction of the threshold dose.

The hazard identification step consists of identifying the potential sources of risk. It involves not only inventorying hazardous agents, but an overall delineation of the scope of the assessment, including the limits of the site area, with consideration of individuals and populations who might be exposed. If possible, this step should also include setting criteria about the quantity and quality of data needed for decision-making. The number of chemicals of concern can be reduced by screening and prioritizing, based on available concentration and toxicity data and relevant standards and limits. Once hazardous substances have been identified, the relationship between dose and effects is determined. This step usually involves research for existing toxicological information about the hazards. Major sources of toxicological information provided by the EPA are the Integrated Risk Information System (IRIS), the Health Effects Assessment Summary Tables (HEAST), and the Environmental Criteria and Assessment Office (ECAO). Due to the complexity of human biological processes and the current level of knowledge, toxicological data are often highly uncertain, and therefore estimated conservatively.

Exposure and ultimately risk requires both the presence or occurrence of a hazard and the presence of a receptor. The goal of the exposure assessment step is to establish ways by which humans may come into contact with the hazards, and to estimate concentrations in ambient media. Establishing potential exposures requires considering pathways by which the hazard can reach humans, and the routes by which they may be exposed. Pathways are typically through fluids—water and air—and include airborne transport and migration through groundwater and surface water. Exposure routes include inhalation, ingestion, contact with the skin, and irradiation. Actual or potential exposures are estimated by measuring or modeling the chemical concentrations in ambient environmental media. In identifying possible exposed population, future as well as current land use should be considered—people may move into currently unoccupied areas.

The purpose of the risk characterization step is to identify who is at risk and what the risks are, to quantify the risk if possible, and to evaluate the acceptability of the risks. To accomplish this, exposure scenarios and toxicity relationships are linked to establish risk estimates. Common risk estimates associated with water-related hazards include probability of cancer, and in the case of non-carcinogens, non-statistical hazard quotients. As part of this step, uncertainties in estimation are analyzed and characterized. When developing results and conclusions, the requirements of the subsequent risk management process should be considered.

4.8.3.2 *Ecological Risk Assessment.*

Ecological risk assessment is for living organisms other than humans, including both plants and animals. Because organisms develop needbased relationships to one-another, effects on one type of organism can affect other types. For this reason, assessments focus on risks, not just to individual organisms, but to systems of organisms, (ecosystems). Identification and characterization of the impacts to ecological systems are typically much more complex processes than for human health. Plants and animals are faced with natural levels of risk. Due to natural and man-made events, risk levels for an organism can change, resulting in altered health or increased mortality in that group or species. This risk then follows the cause-effect chain of relationships, resulting in risk to other types of organisms. Threats to the natural relationships between organisms are *ecological threats,* and sources of ecological risk are referred to as *stressors.* Stressors can be classified as physical, biological, and chemical in nature, and examples include erosion, pests, and pesticides.

Ecological effects which are selected for focus in risk characterization are called *endpoints.* Endpoints can be selected from various levels of ecological organization, including individual organisms, populations, communities, ecosystems, watersheds, and landscapes. Ecological levels of organization are viewed scientifically as nested models, although this concept of ecology has been considered in some cases as overly simplistic for risk assessment (Bartell).

Appropriate problem formulation is crucial to the ability of the risk assessment to achieve its purpose: to provide information for the risk management process. In this step, the nature and scope of the assessment are established. This includes the identification and characterization of hazards (stressors), and the identification of endpoints.

Consideration should be given to the time-related characteristics of the stressor (frequency, duration), spatial characteristics (extent) as well as magnitude of impact. Selecting endpoints should be based on social, political, and economic factors as well as ecological ones. (Bartell). These considerations will ensure the relevance of the assessment to the risk management process. In order to ensure the economy and adequacy of the risk assessment, expectations about the accuracy and precision of the results should be addressed. To complete the problem definition step, the relationships between the stressors and the ecological effects should be established, and a rough work plan, including methods and required data and tools, should be developed.

Once stressors and endpoints are identified, the stress-response is determined. In this step, relationships between exposure, dose, and response are established. The biological/ecological level of organization used to assess the effects of stressors should be based on the ease of measurement and sensitivity to stresses (Bartell).

"Exposure is assessed by determining the mechanisms that bring organisms into contact with the source of the potential ecological risk and by quantifying the frequency, magnitude, and duration of such contact" (Bartell). In the case of water, dose occurs through physical contact, inhalation, ingestion of water, and ingestion of food. Consideration to time variability of both the stressor and the characteristics and behavior of exposed populations should be given to avoid over- or underestimation of effects.

The level of detail and the degree of conservatism required during the exposure assessment will dictate the method of estimating exposure levels. As a simplified first estimate, it can be assumed that receptors are directly exposed to source levels. For instance, fish can be assumed to be exposed to pollutant concentrations in the outlet pipe of a wastewater stream. For a more accurate estimate, a transport model can be used to estimate exposure levels. Possibly the most accurate estimates can be achieved by field measurement of actual exposure levels.

Risk characterization is the integration of exposure information and stress-response relationships to estimate ecological risks. As part of this step, impacts on risk estimates from uncertainties occurring at all phases of the model are addressed. Methods of characterization (in order of sophistication) include screening calculations, joint distributions, hypothesis testing, physical models, and mathematical models.

4.8.4   Risk Management

The risk management process involves the use of all available data, including the results of risk assessments, to take action to control risk levels, and finds relevance in both the regulatory and the corporate setting. Inputs in addition to risk assessments include political, economic, and social requirements and impacts. In the regulatory setting, risk management is directed toward public interests, and activities include research and monitoring, legislation, the establishment of requirements and guidelines, and regulation and enforcement. In the corporate setting, risk management is focused on compliance to regulatory requirements and overall corporate health.

The primary government agency with responsibility for water quality is the Environmental Protection Agency. Other agencies, which may from time to time be concerned with risk and water quality, include the Department of Transportation, the Department of Energy, and the Federal Emergency Management Agency. Research and monitoring activities include environmental monitoring, experimentation, health surveillance, testing and screening, and modeling. Legislative controls often develop from individual or site-specific cases to broad control. Currently, there are legislative controls to address drinking water quality, surface water quality and groundwater quality. Legislative goals are achieved through the establishment and enforcement of standards, which dictate the limiting concentration of specific substances in water.

For corporations, risks are managed through the use of policies, which is in many ways similar to regulatory risk management, and through the implementation of case specific remedies. For each problem, the management sequence typically involves the development and screening of alternatives, the selection and implementation of a remedy, and subsequent monitoring and review. A systematic approach to developing alternatives is invaluable. Following the sequence of risk generation, possibilities include the elimination or reduction of risk sources, controlling release into environmental pathways, preventative measures for exposed populations, and limiting access to exposed populations. Once a list of alternatives is developed, selection of an alternative involves rating their absolute or relative effectiveness. Points of view include economic and legal requirements and benefits, as well as management and operational complications associated with implementation. After implementation, monitoring the effectiveness or progress of the remedy is sometimes required, and when not, is still often invaluable for effective management.

WATER QUALITY MANAGEMENT

CHAPTER 5
ENVIRONMENTAL MANAGEMENT

5.1 INTRODUCTION

Environmental Management is a relatively young discipline having origingated in the last 20 to 30 years. Awareness of the environment has long been with us in a general way, but only in the last 20 to 30 years have risks been defined and subsequently quantified, as people have become aware of the environmental consequences of discharges to the air, land, and water. These risks were then addressed through federal, state, and local regulations.

National emphasis and focus on Environmental Management began with regulations including the Clean Air Act (1970), the Clean Water Act (1972), the Resource Conservation and Recovery Act (1976), the Toxic Substance Control Act (1978), the Comprehensive Environmental Response, Compensation and Liability Act (1980), and the Emergency Planning and Right-to-Know Act (1986). The regulations are single media, are written in a command and control mode for permitting, monitoring, inspection (auditing), and enforcement, and require regulatory compliance.

Management of these risks became known as Environmental Management (EM). Typically, crisis management was the mode adopted by industries and municipalities, as federal regulations could often be unclear and often overlap. The addition of state and local regulations added to the confusion, often causing an organization to respond to each regulation in crisis mode rather than view environmental risks from a long range, all inclusive, multimedia approach. Responsibility for EM usually fell to the engineering department or other technical discipline. EM sometimes evolved through such functions as health and safety and maintenance. Today, EM is either a stand-alone function, or may be combined with other functions depending on the complexity of the risks associated with the process.

In the beginning, Pollution Control was typically the treatment of a discharge after the process (generally known as "end of pipe" treatment). Today, EM's focus to control pollution has progressed to a more proactive approach such as:

- Elimination or reduction of contaminants at the source.

- Increasing process operating efficiencies.

- Recycling and reusing both water and waterborne contaminants.

- Treatment through the use of pollution control devices.

Today, the cost of waste treatment, storage, disposal, possible enforcement penalties, and the goodwill of the public, banks, investors, and others, drive the need for a cost effective Environmental Management System (EMS) that addresses these needs.

5.1.1 The Future of Environmental Management

We can reasonably surmise that EM is, or will be, an integral part of every waste generating enterprise—whether industrial, municipal, or commercial—and as such does, or should, receive oversight from the Board of Directors, Commissioners, etc. We can also reasonably expect that current regulations will evolve with more stringent compliance limits on currently regulated substances and grow with the listing of additional substances as we learn their effect on the environment.

Thus, an EMS should be professional, proactive in its approach, and require a major cost center to effectively prevent problems and control costs related to environmental compliance. Environmental Management is analogous to Quality Management in that prevention of crises and their potential costs are not easily quantified and, therefore, do not lend themselves to simple payback analysis. However, even though a payback is difficult to prove unless there is a noncompliance event, effective quality control and assurance systems continue to make a very positive impact on the operations of an organization.

The structure of an EMS has matured to incorporate standard business concepts and practices similar to quality control and assurance systems such as:

- Develop environmental policy and commit to that policy.
- Determine how the organization presently compares to that policy.
- Set objectives and targets to meet the policy.
- Implement a program to achieve these objectives.
- Monitor and measure its effectiveness to meet the objectives, and take corrective action.
- Review the EMS periodically to improve it and, thereby, improve overall environmental performance.

The goal for any organization structure is continuous improvement, and an EMS designed with the above concepts will achieve continuous improvement. It should be emphasized that management commitment is the foundation of an EMS and must be firm and unwavering. Less than whole-hearted support will reduce the cost effectiveness of an EMS.

As Environmental Management operates under this EMS structure, we can anticipate a reduction in environmental releases to the environment, fewer noncompliance problems, interaction with regulatory authorities moving toward multimedia inspections rather that the current single media, regulatory command and control procedures becoming less adversarial and more user friendly, and marketplace incentives that encourage environmental excellence. EM will continue to become more strategic in an organization's structure and in its marketing approach to gain market share. EM will become another positive influence within the organization rather than a forced response to imposed regulations.

5.1.2 Relationship between Environmental and Quality Management

Quality Management (QM) is customer driven and deals with customer satisfaction and service; therefore, generally speaking EM and QM focus on different stakeholders. However, within the organization, environment and quality have much in common. Both are the responsibility of every employee within an organization regardless of the function assigned. Both are designed to prevent potential problems, both tangible and intangible, that have unknown and potentially serious cost effects for the organization. Management systems for both are similar in content and structure. Each will have a strategy to prevent the unknown and a tactical response to events outside the defined norm. Both are major cost preventors through continuous improvement of operations and operating efficiencies. Also, both EM and QM have the potential to gain competitive advantages in the marketplace for the business. Both standards focus not on performance, but on how that performance is documented, evaluated, and improved. These similarities may one day bring both environmental and quality functions under common management within the organization.

However, at the present time environmental management and quality management function separately in most firms. There are many quality control and assurance programs available depending on the needs of the process, and Total Quality Management (TQM) has evolved as the more widely used concept for quality management. In more recent years, an evolution of TQM is the international standard ISO 9000 which has been accepted worldwide for its common structure from country to country and its third party registration and periodic certification. ISO 9000 and its variations enhance a firm's acceptability in international trade. It is also becoming accepted as an American national standard as international firms require their suppliers to become registered. Familiarity with the guidelines, third party registration and periodic certifications give trading partners added confidence beyond other quality management systems that an organization is truly committed to practicing quality management. Environmental Management is following a similar structure with the introduction of the international standard ISO 14001—Guidelines for an Environmental Management System (EMS).

5.2   ENVIRONMENTAL MANAGEMENT SYSTEM—ISO 14000

The International Organization for Standardization (ISO) was founded in 1946 to develop manufacturing, trade, and communication standards. The American National Standards Institute (ANSI) is the United States member body to ISO. The goals of ISO standards are to facilitate the efficient exchange of goods and services. These goals apply both within national boundaries and internationally. In September 1996, ANSI approved ISO 14001—Environmental Management Systems—Specification with Guidance for Use. It is a voluntary system and the only one of the ISO 14000 series that requires third party authentication. ISO 14001 provides a means for environmental management to become an integral part of an organization's operations. It provides a framework to address:

External Considerations

1. Implementing strategic environmental management.

2. Meeting a world-wide accepted standard for environmental management.

3. Meeting the goals of governments world-wide for enforcement procedures as well as their procurement policies.

4. Satisfying the expectations of stakeholders, including investors, banks, the public, and environmental groups.

Internal Considerations

1. Provides a system that systematically controls the environmental impacts of the process and integrates these controls into all operations of the business system.

2. Provides a system that analyzes the causes of noncompliance and builds prevention and continuous improvement into the organization's overall operations.

3. Creates through training and group actions an awareness of and responsibility for environmental concerns by both line management and employees.

4. Has the potential to reduce costs of storage, disposal, and excess treatment. May also reduce the cost of materials and labor, and has the potential to lessen legal liability and, perhaps, insurance costs.

The ISO 14001 Environmental Management System is well defined and functional. It is consistent in structure so that it can be accepted nationally and internationally, and it provides a framework for companies of all sizes to reduce costs. It requires considerable documentation; however, this can be very advantageous. For instance, as personnel change position or as an employee leaves an organization, the ISO 14001 stays in place and is easily continued without disruption. It is subject to third-party registration and periodic third-party audits, assuring stakeholders that a viable system remains in place. It is a proactive strategy that represents good business practice for any organization.

The ISO 14001 EMS is one of many excellent environmental management systems in place around the world, but it is the only EMS developed for world-wide use by over 100 countries participating in its final guidelines. Further, we can reasonably anticipate its widespread use by organizations operating within national boundaries.

5.3   ENVIRONMENTAL AUDITS

Industry management teams and environmental managers are finding it a very challenging and daunting task to ensure that their facilities are in compliance with today's environmental laws and regulations. This is a complex task that is due not only to the enormous volume of applicable laws and regulations, but also due to the fact that these laws and regulations are growing at a exponential rate with no signs of decline on the horizon.

Industry is finding that the cost of compliance is becoming a significant expense and that the cost of noncompliance can be enormous and is an unacceptable alternative. Heavy fines for noncompliance and civil and criminal liabilities are on the increase. The industrial management teams must make every reasonable effort to ensure that their environmental programs are in place and effective.

5.3.1   Types of Audits

The audit is one of the primary tools to measure how well a facility is complying with current environmental regulations. Today the management team is seeing three basic types of audits: internal, regulatory and customer audits. Internal audits are conducted by in-house personnel or by outside consultants to determine the compliance status of the facilities. The results of the internal audit are used to improve the environmental performance of the facility, to make improvements and modifications to the current environmental programs, to ensure continued compliance with all rules and regulations, to reduce the amount of risk imposed on the facility due to noncompliance, and finally are used as a tool to prepare for scheduled or unscheduled regulatory or other outside audits. The regulatory audit is conducted by a regulatory agency (State, EPA, or local agencies) to determine the compliance status of the facilities. These regulatory audits can be limited to a single area such as air emissions, water or hazardous waste disposal, or they can be multimedia audits covering the entire environmental spectrum.

Unsatisfactory results from the regulatory audit can be very unpleasant, ranging from a simple Notice of Violations to heavy fines and even civil and criminal litigation. Customer audits are becoming more popular as the consumer population becomes aware of the environmental impact of their purchasing decisions. As a result, manufacturers are starting to audit their suppliers environmental programs so that they can demonstrate to the consumer that their products are as environmentally friendly as possible. Unsatisfactory results from the customer audit can be just as unpleasant as the regulatory audit, resulting in lost sales and reduced revenue, an outcome that most management teams can ill afford.

The purpose of this section is to help the management team and environmental manager successfully prepare for regulatory and customer audits in order to prevent or limit any unpleasant results. As with most situations, good preparation is a prerequisite for success. The audits are not something to be left to chance. The efforts put into preparation for the audit and how the audit is handled can make a very positive impact on the results of the audit. This section will also help the management team and environmental manager to develop an Internal Audit Program. The Internal Audit Program is the best tool available to ensure that the existing environmental programs are adequate and that the facilities are in compliance with existing rules and regulations.

5.3.1.1 Regulatory Audits.   Preparations are the key to any successful audit. These preparations must be an ongoing process that is integrated into the daily operations, and not a last minute act of desperation once it is discovered that an audit has been scheduled. In most cases, it is first learned that the regulatory audit is to be conducted is when a call is received from the receptionist that someone from the State Regulatory Agency or USEPA is in the front lobby and wants to see the responsible party. Even if one is fortunate enough to receive prior warning, it will most likely provide only a few days in which to prepare. In today's fast paced work environment, the luxury of being able to lay everything aside in order to concentrate on getting ready for an audit does not exist.

Environmental Compliance is not one person's responsibility. It must be stressed that Environmental Compliance is everyone's responsibility, from the receptionist to the CEO. In preparing for the audit, everyone must understand their responsibilities and the impact their job has on the success of their organizations environmental programs. It takes a team effort to have a truly successful environmental program and it takes a team effort to have a successful audit.

Preparations for a successful audit should start with the receptionist. A written operating procedure should be prepared which provides detailed instructions on how the auditors are to be received and handled. The receptionist should be instructed to be friendly and helpful. The receptionist should be informed as to which personnel should be contacted within the organization. If the primary contact is out of the office, there should be a secondary contact or even a third contact. Since the front lobby is generally the first thing that the auditor will see, it is extremely important that the receptionist give the auditor a very good first impression. You want the auditors to see an organization that is very professional and willing to help.

Documentation is probably the single most important issue of any audit. How the environmental records are kept and organized can often be a direct reflection on the manner in which the environmental programs are maintained. If it is perceived that the records are kept in a lax manner, it may very well be assumed that the environmental programs are also kept in a lax manner. Maintaining the environmental records is an ongoing process that cannot be addressed on an occasional basis or when it is discovered that the auditor has finally arrived for your annual compliance audit. If responsible people have to hunt and search for various permits and records and they are haphazardly filed in different locations or misplaced, it can give the auditor a very bad impression. Poor documentation and record archives can also result in possible Notice of Violations or worse if proof of compliance cannot be provided.

In preparing for an audit, all pertinent records should be maintained in a professional manner and stored in an easily accessible area. They should be well documented, labeled and filed in a logical manner such that the documents are easily retrieved. All the documents or copies of documents should be stored in a central location that is easily accessible during the audit. It is extremely important that during the audit, information that the regulator may request such as permit information, background information used in the preparation of the permit applications, current and past production records, and any pertinent correspondence concerning your environmental programs, can be easily produced.

Good documentation, which is easily retrieved and comprehensive, will often make a very positive and professional impression on the auditor. If the auditor can be shown that the environmental programs are well documented and professionally run, the actual inspection will often not be as intensive as it could be. On the other hand, if during the documentation phase there is a problem in finding the documentation that the regulator has asked for, or the information cannot be produced in a timely manner, or if the records are kept in a haphazard fashion, then the auditor's impression of the environmental programs may be more negative than it should be and may result in a more aggressive physical inspection.

The environmental files should be kept in a dedicated filing cabinet or in a set of well marked three-ring looseleaf binders. The filing cabinet or binders should be kept in an area that is accessible to everyone that may have a need to review the files. Although access to these files is very important, security of these files should also be considered.

In addition to the environmental files, there should be an accurate drawing of the facility layout that shows the location of the various emissions sources, stack locations, storm water outfalls, and wastewater discharge points. The purpose of this layout is to provide an accurate location of the various discharge or emission points during the audit and to show which sources they are related to. It is also a very good idea to label the various discharge points with a unique source identification number so that there is no confusion during the audit. It can be very embarrassing during the audit when the emission point cannot be identified or the identities become confused. More importantly, the inability to properly identify the emission source correctly may damage the credibility and the professional image that is trying to be presented to the auditor. If there is uncertainty about this aspect of the environmental program, in the eyes of the auditor, there now may be questions about the other areas, which may prompt closer investigation.

Although the environmental manager may know exactly what each emission source is for and can find it with his or her eyes closed, the need for an accurate and updated site plan with source identification is still important. The audit may take place when the environmental manager is not available and a less experienced member of the team will be required to escort the auditor. The site map should allow the less experienced team member to escort the audit team and accurately find and describe the various emission sources in a professional manner.

Another very important reason for maintaining an accurate and updated site plan with emission source locations is that it provides the management team with an extremely helpful tool in managing the environmental programs. When there is an operational upset, the site map can help the team members quickly determine the cause of the emissions problem by properly identifying the source. The discharges or emissions can then be brought into compliance either by correcting the problem, isolating the source, or by shutting the equipment down until the problem can be identified and corrected.

In older plants or in industrial plants with various or complex processes which change frequently, there are always pipes and potential discharges into the wastewater streams which are difficult to identify or that are no longer in service. It is extremely important that these potential sources are identified. Often old engineering drawings can be of use in identifying these sources. Dye tests may also be a tool to identify where various wastewater streams discharge. Pipes that have discharge openings into wastewater streams that are no longer in service should be removed or capped to prevent any unauthorized discharges. It is also important to identify all sources of process wastewater to insure that they are being properly permitted and treated as necessary to allow discharge to a publicly owned treatment works (POTW). Boiler and air compressor blowdown are often overlooked as sources of process water discharges. All process wastewaters must discharge to a POTW or have the appropriate discharge permit.

Production records are vital to maintaining compliance with today's operating permits. Most operating permits today specify that a log of process inputs must be kept and maintained for a specified number of years in order to quantify the amount of emissions that a source has generated. The source is considered non-compliant if these documents cannot be located. Although it is not required that these records be kept in the same location as the environmental files, it is recommended that a hard copy be maintained with the Environmental Files. By maintaining a copy of the production records, you can verify on a regular basis that the necessary documentation is being kept. This will eliminate any surprises during the audit. Another benefit in maintaining a copy of production records with the environmental records, is that during the audit, the necessary records can be rapidly found and compliance proven. It can be extremely important during the audit not to have to expend a lot of time hunting people down and searching for the necessary documents.

Maintenance records are also a vital component in today's operating permits. Environmental control equipment must be maintained in good operating condition if it is to be expected to efficiently control emissions and meet the permit specification. Repairs and routine maintenance checks must be documented and these records maintained for a specified number of years. As with the production records, these files are not required to be stored with the environmental files. However, for the same reasons given above for the production records, it is recommended that a hard copy be kept with the environmental files.

It cannot be over-emphasized, how important it is to maintain all the environmental files and supporting documentation in a neat and well organized manner. All files should be clearly labeled so that the desired file can be easily located. The entire management team should be familiar enough with the files so that they can locate specific files during the absence of the environmental manager.

Plant cleanliness and safety are two issues that go hand in hand with the environmental program. As mentioned earlier, first impressions are very important for a successful audit. If a plant is slovenly kept and there are obvious safety problems, the auditor may find that the same attitude has carried over to the environmental programs. This is another area that requires complete teamwork and is an ongoing process. There will not be enough time or money to do a last minute cleanup prior to the audit. The facilities must be cleaned on a daily basis. All spills must be cleaned up immediately and not allowed to accumulate. Waste management and hazardous waste management programs are not within the scope of this book. However, it is extremely important that none of the other environmental programs are overlooked, especially with the increase of multimedia inspections and audits.

It is in the best interest of the facility being audited to make the process as easy on the auditor as possible. On a regular basis, the environmental coordinator should conduct a self-audit. Start by reviewing all current permits. Check that all permits are current and do not require renewal. Review all permit conditions and determine how to prove compliance. Then assemble the necessary documentation and ensure that it is complete and that the required number of years are archived. If calculations are required, be prepared to reproduce those calculations and be able to show the appropriate references. Be able to prove compliance in a very confident and professional manner.

It is important to be prepared to help the auditor understand the process. Provide simple process flow diagrams that will provide the auditor with a basic understanding of each emission source.

Show how the raw and final process materials are stored and how they are transported and handled, typical process rates, description of emissions with typical emission levels, and the type and location of all emission control equipment. Keep the process description as simple as possible, yet with enough information that the auditor can grasp the basic understanding of the processes and discharge sources. During the audit, should the auditor have questions on the process, ensure that his questions are answered quickly and accurately. If the questions can't be answered immediately, tell the auditor that you don't know but will find out. Then, ask another team member to get an answer to the auditor's question as soon as possible, preferably prior to his completion of the audit.

The last issue in making the audit as easy as possible for the auditor is to assemble all the necessary documentation that was discussed above. When the auditor asks for proof of compliance, it should be easy to find the documents that are needed without having to hunt for someone to find the right files or hunt for it in someone's office.

Again, answer all of the auditor's questions as quickly as possible in an accurate and professional manner. The goal is to present an image of a very professional organization that is concerned with maintaining their environmental programs in a very effective manner. Answer the auditor's questions quickly, help him or her understand the process, and demonstrate compliance in a very confident and profession manner.

The regulatory audit is a fact of life. It will happen, and most likely, at a time that is most inconvenient. The audit should be anticipated and prepared for at all times. As mentioned previously, the audit cannot be prepared for once it is announced by the receptionist that the auditor is waiting in the lobby.

A post audit interview to review the preliminary finding should be requested of the auditor at the completion of the inspection. During this brief meeting, the following items should be accomplished:

• Determine the auditor's initial impression of your facility's compliance status.
• Determine if there are any misconceptions concerning your process and operations and correct them.
• Does the auditor require any additional information to complete the audit?
• If additional data is required, determine the time frame for supplying the information.
• Are there any situations that need immediate attention?
• Were any major discrepancies uncovered during audit?
• Were any minor discrepancies uncovered during audit?
• Request a copy of the final audit report.

If possible, the facility's management team should be invited to attend the post audit interview. This will allow the management team to participate in the audit process and to have a basic understanding of the preliminary findings, thereby eliminating any shock from the final audit report. Since the success of the environmental program is a team effort, the team leaders need to understand their responsibilities and their impact on the outcome of the audit and where they can help to make improvements.

Depending on the outcome of the regulatory audit, an official notification of the audit results may or may not be received. If no major or minor problems were found and a final audit report was not requested, no formal notification may be received, depending on the State or agency conducting the audit. However, if problems were found, then a Notice of Violation or other official report will most likely be received. Upon receipt of the audit report, it is imperative to communicate these findings to the appropriate people.

The findings should be reviewed as soon as possible to determine their validity. Was the plant actually out of compliance or was there possibly a misinterpretation of the data or misconception of the process by the auditor. If the Notice of Violation or other findings are believed to be in error, then a letter to the auditing agency should be drafted which describes in detail why it is believed that the facility was actually in compliance.

If the plant is actually in violation of regulations or its permits, then an "Action Plan" must be developed with a reasonable timetable to accomplish the work that will bring the facility back into compliance. The auditing and/or enforcement agency should be notified of the proposed Action Plan and their approval of these plans requested. Some negotiations with the regulatory agency may be required in the development of this Action Plan and the timetable for the completion of the work to ensure compliance with all regulations and permits. In addition to the Notice of Violations, the facility can receive a fine. Keep this in mind while developing the Action Plan for possible use in negotiating a reduction in the amount of the fine or elimination of the fine. Although the potential for a reduction of the fine may not be very large, it is to the plant's advantage to pursue this option.

The Action Plan must be implemented as soon as it has been accepted by the regulatory agency. The established timetable must be adhered to and routine update reports on the progress of work to bring the facility into compliance must be made to the regulatory agency. Any problems in meeting the Action Plan should be dealt with immediately and the regulatory agency informed of any changes or problems.

In addition to establishing the above Action Plan for the correction of existing problems, the existing Management Plans should be reviewed to determine why these problems were allowed to exist and what actions are needed to prevent similar problems from occurring in the future. Updated management plans should be implemented and routine updates provided to the management team on the status of these plans.

The regulatory audit, its findings, and resulting corrective actions should be well documented for future reference. These documents should include at a minimum:

- Preliminary findings
- Immediate corrective actions
- Official inspection report with findings
- Action plans, if required
- Management plans, if required
- Capital expenditure summary
- Results of any post inspection negotiations and agreements
- Completion date of action items in plan

5.3.1.2 Customer Audits.   Customer audits should be treated the same as regulatory audits. The preparations for the customer audit should be no different those for the regulatory audit. The relationship with the customer audit team will possibly be a little more relaxed than what would be expected with the regulatory audit, however, the seriousness and importance of this audit cannot be overstressed.

As with the regulatory audit, the primary goal of the management team and environmental manager is to ensure that the customer audit team leaves the site with the perception that this is a very professional organization that is very serious about its environmental programs. Good preparations are mandatory to ensure a successful customer audit.

5.3.1.3 Internal Audits.   As mentioned earlier, one of the best things that can be done to improve the effectiveness of the Environmental Programs is to implement an internal audit program. The internal audit does several things to assist in the management of these programs. First, it is an early warning system that can point to potential problems before they can become major noncompliance issues. It can identify developing trends that may need to be corrected. The internal audit should focus attention on the importance of the Environmental Programs and how good teamwork is critical for its success. Internal audits can help in the preparation for an audit by a regulatory agency or others outside the immediate organization and provide time to correct any deficiencies. Lastly, an internal audit program can be an extremely effective training tool for your management and work teams.

The first step in the development of the internal audit program is to obtain the commitment from management that they are willing to provide the necessary resources for this program. For the program to be successful, they must be willing to supply the necessary manpower for the audit team and provide for the expenses that may be incurred during the audit process. It should be understood that this is a long-term commitment and not a one-time event. They must have a thorough understanding of why this program is needed and its expectations and goals. Finally, they must be prepared to act on any findings that are uncovered during the audit.

Once management has accepted that there is a need for the internal audit and have agreed to commit the necessary resources in manpower and expenses, then the next step is to develop an *internal audit plan*. This plan should contain the following:

Internal Audit Plan

- Program objectives
- Scope of the audit
- Audit timetable
- Makeup of audit team
- Audit protocols or checklists
- Audit report format
- Procedures for report distribution

The audit plan should be very clear about the specific goals and objectives of the internal audit and what the expectations are from this program. It should be stressed that this is a tool that is to be used to help improve the performance of the organizations environmental programs and is not to be used as a witch hunt or in a punitive manner. One of the primary goals of this program should be to develop the perception of the workforce that the company is very serious about improving its environmental performance and that its input and cooperation is both desired and appreciated and that it can provide information without any fear of reprisals. If the function of the audit program becomes one of finding fault with individuals instead of identifying deficiencies in the program, then the program will rapidly lose the willing cooperation of the workforce which is so critical for the success of this program.

The *scope of the audit* should be specified, establishing what facilities will be included and which programs will be reviewed. This will become extremely important as the audit team begins its preparations for the inspection.

An audit timetable should be determined and promulgated. Once the timetable has been established, it should be adhered to as much as possible. In establishing the timetable, the following items should be included as a minimum:

Internal Audit Schedule

- Establish audit date
- Send out preliminary questionnaire to facility management with required return date
- Audit preparations based on above questionnaire
- Conduct audit
- Preliminary audit report sent to facility management for review
- Final audit report
- Due date for action plan to address non-compliant areas

The audit date should be established in accordance with the frequency of your internal audit program. Once the date has been agreed to, it should not be changed without good reason. The internal audit should be treated as if it were being conducted by an outside agency. If the date is allowed to fluctuate without good reason, the workforce may perceive that the internal audit is of little value or importance.

Once the date has been selected and promulgated, the audit team should develop a *preliminary audit questionnaire* which is then forwarded to the facility management team. The management team should be given a reasonable amount of time to respond to the questionnaire. However, it must be understood by all parties that there is a deadline that must be met and that the questionnaire must be returned by the specified date. The questionnaire should request basic information about the facility so that the audit team can make the necessary preparation for the audit. The following items should be included in the questionnaire as a minimum:

Preliminary Audit Questionnaire
- Facility identification information—name, address, telephone, etc.
- Facility description—area under roof, number of buildings, provide site plan if possible
- Description of area surrounding the facility, population characteristics, nearest residential area, schools, etc.
- Management team—names, organization structure, etc.
- Technical contact
- Brief description of process or processes
- Brief descriptions of air emissions
  - Number of permits
  - Summary of emissions, type of controls, type and quantity of emissions
- Brief description of solid waste stream
  - Type of solid wastes
  - Volume of wastes
  - Method of disposal
  - Name and quantity of hazardous wastes
  - Method of disposal of all hazardous wastes
- Wastewater discharge requirements
  - NPDES permit
  - Description of all wastewater discharges
  - Description of treatment system
  - Description of test methods and required reports
- Stormwater Pollution Prevention Program
  - Is SWPPP Plan up-to-date?
  - NPDES Permit Number and expiration date
- Number of Underground Storage Tanks, description of UST
- Potable water source
- SPCC Plan up-to-date and certified
- Number and description of all spills
- List of all hazardous materials stored on site
- Has the facility received any regulatory notices, violations, or citations
- Are there any PCB or PCB-contaminated equipment or materials on-site now or at anytime in the past
- Has an asbestos survey been conducted at the facility?
- List of all documents to be made available for the audit; permits, permit applications, correspondence, air monitoring reports, water sampling reports, site plans, MSDS, etc.

A brief letter should be attached to the *pre-audit questionnaire,* which will explain how the audit will be conducted and what is expected of the facility being audited. The letter should define who will be conducting the audit, when it will take place, approximately how long the audit will take, what will happen at the conclusion of the audit, and who should be contacted if there are any questions. The letter should explain that the purpose of the pre-audit questionnaire is to help prepare the auditors so that the audit can be conducted in such a manner as to minimize the time on site.

Based on the results of the pre-audit questionnaire, the auditors should develop a set of protocols or checklists which will cover all appropriate federal, state and local regulations that apply to the site. If desired, the protocols may be purchased from an outside source. Regardless of whether the protocols are developed in-house or purchased from an outside source, they must cover all areas of concern. The protocols should adequately cover the various requirements and should list the references so that the auditor can quickly look up the specific regulatory requirements.

Once the protocols have been developed, the audit team should assign the protocol responsibilities so that the individual audit team members can begin to prepare for the audit. The specific scope of the audit should be verified with the entire audit team.

In preparation, the auditors should study their assigned protocols and should review all applicable regulations. Although the protocols should provide a good background on the regulatory requirements, the actual regulations should be reviewed in their entirety prior to the audit. Based on the above research, the auditors should determine the documentation required and develop a list of things to see and questions requiring answers: in short, how he or she will conduct the audit. Good auditors will have their homework completed prior to the day the audit is scheduled so that little time is wasted. Because of the resources in manpower that will be committed to this audit process, it is extremely important that all preparations be completed well in advance of the scheduled date.

On the scheduled date of the audit, a *pre-audit meeting* should be scheduled for all involved personnel. If members of the audit team are from other locations, then the audit team leader should take time to introduce each member of the team. On-site personnel who are responsible for the various areas should now be assigned to assist the audit team members during the audit. The facilities management team should be present at the meeting and it should be prepared to provide its support to the audit process and the audit team. At this time, the audit team leader should go over the proposed scope of the audit and should outline the various needs of the audit team. The audit schedule should be discussed and any conflicts with the production schedule resolved to everyone's mutual satisfaction. Everyone should understand the proposed audit process, what will be expected of them, and a time and place should be established for conducting the post audit meeting.

During the pre-audit meeting, time should be set aside to discuss all pertinent safety issues and concerns that the audit team members need to be aware of during their inspections. Personnel protection equipment should be provided as necessary. If there are any restricted access area, these should be identified and discussed. It may be necessary to schedule an alternate date to review these areas when a shutdown has been scheduled and access does not pose any dangers.

Time should also be taken to answer any questions from either the audit team or from those whose facilities are being audited. Everyone should have a clear picture of what is happening, what the audit will hopefully achieve, and what is expected of each person to ensure that the results of the audit are meaningful and worth the time and effort being expended.

At the completion of the pre-audit meeting, the audit team should immediately start conducting the audit with the assistance of those that have been assigned to escort them. The first step that the auditor will want to take is to conduct a thorough review of all pertinent documentation such as permits, permit applications, production records, training records as applicable, correspondence with regulatory agencies, and other such resources. The documentation should be well-organized and complete without gaps in time or areas. Should such a gap be found, a complete and detailed explanation should be requested. The documentation will often point to areas of weakness or areas of concern that the auditor will want to look at in more detail. Often the operating permits will require that specific records be kept for a specified length of time. The auditor should verify that all required records are readily available and that they cover the required number of years. The records should be reviewed in detail to determine if there are periods of time that exceed the permit limitations. All such discrepancies should be recorded in detail.

Once the documentation has been thoroughly reviewed, the auditor will then conduct an on-site inspection of all areas assigned. During this inspection, the auditor will want to pay close attention to the areas of concern that were found in the documentation review. The auditor should determine how the production records are maintained and what input the operators have in the production logs. The operators should be interviewed to determine their knowledge and understanding of all pertinent operating permit requirements in their assigned areas. Do they understand what is required and what is expected of them and are they satisfactorily meeting these obligations. If not, the auditor should probe to determine reasons why the requirements or regulations are not being met.

As was mentioned earlier, it is extremely important that everyone understand that the purpose of the internal audit is not to find fault with individuals, but to identify problems and shortcomings with the environmental programs. This must be emphasized time and time again if this program is to succeed.

Following the on-site inspection, the auditor should find a quiet space to review his or her notes and findings. Any findings, which are suspected to be in violation of current regulations, should be reviewed and confirmed with the applicable Federal, State, or local regulations. The findings should be very detailed and the applicable source identified for future reference. It may be necessary at this point to go back and review some of the documentation, or to go back into the facilities to confirm the details of the findings. When the audit team has completed their individual assignments, the team leader will then assemble all the auditors in order to review their work and to start putting together the rough draft.

It is important to complete the rough draft while the team is still on-site because the findings are still fresh in the auditor's mind. Once the auditor leaves the site or after several days have passed, the level of detail that the auditor can recall becomes very limited. Also, completion of the rough draft at the conclusion of the audit allows the audit team to go back into the field to confirm any questionable details. But probably the most important reason is that it gets the work completed. Once the auditors are allowed to go back to their regular duties, the demands of their jobs will naturally delay the completion of the audit. Once the rough draft is complete, the audit team should prepare for the *exit interview*.

The team leader should prepare a summary of all findings in the order of major findings followed by the minor findings. The facility's management team and all other responsible personnel should be invited to the closing conference. The team leader should briefly reintroduce the audit team and which areas they covered. The goals, objectives and scope of the audit should again be reviewed to ensure that there is as complete an understanding throughout the organization as possible. The team leader should then provide an overall summary of the audit, emphasizing all areas that were found to be commendable and any significant deficiencies. Then each individual area should be covered and all findings should be reviewed. The audit team should ensure that all findings are based on the receipt of correct and complete information. Questions concerning the findings should be addressed and satisfactorily answered if at all possible. All immediate health or safety problems to be corrected prior to the publication of the final report should be identified at this time. At the conclusion of the meeting, the facility management team should be given a copy of the findings or a copy of the rough draft if at all possible. This will allow the facility to start working on an action plan which will address the various findings of the audit.

The final report should be completed as soon as possible after the audit and sent to the facilities manager for his review and comment. A time limit for his or her comments should be established and adhered to as much as possible. Distribution of the final report should be limited to the facility manager or distributed in accordance with the corporate policies. Other distribution should be considered very carefully due to the legal ramification.

The facility management team should now establish a comprehensive *action plan* with a reasonable and achievable timetable for addressing all discrepancies found during the audit. The action plan and status of all outstanding discrepancies should be reviewed on a regular basis with all appropriate personnel. All management plans should be reviewed and corrected to ensure that existing discrepancies are not reoccurring problems in the future. The final audit report, action plan and associated documentation should be filed for future reference.

5.4   WATER CONSERVATION AND REUSE

Water conservation and reuse are rapidly becoming a necessity for industry and water utilities. *Water conservation* can be defined as actions taken to minimize the amount of water used to accomplish a specific task. *Water reuse* is recycling water through a given process or secondary processes in place of consuming clean unused water.

Industry and water utilities are being forced to consider these options for many reasons. The primary reasons being the increased competition for clean water due to declining water tables, reduced sources of clean waters, and increased demands from both industry and residential growth, all resulting in higher costs for this natural resource.

As the use of water has increased, loading on pretreatment plants, on-site waste treatment facilities, and POTWs has also increased dramatically, resulting in higher operating costs. These wastewater treatment facilities are often operating at their maximum capacity, which can severely limit their capacity to meet discharge permit limits. Industries are often required to pay significant penalties, in the way of surcharges, to their local utilities because their pretreatment plants cannot meet their discharge limits due to high hydraulic loading of the systems. We can also expect that regulatory requirements on discharge permits for individual generators and POTWs will only become more stringent about placing more pressure on the generators to meet their permit limits.

Water conservation and reuse can have tremendous benefits through decreased costs of purchased water and reduced costs for treatment of wastewaters. Prevention of discharge violations as a result of overloaded systems can be a significant inducement for water conservation and reuse. Other savings that can be accomplished through these program are the maximum utilization of existing equipment and the prevention or postponement of constructing new and larger wastewater treatment facilities. The cost of capital for design and construction of these facilities can be very expensive. By implementing water conservation and reuse programs, the decision to expand the treatment facilities can be placed on hold, and the available funds can then be used for expansion or improvements to process equipment.

Another reason for implementing water conservation and reuse programs is the growing public demand for conservation of our natural resources. In order to be a "good neighbor," it will become increasingly important for industries to conserve their use of clean waters and to do their part in reducing the loading on the local POTW.

The first step in developing a water conservation and reuse program is to conduct a site use survey to determine where and how waters are being used. It would be extremely helpful to develop a spreadsheet and/or diagram of the water usage with specific details as shown below:

1. Location and quantity of water usage

2. Temperature requirements

3. Water quality requirements, i.e. pH, hardness and limitations on solid content, must meet clean water standards, etc.

4. Any special process requirements

Once water use has been quantified, the various use points should be prioritized according to cost and how easy it will be to modify the process to minimize water use. When looking for ways to minimize water use, utmost care should be taken such that process quality is not jeopardized. It may be necessary to conduct pilot trials before committing large sums of capital to modifying the process to insure that the process quality is not reduced. In looking for ways to reduce water usage, the following areas are suggested for initial investigation.

Pumps. Facilities with a large number of pumps can consume a significant amount of water for packing cooling and flushing water. Flow monitors should be installed with manual control values to insure that the proper amount of water is being used to protect the equipment, as well as to prevent the use of too much water. Another consideration should be to install solenoid valves that will automatically turn the water on and off with the pump. This again is protection for the equipment to prevent it from running dry and causing damage as well as wasting water when the pump is no longer in use.

Wash Water. Many processes require a fresh water rinse or spray wash to remove unwanted contaminates. Counterflow rinse/washes should be considered to reduce the overall use of fresh water.

Air knives can often be utilized to prevent carry-over of a more contaminated rinse system to a lessor contaminated rinse system, reducing the need for additional makeup water in the rinse system. Low-pressure high volume spray nozzles should be replaced with high pressure, low volume spray nozzles. Solenoid valves should also be considered to automatically shutoff off spray nozzles when not needed.

Blowdown and Makeup Waters. Blowdown and makeup waters are often required to reduce the total suspended and dissolved solids that can build up in a process which can have a negative impact on the boilers, cooling towers, processes, etc. Conductivity meters with controls that will automatically blowdown the system at predetermined levels instead of a manual blowdown can reduce water use as well as chemical treatment requirements.

Heat and Cooling Systems.   Once-through water use such as boilers without condensate return systems and non-contact cooling systems should be eliminated where possible. Air conditioner condensation should be considered for reuse as makeup to cooling towers. Water treatment of boilers and cooling systems should be routinely evaluated to insure the best possible feedwater quality to protect the equipment, as well as to conserve water use through reduced blowdown and makeup waters.

Process Equipment.   Each process should be examined for areas where improvements can be made to eliminate or reduce spills or areas that need routine cleaning. These modifications will not only reduce the amount of water required for cleanup, but will reduce the overall manpower requirements for the process and will improve the safety and possibly improve the work environment.

Water Hoses.   All water hoses should be equipped with nozzles that will automatically close when not being used.

Water Conservation Team.   A water conservation team composed of maintenance, engineering and production personnel should be formed to routinely review water usage in order to identify new sources of water conservation and to evaluate current practices for areas of potential improvements.

Once water usage has been minimized as much as practical through water conservation, the next step should be to consider all possible water reuse points. It will be necessary to quantify the amount of wastewater being generated and the quality of that water. Each waste stream should be identified and evaluated as shown below:

1. Quantity of wastewater being generated

2. Source of wastewater

3. Temperature

4. pH

5. Types and range of contaminants

6. Unit operations available for reuse

Now that water usage requirements have been established along with the quality of water required, water reuse can now be evaluated. Using the spreadsheets and/or flow diagrams, match which waste streams can be reused in which processes. As with water conservation, it will be most helpful to establish a prioritized list by ease of modification, cost of implementation and potential for negative impact on product quality.

Some wastewaters may be able to be reused as is. Others may have to be treated to remove solids, adjust pH, etc. An engineering and economic analysis will have to be conducted to insure that the wastewaters will not impact the quality of the product being manufactured or potentially damage the process equipment and that the savings in water purchase and wastewater treatment can be justified.

Introducing treated wastewater does raise concerns about allowing impurities and contaminates into the processes. Inorganic and organic constituents may continue to build as water is reused until it reaches a point where product quality is affected. It may be necessary to run pilot trials to determine if and how water reuse will affect the quality of the product being made.

WATER POLLUTION
REGULATIONS

CHAPTER 6
REGULATORY STANDARDS

6.1 INTRODUCTION

Water pollution consists of organic or inorganic solids, liquids and/or gases dissolved or suspended in wastewater. These residue by-products of human existence may or may not be compatible with the environment into which they are discharged. Historically, this residue was emptied into nearby streams and dispersed downstream or dumped onto nearby land to be stored or percolate downgradient into the soil. With urbanization and industrialization, these historical means of disposal have become impractical since they negatively affect downstream or down-gradient human inhabitants or the environment. Consequently, civilizations have developed a series of rules, laws, and regulations to protect themselves and their environment.

6.2 HISTORY OF REGULATIONS

The evolution of water control regulations in developed nations has paralleled industrial growth. Rural or agricultural nations still exist with little or no water control regulation. Most efforts at control are similar to those of the United States, which will be used as an example in this Handbook. At the time of this publication, the European Union was developing water pollution control regulations to standardize those of the member nations.[1] Several other nations, notably some Pacific Rim nations, have adopted standards based on those of the USEPA.

The first concern of a developing nation, as was the case in the USA, is typically public health. Water-borne diseases such as amebic dysentery, which affects 50 percent of the population in some developing counties; cholera, which is 30 percent fatal; salmonellosis, which still kills millions of infants each year; and parasites such as schistosomoses, have led most civilizations to exact environmental legislation for survival. The author is personally familiar with villages in which virtually no one remembered a time in their life when they did not have diarrhea.

In the United States, one of the first national efforts at water pollution control was by the U.S. Public Health Service, a division of the Health, Education and Welfare Department, which enacted the Water Pollution Control Act in 1948 and the Water Quality Act of 1965. During this period, federal grants were given to municipalities to improve sewage treatment systems. Most universities taught and most engineers practiced, during that era, a cookbook approach to controlling water pollution. Systems primarily patented or developed by manufacturers were used almost exclusively to treat municipal wastewater, and little concern was given to industrial wastewater which either discharged to a municipal sewerage system or directly to a stream, unless the discharge was unsightly or odorous.

In more recent years, the federal emphasis on wastewater treatment in the USA has encouraged a more rational approach to wastewater control, and academics around the country have begun to realize and communicate that wastewater treatment can be scientifically modeled. The rapid advancement of treatment systems and methods in the 1960's led to the acceptance by the U.S. that pollution could be controlled, and eventually in 1970, to the formation of the Environmental Protection Agency (EPA) as a separate agency of the U.S. government.

The EPA promulgated regulations prolifically during the 1970's and 1980's including the Clean Water Act of 1972, the Federal Water Pollution Control Act Amendments of 1972, as amended in 1977 (also called the Clean Water Act), the Water Quality Act of 1987 and the Spill Prevention Control and Countermeasure Plan (SPCC) 40CFR Part 112 in 1972.

6.3 PHILOSOPHY OF STANDARDS

There are three basic philosophies of environmental regulatory standards: risk-based, performance-based and concentration-based standards. Risk-based standards emphasize the risk to public health and the environment from the discharge of a certain concentration of a pollutant. The theory behind this philosophy is that many pollutants are actually nutrients at a certain low concentration and pollutants at higher concentration (i.e. chromium, zinc, nitrogen, phosphorous, etc). Other pollutants are more toxic in certain receiving waters than in others (i.e. alkaline and hard waters). The risk in a certain environment is therefore more important in setting a standard than the type of treatment for pollutant removal or even the concentration of the pollutant in a discharge. The reader is cautioned to research the literature used to establish standards and to ascertain if they are applicable to a wastewater and a receiving stream before accepting the standards without question.

In Europe, there is generally more emphasis on protection of health and the environment than on regulation performance requirements or concentration limitation. This more rational but more difficult approach has proved successful in several countries, but requires that the general population effectively force industries into protection of the environment through consumer response.

The disadvantage of risk-based standards is that they are very difficult to establish. For instance, a particular wastewater may have many pathways to enter a human, an animal or a plant. These pathways include volatilization and breathing of a pollutant, drinking a pollutant, absorbing a pollutant through skin, or transpiration of a pollutant through roots. These pathways are different for children and adults. The opportunities for these pathways are numerous as are the environmental factors which affect the risk such as temperature, concentration, alkalinity, hardness, wind direction and velocity, etc.

Performance-based standards are popular now in air pollution control and in water pollution control and include the establishment of a maximum, best or generally available or achievable control technology. These standards are set from an analysis of existing industrial treatment systems and have found their way into the establishment of many of the categorical standards now used in wastewater regulations.

The advantage of performance-based standards is that they can be shown to be feasible or practicable and can be imposed for all industries regardless of location. The disadvantage of performance-based standards is that they do not concentrate on the damage to and protection of the environment and are instead limited to economical treatment methods. This theoretically can limit technological advances that other types of regulations encourage. This limitation could be one of the reasons that many technological advances in wastewater are developed in Europe.

Concentration-based standards are typically set in units of mass per volume or mass per time (i.e. ppm, mg/l, lbs/day, or kg/day). The purpose of these standards is to set a level at which the receiving body of water can assimilate or disperse the pollutant without preventing its established use (i.e. water supply, fish and wildlife propagation, irrigation, recreation or navigation). The justification of these standards is typically based on epidemiological studies (water supply and recreation), toxicity tests (fish, wildlife propagation and irrigation), or settleable or floatable solids accumulation (navigation). To prevent the dilution of the concentration of wastewater to meet standards, mass-based (lbs or kg/day) standards are normally established in addition to concentration (mg/l) standards. The advantage of concentration-based standards is that they can be related to literature, and therefore are theoretically more objective. They can also be applied to all industries regardless of their type or location. (i.e. chromium is chromium at a certain valence, regardless of whether it comes from a chrome tannery or a chrome electroplater).

Concentration-based standards may be environmental quality-based or effluent-based. Environmental quality-based standards are theoretically risk-based standards designed to protect the assigned use classification of a receiving body. Effluent-based standards concentrate on the quality of the discharge.

The disadvantage of concentration-based standards is that they typically must incorporate an application or safety factor. For instance, the EPA has set an application factor of 10 for non-cumulative

TABLE 6.1 CFR Water Programs

Program	40 CFR part numbers
Oil Removal	109, 110, 112, 113, 114
Hazardous Substances	116, 117
National Pollutant Discharge Elimination System	122–125
Toxic Pollutant Effluent Standards	129
Water Quality Standards	130, 131
Secondary Treatment Regulation	133
Test Procedures	136
Marine Sanitation	140
Underground Injection	144, 145, 146, 147, 148

pollutants and 100 for pollutants which accumulate in the internal organs of humans or animals. This arbitrary application factor is somewhat indefensible, but certainly some safety factor is needed to protect public health and the environment. A second disadvantage of concentration-based standards is, as explained above for risk-based standards that the environment of the receiving stream is typically not considered but is usually a major factor in toxicity.

6.4 WATER QUALITY CONTROL

Table 6.1 relates the various major water programs developed as a result of the Water Quality Act to the Part of Title 40 of the Code of Federal Regulations (CFR), all under subchapter D, Water Programs:

The Water Quality Act set as its basis that the existing water quality would not be further deteriorated and that overall existing pollution would be reduced.

6.5 SPILL PREVENTION CONTROL AND COUNTERMEASURE PLAN

The 1972 SPCC Plan is detailed in 40 CFR Part 112 was revised July 17, 2002, and applies to the aboveground storage of petroleum products if the total quantity at the facility is over 1320 gallons, and to underground storage of petroleum products if the quantity is above 42,000 gallons. The original plans were required in 1974 while the latest revision requires a new SPCC by February 17, 2006 and implementation by August 18, 2006, and must be amended if a change in storage occurs or if more effective technology becomes available. Plans must be reviewed and re-certified every 5 years. The Plans must include a diagram of the facility with locations of each container, transfer station and connecting pipe. Included is also a list of the type and quantity of all oil containers, 55 gallons or more in capacity. The Plan requires vessel containment, security, countermeasure, spill response and clean up procedures, emergency and reporting procedures, vessel inspection and testing methods, and employee training. The SPCC plans require a method of preventing all releases of petroleum products and responding to controlling any release that occurs.

6.6 FEDERAL WATER POLLUTION CONTROL ACT

In 1974, Section 201 of the Federal Water Pollution Control Act was established requiring urban planning areas to develop a state-approved plan for wastewater pollution control. In 1975, Section 208 of the Federal Water Pollution Control Act required regional coordination of 201 plans.

The Federal Water Pollution Control Act also required that all industries be classified into Standard Industrial Classifications (SIC) and required that the EPA establish categorical standards for each SIC code discharging pollutants. These standards were published starting in 1977 and continue to be revised today. These standards are published in CFR 40 Subchapter N in Parts 401 to 499.

The Standards include pretreatment regulations for wastewater discharging into a Publicly Owned Treatment Works (POTW) and also for direct discharges into streams.

6.7 THE NATIONAL POLLUTION DISCHARGE ELIMINATION SYSTEM

States were required by the Federal Water Pollution Control Act to develop a National Pollution Discharge Elimination System (NPDES) permit program for direct discharges requiring industries to monitor their wastewater discharge by sampling at a permitted frequency, under 40 CFR Parts 122–125.

The Federal Water Pollution Control Act required that states classify all navigable waters into one or more of the following uses:

- Public water supplies
- Propagation of fish and wildlife
- Irrigation
- Recreation
- Navigation

Once these waters were classified into these uses, standards were to be set by the states for these uses. Economics could be used for setting classifications, but not standards.

Acceptability of state standards were determined by the 1976 EPA publication Quality Criteria for Water, U.S. Department of Commerce National Technical Information Service publication PB-263943 under 40 CFR 131. The state-developed criteria had to be equally protective for the designated use as the limits in the publication. These criteria were used to set the NPDES permit levels for both industries and POTWs. In time, the POTWs were required to issue permits for all industrial users of their system. Pre-treatment program pollutant discharge limits were set to prevent the following:

- Passing of the pollutant through the POTW at a level exceeding the POTW NPDES permit
- A negative effect on the POTW treatment efficiency
- The collection of a toxic pollutant in the POTW wastewater sludge at a hazardous level

All industries in their application for a pre-treatment permit have to reveal all chemicals discharged in their wastewater and each is reviewed for these three criteria. If certain pollutants are discharged by an industry into a POTW and these pollutions are not removed by the POTW, the state must establish limits on the POTW for these pollutants.

As the NPDES point source program was established, it was recognized that much of the total pollution entering streams is discharged from non-point sources such as urban areas, industries and agricultural areas. In 1990, Section 405 of the Water Quality Action of 1987, added an NPDES system for urban areas over 100,000 population and for industries. At the time of this publication, cities under 100,000 population and cities adjacent to cities over 100,000 population are being required to apply for an NPDES permit.

This system is similar to the point source permit program and requires reporting and/or control of certain pollutant discharges during stormwater runoff. In 2002, the USEPA required States to do an analysis of all navigable streams and determine the Total Maximum Daily Load (TMDL) that they can assimilate without deterioration. These analyses located specific permitted or not permitted discharges into the streams which caused deterioration. As a result, municipal and industrial NPDES permits have been revised to prevent deterioration.

6.8 STORMWATER POLLUTION PREVENTION PLANS

The Water Quality Act requires all applicable industries and municipalities to develop a Stormwater Pollution Prevention Plan (SWPPP) which includes an analysis of the runoff areas, flow estimations from these areas and the location of all toxic materials stored, processed, loaded, or unloaded in these areas which could leak or spill and be carried by stormwater to a surface water.

The Plans require training and stormwater runoff sampling to monitor discharges. The outfalls selected in this Plan are permitted through the NPDES system and if reportable permit levels are exceeded, a Corrective Action Plan is typically required to eliminate the cause of this exceedance. Municipal SWPPP's incorporate a system to prevent industrial and commercial establishments from adding pollutants which would cause the municipality to exceed the permit levels.

6.9   RESOURCE CONSERVATION AND RECOVERY ACT

In 1972, EPA enacted the Resource Conservation and Recovery Act (RCRA) under 40 CFR 260–267. The goal of the Act was to regulate current hazardous waste problems as opposed to the Comprehensive Environmental Response, Cleanup and Liability Act (CERCLA) of 1980, also known as *Superfund* which had the goal of solving past hazardous waste problems.

Under RCRA, a hazardous waste is defined as a solid waste, which, because of its quality, concentration, or physical, chemical, or infectious characteristics may:

- Cause, or significantly contribute to an increase in mortality or an increase in serious irreversible, or incapacitating, reversible illness, or
- Pose a substantial present or potential hazard to human health or the environment when improperly treated, stored, transported, or disposed of, or otherwise managed.

EPA's four characteristics of a hazardous waste under RCRA are:

- Ignitability at less than 140°F
- Corrosivity (less than a pH of 2.0 or greater than 12.5)
- Reactivity
- Toxic Characteristic Leading Procedure (TCLP)

The TCLP test requires that the solid waste be shredded to maximize its surface area, mixed with water and a buffer, brought to a pH of 4.8 with citric acid and well mixed for 24 hours. This procedure is designed to duplicate the conditions in a landfill where organic degradation produces CO_2 in the presence of water and forms carbonic acid (H_2CO_3), a weak acid with an effect similar to citric acid. Tests have shown that in landfills, the pH is lowered by this action to approximately 4.8. This relatively low pH will leach many toxic organics and inorganics from the solid state and dissolve in water. After the mixing period, the mixture is tested for the 43 specific organic and inorganic elements and compounds shown in Table 6.2. If any of these tests exceed the threshold values, the sample is classified as hazardous.

From this description, it should be apparent that a waste cannot be hazardous, unless it is ignitable, corrosive, or reactive, without being converted to a liquid state. Therefore, the RCRA regulations do affect water pollution control response.

A residual from wastewater treatment, especially if industrial wastewater is involved, must be tested to determine if it is hazardous before storage, transportation or disposal.

6.10   SUPERFUND AMENDMENTS AND REAUTHORIZATION ACT (SARA)

In 1986, the EPA promulgated Title III of CERCLA entitled Emergency Planing and Community Right-to-Know. The Occupational Health and Safety Administration (OSHA) identified and regulated hazardous chemicals present in the workplace for the benefit of employees, while SARA required the reporting for public information of hazardous substances stored and released into the environment.

TABLE 6.2 TCLP Chemicals

Containment	mg/L
TCPL Metals	
Arsenic	5.0
Barium	100.0
Cadmium	1.0
Chromium, Total	5.0
Lead	5.0
Mercury	0.2
Selenium	1.0
Silver	5.0
TCLP Volatiles	
Benzene	0.5
Carbon Tetrachloride	0.5
Chlorobenzene	100.0
Chloroform	6.0
1.2-Dichloroethane	0.5
1.1-Dichloroethylene	0.7
2-Butanone	200.0
Tetrachloroethylene	0.7
Trichloroethylene	0.5
Vinyl Chloride	0.2
TCLP Semi-Volatiles	
Pyridine	5.0
O-Cresol	200.0
M-Cresol	200.0
P-Cresol	200.0
1.4-Dichlorbenzene	7.5
2.4-Dinitrotoluene	0.13
Hexachlorobutadiene	0.5
Hexachloroethane	3.0
Nitrobenzene	2.0
Pentachlorophenol	100.0
2.4.5-Trichlorophenol	400.0
2.4.6-Trichlorophenol	2.0
Hexachlorobenzene	0.13

There are four distinct and separate requirements that call for reporting under Title III:

- Sections 302 and 303
  If an Extremely Hazardous Substance is present in a quantity greater than or equal to its Threshold Planning Quantity, a facility must notify the proper authority and designate a Facility Emergency Coordinator.

- Section 304
  If a Reportable Quantity of an Extremely Hazardous Substance or a CERCLA Hazardous Substance is released and could expose human health or the environment off-site, a facility must notify the proper authority.

- Sections 311 and 312
  If a facility is subject to the requirements of the Hazardous Communications Standard (Worker Right-to-Know), it must submit Material Safety Data Sheets and Tier I and Tier II chemical inventory forms to various agencies.

• Section 313
  If a facility falls into SIC Codes 20 through 39, has 10 or more full-time employees, and has had cumulatively on site the previous year any Title III toxic chemicals in an amount that met or exceeded a specified threshold, it must submit a Toxic Chemical Release Inventory Form (Form R).

6.11   FUTURE REGULATORY STANDARDS

The most accurate method of predicting future wastewater regulatory standards is to understand the history of regulations. The following are suggestions to help prepare for future changes in regulations:

• Look at past trends of regulatory philosophy.

• Regulations have moved from protecting human health to protecting the environment. Therefore more environmental indicators may be required, i.e. the biological tests of streams. Bioassays on fathead minnows and *sera daphnia* may be extended to other biota.

• Risk-based regulations have increased in acceptance.

• An ever decreasing level of permitted pollutant discharges.

• As more synthetic chemicals are developed, more are added to regulations.

• Each industrial classification has seen added controlled pollutants.

• Look at regulation of air and soil.

• Pollution taxes or charges have been imposed.

• Marketable pollution credits have been experimented with.

• Self-reporting programs with criminal penalties have been added.

• Bubble concepts have been tried with several discharges being added together.

• Look at environmental coverage extensions.

• Smaller industries may be included.

• Minimum thresholds may be reduced.

• Agricultural lands may be added to NPDES Stormwater Permits.

• Further control of RCRA treatment, storage and treatment facilities, especially in flood plains, environmentally sensitive areas or geologically fractured areas.

• Regulation of non-hazardous wastes.

• More recycling and reuse requirements.

6.12   REFERENCES

1. "E.C. Environmental Legislation, Water Protection and Management," European Document Research, 1725 K. Street NW, Suite 510, Washington, D.C. 20006, September 1994.

P · A · R · T · IV

WATER POLLUTION CONTROL

CHAPTER 7
WASTEWATER TREATMENT PLANT DESIGN

7.1 INTRODUCTION

7.1.1 Characterization

As explained in Chapter 3, before wastewater treatment is considered, a wastewater must first be characterized as to its flow, its constituents, and the variability of each of these parameters. These constituents can be physically, chemically, or biologically treated. Depending on the constituent, more than one type of treatment may be necessary to remove it from the wastewater.

7.1.2 Final Discharge

A second consideration prior to design is the final point of discharge of the treated wastewater. The ultimate fate of the wastewater can be the air, the water, or the land. As explained previously, pollutants can be changed in state between vapors, liquids, solids to be more compatible with the air, water or land where they are discharged. Regulatory limitations may allow more total pounds of pollutants to be discharged into one of these realms than the others, and this must be a consideration in the treatment decision.

A liquid wastewater can be discharged directly to a stream under a NPDES permit in the U.S. as explained in Chapter 6, or be discharged through a Publicly Owned Treatment Works (POTW) to a stream under a NPDES permit.

7.1.3 Treatability Studies

Chapters 8, 9 and 10 describe treatability studies, which can be run inexpensively to experiment on a bench scale with various methods of treatment in order to optimize the selected method(s). It is recommended that a full-scale treatment design never be attempted on a wastewater without first completing a treatability study, unless the wastewater is identical in character, flow and variability to a previously experienced wastewater. Even relatively simple municipal wastewater not containing industrial wastes can vary in parameters such as strength, temperature, stormwater, distance of the treatment system from the customers, pumping stations, topography, discharge streams, etc.

Industrial wastes may have detergent containing wastewater, emulsified wastewater, highly variable flows, inorganics, organics, domestic wastes or other constituents which will affect the design and require prior testing. Figures 7.1 through 7.3 are photographs of treatability studies on various wastes.

FIGURE 7.1   Sedimentation treatability study.

7.1.4 Pilot Studies

In cases of unknown wastewater character or variability, in very large systems, or in systems incorporating experimental methods, a pilot system can be installed in the field using influent from actual wastewater streams. Such a pilot scale treatment could have a flowrate as small as one gallon per minute. These systems have several advantages such as:

- Flexibility of experimentation with different treatment methods at a relatively low cost
- Operating with the receiving wastewater, which will have the same varying temperatures, flows, and pollutant concentrations
- Allowing future operators to experience operation of the new system on a small scale

In order to be effective, these systems may require several months of operation and should continue over several seasons, especially winter. Figures 7.4 and 7.6 are photographs of pilot systems successfully operated.

7.1.5 Design Considerations

7.1.5.1 General.   It is suggested that this section be used as a checklist for design of a wastewater treatment facility.

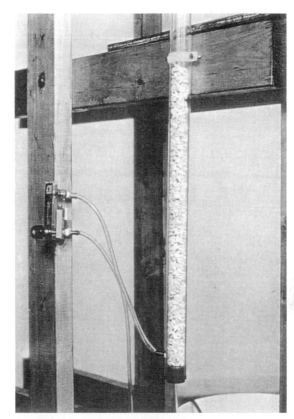

FIGURE 7.2   Air stripping treatability study.

FIGURE 7.3   Biological treatability study.

FIGURE 7.4 Organic chemical company pilot study.

7.1.5.2 Site Characteristics. The characteristics of the site play an important role in selection of a treatment technology and in the design of the chosen system. The following characteristics should be taken in account prior to design:

- Soil and geology
  - Soil type
    The soil type will normally affect design decisions. A rocky site may afford a high bearing strength for supporting structures, but be very expensive to excavate. The sand, loam, and/or clay content of a soil will affect bearing strength, drainage and erosion protection decisions.

FIGURE 7.5 Municipal toxicity removal and sludge reduction pilot system.

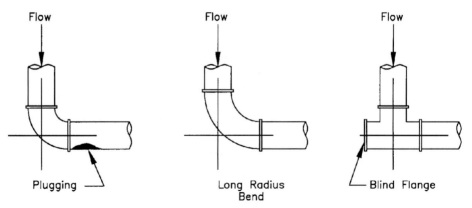

FIGURE 7.6 Methods to minimize plugging.

- Karst terrain
  Karst terrain may include sinkholes, caves and/or pinnacle rocks, which will affect design and drainage
- Site slope
  Ideally, cuts and fills on a treatment site will be balanced to minimize costs. Depending on the soil type, an approximate 20 percent cut overage should be planned to provide sufficient cut volume to overcome compaction and hauling losses. Flat sites may require multiple pumping stations where more steeply sloped sites can make more use of gravity flow between units. The author has designed a biological wastewater treatment system with pumping only for sludge recirculation.
- Drainage
  Final grading should be designed so that the surface flows away from all structures and passageways. Surface water can be conveyed by either ditches, trenches, or underground pipes. Ditches have the advantage of being easy to maintain but require more surface area. They are relatively inexpensive, but allow more wastewater volatilization. Ditch lining should be considered based on soil type and flow velocity with Table 7.1 giving general recommendations.
  Trenches are typically rectangular in shape, are easy to maintain and require less surface area than a ditch. Most trenches must be protected with handrails, grating, or plate covers, which makes maintenance more difficult.
  Underground piping uses the least surface area, but for maintenance purposes structures should not be built above the piping. Pipes are normally less expensive than trenches, but are difficult to maintain and allow little chemical volatilization. It is recommended that underground piping contain a catch basin, junction box, or as a minimum, a clean-out at each grade or direction change to facilitate maintenance. These units should also be placed a maximum of 350 feet apart to facilitate maintenance unless the lines are larger than 8 inches in diameter. This distance is also dependent on pipe cleaning equipment available. Pipes should be sloped sufficiently to allow a self-cleaning velocity if settleable solids are present in the wastewater.

TABLE 7.1 Ditch Lining Recommendations

Mean flow velocity	Type of lining
• 3 FPS and less	• Seeded
• 3–8 FPS	• Sod
• 8–15 FPS	• Riprap or concrete
• Over 15 FPS	• Concrete Paved

- **Flood levels**

  Flood levels must always be considered in design. It is recommended that the tops of all structures be above the 100-year flood plain and preferably a 500-year flood plane. In addition, all motors should be above the 25-year flood plain and all access routes above the 10-year flood plain.

- **Trees**

  The presence of trees in the area can cause leaves to accumulate in facilities.

- **Wind**

  Prevailing winds must be considered due to potential odors and aerosols.

- **Nearby development**

  Adjacent development can affect the type of treatment selected. For example units can be covered to minimize odors or aerosols. The treatment system can also be buried or disguised to be more aesthetically pleasing. Trees or walls can be used to minimize noise, odors or aerosols.

- **Receiving streams**

  Obviously, bodies of water receiving treated wastewater determine the permit parameters, but they can also determine effluent sampling methods and outfall structure design.

7.1.5.3 Layout of Treatment Plant

- **Relation to wastewater sources**

  A central location is ideal to minimize piping but because of odors, land use consideration, or operation, a remote area may be preferred. A downstream location will minimize or can eliminate pumping.

- **Outfall or Effluent Line**

  The outfall or effluent line length can be minimized based on the selected location. The suspended solids in an outfall line should be less than in an influent line. Therefore, settling is less critical, and pipe slopes can be smaller. For maintenance purposes, clean-outs rather than junction boxes can be used on outfall lines to minimize costs.

- **Optimization of operation and maintenance**

  Operation and maintenance both require personnel access and movement. Easy access to all equipment or facilities, which must be operated and/or maintained is important. The following suggestions are made for accessibility:

- **Valves**
  - Handwheel or lever operated valves are more easily operated than those using operating nuts.
  - Handwheels or operating nuts should be extended to an operating surface with extension stems, chains or rods.
  - Operating nuts can be mounted in a valve box and imbedded in a concrete surface to prevent interference with traffic or tripping hazards if the valve is not routinely operated.
  - Screwed fittings can be used for smaller valves (usually 2 inches and less) and should be provided with a nearby union for ease in removal.
  - Access should be provided to valve flange bolts and nuts.
  - Valves containing liquid may need to be heat traced and insulated to prevent freezing.
  - Underground valves should be provided with a valve box to protect the operating shaft.

- **Pumps**
  - All pumps and motors should have at least two feet of access around the unit.
  - All mounting bolts should be accessible.
  - Check pumps for heat tracing and insulation need.
  - Consider pump lubrication and lubrication drainage system (i.e. grease, oil, or water systems).

- **Equipment**
  - All equipment should have at least 2 feet of access around the unit.
  - Manufacturer's drawings should be checked to determine additional access needs for removal of all parts of the equipment.
  - Consider Occupational Safety and Health Act (OSHA) standards for all access platforms.
  - Operating and Maintenance Personnel traffic distance and level changes should be minimized to encourage routine checking and maintenance. Consider a recommended route for maintaining and checking all equipment and make this route convenient and safe.

- **Chemical delivery and usage**
  - Chemical unloading, transfer, storage and delivery should be optimized.
  - Certain chemicals such as lime are relatively insoluble and settling in piping must be eliminated by continuous pumping at 2 feet per second or greater velocity or by transferring the mixture in easily cleaned troughs.
  - Certain chemicals such as ferric sulfate are hydrophilic and tend to absorb moisture and clump. Therefore, storage areas should be dehumidified.
  - For maintenance and operator safety, dust collectors should be provided in loading areas for chemicals such as hydrated lime.
  - Certain chemicals, such as alum, can be purchased in liquid form to minimize freezing problems. Freezing problems should be considered in these cases.
  - Bulk purchase of chemicals requires storage but can lower total costs.
  - Chemical unloading methods should be considered in design of units such as loading ramps and quick disconnect fittings. Other factors such as noise, freezing, settling, etc. should also be considered.

- **Cranes and hoists**
  - The installation and removal of all equipment must be planned for in the design phase. This may require stationary or moveable cranes or hoists.
  - Openings in floors, roofs or walls may need to be provided for equipment installation and removal.
  - Design standards such as doubling static loads to account for moving loads on crane beams should be considered.

- **Piping**
  - Piping distances should be minimized to lower costs and head losses due to friction. This consideration will affect unit location and relation.
  - If piping is coordinated well with structures such as tanks and basins, pipe rack costs can be minimized.

- **Site slope**
  - The slope of the site was discussed above as it relates to cuts, fills and pumping stations. The slope also affects the site layout. Units can be located in series down grade to facilitate gravity transfer between units. A pumping station below the lowest unit can transfer waste or sludge to the top of a parallel chain of units also flowing by gravity.
  - Access to operating platforms, walkways, and valves, etc is affected by site slope. A unit may need to be rotated to allow grade access to a certain side without installing retaining walls.
  - One advantage of buried tanks is that soil below the frost line is typically at about 55°F year around. This temperature can aid or cause heat loss in the summer or heat retention in the winter. Buried steel tanks should be protected with coatings and galvanic anodes to minimize corrosion.

- **Future expansion**
  In most cases, future expansion should be planned for in an initial design. This should include space for one or more unit expansions:

  - Rectangular concrete, steel, and some plastic structures can be expanded by shape duplication and using common walls in order to minimize the use of available land. If such expansion is planned, reinforcing steel and water stops should be exposed and protected for concrete structures, and steel or plastic extensions provided onto which to weld future walls and bottoms for steel or plastic structures.
  - A designer should think of expansion three dimensionally. In some cases, a vessel which is designed for volume instead of surface area can be increased in depth to provide expansion.
  - Consider gravity and pumping ramifications in allowing for expansion.
  - Consider cut and fill balancing for expansions.
  - Leave blind flanges on tees in distribution piping likely to serve future units. The installation of valves in these tees can prevent future shut-downs during construction. These valves should be exercised occasionally to maintain operability.
  - Consider the balancing of flows for expansion, i.e., different pipe lengths cause different friction losses and correspondingly different flows whether a line is pumped or flows by gravity.

7.1.5.4 Material Selection

- **Corrosion resistance**
  - Many wastewaters and wastewater treatment chemicals have a low pH which will corrode certain surfaces. All materials in contact with these chemicals should be selected to resist this corrosion. This can include tanks, pumps, piping, mixers, and valves, etc. In general, concrete, carbon steel, aluminum and bronze can be corroded with low pH liquids and should be substituted for more resistant materials or coated for protection.
  - Many coatings may be designed to resist constant pH levels of around 4.0 and periodic pH levels of around 2.0.
  - It is cautioned that certain gases such as chlorine and hydrogen sulfide can form strong acids when they are condensed in the presence of water vapor.
  - A high pH liquid, especially at elevated temperatures, can cause scaling in piping or equipment.

- **Solvents**
  - Many plastics and fiberglass resins can be dissolved with certain solvents. The most common of these are gasoline, toluene, trichloroethylene, methylene chloride, methyl ethyl ketone, and tetrachloroethylene. When these solvents are present, they should either be removed before contacting affected materials, or the materials should be coated for protection.

- **Abrasion**
  - Solids such as sand and metal shavings can abrade many materials, especially at bends and tees, etc. These critical connections as well as other piping and equipment may need to be protected with abrasive resistant materials such as teflon or monel steel.

- **Material selection charts**
  - The reader is referred Appendix D-9 for the chemical resistance of selected material to certain chemicals.

7.1.5.5 Piping Design.
Piping as a means for transferring wastewater and wastewater treatment chemicals is usually used since it has the advantages over troughs of allowing conventional pumping, containing odors, preventing splashing, and preventing steep slopes or drops in troughs.

- **Velocities**
  One of the most important considerations of piping design is velocity.
  - Minimum velocity
    If a typical liquid contains settable solids, its flow velocity should be greater than 2.0 ft/s to prevent settling or floating in the pipe and its subsequent stoppage. This velocity can be determined more accurately by applying Stoke's Law as explained in Section 8.6.3.
  - Maximum velocity
    At a certain velocity, depending upon the abrasive content, piping material can erode. Even without abrasive solids, the maximum velocity to prevent erosion without pipe lining is 9.0 ft/s.
  - Pipe cleaning
    Piping that carries liquids with a high concentration of solids such as sludge should be provided with a cleanout at each bend and tee. These cleanouts can consist of the replacement of a bend with a tee and replacement of a tee with a cross to provide access to the spots where plugging is most likely. Plugging typically occurs when a velocity is slowed by turbulence and where one stream flows into another. Long radius bends and wyes instead of tees also can help prevent plugging. See Fig. 7.6.
  - Pipe flushing
    When plugging occurs, a pipe under pressure can be flushed with compressed air, compressed nitrogen, or high pressure water as shown in Fig. 7.7.
  - The designer may wish to provide flushing with the flow or against the flow, but normally, flushing will be toward a basin and away from a pump. In the case of a gravity line between two basins, the designer may place two valves in the line, one on either side of the flushing tap.

- **Head loss**
  Head loss is the friction loss of head caused by the velocity in a pipe. The faster the liquid flows, the greater the head loss.

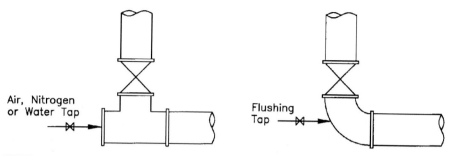

FIGURE 7.7   Methods to flush piping.

- It is recommended that the reader consider head loss in terms of a hydraulic grade line which can be determined by a series of pipes or pressure gages along a stream of water as shown in Fig. 7.8.
    In Fig. 7.8, the slope of the hydraulic grade line (HGL) from point A to point B is called the *entrance loss.* It is the vertical distance a drop of water requires to attain the velocity at point B, starting at a zero velocity at point A. The distance required is determined as h_g in Eq. (7.1).

$$h_g = \frac{v^2}{2g}$$
(7.1)

where g = acceleration due to gravity = 32.2 ft/s² = 980 cm/s²
h_g = entrance loss head in ft or cm
v = the velocity of flow at point B

If the entrance into the pipe is not a simple squared off connection as shown in Fig. 7.8, the entrance losses in Appendix A-1 can be used.
    The velocity at point B is equal to:

$$V = \frac{Q}{A}$$
(7.2)

where V = velocity in ft/s or cm/s
Q = flow in ft³/s or cm³/s
A = area in ft²/s or cm²

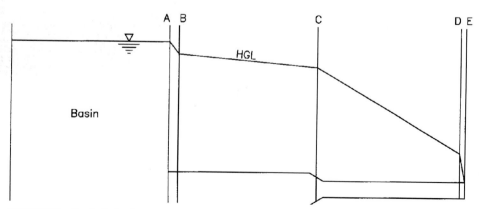

FIGURE 7.8   Gravity hydraulic head line.

The slope of the HGL from point B to point C is determined by Hazen William's formula:

$$h_L = 0.2083 \left(\frac{100}{C} \right)^{1.85} \frac{Q^{1.85}}{d^{4.8655}} \qquad (7.3)$$

where h_L = friction head loss in feet of liquid per 100 feet of pipe
C = constant accounting for surface roughness (see Appendix A-2)
Q = flow in gpm
d = inside diameter of pipe in inches

The slope at the HGL from point C to point D is also determined by Hazen William's formula and is steeper than the HGL from B to C because the C to D pipe is smaller.

The slope of the HGL from D to E is almost vertical dependent upon the outlet condition of the pipe. The HGL must be even with the water surface at the pipe exit, since there is zero head at that point.

Appendix A-3 in the Appendix is a series of charts showing velocities, and head losses for various pipe sizes based on the Hazen Williams formula. Note that the friction or "C" factor varies with the type of pipe. Appendix A-9 gives equivalent pipe lengths for bends, tees, valves and other pipe changes.

- **Pipe supports**
  - Overhead pipes must be supported as per the manufacturer's recommendations to prevent bending. Without the manufacturer's recommendations, it is suggested that pipe racks be spaced to allow a maximum of ½ inch deflection in accordance with Eq. (7.4):

$$\frac{0.0130wL^4}{EI} = \Delta l \qquad (7.4)$$

where Δl = deflection
w = weight of pipe plus liquid
I = moment of inertia
E = modulus of elasticity
L = support spacing

The reader is cautioned that with plastic pipes, I is reduced at higher temperatures.

- Pipes should also be supported laterally to brace against water hammer. The pressure caused by water hammer may be estimated by Eq. (7.5):

$$\text{pressure rise} = \frac{(0.433)a \cdot V}{32.1 \times 1} \qquad (7.5)$$

$$a \text{ for cast iron pipe} = 4000$$

$$a \text{ for steel pipe} = 3600$$

$$a \text{ for ductileiron pipe} = 3550$$

Rule of thumb for simple conduit pressure rise above static equals 60 psi per fps velocity in the line if change in velocity takes place within the critical time C_t. PC_t = time for pressure wave to travel at 4000 fps to end of line and back.

- Because of piping movement, pipes should be attached firmly to all pipe racks and braces.
- **Vertical pipe clearance**
  - The designer should provide a minimum of 8 feet vertical clearance in walkways and greater clearance as appropriate in forklift routes, roadways, and railroads.
- **Freeze protection**
  - Freeze protection should be considered for all overhead pipes in the form of draining, either automatic or manual, or heat tracing and insulation.

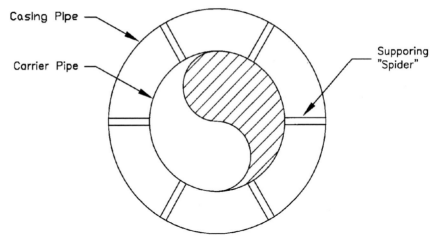

FIGURE 7.9 Concentric double walled pipe.

- Aboveground piping can be freeze protected by double walling the pipe and surrounding the inner pipe with steam, or a hot liquid.
- **Underground piping**
  - The designer should consider galvanic action of the soil on the pipe and provide sacrificial anodes if needed to prevent corrosion.
  - Underground piping has the advantage of resisting freezing but it is more difficult to maintain. It is recommended that junction boxes, or as a minimum, cleanouts, be provided at least every 350 feet.
  - Underground piping should be buried deeply enough to resists surface pressures from vehicles or other surface loads. As a minimum, 30 inches of cover with no vehicular traffic, and 48 inches across traffic should be provided.
  - The designer is cautioned to provide a pipe strength sufficient to resists soil and surface loading pressure.
  - A double walled pipe can be provided underground to allow leak detection if required as shown in Figs. 7.9 through 7.11.

The eccentric pipe requires no support and may be used when forces allow an unsupported interior pipe.

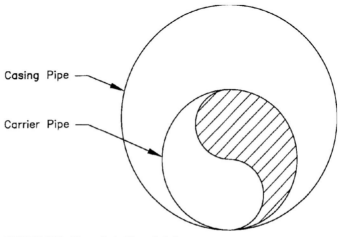

FIGUER 7.10 Eccentric double walled pipe.

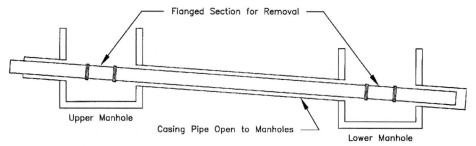

FIGURE 7.11   Section of the double walled pipe.

7.1.5.6   Valve Design.   Valves can be installed in piping to cut off flow, to isolate pumps or equipment, or to throttle flow.

- **Cutoff valves**
    - Cutoff valves include butterfly, ball, globe, gate, plug, mud, check and flap valves. The reader is referred to Fig. 7.12 for drawings of these valves.
    - Butterfly valves include a butterfly with a center axis which can be rotated 90 degrees with an external handle to open, shut or throttle.
    - Butterfly valves can trap solids on the butterfly and prevent complete closure.
    - If used for throttling, butterfly valves can be eroded. The designer is also cautioned that the amount of throttling by a butterfly valve is not directly proportional to the closure.
    - For liquid lines, ball valves work well under one inch in diameter. It is recommended that globe valves be used for piping one inch through 4 inches and gate valves be used over 4 inches.
    - Ball valves can provide full pipe size openings and are one quarter turn open-close. Ball valves can become stuck without routine use or if solids are present in the liquid. Ball valves usually contain screwed or glued fittings.

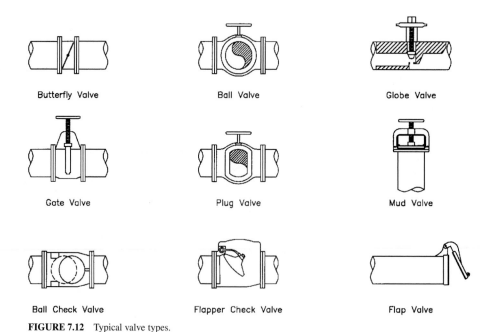

FIGURE 7.12   Typical valve types.

- Globe valves are similar to ball valves in operation and are available in screwed, glued and flanged fittings. They may not have full pipe sized openings. Ball and globe valves are typically open when the handle is perpendicular to the pipe and closed when it is parallel.
- Gate valves have a threaded shaft and gate that is raised and lowered with multiple turns of an operating nut or handwheel. For valves greater than 12 inches in diameter, it is recommended that a gear reducer be provided to ease the force necessary to open and close the valve.
- With excessive velocities or abrasive fluids, the gates of gate valves may erode.
- Gate valves with rising stems can be used to visually observe the percent open, otherwise observation of the valve will not divulge the degree of opening.
- Plug valves are similar to ball or globe valves, but are designed to resist plugging with solids. An eccentric plug valve is especially effective in resisting stoppage.
- Mud valves are vertical plugs similar to bath tub stoppers which can be used to drain a basin. These valves have an extension stem reaching above the water level.
- Check valves include a ball or a flapper usually hinged at the top which will only open one direction. Three types are common, the ball check, which has a ball seating against a collar, and which closes and opens depending on the pressure direction; the spring loaded check, which returns to closed against a seat when the water pressure on one side is greater than the spring pressure; and the weight and lever check which returns to closed against a seat when the water pressure on one side is greater than the pressure caused by the weight. The weight to lever check is the only type of check valve with a resistance that is adjustable.
- Flap valves are usually installed at the ends of lines discharging into a stream or basin and are designed to close when the water pressure in the stream or basin is greater than the pressure in the pipe.
- **Throttling valves**
  - As explained above, butterfly valves can be used for throttling, but except for pure water use, are not as effective as other types. More often used are diaphragm type valves that use a pressure of some type to close a diaphragm over a seat, gradually allowing less flow, but retaining a relatively smooth channel of flow. (See Fig. 7.13.)

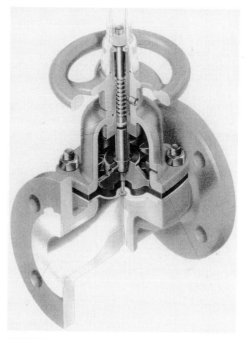

FIGURE 7.13   Diaphragm valve.

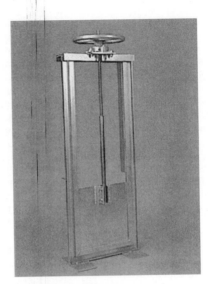

FIGURE 7.14 Slide gate.

7.1.5.7 Gate Design. There are several types of gates to shut off or throttle flow. The two most often used in wastewater treatment are slide gates and sluice gates.

- **Slide gates**

    Slide gates are typically less expensive than sluice gates, are usually constructed of fabricated material, and are not typically as water tight. The better gates will incorporate rubber gaskets around the edges which improve closure when subjected to water pressure. Such a gate is shown in Fig. 7.14. The least expensive slide gates are flat plates that simply slide in imbedded channels as shown in Fig. 7.15. As shown in Fig. 7.14, these gates can be motor controlled or hand wheel operated.

- **Sluice gates**

    Sluice gates typically have machined closure surfaces and are much more expensive than slide gates. Because of machining and/or the use of gaskets, these gates are virtually watertight. These gates can also be handwheel or motor operated. Fig. 7.16 shows typical rectangular gates.

7.1.5.8 Pump Design. Wastewater pumps typically consist of a rotating or reciprocating movement which serves to move wastewater in the direction of the movement. The most common types of pumps are described as follows:

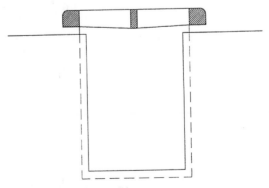

FIGURE 7.15 Alternate slide gate.

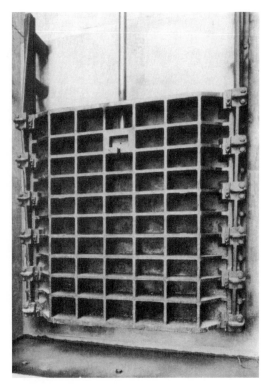

FIGURE 7.16   Sluice gate.

- **Centrifugal pumps**
  Centrifugal pumps are rotating pumps which can have a vertical or horizontal shaft (or a shaft at an angle) and serve to move the wastewater with an impeller rotating within a housing. The impeller consists of blades, shaped and positioned to move the liquid. Figure 7.17 shows a typical centrifugal pump cut away to demonstrate the pumping mechanism.

- **Pump tolerances**
  The closer the tolerance between the impeller and the housing, the higher the efficiency, but the more likely an impeller is to plug or wear and lose its design efficiency. Consequently, wastewater pumps will normally have greater tolerances than pumps transferring water containing no solids.

- **Pump solids passing capacity**
  The pump impeller can be shaped and positioned to enable the pump to pass a solid sphere. Centrifugal pumps are classified by the diameter sphere which can be passed. The designer is cautioned to consider this feature in selecting a pump so that solids entering the pump housing or volute will be passed. Figure 7.18 shows a recessed impeller pump in which the impeller is recessed from the wastewater flow to prevent plugging. Generally, the recessed impeller pumps and those passing larger solids are less efficient than others.

- **Grinder or cutter pumps**
  A pump impeller can be designed to grind, cut or shred solids in a wastewater so that the pumps and the force main, or pipe through which the pump discharges, is less likely to plug. (See Fig. 7.19.)

- **Pump and head curves**
  Centrifugal pumps are rated by flow and head. **Flow** is the flow rate typically expressed in gallons per minute (gpm), millions of gallons per day (MGD) or cubic feet per second (CFS). Smaller

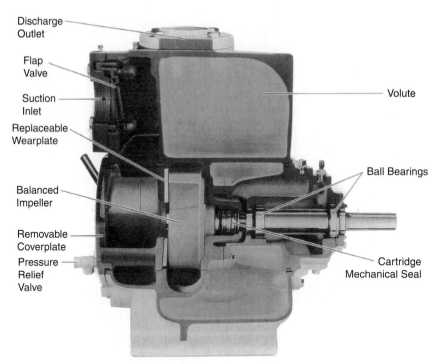

Discharge Outlet

Flap Valve

Suction Inlet

Replaceable Wearplate

Balanced Impeller

Removable Coverplate

Pressure Relief Valve

Volute

Ball Bearings

Cartridge Mechanical Seal

FIGURE 7.17   Centrifugal pump mechanism.

pumps can be expressed in gallons or cubic feet per day (gpd or CFD). **Head** is the height of water column the pump will provide at a certain capacity. The **shutoff head** is the head at which a centrifugal pump will have no capacity, where the impeller turns but will not move the wastewater. Figure 7.20 shows a typical centrifugal pump curve depicting flow, head, horsepower, efficiency, impeller diameter, and net positive suction head.

FIGURE 7.18   Recessed centrifugal pump. (*Source:* Wemco Pumps.)

FIGURE 7.19 Catalogue typical grinder pump.

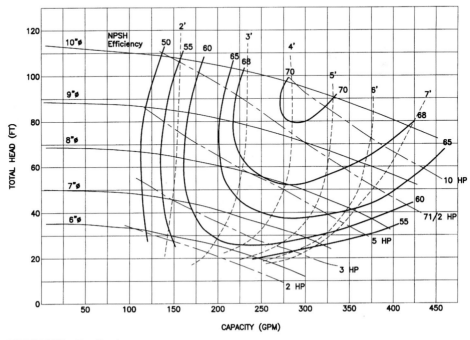

FIGURE 7.20 Centrifugal pump curve.

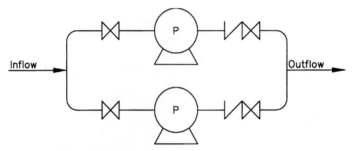

FIGURE 7.21   Parallel pump station.

This curve shows the relationship between pump **flow** and **head.** Some pumps will have a relatively flat curve so that the capacity is increased or decreased drastically with a relatively low change in head. Note that the point at which the impeller capacity/head curve intersects the ordinate at zero flow is the **shutoff head.**

The right ordinate relates the **horsepower** required at various capacities. Note, that as flow increases, horsepower increases. Therefore, there is an overloading capacity beyond which horsepower must be increased to prevent motor damage. The designer should specify sufficient horsepower to allow the pump to operate without it's motor overloading at the maximum possible flow.

The **efficiency** curve relates the pump efficiency, in percent, to capacity and head. It should be noted that there is a point of maximum efficiency for each pump. The maximum efficiency is the center of the concentric efficiency ovals. In Fig. 7.20, the maximum efficiency is greater than 70 percent at a flow of 300 gpm and a head of 95 feet.

The **impeller diameter** curves demonstrate that the flow and head are proportional to the impeller diameter. A given pump housing will accommodate the series of impellers shown on the curve, in Fig. 7.20. The pump housing will accommodate impeller diameters of 6 inches to 10 inches.

The **net positive suction head** (NPSH) curves indicate the suction lift that the pump will provide for each impeller. The suction lift is the vertical distance from the water surface to the impeller center. If a pump is to be mounted above the water surface, a sufficient NPSH must be provided or the pump will not lift the water. As a pump impeller wears the NPSH will decrease.

• **Multiple pump head curves**
  When two centrifugal pumps are pumping in parallel as in Fig. 7.21, their curves will be additive along the flow axis as shown in Fig. 7.22. When two centrifugal pumps are pumping in series as in Fig. 7.23, their curves will be additive along the head axis as shown in Fig. 7.24. The distance between the pumps is important only in that it causes a loss of head. Each pump simply adds the head provided by the pump.

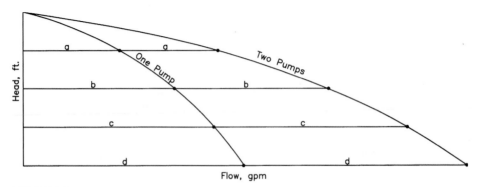

FIGURE 7.22   Parallel pump head curve.

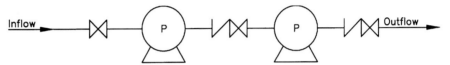

FIGURE 7.23 Series pump station.

- **System head curves**

  A system of piping, fittings and valves can be shown as a curve relating head and flow by applying the Hazen Williams or similar formula to pipe lengths as explained above. Figure 7.25 shows a typical piping system head curve. Because of the exponential nature of the Hazen Williams formula, the curve is concave. Note that a piping system head curve always begins at zero on both axes since there is zero head with zero flow. When the specific piping system head curve is superimposed on a pump head curve, a design point is apparent with a specific head and flow. This point will only change for a given rpm pump curve if the piping system is altered by throttling a valve or changing a route, thereby correspondingly altering the piping system head curve. Figure 7.26 shows a typical system head curve superimposed on a pump curve.

- **Centrifugal pump configurations**

  Centrifugal pumps can have their impellers and their motors submerged and can have the supply water head below or above the center of the pump. These configurations are shown in Figs. 7.27 through 7.30.

  The pump in Fig. 7.27 shows three pumps or stages in series that, in accordance with the above explanation, triples the head with the same flow. If the distance between the highest pump and the

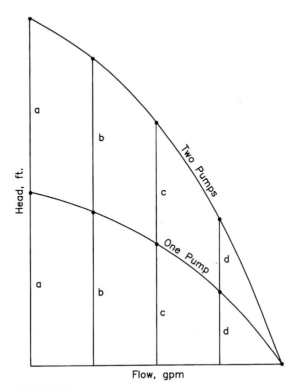

FIGURE 7.24 Series pump head curve.

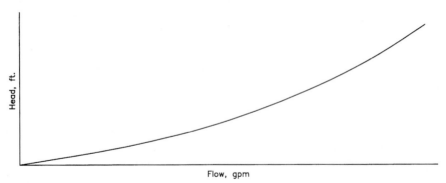

FIGURE 7.25 Piping system head curve.

water level is insufficient, air can be pulled into the pump causing cavitation, resulting in loss of efficiency or pump damage. A pump manufacturer can provide a designer with a minimum submergence. In the absence of a recommendation, at least one foot submergence should be provided. Figure 7.28 shows a manufacturer's drawing of a vertical turbine pump.

Figure 7.29 shows a schematic of a submersible pump. In this case, the motor, the wiring connections, and the drive shaft connection between the motor and the pump must be sealed. Figure 7.30 shows a manufacturer's drawing of a submersible pump. Note the guide pipes and chain, designed to allow the pump to be removed for maintenance. Figure 7.31 shows a positive suction head pump which requires the pump to be mounted below the minimum water level. As in the case of the vertical turbine pump, the manufacturer can provide the minimum submergence depth required.

Figure 7.32 shows a negative head centrifugal pump schematic. Many of these pumps rely on a foot valve or a check valve under the water level to retain water in the suction piping. This keeps them from pumping air and losing their prime or their capacity to transfer water.

A second method of maintaining pump prime is for the pump to fill a chamber with water after a pump cycle. This water allows the pump to begin pumping and pulling water from below. Because a wastewater pump has relative loose tolerances, it will not efficiently pump air. If air can find its way through the pump impeller to the water being pumped, the water will not be drawn into the pump. If there is a water seal provided to prevent this air from entering the suction pipe, a vacuum can be drawn which may serve to pull the water to the pump. This seal can be in the form of water stored in the pump itself or in the discharge pipe. If a complete vacuum is pulled and the weight of the water in the suction pipe is less than the negative vacuum pressure, the water will be lifted. As explained above, many pump curves will provide a net positive suction head (NPSH) which is the maximum length of suction pipe a pump can lift.

Figure 7.33 shows a manufacturer's drawing of a negative head pump.

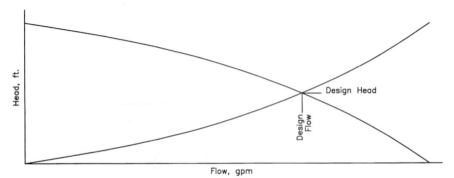

FIGURE 7.26 Pumped piping system head curve.

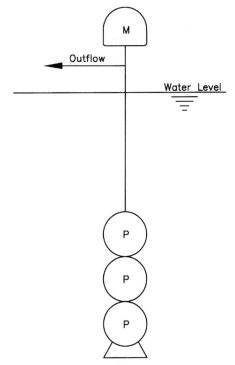

FIGURE 7.27 Submerged centrifugal pump control
(vertical turbine).

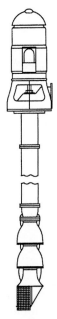

FIGURE 7.28 Vertical turbine pump installation.

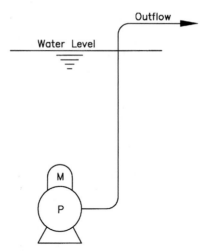

FIGURE 7.29   Submerged centrifugal pump
and motor (submersible)

- **Diaphragm pump**

    A diaphragm pump operates when a flexible diaphragm moved by mechanical or air pressure lifts
    an amount of water. The water is held in its lifted position at the end of the stroke by a ball or check
    valve. This type of pump is a positive displacement pump, with its head only limited by the horse-
    power of the mechanical system or the air pressure. Its capacity is a function of the size of its
    chamber and the frequency of its strokes. Figure 7.34 shows a manufacturer's drawing of an air
    diaphragm pump. These pumps can be used for the transfer of wastewater or sludges.

- **Piston pump**

    A piston pump is a positive displacement pump operating with a plunger or piston moving a col-
    umn of water retained by a check valve. Figure 7.35 shows a manufacturer's drawing of a piston
    pump. These pumps are efficient for the moving of thick sludges.

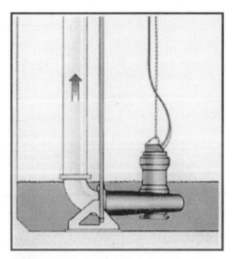

FIGURE 7.30   Submersible   pump   installation.
(*Source*: Sarlin Pump Co.)

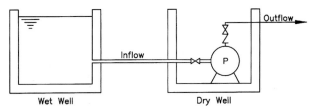

FIGURE 7.31 Positive suction head centrifugal pump.

- **Peristaltic pump**

  A peristaltic pump is a positive displacement pump which moves a column of water by squeezing a tube with a rotating cam. Each squeezing movement of the cam moves the water onward. Figure 7.36 shows a manufacturer's drawing of a peristaltic pump. These pumps are efficient for pumping small quantities of liquids such as for chemical feed operation in wastewater treatment.

- **Progressive Cavity Pump**

  Figure 7.37 shows a manufacturer's drawing of a progressive cavity pump which moves water along a rotating spirally shaped impeller. These pumps are especially efficient in pumping thick solids.

- **Screw Pump**

  Figure 7.38 shows a photograph of a 55,000 gpm screw pump system for transfer of stormwater at the peak of a 10-year storm. Screw pumps are efficient for low head lifts less than about 30 feet, and high flows. Because the Archimedes Screw principal is lifting, rather than pumping the water, they have the advantage of transferring the exact amount of water entering the system, therefore they are a true variable speed pump without a variable speed motor.

- **Constant flow pump systems**

  Many wastewater treatment components, such as physical and chemical treatment systems, operate most efficiently at constant flows. These systems require a pumping system such as the centrifugal pumps or the positive displacement pumps, described above, which are turned on at a low level and off at a lower level. A two-flow system can be arranged with a second pump cutting on at a high level, if the treatment capacity can be doubled to accommodate the flow. Otherwise, the treatment system must be designed for the higher flow.

- **Variable flow pump systems**

  Biological systems operate most efficiently with a constant or slowly changing flow, therefore, variable flow pumps are ideal for these systems. Any of the centrifugal or positive displacement pumping systems described above can be operated as a variable speed system by varying the rpm or stroke frequency. The rpm or the stroke can be varied electrically or mechanically. As explained above, a screw pump will automatically operate as a variable flow pump.

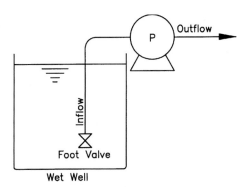

FIGURE 7.32 Negative head centrifugal pump.

FIGURE 7.33   Negative head centrifugal pump installation.

7.1.5.9   *Material and Chemical Handling Design*

- **General**

  Chemical feed systems in wastewater treatment are either gaseous, liquid, slurries or solids. Chemicals can be used to prepare a wastewater for treatment or to actually treat the wastewater through the processes of pH control, oxidation/reduction control, metal precipitation, coagulation, adsorption, gas transfer, nutrient addition or disinfection.

- **Gaseous chemical addition**

  Chemicals in the gaseous form are generally easy to handle since they can be injected into wastewater under pressure without the use of pumps or solids handling equipment. Gases commonly

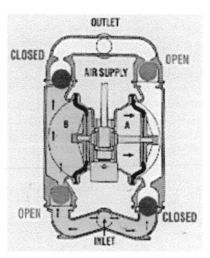

FIGURE 7.34   Air diaphragm pump.

View of SD Pump showing flow

FIGURE 7.35 Piston pump. (*Source:* Williams Machine Co.)

added to wastewater include oxygen, ozone, carbon dioxide, nitrogen, ammonia, chlorine, and sulfur dioxide. Oxygen and ozone are added as an oxidant or an energy source. Carbon dioxide can be used as a pH lowering gas since it forms the weak acid, carbonic acid. Nitrogen is used as a non-combustive pressure source, a purging gas, or as a nutrient source. Ammonia can be added as a nutrient source. Chlorine or other halogens, such as bromine or iodine, are added as oxidants or disinfectants. Sulfur dioxide is added as a reducing gas, normally to minimize chlorine residual.

These gases can be purchased in pressure tanks, usually from 150 pounds to 2000 pounds, and the gases can be fed under the supplied pressure using a valving system, a manometer to measure flow, and an injector as shown in Fig. 7.39. Water may be used as a carrier fluid in which to solubilize the gas and to allow easier conveyance.

The seepex multi-house axial flow perstaltic pump MAP

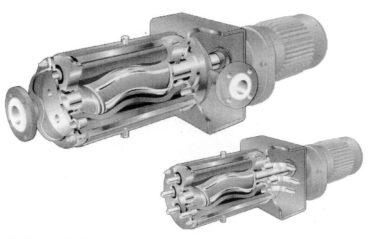

FIGURE 7.36 Peristaltic pump installation.

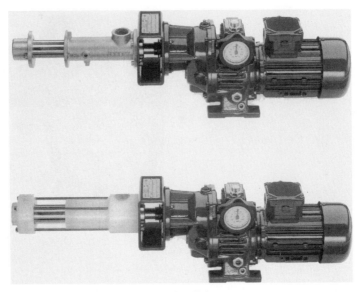

FIGURE 7.37   Progressive cavity pump installation.

FIGURE 7.38   Screw pump.

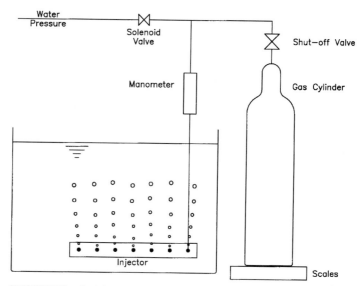

FIGURE 7.39   Gas injection system.

Gases can be fed directly without water solubilization, but are more difficult to handle and disperse.

Ozone can be manufactured on site from oxygen. The characteristics of gaseous chemical used in wastewater treatment can be found in Appendix D-5.

- **Liquid chemical addition**

A liquid chemical consists of either a pure liquid chemical or a soluble solid chemical dissolved in water, which can be added to a wastewater by gravity or under pressure. The liquid can be formed as part of the process by dissolving a solid chemical in water, the liquid can be purchased previously dissolved in water, or the chemical can be purchased and fed as a pure chemical.

In all cases, some pressure, whether by head or pump, must be used to overcome friction, and to allow the transfer of the chemical from its storage location to its point of feed.

Examples of liquid chemicals used in wastewater treatment are acids and bases such as HCl, H_2SO_4, NaOH and $Ca(OH)_2$ for pH control and metals precipitation; metal salts such as liquid alum, $Al_2(SO_4)_3 \cdot 18\ H_2O$, for coagulation; polyelectrolytes for coagulation; and sodium bisulfite, $Na(SO_2)_2$ and sodium hypochlorite, $Na(OCl)$, for oxidation-reduction reactions.

These liquids can be fed by gravity or pressure in proportion to flow, pH, oxidation reduction potential (ORP), or zeta potential, using proper instrumentation. Normally, a gravity system will use a throttling valve which can be automatically controlled in proportion to the variable corresponding parameter. A pressure system normally uses a positive displacement pump variably controlled in proportion to the parameter. An example of a pressure fed chemical feed system is shown in Fig. 7.40.

Examples of the characteristics of liquid chemicals used in wastewater treatment can be found in Appendix D-5.

- **Solid chemical addition**

Chemicals in the solid form can be fed directly into the wastewater, can be dissolved in water prior to feeding and fed as previously described as a liquid, or can be fed as a slurry when insoluble.

Examples of solid chemicals which can be directly fed are lime, $Ca(OH)_2$ for pH adjustment and softening; soda ash, Na_2CO_3 for softening; alum, $Al_2(SO_4)_3$; ferric sulfate, $Fe_2(SO_4)_3$; ferric chloride, $FeCl_3$, ferrous sulfate $FeSO_4$ and polyelectrolytes for coagulation; and calcium hypochlorite $Ca(OCl)_2$ and sodium bisulfate $Na(SO_2)_2$ for oxidation reduction reaction.

The solids can be purchased in powder or granular form in bags, or in bulk in trailers or railcars. Bags are typically loaded into a hopper above a screw or plunger dry chemical feeder. If the chemicals

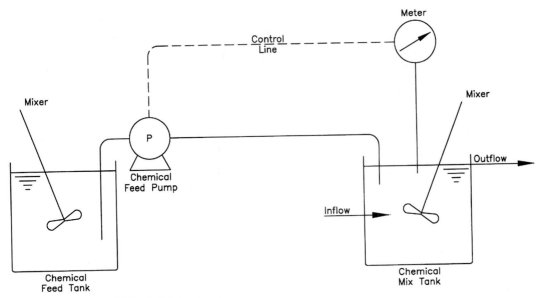

FIGURE 7.40 Pressure fed chemical feed system.

can cause dust discharges during loading, a dust collector may be needed. Bulk chemicals are typically blown from the tank into a hopper. A dry chemical feed system is shown in Fig. 7.41.

With dry chemical feed systems, the powder or granule is dissolved in a mixing chamber below the dry chemical feeder, and it is fed as a dissolved liquid or a slurry. Insoluble chemicals, such as lime, can be transferred as a slurry more easily than dry, but the solids in a slurry will settle in low velocity or turbulent sections, causing stoppage. To prevent this, it is recommended that the slurry be kept mixed and pumped in a loop system with a velocity greater than 2 fps. A slurry system is shown in Fig. 7.42.

- **Design suggestions**
The valves and piping at the chemical feed point should be minimized in length and maximized in size. Certain chemical feed systems may need to be heat traced and insulated to prevent freezing. Hydroscopic chemicals such as ferrous sulfate should be stored in a dehumidified room to prevent clumping.

FIGURE 7.41 Dry chemical feed system. (*Source* Cehm Flow. Inc.)

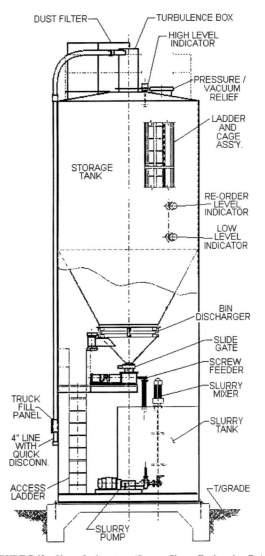

FIGURE 7.42   Slurry feed system. (*Source*: Slogan Engineering Co.)

7.2.   *AUTOMATIC PROCESS CONTROL FOR WASTEWATER TREATMENT FACILITIES*

7.2.1   Background

Process control is the implementation of information-based decision-making for the achievement of specific process characteristics. While an operator manually starting pumps and opening valves is an example of this definition, we will be focusing on automatic process control using mechanical and electrical equipment, with human interaction being limited to a design and supervisory role.

One reason for process control is to regulate processes in order to minimize risks. Additional benefits are that the risks of violating permits (such as discharge permits) are reduced, while the opportunity of meeting other regulatory requirements such as monitoring and reporting is enhanced. Furthermore, many permits have conditions requiring the installation of automatic process control equipment. An example of this type of automatic control is a permitted maximum chlorine residual limit of 0.1 mg/l. This level is not likely to be achieved without automatic controls. Another benefit of automatic process control is financial savings. It is normal to reduce chemical and energy costs by 10 to 20 percent when automatic process controls are incorporated, resulting in a payback period.

7.2.2   Theory

Process control is typically implemented in either a feedback or feedforward manner. In the feedback control scheme, action is taken after monitoring occurs. In feedforward control, action is taken based upon a process model without information regarding the outcome of the action. The difference between these two control schemes is illustrated by the act of driving an automobile. The person (controller) driving the car bases his steering on the direction of the road ahead. Within the driver's mind, there is a model relating amount of steering to the curvature of the road, speed of the car, and other factors. This is an example of feedforward control. If feedback control were to be employed in this example, the driver would base the steering not on the road ahead, but on the road behind the car seen in the rearview mirror.

From the previous example, it may seem that feedforward control is the better approach. However, this is not the case. The key to successful feedforward control is the quality of the process model, which is not accurately known in most real-world cases. One common use of feedforward control is in ratio control of chemical feeds, as is the case with chlorine residual control. In this case, an accurate model involving the ratio of the feeds to the system is known, and control is possible. More often, feedback control is employed. The reasons are many, including the existence of off-the-shelf equipment, a well-developed theory, and the ability to attain good control in most cases which have reasonably quick dynamics.

The need for control arises because of disturbances that affect the process. Disturbances can be external or internal. External disturbances can include influent characteristics such as flow rate, concentration, and composition. Internal disturbances include those generated by recycle pumps, or by sidestreams from other processes, such as sludge processing. Typically, internal disturbances can be controlled.

Feedback automatic process control is typically conceptualized through the use of a process control loop. The control loop (see Fig. 7.43) represents the flow of information in the decision-making process. Sensors gather information on the system, namely the value of the measured variable. The output from the sensor is transmitted to the controller, where the value of the measured variable is compared to a reference value called the setpoint. The controller then computes its controller action using the difference between the actual and setpoint values of the measured variable. The controller action is used as the input to a final control element. The final control element changes the value of the controlled variable. The loop is thus completed.

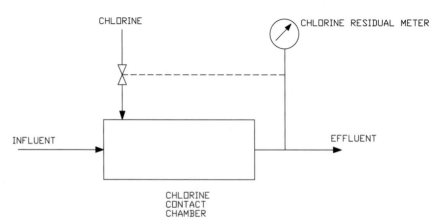

FIGURE 7.43   Process control loop diagram (feedback).

7.2.3 Automatic Control Example

An example of automatic process control is demonstrated in Figure 7.44 where a biological activated sludge system is controlled by a series of Total Organic Carbon (TOC) Analyzers, Dissolved Oxygen (DO) Meters, and Sludge Interface Meters.

In Figure 7.44, the Dissolved Oxygen in the Activated Sludge Basin is analyzed and compared with a pre set valve and the Blowers are controlled from the results. A modification to this system, would be to override the Blower Control with the Effluent TOC reading. For instance, if the effluent TOC, exceeds a pre-set valve, the blowers could be increased in capacity to increase the DO beyond the pre-set valve by overriding the Dissolved Oxygen control system.

The Clarifier Sludge Valve can be controlled to withdraw sludge based on its depth. The sludge depth should never be so high as to allow solids to be carried out of the Clarifier due to excessive velocity as explained in Chapter 8. An alternative, simpler but less effective control is to pace the Clarifier Sludge Valve with a timer, set to open the valve a certain number of minutes an hour.

The selection of Return Activated Sludge verses Waste Activated Sludge can be made using a TOC Analyzer in between the Activated Sludge Basin and the Clarifier. The RAS valve would open when the TOC leaving the Activated Sludge Basin was lower than a pre-set value and the WAS valve would open when the TOC was higher than the pre-set value. Chapter 10 explains how TOC can be used as an indicator of Mixed Liquor Volatile Suspended Solids (MLVSS) and is actually a better indicator of active mass.

The second TOC Analyzer at the effluent of the clarifier could be used to override the RAS and WAS control valves by increasing the RAS when the effluent TOC becomes too low. The relationship between TOC and BOD and COD is explained in Chapter 3.

The practical key to successfully operating this system is to keep the DO probes clean and to acidify the Activated Sludge Basin sample and the Effluent sample before they enter the TOC Analyzer in order to dissolve all solids.

7.2.4 Instrumentation

7.2.4.1 Sensors.   Sensors include instrumentation such as analyzers, flow meters, level meters, temperature meters, and pressure meters. These instruments are dynamic, physical systems. As a

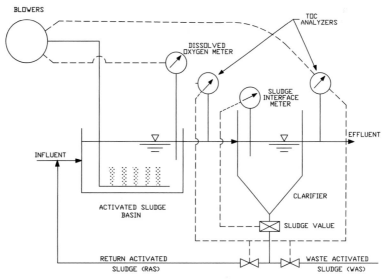

FIGURE 7.44   Automatic process control.

result of the intimate contact and interaction between the process and sensor systems, the sensor produces an output. Since there is a one-to-one mapping between the measured variable and the sensor variable, the process variable is easily obtained from the sensor output.

Wastewater treatment provides a harsh environment for instruments to operate. It is in this harsh environment that they must maintain long-term accuracy. Temperature, moisture, and corrosion shorten instrument life. Due to the electronic nature of most instruments, they must operate within a narrow temperature range of about – 10 to 40°C and a humidity range of 30 to 80 percent. Corrosion can become a problem when hydrogen sulfide, chlorine, sulfur dioxide, and other corrosive chemicals are above certain threshold values.

There are several steps that can be taken in the selection phase to reduce environmental problems. The use of chemical corrosive resistivity charts to choose correct material of construction is often the primary means of extending instrument life. Also, the use of an appropriate enclosure such as National Electrical Manufacturer's Association NEMA 4X enclosure, which is the designation for a watertight, indoor/outdoor, corrosion resistant enclosure.

An electrical instrument performs many tasks. In order to gather and relay the appropriate information, a series of tasks must be performed. First, the variable in question must be converted to an electronic signal. This is the primary sensor task. The electronic signal must be scaled appropriately, amplified, linearized, and cleaned (noise reduction) if necessary. Environmental factors such as temperature and pressure often influence the sensor reading. The effects of temperature and pressure are compensated for in the variable determination.

7.2.4.2 Analyzers. The analyzers of primary interest in wastewater treatment include those that quantify dissolved oxygen (DO), chlorine residual, pH, and suspended solids. The premise behind the DO, Cl, and pH meters is that the amount of current generated between a set of electrodes is related to the amount or level found in a representative sample within an electrolytic cell. The accuracy of the DO and Cl analyzers are about 3 percent, while pH can be measured to within 0.1 units. Suspended solids are measured by quantifying the amount of light or ultrasonic wave transmitted or scattered by the stream. Accuracy varies with the concentration of suspended solids, but a ballpark figure of 5 percent can be expected.

7.2.4.3 Flow Meters. There are a wide variety of materials flowing in a wastewater treatment facility. Solids, liquids, gases, and various combinations of the states must be metered. One of the primary issues of flow measurement in wastewater treatment is flow meter fouling that occurs as a result of solids. There have been many varieties of flow meters developed to handle the extreme compositional variety. For liquids and liquids/solids metering that are conductive, magnetic flow meters are preferred due to their accuracy (about 1 percent), for a wide range of solids concentration, minimal head loss, and their serviceability. For lower solids concentrations, a sonic flow meter may be more economical. If the liquid has a very small solids concentration and a head loss of 5 PSI can be tolerated, a turbine flow meter can be used, which is extremely accurate (0.25 percent). Venturi meters are commonly used for liquid/gas flows. To measure gas flow, either an orifice plate or an averaging pitot is used in combination with a differential pressure measuring device.

The previously mentioned flow meters are all of the closed conduit variety. For high solids concentrations, an open conduit flow meter may be used. Parshall flumes and Palmer-Bowlus flumes are the open conduit flow meters used most often. These devices consist of a specially shaped conduit, which alters the liquid height, in conjunction with a level meter. There are mathematical models which relate flow depth through these devises to flow. Weirs include a shaped plate over which the fluid must flow. Flumes are similar to Venturi flow meters in that they present a known geometry to the approaching flow. These devices are accurate to within 5 percent. The accuracy is severely degraded if the approaching flow is not tranquil and if clearance recommendations are not observed. Flow can also be mathematically modeled using an orifice. Appendices A-11 through A-16 give flows for various weir and orifice configurations.

7.2.4.4 Level. There are two sources requiring level information: open conduit flow meters and tanks. For open conduit flow meters, the most commonly used level measurement methods are floats bubblers, capacitance probes, ultrasonic probes and radar. The floats and capacitance probes require regular cleaning and removal of coated material to maintain their accuracy. Liquid level in a tank can be measured by the conceptually simple method of using a float or floatation device coupled with an indicator. The accuracy of floats is severely impacted by build-up of material coating the surface which changes the buoyancy properties. Alternatively, a non-contacting sonic, ultrasonic, or microwave sensor may be used. These devices measure the return time of waves reflected off the liquid surface. They are not appropriate for use when there is a dense foam or froth on the liquid surface, in which case a radar sensor is preferable. The sonic level meters are also commonly used in conjunction with flumes. An indirect level measurement method is a load cell, or pressure sensor used when the density of the liquid is relatively constant over time.

7.2.4.5 Pressure. Bellows, diaphragms, and Bourdon tubes are pressure sensing elements that are commonly used. If appropriate, isolation diaphragms and purge systems are used. These elements can be used in any wastewater treatment process to measure flow, level, or gauge pressure.

7.2.4.6 Temperature. Process temperature measurement can be accomplished through the use of thermocouples, Resistance Temperature Detectors (RTD), and thermisters, and thermal bulbs. Although the ruggedness of thermocouples and thermal bulbs is beneficial, they are less accurate than RTDs and much less than thermisters.

7.2.5 Controllers

A controller receives information about the process from the sensor. Based upon this information, the controller makes an automatic and intelligent decision regarding the action that must be taken to operate the process. The two most common distinct control units are the automatic continuously regulating controller and the sequential state controller. These individual units can be networked into a distributed control system (DCS) or a supervisory control and data acquisition (SCADA) system. PC-based systems used for automatic process control blur the lines between the different types of controllers. A comparison between today's DCS and SCADA systems shows a convergence of capabilities and the terms are used synonymously. In the very near future, the technology will have merged to the point where it may be accurately termed a DCS/SCADA system.

7.2.5.1 Continuously Regulating Controller. The most common continuously regulating controller is the Proportional-Integral-Derivative (PID) type. A PID controller regulates a process by continuously changing process variables. With a PID controller, feedback control can be successfully implemented in most cases. Tuning of PID controllers has historically been a difficult and time consuming task. Today, however, many PID controllers are auto-tuning, meaning they have the built-in ability to determine the most appropriate constants for the process being controlled. This is not only important during initial setup of the controller, but also, as conditions change or the process characteristics drift, re-tuning of the controller is easily and quickly accomplished.

7.2.5.2 Sequential State Controller. The most commonly used sequential state controller is the Programmable Logic Controller (PLC). The PLC is a discrete event-oriented type controller primarily utilizing sequential logic solving. Depending upon the values of the variables input to the PLC, a specific set of instructions will be executed sequentially. The sequence of steps is programmed into the PLC using ladder logic. The actions taken by the PLC change the state of the system. Common actions taken by a PLC include turning motors and pumps on and off, and opening or closing a valve, etc.

7.2.5.3 Distributed Control System (DCS). The PLCs and PID controllers can be net-worked together in a modular fashion along with any other sensors, as well as information end-users, into a distributed control system (DCS). There are several benefits of having a DCS. First, complete plant information is readily available to plant operators in order to make supervisory control decisions and handle process alarms. Second, all of the information gathered is available for the individual control units to operate. This action necessitates very high speed I/O.

7.2.5.4 Supervisory Control and Data Acquisition (SCADA). Supervisory control and data acquisition (SCADA) systems are quite similar to DCSs in that they both have access to the network of distinct PLC and PID controllers. However, the emphasis in DCS is on time determinism, speed and control capability, while the emphasis is on I/O alarming, logging and viewing in SCADA. Also, since SCADA performs supervisory control (setpoint determination), a DCS could be embedded within a SCADA system. This is not to say that DCS is a subset of SCADA.

7.2.5.5   PC.   The emergence of small, high speed, and low cost personal computers (PC's) has lowered the cost of automatic control systems. With the appropriate software and data acquisition hardware, the PC can perform each of the tasks mentioned previously.

7.2.6   Final Control Elements

Final control elements complete the process control loop by altering one or more process variables. They can either be state-based, such as on/off pumps, or continuously variable, such as control valve position. To manipulate a final control element, a signal is sent to its actuator. An important factor in selecting final control elements is the fail-safe mode. The important question to ask is in what mode should the final control element be in if plant utilities were lost.

7.2.6.1   Actuators.   Actuators are the physical equipment which receive the controller action request and reposition the control valve or turn a motor on or off. These devices receive either a pneumatic or electrical signal and convert it to a desired state. Although pneumatic actuators are rugged, they also require additional air supply equipment; for instance, an air compressor, air dryer, or pressure regulator may need to be installed.

7.2.6.2   Control Valves.   A control valve alters the flow in a pipe. Valve position is a continuous variable which has values ranging from completely closed to completely open. Valves are selected based on appropriate material of construction, valve type (globe, ball, butterfly, diaphragm, plug), flow characteristics (quick opening, linear, equal position), and valve coefficient C_v. The valve coefficient is typically calculated based upon the largest permissible pressure drop. See "Valve Design," Section 7.1.5.6 of this chapter.

7.2.6.3   Motors/Pumps.   Pumps provide the hydraulic head necessary for flow. Motors drive pumps and other mechanical process equipment. There are two basic types of pumps and motors: fixed-speed (on/off) and variable-speed. See "Pump Design," in section 7.1.5.8 of this chapter.

7.2.6.4   Communication/Signal Transmission.   In order for the various instruments of equipment to communicate the process and control information properly, they need to be able to transmit this information and communicate it in a common language.

7.2.6.5   Transmitters.   Transmitters are basically amplifying relays. The output signal produced by many instruments is not strong enough to travel very far. Pneumatic and electrical are the two most commonly used transmitters. Hydraulic transmitters have also been used in the past. Pneumatic transmitters output an air stream with a pressure in the range of 3 to 15 psi that can easily travel anywhere on site. Electrical transmitters output a current in the range of 4 to 20 ma.

7.2.6.6   Signal Converters.   Often two devices need to have a "translator" to convert between different communication standards. For instance, a flow transmitter may output an electrical signal which is being sent to a pneumatically actuated control valve. In order for this to work, the signal must first be converted from 4 to 20 ma to 3 to 15 psi.

Another type of signal conversion that takes places is in the case of data acquisition by computers. The language that computers speak is binary numbers. Not only do computers operate their set of instructions on binary numbers, or rather digitally, but computers input and output must be digital as well. However, transmitters typically send an analog signal between 4 to 20 ma or 3 to 15 psi. In order for the computer to access this information, an analog-to-digital converter is required. The A/D converter samples the analog information stream sent to the transmitter and discretizes it into digital information. Likewise, a digital-to-analog (D/A) converter converts the computers digital information into an analog signal that the other equipment in the control loop can use.

7.2.6.7   Networking.   In order for sensors, PLCs, and PID controllers to function correctly in a DCS or SCADA system, they must have the capability to be properly networked. There are many standards for connectors such as RS-232-C, RS-423-A, RS-422-A, and RS-485. These connection standards differ in transmission length, number of receivers, transmission rate, etc. Network technology from the computer industry has been applied to control system connectivity as well. The control system can be implemented in a TCP/IP-based network. The control units can be in remote locations with information transferred over wireless networks.

7.2.7   Examples of Automatic Wastewater Control

7.2.7.1   pH Control.   One of the most common methods of control is by pH. By definition, pH is the negative logarithm of the hydrogen ion concentration, so that it is a practical measure of the acidity, or lack thereof, of a liquid. This measurement can be accomplished continuously with a pH probe. An automatic control system will typically consist of a probe, a transmitter, a receiver, a signal converter and a controller. The computer can be programmed to give a signal at several low and/or high pH levels. Signals at these pH levels can open and/or close valves on pipes containing acid or base liquids, or turn on or off pumps.

A caution in the use of pH control is that the pH probes can become fouled with solids which affects their accuracy. Therefore, the probes should be cleaned periodically, normally between daily and weekly. Experience will dictate the frequency. Cleaning can be accomplished by removing and cleaning in accordance with the manufacturer's recommendations. Cleaning frequency can be minimized by placing the probe in a self cleaning velocity condition. Any liquid used for this cleaning must not affect the pH reading.

Calibration of pH meters should be done periodically, normally weekly, by comparing the reading with a portable calibrated pH meter. Experience will dictate the required frequency of calibration.

pH control will be explained in more detail in Chapter 9.

7.2.7.2   Flow Control.   Another common method of automation is by flow. Flow is measured using liquid depth in a controlled passage, liquid energy (such as velocity as measured by a propeller or differential pressure), the movement of solids in liquid, or the movement of sound in liquid.

Like pH control, a flow control system typically consists of a measuring device, a transmitter, a receiver, a signal converter and a controller. The controller can actuate pumps, chemicals, dissolved oxygen, or other variables. One practical use of a controller, consists of a combination of flow measuring devises, such as venturi, and a flow control valve. This system is only feasible in a relatively clear liquid, and can be controlled electronically or mechanically.

Any flow measuring devise is subject to corrosion, scaling or solids collection. If the cross sectional area or shape of the devise is altered, the flow reading will be changed and the devise will have to be cleaned, repaired or replaced. For solids accumulation, the solution is periodic cleaning or screening to prevent accumulation. A caution is that screening may remove the solids which should be subsequently treated, if flow measurement is toward the head of a system.

Flow control will be explained in more detail in Chapter 8.

7.2.7.3   Dissolved Oxygen Control.   In aerobic biological wastewater treatment, as explained in Chapter 10, bacteria must have oxygen, as do all animals, as an energy source to break hydrocarbon food (usually measured in COD, BOD or TOC) into CO_2 and H_2O. Since the oxygen concentration level is critical in this process, usually optimal between 0.5 mg/l and 6.0 mg/l, it is imperative that the dissolved oxygen (D.O.) level be routinely measured.

D.O. in the form of air, pure oxygen, or an oxidizing chemical, must be added to the system by passing the gas through the liquid or passing the liquid through the gas. The amount of gas added to the liquid or the amount of liquid sprayed through air can be continuously controlled by D.O. meters in order to optimize the D.O. level, thereby maximizing biological reaction rate and minimizing cost.

A caution in the use of automatic D.O. control, is that the D.O. probes must be cleaned and calibrated periodically to maintain accuracy.

Cleaning can be accomplished by removing and cleaning in accordance with manufacturer's recommendations. Partial cleaning, in order to minimize removal, can be accomplished by assuring that a liquid movement sufficient to hydraulically clean , passes the probe continuously, or at least periodically. The washing liquid must contain the same D.O. as the vessel to prevent affecting the D.O. reading.

Calibration of D.O. meters should be done by comparing the D.O. reading with a calibrated portable D.O. meter periodically. The frequency of calibration should be based on experience with the loss of accuracy.

7.2.7.4   ORP Control.   As explained in Chapter 9, wastewater can be made less objectionable in many cases by using oxidation-reduction chemistry. Instruments have been developed to measure the potential for oxidation or reduction in volts. This measurement is typically used to control oxidation or reduction chemicals to affect the treatment efficiency. Typically, the system will consist of a probe, a transmitter, a receiver, a signal converter and a controller. The controller will usually control on-off valves or a chemical feed pump.

7.2.7.5   Sludge Interface Control.   If settled or floatable solids become too deep in a basin, solids carry over will occur. In order to prevent this, solid depth must be measured periodically. This can be done continuously with an interface measuring devise using ultrasonic and/or turbidity to differentiate between solids of different densities. The signal from this devise can be used to automate solids removal from the basin by opening a valve or operating a pump. The system will typically consist of the interface measuring device, a transmitter, a receiver, a signal converter and a controller.

A simple method to automate solids removal is to control the solids discharge valves or pumps with an adjustable time clock.

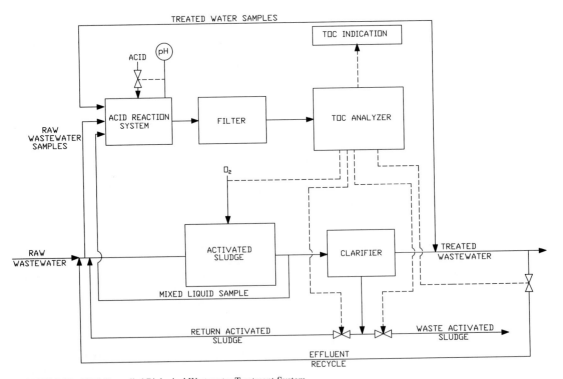

FIGURE 7.45   TOC Controlled Biological Wastewater Treatment System.

7.2.7.6 Total Organic Carbon Control.   Total Organic Carbon (TOC) is a measure of the concentration of a carbon except for CO and CO_2 in a liquid. TOC can be used to measure food and/or microorganisms in any organics removal wastewater treatment system. A TOC analyses is not literally continuous, but does make TOC measurements, up to several samples each minute. The TOC Analyzer will analyze gaseous CO and CO_2 in a sample and then combust the remaining organics, converting them into CO_2 and measure that result. There is a theoretical relationship between TOC and BOD and COD as explained in Chapter 3.

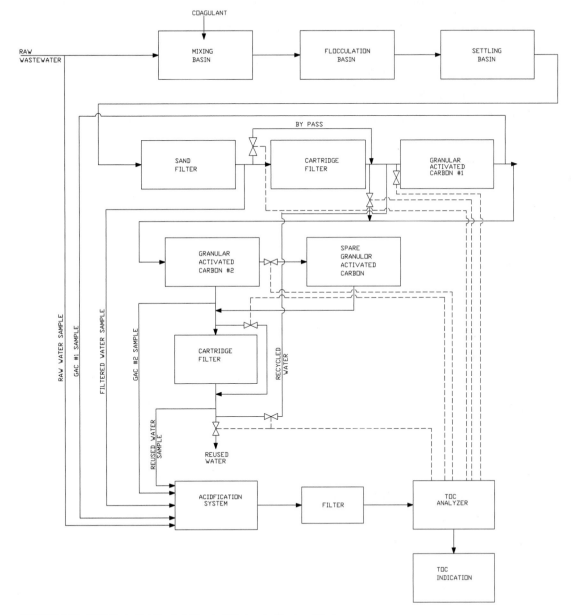

FIGURE 7.46   TOC Controlled Physical Chemical Wastewater Treatment System.

The TOC Analyzer can be connected to a controller through a transmitter, a receiver, and a signal converter. The controller can be used to add air or oxygen, add chemicals, recirculate sludge or waste sludge. A rotator system can be installed to alternate between several sample points. Figure 3.9 shows a photograph of an on line TOC Analyzer at a chemical plant and Figure 7.45 shows a schematic drawing of a TOC controlled biological system. In a chemical or physical wastewater treatment system, a TOC Analyzer can be used to backwash filters or to take a carbon adsorption system off or on line as shown in Figure 7.46. One important use of the system shown in this Figure is the early detection of carbon column breakthrough.

A caution in the use of TOC Analyses is that solids can plug the system. Normally TOC Analyzers are preceded with a cartridge filter to remove these solids, but any of the solids which contain organics then are not analyzed. A solution to this problem is to acidify the solids containing samples before they are filtered and analyzed to dissolve the solids. Then, only inorganic solids will be filtered out.

7.2.7.7   Zeta Potential Control.   In coagulation, as explained in Chapter 9, the change of a solid is typically altered to enhance settleability. A Zeta meter measures this change and can be used with a transmitter, a receiver, a signal converter, and a controller, to control the amount of coagulant for optimization of the process.

CHAPTER 8
PHYSICAL TREATMENT

If physical pollutants are determined to be present in a wastewater, physical treatment methods should be considered. These methods are generalized as follows:

- Liquid equalization
- Solids/liquid separation
- Liquid/liquid separation
- Gas/liquid mixing
- Liquid/liquid mixing
- Solid/liquid mixing
- Temperature control

Physical treatment can also be used prior to chemical or biological treatment to remove a constituent which might cause problems with subsequent treatment. Physical pollutants for the purpose of this Handbook, consist of flow, solids, liquids of varying specific gravities, and temperature.

8.1 EQUALIZATION

8.1.1 Introduction

Equalization is the process of reducing or dampening variations in water or wastewater characteristics. There are actually two processes that fall under this topic: flow equalization and concentration equalization, addressing water quantity and water quality, respectively. Flow equalization is achieved by retaining water during surge periods and releasing water during low flow periods. Concentration equalization is the process of mixing high concentration water with low concentration, yielding intermediate concentrations. Both processes require the routing of the wastewater stream through storage facilities.

Systems can be configured to equalize flow, concentration, or both. Figure 8.1 summarizes the types of systems and the results achieved.

Applications of equalization to industrial and municipal wastewater treatment include the following:

1. Reducing peak flows to downstream treatment processes including stormwater
2. Reducing peak concentrations of key constituents for various downstream treatment processes
3. Optimizing conditions for downstream biological processes, including reducing peak organic loading, reducing and slowing variations in organic loading, reducing peak concentrations of toxic constituents, and raising minimum organic loading and flowrate to maintain biomass

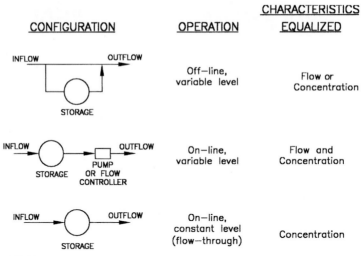

FIGURE 8.1   System configurations.

4. pH control (i.e. to minimize the use of chemicals)

5. Controlling discharge of industrial wastes including stormwater to municipal systems in order to distribute waste loads

In general, the economic value of equalization is that by reducing peak flows and concentrations, the size of more expensive downstream treatment facilities can be reduced.

Depending on the particular application, the criteria for effluent concentrations from the equalization system will include one or more of the following: a minimum and/or maximum outflow, a minimum and/or maximum effluent concentration, and a maximum change in effluent concentration. These criteria will dictate the size of the system needed.

8.1.2   Theory

8.1.2.1 *Flow Equalization.* A common design procedure, presented in the design section of this chapter, involves estimating the required storage capacity to provide a constant outflow. For a flow equalization system that can provide a constant outflow, excess inflow is stored in the basin, and flow shortages are supplemented by flows from the basin. See Fig. 8.2.

Although capacity is usually estimated graphically, a presentation of the theory behind the design technique can be valuable. A practical relationship between flow and the volume of water in the basin can be established by assuming that the outflow is constant and recognizing the principle of volume balance, where the rate of accumulation in the basin is equal to the inflow minus the outflow:

$$\frac{dV}{dt} = Q_{in} - Q_{out} \tag{8.1}$$

By integrating the equation, the relationship for the volume in the basin at any time t is:

$$V - V_0 = \int_0^t Q_{in}(t)dt - Q_{out}\, t \tag{8.2}$$

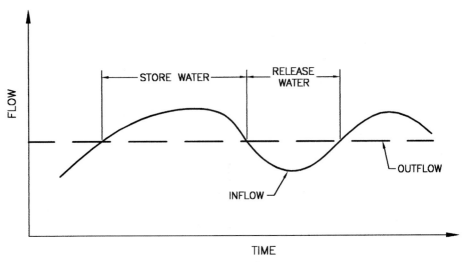

FIGURE 8.2 Flow equalization.

where t = time interval or elapsed time
V = volume at time t
Q = flow
V_0 = volume in the basin at the beginning of the time interval.

It should be noted that the relationship is only valid for non-negative values of V—that is, the basin can never be less than empty. Because inflows cannot typically be represented by mathematical relationships, the integral for inflow is left unsolved. To better relate the theory to the design procedure, the terms are rearranged as follows:

$$V - V_0 = \int_0^t Q_{in}(t)dt - (Q_{out}t - V_0) \tag{8.3}$$

Figure 8.3 shows the relationship graphically, with one line representing the cumulative inflow, and the other the cumulative outflow minus the initial volume of water.

8.1.2.2   Concentration Equalization.   Design techniques for concentration equalization focus on achieving sufficient dampening to reduce peak concentrations of certain constituents, and in the case of biological treatment, to raise low BOD loadings or to prevent large swings in loadings. It is useful, then, to establish the effect of storage capacity on dampening.

In design, concentration equalization systems are usually considered to be completely mixed—that is, influent constituents are instantaneously dispersed evenly throughout the basin. (In reality, this is often not the case—see the "Design" section). For a completely mixed basin with a constant water volume (i.e. a flow-through system), a mass balance can be used to establish a relationship between volume and concentration. For a particular constituent, the rate of change of its mass in the basin is equal to its influent mass rate minus its effluent mass rate:

$$V\frac{dC}{dt} = QC_{in} - QC_{out} \tag{8.4}$$

C = Concentration

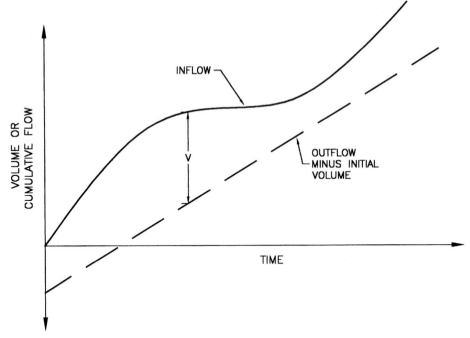

FIGURE 8.3   Cumulative inflow and outflow vs. time.

In many cases where historical or projected flow data are being used to size a system, it is practical to assume that the flow and concentration values at each time interval are constant over the interval. In this case, the equation can be integrated to the following form:

$$C = C_{in} - (C_{in} - C_{out})\, e^{-Qt/v} \tag{8.5}$$

A non-exponential approximation of the mass balance equation has also been applied to design techniques:

$$C = C_0 + \frac{(C_{in} - C_o)Qt}{V} \tag{8.6}$$

Equation (8.6) is a rougher approximation, and it is only valid when the total volume of water passing through the basin, Qt, is less than the volume of water in the basin, V.

8.1.3 Design

8.1.3.1 Flow Equalization Capacity.   Typically, basins are sized graphically using historical or projected flow data. Basins are designed assuming that flow changes are cyclical. In municipal wastewater treatment, the cycle is usually daily (diurnal). In industrial wastewater treatment, flow characteristics are typically dictated by industrial operations unique to each situation.

The design technique consists of developing a graph of the cumulative inflow during the cycle period, with boundary lines above and below.

The boundary lines should be parallel to the cumulative outflow line (that is, their slope should be equal to the outflow) and should just touch the inflow line. If the inflow curve stays on only one side of the effluent line, then there will be only one boundary line drawn, and the effluent line serves as the other boundary line. The lower boundary line is similar to the one in Fig. 8.4: any time, t, the difference between the inflow curve and the lower boundary line represents the volume of water in the basin. The difference between the upper boundary line and the cumulative inflow curve at any time, t, is the remaining capacity in the basin. The vertical distance between the two boundary lines is the total required capacity of the basin.

Problem: It is desired to provide flow equalization for an industrial wastewater stream. Plant discharges are on a weekly cycle. A set of historical flows has been selected for design, as shown below. Determine the required storage capacity of the basin.

Time		Point No.	Flow (MGD)	Cumulative flow (MG)
		0		0.000
Sunday	am	1	0.043	0.022
	pm	2	0.025	0.034
Monday	am	3	0.092	0.080
	pm	4	0.089	0.125
Tuesday	am	5	0.037	0.143
	pm	6	0.057	0.172
Wednesday	am	7	0.268	0.306
	pm	8	0.19	0.401
Thursday	am	9	0.164	0.483
	pm	10	0.202	0.584
Friday	am	11	0.154	0.661
	pm	12	0.164	0.743
Saturday	am	13	0.039	0.762
	pm	14	0.034	0.779

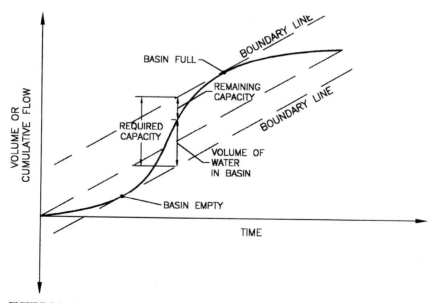

FIGURE 8.4   Flow equalization capacity design.

Solution: It is assumed that the system will discharge a constant flowrate. The cumulative inflows and outflows are graphed. Boundary lines are drawn parallel to the outflow line. The vertical difference between the boundary lines is measured, and the required storage capacity is 237,000 gallons. See Fig. 8.5.

8.1.3.2 Concentration Equalization Capacity.   With adequate flow and concentration data, a system can be sized by developing a model of the effluent concentration from a completely mixed basin of a given size. The model results can then be analyzed, and the basin volume can be adjusted until the results meet the design criteria.

Assuming that influent flow and concentration are constant, Eq. (8.5) can be applied to each time interval, with C_0 equal to the value of C from the previous time interval. For the first time interval, C_0 can be set at the average influent concentration or thereabouts. See Fig. 8.6.

The model is easier to understand through an example.

Design example

Problem: Given the following data, determine the required capacity of a completely mixed, flow-through equalization basin that will reduce the peak concentration to 800 mg/L.

Day	Average flow (gpm)	Average influent concentration (mg/L)
1	45	340
2	85	412
3	20	600
4	24	1005
5	90	1200
6	78	815
7	95	735
8	19	210
9	20	280
10	22	250

Solution: Using Eq. (8.5), a spreadsheet is developed that calculates the effluent concentrations at the end of each day. The basin volume is then adjusted until a peak concentration of 800 mg/L is achieved. The required basin volume is 240,000 gallons.

Upon completion, the spreadsheet contains the following relationships:

	A	B	C	D
	Day	Average flow (gpm)	Average influent concentration (mg/L)	Effluent concentration at end of day (mg/L)
1		Basin volume (gal)	240000	
2				
3				
4				
5	1	45	340	400 = +C5-(C5-D4)*EXP(-B5* (A5-A4)*1440/$C$1)
6	2	85	412	= + C6-(C6-D5)*EXP(-B6*(A6-A5)*1440/$C$1)
7	3	20	600	= + C7-(C7-D6)*EXP(-B7*(A7-A6)*1440/$C$1)
8	4	24	1005	= + C8-(C8-D7)*EXP(-B8*(A8-A7)*1440/$C$1)
9	5	90	1200	= + C9-(C9-D8)*EXP(-B9)(A9-A8)*1440/$C$1)

	A	B	C	D
	Day	Average flow (gpm)	Average influent concentration (mg/L)	Effluent concentration at end of day (mg/L)
10	6	78	815	= + C10-(C10-D9)*EXP(-B10*(A10-A9)*1440/$C$1)
11	7	95	735	= + C11-(C11-D10)*EXP(-B11*(A11-A10)*1440/$C$1)
12	8	19	210	= + C12-(C12-D11)*EXP(-B12*(A12-A11)*1440/$C$1)
13	9	20	280	= + C13-(C13-D12)*EXP(-B13*(A13-A12)*1440/$C$1)
14	10	22	250	= + C14-(C14-D13)*EXP(-B14*(A14-A13)*1440/$C$1)

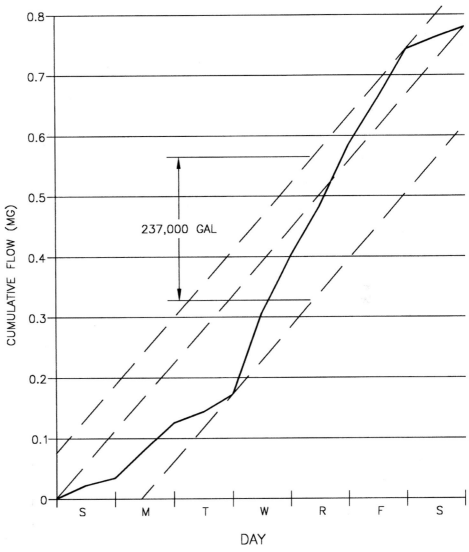

FIGURE 8.5 Flow equalization capacity design example.

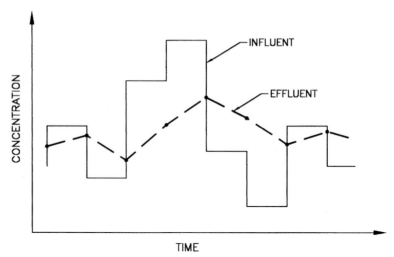

FIGURE 8.6   Concentration equalization capacity.

See Fig. 8.7

The results are tabulated below, along with a graph of the influent and effluent concentrations.

Day	Average flow (gpm)	Average influent concentration (mg/L)	Effluent concentration at end of day (mg/L)
			400
1	45	340	386
2	85	412	396
3	20	600	419
4	24	1005	498
5	90	1200	791
6	78	815	800
7	95	735	772
8	19	210	711
9	20	280	662
10	22	250	611

8.1.3.3   Combined Systems.   When it is desired to attenuate both flow and concentration, a single basin with a level-controlled pumping system can be used. See Fig. 8.8.

Concentrations are dampened by setting the pump shutoff level to provide a minimum volume of water in the basin. The upper portion stores excess inflows and releases them during low-flow periods.

A simple design technique can be used to size such a system by treating it as two basins stacked on top of each other, the bottom one being for concentration equalization, the top one for flow. The total basin capacity required can be determined by estimating the individual capacities, and then summing them.

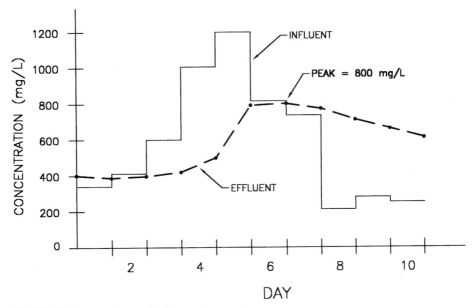

FIGURE 8.7   Concentration equalization capacity example.

An alternate design technique, similar to that for concentration equalization alone, consists of developing a model of the effluent flows and concentrations, and adjusting high and low water levels until satisfactory results are achieved. Such a model can be useful when variations in flows and concentrations are related. An example of this is when high flows are associated with low concentrations, and vice versa. Output from the model could also be used as data for design of downstream treatment process.

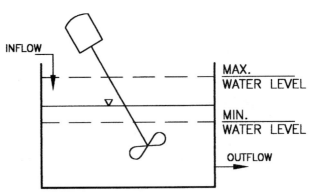

FIGURE 8.8   Level-controlled system.

Assuming influent concentrations and influent and effluent flows are constant, a relationship for the effluent concentration from a completely mixed basin is:

$$C = C_{in} - (C_{in} - C_0)\left(\frac{V}{V_0}\right)^{Q_{in}/(Q_{in}-Q_{out})}$$

(8.7)

where the volume of water in the basin, V, is:

$$V = (Q_{in} - Q_{out})t + V_0$$

(8.8)

These relationships are only valid for non-negative volumes.

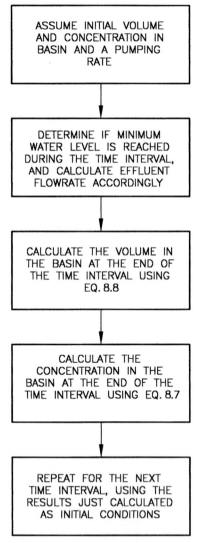

FIGURE 8.9   Calculation of effluent flow and concentrations flowchart.

Because the basin level controls effluent flows, a model must include a decision-making process. A flowchart of such a model is shown in Fig. 8.9. This flowchart could be used to set up a program or spreadsheet that calculates effluent flows and concentrations.

8.1.4 Equipment Installation and Operation

Equalization systems are typically near the front of the treatment flowpath. Due to their simplicity, equalization systems can often be combined with other treatment processes, such as pH adjustment and aeration. System volumes vary from a few thousand gallons to several million gallons, and accordingly, basin designs will vary. Small and medium sized systems can utilize circular or rectangular basins of steel or concrete. Very large systems are often configured as earthen lagoons, with floating agitators.

In industrial applications, wetted surfaces of the basin may need to be lined or painted to prevent corrosion. Where leakage is possible, basins may need to be equipped with secondary liners or containment to prevent contamination. Liner selection should be made with consideration to wastewater constituents.

Most systems will need to provide mixing. For concentration equalization, it is very important to provide adequate mixing to disperse constituents. For flow equalization, mixing is necessary only to prevent the deposition of solids in the basin. Sometimes no mixing is used, and the basin is dredged or cleaned periodically. The shape of the basin will affect mixing efficiency. A circular shape is well suited for mixing. For a rectangular basin used for concentration equalization, it is best to use a length to width ratio close to 1.0. It is also helpful to eliminate corners or pockets that may act as dead zones. Common mixing equipment choices are: vertical (floating) or horizontal (through-wall) impeller-type mixers, jet mixers utilizing pumps, and aerators.

Most equalization systems can be set up to operate automatically. Variable-level systems are typically set up with a level sensor/controller that starts and stops the pump(s) on maximum and minimum water level. Mixers and automatic pumping systems should be checked regularly for proper operation.

8.1.5 Design Hints

- Size to equalize flow should be greater than flow cycle.
- Mixing with diffused air:

  Partial mixing: 4 to 6 scfm/1000 ft$_3$

  Complete mixing: 20 to 30 scfm/1000 ft$_3$
- Mixing with surface aerators:

  Partial mixing: 0.1 HP/1000 g

  Complete mixing: 0.5 HP/1000 g
- Mechanical Mixing

  $P = kpn^3D^5$
- Design mixers to prevent solids accumulation or provide for floatable and settleable solids removal.
- Length to width to depth ratio for around 1:1:1.
- Basin shape approaching a sphere (ideal shape) is preferred.

  See paragraph 8.10.3.5 for a definition of units

  Figure 8.10 shows a flow concentration equalization basin at an Organics Chemical Plant. The pH Control System is shown in the upper right. Figure 8.11 shows a lagoon type flow/concentration equalization basin which is inadequately mixed resulting in floating scum.

FIGURE 8.10   Flow/concentration equalization basin.

FIGURE 8.11   Inadequately mixed lagoon equalization basin.

8.2 SCREENING

Screening is a step used to separate inorganic solids from a liquid by particle size. The purpose of screening is to eliminate these inorganic solids by the least expensive method for separate disposal, either for economic reasons or to prevent plugging of subsequent treatment stages.

There may be a need to wash removed screenings with a spray system to minimize odors.
The simplest form of solids separation is grids or sieves of a spacing less than the diameter of the solid being removed. The solids caught on these screens must be removed, manually or automatically, through a physical or hydraulic system. Screening systems do little to affect the BOD of a wastewater since the solids caught are usually inorganic and would not be caught in a BOD sample even if they were organic. The screenings should be limited to inorganics as much as possible to minimize odors. Examples of screens are as follows:

- Bar screens

  Bar screens are parallel sloping bars that trap the solids and are normally cleaned manually with rakes or mechanical devices.

  Design suggestions are as follows:
  - Bar screen spacing—1 inch (2.5 cm)
  - Total area of openings under minimum water level—2 times perpendicular area of sewer line
  - Slope of bars—30° to 45° from horizontal
  - Velocity through openings under minimum water level

    1.25 to 2.5 fps

    (0.4 to 0.8 mps)

- Provide automatic by-pass in case of plugging

See Fig. 8.12 for the design of a typical bar screen.
The design calculations for this design are as follows:

1. Size of open area = 2 times sewer area

   24 in. ϕ pipe = 1 ft. radius pipe

   Area = $\pi (1)^2$

   Area = 3.14 ft^2 Open Area = Area × 2 = 6.28 ft^2 = 2 ft. high × 3.14 ft. wide

2. 1 in. openings

   $\dfrac{3.14\ \text{ft.}}{1\ \text{in.}} = 37.68$

   ∴ 38 openings, 39 bars

   39 × $\frac{1}{4}$ in. = 9.75 in. = 0.81 ft.

   Total width = 0.81 ft. + 3.14 ft. = 3.95 ft. wide, say 4 ft. 0 in. width

3. Flow rate = 5.0 MGD = 7.74 cfs

   7.74 cfs ÷ 6.28 ft^2 = 1.23 fps

   ≈ 1.25 fps, therefore OK

4. Bar length using 30° angle

   tan 30° = 24 in./length

   Length = 24 in./0.57735 = 41.57 in.

   = 3.46 ft., say 3 ft. 6 in. length

8.2.1.1 Self Cleaned Bar Rack. A bar screen mechanically cleaned is shown in Fig. 8.13

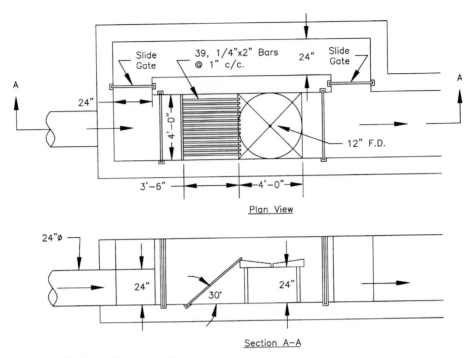

FIGURE 8.12   Typical bar screen design.

8.2.1.2 Traveling Screen.   A hydraulically cleaned rotating mesh screen is shown on Fig. 8.14. These screens can be either horizontal or vertical.

8.2.1.3 Fine Screen.   Fine screens are mesh screens that normally have openings from 15 μm (fine) to 2.5 mm (coarse) in size. These screens are typically cleaned with a spray system but have little BOD reduction.

FIGURE 8.13   Self cleaned bar rack.

FIGURE 8.14   Traveling screen.

A vibrating screen is efficient for lint removal and is shown in Fig. 8.15.

A rotating screen is cleaned in each rotation and is shown in Fig. 8.16.

A wedge wire screen contains sloped louvers using surface tension of the water to separate it from the solids as shown in Fig. 8.17. The separated solids slide down the outside slope. See also Fig. 8.18.

FIGURE 8.15   Vibrating screen. (Sweco)

FIGURE 8.16   Rotating screen. (Sweco)

8.3 SHREDDING

An option or addition to removing larger solids through screening is to shred the solids so they will pass into the wastewater system and be removed through a further process such as flotation or sedimentation. Shredding is especially effective for larger organic solids that will easily settle or float. These solids would cause odors and further increase disposal problems if removed by a screen. Shredding will also prevent clogging of subsequent equipment. A shredder consists of a series of shredding blades acting against a flat or circular bar screen with positive pressure against the screen provided by the water flow. The object must be shredded to a size smaller than the bar screen opening in order to pass through the screen. Typically bar spacing will be approximately $1/4$ inch (6mm).

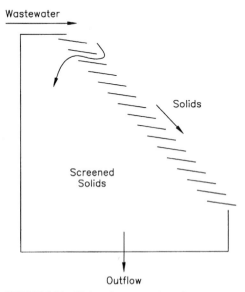

FIGURE 8.17   Wedge wire screen schematic.

FIGURE 8.18   Wedge wire screen installation. (Bauer)

The design criteria includes the hydraulic capacity, the strength of the bars to prevent breakage, the hardness of the blades to minimize wear, the means for removal of solids not shredded, and automatic by-passing so that operation of downstream units can continue if the bars are plugged. Normally, the by-pass system is simply an overflow system for high levels of water.

Figures 8.19, 8.20 and 8.21 show common commercially available systems for shredding.

8.4   *GRIT REMOVAL*

8.4.1   General

The purpose of a grit removal system is to remove inorganic settleable solid particles known as grit. Grit can be sand, small pebbles or industrial waste such as filings, turnings, etc. Generally an inorganic solid that is considerably denser than water can be removed with a grit chamber. Grit chambers are typically placed after screens, but ahead of pumps and other equipment which could be damaged by the abrasiveness of grit. They can be placed either before or after shredders depending on whether there is more probability of unshredded particles plugging the grit chamber or the grit damaging the shredder.

8.4.2   Types of Grit Removal

The types of grit chambers discussed in this chapter are longitudinal chambers, centrifugal chambers and aerated chambers.

8.4.3   Design

A grit chamber, as any sedimentation basin, works best when inlet flow is uniform. This is especially critical for a grit chamber because, if the flow is too large, grit will not be removed, and if the flow

FIGURE 8.19 Comminutor shredder. (Franklin Miller)

FIGURE 8.20   Barminutor for shredder. (Chicago Pump Company)

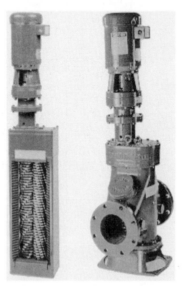

FIGURE 8.21   Muffin monster shredder.
(JWC Environmental)

is too small, organic solids will settle with the grit, causing odors and making disposal more expensive.

8.4.3.1   Longitudinal Grit Chamber.   The velocity of flow through a longitudinal grit chamber should be 0.5 to 1.0 ft/s (0.15 to 0.30 m/s). The detention time in a longitudinal grit chamber should be $1/2$ to 1 minute. In order to limit velocity variations, the following methods can be used:

- Equalization prior to the grit chamber
- In a long, narrow chemical, an effluent device can be used to minimize the velocity variation. In accordance with Eq. (8.9), if the depth of flow is doubled as the flow doubles, the cross-sectional area will also double, allowing the velocity to remain the same. This can be accomplished with an effluent weir that has the characteristic of changing the depth equally with changing flow.

$$Q = V/A \tag{8.9}$$

where Q = flow in ft^3/s or m^3/s
$\quad\quad\ \ V$ = velocity in ft/s or m/s
$\quad\quad\ \ A$ = cross-sectional area in ft^2 or m^2

This is approximated with a long rectangular channel as shown in Fig. 8.22, where $Q_2 = 2Q_1$.

The flow depths are not necessarily proportional to the depths toward the drawdown at the discharge point, but the settling should have taken place before that point.

Settled grit in the relatively shallow channel must be removed manually or automatically with regularity to prevent its affecting the flow.

8.4.3.2   Centrifugal Grit Chamber.   A centrifugal grit chamber typically has a submerged tangential side entrance pipe and a surface exit pipe. The tangential velocity forces the solids to the walls and then to the bottom for removal from a centrally located bottom exit pipe. The slower moving solids in the center of the basin also will settle. The basin should have a surface area of approximately 2000 gpd/ft^2 and the lower side slopes should be at least 60° from the horizontal as shown in Fig. 8.23.

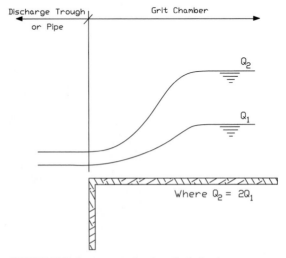

FIGURE 8.22   Long rectangular channel grit chambers.

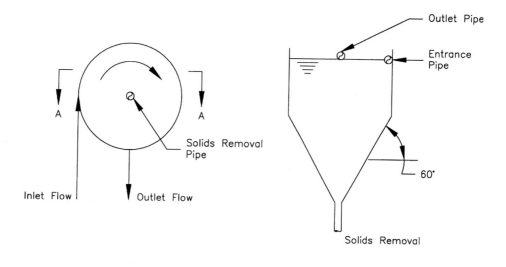

FIGURE 8.23   Centrifugal grit chamber.

8.4.3.3   Aerated Grit Chamber.   The addition of coarse bubbles to a grit chamber will reduce the specific gravity of the wastewater so that the settling velocity of the grit is increased. The organic solids will tend to be suspended by the air bubbles and will not settle. Figure 8.24 shows an Aerated Grit Chamber installation.

The air flow should be adjustable or sized so that the water flow across the bottom is approximately 1 to 2 ft/s (0.3 to 0.6 m/s) so that the grit is washed but not suspended.

The total hydraulic detention time should be 5 to 10 min. and the air flow approximately 3ft³/min./l.f. of tank width (0.28m³/min./m).

8.5   SEDIMENTATION

8.5.1   Introduction

Sedimentation and flotation are the most basic and generally the most feasible of wastewater treatment processes. This section will concentrate on sedimentation as a method for separating solids from liquids. Some of the theory and practical recommendations discussed will also apply to the settling of a dense liquid from a less dense liquid. Terms used synonymously with sedimentation are clarification and settling.

8.5.2   Theory

8.5.2.1   General.   The purpose of sedimentation is to remove settleable solids from liquids. If solids do not naturally settle completely, it is normally more efficient to float the solids. A quick visual jar test will indicate the best solution to sedimentation/flotation, so that a system will not experience the inefficiency of competing with gravitational forces. See Figs. 8.25 and 8.26.

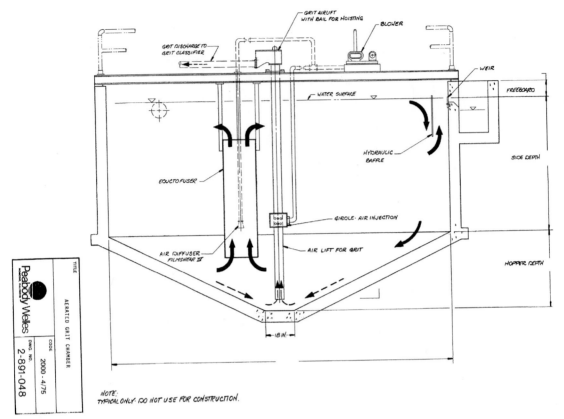

Labels in figure:
GRIT AIRLIFT WITH BAIL FOR HOISTING
BLOWER
GRIT DISCHARGE TO GRIT CLASSIFIER
WEIR
WATER SURFACE
FREEBOARD
HYDRAULIC BAFFLE
SIDE DEPTH
EDUCTO FUSER
GIRDLE - AIR INJECTION
HOPPER DEPTH
AIR DIFFUSER FILMSHEAR II
AIR LIFT FOR GRIT
18 IN.

TITLE
AERATED GRIT CHAMBER
CODE 2000 - 4/75
DWG. NO. 2-891-048
Peabody Welles

NOTE:
TYPICAL ONLY - DO NOT USE FOR CONSTRUCTION.

FIGURE 8.24   Aerated grit chamber.

The design parameters for sedimentation include the specific gravity of the liquid, the specific gravity of the settling particle, the viscosity of the liquid, and the particle characteristics such as shape, diameter, concentration, and surface type. If particles are spherical, of uniform diameter and specific gravity, and are discrete (non-touching), then settling velocity can be determined by Stoke's Law. Particles that interact with each other, either flocculently or compressively, will have a hindered settling rate that must be determined experimentally. Stoke's Law cannot be used for these types of settling since it will predict a greater settling velocity than the actual settling velocity. Increasing the specific gravity of the particle, reducing the specific gravity of the liquid, raising the temperature of the liquid, or reducing the forward flow velocity of the liquid can accelerate sedimentation rates. Figure 8.27 shows the relationship of the three most common variables in sedimentation.

Discrete Settling as determined by Stoke's Law is indicated by point 1 in Fig. 8.27, with homogeneous, dilute, non-flocculent particles. Unfortunately, there are limited actual instances of this type of settling. The closest examples are sand, metal filings, and precipitation of dilute metal salts or hydroxides. Only this type of settling can be accurately modeled by Stoke's Law.

Point 2 in Fig. 8.27 demonstrates homogeneous, concentrated, non-flocculent particles. In this case, the particles can interact with each other as hindered settling. Examples are concentrated sand, metal filings and metal salt hydroxide solutions. Point 6 demonstrates heterogeneous, concentrated, non-flocculent particles. Examples are concentrated sand, metal dust or filings, and metal salts or

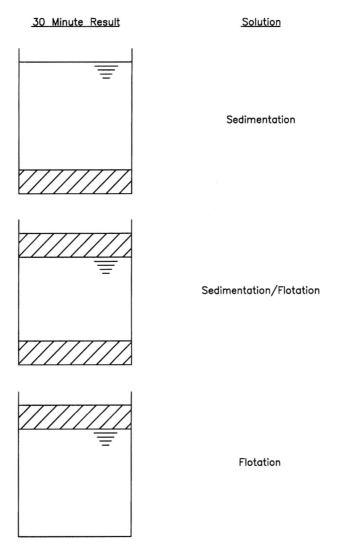

30 Minute Result   **Solution**

Sedimentation

Sedimentation/Flotation

Flotation

FIGURE 8.25 Sedimentation/flotation selection.

hydroxides with a low uniformity coefficient. Points 2 and 6 demonstrate particle characteristics requiring settling rates determined by experimentation and known as **Concentrated Settling.**

Flocculent settling is the condition demonstrated at points 4 and 8 in Fig. 8.27 depending on homogeneity. Flocculent settling rates must be determined by experimentation.

Compression settling occurs with concentrated flocculent particles and is demonstrated at points 3 and 7 in Fig. 8.27. Compression settling must be determined by experimentation.

These types of settling are described as follows in more detail with recommended models and tests for design.

Zone settling may occur in flocculant, homogeneous, concentrated conditions (point 4) when particles, usually due to coagulation as explained in Chapter 9, tend to adhere to one another in a zone and are of such uniformity that the liquid above the zone is clear and a clear interface can be seen

FIGURE 8.26 Photograph of sedimentation/flotation selection treatability study; food processing plant.

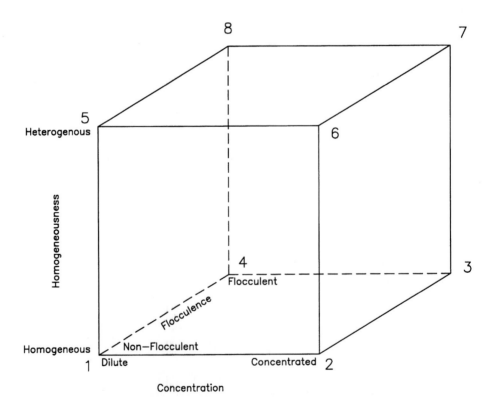

FIGURE 8.27 Sedimentation variables.

between the clear water and the sediment. This condition requires basin sizing by experimentation as explained below.

If the water surface after a few minutes settling has a layer of very clear liquid and the solids interface drops with time, zone settling is present. If the surface area is still turbid after a few minutes and there is no defined interface, flocculant settling is present.

8.5.2.2 Discrete Settling. Discrete settling occurs when there is no variation in size, shape, surface characteristics or density between particles. Particles tend to settle parallel to each other without interference between particles and with uniform velocity so that particles do not interact vertically or horizontally. In this case, particles settle in accordance with Stoke's Law as discussed below.

If observation indicates that particles settle at a uniform velocity between the surface and the sludge interface near the bottom, discrete settling should be present. Because discrete settling is the only type of settling which can be modeled, it is recommended that Stoke's Law be used to calculate the rate of settling.

Stoke's Law can be stated as follows:

$$v_s = \frac{gd^2 (p_1 - p_2)}{18u} \qquad \text{For } 10^{-4} < N_R < 1 \tag{8.10}$$

where v_s = rate of settling, cm/s
  g = acceleration due to gravity = 980 cm/s^2
  d = particle diameter in cm
Note: If Stoke's Law is used as an approximation for the velocity of settling for spherical particles of different but relatively consistent diameters (high uniformity coefficient), the mean grain diameter = d.
  N_R = Reynold's Number
  p_1 = density of particle in g/cm^3
  p_2 = kinematic density of fluid in g/cm^3 (See Ap. C-4)
  u = viscosity of fluid in poises (g/cm-s) (See Ap. C-4)

$$N_R = Vd/v_s \tag{8.11}$$

where V = liquid flow velocity in cm/s
  d = particle or mean grain diameter, cm
  v_s = rate of settling, cm/s
Note: Reynold's Number is related to the coefficient of drag.

The theory of discrete settling is that v_0 is the vertical particle velocity of diameter d_0, that is barely 100 percent removed as shown in Fig. 8.28.

The path of v_0 represents the highest elevation at which a particle of diameter d_0 will be found at time t. Since settling velocity equals depth of settling with time:

$$v_s = \frac{h_0}{t_0} \tag{8.12}$$

where v_s = settling velocity in ft/sec or m/sec
  h_0 = total depth at time t_0, in inches, ft, or m
  t_0 = total time to settle to depth h_o

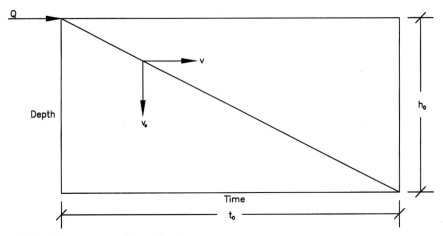

FIGURE 8.28 Discrete settling, uniform d.

Total time, t_0, in a basin also equals the volume of the basin divided by the flow through the basin.

$$t_0 = \frac{V}{Q} \tag{8.13}$$

where V = basin volume in ft³ or m³
$\quad\quad Q$ = flow in ft³/s or m³/s

Substituting Eq. (8.13) for t_0 in Eq. (8.12):

$$v_s = \frac{h_0}{V/Q} \tag{8.14}$$

since

$$V = LWh_0 \tag{8.15}$$

where L = length of basin in ft or m
$\quad\quad W$ = width of basin in ft or m
$\quad\quad h_0$ = basin depth in ft or m

Substitute Eq. (8.15) for V in Eq. (8.14):

$$v_s = \frac{h_0}{\frac{LWh_0}{Q}} \tag{8.16}$$

This simplifies to:

$$v_s = \frac{Q}{LW} \quad (8.17) \quad = v_s = \frac{Q}{A} \tag{8.18}$$

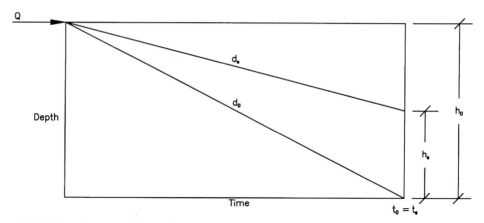

FIGURE 8.29   Discrete settling, variable d.

where A = the surface area of the basin in ft^2 or m^2

Therefore, the efficiency of discrete settling can be expressed in terms of surface loading rate, Q/A, in gal/min/ft^2 or m^3/s/m^2.

See Fig. 8.29. The efficiency of removal of a particle of diameter, d_0, in a variable diameter discrete settling basin can be determined as follows:

Since

$$h_0 = v_0 t_0 \tag{8.12}$$

$$\frac{h_s}{h_0} = \frac{v_s t_s}{v_0 t_0} \tag{8.19}$$

where h_s = depth at t_0 for particle d_s
    h_0 = depth at t_0 for particle d_0
    v_s = settling velocity of particle d_s
    v_0 = settling velocity of particle d_0
    $t_0 = t_s$ = detention time

Since $t_0 = t_s$, these units cancel giving:

$$\frac{h_s}{h_0} = \frac{v_s}{v_0} \tag{8.20}$$

Inserting v_0 in Eq. (8.18):

$$\frac{h_s}{h_0} = \frac{v_s}{\frac{Q}{A}} = f \tag{8.21}$$

where f is the efficiency of removal of a particle of v_s settling velocity.

Using the previous discrete settling theory, one can see that installing equally sized horizontal compartments in a settling basin can increase the settling efficiency, but installing vertical baffles will not affect the settling. See Fig. 8.30.

Since: $A_1 = A_2 = A_3 = A_4$, and $Q = Q_1 + Q_2 + Q_3 + Q_4$, or $A_1 = 1/4\ Q$,

$$v_1 = \frac{Q}{\dfrac{4}{A}} \tag{8.22}$$

where v_1 = settling velocity in Compartment 1, in ft/s or m/s.

Substituting Q in Eq. (8.18) for Q in Eq. (8.22):

$$v_1 = v_s \frac{A}{\dfrac{4}{A}} \tag{8.23}$$

Canceling the areas:

$$v_1 = \frac{v_s}{4} \tag{8.24}$$

Substituting h_0/t_0 in Eq. (8.12) and likewise, h_1/t_1 for v_s and v_1 in Eq. (8.24):

$$\frac{h_1}{t_1} = \frac{h_0}{\dfrac{t_0}{4}} \tag{8.25}$$

Since,

$$h_0 = 4h_1 \tag{8.26}$$

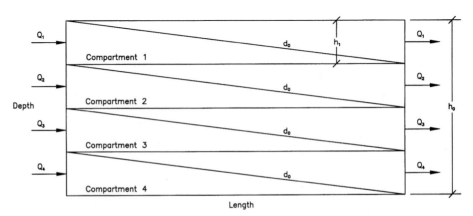

FIGURE 8.30   Multiple horizontal compartment settling.

Substitute $4h_1$ in Eq. (8.26) for h_0 in Eq. (8.25):

$$\frac{h_1}{t_1} = \frac{4h_1}{\dfrac{t_0}{4}} \tag{8.27}$$

where h_1 = depth of compartment 1 in ft. or m.
Simplify:

$$\frac{h_1}{t_1} = \frac{h_1}{t_0} \tag{8.28}$$

Dividing each side by h_1:

$$t_1 = t_0 \tag{8.29}$$

This relation indicates that we can remove a particle with a settling velocity, v_0 in time t_0 in each of the four compartments giving us four times the efficiency of a basin without compartments. See Fig. 8.31.

For a basin of similar size, divided into four vertical compartments, the detention time through each compartment is $t_0/4$ and the velocity is the same as in a non-compartmentalized basin. Therefore, in accordance with Eq. (8.14):

$$v_0 = \frac{h_0}{V/Q} = v_1 = \frac{h_1}{V/4Q} \tag{8.30}$$

or, transposing units:

$$h_1 = \frac{h_0 V/4Q}{V/Q} \tag{8.31}$$

Canceling the V and Q:

$$h_1 = \frac{h_0}{4} \tag{8.32}$$

or a particle of d_0 diameter will settle only $1/4$ of the depth in each of the four compartments.

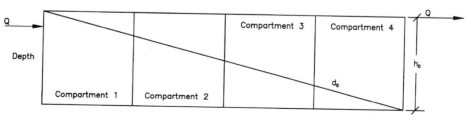

FIGURE 8.31 Multiple vertical compartment settling.

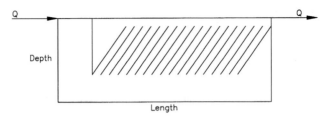

FIGURE 8.32 Inclined plate separator schematic.

8.5.2.3 Inclined Plate Settling. The reason for the depth in discrete settling is for storage capacity and not efficiency. Since it would be difficult to remove sludge from horizontal compartments, the practical application of this theory is inclined plate separators. See Fig. 8.32.

The settling velocity of a particle in each plate is shown in Fig. 8.33.

$$v = \text{the forward velocity} = Q_0/A \tag{8.33}$$

where

$$Q_0 = Q/\text{no. spaces} \tag{8.34}$$

where Q = total basin flow
  Q_0 = flow through one space
  v_s = particle settling velocity
  v_0 = particle settling vector

It can be seen in Fig. 8.31 that a particle of diameter, d_0 will strike the bottom of the inclined plate at a reduced time compared to that in a non-compartmentalized basin.

In order for the particles to be removed automatically from the bottom surface of the inclined plate, the plate slope must exceed the **angle of repose**, θ of the wet sludge. Many particles will have an angle of repose of 60°. Flocculent particles will tend to have a greater angle of repose.

Figures 8.34 through 8.36 and Figs. 8.39 show settling in demonstration tubes with one vertical and one inclined at an angle of 60° with the horizontal at times 0, 5 seconds and 10 seconds.

Typical inclined plate sedimentation basins are shown in Figs. 8.37 and 8.38.

Since discrete settling efficiency is a factor of surface area only, inclined plate settling efficiency is computed by projecting the length and width of a plate to a horizontal plane and totaling the horizontal area of all plates.

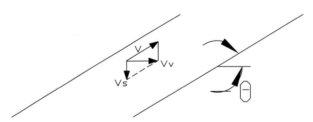

FIGURE 8.33 Inclined plate settling velocities.

 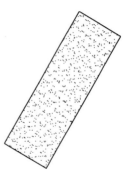

FIGURE 8.34 Settling, time 0.

 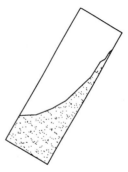

FIGURE 8.35 Settling, time 5 seconds.

 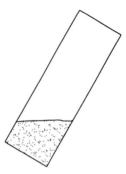

FIGURE 8.36 Settling, time 10 seconds.

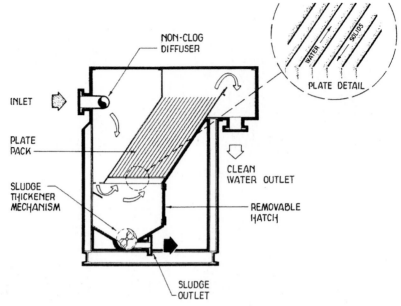

FIGURE 8.37 Inclined plate installed. (Great Lakes Environmental)

The equivalent area of plate P, is the horizontal length L, times the plate width, W.

$$\text{Total plate area} = (L_1 + L_2 + L_3 + L_n)\, W \qquad (8.35)$$

Any surface area not covered by plates should be added to the plate surface area to compute the total surface area.

FIGURE 8.38 Inclined plate photograph.

FIGURE 8.39   Inclined plate separator efficiency.

If the flow through a basin is 100,000 gal/day, the settling basin capacity is expressed as:

$$\text{Capacity} = \frac{100{,}000 \text{ gal}}{(L_1 + L_2 + L_3 + L_n)W \text{ ft}^2} \tag{8.36}$$

8.5.2.4 Flocculent Settling.   Flocculent settling occurs when particles flocculate or aggregate, and agglomerate or cluster. As particles cluster, they settle faster, since their diameter is greater (see Stoke's Law (8.10). Heavier particles with high settling velocities overtake and coalesce with smaller, lighter particles. This opportunity increases with depth.

If observation indicates that particles increase in velocity between the surface and the sludge interface and the liquid near the surface never becomes completely clear, flocculent settling is occurring. The smallest particles are the slowest to begin to settle and keep the surface area turbid.

Practical examples of flocculent settling are primary settling before treatment and the upper portions of secondary settling after biological treatment or with a coagulant. Many industrial sludges will exhibit flocculent settling.

8.5.2.5 Zone Settling.   Zone settling occurs when the particle concentration is so great that the particles touch each other and maintain their spatial relationship, settling as a mass or matrix. The mass usually exhibits a definite interface between the solid mass and the clarified liquid above. The mass settles more slowly than discrete particles because the upward velocity of the water being displaced is so great that it reduces the effective downward velocity of the mass.

If observation indicates that particles decrease in velocity between the surface and the bottom, and the liquid near the surface above the interface is clear, then zone settling is occurring.

8.5.2.6 Compression Settling. Compression settling occurs when the particle concentration is so great that the particles rest on each other and mechanically support each other. The weight of the particles above slowly compresses the lower layers, squeezing out the water. Settling can be slightly accelerated by slow stirring to break up bridging between the particles.

Practical examples of compression settling are the lower levels of secondary settling and sludge thickening.

8.5.2.7 The Effect of Basin Shape on Settling. Figures 8.40 and 8.41 show a rectangular and a circular settling basin.

In a circular settling basin, the flow velocity, v is reduced as the flow approaches the periphery in a center feed clarifier, and is increased in a peripheral feed clarifier. Since a reduced flow velocity is needed to settle the smaller particles, more efficient use is made of the settling basin volume when the larger particles are settled first. For this reason, most designers prefer a center feed clarifier.

In Fig. 8.42a, with the same settling velocities, v_s, the decreasing velocities for the center feed clarifier allow steeper settling slopes and therefore more settling efficiency toward the periphery. See also Fig. 8.43.

8.5.2.8 The Effect of Velocity on Settling. For simplicity, the effects of velocity, entrance conditions, and exit conditions on settling will be demonstrated for discrete settling. Particles experiencing other types of settling will be similarly affected. A particle will have a flow velocity v_F along its path as follows and as shown in Eq. (8.37).

$$Q = v_F A \tag{8.37}$$

where $Q_{M^{3/4}}$ = flow in ft³/s, m³/s
v_F = flow velocity of ft/s, m/s
A_{M^2} = cross sectional area in ft², $M^2 = Wh$

FIGURE 8.40 Rectangular settling basin. (HiTech Environmental)

FIGURE 8.41   Circular settling basin. (HiTech Environmental)

or

$$v_F = Q/Wh \qquad (8.38)$$

All particles will have a settling velocity, v_s vertically downward because of gravity. The velocity in the direction of flow will cause a vector, v_v in the direction of settling.

Depending on the relative magnitudes of the flow velocity v_F and the settling velocity, v_s, the settling particle can settle to the bottom of a basin, or it will be carried out the effluent end.

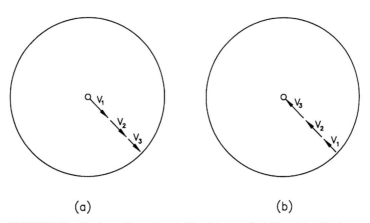

FIGURE 8.42   Circular settling basin velocities. (*a*) center feed, (*b*) peripheral feed.

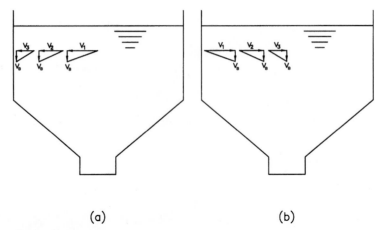

FIGURE 8.43   Circular settling basin velocities, cross-sections. (*a*) center feed, (*b*) peripheral feed.

In most practical cases, the wastewater will have particles with varying sizes, shapes and specific gravities which will all be carried along with the same flow velocity, v_F. In these cases, the smaller, less dense particles are those which are critical in the design of the settling basin. See Figs. 8.44 and 8.45.

8.5.2.9   *The Effect of Entrance Conditions on Settling.* If all flow enters a settling basin at one point, the particles must be distributed across the entire basin width and depth before the flow velocity, v_F is minimized. See Fig. 8.46.

The velocity at point *B* according to Eq. (8.38) will be approximately $Q/W_B h_B$ and the velocity at point *A* will be $Q/W_A h_A$. Depending on the efficiency of distribution, the velocity at point *A* could be many times faster than the velocity at point *B*, with a corresponding vector velocity v_{sA} much greater than v_{sB}. The particles at point *A* will therefore be more difficult to remove.

The solution to this dilemma is to effect a rapid dispersion at the basin entrance of all particles, both vertically and horizontally. Ideally, there will be a complete distribution across the entire entrance cross-sectional area. Since this is impractical, it may be approached by a series of small openings. The smaller and fewer the openings, the higher the velocities through the openings, but the more uniform these velocities will be.

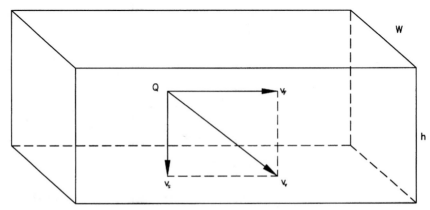

FIGURE 8.44   Settling basin particle velocities.

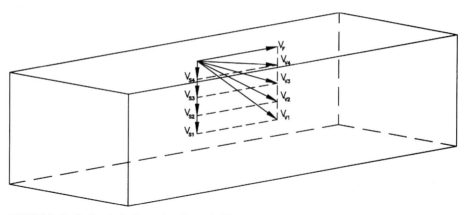

FIGURE 8.45   Settling basin flow and settling velocities.

The uniformity of the velocity through each opening is a factor of the flow Q, which passes through the opening. This flow in turn is a factor of the friction loss of the flow between its source and the opening. Figure 8.47 shows a piping distribution system at the entrance to a basin. Figures 8.71 and 8.73 show an inlet distribution wall using concrete blocks with some blocks laid on their side to allow openings. Note that the open blocks toward the center are farther apart than those toward the sides in under to compensate for the greater friction loss in the flow to the side ports.

Referring to Fig. 8.47, because the distance from A to points B and D is longer than from A to point C, there is more friction in a similarly sized pipe from A to B or D than from A to C. Therefore, there is less flow from A to B or D than from A to C, in accordance with the Hazen-Williams Equation for closed channel flow:

$$Q = 1.318A \times R^{0.63}S^{0.54} \tag{8.39}$$

Where Q = Flow in cfs
  A = Cross sectional area in ft^2
  R = Hydraulic Radius
    $R = A/P$
  P = Wetted Perimeter in ft
  S = Head loss in ft. per foot of pipe

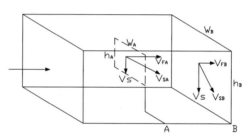

FIGURE 8.46   Settling basin entrance conditions.

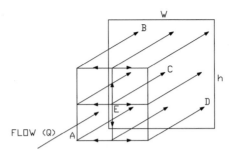

FIGURE 8.47 Settling basin entrance piping.

The solution to balance flows through the inlet points is to alter the pipe length or pipe sizes which will balance the friction losses and therefore the flows. For example, if the length of *E-C* in Fig. 8.47 equals the length of *E-B*, and the fittings are similar, the friction will be identical and therefore the flow distribution will be equal. A solution other than making the pipes identical is to increase the size of pipe *E-B* so that its flow is equal to the flow through the shorter but smaller pipe *E-C*.

This method of flow distribution can be accomplished in a less expensive, but adequate manner, by installing a distribution chamber followed by openings in a wall such as shown in Fig. 8.48 below.

More detailed information concerning piping hydraulic is included in section 7.1.5.5, "Piping Design."

8.5.2.10   The Effect of Exit Conditions on Settling.   The same situation occurs at the exit of a basin as does at the inlet, in that the closer to the exit, the higher the exit velocity (the opposite situation as depicted in Fig. 8.46). If the velocity in an exit pipe is 2 ft/s (a non-settling velocity as determined by Eq. (8.10)), the velocity at a distance from the exit pipe equals the flow divided by the cross sectional area affected by the velocity. Theoretically, this area is in the form of a hemisphere

FIGURE 8.48 Settling basin flow distribution wall.

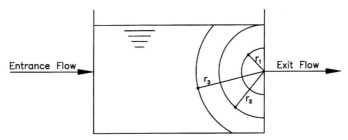

FIGURE 8.49   Settling basin exit velocity effect.

with all points in the surface equidistant from the exit pipe, as shown in Fig. 8.49. The exit velocity effect is equal to:

$$v_e = \frac{Q}{2\pi r^2} \tag{8.40}$$

where v_e = the exit velocity effect in fps, Mps
Q = the entrance flow in cfs, m^3/s
π = 3.1416
r = the radius of the hemisphere in feet, m

If the velocity at any point before the exit pipe is greater than the settling velocity of a particle, the particle will tend to be drawn to the exit pipe rather than settling.

To prevent this problem, the exit velocity must be minimized and ideally limited to the settling velocity of the critical particle. A rectangular weir with a length of 10,000 gpd/linear ft. has been demonstrated to allow a slow enough velocity to prevent particle carryover.

8.5.2.11   The Effect of Float Skimming on Settling.   Any solid or liquid particle which has a specific gravity less than water will rise during quiescent conditions, typically in accordance with Stoke's Law (Eq. (8.10)). Floatable solid particles are typically called *scum* and floatable liquid is called *float*. If these particles are removed from the surface with a velocity greater than the settling velocity, normally settleable particles can be resuspended and prevented from settling. The theory of this velocity is similar to that depicted in Fig. 8.49 for exit conditions. In accordance with section 8.5.2.10, the exit velocity effect from scum skimming should be minimized by increasing the cross-sectional area of the scum skimming device.

8.5.2.12   The Effect of Heavy Liquid or Sludge Removal on Settling.   The velocity at which heavy liquid or sludge is removed causes a velocity effect similar to that described above for the float skimming effect. Because heavy liquids and sludge are typically removed from the bottom of a basin, this velocity effect tends to increase particle-settling velocities. But, since the velocity effect will move water more easily than it will move a solid particle (the velocity of which is reduced due to drag of the particle through the viscosity of the liquid) or a heavier liquid, it will pull water through the heavier liquid or sludge, thus diluting the heavy liquid or sludge with water. Since one of the possible purposes of settling is to concentrate heavy liquid or sludge, this effect may not be acceptable. It causes more volume of heavy liquid or sludge to handle.

8.5.3   Samples Needed

8.5.3.1   Quantity of Samples.   In order to run the treatability studies recommended below, a proper sample of the wastewater must be taken. It is suggested that at least twice the capacity of the treatability study settling unit be gathered. For instance, if a 4-inch diameter and 4-foot high (the minimum recommended) settling unit is used, about 5 gallons of sample should be taken. If a 6-inch diameter and 8-foot high (the maximum recommended) settling unit is used, then about 24 gallons is needed.

8.5.3.2   Quality of Sample.   As explained above under section 8.5.2 "Theory," solid particles settle at different rates depending on their diameter and density, and therefore require different surface loading rates. The sample required for running sedimentation treatability studies must be representative of the suspended solids in the wastewater. If a representative grab sample cannot be taken, a sample with the slowest settling (smallest and/or less dense) particles should be obtained. The sample should be taken from a stream moving at least 2-fps velocity so representative solids will not settle in the conduit and, therefore, be missed. This sampling point could be a tap on a pipe under pressure, which should be well mixed and representative.

If the samples contain bacteria that could break down biodegradable solids, the sample should be iced or refrigerated to preserve the sampled conditions.

In order to optimize settling, many wastewaters are coagulated and/or flocculated. In those cases, the "floc" can be destroyed by turbulence. Therefore care must be taken to keep the sample as quiescent as possible. Before testing or removing an aliquot of the sample for testing, the sample must be re-mixed at an energy rate adequate for particle suspension but low enough to prevent particle shear. Slow hand mixing with a rod or paddle is usually adequate.

8.5.4   Treatability Studies

8.5.4.1   General.   There are two possible goals or purposes for settling: to clarify the liquid, or to concentrate the settled sludge. As explained above, a characteristic of flocculent settling is that the liquid surface does not immediately become clear, while in zone settling it does. The selected purpose will dictate the type of treatability.

The goal of a treatability study is to size the surface area, for a circular basin, the length and width, for a rectangular basin, and the depth of a settling basin. Recommended treatability studies for the four types of settling are as follows.

8.5.4.2   Discrete Settling.   If the particle settling velocity can be computed from Stoke's Law, Eq. (8.10), by knowing the critical particle diameter and density, and fluid density and viscosity, no treatability studies are needed to size a settling basin for discrete settling.

As an option, since discrete particles settle independently, the velocity can be measured visually with a settling column. The most accurate column for these tests should be 6-inch (15 cm) diameter, 10 feet (3m) long with taps at 2-foot (60-cm) intervals. The minimum size column recommended is 4-inch (10-cm) diameter, 4 feet (1.2 M) long with taps at 1-foot (30-cm) intervals.

If visual observation of particle settling is impractical, a test can be run using total suspended solids measurements as follows:

- Fill settling column with wastewater at time, $t = 0$.
- Measure suspended solids from each tap each period of time, t_n.
- Plot the results as shown in Fig. 8.50.

This plot will indicate clarity in the form of suspended solids after each time tested and at each depth, h. The depth is important because of the effluent structure and its exit velocity effect on settling efficiency at various depths.

8.5.4.3   Flocculent Settling.   Since the opportunity for coalescing increases with depth, flocculent settling efficiency, unlike discrete settling, depends on depth. There is no model or mathematical relationship to determine the effect of flocculation on sedimentation. Settling column analyses are required to evaluate this effect.

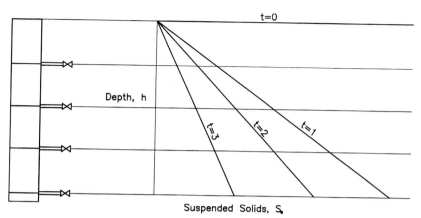

FIGURE 8.50 Discrete settling treatability.

A settling test can be run using suspended solids measurements as follows:

- Fill settling column with wastewater at time, $t = 0$.
- Measure suspended solids from each tap each period of time, t_n.
- Plot the results as shown in Fig. 8.51.

This plot will indicate clarity in the form of suspended solids after each time, t at each depth h. The depth is important because the velocity increases with depth and because of the exit velocity effect on settling efficiency.

8.5.4.4 Zone Settling. In zone settling, the rate of settling decreases with depth, therefore the depth of the basin is important in the test. Since clarity occurs at the surface almost immediately after settling begins, suspended solids in the settling zone is not a meaningful test to determine zone settling efficiency. The following settling test is recommended for zone settling conditions:

- Fill settling column with wastewater at time, $t = 0$.
- Record upper sludge blanket interface at each period of time t_n and when as the sludge blanket velocity begins slowing, measure sludge suspended solids from the bottom at time t_n.

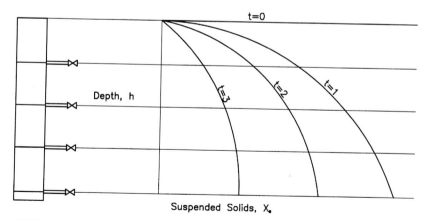

FIGURE 8.51 Flocculent settling treatability.

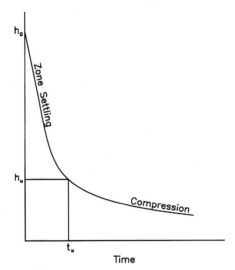

FIGURE 8.52 Zone settling treatability.

- Plot the results as shown in Fig. 8.52.
- Calculate time to reach a designed sludge concentration as follows:
  - Construct a horizontal line at H_u which corresponds to the designed underflow suspended solids concentration, C_a.
  - Construct a vertical line from the intersection of the horizontal line H_u and the settling curve to the ordinate of the graph to read t_a, the time required to reach the desired underflow concentration.
  - Combining Eq. (8.12), $V_0 = h_0/t_0$ and (8.18), $V_0 = Q/A$ from discrete settling, the following relationship develops:

$$A = \frac{Q}{h_0/t_0} = \frac{Qt_0}{h_0}$$

(8.41)

where A = surface area required for desired sludge concentration, ft^2, m^2

Q = flow rate into tank, ft^3/s, m^3/s

h_0 = initial height of sludge blanket interface in column, ft, m

t_u = detention time to reach desired sludge underflow concentration, s

8.5.4.5 Compression Settling. The test described above for zone settling should also be used for compression zone. Stirring should be investigated to improve sludge concentration. Stirring should move at a peripheral velocity of about 1 foot per minute to prevent resuspension of solids. Figure 8.53 shows a method successfully used for this purpose.

An electric clock motor can be used to obtain the desired rpm.

8.5.5 Design

8.5.5.1 General. The variables for design of a settling basin are as follows:

- Batch or flow through
  - Shape
  - Surface area

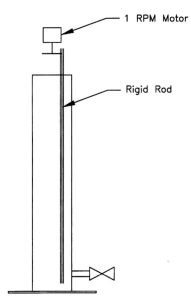

FIGURE 8.53 Compression settling treatability study stirring mechanism.

- Length and width
- Depth
- Detention time
- Entrance conditions
- Exit conditions
- Scum removal facilities
- Sludge removal facilities

The most common types of settling basins are shown in Figs. 8.54 through 8.64. The variables and design criteria for each of these types are recommended below starting with section 8.5.5.2.

8.5.5.2 Settling Tank Design.   The most basic design decision is between a batch, or fill and draw system, and a flow through system. See Fig. 8.54 for batch systems and Figs. 8.55 through 8.64 for flow through systems.

- Batch system: The advantages and disadvantages of a batch settling basin are as follows:
  - Advantages
  - The settling time is variable so efficiency can be optimized even though particle characteristics may vary between batches.
  - With a dual system, one basin can be filled while the other is settling to duplicate the performance of a flow through system.
  - The sludge can be removed before or after the effluent is removed to optimize efficiency.
  - This basin can be circular or square.

- Disadvantages
  - Much more labor is required to operate a batch system since scum, sludge and clarifier liquid must be withdrawn manually.
  - This system is more sensitive to operational error especially at the scum-liquid and liquid sludge interfaces.
  - Variable level drawoffs are prone to mechanical failure.

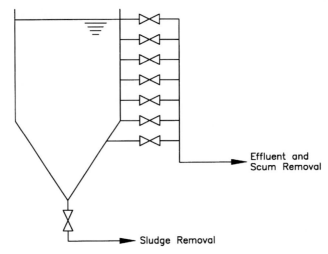

FIGURE 8.54   Batch settling basin with header exit system.

- Flow-through system: The advantages and disadvantages of a flow-through system are as follows:
  - Advantages
    - The sludge can be removed continuously or intermittently, manually or automatically.
    - The basin can be circular or rectangular.
    - Less operator time is required.
    - No variable level draw offs are required.

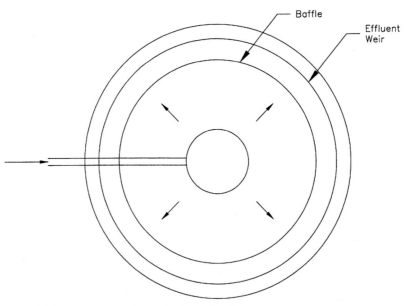

FIGURE 8.55   Circular center feed settling basin, plan view.

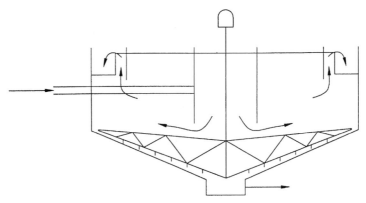

FIGURE 8.56   Circular center feed settling basin, section view.

- Disadvantages
  - The constant settling time of a flow through system prevents the flexibility to most efficiently settle particles of varying characteristics.

8.5.5.3 Sequencing Batch Reactor.   A modification of a batch settling system is a sequence batch reactor (SBR). In a sequencing batch reactor, sludge is settled in a batch process, the clarified liquid drawn off above the sludge-liquid interface, and the sludge retained for the activated sludge process in the same vessel or in a separate vessel. Two vessels are used in the system so that raw wastewater can be accumulated in one tank while treatment proceeds in the second tank.

In a two basin system, as shown in Figure 8.65 the sedimentation basin is used only for settling and the sludge is drawn from it into a separate vessel for activated sludge. The advantage of this

FIGURE 8.57   Circular center feed settling basin, photograph.

FIGURE 8.58 Circular center feed settling basin, photograph.

system is that a hopper bottom settling basin can be used to more efficiently concentrate the sludge and that untreated wastewater can be continuously pumped into the aeration basin. Upon reaching the optimum aeration detention time, the activated sludge is pumped to the settling basin, while the activated sludge basin continues to be filled.

A one tank sequencing batch reactor is shown in Fig. 8.66.

In this system, the wastewater is pumped into a flat bottom activated sludge basin which retains sludge from the previous settling cycle. In order to continuously fill the activated sludge basin, a second holding tank or redundant system is required which means two tanks must be used just as in the

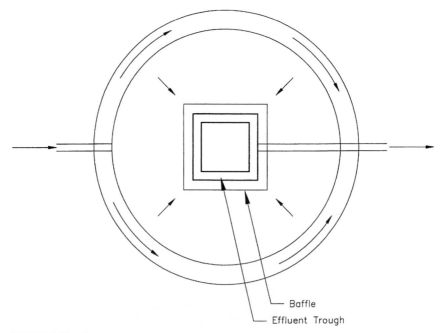

Baffle

Effluent Trough

FIGURE 8.59 Circular peripheral feed settling basin, plan view.

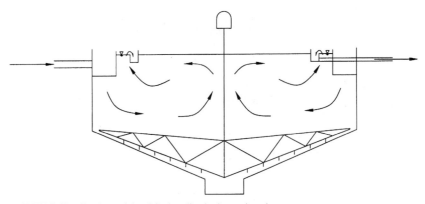

FIGURE 8.60   Circular peripheral feed settling basin, section view.

previously described system. A hopper is impractical for this type of operation which eliminates the advantage which a hopper has of sludge concentration. The advantages and disadvantages of an SBR are as follows:

Advantages:
- Batch operation allows more flexibility to treat varying influent condition.
- Can remove NH_3 and total nitrogen efficiently
- Eliminate clarifier
- Operation can be automated

Disadvantages:
- Requires two tanks
- Does not provide for easy floatable removal
- Mechanical liquid removal system can fail or become plugged with grease.

FIGURE 8.61   Circular peripheral feed settling basin, photograph.

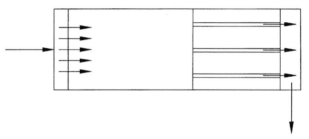

FIGUER 8.62 Rectangular settling basin, plan view.

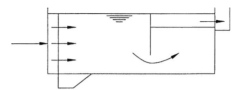

FIGURE 8.63 Rectangular settling basin, section view.

FIGURE 8.64 Rectangular settling basin, photograph.

OMNIFLO SBR Cycle
Anoxic Fill Phase
The reactor is filled with wastewater, Fill can be aerated, anoxic, or a combination of aerated and anoxic. Biodegradation is initiated.

During Anoxic Fill Influent is distributed throughout the settled sludge through the influent is not diluted by mixing, making biological nutrient removal much more reliable.

INFLUENT

React Phase
Influent flow is diverted to another reactor. Aeration continues in the full reactor until complete biodegradation is achieved; mixed liquor is drawn through the IDSC and used as motive Liquid for the jet aerator.

RECIRCULATED
MIXED LIQUOR

Settle Phase
The aerators are turned off and perfect quiescent conditions allow the biomass to settle, leaving the treated supernatant above.

Decant Phase
Treated effluent is removed from just below the liquid surface by the Jet Tech Floating Solids Excluding Decanter.

EFFLUENT

Idle/Waste Sludge Phase
The reactor waits to receive flow. Settled sludge is drawn through the IDSC and pumped to an aerobic digester. The jet motive liquid pump is utilized as a waste sludge pump.

WASTE SLUDGE

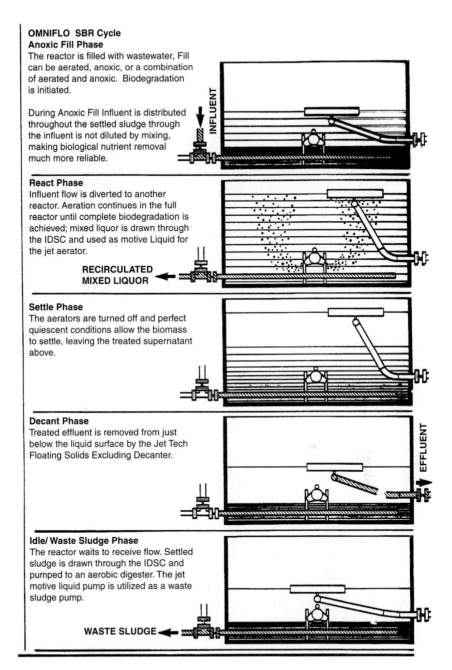

FIGURE 8.65 Sequencing batch reactor, single basin. (Jet Tech)

FIGURE 8.66

8.5.5.4 Shape.   The three most common shapes for the surface of a settling basin are circular, square and rectangular. Shapes are discussed for batch and flow through systems as follows:

- Batch systems

  To optimize the efficiency of a batch settling system, i.e., minimize the detention time, the length of the settling route of a particle should be minimized. If every particle takes the same time to settle, the settling time will be minimized. If the sludge settles onto a basin bottom the same shape as the surface, all particles will settle with the same time. The problem with a basin bottom the same size and shape as the surface is that there is no means to concentrate or remove the sludge. In a batch system, sludge concentration and removal is typically accomplished with a hopper bottom that concentrates the sludge by forcing it into a smaller cross-sectional area and allows single point of removal. A hopper bottoms system is shown in Fig. 8.54.

  A batch system with a conical hopper bottom has as its most efficient surface shape a circle. A batch system with a pyramidal shaped bottom has a square as its most efficient surface shape. A sequencing batch reactor can be of any shape since it provides no sludge concentration. A pyramidal shape in steel is generally less expensive that a conical shape since simple triangles are cut instead of a cone shape.

- Flow-through systems
- Circular center feed

  Figure 8.55 through 8.58 show circular center feed settling basins, the most common configuration for circular settling basins. The advantages and disadvantages of a circular center feed settling basin are as follows:

  Advantages:
  - The flow velocity decreases as a particle approaches the effluent, allowing the smaller particles to settle more efficiently.
  - The peripheral effluent weir maximizes the length and minimizes its exit velocity.
  - The sludge transfer distance to the center sludge hopper is minimized. Most sludge settles quickly toward the center feed.

- The sludge rake and scum skimmer are simple and efficient.
- The scum transfer distance is minimized. Most scum rises slowly and is caught before the peripheral weir.

Disadvantages:

- The center baffle doesn't distribute the entrance velocity well.
- Scum can collect behind the center baffle.
- The center sludge hopper location is far from the edge, normally resulting in a long, inaccessible, underground pipe.
- No common walls can be used between basins, thereby maximizing concrete use and space.
- Circular peripheral feed

Figures 8.59 through 8.61 show circular peripheral feed settling basins. The advantages and disadvantages of a circular peripheral feed settling basin are as follows:

Advantages:

- A peripheral entrance trough can be designed to distribute the inlet flow evenly around the periphery.
- The scum collection is efficient and does not normally collect behind the influent baffle.
- The center sludge hopper location is far from the edge.
- The sludge rake and scum skimmer are simple and efficient.

Disadvantages:

- The flow velocity increases as a particle approaches the effluent.
- The center effluent weir cannot be as long as a peripheral weir, thus causing the exit velocity to be increased.
- The sludge transfer distance to the center sludge hopper is maximized.
- No common wells can be used between basins.
- Rectangular

    Figures 8.62 through 8.64 show rectangular settling basins. The advantages and disad-vantages of a rectangular basin are as follows:

Advantages:

- The inlet flow can be distributed fairly evenly across the influent end of a basin since the width is normally restricted.
- The flow velocity is uniform along the length.
- The sludge transfer distance to the head of the basin or to the side is minimized.
- The scum transfer distance to the side of the basin is minimized.
- All scum collected can be easily removed and normally does not collect behind the entrance weir.
- Common walls can be used for adjacent basins, minimizing concrete and total space used.

Disadvantages:

- The effluent weir is relatively short, increasing exit velocity.
- The sludge collection mechanism is complex.
- Sludge must be scraped a long distance from the bottom at the exit and to the sludge hopper pit.

8.5.5.5   Surface Area.   The surface area in square feet (meters) should be determined as explained under section 8.5.4, "Treatability Studies."

If it is impractical to conduct treatability studies because samples of the wastewater are not available or for some other reason, experience has shown that the following criteria can be used for general design values or for checking the results of treatability studies (see Table 8.1).

The "Recommended Standards for Wastewater Facilities," known as the "Ten State Standards," make the surface area recommendations for the "design, review and approval of plans and specifications for

TABLE 8.1   Surface Overflow Rate for Secondary or
Flocculant Settling and Precipitation

Minimum surface area	200 gpd/ft^2 (0.094 L/s/m^2)
Maximum surface area	1200 gpd/ft^2 (0.564 L/s/m^2)

wastewater collection and treatment facilities" (primarily used for domestic sewage design) shown in
Table 8.2.

8.5.5.6   Length, Width and Depth

- Minimum length
  The Ten State Standards recommend that the minimum length from inlet to outlet be 10 feet to prevent short circuiting, but in some small industrial applications, shorter basins may be effectively
  used, with care taken in the design of the entrance and exit conditions.

- **Length to width ratio**
  In order to prevent short-circuiting, it is recommended that the length of a rectangular basin be 4
  times the width. In addition, the width of a rectangular basin is typically limited to about 20 feet
  due to the structural design of the sludge removal mechanism.

TABLE 8.2

Ten state standards		
Primary settling tanks		
Design average overflow rate	Design peak hourly overflow rate	
1000 gpd/ft^2	1500–3000 gpd/ft^2	
0.47 l/s/m^2	0.71–1.42 l/s/m^2	
Intermediate settling tanks		
Design peak hourly overflow rate		
1500 gpd/ft^2		
0.71 l/s/m^2		
Final settling tanks		
	Design hourly overflow rate	Design peak solids loading rate
Activated sludge	1200 gpd/ft^2	50 lb/day/ft^2
	0.56 l/s/m^2	245 kg/d/m^2
Extended aeration	1000 gpd/ft^2	35 lb/day/ft^2
	0.47 l/s/m^2	171 kg/d/m^2
Single stage nitrification	1000 gpd/ft^2	35 lb/day/ft^2
	0.47 l/s/m^2	171 kg/d/m^2
Two stage nitrification	800 gpd/ft^2	35 lb/day/ft^2
	0.38 l/s/m^2	171 kg/d/m^2
Attached growth	1200 gpd/ft^2	50 lb/day/ft^2
	0.56 l/s/m^2	245 kg/d/m^2

Great Lakes-Upper Mississippi River Wastewater Committee, 1997. *Recom-mended Standards for
Wastewater Facilities,* Health Education Services, Albany, NY, 1997

- **Depth**

  As explained above in section 8.5.2 "Theory," the only purpose for the depth of a settling basin for discrete settling is to store the sludge. For other types of settling, the depth is of design importance and must be determined by treatability studies as explained in section 8.5.4.

  It should be noted that the quicker the sludge can be removed from the bottom of the basin, the shallower the basin can be, since sludge storage is not as critical. The drawbacks of quick sludge removal are as follows:

- No opportunity for sludge concentration
- Fast moving sludge removal mechanism may resuspend sludge

For biological or flocculent settling, it is recommended that the minimum side water depth be 10 feet. The Ten State Standards depths are shown in Table 8.3.

8.5.5.7   Detention Time.   Detention time should be determined by treatability studies in accordance with section 8.5.4. For biological flocculent or zone settling, a 4-hour detention time has been successfully used for a typical design value.

8.5.5.8   Entrance Conditions

- **General**

  Section 8.5.2.9 describes the theory of entrance conditions on settling. The ideal entrance design is one that distributes flow uniformly across the entire width and depth of a basin and dissipates the inlet velocity. A precaution in designing entrance conditions is that any structure that allows a flow velocity of less than 2 ft/s can allow settling and flotation within the structure. This may cause flow, operation and maintenance problems.

- **Circular center feed basins**

  In a circular, center feed settling basin, the flow is normally transferred by pipe through the outside wall to a center chamber as shown in Figs. 8.55 and 8.56. The purpose of this center wall is to distribute the flow peripherally, vertically, and horizontally, thereby reducing the velocities and promoting quiescent conditions. The well can also be used for mixing with chemicals and/or flocculation.

  In order to accomplish the flow distribution, three purposes must be accomplished:

- Dissipation of the inlet velocity
- Distribution of the flow horizontally and vertically
- Avoidance of accumulation of scum in the well

The following are various recommendations to accomplish these three purposes (see Fig. 8.67).

$$\phi = \text{diameter}$$
$$h = \text{side water depth}$$
$$v = \text{inlet flow velocity}$$

TABLE 8.3   Ten State Standards Recommended Depths

Type of sludge	Side water (feet)	Depth (meters)
Primary	7	2.1
Secondary following Suspended growth*	12	3.7
Secondary following Attached growth*	10	3.0

*Greater depths recommended for basins in excess of 4000 ft^2 (372 m^2), 70-feet (21 m) diameter and for nitrification plants.

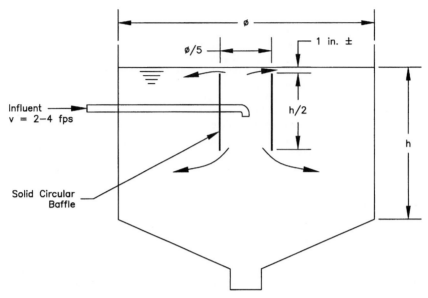

FIGURE 8.67 Center circular feed clarifier entrance design.

The small, one-inch clearance between the top of the solid circular baffle and the water surface is to allow scum to move from behind the baffle. This space does allow a potential for surface short-circuiting but this normally is minimized with an effluent baffle as described below under Exit Conditions.

A second method of distributing entrance flow is to use a perforated circular baffle as shown in Fig. 8.68.

The bottom of the baffle must clear the sludge removal mechanism, and it must be solid so no influent flow is carried through the sludge accumulation area. The inlet well must be smaller in the design to prevent sludge settling in the well. As shown in Fig. 8.66, the bottom of the entrance well

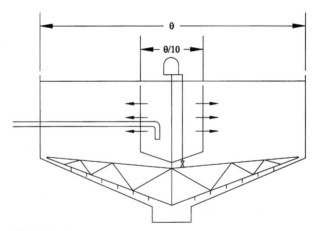

FIGURE 8.68 Circular center feed clarifier alternate entrance design.

can be sloped, with a valve to discharge accumulated sludge periodically. The perforated openings in the baffle should be sized to prevent velocities through the holes greater than one foot per minute using the formulae below to size the openings.

- **Circular peripheral feed basin**
  - Entrance conditions

In a circular, peripheral feed settling basin, the flow is normally distributed to the periphery, through a trough under a baffle. The trough must be designed to distribute the flow uniformly around the periphery. If a distribution trough is used, holes in the trough should be sized, as explained above, with less than 1-fps velocity. They should also be spaced wider apart near the entrance pipe to allow an even distribution.

The flow through each opening can be computed using the orifice flow formula as follows:

$$Q = 0.61 \, A \, (2gh)^{1/2} \tag{8.42}$$

where Q = flow in ft^3/s
  A = cross sectional area of orifice in ft^2
  g = acceleration due to gravity = 32.2 ft/s^2
  h = head or distance of water level over centerline of orifice in ft

The head, h, varies in accordance with the head or friction loss between the entrance point and the orifice in question using Hazen-Williams' Equation (8.44) for closed channel flows, such as in pipes, and Manning's Equation (8.45) for open channel flows, such as in troughs.

$$f = 0.2083 \left(\frac{100}{C} \right)^{1.85} \times \frac{Q^{1.85}}{d^{4.8655}} \tag{8.43}$$

where Q = flow in ft^3/s
  f = friction head in feet of liquid per 100 feet of pipe
  d = inside diameter of pipe in inches
  C = surface roughness constant

Note that Eq. (8.43) is a rearrangement of Eq. (8.39). Appendices A-2 through A-4 are tools that can be used to compute head loss in pipe with respect to the Hazen-Williams' Equation.

Manning's Equation (Eq. (8.44)) is an approximation of Kutter's Formula as follows:

$$V = \frac{1.86}{n} R^{2/3} S^{1/2} \tag{8.44}$$

where V = flow velocity
  S = the channel slope in ft/ft
  n = the roughness coefficient as shown in Table 8.4
  R = the hydraulic radius = A/P

where P = the wetted perimeter of the open channel in ft
  A = the cross sectional area of the water flowing in the open channel in ft^2

Nomographs are readily available in hydraulics textbooks for Manning's Equation. See Appendix A-5 for Kutter's formula nomograph and Appendix A-6 is for the relation between discharge, velocity and percentage depth of flow. See Appendix A-7 for a Manning's Equation nomograph. To achieve a relatively even distribution of entrance flow, the trough holes should be spaced no more than 2 feet apart.

TABLE 8.4 Roughness Coefficients for Manning's Equation

Surface	Value of n
Ductile or cast iron, unlined	0.013
Ductile or cast iron, lined	0.010
Steel	0.011
Concrete poured with steel forms	0.012
Concrete poured with smooth wood forms	0.012
Concrete poured with rough wood forms	0.015

A baffle can be used for entrance flow distribution placed approximately one foot past the orifice holes to further direct the flow downward and prevent short circuiting, as show in Fig. 8.67.

If a baffle is used instead of orifice holes for flow distribution, the flow will be less evenly distributed and will tend to enter the basin close to the entrance pipe. The best solution is to design the baffle so that it is closer to the outer wall near the entrance, rather than at the far side of the basin. See Fig. 8.67 for this type of design.

As in the case of the center feed clarifier, the baffle depth should be approximately one half the side water depth. The distance between the baffle and the outer wall should be calculated using Manning's Equation (8.44).

- **Rectangular Basins**

As explained in section 8.5.2.9 "The effect of entrance conditions on settling," a vertically and horizontally even distribution of entrance flow is ideal. The options for this distribution are as follows:

- Piping network (Fig. 8.47)
- Trough and baffle (Fig. 8.71)
- Trough and perforated well (Figs. 8.72 and 8.73)

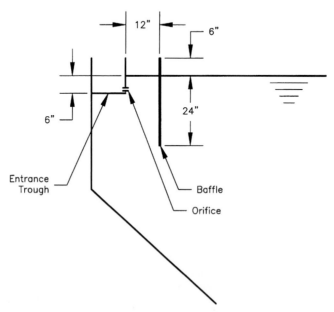

FIGURE 8.69 Peripheral feed circular clarifier office entrance design.

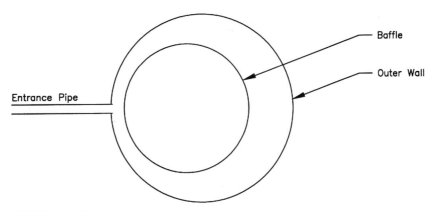

FIGURE 8.70   Circular peripheral feed clarifier baffle entrance design.

As in the case of the circular, peripheral feed clarifier, the distance between the baffle and the entrance wall should be calculated using Manning's Equation (8.44). The baffle depth should be approximately half the side water depth.

Note that the openings in the perforated wall are farther apart as they get closer to the entrance pipe. As in the case of the circular clarifier, the design should be as follows:

- Flow-through openings Eq. (8.42)
- h in Eq. (8.42) is from open channel flow calculations (8.44)

The perforated wall should have open sections or sections that can be opened at the bottom to allow movement of settled solids from the entrance chamber to the basin to facilitate solids removal.

8.5.5.9   Exit Conditions

• General

In a typical settling basin, there are three phases of liquids and/or solids; the scum or float, the clarified liquid, and the sludge or heavy liquid. The exit conditions should separate these three phases to minimize the effect of the interfaces. In other words, as one phase is withdrawn, the amount of the adjacent phase removed should be minimized. In this section, emphasis will be placed on the

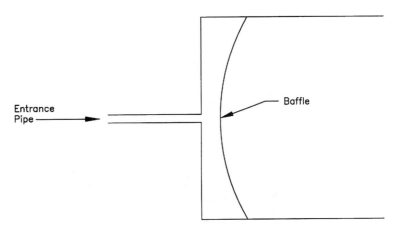

FIGURE 8.71   Rectangular settling basin entrance trough baffle.

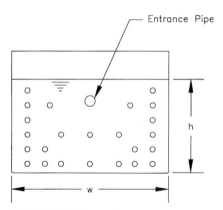

FIGURE 8.72   Rectangular settling basin entrance trough and perforated wall (looking upstream).

removal of the clarified liquid. The removal of the scum or float is explained below under "Float removal facilities" and removal of the sludge or heavy liquid is explained "in under sediment removal facilities."

Section 8.5.2.10 describes the theory of exit conditions on settling. As explained, in a continuous flow basin the linear feet of exit weir length should be less than 10,000 gpd/ft. The depth of flow over a one foot wide weir at a flow of 10,000 gpd, according to Francis, Eq. (8.45) is 0.34 inches.

$$Q = 3.33(L - 0.2\,h)\,h^{1.5} \tag{8.45}$$

FIGURE 8.73   Photograph of rectangular settling basin perforated wall.

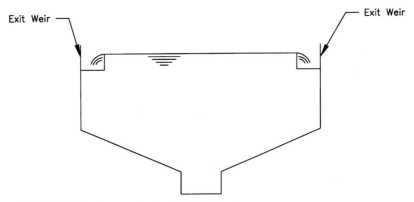

FIGURE 8.74   Circular center feed basin inside exit weir.

where Q = flow in ft³/s
   L = length of weir in ft
   h = head on weir in ft

This recommendation means that an exit weir must be provided that will keep the depth of flow over the weir to less than 0.34 inches. This gives a velocity over the weir of 0.55 ft/s, in accordance with Eq. (8.18).

   For a circular centerfeed clarifier, as long as the surface loading rate on the clarifier is less than 1256 gpd/ft², the exit weir overflow rate will be less than 10,000 gpd/ft. Therefore, normally a sufficiently long exit weir is provided with a circular center feed basin using a peripheral weir. The two most common types of peripheral exit weirs are shown in Figs. 8.74 and 8.75.

   The level of the exit weir, of course, sets the water level in the basin.

• **Effect of the settling of the clarifier structure into the soil**
   A clarifier loaded with water has a weight of 62.4 lb/ft³ of liquid plus the weight of the basin bottom and any other structural or mechanical weights. Therefore a 10-foot deep basin will weigh 624 lbs. for each square foot of bottom area plus structural and mechanical weights. This loading

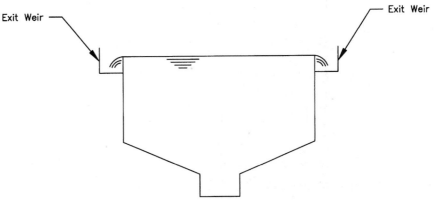

FIGURE 8.75   Circular center feed basin outside exit weir.

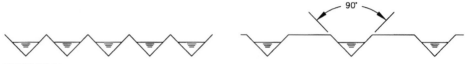

FIGURE 8.76   *V* notch overflow weirs.

could be sufficient to cause slight differential settling of the soil beneath the clarifier. Since the maximum depth of flow over a weir is 0.34 inches, if one side of a clarifier settles more than 0.68 inches, the flow will be linearly diminished farther from the point of settling until there is no flow over the weir opposite the point of settling. With even a smaller amount of settling, the flow will vary based on its depths over the weir. This will increase the flow velocity over the weir at the point of greatest settling, causing particle overflow in these areas.

In order to alleviate this potential problem, *V* notch or semicircular weirs can be used to minimize the differential overflow depth and therefore the overflow velocity. See Figs. 8.76 and 8.77.

The flow velocities through a 90° *V* notch weir can be computed using Eq. (8.46) along with Eq. (8.18):

$$V = Q/A \qquad (8.18)$$

$$Q = 2.4381 \, h^{2.5} \qquad (8.46)$$

where Q = flow in ft^3/s
h = head of water above apex of notch in ft

The flow velocities through a semi circular weir can be computed using the orifice Eq. (8.42): $Q = 0.61A \, (2 \, gh)^{1/2}$ along with Eq. (8.18)) with the flow from Eq. 8.42 divided by 2.

The further apart are the notches, the deeper the flow in each, therefore the higher the velocity through each but the more equalization of velocity with differential basin settling will be realized.

It is recommended that when notches are designed, the horizontal water surface over-flowing the weir still not exceed 10,000 gpd/ft.

• **Circular peripheral feed basin**
Since a peripheral feed basin cannot maximize its exit weir length, other configurations must be developed such as those shown in Figs. 8.78 through 8.80.

The total weir length is recommended to be less than 10,000 gpd per foot. The effective weir length is maximized by allowing water to overflow from the middle of two weirs and/or installing multiple parallel weirs. The solid particles should not be discharged over the weir if the basin is designed in accordance with criteria presented in this Handbook. See Fig. 8.81.

• **Rectangular basins**
One weir across the end of a basin designed at 10,000 gpd/ft would require a surface loading rate of only 25 gpd/ft^2 at a 4:1 length to width ratio, as recommended.

If both sides of this weir are used, the loading rate would still be only 50 gpd/ft^2. To reach the more reasonable surface loading rates, there must be multiple overflow weirs, as shown in Fig. 8.82.

If the baffle is placed 3/4 of the distance from the entrance, there should be 6 weirs to obtain the 10,000 gpd/ft overflow ratio at a 1000 gpd/ft^2 surface loading rate and 3 weirs for a 200 gpd/ft^2 surface loading rate. Possible configurations for these examples are shown in Fig. 8.83.

FIGURE 8.77   Semicircular overflow weirs.

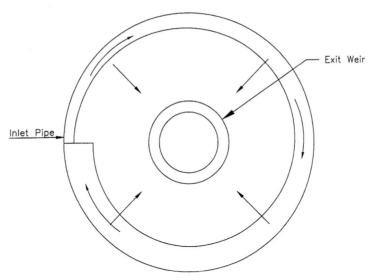

FIGURE 8.78   Circular peripheral feed basin, circular exit weirs.

The baffle can possibly be placed as far forward as one-half the distance from the influent, depending upon the particle settling slopes.

8.5.5.10   *Float Removal Facilities*

- **General**

  Section 8.5.2.11 describes the effect of scum or float skimming on settling. The float can be skimmed continuously or intermittently and manually or automatically. As in the case of the exit

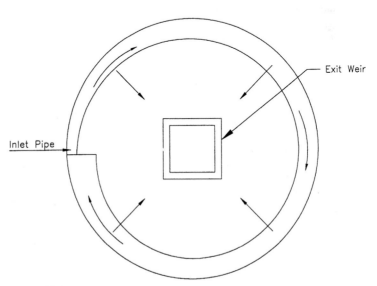

FIGURE 8.79   Circular peripheral feed basin, square exit weirs.

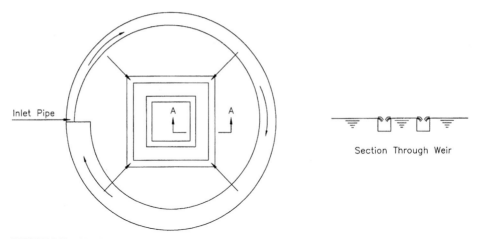

FIGURE 8.80 Circular peripheral feed basin, double square exit weirs.

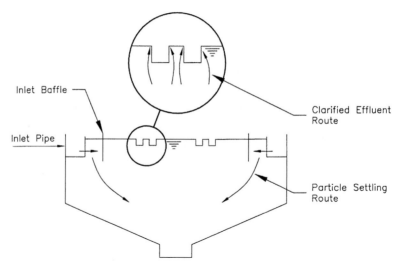

FIGURE 8.81 Circular peripheral feed basin particle flow.

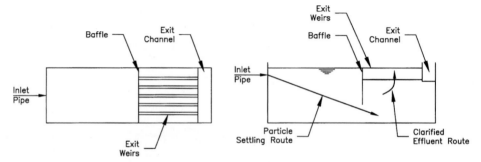

FIGURE 8.82 Rectangular basin exit weirs.

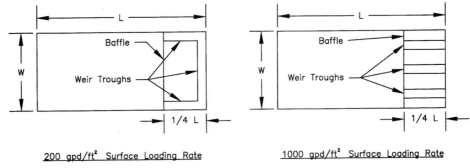

200 gpd/ft² Surface Loading Rate   1000 gpd/ft² Surface Loading Rate

FIGURE 8.83 Rectangular basin weir designs.

conditions explained in section 8.5.5.9, the depth of flow over the skimmer inversely affects the velocity. An excessive velocity will draw settling particles over the skimmer. The design approach to compute this velocity is to use equation: $Q = 3.33 (L - 0.2h) h^{1.5}$ and $V = Q/A$. The skimmer should be located near the exit structure to maximize the distance between the skimmer and the settling sludge particles. The float is drawn at a much lower flow than the flow through the settling basin but the float removal velocity should still be limited to 0.55 fps or 10,000 gpd/ft².

- **Continuous skimming**
  Ideally, in a flow-through system, float should be continuously skimmed so that the exit velocity is minimized. The problem in most applications is that since the length of the skimmer must be maximized to minimize the exit velocity, the depth of flow over the skimmer will be very small for continuous skimming. This small depth can cause the surface to become plugged with solids of larger diameter than the skimming depth. If the wastewater is very homogeneous and without larger floating particles, or if the float or scum amount is larger, continuous skimming is a possibility. Figure 8.84 shows a skimmer for a circular basin.

  Because of the flow velocity, the scum will tend to accumulate at the surface of the skimmer if the skimmer is located near the basin effluent. With very low flow velocities and/or high winds, the

FIGURE 8.84 Circular basin skimmer. (Hi Tech Environmental)

scum can be moved to other parts of the basin. Methods of directing scum to the skimmer are as follows:

- Spray system
- Circular scraper (see Fig. 8.85)
- Rectangular Scraper (see Fig. 8.86)

The surface velocity caused by movements of the float should be lower than 2 ft/min to maintain settling and flotation of particles.

- **Intermittent skimming**

  Normally, scum or float is allowed to accumulate to a thickness that is easier to skim and then withdrawn intermittently. This can be done manually by opening a skimmer or moving the position of a skimmer so it will accept float, or automatically with a timer. In the case of circular settling basins, the skimmer is effactually activated once a cycle (see Fig. 8.84).

- **Batch settling basin skimming**

  In a batch settling basin, the float is necessarily skimmed intermittently after it accumulates. It is normally removed prior to the clarified liquid removal so that it will not interfere with clarified liquid or sludge removal. Float can be removed using a floating skimmer as the clarified liquid or a sludge is being removed, but this complicates the mechanical facilities.

  Experience has shown that these floating variable level skimmers tend to collect floating solids which affect their balance and/or movement.

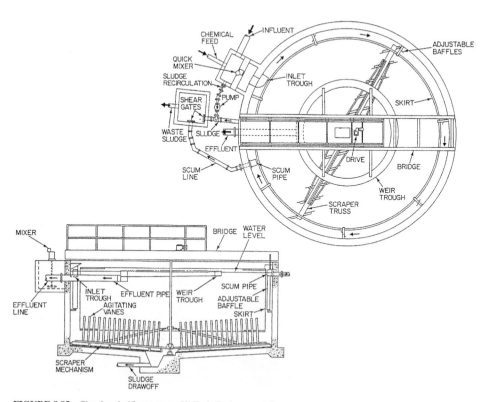

FIGURE 8.85   Circular clarifier scraper. (Hi Tech Environmental)

FIGURE 8.86 Rectangular clarifier scraper.

Normally a skimmer is used which can be activated by opening a valve or lowering the skimmer level when needed. This can be a circular slotted pipe or any sort of movable gate. The simplest type of skimmer is simply a stack of thin metal plates which are removed as needed to allow surface skimming (see Fig. 8.87).

8.5.5.11 Sediment Removal Facilities

- **General**
  Section 8.5.2.12 described the effect of sludge removal on settling. Sediment may consist of heavy liquids and/or solids or sludge and must be removed slowly enough to prevent the concurrent removal of water. This section will discuss mobile or stationary and continuous or intermittent sediment removal facilities for flow through and batch settling basins.

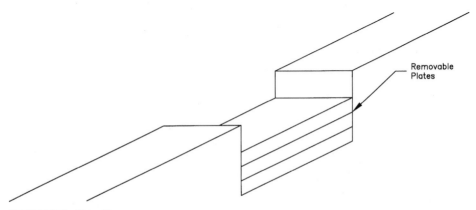

Removable Plates

FIGURE 8.87 Plate skimmer.

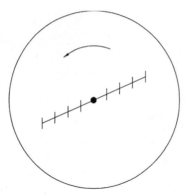 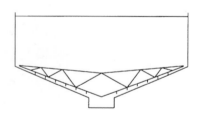

FIGURE 8.88   Circular settling basin rakes.

It is recommended that the sludge removal velocity be less than the particle settling velocity unless the sludge is accumulated in a sump and prior to removal.

- **Mobile sediment removal facilities**

  Removal facilities can be mobile or static and can use rakes or suction to remove the sediment. Mobile facilities in a circular basin will typically rotate from a center shaft and the rake system will push the heavy liquid or sludge along the bottom toward a center sludge well such as shown in Fig. 8.88.

  This moving rake should rotate with a peripheral speed of less than 2 ft/m to prevent sludge or liquid re-suspension. The rake can either move continuously or intermittently, and with manual control or automatic control, with a timer. The center shaft can also be used to drive a float or scum removal rake.

  A suction system is also rotated from a center shaft, but will use openings in a pipe or pipes to suck or draw the sludge into the removal system. This suction can be caused by a vacuum pump, an eductor, or simply a water level difference (see Figs. 8.89 and 8.90).

  The water level difference can be controlled as shown in Fig. 8.90 with a telescopic valve or with a gate system. As long as the water level leaving the sediment receiving pipe is lower than the water level in the basin itself, the higher basin water level will push the sediment through the suction pipe to and out of the sediment receiving pipe. The lower the outlet of the sediment receiving pipe, the faster the sediment will be removed. In case of the sludge, if it is removed too slowly, it will accumulate in the settling basin and will also settle in the sludge removal piping and eventually plug them. If the sludge is removed too quickly, as explained above, water will be drawn out with the sludge. Because of the risk of pipe plugging, it is recommended that sludge carrying pipes be a minimum of 4 inches in diameter. It is also recommended that the sludge flow in the pipes be greater than 2 fps to prevent settling. Since this velocity requires 80 gpm of flow (Eq. 8.39 or Appendix A-5 for a 4-inch pipe), a small system must use intermittent sludge removal to prevent pipe pluggage. It is also recommended that each bend in the sludge piping be provided with a clean out or tee and, preferably, a tap for air or water pressure cleaning (see Fig. 8.90).

  In designing a piping system for heavy liquid or sludge, the friction factor used in Eqs. 8.39 or 8.43 must be increased to compensate for the viscosity. See Table 8.5 for sludge concentration correction factors.

  Heavy liquid is much less likely to plug a piping system and can be removed in piping smaller than 4 inches in diameter. The clean out systems are also less critical.

  Moving rakes in a rectangular basin should move a speed of less than 2 ft/min to prevent sludge or heavy liquid re-suspension. These rakes typically move sludge to a sludge sump as shown in Figure 8.92.

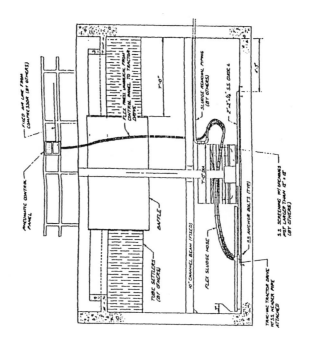

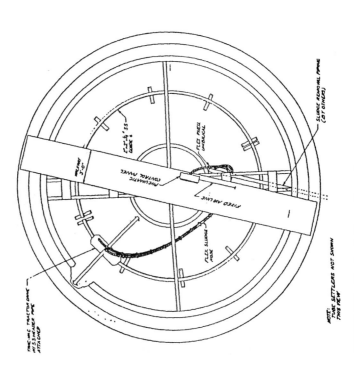

FIGURE 8.89 Circular clarifier suction system.

FIGURE 8.90   Circular clarifier suction removal. (Hi Tech Environmental)

The flights consist of chemically resistant boards moved by chains that are attached to a drive sprocket. This flight system can be modified to push float to a skimmer as shown in Fig. 8.93.

Note that the sludge sump should be at the entrance end of the basin, because the majority of the sediment will be accumulated on that end.

Structurally, this sort of sludge removal system is limited to approximately 20 ft in width.

The flights, chains, gearing and motors must be designed to overcome the torque caused by the weight of the sediment. It is recommended to compute torque when scraping non-flocculant sediment.

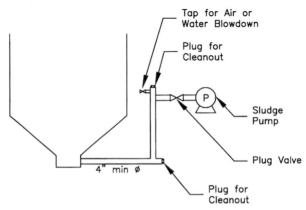

FIGURE 8.91   Recommended sludge removal piping.

TABLE 8.5 Sludge Concentration Correction Factors

Sludge Concentration, %	Multiply Friction Factor By:
1	1
2	1.5
3	2
4	2.5
5	3
6	4
7	5
8	7
9	9
10	12

- **Stationary sediment removal facilities**

  In order to completely remove sediment from the bottom of a basin, a stationary removal system may be used. Heavy liquid sediment will move horizontally to a removal point. If the heavy liquid layer is relatively thin, however, the higher viscosity of the heavy liquid will allow the less viscous water above to move through the heavier liquid. In the case of the more viscous sludge, there is even more tendency for water to be drawn through the sludge. Consequently, sludge must be moved by gravity to all draw-off points.

  It is recommended that the side slopes of a hopper bottom be greater than 1:1or 45°, when heavy liquid is removed in a stationary skimmer (see Fig. 8.94).

  In a sludge removal system, the hopper bottom should have a 2:1or 60° from horizontal slope to prevent solids accumulation on the sides (see Fig. 8.95).

  In the middle of the 20th century, Howard K. Bell developed a "Bell Bottom" system consisting of multiple hopper bottoms as shown in Fig. 8.96.

  All stationary systems require sludge removal lines with the velocity constraints for sludge and heavy liquid recommended above.

- **Continuous sediment removal systems**

  In a flow-through basin, sludge accumulates continuously, and ideally, it should be removed continuously. As noted previously, continuous sludge removal may be impractical in small systems. The advantage of a continuous system is that sludge can be constantly recirculated to another treatment system such as activated sludge, to prevent slugging the system. With a precipitation system, such as metals removal, sludge is discharged directly to a thickener or a solids concentration system, and intermittent sludge removal is satisfactory.

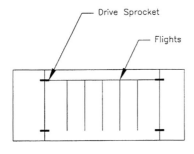 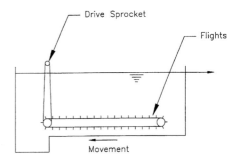

FIGURE 8.92 Rectangular clarifier scraper.

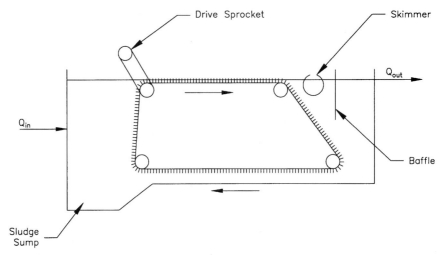

FIGURE 8.93 Rectangular basin scraper/skimmer system.

FIGURE 8.94 Hopper bottom for heavy liquids.

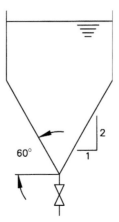

FIGURE 8.95 Hopper bottom for sludge.

Continuous sediment removal systems can be mobile or stationary, but are most often mobile in larger systems. In these systems, the sludge is gathered and transferred continuously over a large area and concentrated into a small hopper for removal.

- **Intermittent sediment removal systems**
  In systems where the velocity of sludge removal required to prevent sludge settling in the piping (usually 2 fps) can cause clear liquid to be pulled through the sludge, the solution is typically an intermittent withdrawal. This withdrawal can be manual or automated.

  A manual withdrawal consists of opening the sludge withdrawal valve or turning on the sludge withdrawal pump when the sludge in the basin (batch systems) or hopper (flow through systems) reaches a critical level. If the clarified liquid is clear enough, the top of the sludge blanket can be visually observed. When this cannot be done, a "sludge judge" such as in Fig. 8.97 can be used.

  The "Sludge judge" is inserted into the settling basin slowly to prevent disturbing of the sludge. The sludge is then allowed to reach its level in the clear plastic pipe. The sludge judge is then capped by shutting off a valve or sealing the upper end by hand and the pipe is removed. This is equivalent to lifting soda out of the glass by putting your thumb over the end of the soda straw. The sludge depth is then visually measured.

  The sludge should be removed with at least a 2 fps velocity in the withdrawal pipe, but at a velocity in the settling basin that will not pull clarified water with the sludge.

  Intermittent sludge withdrawal can be accomplished with a timer. A timer interval should be selected to ensure sludge will not accumulate to a high level that causes it to flow over the effluent weir, or

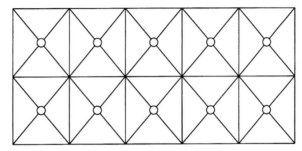

FIGURE 8.96 Bell bottom sludge removal system.

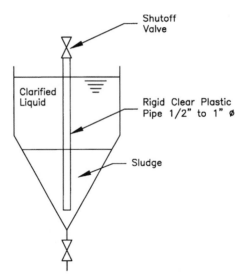

FIGURE 8.97 Sludge judge.

to a low level through which water can be drawn. Variable settleable solids entering a basin will accumulate at different rates and will affect the draw off interval. Intermittent sludge withdrawal can also be accomplished using a sludge interface locator to trip the withdrawal valve or pump.

8.5.6 Settling Basin Design Hints

- Secondary Surface Loading Rate: 200 to 1200 gpd/ft$_2$
- Primary Surface Loading Rate: 1500 to 3000 gpd/ft$_2$
- Minimum depth for biological or flocculant settling or precipitation: 10 feet
- Minimum depth for primary settling: 7 feet
- Rule of thumb detention time: 4 hours (varies)
- Minimum velocity in entrance and exit piping: 2 fps
- Depth of circular center feed influent baffle: h/2
- Depth of effluent baffle 24″ to 48″.
- Minimum length of effluent weir: 10,000 gpd/ft.
- Length to width ratio of rectangular basin: 4:1
- Maximum typical width of rectangular basin: 20 ft
- Maximum sludge scraper velocity: 2 ft/min.
- Minimum sludge removal piping size: 4″

8.6 FLOTATION

8.6.1 Introduction

Flotation is a process used to separate solid or liquid particles from a liquid phase by gravity, perhaps aided by coagulation, flocculation or dissolved air.

A dissolved, soluble or emulsified chemical or a fat, oil, or grease that is of lower specific gravity than water must be chemically converted to a filterable chemical, or it must have its emulsion broken before flotation. Chapter 9 describes the chemical conversion of soluble metals to insoluble hydroxides or sulfides.

8.6.2 Emulsion Breaking

The floatability of a wastewater should be determined before treatability studies or design is considered. When the chemical composition of a wastewater is analyzed, all chemicals with a specific gravity less than water can be floated theoretically. If a quiescent sample of the wastewater does not stratify into layers of water and lighter chemicals, as determined visually, or by tests, there is a solubilization or an emulsion of the chemicals. Depending on the type of emulsion, there are several methods of breaking the emulsion. It may aid in the selection of the method to know the way the emulsion was formed. In this section of the Handbook, the discussion will be limited to fat, oil and grease (FOG) emulsions since they constitute a vast majority of emulsions encountered. Examples of emulsion formation are as follows:

- **Energy**: Mixing, pumping and other examples of adding energy to a wastewater may accidentally or purposefully, mechanically emulsify FOG.

- **Emulsifiers**: Certain organic chemicals such as soaps and detergents, also called saponifiers, will lower the surface tension of water so that a lower specific gravity FOG will become chemically emulsified and will not separate by gravity. This is done deliberately in many industrial applications in order to dilute an oil with water without stratification. Examples are cutting oils, lubrication oils, mineral oil, and vegetable oil.

- **Heat**: Some organic chemicals can become emulsified at high or low temperatures.

In order for emulsified FOG to be removed from a wastewater, the emulsion must be broken. The common methods for de-emulsification are:

- **Acid cracking**: The most common method of breaking an emulsion is **acid cracking**, where the pH of the wastewater is lowered to 1.0 to 2.0 while mixing, followed by gravity flotation. The success of this method can be easily determined in the field or in a laboratory by mixing several beakers of wastewater at different pH levels (from 1.0 to 5.0) and comparing floatability. Acid cracking may be more effective in conjunction with other methods. The author has used acid cracking in several cases of vegetable oil emulsions.

- **Heating or cooling**: Depending on the method of forming the emulsion, heating or cooling may be effective in breaking an emulsion. This can be accomplished with pH adjustment or the addition of a coagulant, or without these additional methods. Most typically, heating to between 100 and 150°F (38 to 66°C) can be effective.

- **Coagulants**: A metal salt such as alum ($Al_2(SO_4)_3$), ferrous sulfate (Fe SO_4), ferric sulfate ($Fe_2(SO_4)_3$) or ferric chloride ($FeCl_3$) or a cationic, anionic or non-ionic polymer may serve to break the emulsion. The author has used alum to break detergent/oil mixtures and polymers to break cutting oil emulsions.

- **Combination of pH adjustment and coagulant addition**: Some emulsions cannot be effectively broken using any of the above three methods alone. These types of emulsions require pH adjustment and coagulant addition. In this method, the pH of the wastewater is lowered to approximately 2.0, then raised to a neutral pH or higher depending on the wastewater. The point of coagulant addition will vary depending on the wastewater, but is normally at the lowest or near neutral pH.

8.6.3 Design of Gravity Floatables Separation

Section 8.5.2 described in detail Stoke's Law for the gravity separation of discrete solids from a wastewater. The same theory applies to light chemicals and FOG removal since as free liquids they are theoretically of spherical uniform shape and will discretely float.

- **Design steps**
- *Step 1*

For the design as presented in this chapter, Stoke's Law is as follows:

$$v = \frac{gd^2(p_1 - p_2)}{18\mu}$$ (8.47)

where v = rate of flotation, cm/s
g = acceleration due to gravity = 980 cm/s²
d = particle diameter, cm
p_1 = particle density, g/cm³
p_2 = kinematic fluid density, g/cm³
μ = fluid viscosity, g/cm-s

Note: see Appendix C-4 for the density and viscosity of water at various temperatures.
For a typical petroleum oil globule diameter of 0.015 cm, Eq. (8.47) reduces to:

$$v = \frac{0.01225(p_1 - p_2)}{\mu}$$ (8.48)

where v = rate of flotation, cm/s

$$v = \frac{0.0241(p_1 - p_2)}{\mu}$$ (8.49)

where v = rate of flotation, ft/m

For other fats, oils and grease, the globule diameter should be determined experimentally.
The first step in the design of a gravity flotation separator is therefore to compute the rate of flotation in cm/s or ft/m using Eqs. (8.47), (8.48), or (8.49).

- *Step 2*

Assign the maximum horizontal flow velocity as follows:

$$V_L = 15 \, v \text{ or } 3 \text{ ft/m} = (1.52 \text{ cm/s}), \text{ whichever is smaller}$$ (8.50)

- *Step 3*

Compute the minimum vertical cross-sectional area $H \times W$ as follows:

$$HW = \frac{Q}{V_L}$$ (8.51)

where HW = minimum vertical cross sectional area or basin depth (H) times basin width (W) in ft²(m²)
Q = flow in ft³/m or m³/m
V_L = maximum horizontal velocity in ft/m (M/m)

Note: 1 cm/s = 0.6 M/m

- *Step 4*

Assign the minimum depth to width ratio of 0.3:

$$\frac{H}{W} = 0.3 \tag{8.52}$$

where H = basin depth in ft (m)
W = basin width in ft (m)

Note: In certain cases, H/W can be increased to an absolute maximum of 0.5.

- *Step 5*

Compute the minimum surface area as follows:

$$LW = F\frac{Q}{v} \tag{8.53}$$

where LW = minimum surface area, length of (L) times width (W) in ft^2 (m^2)

$$F \propto \frac{V_L}{v} = \quad \text{turbulence and short circuiting factor} \tag{8.54}$$

see Step 1 for v
see Step 2 for V_L
see Fig. 8.98 for F

- *Step 6*

Compute the basin length as follows:

$$L = F\frac{V_L}{v}d \tag{8.55}$$

where L = basin length, ft (M)
F = turbulence and short circuiting factor, see Fig. 8.95
V_L = horizontal velocity in ft/m (M/m)
v = rate of flotation, ft/m (M/m)
d = particle diameter

- *Step 7*

Assign basin depth, H. As previously discussed, basin efficiency is independent of basin depth. The purpose of basin depth is for the storage and removal without affecting flotation efficiency of float-ables and settleables.

It is recommend that a flotation basin be no less than 3 feet depth and preferably 4 feet. Practical maximum depth is 8 to 10 feet.

- *Step 8*

Compute basin width, W, as follows:

$$\frac{H \text{ from step 3}}{H/W \text{ from step 4}} = W \text{ in ft (m)} \tag{8.56}$$

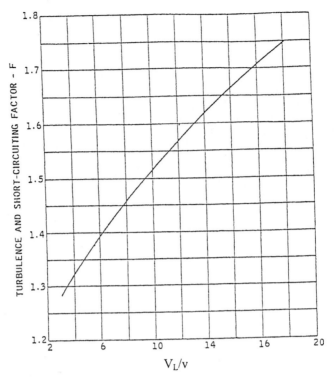

FIGURE 8.98   Recommended values of *F*.

It is recommended that a basin be no wider than 20 feet because of mechanical equipment, baffle limitations and wind action.

- *Step 9*
Compute basin length, *L*, as follows:

$$\frac{LW \text{ from Step 5}}{W \text{ from Step 7}} = L \text{ in ft (m)} \tag{8.57}$$

- **Design recommendations and hints**
The same design factors as explained in the sedimentation section apply to flotation. These recommendations are summarized as follows, but the reader is referred to sedimentation design section for details.
  - Distribute inflow across the entire width and depth of the basin by installing inlet baffles with sized openings.
  - Design a floatables removal system to maximize concentration of floatables and minimize water removal.
  - Design an under flow baffle(s) following floatables removal system to collect floatables. This baffle must be deep enough to prevent pulling of the floatables under the baffle. It is recommended that the velocity of flow under the baffle be limited to 1 ft/m for this reason.

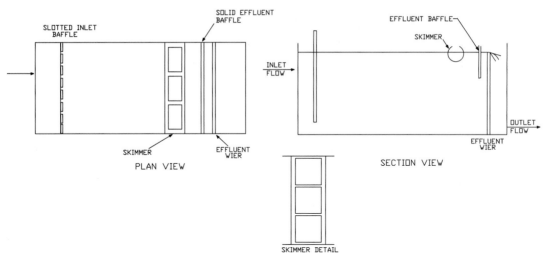

FIGURE 8.99   Flotation separator design.

- Design an overflow weir to set the water level. The weir length should be sufficient to prevent the pulling of floatables under the baffle, preferably greater than 10,000 gpd/L.F. Also the clearance of the baffle and the effluent weir should allow flow velocity to be limited to 1 ft/m.
- If solids or heavy chemicals can accumulate in the bottom, a solids or heavier removal systems should be designed.
- The design loading rate should be a maximum of 2000 gpd/ft^2 (56 L/m/M^2).
- Figure 8.99 shows a typical flotation or separator design. Figures 8.100, 8.101 and 8.102 show photographs of a flotation separator designed in accordance with the above recommendations. Figures 8.102 and 8.103 show solids removal systems for flotation separators.

FIGURE 8.100   Flotation separator photograph.

FIGURE 8.101   Flotation separator entrance photograph.

FIGURE 8.102   Flotation separator skimmer photograph.

FIGURE 8.103   Scum skimmer.

8.6.4   Design of Dissolved Air Flotation Separation

8.6.4.1   Introduction.   One method of increasing the efficiency of flotation is to reduce the specific gravity of the floatables by attaching very small bubbles of air to them using Dissolved Air Flotation (DAF). In this system, a side stream of the liquid is super-saturated under pressure with dissolved air so that the movement of the air bubbles will carry the floatables vertically upward where they can be removed.

A variation of DAF is vacuum flotation which involves saturating the wastewater with air and applying a partial vacuum to cause the dissolved air to be released from solution as minute bubbles.

8.6.4.2   Treatability Studies.   A laboratory unit as shown in Fig. 8.104 can be used to efficiently design a DAF system.

- *Step 1*
Fill the pressure chamber 80 percent full of wastewater.

- *Step 2*
Pressurize the pressure chamber with a hand pump to 40 to 50 psig.

- *Step 3*
Quickly open the petcock.

- *Step 4*
Wastewater will rapidly transfer to the cylinder.

- *Step 5*
Record the detention time and surface loading rate to reach the desired FOG concentration in the cylinder.

- **Design**
  - Air-to-solids ratio should be between 0.005 and 0.060 ml air/mg solids with 0.02 a good design value.
  - Solids loading rate should be less than 2 lb/ft^2/hr (10 kg/M^2/hr).

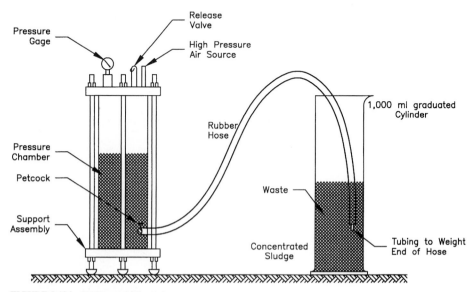

FIGURE 8.104   DAF treatability study.

- Hydraulic loading rate should be less than 0.8 gpm/ft^2 or 1152 gpd/ft^2 (32 L/m/M$_2$).
- Air injection should be at 40 to 50 psig (276–B 345 kilopascals).

Note that this hydraulic loading rate is much greater than that of sedimentation.

- The inlet distribution suggestions made in section 8.5 for circular or rectangular clarifiers also apply to DAF in that the inlet flow should be distributed at a low velocity across the entire depth and width of the inlet cross-section. In flotation, the baffling arrangement should be designed to force particles slowly upward instead of downward.
- The floatables removal system should be designed to maximum the concentration of floatables and minimize the water removed. Some of these systems are described in section 8.5.5.10 and shown in photographs in Fig. 8.102 and Fig. 8.103.
- DAF systems have successfully incorporated an air eduction system based on Bernoulli's principal instead of a pressure tank.

One additional method used for DAF floatables removal is a variable level system where the water level is increased by closing the outlet valve and allowed to overflow a weir. Such a system is shown in the photograph in Fig. 8.105.

- As in the case of a gravity flotation basin, a solid baffle must follow the floatable removal system to trap the floatables. Again, the baffle should have the flow velocity under the baffle and between the baffle and the effluent weir, limited to 1 ft/m.

FIGURE 8.105 DAF system.

• The overflow weir should be long enough to minimize the outlet velocity. Preferably less than 10,000 gpd/L.F. of weir length.

8.7 FILTRATION

8.7.1 Introduction

Filtration is a process that uses voids in paper, fabric or between granular solids, to strain settleable solids from a wastewater. There are four primary mechanisms which aid filtration:

• Straining

• Sedimentation

• Adsorption

• Vander Waal's force

8.7.1.1 Straining.   Straining is the primary process used in filtration and the only process used in a cartridge type filter using paper or fabric filter media. Initially, only particles larger than the filter media pore size will be removed by straining. This allows a designer to easily determine removal efficiencies of a filter using straining. The first solids collected on a filter media form a layer of particles with larger than pore size particles next to the media. With this layer, gradually smaller sized particles are trapped between the particle voids, which are smaller than the particles. This layer will remove smaller sized particles as it increases in depth. If biological particles are present, this layer, known as a *schmutzdecke,* can also act as a biofilter. In either case, the build-up of smaller particles not only increases the filter efficiency, but increases the pressure loss through the filter. In the case of a granular filter, this layer causes most of the filtration activity to occur on the surface with little or no efficiency beneath the surface. The only purpose then of the remaining granular depth is to support the surface media and provide a safety factor.

8.7.1.2 Sedimentation.   In granular filters, each void area consists of a small sedimentation basin. If velocities through the voids are low enough, the amount of settling in these voids can be significant. This efficiency is theoretically able to be modeled, but is typically ignored when computing overall filter efficiency because of the normally high flow velocities through the voids.

8.7.1.3 Adsorption.   Since biological particles tend to grow on the surface of solids, as in an attached growth biological system, granular filters can have some adsorptive removal efficiency. As mentioned above, the *schmutzdecke* also acts as a biofilter. Again, since adsorption efficiency levels are low, especially in a rapid sand filter that is backwashed often, they are usually ignored in design calculations.

8.7.1.4 Vander Waal's Force.   Vander Waal's Force is the relatively weak gravitational attraction between molecules because of the electric polarization induced in each of the particles by the presence of other particles.

8.7.2 Types of Filters

8.7.2.1 Cartridge Filter.   A cartridge filter usually consists of a series of vertical bags supported in a canister in order to maximize the filter surface area. The media is typically paper or fabric. Paper media is disposable with the paper and filtered material disposed of as a solid waste when the pressure loss through the filter reaches a maximum level of about 8 feet of head or 3.5 psig. Theoretically, a fabric bag can be backwashed internally or removed and washed.

8.7.2.2 Pressure Filter.   A pressure filter is a granular filter which uses the pressure from a pump or a high head to force the water through the media. The head must be greater than the expected

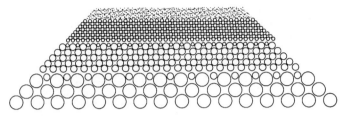

FIGURE 8.106 Filter media stratification.

maximum head loss through the filter. A pressure filter is typically fed through the top and back-washed from the bottom.

8.7.2.3 Gravity Filter. A gravity filter is a granular filter which uses only the pressure of the water above the media to force the water through the media. Theoretically, a gravity, or for that matter, a pressure filter can be upflow or downflow, but because of particle stratification after backwashing, a downflow, up-backwash filter is more efficient. Stoke's Law dictates that larger particles settle faster, therefore after backwashing, the larger particles accumulate on the bottom as shown in Fig. 8.106.

During filtration, these particles will retain their positions, which unfortunately are opposite that of maximum efficiency. If the larger particles were on top, more of the depth of the filter could be used in a downflow filter. If the filter were used in an upflow direction, the backwashing would have to occur downflow, with no ability to expand the media, resulting in inadequate cleaning. To counteract this stratification tendency, two methods have been used:

- **Particle Uniformity**
  If all filter particles were exactly the same size, they would not stratify after backwashing. This is impractical for two reasons: the difficulty of screening filter media that accurately, and the breaking up of media during backwashing, forming finer media. Theoretically, synthetic media can be manufactured with a constant uniformity coefficient. Screened media is typically specified with a uniformity coefficient (U.C.) as shown in Table 8.6 in order to minimize the particle size variation.

TABLE 8.6 Comparison of Rapid and Slow Sand Filters

	Slow sand	Rapid sand
Rate of filtration	1–10 MGAD	100–200 MGAD 125 MGAD = 2 gpm/ft^2 (81 L/min/m^2)
Depth of bed	12 in. (0.3 m) gravel, 42 in. (1.0 m) sand initial, 24 in. (0.6 m) final	18 in. (0.5 m) gravel 30 in. (0.75 m) sand
Sand size	d_{10} = 0.24–0.35 mm u.c. = 2–3	d_{10} = 0.45 mm and higher u.c. = 1.5 and lower
Sand distribution	Unstratified	Stratified
Underdrainage	Spilt tile laterals	Perforated pipe laterals
Loss of head	0.2 ft. (0.06 m) initial, 4 ft. (1.2 m) final	1 ft. (0.3 m) initial, 9 ft. (2.7 m) final
Time between cleaning	20–60 days	12–40 hours
Method of cleaning	Scrape off surface	Backwashing
Amount of washwater	0.2%–0.6% of filtered water	1-4-6% of filtered water
Cost of construction	Higher	Lower
Cost of operation	Lower	Higher
Depreciation of plant	Lower	Higher

- **Multi Media Filters**

  If some of the filter particles are of a material with a lower density, the particles of that material will stratify within themselves, allowing multiple bands of filter media, thus using more of the depth and increasing efficiency and minimizing backwashing.

  Dual media has been successfully used with the upper media of an anthracite coal with a specific gravity of approximately 1.6, much less than sand, and the lower media of sand.

  Triple media has been successfully used with the upper layer being anthracite coal, the center media of sand, and the lower media of garnet, a semi-precious stone.

8.7.3 Sand Filter Classification

Sand filters can be classified into slow and rapid sand filters. Typically, slow sand filters periodically have their surface removed and replaced, and rapid sand filters are backwashed to clean. Table 8.6 is a comparison of the two types of filters.

8.7.4   Filter Design and Design Hints

- **Hydraulics of filtration**

  The head loss through a filter can be calculated as follows:

$$\frac{h}{H} = \frac{5}{g} - \frac{uv(1-F)^2}{F^3} - \frac{S^2}{d} \tag{8.58}$$

where h = head through filter in ft (M)
   H = depth of filter bed
   G = acceleration due to gravity = 32.2 ft/s^2 (980 cm/s^2)
   u = kinematic viscosity of water in ft^2/s (Ns/M^2)
   v = flow velocity ft/s (M/s)
   F = porosity = volume of voids/volume of sand bed (approximately 0.45)
   S = shape factor
      spherical grains 6.0
      worn grains 6.4
      sharp grains 7.4
      crushed grains 8.5
   D = grain diameter in ft (M)

  (See Appendix C-3 and C-4)

- **Backwashing**

  - It is recommended that the rise rate through the horizontal surface area of the filter be 24 to 26 in/min (0.6 m/min) or 18.75 gpm/ft^2 (760 L/min/m^2).
  - When backwashing, the media bed should be expanded from 30 to 50 percent of its settled depth at that rate to achieve sufficient backwashing.
  - The predicted expanded depth can be calculated as follows:

$$H_e = \frac{H(1-F)}{1 - \left(\dfrac{V}{V_s}\right)^{0.22}} \tag{8.59}$$

where H_e = expanded sand depth
H = unexpanded sand depth
F = porosity = volume voids/volume bed (approximately 0.45)
V = approach velocity = Q/A
Vs = settling velocity of particles

- The backwash time is normally from 5 to 20 minutes in length.

- *Backwash frequency*
  - The head loss through a clean filter should be 4 to 6 in (10–15 cm).
  - The maximum permissible head loss is 8 to 10 ft (2.5 to 3.0 m).
  - Backwashing can be accomplished with a pump or an elevated storage tank.

- *Filter loading for sand filters*
  - Standard rate filters can be loaded at up to 2 gpm/ft² (81 L/min/m^s).
  - Iron removal filters can be loaded up to 3 gpm/ft² (122 L/min/m²).
  - High rate filters (usually with multimedia) can be loaded at up to 4 to 6 gpm/ft² (162 L/min/m²).
  - Declining rate filters can be loaded at up to 6 gpm/ft² (244 L/min/m²).

- *Filter troughs*
  - Filter troughs should be a minimum of 3 ft (1 m) from sides.
  - Tops of troughs should be a maximum of 30 inches (0.75 m) from filter surface and a minimum of 12 inches more than the expanded filter bed.
  - Bottoms of troughs should be above the expanded filter media.

- *Filter media for relatively clear liquid*
  - Sand depth should be approximately 30 inches (0.75 m).
  - Sand effective size should be 0.35 to 0.55 mm.
  - Sand uniformity coefficient should be less than 1.70.
  - Anthracite in a dual media filter should be a maximum of 20 inches (0.5m) in a 30-inch (0.75 m) bed.
  - Anthracite should have an effective size of 0.8 to 1.2 mm.
  - Anthracite should have a uniformity coefficient of less than 1.85.

- *Filter Media Support Systems*
  The filter media must have a support system with openings smaller than the media to prevent wash through of the media. Since for sand media this would require a very small opening, subject to stoppage, there are several methods used as follows:
  - Slotted plastic pipe or a rigid plastic pipe (such as PVC) header system can be slotted on the lower half with slots smaller than the grain size (see Fig. 8.107).
  - Glass beads (see Fig. 8.108). Glass beads with void sizes smaller than sand grain sizes can be used to support the sand.
  - The glass beads can then be supported with a perforated plate with holes smaller than the glass beads.
  - Gravel (see Fig. 8.109).
  - River gravel graded from small to large, top to bottom, can be used to support sand filter media, with voids in each size being smaller than the next higher grain size.

- *Filter underdrain systems*
  - Filter underdrain systems should be designed to prevent stoppage and allow even distribution of backwash water.
  - The total area of filter underdrains (perforations) should be approximately 0.003 times the total filter surface area.
  - The plenum area under the filter bottom should be sufficient to carry twice the maximum filter design flow.

- *Filter surface wash*
  - Since the filter surface typically plugs with solids as explained above, it is easier and quicker to clean if it is broken up before backwashing begins. This is normally done with a surface wash or an air wash as follows.

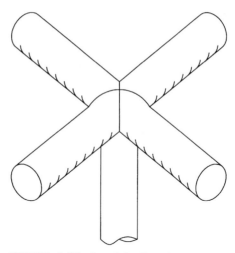

FIGURE   8.107   Stotted pipe filter midea support.

FIGURE 8.108   Photograph of glass beads filter media support.

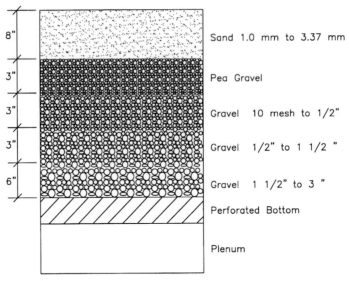

FIGURE 8.109   Gravel filter media support.

- Surface Wash

  A rotating surface wash system can be run for approximately 2 minutes before back-wash begins to break up the surface. The surface wash water pressure should be 45 to 75 psi (310 to 520 kPa).

- Air wash

  Air can be pumped into the plenum below the filter bottom for approximately 5 minutes before backwashing begins in order to break up the media surface. The air will follow the same route as the backwash water and should be provided at 2.0 to 4.0 cfm/ft^2 (0.6 to 1.2 m^3/min/m^2). An option to this type of air wash is to inject air directly into the sand just above, or in place of, the filter surface wash system.

8.7.4.1 Filter Difficulties

- **Negative head**

  If water tends to leave a filter faster than it is supplied, a negative head can result causing the release of gases from the water and subsequent air binding or bubbles of air.

- **Mud balls**

  Grains of sand can stick together because of organic adsorption in sizes from peas to greater than 2 inches (5 cm) in diameter. Like air bubbles, mud balls can cause uneven filtering and backwashing. Mud balls can also cause cracks to develop, allowing deeper penetration or short circuiting of unfiltered water.

  Mud balls will normally be broken during adequate surface washing.

8.7.4.2   Diatomaceous Earth Filters

- Diatomaceous Earth (D/E) Filters were developed during the Second World War by the

  U.S. Army to remove water borne cysts of hystolytica and shistosomes, which are small enough to enter a swimmer's body through the skin.

- Diatomaceous earth consists of the inert skeletal remains of diatoms deposited in certain areas of the world including California, USA.

- D/E filters should produce a filtrate of less than 0.1 ppm turbidity.
- Design
  - Filter consists of a septum or center core porous enough to permit maximum flow while supporting filter aid.
  - Helically wound wire, porous rubber and other porous materials have been used for the septum.
  - Filter aid is required at the rate of approximately 1000 g water/lb (8.33 L/kg).
  - Filter aid is applied at approximately 0.10 lb/ft^2 of septum area (5 kg/m^2).
  - If a run is stopped, backwash and filter aid replacement must occur.
  - Backwashing and replacement is required approximately each 2 hours.
  - Filter aid can be continuously applied in a slurry form to form the filter cakes.

8.7.4.3 Examples of Filters

- **Pressure filters** are shown in Figs. 8.110, 8.111, and 8.112.
- **Gravity filters** normally operate as constant level, variable level and variable head filters.
- **Constant level filters**
  A constant level filter will have a rate of flow controller on its discharge line to control the rate of flow to a pre-set amount. As the filter becomes plugged and the rate tends to be reduced, the rate of flow controller will open, allowing more flow. In effect, the friction loss through the filter, plus the friction loss through the rate of flow controller equals the head over the filter. A constant level filter will usually have a pressure differential measurement system with one sensor above the sand and one in the plenum under the filter floor. When the pressure loss exceeds a certain amount, the filter is manually or automatically backwashed.

  This type of filter is shown in Fig. 8.113.

  A second type of constant level filter is shown in Fig. 8.114, which uses an effluent weir with a level high enough to exceed the total head loss through the filter.

- **Variable level filter**
  A variable level filter is shown in Fig. 8.115. This filter allows the water level (and head pressure) to increase until it reaches a maximum level. Once the maximum level is reached, the filter is manually or automatically backwashed.

FIGURE 8.110 Typical pressure filter.

FIGURE 8.111 Photograph of large pressure filter.

- **Low media depth filter**

Since only the top fraction of a filter actually operates during a cycle, if a filter is back-washed often enough, it need be only a few inches thick. This theory is used in the low media depth filters as shown in Fig. 8.116. In this filter, each section of filter (1 or 2 feet wide) is continuously and sequentially backwashed. The major advantage of this filter is the minimum head loss required.

FIGURE 8.112 Photograph of small pressur filter. (Process Efficiency Products)

FIGURE 8.113   Photograph of a constant level rapid sand filter.

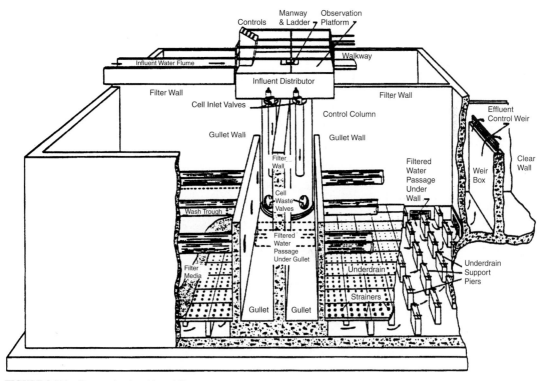

FIGURE 8.114   Constant level rapid sand filter.

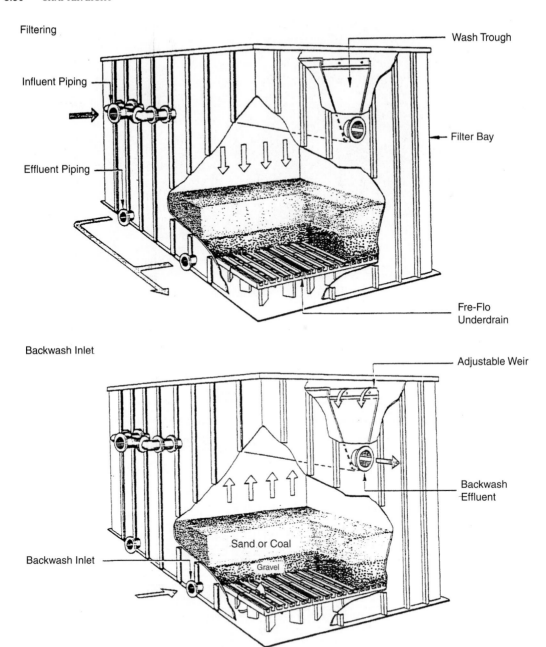

FIGURE 8.115   Variable level rapid sand filters.

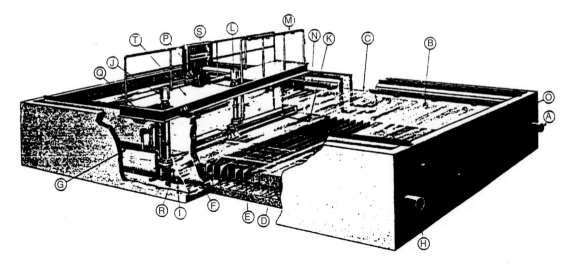

A. Influent line.
B. Influent parts.
C. Influent channel.
D. Compartmented filter bed.
E. Sectionalized under-drain.
F. Effiuent and backwash parts.
G. Effiuent channel.
H. Effiuent discharge line.
I. Backwash valve.
J. Backwash pump assembly.

K. Washwater hood.
L. Washwater pump assembly.
M. Washwater discharge pipe.
N. Washwater trough.
O. Washwater discharge.
P. Mechanism drive motor.
Q. Backwash support retaining springs.
R. Pressure control springs.
S. Control instrumentation.
T. Traveling backwash mechanism.

FIGURE 8.116   Low media depth rapid sand filter.

8.8   MEMBRANE FILTRATION

Membrane filtration is a separation process in which one or more bulk phases is separated from another phase(s) through the use of a membrane. A membrane used in wastewater treatment is a thin solid media that is characterized by the size of its pores (openings). Membranes can be described as macroporous, microporous, or semi-permeable. Macroporous membranes are membranes that allow the transport of a bulk liquid and dissolved constituents by hydrodynamic flow. Due to the tortuous path created by the membrane structure, the hydrodynamic flow is increased through the application of pressure. The transport of the liquid across a microporous membrane in wastewater treatment is accomplished through both hydrodynamic flow and sorption/diffusion. In microporous membranes, the transport across the membrane is due to the application of pressures to the feed side of the membrane. Semipermeable membrane filtration involves permeation of a liquid through a thin barrier, or membrane, by sorption and diffusion and is only accomplished through the application of high pressure to the feed side of the membrane.

8.8.1   Categories of Membrane Filtration

The term *membrane filtration* covers a wide array of treatment processes and applications. Membrane filtration is divided into four categories based on the effective pore size of the membrane as shown in Table 8.7 (Ho and Sirkar, 1992).

TABLE 8.7 Membrane Pore Sizes

Membrane type	Minimum pore size (nm)	Maximum pore size (nm)
Microfiltration	200	10000
Ultrafiltration	1	20
Nanofiltration	0.5	2
Reverse osmosis	0.1	1

Microfiltration is capable of removing relatively large colloidal and suspended solids from wastewater. A common use of microfiltration is the removal of suspended solids too small to be removed by a conventional media filter. Ultrafiltration is used in applications which require greater colloidal solids removal and/or removal of large molecular weight organics. Two common applications for ultrafiltration are the treatment of food processing wastewater and the separation of fats, oils, and grease from wastewater. Nanofiltration is capable of removing smaller molecular weight organics and a portion of the dissolved solids in a wastewater, while reverse osmosis is used to remove dissolved solids from a wastewater.

8.8.1.1   Reverse Osmosis.   Reverse osmosis is a membrane separation process, which, through the application of pressure, reverses the natural phenomenon of osmosis. Osmosis is the flow of water from an area of low ionic concentration to an area of high ionic concentration. Under an applied pressure, water is forced through a semi-permeable membrane from an area of high concentration to that of low concentration. The semi-permeable membrane rejects the solutes and suspended matter in the water while allowing the clean water to pass through. The effective pore sizes of reverse osmosis membranes are typically on the order of 2 to 5 angstroms. This pore size allows the passage of only molecular water. In order for reverse osmosis to take place, the applied pressure must be greater than the naturally occurring osmotic pressure. Osmotic pressure is a characteristic of the solution and is a function of the molar concentration of the solute, the number of ions formed when the solute dissociates, and the solution temperature. Osmotic pressure is calculated in Eq. (8.61) later in this chapter.

Reverse Osmosis (R.O.) has been used successfully in electroplating operations to lower dissolved solids concentration from over 1300 Specific Conductivity in micromhos/cm to less than 125 micromhos/cm. A well designed reverse osmosis system will allow a permeate (treated water) of 5 to 10 percent of total water treated. This means that 90 to 95 % will not be treated. Increasing the temperature has been shown to increase percent recovery. The water entering an R.O. system should have an SDI (silt density index) of less than 3.5 to protect the membrane from fouling.

8.8.1.2   Nanofiltration.   Nanofiltration (NF) uses membranes that have a larger effective pore size than reverse osmosis membranes. This results in lower dissolved solids rejection rates or if the pores are large enough, a complete passage of dissolved ions through the membrane. The osmotic pressure equation (8.61) still applies to NF, but the osmotic pressure difference across the membrane will be lower or may be zero. The osmotic pressure differential across the membrane is taken into account in the equation for water transport through the membrane.

8.8.1.3   Ultrafiltration.   Ultrafiltration (UF) uses membranes that have effective pore sizes several orders of magnitude larger than reverse osmosis. UF units are capable of separating large molecular weight organics from wastewater. UF units can also be used as a pretreatment option for nanofiltration or reverse osmosis. In the case of food processing, UF units can be used to separate proteins and carbohydrates from the wastewater. The proteins and carbohydrates may then be reused in the process or sold as a by-product. Another use of UF units is the separation of emulsified fats, oils, and grease from wastewater.

8.8.1.4   Microfiltration.   Microfiltration (MF) uses microporous membranes that have effective pore sizes much larger than UF membranes. Flow through a microporous membrane can occur without the application of pressure on the feed side of the membrane, but in most wastewater applications, a small pressure difference across the membrane produces significant increases in flow. Microfiltration is most often used to separate suspended and colloidal solids from wastewater.

8.8.1.5 *Membrane Filtration in Wastewater Treatment.* Membrane filtration can be used in a variety of applications for the treatment of wastewater. Membrane filtration represents a treatment technology that is capable of separation not possible in classic wastewater treatment. As the world's environmental concerns over clean water supplies heighten, the application of membrane filtration increases. The primary use of membrane filtration technologies is to provide treatment of the wastewater to such a degree that it may be recycled back to a process (for industrial facilities) or used in other applications such as irrigation, groundwater recharge, or as drinking water. This use of membrane filtration technology is especially common in areas where clean water is scarce, such as in dry, arid climates, especially for removing salt from seawater.

In biological treatment processes, membrane filtration can be used as a tertiary filtration process following secondary clarification. Membrane filtration can also be used in a variety of physical/chemical treatment systems such as FOG removal, organics removal, heavy metals removal, and process water reuse.

8.8.2 Application of Membrane Filtration

Membrane processes can provide ultra pure water for industrial uses, removal of speciated pollutants such as nitrogen and phosphorous or other nutrients, industrial pretreatment before release to a POTW, organics recovery, recovery of suspended or dissolved solids, recovery of water for recycling/reuse, or tertiary treatment prior to discharge to receiving water. Membrane filtration has been successfully used in all of the above mentioned applications. In order to determine the applicability of a membrane filtration process for wastewater treatment, the economics should be evaluated on a cost-benefit basis. Membrane filtration has a high capital and operating cost, but in many cases is the only treatment option that will actually provide the needed results.

8.8.2.1 *Raw Wastewater Analysis.* In order to begin to design a membrane process, a detailed analysis of the wastewater to be treated should be performed. A detailed analysis would include the following constituents:

- Biochemical oxygen demand (BOD_5)
- Chemical oxygen demand (COD)
- Total suspended solids (TSS)
- Total dissolved solids (TDS)
- Fats, oils, and grease (FOG)
- Dissolved anions (chlorides, fluorides, sulfates, etc.)
- Dissolved cations (sodium, calcium, magnesium, potassium, etc.)
- Organics
- pH
- Temperature
- Compounds of interest

The results of the above analysis will help determine:

- The appropriate pretreatment for the membrane unit
- The appropriate membrane configuration
- The appropriate membrane material
- The required pollutant removal efficiency

Biochemical Oxygen Demand and Chemical Oxygen Demand: The BOD_5 and COD concentrations are used in determining the required removal of oxygen demanding components of the wastewater. Also, biological fouling of the membrane can occur with BOD present in the wastewater.

Total Suspended Solids: Microfiltration and ultrafiltration are used to remove suspended solids from wastewater, so an accurate determination of the TSS concentration of the wastewater is necessary to size these systems. Another reason the TSS concentration of the wastewater is important is that suspended solids are a major cause of plugging of membrane systems. In reverse osmosis, nanofiltration and certain ultrafiltration applications, the TSS in the wastewater must be removed prior to the membrane unit. TSS is typically removed using either gravity settling and/or media filters.

Total Dissolved Solids and Dissolved Ions: The TDS concentration is very important when sizing nanofiltration and reverse osmosis units. The primary application for RO systems is the removal of the TDS from the wastewater while NF systems can be designed for partial removal of TDS. Certain dissolved ions in the wastewater can cause scaling, so an accurate determination of the dissolved ions must be performed.

Fats, Oils, and Grease: A primary application of ultrafiltration system is the removal of FOG from wastewater, therefore, an accurate determination of the FOG concentration is necessary for sizing UF systems. Also, FOG can cause fouling in other membrane systems, primarily NF and RO units.

pH: The pH of a wastewater is very important in selecting membrane material. Certain membrane material is not resistant to wide pH ranges, therefore, a misapplication of membrane material may cause corrosion resulting in failure of the membrane unit. Another reason for the importance of knowing the pH of a wastewater prior to system design is that several chemical reactions that form precipitates are pH dependent. If precipitates are formed in the membrane unit, scaling will form and reduce the life of the membrane.

Organics: A complete organic analysis may be necessary to determine if BOD or COD removal is required. An accurate determination of the organic constituents present will aid in the selection and sizing of the membrane system.

Temperature: All membrane materials have an acceptable operating temperature range. If prolonged operation outside of the range occurs, membrane life will be reduced.

Compounds of Interest: Depending on the purpose of the membrane filtration system, other compounds may be of interest and require independent testing.

8.8.2.2   Membrane Fouling.   Membrane fouling is one of the largest problems in the application of membrane processes to wastewater treatment. Fouling occurs when material is trapped within the pores or at the surface of a membrane causing a reduction in membrane performance. Five types of membrane fouling have been identified: membrane scaling, fouling by metal oxides, plugging, colloidal fouling, and biological fouling.

Membrane scaling: Membrane scaling is the result of precipitation of dissolved salts onto the membrane surface. Reverse osmosis processes typically concentrate dissolved salts by a factor of 2 to 4 times (Montgomery, 1985). When initial feedwater dissolved salt concentrations are very high, solubility limits may be exceeded when the water is further concentrated. This results in precipitation of the dissolved salts. As previously stated, the scales increase the osmotic pressure at the membrane surface which reduces water flux and salt rejection rates.

Metal oxide fouling: Metal oxide fouling occurs when soluble compounds in the feedwater are oxidized in the membrane system forming insoluble compounds that can deposit onto the membrane surface. These deposits block the membrane pores and reduce water flux rates. The most common metal oxides formed are those of iron and manganese. Iron fouling is most prevalent and is caused by the oxidation of the soluble ferrous ions to ferric ions and the subsequent precipitation as ferric hydroxides.

Plugging: Plugging is caused by mechanical filtration of suspended solids in the feedwater which are larger than the flow passages in the membrane unit. The problem of plugging is most prominent in hollow fiber membrane configurations and least prominent in tubular membrane configurations due to the relative sizes of the flow channels in the membrane unit. Plugging in tubular membrane designs is possible only when the flow channels of the membrane are smaller than the suspended particles in the feedwater. The flow channels in the NF unit are orders of magnitude larger than this, so plugging should not be a problem.

Colloidal fouling: Colloidal fouling is caused by the entrapment of colloidal particles on the membrane surface. The build-up of entrapped colloids reduces the water flux and solids rejection rates of the membrane.

Biological fouling: Biological fouling refers to microbial adhesion and growth onto the membrane surface. This build-up of microbial growth can decrease water flux rates, increase transmembrane and differential pressures in the system, and decrease solids rejection by the membrane.

Concentration polarization: Concentration polarization is the build-up of a boundary layer of more highly concentrated solute on the membrane surface than in the bulk solution. This occurs as water permeates through the membrane surface and leaves a more concentrated solute layer, which must diffuse back into the bulk solution. The more concentrated solute layer increases the osmotic pressure at the membrane surface, which causes a decrease in water flux and dissolved solids rejection. If the concentration of sparingly soluble salts at the membrane surface exceeds their solubility limits, precipitation or scaling will occur at the surface (Amjad, 1993).

Most fouling can be eliminated by pretreatment of the feedwater and cleaning of the membranes. If fouling is probable even after pretreatment or pretreatment is not an option, the regularity of the fouling must be determined. In order to use membrane processes in which fouling is an issue, cleaning costs and membrane replacement costs need to be incorporated into the economic analysis of a membrane process as a treatment alternative.

8.8.2.3   Membrane Bioreactors (MBR).   An MBR combines biological treatment with a membrane system to provide a very high level of organical suspended solids removal. This system is described further in Section 10.

8.8.3   Treatability Study

The most accurate method for sizing membrane processes is to conduct a treatability study. In the study, a bench scale membrane system is used to treat small quantities of the wastewater of concern. Typically, treatability studies are conducted by a membrane systems manufacturer, an engineering firm, or a research institute. The goal of a membrane process treatability study is to obtain the following information:

• Appropriate pretreatment
• Inlet pressure requirements (psig)
• Treated water flux through the membrane (gpd/ft^2)
• Rejection rates of contaminants
• Appropriate number of stages
• Suitable membrane material
• Appropriate membrane configurations

Pretreatment, suitable membrane material, and appropriate membrane configuration are typically investigated prior to commencing laboratory experiments.

8.8.3.1   Pretreatment.   Appropriate pretreatment can be selected based on the lab analysis of the wastewater constituents, but a treatability study proves the suitability of the chosen pretreatment

TABLE 8.8 Pretreatment Selection

Wastewater constituent	Membrane process	Pretreatment technology
Total suspended solids	UF, NF, RO	Gravity settling, media filtration, cartridge filtration
Organics	NF, RO	Activated carbon adsorption
Chlorine	MF, UF, NF, RO	Activated carbon adsorption
Scalants (calcium, magnesium, metal hydroxides, metal oxides, etc.)	MF, UF, NF, RO	Precipitation, filtration, chemical addition, ion exchange
Fats, oils, and grease	NF, RO	Gravity separation, emulsion breaking, ultrafiltration
High/low pH values	MF, UF, NF, RO	pH adjustment
Organic matter	MF, UF, NF, RO	Gravity separation, filtration, chlorination

options. A treatability study also allows for the differentiation of similar pretreatment options. Table 8.8, below, can be used to select pretreatment options.

Selection of proper wastewater pretreatment prior to the membrane filtration unit is one of the most important design aspects of membrane filtration due to the high probability of membrane fouling without proper pretreatment.

8.8.3.2 Membrane Materials.   Another important aspect in the design of a membrane filtration system is the membrane material. The ideal membrane material has the following qualities: high selectivity, high permeability, mechanical stability, temperature stability, and chemical resistance. Membrane materials may be classed in three broad categories: modified natural products, synthetic products, and inorganic products (Rautenbach and Albrecht, 1989).

The modified natural product membranes consist of such materials as cellulose acetate, cellulose acetobutyrate, and cellulose regenerate. Cellulose acetate (CA) is one of the oldest membrane materials in use today. The major drawbacks of this material are its poor temperature and chemical stability.

The synthetic product membranes consist of such materials as aromatic polyamide, polysulfone, polyethylene, polypropylene, and polyfuran. Synthetic product membranes are very versatile and may have excellent temperature stability and chemical resistance. Table 8.9, below, gives a general comparison of cellulose acetate and various synthetic product membranes (Ho & Sirkar, 1992; Rautenbach & Albrecht, 1989).

The inorganic materials used in membrane construction consist of oxides of aluminum, silver, titanium, and zirconium. The chemical resistance and temperature stability of the inorganic membranes are phenomenal, but currently, the major drawback with inorganic membranes is the mechanical instability of the membranes. They can be very brittle and therefore difficult to produce and transport.

TABLE 8.9 Membrane Selection

Material	pH range	Maximum temperature at pH = 7(°C)	Chlorine resistance	Solvent resistance
Cellulose acetate	4.5–9	55	Good	Poor
Polyamide	3–12	80	Poor	Good
Polysulfone	0–14	80	Good	Good
Polyacrylonitrile	2–12	60	Good	Poor
Polyfuran	2–12	90	Poor	Good

TABLE 8.10 Membrane Configurations

Configuration	Simplicity of flow path	Resistance to mechanical damage	Ease of mechanical cleaning	Power consumption (kwh/ft^2)
Tubular	Good	Good	Good	Poor
Plate and frame	Poor	Good	Fair	Good
Spiral wound	Fair	Poor	Poor	Good
Hollow fiber	Fair	Good	Poor	Good

8.8.3.3 Membrane Configurations. The configuration of the membrane module plays an important role in the performance of the membrane and its application. Membranes can be configured in four designs: hollow fiber, spiral-wound, tubular, and plate and frame systems. Hollow fiber and spiral wound membranes are the most popular configurations because of the large membrane surface area per volume thata they provide. The major disadvantage of such configurations is the high probability of fouling due to suspended particles in the wastewater. Fouling is a problem for these designs because the openings for the water to pass through are relatively small. This can be compensated for by adding pretreatment steps to remove the suspended solids.

Tubular designs were some of the earliest practical membrane treatment devices. The tubular configuration provides large well-defined flow passages and high flow velocities. This lowers the tendency of the membrane to foul and makes the membrane easier to clean. The disadvantage of such a design is the low membrane surface area to volume ratio and potential to leak between membrane leaves. Plate and frame configurations use flat sheet membranes and are of similar concept to the plate and frame filter press. Table 8.10 gives a general comparison of the different membrane configurations seen in wastewater treatment (Ho & Sirkar, 1992; Rautenbach & Albrecht, 1989).

8.8.3.4 Pressure Requirements. The inlet pressure requirements will vary depending on the effective pore size of the membrane. Table 8.11 lists inlet pressure requirements for the different membrane processes are (Ho and Sirkar, 1992).

The inlet pressures shown above vary due to desired recovery rates, desired rejection rates, dissolved solids concentration, amount of fouling, amount of membrane compaction, and the effects of concentration polarization.

8.8.3.5 Water Flux. A treatability study will enable actual determination of a membrane's treated water flux during treatment of the wastewater. The flux in gallons per day per square foot of membrane surface determines the number of membrane modules needed for treatment. The flux results also help determine the appropriate membrane configuration in order to minimize membrane modules.

The major operational parameters of concern in membrane filtration design are the treated water flux, water recovery rates and constituent rejection rates. The treated water flux and constituent rejection rates are determined through a treatability study in which the pretreatment technologies, membrane material, feed pressure, and the number of stages are varied.

TABLE 8.11 Membrane Operating Pressures

Membrane type	Minimum pressure differential (psi)	Maximum pressure differential (psi)	Average pressure differential (psi)
Microfiltration	1	50	20
Ultrafiltratio	15	600	168
Nanofiltration	50	600	211
Reverse osmosis	200	1200	514

The flux of treated water through a reverse osmosis membrane is described by the following equation (Amjad, 1993):

$$Q_p = \frac{K_w \times (\Delta P - \Pi) \times A}{\tau} \qquad (8.60)$$

where Q_p = water flow rate across the membrane (L/min)
K_w = membrane permeability coefficient (L/(kPa min m))
ΔP = hydraulic pressure differential across the membrane (kPa)
Π = osmotic pressure differential across the membrane (kPa)
A = membrane surface area (m²)
τ = membrane thickness (m)

The membrane permeability coefficient can be obtained from the membrane manufacturer. For dilute solutions, the osmotic pressure differential across the membrane is approximated using the van't Hoff Equation (Amjad, 1993):

$$\Pi = v_i \times c_i \times R \times T \qquad (8.61)$$

where Π = osmotic pressure (kPa)
v_i = the number of ions formed if the solute dissociates
c_i = molar concentration of the solute (mol/L)
R = universal gas constant (kPa L)/(mol K))
T = temperature of the solution (K)

For microfiltration and ultrafiltration applications, the osmotic pressure differential across the membrane can be approximated to be zero.

Typical treated water flux rates from the different membrane types are shown in Table 8.12 (Amjad, 1993; Ho & Sirkar, 1992; Rautenbach & Albrecht, 1989; Belfort, 1989).

Water flux through a clean membrane will decrease with time due to compaction of the membrane. Compaction is caused by creep deformation of the membrane over time and is dependent on the membrane material, applied pressure, and feedwater temperature. Creep deformation is directly proportional to the applied pressure and temperature. Compaction tightens the membrane's rejection layer and decreases the water flux across the membrane. A log-log plot of water flux versus time for a given temperature and pressure produces a straight line (Amjad, 1993). Compaction data for a given membrane type can be supplied by the membrane manufacturer and is used to predict membrane performance at a future time. With this data, membrane systems can be designed with such a capacity to offset the decrease in water flux due to compaction.

8.8.3.6 Rejection Rate.   Rejection rates of a membrane unit are measured as the percent of the constituent of concern removed in the product water. The equation which describes rejection is as follows (Montgomery, 1985):

TABLE 8.12 Membrane Flux Rates

Membrane filtration type	Minimum flux rate (gal/day/ft²)	Maximum flux rate (gal/day/ft²)	Average flux rate (gal/day/ft²)
Microfiltration	300	3500	1000
Ultrafiltration	12	490	150
Nanofiltration	17	290	28
Reverse osmosis	2	62	23

$$R_i = \left(1 - \frac{C_i}{C_f}\right) \times 100 \tag{8.62}$$

where i = constituent of concern
  C_i = concentration of constituent in the treated water (mg/L)
  C_f = concentration of constituent in the feedwater (mg/L)
  R_i = Rejection rate

Rejection rates and treated water flux rates determine the number of stages needed in a full-scale membrane filtration system. Multiple stages are used when the application calls for either very high rejection rates, or very high water flux rates, or a combination of the two. In a multiple staged system, either the concentrate or the filtrate from one stage is used as the feed for another membrane unit. The use of multiple stages in a membrane filtration system allows the designer to achieve high rejection and water flux rates that would be impossible from a single stage filtration system.

The treatability study is a crucial step in designing a full-scale membrane filtration system. The treatability study allows a greater optimization of the treatment process prior to any large capital investment in a full-scale system. Without the parameters determined in a treatability study, an accurate full-scale design of the membrane filtration system is impossible.

8.8.4 Design of Full-Scale Membrane Filtration Systems

8.8.4.1 Sizing of Membrane Units.   Once an accurate analysis of a wastewater's constituents is performed and a treatability study is completed, design of a full-scale membrane treatment system can begin. The first step in the full-scale design of a membrane unit is the determination of the total membrane surface area required. This can be calculated using the following equation:

$$A_m = F \times Q_t \tag{8.63}$$

where A_m = required membrane surface area (ft^2)
  F = treated water flux (gpd/ft^2)
  Q_t = required treated water flowrate (gpd)

For microfiltration and ultrafiltration applications, the required treated water flowrate can be approximated by the incoming raw wastewater flowrate. For nanofiltration and reverse osmosis applications, a certain quantity of the raw wastewater will not pass through the membranes. The quantity of raw wastewater that does not pass through the membranes contains high levels of dissolved solids and must be disposed of or further treated. The required treated water flowrate can be calculated for NF and RO applications using the following equation:

$$Q_t = Q_r = R_w \tag{8.64}$$

where Q_t = required treated water flowrate (gpd)
  Q_r = raw wastewater flowrate (gpd)
  R_w = recovery efficiency (%)

The recovery efficiency is determined through an economic analysis comparing the cost of final concentrate treatment or disposal and the increase capital and operating cost of a larger membrane unit.

Once the total membrane surface area required is determined, the next step in the design of a full-scale system is to select the appropriate pretreatment alternatives. As previously stated, pretreatment of the wastewater prior to filtration through the membrane unit is, in most cases, absolutely necessary. Without proper pretreatment of the wastewater, a membrane treatment system will require significant operation and maintenance costs. Once the proper pretreatment technologies are selected, each step must be properly sized. For design procedures and criteria for each treatment technology, please refer to the respective chapters of this book.

8.8.4.2 Equipment.   The equipment necessary for the installation of the full-scale system includes:

1. Pretreatment equipment
2. Transfer and feed pumps
3. Membrane modules
4. Feed, permeate, and concentrate storage tanks
5. Piping and valving
6. Instrumentation and controls

Figure 8.117 gives a schematic of a typical membrane system.

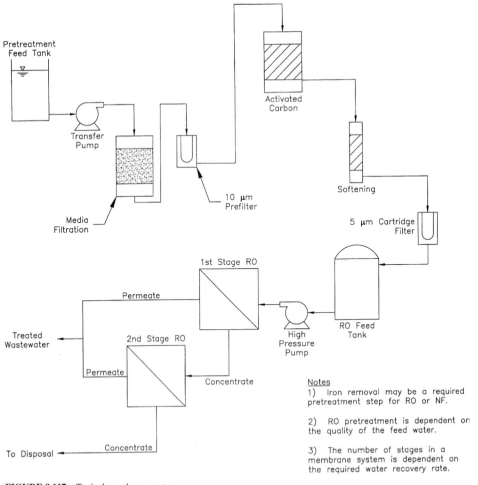

FIGURE 8.117   Typical membrane system.

As stated previously, the pretreatment portion of the system can vary based on the feedwater constituents and desired treatment of the wastewater. The material of construction and all equipment must be compatible with the constituents in the wastewater. In most applications, oxidation of iron in the materials of construction should be avoided at all cost due to the fouling potential of iron hydroxides and iron oxides. Therefore, it is advisable to select materials of construction that will not corrode from the wastewater, the elements (if outside), or time.

The instrumentation and controls include level controls on the feed, permeate, and concentrate storage tanks. Also, pressure gauges are necessary on pretreatment media filtration, and the membrane feed pump.

8.8.4.3 Cost of Membrane Filtration.   Broad estimates of membrane filtration systems are difficult. The difficulty results from the contributions to cost of the water recovery and pollutant rejection rates. The most expensive units will deliver both high water recovery and pollutant rejection. While the least expensive units will give low water recovery and low pollutant rejection rates. From an economic analysis, it is usually more feasible to have a system that provides either high water recovery or high pollutant rejection rate, but not both. Of course, certain applications will dictate the need for both high water recovery and high pollutant rejection. Table 8.13 provides a wide range of costs, which reflect the range of efficiencies of the individual membrane systems (Wiesner, et al, 1994; Ho & Sirkar 1992; Rautenbach & Albrecht, 1989).

The reverse osmosis and nanofiltration costs in the table above are based on a 1 MGD system. Modifying the costs for lower flowrates will increase the above unit costs using the following equation:

$$C_i = C_{1MGD} \times \left(\frac{Q}{1MGD} \right) 1 - n \tag{8.65}$$

where C_i = unit cost at a given flowrate ($/gpd)
C_{1MGD} = unit cost at 1 MGD ($/gpd)
Q = actual flowrate of system (MGD)
n = 0.6 for normal unit process
n = 0.75–1.00 for RO and NF systems
n = 0.52 for system up to 100,000 gpd

8.8.5 Installation

The installation of a membrane system may be conducted by the supplier, sub-contracted by the supplier or sub-contracted by the purchaser. In any case, most suppliers have the personnel to supervise installation. The supervision can usually be covered in the purchase cost of the system.

The degree of complexity of the installation will depend on the flowrate of wastewater to be treated, the degree of automation of the system, and the required pretreatment technologies. For

TABLE 8.13   Membrane Filtration Costs

	Low ($/gpd)	Average ($/gpd)	High ($/gpd)
Reverse osmosis	$2.26	$4.50	$6.800
Nanofiltration	$0.54	$1.35	$2.10
Ultrafiltration	$0.40	$1.60	$2.80
Microfiltration			
Metallic membranes	$0.23	$1.60	$3.00
Ceramic membranes	$0.23	$1.30	$2.30
Polymeric membranes	$0.16	$1.20	$2.20

smaller applications, the entire membrane system can be skid mounted with the majority pre-assembled at the manufacturing facility. The skid-mounted system can include all pretreatment equipment, piping and valving, and associated tankage. In this case, the inlet wastewater feed, outlet clean water discharge and the outlet concentrate discharge must be connected up to the skid mounted system.

For larger applications, installation of the system would become more complex. The treatment of a large wastewater flow or a flow with a high contaminant removal requirement may require more than one stage. Multiple stages require an increase in the number of membrane modules, the number of feed pumps, and the amount of piping, and valving. For larger flows, the pretreatment technologies would also be a large part of the overall system installation.

8.8.6   Operation

Membrane treatment systems can be run continuously without operator attention if designed with the correct controls and instrumentation. Regardless of the level of automation of the system, membrane cleaning and routine maintenance on the pumps is necessary with any size unit. Depending on the level of fouling, membranes will require cleaning. The cleaning can be accomplished by flushing clean water through the system or by flushing a chemical cleaning solution through the membranes. In either case, membrane cleaning involves turning all or a part of the system offline. This will require opening and closing inlet valving as well as potentially changing outlet valving. In industrial applications, membrane systems may only run during operation of the facility. In this case, the membranes may be cleaned before or after a day of operation. Membrane systems in municipal application, in which constant operation is required, will need to be taken offline while the membranes are cleaned. If the system permits, only the portion of the system that is being cleaned needs to be taken offline.

MEMBRANE FILTRATION REFERENCES

Amjad, Z., *Reverse Osmosis: Membrane Technology, Water Chemistry, and Industrial Applications,* Van Nostrand Reinhold, New York, New York, 1993.

Belfort, G., Synthetic Membrane Processes: Fundamentals and Applications, Academic Press, Orlando, FL, 1984.

Ho, W. S. W. and Sirkar, K. K. ed., *Membrane Handbook,* Van Nostrand Reinhold, New York, New York, 1992.

James M. Montgomery, *Water Treatment Principles and Design,* John Wiley & Sons, New York, New York, 1985.

O'Shaughnessy, W, Clark, W. Lizotte, R and Mikotel, D, "Fulfilling a Vision", Industrial Wastewater Jan, Feb, 1997.

Rautenbach, R. and Albrecht, R., *Membrane Processes,* John Wiley & Sons, New York, New York, 1989.

Wiesner, M. R., Hackney, J., Sethi, S., Jacangelo, J. G., Laine, J. M., "Cost Estimates for Membrane Filtration and Conventional Treatment," *JAWWA,* Vol. 86, No. 12, pp. 33–41, 1994.

8.9   TEMPERATURE CONTROL

8.9.1   Introduction

The temperature of a wastewater effects the rate of settling and flotation, the rate of chemical reaction, and the rate of biological activity. The rate of settling, flotation and filtration increases since the viscosity of the water decreases. Most chemical reactions including volatization increase rates with increasing temperature. Biological reaction rates approximately double for each 18°F (10°C) the liquid temperature rises. Wastewater may need to be cooled if discharge permits require a maximum discharge temperature. The only effective method for controlling the temperature of a wastewater is to transfer heat to the wastewater from some other medium. Wastewater temperature manipulation can be accomplished through several techniques.

8.9.2 Design

- Hot water is normally used to raise the temperature of a wastewater. A gas-fired boiler can be used to heat water which is then run through the shell of a heat exchanger to raise the temperature of the wastewater flowing through the tubes. Direct steam injection is also used in some circumstances to heat the wastewater.
- Chilled water including well water can likewise be used in a heat exchange to reduce wastewater temperature.
- Refrigerants such as freon can be used to lower the temperature of a wastewater in a refrigeration-type system.
- A louvered-type cooling tower or a spray system can be used to lower the water temperature to the temperature of the ambient air.

8.10 MIXING

8.10.1 Introduction

Mixing of a wastewater can be used for the following purposes:

- Where a gas, a solid, or a liquid must be completely intermingled with the wastewater in the case of chemical addition
- Where the solids in a liquid must be kept in suspension
- Where a suspended growth aerobic or anaerobic system must have intimate contact between food and microorganisms
- Where air or oxygen must be mixed with wastewater in a suspended growth aerobic system

8.10.2 Methods of Mixing

Mixing can be accomplished by any method of adding energy to the wastewater. Examples are:

- Hydraulic jumps in channels (Fig. 8.118)
- Venture flumes (Fig. 8.119)
- Fittings in pipeline (Fig. 8.120)
- Volutes of pumps (Fig. 8.121)
- Static mixers (Fig. 8.122)
- Paddles (Fig. 8.123)
- Turbines (Fig. 8.124)
- Propellers (Fig. 8.125)
- Gas movement through the liquid (Fig. 8.126)

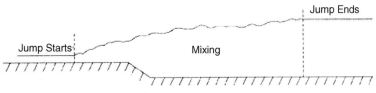

FIGURE 8.118   Hydraulic jump mixing.

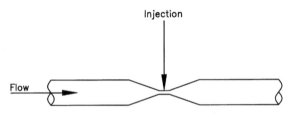

FIGURE 8.119   Venturi mixing.

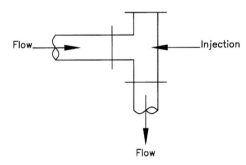

FIGURE 8.120   Pipeline fitting mixing.

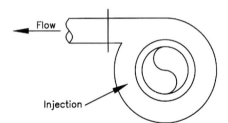

FIGURE 8.121   Pump volume mixing.

FIGURE 8.122   Static mixer. (Westfall Manufacturing Co.)

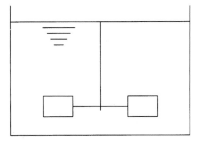

FIGURE 8.123   Paddle mixer.

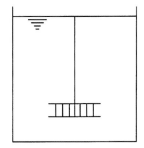

FIGURE 8.124   Turbine mixing.

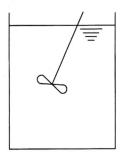

FIGURE 8.125   Propeller
mixing.

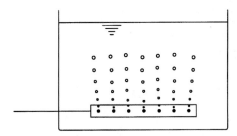

FIGURE 8.126   Gas mixing.

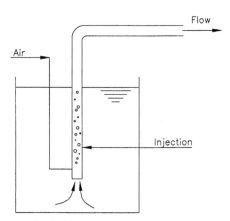

FIGURE 8.127   Air lift pump mixing.

- Air lift pumps (Fig. 8.127)
- Water eductor (Fig. 8.128)

8.10.3 Design of Mixers

8.10.3.1 Detention Time.   Total mixing detention time for chemical addition should be between $1/_2$ and 5 minutes. Less time is inadequate for mixing because of hydraulic inefficiency. More time requires unnecessary energy and promotes short circuiting. The lower the energy per detention time, the greater the detention requirement, but detention times greater than 5 minutes require extensive baffling to prevent short circuiting. This extensive baffling tends to approach a plug flow situation, which again prevents adequate mixing.

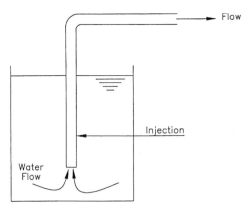

FIGURE 8.128   Water eductor mixing.

In the case of solids suspension and suspended growth biological systems, the detention time in the tank basin is usually dictated by some other parameter. Therefore, the amount of mixing required in these situations will vary greatly depending upon application.

8.10.3.2   Rotational Speed

• Large blenders normally rotate between 25 and 100 rpm, but may promote blending and subsequent short circuiting.
• Propellers normally rotate up to 2000 rpm.

8.10.3.3   Vortexing

• Vortexing occurs when the wastewater swirls rather than mixes. The fluid tends to circulate at a similar velocity as the propeller or paddle and decreases the effectiveness of mixing.
• An indication of vortexing is the presence of a whirlpool.
• Vortexing can be prevented by
  • Mounting propeller shaft vertically, but off center
  • Mounting propeller shaft at an angle (see Fig. 8.129).
  • Installing vertical baffles to break up the vortexing

8.10.3.4   Power Calculation.   The velocity gradient, G is an indication of the energy imparted in mixing as follows:

$$G = \sqrt{\frac{P}{uV}} \qquad (8.66)$$

where G = velocity gradient sec^{-1}
$\quad\quad\ P$ = power required in ft.-lbs/s (watts)
$\quad\quad\ u$ = dynamic viscosity in lb-s/f^2 (N-s/m^2)
$\quad\quad\ V$ = Basin volume ft^3 (m^3)

See Appendices C-3 and C-4.

FIGURE 8.129   Angle mounted propeller mixer.

The velocity gradient G, can vary between 20 and 500 for a typical wastewater applications. Note that in the exponential relationship, power is increased logarithmically with G; therefore slight changes in G dramatically affect the power requirement. As a rule of thumb, the energy gradient should range as follows:

- Chemical addition: $G = 250-500$ (depends upon chemical type and state)
- Solids suspension: $G = 100-250$ (depends on characteristics of the solids)
- Suspended growth biological growth systems: $G = 100-200$ (depends on basin configuration)
- Flocculation: $G = 30-50$

8.10.3.5   Propeller Design.   The power computed above should be used in Eq. (8.67) to compute an unknown rotational speed, propeller diameter or propeller-type when the remaining parameters are selected.

$$P = kpn^3 D^5 \qquad\qquad (8.67)$$

where P = power in ft-lbs/sec (watts)
k = blade constant (see Table 8.13)
p = mass density of fluid in slugs/ft^3 (kg/m^3)
n = rotational speed in rev/sec
D = impeller diameter or ft (m)

See Appendices C-3 and C-4.

- A small impeller of high speed gives high turbulence and is best for dispersing of gases or chemicals in relatively small amounts and for causing particle shear.
- A large slow impeller gives low turbulence and is best for mixing of large flows of chemicals or the blending of two streams.
- Mixing reaction rates proceed corresponding to constituent concentrations as follows:

$$t = \frac{V}{Q} = \frac{C_0 C_t}{dc/dt} \qquad\qquad (8.68)$$

where t = detention time in sec.
V = volume in ft^3/M^3
Q = *flow in ft^3/s* (M^3/s)
C_0 = initial concentration
C_t = concentration at time, t
dc/dt = rate of change in concentration of Ct from a first order curve determined from lab experiments.

8.10.4   Design Hints

- Detention time 0.5 to 5 minutes
- Chemical Addition: $G = 250-500$
- Solids Suspension: $G = 100-250$
- Suspended Growth Biological Systems: $G = 100-200$
- Flocculation: $G = 30-50$
- Blending: rpm $= 25-100$

TABLE 8.14 Mixer Relative Powered Speeds for suspension
Conditions (Oldshue and Herbst, 1992)

Criterion	Speed Ratio	Power Ratio
On Bottom Suspension	1.0	1.0
Off – Bottom Suspension	1.7	5.0
Complete Uniformity	2.9	25.0

- Mixing: rpm <2000 rpm
- A paddle mixer in a vertical, cylindrical, flat bottom vessel should have an outside diameter of 1/3 of the vessel diameter and be approximately 1/3 of the vessel diameter off of the bottom.
- Vortex baffles in a vertical, cylindrical, flat bottom vessel should be about 1/12 of the diameter of the vessel in length, and a minimum of four baffles should be used.
- Bottom corner fillets can be used to minimize quiescent areas.
- Table 8.14 gives the relative power of speeds of a mixer for various solids suspension conditions:

In Table 8.14, On-Bottom Suspension is when all solid particles are in motion without substantial vertical movement. Off-Bottom Suspension is when particles have vertical movement with larger particle close to the bottom and smaller particles distributed to the surface. Complete Uniformity is where there is a completely mixed condition of all solids particles throughout the basin.

8.11 EVAPORATIVE TREATMENT

8.11.1 Evaporative Treatment Theory

Evaporative treatment is a physical process that utilizes heat to evaporate wastewater. Wastewater evaporation can be accomplished using heat from one of several sources: natural gas, LP gas, fuel oil, steam, or electricity. Evaporator configurations include forced circulation, vertical tube (rising and falling film), and calandria-type systems.

8.11.1.1 Forced Circulation. The wastewater in a forced circulation evaporator is pumped through heat exchanger tubes where heat is added to the liquid. From the heat exchanger, the wastewater enters the vapor body where the evaporation occurs as shown in Fig. 9.130.

8.11.1.2 Rising Film. The wastewater in a rising film evaporator is heated in vertical heat exchanger tubes and evaporated above the tubes in the vapor body. This type of evaporator is used with wastewaters that do not have a high TDS concentration because the dissolved solids will form scale on the tube once the water is evaporated. A rising film evaporator is shown in Fig. 9.131.

8.11.1.3 Falling Film. The wastewater in a falling film evaporator is heated in vertical heat exchanger tubes and evaporated in a separate vapor body below the tubes. This type of evaporator is used with wastewaters that have relatively high dissolved solids concentration because the dissolved solids can be removed from the bottom of the vapor body after the water is evaporated. A falling film evaporator is shown in Fig. 9.132.

8.11.1.4 Calandria. The wastewater in a calandria-type evaporator is recirculated through vertical heat exchanger tubes using the differences in specific gravity of the incoming wastewater and the heated wastewater. Calandriatype evaporators can be used for wastewaters with high TDS due to the configuration of the unit. Figure 9.133 shows a typical calandria-type evaporator.

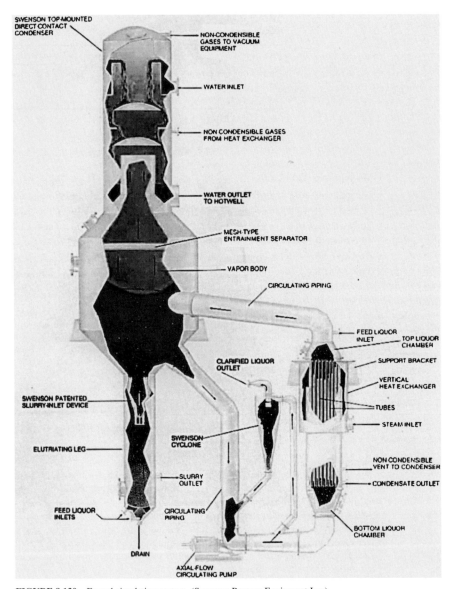

FIGURE 8.130   Forced circulation system. (Swenson Process Equipment Inc.)

8.11.1.5   Application of Evaporative Treatment.   The most common uses for evaporative wastewater treatment technologies are for small batch generation of wastewater, wastewaters that contain valuable constituents that need to be recycled, wastewaters that are extremely difficult to treat conventionally, or for applications in which zero discharge is desired.

Evaporative technologies for wastewater treatment have the following advantages:

• Greatly reduces quantity of waste requiring disposal
• Provides zero discharge of wastewater

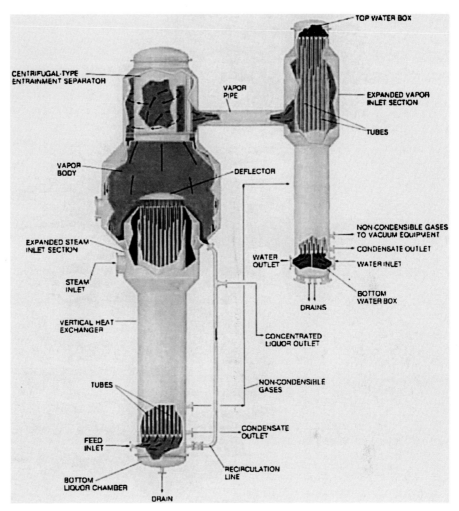

FIGURE 8.131   Rising film evaporator.(Swenson Process Equipment Inc.)

- Recovers valuable constituents in wastewater
- Reduces environmental liability
- Treats wastewaters that are difficult to treat

    Disadvantages of evaporative technologies are:

- High operating cost associated with heating water
- Potential maintenance problems with salt accumulation and disposal
- High capital cost
- Must still dispose of concentrated residual

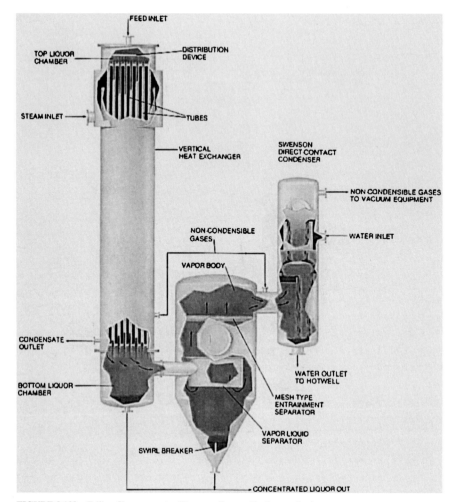

FIGURE 8.132 Falling film evaporator.(Swenson Process Equipment Inc.)

8.11.2 Treatability Study

A treatability study for an evaporative treatment process is a fairly simple procedure. The main goal of the study is to determine evaporation rate as a function of temperature as well as the quantity of sludge generated during the evaporation process.

Both of these criteria can be determined by placing a certain quantity of the raw wastewater in a glass beaker and heating the beaker with a hot plate. While the wastewater is being heated, the temperature of the wastewater and volume of wastewater remaining should be measured with time. The temperature of the wastewater should be measured at the boiling point and again during boiling.

Once the wastewater has been evaporated, the weight and volume of sludge remaining should be measured. A complete analysis of all sludge constituents should be performed enabling the engineer to determine what constituents evaporated with the wastewater and what remained in the sludge. Also, a TCLP should be performed on the sludge to determine its hazardous classification for landfill disposal.

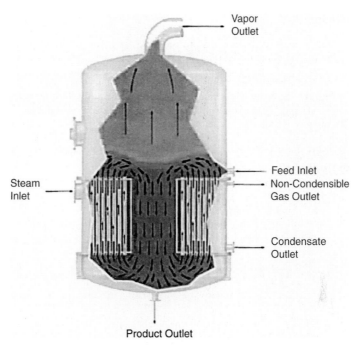

FIGURE 8.133   Calandria-type evaporator. (Swenson Process Equipment Inc.)

8.11.3   Design of Evaporative Treatment Processes

The design of evaporative treatment processes involves sizing the evaporator and related equipment for a certain evaporation throughput. Evaporators can be batch treatment or flow-through systems. For both types of systems, evaporator manufacturers typically size the evaporator system based on information that they receive from the treatability study such as the wastewater's boiling point, constituents evaporated, and the amount of sludge produced.

The number of calories required to bring the wastewater to boiling can be computed by using the definition of a calory as the heat required at one atmosphere to raise one gram of wastewater one degree centigrade. Likewise, a B.T.U. is defined as the amount of heat required to raise the temperature of one pound of water one degree Fahrenheit. The calories or BTU's required to hold wastewater at its boiling point, is equal to the heat loss.

8.11.4   Installation

Small scale evaporation systems are easily installed with the majority of the equipment assembled (by the manufacturer) prior to delivery. The only equipment that would require installation in this case would be the liquid transfer system to the evaporator. In some cases, this would require a sump with level controls, a transfer pump and in some cases (for batch treatment), a holding tank.

Larger scale systems would require more complex liquid transfer systems as well as footing installation for the system. Depending on the size, the evaporator can be delivered in one or more pieces. When the evaporator size requires delivery of several pieces of the system, sufficient time must be allowed for installation. Typically, for medium size systems, the evaporator can be installed in 1 to 2 weeks with larger systems taking longer.

Most evaporator applications in wastewater treatment are smaller in scale where the installation of the system could take a few days.

8.11.5 Operation

Operation of an evaporator may require a significant time commitment depending on the size of the unit and/or the amount of sludge created. The more inexpensive small scale evaporators do not provide a quick, easy method for sludge removal. To minimize operating cost, equipment must be purchased with the evaporator for easier sludge removal. For complete manual systems, the capital cost will be relatively low compared to those in which certain functions are automated. Before buying a small-scale evaporator, the economics of the increased capital cost over the decreased labor cost must be examined.

8.12 REFERENCES

Calgon Carbon Corporation, "Basic Concepts of Adsorption on Activated Carbon."

Cooper, C. D., and F. C. Alley, *Air Pollution Control: A Design Approach* (Prospect Heights, ILL.: Waveland Press, Inc., 1986).

Dobbs, Richard A. and Jesse M. Cohen, *Carbon Adsorption Isotherms for Toxic Organics,* EPA-600/8-80-023, 1980.

Dow Chemical Company, Dowex Monosphere Resins.

Knabel, Kent S., "For Your Next Separation Consider Adsorption," *Chemical Engineering,* November 1995, pp. 92–102.

J.Y. Oldshue and N.R. Herbst, A Guide to Fluid Mixing, Lightnin, 1992

Perry's Chemical Engineers Handbook, McGraw-Hill, Inc., 6th ed., 1984.

Speitel, Gerald E. Jr., University of Texas at Austin, notes, 1990–1.

Swindell-Dressler Company, *Process Design Manual for Carbon Adsorption,* for the EPA Technology

Transfer, Program #17020 GNR, 1971.

Yaws, Bu, and Nijhawan, "Adsorption-Capacity Data for 283 Organic Compounds," *Environmental Engineering World,* May–June 1995.

CHAPTER 9
CHEMICAL TREATMENT

9.1 INTRODUCTION

In wastewater treatment, the borders between physical, chemical, and biological treatment are sometimes blurred, because theoretically, most actual processes involve a combination of two or even all three of the general types of treatment. The purpose of the separation in this handbook is to categorize the general types of treatment and to demonstrate that the types of treatment must be compatible with the characterization of the wastewater.

In general, chemical treatment is used when a pollutant in wastewater must be altered or manipulated chemically in order to reduce the pollutant's concentration in the effluent. The most common types of chemical treatment are:

- pH control
- Chemical oxidation/reduction
- Metals precipitation
- Coagulation/flocculation
- Stripping
- Adsorption

9.2 pH CONTROL

pH Control is a frequently occurring neutralization process in industrial wastewater treatment system. Also, there is a need for pH control in municipal systems that have major industrial discharge.

9.2.1 pH Control Theory

pH is the chemical term used to quantify the concentration of the hydrogen ion in the liquid. The letter "p" symbolizes the negative logarithm of the ion concentration, and the letter "H" symbolizes the hydrogen ion, chemically designated $[H^+]$, in terms of moles per liter. The term pH therefore designates the negative logarithm of the hydrogen ion concentration in moles per liter.

Since water, H_2O, is poorly ionized at approximately one molecule per 500,000,000, a constant can be designated for the relationship of hydrogen ions and hydroxide ions versus water ions:

$$K_{eq} = \frac{[H^+][OH^-]}{[HOH]}$$

The number of unassociated water ions is extremely large compared to the number of unassociated hydrogen or hydroxide ions. Therefore, $[HOH]$ is assumed to be a constant, K_1, and:

$$K_{eq} = \frac{[H^+][OH^-]}{K_1}$$

Therefore:

$$K_{eq} \times K_1 = [H^+][OH^-] = K_w = 10^{-14}$$

Since it has been found that K_w equals 10^{-14} moles/liter:

$$K_{eq} \times K_1 = K_w = 10^{-14}$$

then:

$$[H^+][OH^-] = K_w = 10^{-14}$$

If there is the same number of hydrogen ions as hydroxide ions in a water, then:

$$[H^+] = [OH^-] = 10^{-7} \text{ moles/liter}$$

Therefore, neutralized water with the same number of hydrogen and hydroxide ions has a pH of 7 since pH is the negative logarithm of the hydrogen ion concentration in moles/liter. Conversely, the pOH, or the negative logarithm of the hydroxide ion concentration is 7 for neutral water.

An example of the practical use of the pH calculation is given as follows:

Example 8.1

Problem: Calculate the pH of a 365 mg/L solution of HCl.

Solution: By definition, a one molar solution contains one mole per liter. One mole, by definition, is the molecular weight in grams. The Periodic Table (Table 3.3) shows molecular weights. The molecular weight of HCl is:

$$H = 1 \text{ g/mol}$$
$$+ Cl = 35.45 \text{ g/mol}$$
$$\overline{HCl = 36.5 \text{ g/mol}}$$

Therefore, 36.5 g/L is a one molar (M) solution of HCl and 365 mg/L or 0.365g/L is a 0.01 M solution or 1.0×10^{-2} moles/L. Strong acids, such as HCl, are considered to dissociate or ionize completely in a dilute solution. Therefore:

$$[H^+] = 1.0 \times 10^{-2} \text{ moles/L}$$

$$pH = \frac{1}{\log[0.01]} = -\log[0.01]$$

$$pH = \log 10^{-2}$$

$$pH = 2$$

Or, the pH of a 365 mg/L solution of HCl is 2.

Note that the relationship between adjacent pH units is logarithmic. Therefore, if one liter of HCl with a pH of 1.0 is added to one liter of pure water with a pH of 7, the following occurs:

$$\text{Acid: pH} = 1.0$$

$$pH = 1.0 = \log[10] = \log\left[\frac{1}{0.1}\right]$$

Therefore:

$$[H^+] = 0.1 \text{ or } 10^{-1} \text{ moles/L}$$

$$\text{Water: pH} = 7.0$$

$$pH = 7.0 = \log[10,000,000] = \log\left[\frac{1}{0.0000001}\right]$$

Therefore:

$$0.0000001 + 0.1 = 0.1000001 \text{ moles/2 liters}$$

$$[H^+] = 0.05000005 \text{ moles/L}$$

$$pH = \log\left[\frac{1}{0.05000005}\right] = 1.3$$

If 10 liters of water are added, the hydrogen ion concentration is:

$$10 \times 0.0000001 + 0.1 = 0.100001 \text{ moles/11 liters}$$

$$[H^+] = 0.0090909 \text{ moles/L}$$

$$pH = \log\left[\frac{1}{0.009090901}\right] = 2.04$$

Therefore, 10 liters of water at neutral pH will raise the pH of the mixture from 1.3 to approximately 2.04. Likewise, if 100,000 gallons of water are added:

$$1000,000 \times 0.0000001 + 0.1 = 0.11 \text{ moles/100,001 L}$$

$$[H^+] = 0.0000011 \text{ moles/L}$$

$$pH = \log\left[\frac{1}{0.0000011}\right] = 5.96$$

Therefore, it will take 1,000,000 gallons of pure water to raise the pH of one mixed gallon of HCl with a pH of 1.0 to a pH of approximately 7.0.

This example demonstrates graphically why a gallon of concentrated acid cannot be dumped into a running sewer line and be expected to neutralize before the sewer reaches the stream, a city sewer, or a pH meter. It would need to be mixed with 1,000,000 gallons of water to do so. Even if that much water were available, it is unlikely that there would be complete mixing of the entire amount with the one gallon of acid.

With complete mixing, acid or base molecules are dissociated completely, and the pH change is instantaneous, as in the case of a dilute strong acid or base. Therefore, the mixing theory in Chapter 8, "Physical Treatment," applies to pH control. High energy and not a long detention time is critical for mixing and consequently, pH control. Indeed, without adequate mixing, pH control with large basins will be less effective than with smaller basins. The photographs in Figs. 9.1 and 9.2 show a large and small pH basin for a flow of 300 gpm. The small basin replaced the larger one and was much more effective in pH control because of short-circuiting due to inadequate mixing in the large basin. Occasionally, a second or even a third small pH basin in series is needed to control radical pH swings, but will still be less expensive than a larger basin.

FIGURE 9.1   Large inefficient pH basin.

9.2.2 Design

The parameters that must be considered in the design of a pH control basin are as follows:

- The correct number of acid and base ions must be added to the wastewater to neutralize the exist-ing acid or base ions to the level required by the permit or for subsequent wastewater treatment steps.
- The pH basin should have an adequate size to assure complete mixing without excess lag.
- The inlet flow should be baffled so that it is distributed evenly vertically and horizontally to avoid short circuiting as explained in Chapter 8.
- The outlet flow velocity should be minimized with baffles or weirs as explained in Chapter 8.
- Enough energy must be added for mixing to completely mix the basin in accordance with Chapter 8.
- The basin should be shaped so that mixing efficiency is optimized.
- In a feed back system, as explained below, the lag must be less than the time of change of one pH unit.
- In a feed forward system, as explained below, a final alarm or subsequent pH adjustment must be made to assure success.
- The rate of acid or base addition must be quicker in the form of hydrogen or hydroxide ions, than the rate of change of pH.
- Enough acid or base must be added over this time period to neutralize the wastewater to levels allowed by permit or subsequent wastewater treatment steps.

Each of these design parameters will be discussed below.

9.2.2.1 Size.   It is recommended that the pH basin have a minimum detention time of 30 seconds and a maximum detention time of five minutes at peak of flow.

FIGURE 9.2   Small efficient pH basin.

9.2.2.2   Inlet Flow.   It is recommended that the flow be distributed evenly using baffles or openings in a wall as explained in paragraphs 8.5.5.9 and Figures 8.71 and 8.72. The baffle or wall should prevent the short circuiting of the flow from the entrance pipe by positioning a portion of the baffle or wall without an opening directly opposite the inlet pipe. The distribution should also be controlled as explained in Chapter 8 so that inlet flows are balanced across the entire width and depth of the basin. See the note below concerning small basins.

9.2.2.3   Outlet Flow.   It is recommended that the outlet flow velocity be minimized in accordance with paragraphs 8.5.2.10 and 8.5.5.9 and Figure 8.74 through 8.83.

In small pH basins or where the width is less than 3 times the pipe diameter, adequate mixing can be realized by entering the basin at the bottom and exiting at the top so that the flow passes directly through a mixer.

9.2.2.4   Mixing.   It is recommended that mixing be in accordance with Section 8.10 including the use of formulas (8.66) and (8.67). It is cautioned that the propeller rotation and direction be coordinated with the basin configuration in order to optimize the turbulence provided by the energy.

9.2.2.5   Basin Shape.   The optimum basin size is one which minimizes the distance between the source of energy and all points in the basin. Therefore the optimal shape is a sphere. Since this shape is impractical, the best compromise is typically a cube with the energy source in the center.

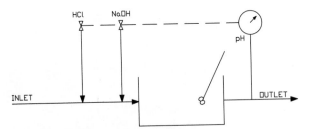

FIGURE 9.3   Feed Back pH Control System

9.2.2.6   *Feed Back pH Control Systems.* During the detention time between pH sensing and acid or base addition, if the inlet pH changes, the chemical added may not be appropriate for the condition. For instance, if the detention time of the basin is one hour, the outlet water which is tested by the pH meter entered the basin more than one hour before it was tested. The water entering the basin at the time of the test may not be of the same pH. For example, if a pH of 11 is sensed and acid is called for, then the inlet waste water drops to 4.0, acid will be added to an acidic condition. Even with a completely mixed basin, only a small portion of the inlet water reaches the effluent when the pH is sensed. Theoretically, the pH does not completely reach the influent level until the entire detention time is passed.

The best application of a feed back pH control system is when pH changes to a higher or lower pH unit in more time than the detention time of the basin. If this does not occur, the solutions need concentration equalization prior to the basin or a different kind of pH control. Figure 9.3 shows a Feed Back pH Control System.

9.2.2.7   *Feed Forward pH Control Systems.* One solution to the lag problem of a feed back pH control system is a feed forward system. In this system, the pH is monitored before the inlet of the basin, thus sensing the pH change as it occurs. The feed of the acid or base is downstream of the pH meter and either immediately before the basin or at its inlet point. It is cautioned that there must be positive liquid movement from the meter to the feed points. The author has observed many poorly functioning systems with the pH meter at the acid and base feed the basin. In this case, the mixing pattern must be relied upon to prevent the pH meter from sensing the acid or base being fed rather than the inlet pH.

The major weakness of a Feed Forward system is the lack of a final pH adjustment. The amount of acid or base fed must be assumed before feeding. The control system is very seldom adequate to match the pH need with acid or base addition because of the logarithmic relationship of the pH need and the very different relationship between pumps, valves, and feed rates. Figure 9.4 shows a Feed Forward pH Control System.

9.2.2.8   *Multiple Basins.* There are two workable solutions to the previously described problems.

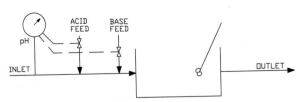

FIGURE 9.4   Feed Forward pH Control System

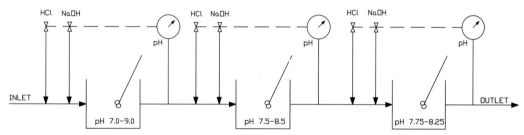

FIGURE 9.5 Multiple Basins pH Control System

In either case, two basins can be used with less effectiveness, but less cost than three basins. The author has designed an effectively operating three tank pH control system from an electro-plating system serving more than one hundred plating tanks and one in a yeast plant with radical pH swings.

As the figures demonstrate, each subsequent basin can have more closely controlled pH ranges. One existing pH basin can be split into three separate basins, but according to the mixing theory expressed in section 8.9, each basin should approach a cube in shape. Figure 9.5 shows a multiple pH basin System.

9.2.2.9 Constant Acid and Base Feed Systems. If a constant feed pump or on/off valve is provided for pH control, a system will add acid or base at preset levels, as shown in Fig. 9.3.

As can be seen from Fig. 9.3, in a rising pH condition, the pH will continue to rise even after acid is added at point A until sufficient hydrogen ions are present to begin a lowering of the pH. The maximum pH at point B must be lower than that required by the system. When pH is lowered to point C, the acid ceases being fed. If acid is overfed or if inlet pH has dropped and the pH continues to fall, base will be added at point D. Again, an additional pH drop will occur to point E. Therefore, point D, as well as point A, must anticipate the subsequent pH change before correction. As the pH is corrected by the base and begins rising, the base feed is turned off at point F. Note that points C and F must be far enough apart to prevent the feed of acid to counteract the base feed, or vice versa. It is suggested that the pH difference between points C and F be at least twice the larger of the pH difference between points A and B or points D and E.

A further complication in the use of a constant acid or base feed system is that the rate of addition of hydrogen or hydroxide ions must be greater at the higher or lower pH levels, than when the pH approaches closer to neutral. Consequently, there is a tendency to overfeed acid or base, requiring compensation.

9.2.2.10 Variable Acid and Base Feed Systems. If a control valve or pump rate can be set proportional to the pH requirement, the efficiency of the system can be increased. The difficulty of this operation is the logarithmic pH relationship versus the valve or pump flow characteristics. The only method that accurately accomplishes this relationship is the use of a rate of flow controller with the rate set logarithmically by pH required.

A second method of variably adjusting pH is to feed two different concentrations of acid and/or base. In this case, at the beginning of a feed cycle, at a low or high pH, the concentrated base or acid will be fed to allow the quicker transfer of more hydroxide or hydrogen ions. When the pH reaches a higher (or lower) point, the less concentrated base or acid will be fed. Figure 9.6 shows a Multiple Concentration pH Control System.

A two basin feed back (or feed forward) system can be designed in accordance with the pH Titration curve shown in Figure 9.7 by feeding a concentrated base in the first basin to raise the pH from 2.0 to 3.0, followed by a dilute base to raise the pH from 3.0 to 12.0. This dilute base addition prevents the nearly vertical pH rise caused by feeding only a concentrated base. The logarithmic nature of pH causes the steep rise which in many cases will cause overshooting of the desired pH.

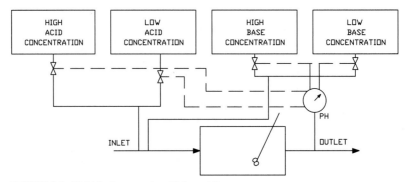

FIGURE 9.6 Multiple Concentration pH Control System.

9.2.2.11 Design Suggestions. The design of an efficient pH control system should adhere to the following suggestions:

- In a feed back system, place the pH meter so that it will read the pH of the wastewater after the acid or base is completely mixed with the wastewater.
- In a feed forward system, place the pH meter so that it will read the pH of the wastewater well before the acid and base are added and at a point where the acid or base cannot possibly reach the meter.
- Design the mixer in accordance with section 8.9 and place it in the center of a chamber which is as close to cubical or cylindrical as possible, with baffle or mixer placement located to prevent vortexing.
- Allow between 0.5 and 5 minutes detention time in the basin.

9.3 CHEMICAL OXIDATION/REDUCTION

9.3.1 Theory

Oxidation-reduction reactions, or "redox" reactions, are fundamental parts of wastewater treatment. All forms of chemical wastewater treatment as well as biological wastewater treatment incorporate redox reactions. Compounds containing carbon, sulfur, nitrogen, oxygen, and metals with more than one oxidation state, such as chromium, iron, and manganese encounter processes which change the oxidation state, and thus substantially alter their chemical properties. This change in oxidation state is the basis for chemical removal of organics, nitrogen, sulfur, and metals from wastewater.

The processes of biological treatment also depend upon redox reactions catalyzed by microorganisms. Disinfection and oxidation with such chemicals as chlorine, chlorine dioxide, permanganate, hydrogen peroxide, oxygen, ozone, and sulfur dioxide are also example of redox reactions. Oxidation-reduction reactions are also the basis for BOD and COD tests used to evaluate wastewater. Biological treatment is covered in Chapter 9 of this Handbook and disinfection is covered later in this chapter. The basic fundamentals of redox reactions and some specific applications will be covered here.

Chemical equations are necessary for the understanding of redox reactions and how they can be used to treat a specific waste stream. The reactants and products from the system, which is in equilibrium, must be presented in a balanced equation. Redox reactions consist of two parts. The first part is the *oxidation* reaction. The oxidation of a substance occurs when it loses or donates electrons. An alternate definition for oxidation is the combination of a compound with oxygen or an increase in oxidation state. The second part is the *reduction* reaction. The reduction of a substance occurs when a substance gains or accepts electrons. An alternative definition for reduction is the decrease in oxidation state.

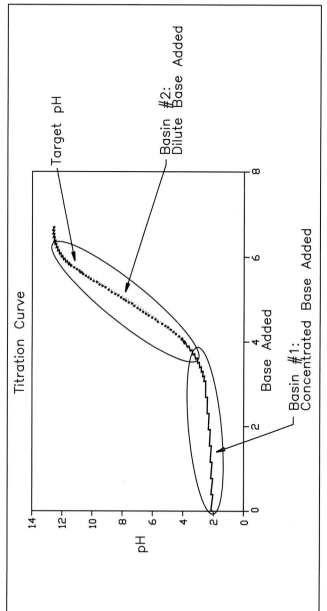

FIGURE 9.7 Dilute concentrated base feed.

9.3.1.1 Stoichiometry. Oxidation-reduction reactions must be represented by balanced chemical equations. The reaction may be considered in terms of two "half" reactions. One half reaction is an oxidation process where the ion looses electrons from a reactant to form an oxidized product. The other half reaction is a reduction process where the reactant gains electrons to form a reduced product. To balance the equations, the number of electrons exchanged must be the same.

Consider the reduction of Hexavalent chromium to Trivalent chromium by ferrous iron in acid solution. The reduction reaction is:

$$6e^- + Cr_2O_7^{2-} + 14H^+ \rightarrow 2Cr^{3+} + 7H_2O \tag{9.1}$$

1. The oxidation state of chromium in the dichromate is reduced from $+6$ to an oxidation state of $+3$.
2. The process requires six electrons.
3. The hydrogen and oxygen retain the same oxidation state on both sides of the half reaction.
4. The equation is balanced with respect to charge (net charge is $+6$ on both sides) and mass (net 7 O's, 14 H's, and 2 Cr's on both sides).

The practical purpose of this reaction in wastewater chemistry is that trivalent chromium is insoluble at certain pH levels and will settle, unlike hexevalent chromium, which is soluble at all pH levels.

Chromium compounds in natural environments tend to become reduced to the trivalent state, since environments are naturally reducing. There is no "natural" oxidizer except for oxygen itself, which is a very weak oxidizer, but most sulfur compounds are reducers.

The reduction of divalent iron to trivalent iron is a reduction reaction:

$$Fe^{2+} \rightarrow Fe^{3+} + 1e^- \tag{9.2}$$

The practical purpose of this reaction in wastewater chemistry is that divalent ferrous iron is soluble at all pH levels. This is the natural or reduced state of iron which is found in groundwater. Trivalent iron is insoluble at pH levels above about 4.0.

In Equation 9.2, the oxidation state of iron is increased from $+2$ to $+3$ with the loss of one electron. The half cell is balanced with respect to charge and mass.

The net reaction, a combination of the two half reaction requires the loss and gain of six electrons is as follows:

$$\text{oxidation:} \quad 6Fe^{2+} \rightarrow 6Fe^{3+} + 6e^- \tag{9.3}$$

$$\text{reduction:} \quad 6e^- + Cr_2O_7^{2-} + 14H^+ \rightarrow 2Cr^{3+} + 7H_2O \tag{9.4}$$

$$\text{net reaction:} \quad 6Fe^{2+} + Cr_2O_7^{2-} + 14H^+ \rightarrow 6Fe^{3+} + 2Cr^{3+} + 7H_2O \tag{9.5}$$

The balanced equation permits an evaluation of the following relationships:

The equivalent weight: The equivalent weight of $K_2Cr_2O_7$ is the molecular weight divided by 6, the number of electrons in the reaction:

$$Eq \ Wt = \frac{294.189}{6} = 49.0315$$

Concentrations: The weight of $FeSO_4$ required to reduce 1 kg of $K_2Cr_2O_7$ is

$$Wt = 1 \text{ kg } Cr_2O_7 \left(\frac{1000 \text{ equivalents}}{49.0 \text{ kg } K_2Cr_2O_7} \right) \left(\frac{151 \text{ grams } FeSO_4}{1 \text{ equivlant}} \right) = 3000 \text{ grams}$$

One useful concept in oxidation-reduction reactions is the redox potential. Associated with each half reaction is a standard potential, the voltage of the half reaction when each reactant or product is at "unit-activity" and 25°C. For example, the potential for the divalent iron oxidation to trivalent iron is $E_0 = -0.77$ volts. The potential for the reduction of trivalent iron to divalent iron is $+0.77$ volts.

The standard electrode potential for the cell is the sum of the oxidation and reduction potentials: For the reaction represented by Eq. (9.5), the standard electrode potential for the cell is

$$E_0 \text{ Cell} = E_0 \frac{Cr^{6+}}{Cr^{3+}} + E_0 \frac{Fe^{2+}}{Fe^{3+}} = +1.33 \text{ volts} + (-0.77) \text{ volts} = 0.56 \text{ volts} \qquad (9.6)$$

If E_0 is positive and all activities are unity, the reaction will proceed spontaneously as written. The standard electrode potential values as E_0 for the reduction reaction half-reaction for several elements relative to water chemistry are tabulated in Table 9.1.

The discussion so far considers standard half-reactions at standard state. The conditions assume unit activity for all reactants and products, that is 25°C, 1 atm pressure, and concentration 1 mole/l. To determine the effect of reactant and product concentrations, the Nernst Equation is used. The simplified form of this equation is:

$$E = E_0 - \frac{0.59}{n} \log Q \qquad (9.7)$$

TABLE 9.1 Electrode Potentials of Common Reduction Half Reactions

Reduction half reaction	E_0 (Volts)	Reduction half reaction	E_0 (Volts)
$Li^+ + e^- \rightarrow Li$	−3.05	$Sn^{4+} + 2 e^- \rightarrow Sn^{2+}$	0.15
$K^+ + e^- \rightarrow K$	−2.93	$Cu^{2+} + e^- \rightarrow Cu^+$	0.15
$Ba^{2+} + 2 e^- \rightarrow Ba$	−2.90	$Cu^{2+} + 2 e^- \rightarrow Cu$	0.34
$Ca^{2+} + 2 e^- \rightarrow Ca$	−2.87	$Cu^+ + e^- \rightarrow Cu$	0.52
$Na^{2+} + e^- \rightarrow Na$	−2.71	$I_2 + 2 e^- \rightarrow 2I^-$	0.53
$Mg^{2+} + 2 e^- \rightarrow Mg$	−2.37	$Mn^{7+} + 3 e^- \rightarrow Mn^{4+}$	0.58
$Al^{3+} + 3 e^- \rightarrow Al$	−1.66	$Fe^{3+} + e^- \rightarrow Fe^{2+}$	0.77
$Mn^{2+} + 2 e^- \rightarrow Mn$	−1.18	$Hg_2^{2+} + 2 e^- \rightarrow 2 Hg$	0.79
$Zn^{2+} + 2 e^- \rightarrow Zn$	−0.76	$Ag^+ + e^- \rightarrow Ag$	0.80
$Cr^{3+} + 3 e^- \rightarrow Cr$	−0.74	$Br_2 + 2 e^- \rightarrow 2Br^-$	1.07
$Fe^{2+} 2 + e^- \rightarrow Fe$	−0.44	$Mn^{4+} + 2 e^- \rightarrow Mn^{2+}$	1.21
$Cr^{3+} + e^- \rightarrow Cr^{2+}$	−0.41	$O_2 + 4H^+ + 4 e^- \rightarrow 2H_2O$	1.23
$Cd^{2+} + 2 e^- \rightarrow Cd$	−0.40	$Cl_2 + 2 e^- \rightarrow 2Cl^-$	1.36
$Co^{2+} + 2 e^- \rightarrow Co$	−0.28	$Au^{3+} + 3 e^- \rightarrow Au$	1.50
$Ni^{2+} + 2 e^- \rightarrow Ni$	−0.25	$H_2O_2 + 2H^+ + 2 e^- \rightarrow 2H_2O$	1.23
$Sn^{2+} + 2 e^- \rightarrow Sn$	−0.14	$Co^{3+} + e^- \rightarrow Co^{2+}$	1.82
$Pb^{2+} + 2 e^- \rightarrow Pb$	−0.13	$F_2 + 2 e^- \rightarrow 2F^-$	2.87
$H^+ + 2 e^- \rightarrow H_2$	0.00		

Source: http: // fandango.austin.cc.tx.us/rvsmthsc/chem/chem-Standard.html.

Table 9.2   Chemical Reduction/Oxidation (REDOX)

Compound	Common Form	Redox Process	Treatable Form
Fe	Fe^{2+}	Oxidation	$Fe^{3+} - Fe(OH)_3$
Mn	Mn^{2+}	Oxidation	$Mn(IV) - MnO_2$
Cr	$Cr(VI)$	Reduction	$Cr^{3+} - Cr(OH)_3$
Cyanide	CN^-	Oxidation (2 Steps)	$CNO^- - CO_2, N_2$
Chlorinated VOCs	TCE	Reduction	Ethylene
Toxic/Refractory Organics	Phenolics	Oxidation	CO_2, other organics

where: $K = 0.059 = 2.3 \times RT/F$
R = gas law constant
T = absolute temperature
F = faraday constant
E_0 = standard electrode potential
E = electrode potential
Q = equilibrium quotient

The following Table 9.2 summarizes several chemical species which exist in more than one oxidation state and are more easily treated after oxidation or reduction.

9.3.1.2   Hexavalent Chromium Reduction.

As explained above, hexavalent chromium can be reduced to the less soluble trivalent chromium which will form chromium sulfate which is insoluble as chromium hydroxide at a pH of around 8.5.

The recommended procedure is to lower the pH to 2.5 with sulfuric acid and reduce at an ORP level of −300 mv with sodium metabisulfite or another reducing agent. See Formula 9.8 for this reaction.

$$4H_2CrO_4 + 3Na_2S_2O_5 + 3H_2SO_4 \rightarrow$$
$$3Na_2 + SO_4 + 2Cr_2(SO_4)_3 + 7H_2O \tag{9.8}$$

This reaction should take around 3 hours. Following this reduction, the hydroxide precipitation will occur if the pH is raised to approximately 8.0 to 8.5. (See Figure 9.8).

Other reducing agents are shown in Table 9.3.

9.3.1.3   Cyanide Oxidation.

Cyanide (CN) is a molecule consisting of carbon and nitrogen ions and is very toxic at pH levels below 7.0 as hydrogen cyanide (HCN) gas. Even at a pH of 9.0, the HCN is approximately 58% of the total cyanide.

CN will form metal cyanide complexes with zinc, cadmium, and copper which can dissociate and release free CN. Zinc, copper, and cadmium cyanide are more toxic than an equal concentration of sodium cyanide. Temperature increase also results in increased toxicity of two to three times, per 10°C, as does reduced dissolved oxygen content. The most practical way of removing CN is to oxidize the CN is a two stage reaction as follows:

First Stage

- Raise pH to 10 to 11.5 with NaOH
- Add oxidizing agent such as Cl_2 gas or sodium hypochlorite solution to an ORP of +350 to +400mv
- Mix for 5 to 15 minutes
- The following reaction occurs:

$$NaCN + 2NaOH + Cl_2 \rightarrow NaCNO + 2NaCl + H_2O \tag{9.9}$$

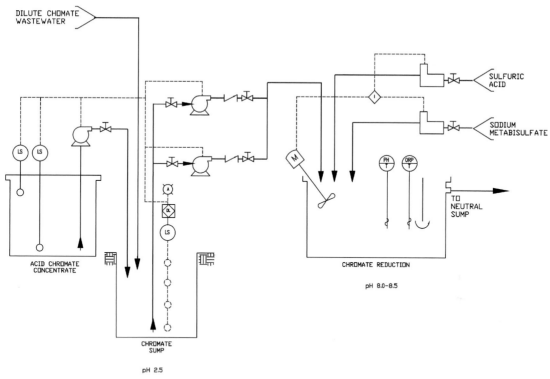

FIGURE 9.8   Hexavalent chromium reduction.

Second Stage
- Lower pH to 7.5 to 8.0 with acid
- Add oxidizing agent to an ORP of +600 mv
- Mix for approximately 60 minutes
- The following reaction occurs:

$$2NaCNO + 4NaOH + 3Cl_2 \rightarrow 6NaCl + 2CO_2 + N_2 + 2H_2O \qquad (9.10)$$

Figure 9.9 shows a recommended flow diagram for oxidation of CN to CO_2 and N_2 gases.

Table 9.3   Reducing Agents

Elemental Iron:	Fe°
Ferrous sulfate	$FeSO_4$
Sulfur dioxide	SO_2
Sodium sulfite	Na_2SO_3
Sodium bisulfite	$NaHSO_3$
Sodium thiosulfate	$Na_2S_2O_3$
Sodium hydrosulfite	$Na_2S_2O_4$
Sodium metabisulfite	$Na_2S_2O_5$

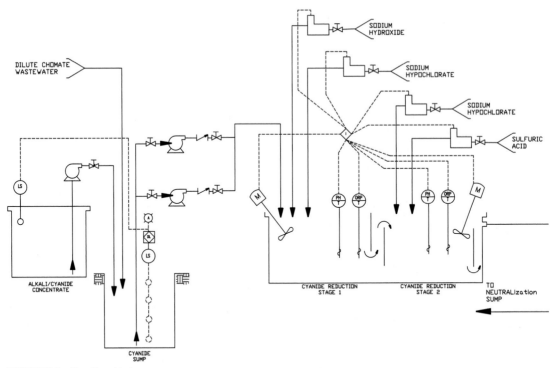

FIGURE 9.9 Cyanide oxidation.

9.3.1.4 Other Oxidation Methods. Other oxidizing agents are shown in Table 9.4.

One caution in the use of chlorine as an oxidizing agent for CN is that under certain conditions it can form highly toxic cyanogen chloride.

Table 9.5 shows the results of the chemical oxidation of various compounds by chlorine, potassium permanganate and ozone in terms of half-lives.

9.3.2 Design

Many of the considerations for pH control apply in redox systems. In redox systems, an electrode called an oxidation/reduction potential (ORP) sensor is used to monitor the electrode potential of the solution. The ORP electrode is useful to detect the concentration of the reactants and the products in

TABLE 9.4   Oxidizing Agents

Chlorine	Cl_2
Ozone	O_3
Sodium hypochlorite	$NaOCl$
Calcium hypochlorite	$Ca(OCl)_2$
Hydrogen peroxide	H_2O_2
Potassium Permanganate	$KMnO_4$

TABLE 9.5   Oxidation Half-Life of Various Organic Compounds

Compound	Chlorine	Permanganate	Ozone
Acetophenone	26 d	43 d	25 min
Benzaldehyde	>3.2 d	36 min	28 min
Benzothiazole	8.2 min	>5.8 d	22 min
1, 2-Bis (2-chloroethoxy) ethane	>20 d	67 d	50 min
Bis (2-chloroethoxy) ethane	>20 d	15 d	21 min
Borneol	1.4 d	7 d	53 min
Camphor	>3.2 d	>5.8 d	>12 min
p-Dichlorobenzene	>4.2 d	>22 d	N/A
p-Nitrophenol	2.1 h	1.1 d	2 min
Methyl-m-toluate	>20 d	22 d	5.5 min
p-Tolunitrile	>20 d	28 d	6.4 min
Diacetone-L-sorbose	100 d	>14 d	2.8 min
Diacetone-L-xylose	>15 d	>14 d	2.3 min
Toluene	N/A	N/A	2.8 min
Ethylbenzene	N/A	N/A	2.8 min
1, 2, 3-Trimethylbenzene	N/A	N/A	1.9 min

Source: "Wastewater TryOzone", Sheng H. Cin and Kuo C. Yeh, Chemical Engineering, May 1993.

the solution. For example, an ORP electrode in a solution containing potassium dichromate may be controlled by the Cr_2O_7 half reaction and be quite high. The electrode potential of a solution containing little Cr_2O_7 and divalent iron will be controlled by the Fe^{2+} half reaction and be lower.

The diagram shown in Fig. 9.10 illustrates how the ORP would vary during a neutralization reaction.

9.3.3   Design Hints

Chemical oxidation/reduction systems should be designed using the following suggestions:

- Select reactants that do not cause undesirable side reactions or may be difficult to treat
- Hydrogen peroxide may be used as an oxidizing agent or a reducing agent, the reaction products being water or oxygen
- Sulfur dioxide, sodium thiosulfate, or similar oxidized forms of sulfur are frequently used to reduce oxidized forms of chromium and chlorine
- Chlorine or sodium hypochlorite, as well as potassium permanganate solutions are used to oxidize organic compounds. Sodium hypochlorite is used to oxidize cyanide.

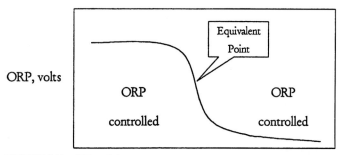

FIGURE 9.10   ORP variation.

- Redox reactions must chemically reach completion so that the detention time is variable and should be determined by treatability studies. The minimum detention time should be five minutes at peak flow.

- Mixing for redox reactions should be provided in accordance with Chapter 8 recommendations.

9.4 METALS PRECIPITATION

9.4.1 Theory

Wastewater containing dissolved metals must be treated to reduce the concentration of metals below the toxicity threshold for organisms potentially exposed to the wastewater. Practically, this form of wastewater treatment usually involves reducing the metal concentration from an initially high level (C_0) to a target concentration (C_f), established by a regulatory requirement. One practical approach is the conversion of soluble metal ions to insoluble metal salts, with subsequent removal of solids by gravity settling, filtration, centrifugation, or similar solids-liquids separation techniques. This section describes commonly found wastewater conditions and proven techniques based on precipitation to acceptably reduce the metal content.

Metals found in wastewater solutions may be cations or anions. Cations are positively charged forms of elements which may be precipitated as an insoluble salt at optimum conditions, usually an optimized pH level. The most common precipitating salts for metal cations are hydroxides, sulfides, and carbonates. Copper (Cu^{2+}), iron (Fe^{3+}), lead (Pb^{2+}), and cadmium (Cd^{2+}) are examples of metal cations that can be removed by hydroxide, sulfide, or carbonate precipitation. Anions are negatively charged forms of metals, which may be precipitated by selected cations or be converted to cations by reduction reactions prior to precipitation. Chromate ($Cr_2O_7^{2-}$), and permanganate (MnO_4^-) are examples of metallic anions.

A general picture of the nature of precipitation can be shown by solubility curves as shown in Fig. 9.11. In the event the concentration of an insoluble salt is low and the solubility product of the system is less than k_{sp}, the metal ions and the precipitating anion remain in solution. When the concentrations of cation and anion are sufficiently high, the solubility of the salt in the liquid is exceeded and precipitation occurs. Initially, the supersaturated solution is formed and there is a period of time in which no reduction of the dissolved salt occurs. This so-called induction period represents the time for nuclei to form. The nucleation event may be hastened by the mechanical action of vibrating the solution or the introduction of a "seed crystal". During the nucleation period, the appearance of an amorphous phase is evident with a reduction of the dissolved constituent in the solution. The precipitate in the amorphous phase does not have a well organized crystal structure. The particles are often small and more soluble than larger crystals that appear later on. The amorphous crystals are not stable and slowly transform into larger crystals representative of the more stable solid phase.

The precise details of crystallization, crystal growth, and establishment of equilibria for a given precipitation system may be complicated by many factors including temperature, presence of other dissolved or suspended matter, and even the container and the degree of agitation.

The solubility curves illustrated in Fig. 9.11 have been found useful to predict the behavior of metals towards optimum removal under a range of experimental conditions.

The solubility of the metals under conditions to precipitate metal hydroxides, sulfide, and carbonate is greatly influenced by the pH of the solution. At low pH, the concentration of metal ions is highest. The reason for this can be explained by the example below.

The precipitation reaction described by the equations

$$M^{2+} + 2OH^- \rightarrow M(OH)_2 \quad \text{(solid)} \tag{9.8}$$

$$M^{2+} + S^{2-} \rightarrow MS \quad \text{(solid)} \tag{9.9}$$

An equilibrium exists between the metal cation, hydroxide, sulfide or carbonate ions, and the insoluble metal hydroxide or sulfide.

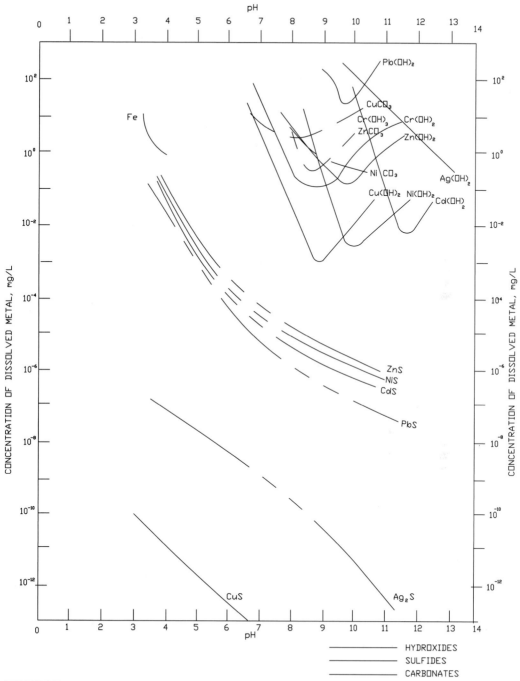

FIGURE 9.11 Solubility Curve.

The equilibrium reaction for divalent cations is described by the equilibrium constant in the equation:

$$[M^{2+}][OH^-]^2 = K_{sp} \tag{9.10}$$

where: K_{sp} = solubility product constant
   $[M^{2+}]$ = concentration of soluble cation
   $[OH^-]$ = concentration of soluble anion

Table 9.6 shows solubility products for various compounds.

In solutions containing sodium hydroxide, a common material to precipitate metal hydroxides, the hydroxide ion concentration is adjusted by incrementally adding excess NaOH with mixing and careful recording of the pH of the solution. Initially, the hydroxide is required to neutralize acid associated with the solution. At a point where the product of the metal ion concentration and the hydroxide ion concentration is equal to the value of K_{sp}, the metal hydroxide precipitates. The addition of additional sodium hydroxide results in the precipitation of additional metal hydroxides. At any condition, the equilibrium is described by Eq. (9.10). The product of $[M]$ and $[OH^-]$ is equal to the value of K_{sp}.

Using zinc as an example, K_{sp} for Zn(OH)$_2$, the pK = 17.2 and $K_{sp} = 6.3 \times 10^{-18}$. Table 9.7 shows how zinc can be precipitated from a plating bath waste stream.

Table 9.8 identifies representative ionic metal species.

9.4.1.1 Metal Hydroxides. Metal ions react with lime (Ca(OH)$_2$), caustic (NaOH), or other sources of hydroxide ions to form insoluble metal hydroxides. A typical reaction for an acid zinc chloride electroplating bath or rinse wastewater is:

$$Zn_2 + 2Cl^- + 2Na^+ + 2OH^- \leftrightarrow Zn(OH)_2 \text{ (solid)} + 2Na^+ + 2Cl^- \tag{9.11}$$

One measure of the solubility of a slightly soluble metal hydroxide is the solubility product constant, an equilibrium constant defined by the equation:

TABLE 9.6   Solubility Product Constants at 25°C

Solid	pK$_{sp}$	Solid	pK$_{sp}$
Fe(OH)$_3$ (amorph)	38	BaSO$_4$	10
FePO$_4$	17.9	Cu(OH)$_2$	19.3
Fe$_3$(PO$_4$)$_2$	33	PbCl$_2$	4.8
Fe(OH)$_2$	14.5	Pb(OH)$_2$	14.3
FeS	17.3	PbSO$_4$	7.8
Fe$_2$S$_3$	88	PbS	27
Al(OH)$_3$ (amorph)	33	MgNH$_4$PO$_4$	12.6
AlPO$_4$	21	MgCO$_3$	5
CaCO$_3$ (calcite)	8.34	Mg(OH)$_2$	10.7
CaCO$_3$ (aragonite)	8.22	Mn(OH)$_2$	12.8
CaMg(CO$_3$)$_2$ (dolomite)	16.7	AgCl	10
CaF$_2$	10.3	Ag$_2$CrO$_4$	11.6
Ca(OH)$_2$	5.3	Ag$_2$SO$_4$	4.8
Ca$_3$(PO$_4$)$_2$	26	Zn(OH)$_2$	17.2
CaSO$_4$	4.59	ZnS	21.5
SiO$_2$ (amorph)	2.7		

Source: Snoeyink and Jenkins 1980.

TABLE 9.7   Zn Removal by Precipitation

Condition	Concentration of Zn Zn(OH)$_2$	pH of the solution
Initial conditions: 50 mg/l zinc plating bath in 0.5 M HCl, total volume 100 liters	[Zn] = 50 mg/l × 1 mmole/65 mg = 0.77 mmolesO/l	pH = −log [H] = −log 0.50 pH = 0.30
Add sodium hydroxide to adjust pH to 7.0. Assume no volume change.	[Zn] × [OH]^2 = 0.77 × 10^{-3} × (1.0 × 10^{-7})^2 = 7.7 × 10^{-18} Because product > k$_{sp}$, precipitation of Zn(OH)$_2$ is possible. [Zn] = K$_{sp}$/[OH]^2 = 6.3 × 10^{-18}/ (10^{-7})^2 = 6.3 × 10^{-4} moles/ 1 Zn = 6.3 × 10^{-4} × 65 mg/10^{-3} moles = 41 mg/l Therefore about 10 mg/l Zn removed.	pH = 7.0 [OH] = 1.0 × 10^{-7}
Add additional NaOH to adjust pH to 9.5	[Zn] = K$_{sp}$/[OH]^2 = 6.3 × 10^{-18}/ (10^{-5})^2 = 6.3 × 10^{-8} moles/ 1 Zn = 6.3 × 10^{-8} × 65/10^{-3} moles = 4.1 × 10^{-3} mg/l About 50 mg/l Zn removed and 0.004 mg/l left in solution	pH = 9.0 [OH] = 1.0 × 10^{-5}

If the chemical reaction follows the form,

$$n(\text{Cation}) + m(\text{Anion}) \rightarrow \text{precipitate} \tag{9.12}$$

then

$$K_{sp} = [\text{cation}]^n\,[\text{anion}]^m \tag{9.13}$$

Several values for K_{sp} can be found in Table 9.6 in the previous section.

There are many metals ions that are not reactive with hydroxide ions, and little or no metal hydroxide precipitation is possible. Sodium, potassium, magnesium, and calcium form soluble forms with hydroxide.

There are several metals that initially form insoluble solids with hydroxide, but with very high concentration of hydroxide (high pH), the metal is soluble. The reaction of precipitated zinc hydroxide with excess sodium hydroxide produces soluble forms of hydrated zinc anions.

$$Zn(OH^-)_2 + 2Na^+ + 2OH^- \leftrightarrow 2Na^+ + Zn(OH)_4^{2-} \tag{9.14}$$

TABLE 9.8   Ionic Species

Metal	Ionic form	Insoluble form
Copper	Cu^{2+}	Cu (OH)$_2$ or CuS
Chromium	Cr^{+3} (trivalent chrome)	Cr(OH)$_3$ or Cr$_2$S$_3$
Hexavalent chromium	Cr$_2$O$_7$$^{-2}$ or CrO$_4$$^{-1}$ (hexavalent chromium)	Ag$_2$CrO$_4$
Zinc in acidic solution	Zn^{2+}	Zn(OH)$_2$ or ZnS
Zinc in basic solution	Zn(OH)$_4^{2-}$	Zn(OH)$_2$ or ZnS
Lead	Pb^{2+}	Bo or PBS, or Pb(OH)$_2$

The generalized hydroxide precipitation formula is as follows where M^{tt} is the divalent metal: with lime;

$$M^{tt} \{CO_3, SO_4, Cl_2\} + Ca(OH)_2 \rightarrow M(OH)_2 \downarrow + Ca^{++} \{CO_3 \downarrow, SO_4 \downarrow, Cl_2\}$$

With caustic;

$$M^{tt} \{CO_3, SO_4, Cl_2\} + 2Na(OH) \rightarrow M(OH)_2 \downarrow + Na_2 \{CO_3 \downarrow, SO_4 \downarrow, Cl_2\}$$

There is less sludge produced with caustic precipitation, and caustics easier to handle than lime, but lime enhances sedimentation, removes sulfates, and is less expensive.

The generalized sulfide precipitation formula is as follows:

$$M^{++} \{CO_3, SO_4, Cl_2\} + \{Na_2, NaH, Mg, H_2, Fe\} S \rightarrow MS \downarrow$$
$$+ \{Na_2, NaH, Mg, N_2, Fe\} \{CO_3 \downarrow, SO_4 \downarrow, Cl_2\}$$

The generalized carbonate precipitation formula is as follows:

$$M^{++} \{SO_4, Cl_2\} + Na_2CO_3 \rightarrow MCO_3 \downarrow + Na_2 \{SO_4, Cl_2\}$$

9.4.1.2 Metal Sulfides.   Heavy metals react with hydrogen sulfide over a wide pH range to form insoluble metal sulfides. The reaction of zinc is typical:

$$Zn^{2+} + H_2S \leftrightarrow 2H^+ + ZnS \text{ (solid)} \tag{9.15}$$

The reactant providing sulfide may be sodium sulfide (Na_2S), sodium hydrogen sulfide (NaHS), or slightly soluble ferrous sulfide (FeS).

9.4.1.3 Comparison of Metal Hydroxide Sulfide and Carbonate Precipitation.   Metal hydroxide precipitation is more commonly practiced than metal sulfide or carbonate precipitation. The prevalence of hydroxide over sulfide and carbonate precipitation is due to the fact that less expensive reagents like lime or caustic can be used as opposed to sodium sulfide or sodium carbonate, or sodium bicarbonate. Odor and handling of potentially toxic gas (H_2S) is generally a disadvantage for sulfide systems.

The distinct advantage of sulfide systems is the ability to produce significantly lower residual metal ion concentrations in treated wastewater. Figure 9.11 compares the solubility curves for Cu, Zn, Cd, and Pb when the metals are precipitated from a caustic solution (NaOH), a sodium sulfide solution, or sodium carbonate. In Fig. 9.11, the pH of the treatment solution is varied over a wide range (pH 2 to pH 13). The concentration of metal in sulfide systems is typically orders of magnitude lower than the hydroxide system.

Metal sulfides also permit precipitation in complicated solutions containing EDTA or other chelating agents which always cause incomplete precipitation problems in hydroxide systems. There is a danger of toxic H_2S being produced in sulfide precipitation.

In general, carbonate precipitation is the most expensive and is difficult to control since it is easy to overdose.

Sludge production from the processes is compared for certain metals in Table 9.9. Table 9.10 lists the advantages and disadvantages of Iron versus alum.

TABLE 9.9 Comparison of Precipitation Techniques—Sludge Production

Method	Metal	Sludge Production
Hydroxide-Lime	$CuSO_4$	3.68 lb Sludge/lb Cu
Hydroxide-Lime	$PbSO_4$	1.82 lb Sludge/lb Pb
Hydroxide-Caustic	$CuSO_4$	1.54 lb Sludge/lb Cu
Sulfide	$CuSO_4$	1.50 lb Sludge/lb Cu
Carbonate	$PbSO_4$	1.29 lb Sludge/lb Pb

9.4.1.4 Use of the Solubility Product.   Use Ksp when you have a slightly soluble electrolyte as in heavy metals removal or other types of coagulation.

- When a poorly ionizable salt (slightly soluble electrolyte) is dissolved in water, at some point no more will dissolve.

- When you have this equilibrium between a solid and a liquid, you can consider the solid a constant since it is in equilibrium and will not further accumulate.

- If the product of the ionic molar concentration is less than K_{sp}, it is super saturated and crystals will form and precipitation will occur.

9.4.1.5 Interference from Complexing Agents.   Complexing Agents may be used with metals in industrial applications. These combine with metals to form more soluble metal complexes that are stable in water and can be more difficult to treat. Examples include:

- Citrates
- Ethylenediaminetetraacetic Acid
- Tartaric Acid
- Tiron
- Polyphosphates (Inorganics)

    Figure 9.12 shows the effect on solubility of Nickel with a complexing agent. The effect of complexing agents may be overcome with a variety of techniques. Examples include:

- Destruction of the complex forming agent with chemical oxidation
- Precipitation of less soluble metal compounds

TABLE 9.10 Advantages and Disadvantages of Iron vs. Alum

Advantages	Disadvantages
Heavier, denser floc	Must use lime
Better settling	Close control is necessary
Works over a wider pH range than alum	More expensive
More effective in color removal	More corrosive than alum
	Ferous is hydrophillic (water loving)

Source: "Wastewater TryOzone", Sheng H. Cin and Kuo C. Yeh, Chemical Engineering, May 1993.

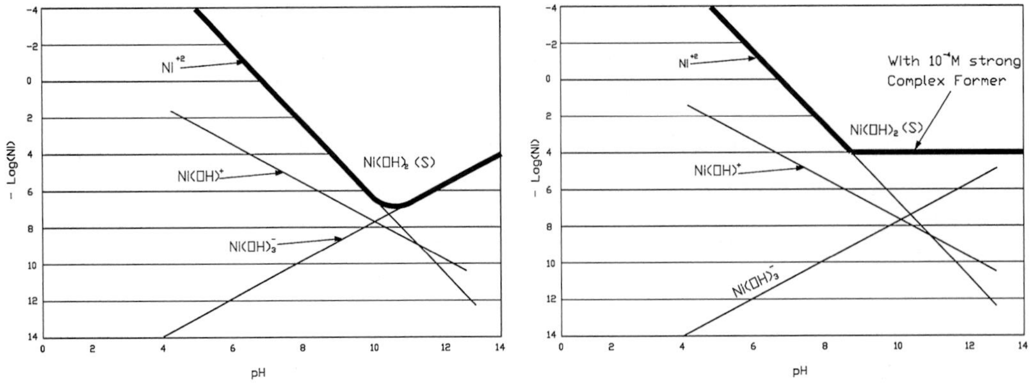

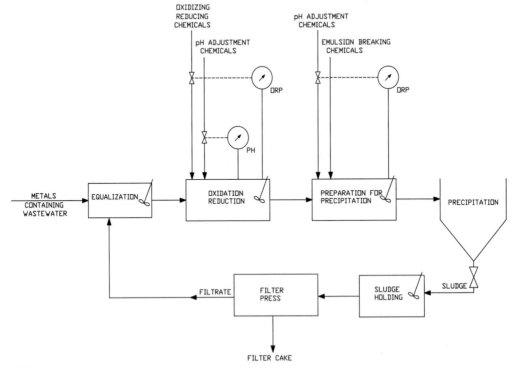

FIGURE 9.12 Nickel with Interference from Complexing Agents

• Addition of other metals to compete for the complex forming agent. Examples include Calcium and Ferrous Iron

$$Ni\text{-}Complex + Ca^{2+} \rightleftarrows Ni(OH)_2 + Ca\text{-}Complex$$

$$Ni\text{-}Complex + Fe^{2+} \rightleftarrows Ni(OH)_2 + Fe\text{-}Complex$$

9.4.1.6 Metals Removal. A typical Metals Removal system is shown in Figure 9.12a.

FIGURE 9.12a Metals removal system.

TABLE 9.10a pH Requirements for Heavy Metal Removal by Lime and Ozonation

Metal	Lime	Lime and Ozonation
Aluminum	-	7.0
Cadmium	10.0	10.0
Chromium	9.0	7.0
Cobalt	11.0	9.0
Copper	9.0	7.0
Iron	4.0	4.0
Lead	-	5.1
Manganese	9.0	6.0
Nickel	11.0	6.0
Silver	11.0	10.0
Zinc	10.0	9.0

Source: "Wastewater TryOzone", Sheng H. Cin and Kuo C. Yeh, Chemical Engineering, May 1993.

9.4.1.7 Design Hints. It is recommended that the following be considered in designing a metals precipitation system:

- The pH control system be designed in accordance with the recommendations listed above with special caution that with hydroxide precipitation for metals with steep "U" shaped solubility curves, multiple basins or some other pH control system may be required.
- The hydraulic detention time in a metals precipitation system should normally be from $1/2$ to 5 minutes at peak flow.
- Mixing should be done in accordance with the recommendations in Chapter 8.
- Precipitated metals sludge may be toxic and must be disposed of as such.
- Precipitated metals sludge in a flow through system must be removed while under the liquid head. Removal should be slow enough to prevent carry through of liquid and fast enough to prevent plugging. This sludge is normally only from 1 to 2% solids.
- Precipitated metal sludge in a batch system can be removed while under the liquid or after the liquid is drawn off. If liquid is drawn off, draw off should be slower as the draw off tap approaches the liquid/solid interface to prevent solids carryover. The sludge in the first case will be about 1 to 2% solids.
- A flow through basin should be designed with baffling in accordance with the recommendations in Chapter 8.
- The system should be completely mixed with added chemicals in accordance with the recommendations in Chapter 8.
- Ozone precipitation can be used in conjunction with lime, to allow lower pH levels than with lime alone. Table 9.10a shows the results.

9.5 COAGULATION AND FLOCCULATION

Coagulation and flocculation are the methods used for aggregating suspended solids into larger and perhaps denser particles that will settle more quieely or become more filterable. For the purposes of this discussion, coagulation is the addition of chemicals to water to destroy or reduce repulsive forces and induce particle agglomeration. Flocculation is the physical process of promoting particle contact to facilitate the agglomeration to larger settleable floc. In practice, these processes are usually accomplished using two different tanks in series. The first tank is a rapid mix tank, into which the coagulant dose is added and the particles become destabilized. The second tank is the flocculation tank, which is a slow mix tank to promote particle collision and aggregation.

9.5.1 Coagulation Theory

Colloids are particles that are less than one micron (10^{-6} m) in size that remain suspended in water. Some colloids are stable thermodynamically (energetically), making them more difficult to coagulate. These are usually ordered structures such as soap and detergents molecules, proteins, starches, and large polymers. Others colloids, however, are not stable thermodynamically, and can therefore coagulate more readily. Examples of these include such particles as microorganisms, algae, organic particles (color), clay, and metal oxides. In coagulation processes, the terms stable and unstable usually refer to the rates at which thermodynamically unstable particles coagulate. Therefore, in practice, stability has a kinetic connotation, not a thermodynamic one. Stable particles coagulate slowly, and unstable particles coagulate quickly. Therefore, the goal of coagulation is to decrease particle stability.

In addition to being very small, colloids also have a very high specific surface area, defined as surface area of the particle per volume of the particle. This contributes to the fact that most colloids also have negatively charged surfaces that cause mutual repulsion. Different processes induce this charge on colloids in solution. Examples of such processes include:

- Redox reactions involving hydrogen ions and the acceptance or donation of protons
- Redox reactions involving solutes in the water other than hydrogen ions
- Imperfect molecular structure within the molecule itself

Even though the surface charge of a colloid is negative, a colloidal suspension has no net electrical charge. This is the result of an ionic cloud surrounding the particle. The ions forming the portion of the cloud closest to the colloid itself are positively charged. This combination, a strong surface charge (negative) and an ionic cloud (positive) results in a stable suspension with no net charge. This is shown in Fig. 9.13.

As can be seen in Fig. 9.13, the total electric potential around the particle is the Nernst potential, and this is equal to the potential of the particle itself. The negatively charged particle has a strong layer of positively charged ions around it. This is called the Stern layer. The electrostatic potential across this layer is consequently called the Stern potential. It is at the extent of the Stern layer that the particle can be forced to move. The potential from the outside of the Stern layer to the outside of the total electric potential surrounding the particle is called the Zeta potential. If particles are placed in an electrolyte solution and an electric current is passed through the solution, the particle, depending on its surface charge, will attract to one or the other of the electrodes. As has been noted, most colloids are negatively charged and they will therefore move to the positive electrode. This potential for a particle to move in a solution is the Zeta potential. It is this Zeta potential that must be overcome to effectively coagulate particles.

When colloidal particles that are similar approach each other, their diffuse layers, the Zeta layers, begin to overlap and interact. This interaction results in a repulsive force between the particles. Also acting on the particles are van der Waals forces. Van der Waals forces are attractive forces that act on all particles and between all particles. They arise from dipole interaction between the colloids and water. An attractive potential energy arises from these attractive van der Waals forces. The difference between the Zeta potential and the potential caused by these results, is a net potential energy acting upon particles in water. When a colloidal suspension will not settle or coagulate, the Zeta potential cannot be overcome by van der Waals forces, and chemical coagulation is necessary to destabilize the particles. Figure 9.14 shows this graphically.

Chemical coagulation can destabilize particles in the following different ways:

- The first method is by compression of the outer electron cloud by ions in solution. In this method of destabilization, the ions do not react directly with the colloids, but only act upon the colloids electrostatically. This works by ions such as Na^+, Ca^{2+}, or Al^{3+} being added to the solution. The

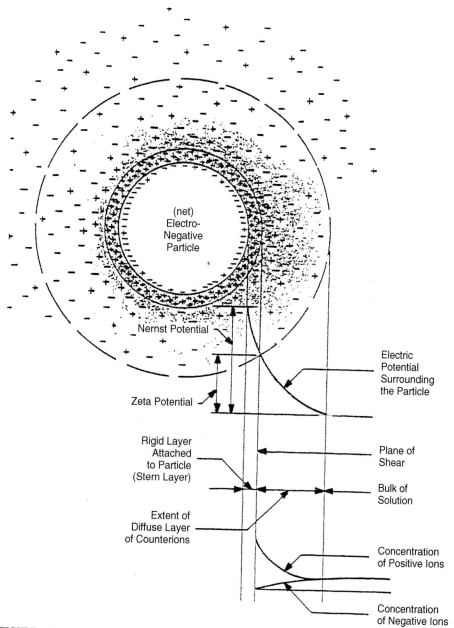

FIGURE 9.13 Negatively charged colloid with positive ionic closed. Nernst, stern, and zeta potential. Sources: Amirtharajah and O'Melia. 1990.

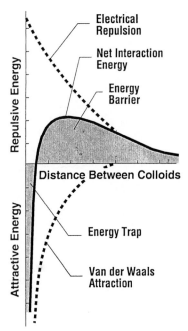

FIGURE 9.14 Forces Acting on colloidal particlae *Source* www.zeta-meter.com/ 5min. pdf

negatively charged colloids attract the positively charged ions. As the counterions approach the colloids, the other positive electrons in the Zeta layer move away to maintain electrostatic neutrality, thus decreasing the layer thickness and range of repulsive interaction.

• The second method of destabilization is adsorption and charge neutralization. In this case, ions are added to the water in the form of hydrophilic coagulants or positively charged ions that adsorb to the particle surface, and thus neutralize the charge. If this process is pushed past neutralization, the effective charge of the particle can be changed, allowing for repulsive forces to once again increase.

• The third method of destabilization is enmeshment in a precipitate. This is accomplished when metal salts are added to a water and precipitate as metal hydroxides. Colloids can become enmeshed in the formations or join after they are formed. This has been termed sweep floc.

• The fourth method of chemical destabilization is adsorption and interparticle bridging. This method usually occurs when synthetic, long chained polymers are used to adsorb to the surface of colloids. The polymers accomplish destabilization when they adsorb to several colloids and thereby, bind them together by interparticle bridging.

These methods of destabilization can act individually or in conjunction with one another, but all require adding counter ions or charges to coagulate. The greater the ionic strength of the counter ion, the greater the destabilization effectiveness. This is shown in Fig. 9.15. The design of a coagulation treatment system requires testing to determine the best coagulation option.

9.5.2 Coagulation Design

As noted earlier, chemical coagulation is the addition of chemicals to water in order to destroy or reduce repulsive forces and induce particle agglomeration. A few of the more commonly used coagulants are aluminum salts, iron salts, and organic polymers.

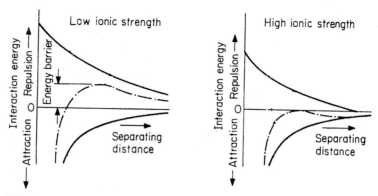

FIGURE 9.15   Ionic strength and coagulant effectiveness. Source: Amirtharajah and O'Melia, 1990.

9.5.2.1   Aluminum Salts.   The most common aluminum salt used for coagulation is alum. Alum incorporates several of the destabilization methods mentioned previously. Alum has the chemical form:

$$Al(SO_4)_3 \cdot 18H_2O \tag{9.16}$$

In practice, however there are usually only about 14 to 16 water (H_2O) molecules, giving an average molecular weight of approximately 600g/mol.

When alum is added to water, it reacts with the alkalinity in the water according to the following reaction,

$$Al_2(SO_4)_3 \cdot 14H_2O + 3Ca(HCO_3)_2 \rightarrow 3CaSO_4 + 2Al(OH)_3 + 6CO_2 + 14H_2O \tag{9.17}$$

Alum also undergoes hydrolysis according to the following reaction:

$$Al_2(SO_4)_3 \cdot 14H_2O + 6H_2O \rightarrow 2Al(OH)_3 + 6H^+ + 3SO_4^{2-} + 14H_2O \tag{9.18}$$

In practice, the above reactions actually produces aluminum oxyhydroxo complexes, or micro flocs that crowd the colloids. These microflocs are positively charged, and they neutralize the charge in colloids and attract negatively charged colloids.

The optimum pH for alum coagulation is 5 to 7. This optimum is lower for soft, colored waters and higher for hard waters. Alum naturally lowers the pH of the water to which it is added, depending upon how much alkalinity is initially present. With the production of CO_2, the pH drops as CO_2 is converted to H_2CO_3, carbonic acid. A base such as lime may have to be added to maintain the optimum pH range. If the solution has high alkalinity, such $Ca(HCO_3)_2$, the pH adjustment probably will not be necessary, otherwise, lime or a base should be added to keep the pH around 6.

9.5.2.2   Iron Salts.   The most common iron salts used for coagulation are copperas (ferrous sulfate), ferric chloride, and ferric sulfate. Copperas, or ferrous sulfate, is a coagulant that must be accompanied by lime to maintain the proper pH level, and therefore adequate hydroxide ion concentration, for coagulation to occur. When the lime is added as CaO to the water first, the chemical reactions are as follows:

$$CaO + H_2O \rightarrow Ca(OH)_2 \tag{9.19}$$

and adding the ferrous sulfate:

$$FeSO_4 \cdot 7H_2O + Ca(OH)_2 \rightarrow Fe(OH)_2 + CaSO_4 + 7H_2O \qquad (9.20)$$

resulting in soluble ferrous hydroxide, $Fe(OH)_2$, initially, but continuing:

$$4Fe(OH)_2 + O_2 + 2H_2O \rightarrow 4Fe(OH)_3(s) \qquad (9.21)$$

producing insoluble ferric hydroxide, $Fe(OH)_3$. This ferric hydroxide accomplishes coagulation by the mechanisms discussed previously.

If the lime is added after the copperas, the copperas reacts with the natural alkalinity, $Ca(HCO_3)_2$, as follows:

$$FeSO_4 \cdot 7H_2O + Ca(HCO_3)_2 \rightarrow Fe(HCO_3)_2 + CaSO_4 + 7H_2O \qquad (9.22)$$

and adding the lime, as in equation (9.20), the following then occurs:

$$Fe(HCO_3)_2 + 2Ca(OH)_2 \rightarrow Fe(OH)_2 + 2CaCO_3(s) + 2H_2O \qquad (9.23)$$

and finally:

$$4Fe(OH)_2 + O_2 + 2H_2O \rightarrow 4Fe(OH)_3(s) \qquad (9.24)$$

In this second scenario, if lime is added after the copperas, twice as much lime must be added and more sludge is produced, and consequently, must be disposed. *Therefore, it is much more economical to add the lime before the ferrous sulfate.*

There are several advantages of the iron and lime process over alum, but there are also some disadvantages. These include:

- Standard size granular ferrous sulfate tends to cake and arch in dry feed hoppers or storage bins.
- Slower response than alum
- Adds hardness, may need to be stabilized
- Dosing of two chemicals more difficult than one.

Other iron salts include ferric chloride, and ferric sulfate. These coagulants undergo reactions similar to the reactions mentioned for copperas. There are also similarities for the advantages and disadvantages of these and ferrous sulfate.

Table 9.9 shows the advantages of ferrous coagulation compared to aluminum coagulation.

9.5.2.3 Coagulant Aids. Coagulant aids are subtances added to water in addition to the primary coagulant to enhance the coagulative process. One such aid that has already been discussed is lime. The addition of lime is necessary to control the pH when either the wastewater is not initially at the optimum pH range or the coagulant dose drops the pH below the optimum range. Another coagulant aid is bentonite clay, which is used to add weight to coagulated colloids and to form a nucleus to begin agglomeration. The addition of Bentonite clay increases the number of negatively charges particles and coagulates with the colloids to form a heavier, denser floc. Also used as a coagulant aid is soda ash, $NaCO_3$. It is easier to use than lime, it works in the same manner, but not as well. Another aid to coagulation is recycled sludge from previous coagulation. This sludge acts in the same way as bentonite clay, as a weighting agent.The author has successfully used a slip stream from the Return Activated Sludge in a biological treatment plant to coagulate the flow to the clarifier. This idea can also be used to coagulate chemical wastewater with a return of sludge from the clarifier.

Silica gel or activated silica can also be added as a coagulant aid. These anionic colloidal sol of silica particles lower zeta potential of other colloids and also act as a polymer for cross-linkage. Similarly, organic polyelectrolytes, when used with other coagulants, or when used with pH adjustment, greatly aid in coagulation. These high molecular weight, long chain polymers provide cross-linkage or bridging and adsorption, and they can have a negative, (anionic), positive, (cationic), or neutral (nonionic) charge. For example, a positively charged polymer attracts negatively charged colloids and bridges the particles. This combined agglomeration then attracts another long chained polymer which then attracts more negative colloids, and this continues forming a settlable floc.

Emulsion and dispersion polymers are coated with oil. Polymer activation requires the oil to be "scrubbed" from the polymer, exposing the polymer to water. Scrubbing is accomplished with sufficient mixing energy. Insufficient mixing causes polymer gelling and maximum activation is unattainable. Over mixing following activation can cause polymer damage, defeating the activation process. Surfactant-rich polymer solution can be provided to aid in chemically scrubbing the polymer.

The ideal mixing for emulsion and dispersion polymers is an initial high, non shearing mixing energy directly at the point of polymer/water contact to scrub the oil from the polymers, followed by high turbulent mixing to complete the dilution, distribution, and hydration processes.

Solution or Mannich type polymers are pre-activated or extended and are very viscous. This type of polymer does not need the initial mixing but is highly susceptible to damage from excessive mixing energy.

In addition to chemical coagulants, electrocoagulation is a different coagulation alternative. Electrocoagulation consists of the use of an AC current to neutralize the natural surface charge on a particle, resulting in flocculation and settling. An example of how this can work can be demonstrated by a system developed by Vanguard Environmental in Houston, Texas. The parameters for the current are a 30-second detention time at 25 volts, 120 amps, and 60 hertz. This prototype runs at 120 gallons for 30 to 50 cents/1000 gallons. The limitaions are that it will not work when the pH is greater than 10.0, and does not work with soluble organics.

9.5.2.4 *Coagulant Determination.*   There are several factors that affect coagulation. Some of the more critical factors include:

- pH
- Mixing effects
- Zeta potential
- Coagulant dosage
- Colloid concentration (turbidity)
- TOC or (color)
- Anions or cations in solution
- Temperature

It is necessary to determine the ideal range for each of these factors if chemical coagulation is to be used in treatment of the wastewater. The most commonly used method for coagulation control is the jar test. Jar tests are used to determine:

- Type of coagulant
- Coagulant dosage
- If a coagulant aid is required and the dosage of such an aid
- Determination of an optimum pH
- Determination of point of addition of pH adjustment chemicals and coagulant aids
- Determination of the dilution of coagulant
- Optimization of mixing energy and time for rapid mix and slow mix

Jar tests can be conducted by rapidly mixing the coagulant and the wastewater in a large beaker or jar for 1 minute, then slowly mixing the solution for 15 minutes, and then allowing the floc to settle for 45 minutes. The beginning parameter and final parameter (COD, TSS, Turbidity, etc.), can be measured or visual inspection can be used to determine effectiveness. An example of typical jar test apparatus can be seen in Fig. 9.16. This mechanical mixing apparatus more closely duplicates a full scale mixing tank since it uses blades instead of magnets. The author typically designs a full scale mixing tank as close to the configuration of a 1000 ml beaker and the mixer blades as possible, when jar tests are used to routinely adjust the coagulant doseage.

When chemical coagulants or aids are added to a wastewater, there are physical criteria that must be considered. The physical state of the coagulant will determine the type of coagulant delivery system. Liquid coagulants generally work better with automated feeding systems, but solid coagulants can easily be added manually. Lime can often cake and be difficult for operators to work with. It should be determined, if lime is used as a coagulant aid, whether it should be slaked or unslaked. Alum is often used as a liquid coagulant. In this form, it is inexpensive, but very corrosive. In all cases the amount of chemical addition should be proportional to flow.

9.5.2.5 Coagulant Rapid Mixing.

9.5.2.5 Coagulant Rapid Mixing.   Although mixing is covered elsewhere in this Handbook, it will be touched on briefly here, in conjunction with coagulation and flocculation. When coagulants are added to water in the form of iron or aluminum salts, two mechanisms occur:

1. Adsorption of aluminum or iron hydrolysis species on colloids causing charge neutralization

2. Sweep coagulation with the precipitating hydroxide

To facilitate the first mechanism, turbulent rapid mixing is necessary. The adsorption of coagulant onto the colloids occurs very quickly. Rapid dispersion is necessary to ensure that the hydrolysis species adsorb onto the colloids. For the hydroxide precipitation, the second mechanism, immediate dispersion is not as critical and the mixing intensity does not have to be as high. After the rapid mix to encourage coagulant-colloid contact, a slow mix, or flocculation, is required to encourage aggregation of the destabilized particles.

FIGURE 9.16   Jar test apparatus.

9.5.2.6 Design Hints.   With the exception of specific mixing requirements described above, mixing time for coagulation should be between $\frac{1}{2}$ and 5 minutes at peak flow.

• Mixing energy should be in accordance with Chapter 8 will caution as to the shearing and dispersion recommendations listed above.

• Associated pH control should be in accordance with the above recommendations.

• Chemical Feed equipment should be selected to avoid stoppage, corrosion, and dust problems. Generally liquid feed is easier and requires less capital cost but liquid chemicals are typically more expensive than solids.

9.5.3 Flocculation Theory

The purpose of flocculation is to bring particles together to form well settling particles. The rate of aggregation is dependent upon the rate of interparticle collisions. When particles aggregate, hydrodynamic shear forces in the water can cause the aggregations to break-up. Aggregation and break-up can occur simultaneously.

Theory: Particle collisions can occur through three different mechanisms:

• Brownian diffusion

• Differential settling

• Fluid shear

Brownian diffusion is often referred to as *perikinetic flocculation* and fluid shear is often called *orthokinetic flocculation*. There are various models used to describe the types of flocculation, as well as other models to describe the effects of varying particle size upon the aggregation of particles.

Brownian diffusion, or perikinetic flocculation, is the random motion of particles caused by contact with water molecules. The thermal energy of the water is the driving force behind this mechanism. Fluid shear, or orthokinetic flocculation, is particle contact caused by fluid movement in currents. The particle follows fluid flow and subsequently contacts other particles also following the flow of the fluid. The fluid motion can be described by the velocity gradient, G, which is the change in velocity over a corresponding distance, $^{dv}/_{dz}$, and is expressed in units of inverse time (1/time). The third type of flocculation, differential settling, is based on particle settling caused by gravity. As particles settle, they collect other particles and agglomerate to further enhance settling and particle contact. Each of these mechanisms plays a role in flocculation. Brownian diffusion and differential settling are based on the thermal energy of the fluid and gravity/quiescent conditions, respectively. Because the thermal energy of water cannot be changed, Brownian diffusion cannot be designed, and because differential settling is not a significant factor in flocculation, completely quiescent conditions are not beneficial. Fluid shear can be induced, however. Fluid motion fields can be generated to produce orthokinetic flocculation.

9.5.4 Flocculation Design

Caution must be taken when inducing fluid shear, because, if the mixing becomes turbulent, then particle dispersion will occur. The goal of fluid shear/orthokinetic flocculation is to increase particle collision, and thereby encourage particle agglomeration. There are several factors that are key to this process, and the most important are:

1. Time
2. Number of particles
3. Size of particles
4. Velocity gradient, G, in the fluid

Factors 1 and 4 can be controlled in the design process.

In accordance with the mixing recommendations in Chapter 8, the velocity gradient in accordance with Formula (8.66) is $G = \sqrt{p/uv}$.

Each mixing basin design should be checked with this formula using u from Appendix C and a G of 250 to 500 s^{-1}.

9.5.5   Coagulation and Flocculation Design Suggestions

9.5.5.1   Coagulation

- Rapid mix to get max dispersion
- $G = 250 - 500$/sec in rapid mix chamber
- Rapid mix detention time $= 0.5$ to 5 minutes
- It is important to make sure units cancel in design equation

9.5.5.2   Flocculation

- $G = 10$ to 75/sec
- $GT = 10^4$ to 10^5, where T is basin detention time
- Maximum tip speed $= 3$ ft/sec
- Maximum paddle area $= 25$ percent of the basin cross-sectional area in plan view
- Cubical or cylindrical basin is best
- Use baffle extending 10 percent into basin to reduce swirl and distribute turbulence evenly
- If G is high flocculate for short time to produce small dense floc that settles well
- If G is low, flocculate for a longer detention time to produce larger fluffy floc that settles well
- Tapered flocculation is best if possible, gradually decreasing G

9.6   DISINFECTION

Disinfection is the process of removing all bacteria from wastewater. Sterilization is the process of removing all organisms. This Handbook will concentrate on methods of disinfection. The purpose of disinfecting wastewater containing domestic sewage is to reduce to a safe level the hazards of infectious disease in receiving waters. The practice of disinfection has greatly reduced waterborne disease outbreaks since 1980 in many areas of the globe.

9.6.1   Disinfection Theory

Disinfection can theoretically be accomplished by physically removing bacteria, but because of their microscopic size, normally bacteria are destroyed physically or chemically.

The most commonly practiced method of disinfection worldwide is chlorination. The toxic effect of total residual chlorine on fresh water organisms has been confirmed at low concentrations. Residual chlorine or bromine can combine with organics in a receiving water to form halogenated organic compounds which may be carcinogenic to humans. In addition, residual chlorine or bromine above 0.01 mg/l can be toxic to fresh water fish and other aquatic life (see Chapter 13). Halo-organic compounds which have been identified in waters in the United States are shown in Table 9.11.

Dechlorination through an oxidation-reduction reaction greatly reduces or eliminates the toxicity caused by residual chlorine. Brominated effluents can also be highly toxic but have a shorter half-life. Ozonated and ultraviolet effluents have shown no toxicity effects.

TABLE 9.11   Halo-organic Compounds Identified in Drinking Water in the United States

1. acetylene dichloride	30. dibromodichloroethane
2. aldrin	31. 1, 4-dichlorobenzene
3. atrazine	32. dichlorodifluoroethane
4. (deethyl) atrazine	33. 1, 2-dichlorobenzene
5. bromobenzene	34. dichloroethyl ether
6. bromochlorobenzene	35. dichloromethane
7. bromodichloromethane	36. dieldrin
8. bromoform	37. heptachlor
9. bromoform butanal	38. heptachlor epoxide
10. bromophenyl phenyl ether	39. 1,2,3,4,5,6,7,7-heptachloronorbornen
11. carbon tetrachloride	40. hexachlorobenzene
12. chlordan (e)	41. hexachloro-1, 3-butadiene
13. chlorobenzene	42. hexachlorocyclohexane
14. chlorodibromomethane	43. hexachloroethane
15. 1, 2-bis-chloroethoxy ethane	44. methyl chloride
16. chloroethoxy ether	45. octyl chloride
17. bis-2-chloroethyl ether	46. pentachlorobiphenyl
18. b-chloroethyl methyl ether	47. pentachlorophenol
19. chloroform	48. 1,1,3,3-tetrachloroacetone
20. chlorohydroxy benzophenone	49. tetrachlorobiphenyl
21. bis-chloroisopropyl ether	50. tetrachloroethane
22. chloromethyl ether	51. tetrachloroethylene
23. chloromethyl ethyl ether	52. trichlorobenzene
24. m-chloronitrobenzene	53. trichlorobiphenyl
25. 3-chloropyridine	54. 1,1,2-trichloroethane
26. DDE	55. 1,1,2-trichloroethylene
27. DDT	56. trichlorofluoromethane
28. dibromobenzene	57. 2,4,6-trichlorophenol
29. dibromochloromethane	

9.6.2   Physical Disinfection

Screening methods of removing bacteria are ineffective because of the small size of the bacteria. Only bacteria which adhere to solid particles large enough to be removed by the screening or filtering device will be removed with the total suspended solids. Consequently less than 5 percent of bacteria will be removed with a coarse screen, 15 percent with a fine screen, 20 percent in a grit chamber, and 30 percent with primary sedimentation. Boiling wastewater for 10 to 20 minutes is another physical method effective for disinfection.

9.6.3   Chemical Disinfection

Chemicals used for disinfection include halogens such as chlorine, bromine and iodine, metallic ions such as silver or copper as well as ozone, hydrogen peroxide, acids, bases, and detergents. The relative oxidation power compared to chlorine at 1.0 is shown in Table 9.11a.

• Chlorination

Liquefied chlorine gas is available in 150- and 2000-pound cylinders in the United States. When gaseous chlorine is added to water, two reactions take place: hydrolysis and ionization.

$$Cl_2 + H_2O \Leftrightarrow HOCl + H^+ + Cl^- \tag{9.25}$$

TABLE 9.11a   Oxidation Power

Relative Oxidation Power (Cl = 1.0)	
Fluorine	2.23
Hydroxyl Radical	2.06
Atomic oxygen (singlet)	1.78
Ozone	1.52
Hydrogen Peroxide	1.31
Perhydroxyl Radical	1.25
Permanganate	1.24
Hypobromous Acid	1.17
Chlorine Dioxide	1.15
Hypochlorous Acid	1.10
Hypoiodous Acid	1.07
Chlorine	1.00
Bromine	0.80
Hydrogen Peroxide	0.64
Iodine	0.54
Oxygen	0.29

In this hydrolysis reaction, the stability constant can be computed as:

$$K = \frac{[HOCl][H^+][Cl^-]}{[Cl_2]} = 4.5 \times 10^{-4} \text{ at } 25°C \tag{9.26}$$

$$HOCl \Leftrightarrow H^+ + OCl^- \tag{9.27}$$

In this ionization reaction, the stability constant is:

$$K_i = \frac{[H^+][OCl^-]}{[HOCl]} = 3.7 \times 10^{-8} \text{ at } 25°C$$

$$K_i = 2.0 \times 10^{-8} \text{ at } 0°C \tag{9.28}$$

The free available chlorine equals the total of OCl^- and $HOCl$. Since $HOCl$ is 40 to 80 times as effective in disinfection as OCl^- (see Fig. 9.17a), the ratio of the two is very important. Note that chlorine is much more effective below a pH of 6.0 than above a pH of 8.0.

• Chloramines

Ammonia in wastewater will combine with chlorine to form chloramines as follows:

$$HOCl + NH_3 \rightarrow H_2O + NH_2Cl \text{ (monochloramine)} \tag{9.29}$$

$$HOCl + NH_2Cl \rightarrow H_2O + NHCl_2 \text{ (dichloramine)} \tag{9.30}$$

$$HOCl + NHCl_2 \rightarrow H_2O + NCl_3 \text{ (trichloramine)} \tag{9.31}$$

The type of chloramine depends on the pH, the ammonia concentration, and the temperature. Normally monochloramines will be formed at a pH greater than 8.5, dichloramines between pH 4.8 and 8.5, and trichloramines below pH 4.8.

When chorine is added to water containing reducing agents and ammonia, residuals develop which yield the curve in Figure 9.17b.

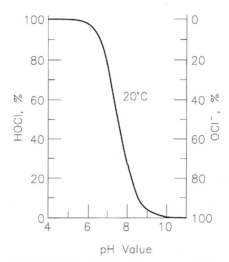

FIGURE 9.17a Relationship between HOCL and OCl⁻.

Chlorine first reacts with reducing agents present in water and develops no appreciable residual (0-A in Fig. 9.17). The dose at point A is therefore that which is required to meet the demand exerted by reducing agents. These include nitrites, ferrous ions, and hydrogen sulfide.

Chloramines are formed when chlorine is added in excess at point A. Mono and dichloramines are considered together since there is little control over which is formed.

When all the ammonia has been reacted, a free available chlorine begins to develop (point B). As free available chlorine residual increases, the previously produced chloramines are oxidized. This results in the creation of oxidized nitrogen compounds such as nitrous oxide, nitrogen and nitrogen trichloride, which in turn reduce the chlorine residuals as seen on the curve from point B to the break-point.

- Breakpoint chlorination

When all chloramines are oxidized, additional chlorine added creates an unequal residual known as the breakpoint (that limit beyond which all residual is free available chlorine).

Stoichiometrically, ammonia can be removed through breakpoint chlorination by feeding 7.6 parts chlorine per part of ammonia. Practically, 8 to 10 parts chlorine must be used.

As in any chemical reaction, the available chlorine must contact all ammonia molecules for effective ammonia reduction. If ammonia exists within suspended solids particles, complete association may not occur. Therefore, these particles should be removed through sedimentation or filtering before applying breakpoint chlorination.

- Chlorine Compounds

The two most common compounds of chlorine used for disinfection are calcium hypochlorite and sodium hypochlorite.

Calcium hypochlorite manufactured under the brand name HTH, reacts in water as follows to disinfect:

$$Ca(OCl)_2 + H_2O \rightarrow 2HOCl + Ca(OH)_2 \tag{9.32}$$

Calcium hypochlorite is produced as a white powder or tablet and is completely soluble in water. It is available in up to 70 percent available chlorine and can cost as much as twice chlorine gas.

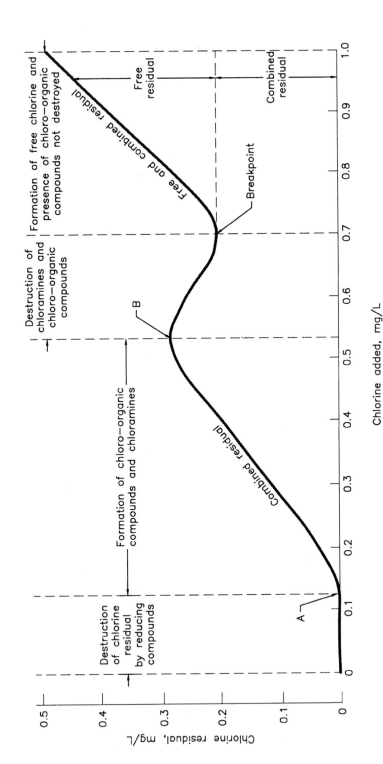

FIGURE 9.17b Generalized curve obtained during breakpoint chlorination.

9.36

Sodium hypochlorite, manufactured as bleach under the brand name Clorox, reacts in water as follows as a disinfectant:

$$NaOCl + H_2O \rightarrow HOCl + NaOH \tag{9.33}$$

Sodium hypochlorite is produced in a clear liquid form and is completely soluble in water. It is from 5% available chlorine to 15% available chlorine and can decompose with exposure to light and heat, especially at higher concentrations. The 15 percent concentration is the maximum practical for stability reasons. Sodium hypochlorite can cost twice as much as chlorine gas.

• Ozone

Ozone has a high oxidation potential and is considered an effective alternative to chlorination since it forms no toxic products when combined with organics. Ozone is 13 times more soluble in water than oxygen, but it is difficult to dissolve more than a few milligrams per liter. Its solubility in water decreases as the temperature drops (see Fig. 9.18).

Ozone decomposes in water to form molecular oxygen as follows:

$$O_2 + H_2O \rightarrow HO_3^+ + OH^- \tag{9.34}$$

$$HO_3^+ + OH^- \rightarrow 2HO_2 \tag{9.35}$$

$$O_3 + HO_2 \rightarrow HO + 2O_2 \tag{9.36}$$

$$HO + HO_2 \rightarrow H_2O + O_2 \tag{9.37}$$

Ozone is typically produced on site by applying an electrical discharge across oxygen or air. Up to 6 kilowatt hours of power is required to generate one pound of ozone from pure oxygen and twice that amount is required to generate ozone from air (see Fig. 9.19). As a comparison, chlorine can be produced using approximately 1.3 kilowatt hours of electricity per pound.

Because ozone is a corrosive gas, care should be taken when specifying materials for the generator. Metallic parts should be stainless steel and scales should be either polytetrafluoroethylene or chlorosulfonated polyethylene. Rubber and synthetic rubber products are deteriorated by ozone. An ozonization system consists of:

• Ambient air or oxygen feed gas treatment unit.

• Ozone Generator

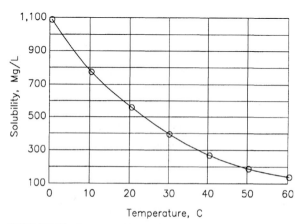

FIGURE 9.18   Ozone solubility in water.

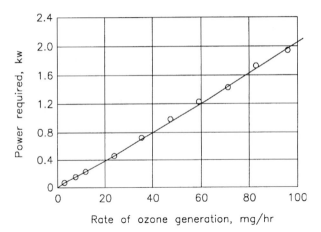

FIGURE 9.19 Ozone power consumption.

• Ozone/water contact mechanism.
• Residual ozone destruction unit.

The feed gas treatment unit should remove all particles including aerosols, moisture, and hydrocarbons, greater than 0.3 μm and 95% of those larger than 0.1μm.

Nearly all commercial ozone generators employ the corona-discharge method and have outputs of 1 to 2 weight percent ozone from ambient air and 2 to 4 % from oxygen. Elective consumption is usually between 15 and 26 kw hr/kg O_3 for an air feed generator.

The ozone/air mixing basin may use static mixers, diffusers, injectors, or packed columns. Diffusion, the most common method, can introduce 1 to 1.5 kg O_3 per kw hr consumed.

Since O_3 is corrosive to metals and harmful to humans in high concentrations, it should not be allowed to escape from the generator or contact vessel without being treated. OSHA requires that O_3 concentrations be less than 0.1 ppm before venting. O_3 destruction units can be catalytic, thermal, thermal-catalytic, or activated carbon.

• Ultraviolet (UV) radiation
Ultraviolet light is absorbed by organic molecular components, essential for the cells biological activity. The excitation of the molecules causes disruption of unsaturated bonds that produces a progressive lethal biochemical change. For most bacterial species, the lethal effect is a function of the wavelength and is greatest between 2500 and 2600 angstroms (A).

For UV radiation to be an effective biocide, the energy dosage must reach the organism. The efficiency of penetration of a bacterium in water is affected by turbidity, color, and the concentration of organic compounds.

• Bromine
Bromine is a liquid at atmospheric pressure and is safer to handle than chlorine. As with chlorine, the amine form is produced when ammonia is present, and bromine can be used to remove ammonia with the breakpoint process.

Bromine, hypobromous acid (HOBr), and monobromamine are considered nearly equal in bactericidal properties and are essentially equal to free chlorine at comparable pH. Whereas chlorine effectiveness decreases with increasing pH, bromine is most effective at high pH. This indicates potential use for use with ammonia stripping and phosphorous removal with lime.

Bromine costs approximately 3.5 times as much as chlorine.

- Iodine
Commercial iodine is a nonmetallic solid and has a dense, dark, shiny metallic appearance. The vapor pressure of iodine at 25°C is only 0.31 mm Hg, therefore it is much less volatile than chlorine with a vapor pressure of 5300 mm Hg and bromine at 215 mm Hg. Iodine does not form iodamine with wastewater. The cost of iodine disinfection is approximately 18 times more than chlorine.

- Bromine chloride
Bromine chloride is a chemical disinfectant similar to chlorine in efficiency. It hydrolyzes in water to form hypobromous acid (HOBr) as follows:

$$BrCl + H_2O \rightarrow HOBr + HCl \tag{9.38}$$

The hydrolysis constant for this reaction is:

$$\frac{[HOBr][H^+][Cl^-]}{[BrCl]} = 2.97 \times 10^{-4} \text{ at } 0°C \tag{9.39}$$

When hypobromous acid reacts with ammonia, bromamines are formed which are superior to chloramines as a bactericide as follows:

$$NH_3 + HOBr \rightarrow NH_2Br + H_2O \text{ (monobromamine)} \tag{9.40}$$
$$NH_2Br + HOBr \rightarrow NHBr_2 + H_2O \text{ (dibromamine)} \tag{9.41}$$
$$NOBr \rightarrow NBr_3 + H_2O \text{ (tribromamine)} \tag{9.42}$$

- Chlorine dioxide (ClO_2)
Chlorine dioxide is a strong oxidant which is unstable and highly corrosive. It is usually generated from the reaction between a sodium chloride solution and chloride in water as follows:

$$2NaClO_2 + Cl_2 \rightarrow 2ClO_2 + 2NaCl \tag{9.43}$$

The cost of producing ClO_2 is as much as 13 times the cost of producing Cl_2.

- Lime
Lime treatment at a pH of 11.5 to 12.0 can be effective with a detention time of 30 minutes at a temperature as low as 1°C.

- Ionizing radiation
Ionizing radiation from cobalt-60, cesium-137, electron accelerators, reactor loops, fuel elements, and mixed fission products have been successfully used for sterilizing foods and have potential for wastewater disinfection.

- Low pH
Exposure of microorganisms to extremes in hydrogen ion concentration has some potential for disinfection. Escherichia coli exposed to a pH of 1-2 for one hour have only a 71 to 80 percent loss in viability. Also, enteric bacteria survive the extreme low pH of the stomach before entering the small intestine. Therefore, for low pH to be an effective disinfectant, long detention periods must be used.

- Photon induced reaction
Photons in the far ultraviolet spectrum in the presence of ozone can provide disinfection capabilities.

9.6.4 Dechlorination Theory

When chlorine or its products are used for disinfection, if chlorine residual is included in a discharge permit, dechlorination will likely be required. It is almost impossible to balance chlorine demand with chlorine feed to the point that residuals are consistently lower than permit limits.

- Sulfur dioxide (SO_2)

  Chlorine, being a strong oxidizer, can be removed with the reducer SO_2 or its products, sodium sulfite (Na_2SO_3), sodium bisulfate ($NaHSO_3$), sodium metabisulfate ($Na_2S_2O_5$), or sodium thiosulfite ($Na_2S_2O_3$) in an oxidation-reduction reaction.

  SO_2 is commercially available as a liquefied gas and is much more soluble in water than chlorine (1.0 lb/gal. At 60°F).

  The dechlorination reaction of SO_2 with free and combined chlorine is almost instantaneous; therefore contact chambers can be replaced by complete mixing at the point of addition.

  SO_2 can be added at a stoichiometric ratio of 0.9 to 1.0 to remove chlorine. Sodium meta bisulfate addition should be 1.34 to 1.0 and sodium bisulfate 1.46 to 1.0. SO_2 costs approximately the same as chlorine and causes no toxicity in receiving waters. Dechlorination occurs as follows: With chlorine or chlorine products:

$$SO_2 + H_2O \rightarrow HSO_3^- + H^+ \tag{9.44}$$

$$HSO_3^- + HOCl \rightarrow Cl^-\ SO_4^{-2} + 2H^+ \tag{9.45}$$

$$\text{Therefore: } SO_2 + HOCl \rightarrow Cl^- + SO_4^{-2} + 3H^+ \tag{9.46}$$

With chlormines:

$$SO_2 + H_2O \rightarrow HSO_3^- + H^+ \tag{9.47}$$

$$NH_2Cl + HSO_3 + H_2O \rightarrow Cl^- + SO_4^{-2} + NH_4^+ + H^+ \tag{9.48}$$

$$\text{Therefore: } SO_2 + NH_2Cl + 2H_2O \rightarrow Cl^- + SO_4^{-2} + NH_4^+ + 2H^+ \tag{9.49}$$

With chlorine dioxides:

$$SO_2 + H_2O \rightarrow H_2SO_3 \tag{9.50}$$

$$H_2SO_3 + 2ClO_2 + H_2O \rightarrow 5H_2SO_4 + 2HCl \tag{9.51}$$

- Activated carbon

  Activated carbon can be used to remove residual chlorine but its adsorption sites are also available to most organics. Therefore if residual organics are present in a wastewater, their demand must be considered in computing carbon use requirements.

  Chlorinated amines, free chlorine, and chlorinated organics will all be successfully adsorbed on granular activated carbon.

- Hydrogen peroxide

  Even though hydrogen peroxide is a strong oxidant (see Table 9.11), it can react with chlorine at a ratio of 0.48 to 1.0 to remove the residuals as follows:

$$OCl^- + H_2O_2 \rightarrow Cl^- + H_2O + O_2 \tag{9.52}$$

9.6.5 Safety and Storage

- Chlorine

  Liquid chlorine is a hazardous chemical and chlorine gas is toxic and can cause death by suffocation. It irritates the respiratory tract mucous surfaces and the skin. The U.S. Occupational Safety and Health Act (OSHA) regulations have set the IDLH (in danger of life and health) level of chlorine in air at 1.0 ppm. The NIOSH standard for 15 minutes time weighted average exposure is 0.5 ppm and the 8-hour average is 1.0 ppm.

  Liquid chlorine vaporizes at atmospheric pressure and ambient temperatures. The gas is 2.5 times as heavy as air and will therefore accumulate in low areas. Fans in chlorine feed areas must be at floor level.

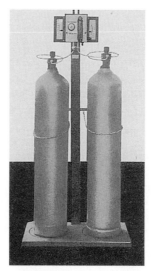

FIGURE 9.20   150 lb. cylinders with chlorinator and scale. (Advance Chlorination)

Liquid chlorine should be stored out of direct sunlight in a way that prevents accidental release. All 150 lb. cylinders should be stored upright and held in place with cables or chains. Ton cylinders should be stored horizontally on saddles. See Figs. 9.20 and 9.21.

Gaseous chlorine will combine quickly with condensed water to form the strong acid HCl, which is extremely corrosive. Therefore all materials which can be exposed to chlorine in a moist atmosphere must be chlorine resistant.

• Calcium hypochlorite and sodium hypochlorite

Since chlorine will readily volatilize from the hypochlorites, the gaseous chlorine precautions must still be observed, especially when removing caps from containers.

FIGURE 9.21   Ton cylinders and scale.

Since neither hypochlorite is completely stable, during long-term storage, the chemical will lose its effectiveness through volatilization. Fifteen percent sodium hypochlorite will lose 50 percent of its strength in 100 days when stored at 75°F. Sodium hypochlorite must be protected from freezing.

Because of the chlorine gas volatility from hypochlorites, the chemicals are very corrosive and material selection is critical. Calcium hypochlorite should be stored in a cool dry location and sodium hypochlorite should also be kept cool to minimize volatilization.

• Sulfur dioxide(SO_2)

Sulfur dioxide is a hazardous, highly corrosive, and irritating gas that causes skin, eye and mucous surface burns. The vapor pressure of SO_2 (35 psi at 70°F) is less than Cl_2 (90 psi at 70°F), therefore it is less volatile and will cause fewer leakage problems.

• Ozone

Ozone is a toxic gas with a maximum recommended 8 hour average exposure concentration of 0.1 ppm. The presence of ozone is easily detected by odor above 0.02 ppm. Exposure to 1000 ppm of ozone for 30 seconds will be mildly irritating, but an equivalent exposure to chlorine may be fatal.

• Bromine

Gaseous bromine is safer to handle than chlorine, but does produce irritating gases. Liquid bromine can cause severe burns.

• Iodine

Iodine can cause severe burns.

• Bromine chloride

Safety precautions for bromine chloride are the same as for chlorine.

• Chlorine dioxide

Sodium chlorite from which chlorine dioxide is generated is highly explosive.

• UV radiation

The operation of a uv system can produce ozone with its safety concerns.

9.6.6 Design

• Chlorine feed rates

Table 9.12 shows recommended minimum chlorine residuals for disinfection based on studies reported by the U.S. Public Health Service.

For domestic wastewater disinfection, dosage rates of chlorine of 8–15 mg/L are usually required. For H_2S corrosion control, the chlorine dosage rate is 2–10 mg/L and for algae control is 1 to 10 mg/L.

The Ten State Standards recommends the following chlorine dosage rates for disinfection of normal domestic sewage as shown in Table 9.13

TABLE 9.12   Recommended Minimum Chlorine Residuals for Bacterial Disinfection of Water

pH Value	Min.Free Available Cl_2 Residual after 10 min. (mg/L)	Min. Combined Available Cl_2 Residual after 10 min. (mg/L)
6.0	0.2	1.0
7.0	0.2	1.5
8.0	0.4	1.8
9.0	0.8	Not Applicable
10.0	0.8	Not Applicable

Note that this Table is based on water and not wastewater. More chlorine residual will be required in wastewater with suspended solids which can contain bacteria that can take up to 45 minutes to disinfect.

TABLE 9.13 Ten State Standard Chlorination Recommendations

Type of Treatment	Dosage (mg/L)
Trickling Filter Plant Effluent	10
Activated Sludge Plant Effluent	8
Tertiary Filtration Effluent	6
Nitrified Effluent	6

- Chlorine and SO_2 Feed Methods

Chlorine can be withdrawn from a 150 lb (68 kg) cylinder at a maximum rate of approximately 40 lb/d (18/kg/d), and from a ton (907 kg) cylinder at 450 lb/d (205 kg/d).

SO_2 can be withdrawn from 150 lb cylinders at 30 lb/d and from ton cylinders at 350 lb/d. Higher rates for chlorine or SO_2 require evaporators. See Figs. 9.20 and 9.21.

Chlorine and SO_2 are typically fed through a chlorinator or sulfonator which consists of a water supply, an educator system and a manometer. The water is carried at pressure past the educator fitter, causing a partial vacuum that allows the evaporated chlorine or SO_2 to volatilize from its liquid storage tank. The gas is mixed with the water and carried through the manometer to measure the flow. The chlorine used is typically measured using a scale by weight loss from the cylinders.

Chlorinators and sulfonators can be cylinder mounted (Fig. 9.22), wall mounted (Fig. 9.23), or freestanding (Fig. 9.24).

FIGURE 9.22 Cylinder-mounted chlorinator. (Advance Chlorination)

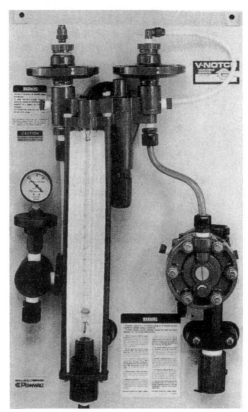

FIGURE 9.23   Wall-mounted chlorinator.

FIGURE 9.24   Freestanding chlorinator.

• Contact time

A disinfectant will work in accordance with Chick's Law as follows:

$$\frac{N_t}{N_0} = e^{-kt} \qquad (9.53)$$

or

$$\text{In} \frac{N_t}{N_0} = -kt \qquad (9.54)$$

where: N_t = number of organisms at time, t
N_0 = number of organisms at time 0
k = constant, time^{-1}
t = contact time

N_t and N_0 are determined by fecal coliform tests and k is determined from the plot of N_t and N_0. Since realistically, organism disinfection efficiency varies with time, the following formula is commonly used:

$$\text{In} \frac{N_t}{N_0} = -kt^m \qquad (9.55)$$

plotted as shown in Fig. 9.25 where m is a constant. If less than 1.0, the disinfection rate decreases with time; if more than 1.0, the disinfection rate increases with time.

Figure 9.26 shows the plot of Eq. (9.55) on log-log paper as per Eq (9.56):

$$\text{Log}\left(-\text{In}\frac{N}{N_0}\right) = \log k + m\log^t \qquad (9.56)$$

It is recommended for adequate reaction time and penetration of total suspended solids, that a detention time of between 30 and 45 minutes must be provided with a plug flow system such as shown in Fig. 9.27.

• Safety design considerations for chlorination

• All chlorination facilities should be placed in a separate room with an outside opening door.

• A fan should be installed at the floor level which automatically turns on when the door to the chlorination room is opened.

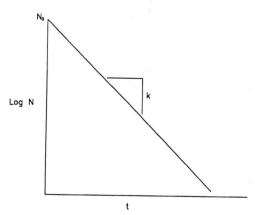

FIGURE 9.25   Organism survival with disinfection.

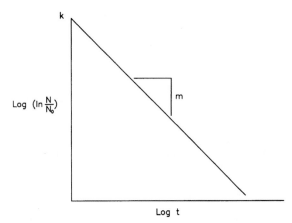

FIGURE 9.26 Organizm survival rate with disinfection varying with time.

- All cylinders should be chained or anchored to prevent movement.
- uv disinfection system

  A uv system is dependent on a constant water level for effective disinfection and to assure the immersion of the molecules. Figure 9.28 shows an example of a uv system with an automatic level controller to assure constant water level.

  Normally about 10 seconds of detention time is needed for adequate disinfection. The uv modules are typically $1/2$ inches in diameter, 4 to 8 feet long, 3 inches center-to-center and 65 watts per module.

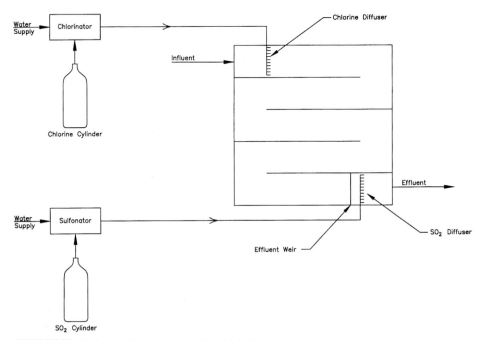

FIGURE 9.27 Typical chlorination/dechlorination disinfection system.

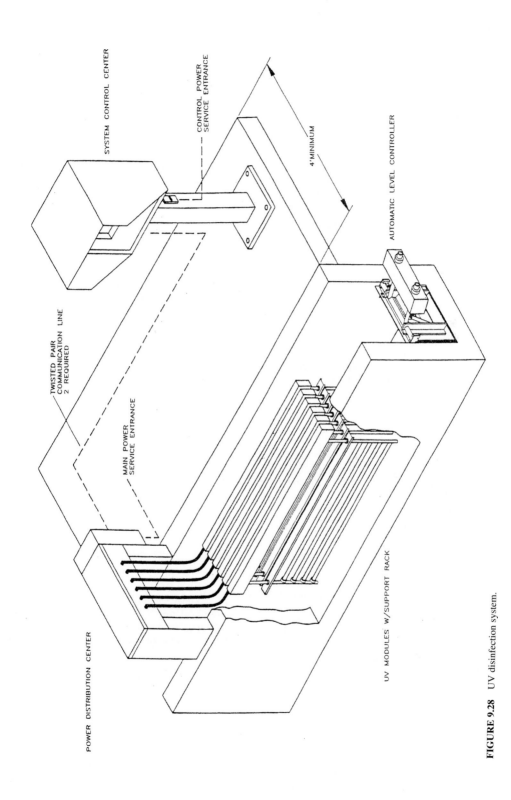

SYSTEM CONTROL CENTER

CONTROL POWER
SERVICE ENTRANCE

4' MINIMUM

AUTOMATIC LEVEL CONTROLLER

TWISTED PAIR
COMMUNICATION LINE
2 REQUIRED

MAIN POWER
SERVICE ENTRANCE

UV MODULES W/SUPPORT RACK

POWER DISTRIBUTION CENTER

FIGURE 9.28 UV disinfection system.

9.7 AIR STRIPPING

9.7.1 Theory

Air stripping can basically be described as the engineered transfer of contaminants from an aqueous phase to the vapor phase—in other words, the transfer of a contaminant from water to air. This can be accomplished using a variety of different methods. More specifically, air stripping treatment technology is the counter current flow of a clean air stream against a "contaminated" wastewater stream. Based upon the contact time and packing surface area, removal of the contaminant from the wastewater stream is accomplished. It is an effective treatment technology for the removal of many organics (especially volatile organics) as well as for ammonia and a few other inorganic chemicals.

Volatiles can be stripped by blowing the gas (usually air or stream) through the liquid, with the liquid flowing through a basin, or by allowing the liquid to flow by gravity down a packed tower, with the stripper gas flowing upward. In either case, the stripping efficiency is proportional to the gas/liquid transfer efficiency, which in turn is proportional to the gas/liquid interface areas. This area is maximized in a basin by maximizing the number of gas bubbles and/or minimizing their size for the volume of gas pumped. In a packed tower, this area is maximized by maximizing the number of liquid droplets and/or minimizing their size for the volume of liquid.

9.7.1.1 General Theory. Air stripping efficiency is governed by the rate of gas transfer of a compound from a liquid. The relationship between the liquid phase concentration and the gas phase concentration can be expressed most readily according to Henry's Law. In its most basic form, Henry's Law expresses the equilibrium partitioning between a liquid phase concentration and a gas phase concentration. Henry's Law can be expressed according to the following two equations:

$$H = \frac{p_{vap}}{C_L} \tag{9.57}$$

where H = Henry's constant (moles/L atm)
p_{vap} = the partial pressure of a pure compound (atm), and
C_{sat} = the saturation concentration of the pure compound in the liquid phase (moles/ or mg/L)

$$H = \frac{C_G}{C_L} \tag{9.58}$$

where H = Henry's constant (dimensionless)
C_G = gas phase concentration (identical units)
C_L = liquid phase concentration (identical units)

H can also be viewed as a measure of a compound's "volatility." The reader should be aware that Henry's constant is commonly reported in three different sets of units. Values for a Henry's constant will be reported as partial pressure (in atmospheres), in mass/volume units (atm m³/mole), or as a dimensionless number (volume liquid/volume gas). The three forms are related according to the following conversions:

$$H_{atm} = H_{dimensionless} \, (4.57T) \tag{9.59}$$

$$H_{atm} = H \left(\frac{atm \cdot m^3}{mole} \right) \times 55600 \tag{9.60}$$

where H_{atm} = Henry's constant, with units in atmospheres
T = temperature, degrees Kelvin

Henry's constant can be experimentally determined by placing a known volume of a pure chemical in an enclosed chamber (pressure 1 atm.), with a known volume of pure water. Once in equilibrium, both p_{vap} and C_{sat} can be measured. Because this experiment is not easy to perform, it is recommended that published Henry's constant data be consulted whenever possible. Values for Henry's constant are presented in a variety of handbooks (Montgomery, Perry), and can be found in a survey of current technical literature. Table 9.14 lists the Henry's constants for several common chemicals of concern. Again, please remember that Henry's constant is not a *measure* of whether or not a chemical can be stripped, but is rather an indication of a chemical's tendency to be removed by stripping.

Generally, a high Henry's constant indicates a high tendency towards stripping, while low constants indicate a low tendency towards stripping. As can be seen in Eq. 9.70, the stripping efficiency of a system is inversely proportional to the absolute pressure of the system. Compounds with low Henry's constants may therefore require high gas-to-liquid ratios, mixing, increased surface area, and or recirculating configurations through a particular treatment process to facilitate adequate removal of the chemicals of concern. These adjustments are described more later in the chapter.

The Stripping Factor (S, sometimes seen as R) is a normalized measure of the tendency of a compound to be removed by air stripping under a given set of conditions. For optimum removal, larger

TABLE 9.14   Henry's Constants (7)

Compound	Henry's constant (atm. mol/m^3)
"Strippable"	$> 10^{-3}$
Carbon disulfide	24.3
Vinyl chloride	2.8
Hexane	1.2
Carbon tetrachloride	3.0×10^{-2}
Tetrachloroethene	1.5×10^{-3}
Methylene chloride	2.2×10^{-3}
o-Xylene	5.3×10^{-3}
Benzene	5.5×10^{-3}
Toluene	6.7×10^{-3}
Ethylbenzene	8.8×10^{-3}
Trichloroethylene	9×10^{-3}
"Not very strippable"	$<10^{-3}$ and $>10^{-5}$
Ammonia	2.9×10^{-4}
Naphthalene	4.6×10^{-4}
1-2 Dichloroethane	9.8×10^{-4}
Ethylene dibromide	7×10^{-4}
"Essentially non-strippable"	$<10^{-5}$
Phenanthrene	3.9×10^{-5}
Acetone	4.0×10^{-5}
Methyl ethyl ketone	4.7×10^{-5}
Pentachlorophenol	2.8×10^{-7}
Diethyl phthalate	8.5×10^{-7}

S is desired (and *S* must be greater than 1 to achieve 100 percent *theoretical* removal) and in practice is typically between 2 and 10.

$$S = H \frac{Q_G}{Q_L} \tag{9.61}$$

where S = stripping factor
H = Henry's constant (dimensionless)
Q_G = gas flow rate (volume/time)
Q_L = liquid flow rate (volume/time)

The mathematical expression Q_G/G_L is also commonly referred to as the Air-to-Water Ratio (A/W). The less volatile a chemical is (lower H), the greater the air- to- water ratio must be (more air flow) to achieve the desired removal efficiency.

9.7.1.2 Factors Affecting Air Stripping. **Temperature:** Increasing temperature results in increased stripping removal of the chemicals of concern from the wastewater. This applies to both increased liquid temperatures as well as to increased gas temperature. Higher temperature is the only reason that steam is used instead of air in stripping.

$$\log H = \frac{-\Delta H^O}{RT} + k \tag{9.62}$$

where ΔH° = heat of enthalpy (kcal/kmol)
R = universal gas constant
T = temperature (degrees Kelvin)
k = constant

These values are available in many handbooks of chemical properties (such as Perry's).

9.7.1.3 Design Equations. The design height of the packing material in a stripping tower can be calculated from the height of a transfer unit (HTU), multiplied by the number of transfer units (NTU). NTU is a function of removal efficiency and the stripping factor.

$$h_T = HTU \times NTU \tag{9.63}$$

HTU characterizes the efficiency of the mass transfer from the liquid phase to the gas phase (as HTU decreases, the efficiency increases), and NTU characterizes the difficulty of removing the chemical from the liquid (as NTU increases, the efficiency increases). The terms HTU and NTU are defined as follows:

$$HTU = \frac{Q_L}{(K_{La})A} \tag{9.64}$$

where Q_L = liquid flow rate
K_{La} = overall mass transfer coefficient (time^{-1})
A = cross-sectional area of the tower

K_{La}, the overall mass transfer coefficient, generally cannot be theoretically determined for a transfer "system," but must be experimentally determined. K_{La} can be experimentally determined according to the following equation (integral method, non-linear regression):

$$C_L = C_{Ls}(1 - e^{(K_{La})^t}) - C_{Loe}{}^{(K_{La})^t} \tag{9.65}$$

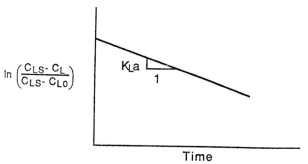

FIGURE 9.29   Integral method.

where C_L = liquid phase concentration at a given time
C_{Ls} = liquid phase, saturation concentration
t = time
C_{Lo} = liquid phase, initial concentration (time = 0)

This equation is expressed graphically in Figs. 9.29 and 9.30. The above equation assumes that C_{Ls} is known for the system being studied. If C_{Ls} is not known, or if instrumentation is available to allow rapid and frequent (essentially continuous) measurement of C_L, the differential method may be easier. The differential method is graphically shown in Fig. 9.31.

It should be noted that some transfer systems have been well studied, and that sufficient data does exist in literature so that K_{La}, for a system being investigated, can be correlated to the existing data. K_{La} is affected by gas flow rate, size of air bubbles (for diffuser systems), mixing intensity, physical properties of chemical being removed, and water quality. Because of these many variables, air stripping is as dependent upon engineering factors as much as upon chemical factors.

In order to facilitate the transfer of the chemical from the liquid phase to the gas phase, it is best to have HTU be as small as possible. As noted in the discussion above, HTU is impacted by several factors, including the packing material. It can be used to compare the removal efficiencies of different packing materials, by holding all other parameters (liquid loading rate, influent concentration, air flow, etc.) equal.

$$NTU = \frac{S}{S-1} \ln\left[\frac{1}{S} + \left(1 - \frac{1}{S}\right)\frac{C_{LI}}{C_{LE}}\right] \tag{9.66}$$

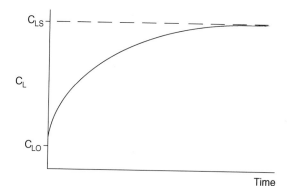

FIGURE 9.30   Integral method with non-linear regression.

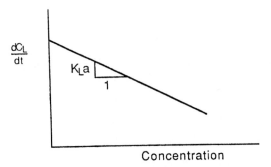

FIGURE 9.31 Differential method.

where S = stripping factor, defined above in Eq. (8.74)
C_{LI} = Influent chemical concentration
C_{LE} = Effluent chemical concentration

Note, this equation assumes that the influent gas concentration is 0, which is typically the case.

The preceding calculations are somewhat involved, and require a significant amount of experimentation to determine several of the required values. Fortunately, a number of computer programs exist to perform these calculations.

9.7.2 Applicable Chemicals

Air stripping may be effectively applied to the treatment of wastewater containing many volatile organic compounds, and is often used for ammonia removal. Table 9.14 lists organics effectively removed by air stripping. In general, chemicals with a Henry's constant greater than 10^{-3} are considered readily strippable. Chemicals with a Henry's constant less than 10^{-5} are considered non-strippable. Chemicals with a Henry's constant between these two values may be stripped under the appropriate conditions. However, please remember that Henry's constant is not a measure of the stripping efficiency of a compound. Even compounds considered non-strippable may be removed by stripping under the proper conditions (higher temperature, A/W, etc.) (3, 4, 6, 8).

9.7.3 Ammonia Stripping Design Example

The author designed the air stripping system shown in Figure 9.32 to remove ammonia from a wastewater. In this case, an ammonium compound was used in a process to solubilize copper. In order to precipitate the copper to meet permit limits, the ammonium compound had to be removed. A pilot study, consisting of a 55 gallon drum, a steam line, an air line, and pH control chemicals was installed to test several temperatures, air rates, and pH levels. The study showed that a temperature of 105°F was optimal. Even the lowest air rate in this test was sufficient. A pH of 11.5 was optimal.

The full-scale design was based on the results of this pilot study and was successful in removing the ammonium compound and upon precipitation, the copper. It should be noted that the gaseous ammonia (NH_3) to liquid ammonium (NH_4^+) ratio is depended on pH, with all NH_3 existing at high pH levels (above about 10.5) and all NH_4 existing at low pH levels.

9.7.4 Treatability Studies

It is rare that a wastewater stream performs "ideally," primarily because most wastewaters do not contain pure concentrations of a compound. Frequently, a wastewater will contain multiple contaminants

FIGURE 9.32 Air stripping system.

that will interact in complex relationships. Other waste stream characteristics such as flow rate, pH, solids content, and temperature can affect air stripping. It is therefore strongly recommended that prior to deigning an air stripper, the designer should conduct a bench-scale treatability study in the lab. The goal of the study is to generate the liquid and gas loading rates and the column removal efficiency.

Following is an example of a small-scale air stripping treatability study.

A computer modeling package, *AirStrip Plus,* is used to determine initial operating parameters for the treatability study. *AirStrip Plus,* released by Choice Computing in 1995 as an upgrade to *AirStrip,* was developed to assist the designing of air stripping equipment. A database of chemicals and physical constants, including Henry's Law constants, is included with the software.

The *AirStrip Plus* program is used to develop a range of air-to-water ratios as a starting point for the treatability study. The initial calculations are performed to determine an acceptable air-to-water ratio based upon the influent and effluent concentrations (a function of column height, diameter, and the packing material). It is recognized that this design is somewhat conservative, due to wall effects, lower wetted surface area, and the size of the packing material.

For the treatability study, the stripping column consists of a 1-inch diameter clear PVC column 19 inches high that is packed with 16 inches of ceramic saddles and contains 8 circular baffles protruding 1/16 of an inch into the column and spaced at 2 inch intervals. The circular baffles are placed in the column to minimize channeling along the sides of the column. The wastewater flow rate into the column is maintained at a constant flow rate for each treatability test run. The wastewater is supplied to the top of the stripping tower by gravity flow, with the flow rate being adjusted according to an adjustable flow meter. Air is supplied to the bottom of the column with an aquarium air pump, or similar low flow air pump. Because the air is supplied at a constant flow rate from the pump, the water flow rate is adjusted to vary the air- to -water ratio (A/W). Typical treatability studies conducted utilized liquid flow rates that varied between 5 and 30 ml/min. As mentioned above, the wastewater flows into the top of the column by gravity, and discharges out the bottom. The direction

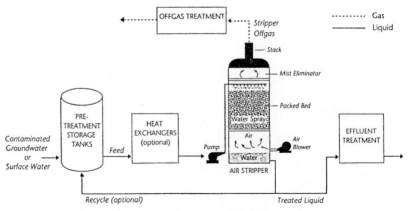

FIGURE 9.33   Schematic of air stripping system.

of the air flow is counter current to the water—into the bottom of the column and out the top. A typical stripping column used in treatability studies is shown in the schematic shown in Fig. 9.33.

9.7.5   Design Process

In reality, the normal design process involves contacting one or more manufacturers of air stripping equipment, providing them with sufficient information regarding liquid loading rates, wastewater characteristics, and effluent standards, and then the manufacturer responds back with a proposed design. This approach is admittedly a "black box" approach. Because of this, we are providing general design approaches for three types of column design, so that the reader is familiar with what takes place in the manufacturer design process.

There are three types of columns: packed, low profile, and high efficiency stripping towers.

9.7.5.1   Packed Column Design.   A packed column stripping tower consists of a cylindrical column, a support plate, a liquid distribution system at the head of the column, and an air supply at the base of the column (for countercurrent flow).

Most design manufacturers have developed experimentally derived removal efficiency curves that form the basis of packed tower column design. By providing flow rates (air and liquid), loading rates, effluent concentrations (air and liquid), and packing materials, the empirical curves are used to size the columns. Many manufacturers have online RFQ submittal forms (such as Delta Cooling, at *www.deltacooling.com*). Treatability and pilot studies are used for chemicals for which design information does not exist. Generally, there are 4 decision variables that must by selected. These include: the height of the packing material (h_T); cross-sectional area of the column (A); stripping factor (S); and the packing material.

Figure 9.34 shows a steam stripper at a chemical plant. Figure 9.35 shows a 300 cfm air stripper for toluene and benzene at a chemical plant.

An example of the general design approach used to determine the empirical curves used by the manufacturers is as follows:

1. Select packing material
2. Select Q_G (or S)
3. Select acceptable ΔP (a function of the packing material and gas loading rate)
4. Calculate A
5. Calculate Q_L/A (the liquid loading rate)

FIGURE 9.34 Steam stripper at a chemical plant.

6. Calculate K_{La}
7. Calculate h_T

This is an iterative process. For most strippable chemicals, A/W should be on the order of 40:1. Packed towers are generally design with a 10:1 height-to-diameter ratio.

9.7.5.2 Low Profile Tower Design. Low profile air strippers consist of a number of rectangular trays (1 to 5) that operate based upon the plug flow of the influent wastewater along a long, narrow channel, which is subjected to countercurrent air flow. Low profile strippers offer a number of benefits over other stripping column designs. Specifically, the more compact design (typically less than 10 feet in height) allows the stripping unit to be installed indoors, protecting the unit from freezing or other climate related problems. The relatively low height also allows easier access to the interior of the unit for cleaning and other maintenance. Typical operating rates are 0.5 to 550 gpm liquid flow rate and 150 to 2400 cfm air flow rate.

As with the design of packed column air strippers, the design of low profile air strippers is based upon empirically derived performance curves. By providing flow rates (air and liquid), temperature

FIGURE 9.35 Air stripper for toulene and benzene.

(liquid and air), effluent concentrations (air and liquid), and other water quality parameters, the performance curves are used to determine the number of trays required for the desired removal efficiency. Many manufacturers have online RFQ submittal forms (such as North East Environmental Products, manufacturer of the Shallow Tray® low profile stripper, at *www.neepsystems.com*). Figures 9.36 and 9.37 show examples of two low profile Shallow Tray air strippers and Figure 9.38 shows an air stripper for methylene chloride and tetrachloroethene installed on a groundwater remediation system.

9.7.5.3 High Efficiency Air Strippers.   In addition to the traditional packed tower and low profile air strippers, alternative stripping technologies can use high pressure or vacuum conditions (commonly referred to as "high efficiency air strippers"). Most commercial stripping systems operate at atmospheric to slightly positive pressure conditions as in the case of a gas bubbled through a water column. The GDT™ and VacGDT™ processes (developed commercially by GDT Water Process Corp., Phoenix, AZ *www.got-h20.com*) are specific applications of pressure and vacuum systems.

In the GDT™ process, detailed in Fig. 9.39, with an example shown in Fig. 9.40, the liquid flow to be treated enters an injector under pressure. This pressure is typically 30 to 60 psig, dependent on well pump and piping pressure drop. The available pressure sets the motive flow rate through the

FIGURE 9.36 Shallow Tray Model 1300. *Courtesy of Northeast Environmental Products.*

process and establishes suction capacity for the stripping gas to be used. The injectors aspirate the stripping gas under vacuum conditions, where the stripping gas expands into the liquid, resulting in violent small bubble gas and causing significant liquid mixing. The dynamic mixing that occurs at the injector allows volatile partitioning and rapid equilibrium attainment in the two-phase flow leaving the injector. The flow then travels to a degas separator for rapid volatilized entrained gas removal.

As the entrained gas/water mixture enters the degassing separator, it is accelerated to a velocity that exerts 4 to 10 times gravity in a lateral force creating a water film at the separator wall and a gas vortex at the central, gas extraction core. This journey of a few seconds has the ability to extract 98 percent of the entrained gases within the water stream.

In the VacGDT™ process, operation is under negative pressure conditions to increase partitioning of volatile compounds. The VacGDT process schematic is detailed in Fig. 9.41 with a typical unit shown in Fig. 9.42. The system injects the liquid at a pressure as low as 5 to 15 psig, equivalent to the static head of most atmospheric storage tanks.

A degas relief valve controls the flow of separated gas under the vacuum created by a vacuum pump designed for the specific gas-to-liquid ratio needed to attain the desired volatile removal rate.

FIGURE 9.37 *Shallow Tray Model 41200.* Courtesy of Northease Environmental Products.

FIGURE 9.38   Photograph of air stripper.

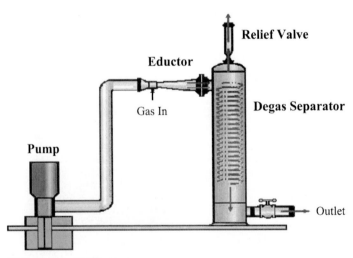

FIGURE 9.39   GDT™ stripping process.

FIGURE 9.40 Example of high efficienc air stripping application (GDT Water Process Corp).

A pump at the outlet of degas separator boosts the pressure from the vacuum condition within the process to the desired pressure for further treatment or distribution.

9.7.5.4 Stripping Tower Media. The key element in air stripping performance design is the packing material selected for use. The different types of commercially available packing are designed to maximize the wetted surface area (for increased air to water contact time), while minimizing pressure drop through the media. This increases the contact with the air stream, and therefore maximizes the removal of the contaminant from the wastewater to air stream. There is therefore no single ideal media for all applications. Packing media is most often ceramic, plastic, or metal alloys. Many types of media exist, with varying impacts on cost, pressure drop, wetted surface, and corrosion resistance. The packing media is most often "dumped" into a column resulting in a random media arrangement, although "arranged" sections of media may be used if necessary. Common packing media shapes include several vendor specific ring shapes, berl saddles, pall rings, tellerettes, and other assorted designs. In the past few years, several manufacturers have developed spiral wound or tube-type stripping media using the thin film theory of gas/liquid transfer, rather than maximizing gas/liquid surface area by maximizing the number or minimizing the size of the droplets. This media provides the benefit of a higher air flow rate, which can be useful for compounds with low Henry's constants (such as semi-volatile organics and ammonia). Perry's Chemical Engineers Handbook is a good source of information for comparing both structural and random packing materials. (3, 2, 1, 5)

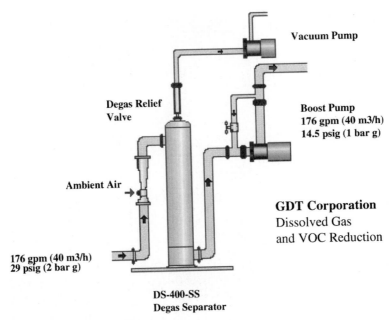

Vacuum Pump

Degas Relief Valve

Boost Pump
176 gpm (40 m3/h)
14.5 psig (1 bar g)

Ambient Air

GDT Corporation
Dissolved Gas
and VOC Reduction

176 gpm (40 m3/h)
29 psig (2 bar g)

DS-400-SS
Degas Separator

FIGURE 9.41 The VacGD™ process.

9.7.6 Design Process—System Design

In reality, the design process for air stripping is a combination of preliminary estimates, bench scale treatability studies, and consultation with stripping tower vendors (either packed tower, low profile, or high efficiency; see section 9.7.5). Preliminary estimates are typically based upon the designer's experiences, and upon general design "rules-of-thumb." However, information is presented below on the equations behind the design, so that, as the designer, you can have an overall grasp of the theory.

9.7.6.1 Preliminary Estimates. Rules of thumb for design parameters are given in Table 9.15.

9.7.7 Design Hints

- Add interior "donut" baffles about every five feet of tank wall to break up wall short circuiting.
- Stripper packing can be of any appropriate shape or material which will break up the wastewater into small droplets.

TABLE 9.15 Design Parameters (Ammonia Removal)

Parameter	VOC removal[1]	Ammonia removal[2]
Liquid loading rate	15–45 gal/min/ft^2	1–2 gal/min/ft^2
A/W	10–50	
Airflow rate		300–500 ft^3/gal
Packing depth	10–20 ft	
Wastewater pH		10.8–11.5
Air pressure drop	0.061–0.068 in. H$_2$O/ft^2	0.015–0.019 in. H$_2$O/ft^2

FIGURE 9.42   Example of vacuum enhanced high efficiency air stripping (GDT Water Process Corp).

• The stripper column may need to be insulated to minimize heat loss. An acid or oxidant flush in system can be installed if scaling, bacterial, or algal growth occurs.

9.8   *ADSORPTION OF AQUEOUS COMPOUNDS*

One of the traditional techniques for controlling the release of organic compounds is adsorption. Adsorption is primarily used for the reduction of organic discharges, though metals removal has been demonstrated in some wastewater applications. Adsorption is a mass transfer process that can generally be defined as the accumulation of material at the interface between two phases. For example, a contaminant in a fluid passing through a bed of carbon is transferred from the liquid phase to the surface of the carbon. More generally, chemicals in the liquid phase preferentially accumulate on an unsaturated solid surface causing the chemical to be removed from the liquid phase. The material upon which the chemical is adsorbed (e.g. carbon) is known as the adsorbent. The material that is adsorbed (typically the contaminant) is known as the adsorbate.

9.8.1 Adsorption Theory

Adsorption is a thermodynamic system in which the various components are striving for equilibrium. The process of adsorption occurs in both steady-state and unsteady-state conditions. The primary force driving the interaction between the adsorbate and the adsorbent is the electrostatic attraction and repulsion between molecules of the adsorbate and the adsorbent. These driving forces can be either physical or chemical.

9.8.1.1 Physical Adsorption. Physical adsorption is a result of intermolecular forces that interact between the adsorbate and the adsorbent. These physical electrostatic forces include the van der Waals force, consisting of weak attraction and repulsion through dipole-dipole interactions and dispersing interactions, and hydrogen bonding. Dipole-dipole interactions are the result of polar compounds orienting themselves so that their charges result in a lower combined free energy. Dispersing interactions are the result of attractive forces between electrons and nuclei of molecular systems. If the molecules come too close to each other, repulsive forces can push the molecules apart. Hydrogen bonding is a special case of dipole-dipole interaction in which the hydrogen atom in a molecule has a partial positive charge, attracting another atom or molecule with a partial negative charge. For liquid phase systems, the van der Waals force is the primary physical force driving adsorption. Physical adsorption is a readily reversible reaction and includes both mono- and multilayer coverage. Because physical adsorption does not involve the sharing of electrons, it generally has a low adsorption energy, and is not site specific. The heat of adsorption for the reaction is on the order of 40 Btu/lb. per mole of the adsorbate. When the intermolecular forces between a chemical molecule in a liquid stream and a solid (the adsorbent) are greater than the forces between the molecules of the liquid stream, the chemical is adsorbed onto the adsorbent surface.

9.8.1.2 Chemisorption. Chemical adsorption (chemisorption), like physical adsorption, is also based upon electrostatic forces. The mechanisms of chemical adsorption are similar to those of physical adsorption, yet are often stronger (approaching the adsorption energies of chemical bonds). Chemical adsorption is produced by the transfer of electrons and the formation of chemical bonds between the adsorbate and the adsorbent. It may be an irreversible reaction and have high adsorption energies. The heat of adsorption is significantly greater than for physical adsorption, ranging from 80 to 400 Btu/lb. mole. It is not unusual for the adsorbate to have chemically changed due to the reaction. Chemical adsorption involves only monolayer coverage, and is a site specific reaction, occurring at specific functional group locations. Functional groups are distinctive arrangements of atoms in organic compounds that give that compound its specific chemical and physical properties.

9.8.1.3 Specific Adsorption. Some reactions have adsorption energies that are higher than physical adsorption, but lower than chemical adsorption. Though not common, these interactions are referred to as "Specific Adsorption." Specific adsorption involves an interaction with a specific functional group on the adsorbent surface, but it does not result in the formation of a true chemical bond. The primary indication of specific adsorption is that the adsorption energies are between chemical (strong) and physical (weak).

9.8.1.4 Factors Affecting Adsorptive Capacity. As a generalization, the adsorptive capacity of a given adsorbent material is proportional to the surface area available. However, there are a number of other factors that affect adsorption. In selecting the appropriate adsorbent material for a particular application, several factors must be considered and balanced. These design variables are briefly described in the following sections.

- **Polarity**
  For activated carbon, non-polar chemicals are preferentially adsorbed because activated carbon is itself non-polar. Polarity is influenced by both physical and chemical forces such as dipole-dipole interactions, dispersing interactions, and hydrogen bonding. Polarity can be a significant factor with some of the adsorbent resins.

• **Charge**
Adsorption of charged (ionized) chemicals is less significant than the adsorption of uncharged particles.

• **Molecular weight**
In general, larger molecules are better adsorbed, unless the size of the molecule is greater than the diameter of micropores within the carbon particle.

• **Temperature**
The temperature of a system is especially important when considering the adsorption of volatile organic chemicals. Adsorption capacity is inversely proportional to the temperature, increasing as temperature decreases. This principle is based upon Gibbs free energy. Gibbs free energy (G) is a measure of the spontaneous change in a system. The change in the Gibbs free energy of a system (ΔG), at constant temperature and pressure, is one measure of the "adsorbability" of a compound. When ΔG is <0, a system reacts spontaneously and adsorption occurs. A compound that does not "react," or adsorb, at a given temperature (ΔG > 0), may adsorb at a lower temperature (if G < 0). However, as the temperature decreases, the kinetic reaction rates also decrease. In practice, temperature does not play a significant role in adsorption efficiency for most waste streams.
Adsorption is an exothermic reaction. As the zone of adsorption moves through an adsorbent bed, the temperature of the bed increases and heat is released to the liquid. Likewise, when the liquid leaves the area of adsorption activity, the liquid transfers the heat back to the bed. For certain chemicals and adsorbents, the heat transfer can be significant. In physical adsorption situations, the amount of heat released during the adsorption process is approximately equal to the latent heat of condensation of the adsorbate and the heat of wetting of the adsorbent by the adsorbate. During chemical adsorption, the heat released is approximately equal to the heat of chemical reaction. The temperature differential during the operation of the bed can be estimated as follows:

$$\Delta T = \frac{6.1}{(C_p/C_i) \times 10^5 + 0.51(C_A/q_e)} \tag{9.67}$$

where ΔT = temperature rise (°F)
  C_p = heat capacity of air (Btu/ft^3 °F)
  C_i = influent concentration of the adsorbate (ppm)
  C_A = heat capacity of the adsorbent (Btu/ft^3 °F)
  q_e = equilibrium loading of the adsorbent (lbs./100 lbs.)

• **Surface area**
Physical characteristics are very important in selecting adsorbent materials. One of the major considerations when selecting the adsorbent is the surface area of the material. Adsorbent materials used for wastewater treatment are highly porous. The majority of the surface area of an activated carbon particle is provided by the pore structure. Another surface area property affecting adsorption is the distribution of pore size diameters. The larger adsorbate molecules can only adsorb in the larger diameter pores. Two carbon particles with different pore size distributions will exhibit different adsorption performance. Table 9. 16 lists typical surface areas of different activated carbons. Determining the surface area of a unit volume of activated carbon using iodine adsorption is typically accomplished by the Brunauer-Emmett-Teller (BET) method.

A characteristic related to particle surface area is adsorbent particle diameter. The particle diameter influences the rate of adsorption. Smaller diameter particles such as powdered activated carbon (PAC) have a shorter diffusion path resulting in more rapid adsorption. The mass transfer rate increases in inverse proportion to the particle diameter (d3/2). The internal adsorption rate increases in inverse proportion to d2.

The use of smaller diameter adsorbent particles also results in a higher pressure drop across the adsorbate bed than exhibited with larger diameter particles. As the Reynolds number (the measure

TABLE 9.16 Typical Surface Areas of Activated Carbon

Typical surface areas of activated carbon		
Carbon Base	Base material	Surface area (m^2/g)
PCC SGL	Bituminous coal	1000–1200
PCC BPL	Bituminous coal	1000–1200
PCC RB	Bituminous coal	1200–1400
PCC GW	Bituminous coal	800–1000
Calgon Filtrasorb 300		950–1050
Calgon Filtrasorb 400		1000–1200
Columbia CXA/SXA	Coconut shell	1100–1300
Columbia AC	Coconut shell	1200–1400
Columbia G	Coconut shell	1100–1150
Darco S51	Lignite	500–550
Darco G60	Lignite	750–800
Darco KB	Wood	950–1000
Hydro Darco	Lignite	550–600
Nuchar Aqua	Pulp mill residue	550–650
Nuchar C	Pulp mill residue	1050–1100
Nuchar (various)	Pulp mill residue	300–1400
Norit (various)	Wood	700–1400

Source: Calgon Carbon Corporation, "Basic Concepts of Adsorption on Activated Carbon," p. 2.

of laminar or turbulent fluid flow) for the waste stream increases (becomes more turbulent), the pressure drop across the bed increases. Minimal pressure drop occurs with adsorbent particles that are uniform in size and are spherical.

- **Pore size distribution**
  Pore size distribution is a measurement of the percent of the space of a particle occupied by micropores (pore diameter <2 Å), mesopores (pore diameter >20 Å and <500 Å), and macropores (pore diameter >500 Å). Figure 9.43 illustrates a portion of a typical particle cross section. A molecule cannot penetrate into a pore smaller than a given minimum diameter (depending upon the size of the molecule). This process screens out larger molecules, and allows smaller molecules to penetrate further into the adsorbent particle or to adsorb in the smaller diameter pores.

- **Other factors**
  In addition to the above-mentioned parameters which impact adsorption, other factors that may be significant to the design of an adsorption control system include resistance to flow, as determined by the size and shape of the adsorbent particles, as well as the adsorbent bed depth, waste stream velocity, and desired removal efficiency.

9.8.2 Applicable Chemicals

Activated carbon may be effectively applied to the treatment of wastewater containing many organic and some inorganic contaminants. Table 9.17 lists organics effectively adsorbed by activated carbon. Alcohols, amines, glycols, ketones, many acids, and low molecular weight organics are not effectively adsorbed by activated carbon. Appendix G contains carbon isotherms, adsorptive capacities and carbon doses for various chemicals as determined by experimentation as explained below.

9.8.3 Isotherm Generation

The adsorption of a compound onto any adsorbent can be described mathematically by an "adsorption isotherm." Simplified, the quantity of adsorbate that can be bound up on a particular adsorbent is a

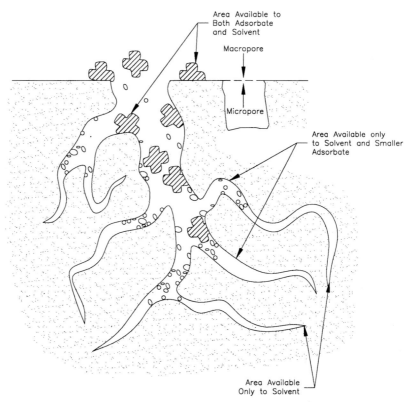

FIGURE 9.43 Concept of molecular screening in micropores. (Cooper and Alley)

function of the concentration of the adsorbate and the temperature. Isotherms are a measure of the capacity of an adsorbent as a function of the concentration of the adsorbate in a feed stream. The key to determining the capacity and life of an adsorbent is the development of adsorption isotherms. Treatability studies provide the most accurate means of determining the specific adsorption properties of a waste stream for a given adsorbent. Experimental data can be determined on a volumetric basis, gravimetrically, or chromatographically. According to the volumetric method, an adsorbent is exposed to several adsorbate concentrations, and a mass balance is used to determine the adsorbed concentration. The volumetric method is the most common method used because of its low cost and straightforward procedures. A typical volumetric experimental procedure is described below in Sec. 8.13.3.3.

In its most general form, the equilibrium concentration of an adsorbate can be expressed as follows:

$$q^* = f(C_o, T) \tag{8.68}$$

where q^* = the equilibrium concentration in moles/unit weight of adsorbent
$\quad C_o$ = Influent concentration
$\quad T$ = temperature adsorbate in a feed stream

Experimental data derived from adsorption isotherm experiments will typically fit one of five different curves. The five types of curves are shown in Fig. 9.44, where n_M is defined as either the monolayer or maximum loading, C is the adsorbate concentration, and C_{sat} is the saturation concentration of the adsorbent. Type I adsorption is convex upward throughout the curve, and is considered

TABLE 9.17 Classes of Organic Compounds Amenable to Adsorption of Activated Carbon

Classes of organic compounds amenable to adsorption of activated carbon
Aromatic Solvents
Benzene, Toluene, Xylene, Nitrobenzenes
Polynuclear Aromatics
Naphthalene, Biphenyls, Acenaphthalene, Benzopyrenes
Chlorinated Aromatics
Chlorobenzene, Toxaphene, DDT, Aldrin, PCBs
Phenolics
Phenol, Cresol, Resorcinol
High Molecular Weight Aliphatic Amines and Aromatic Amines
Aniline, Toluene Diamine
Surfactants
Alkyl Benzene Sulfonates
Soluble Organic Dyes
Methylene Blue, Textile Dyes
Fuels
Gasoline, Kerosene, Oil
Chlorinated Solvents
Carbon Tetrachloride, Perchlorethylene
Aliphatic and Aromatic Acids
Tar Acids, Benzoic Acids

Source: Calgon Carbon Corporation, Speitel.

favorable for adsorption. Type II adsorption is concave upward throughout the curve and is considered unfavorable for adsorption. Types III, IV, and V adsorption curves have multiple inflection points and therefore both concave and convex portions. Type IV and V curves can exhibit hysteresis, a condition where desorption occurs along a different isotherm than adsorption. Hysteresis only becomes an item of concern to the design when the system in question will have built-in regenerative capability. A sixth condition that may be observed is a linear adsorption curve, as modeled by the Langmuir equation, described below in Sec. 9.8.3.1.

Based upon the plotted data, equations are fitted to the data to determine the best predictive model. Once the isotherm that most accurately fits the data has been determined, a user may accurately predict, based upon the concentration of the adsorbate in the wastewater, the amount of adsorbent requires to "capture" a given amount of the adsorbate.

9.8.3.1 Single Component Isotherms.

Several isotherms have been developed for use in modeling the adsorption equilibrium. While numerous adsorption isotherms exist and are useful, the two most common equations used to describe the equilibrium between a surface (adsorbent) and a chemical in solution (adsorbate) are the Langmuir and Freundlich adsorption equations. A number of other equations have been developed either as extensions of the equations mentioned above, or as new adsorption equations altogether. Among the other single component equations encountered, although not presented here, are the BDDT (Brunauer, Deming, Deming, and Teller), the Langmuir-Freundlich, Unilan, Toth, and three "Dubinin" equations. The more complex equations will typically more accurately model the adsorption data observed, though they necessitate the collection of more data, which can be an expensive and time-consuming process. These other methods all concentrate on various aspects of adsorption isotherms that exhibit different levels of accuracy over various ranges of data. In choosing an equation, a designer should tend towards the simplest equation that can account for the "non-ideal" conditions observed in the data. In many cases, either the Langmuir or Freundlich equations are sufficient. For additional information on these other isotherm equations, see Knaebel, 1995.

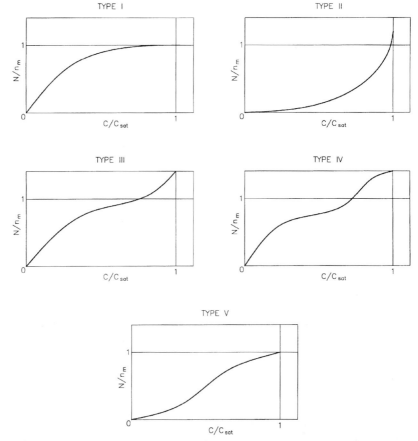

FIGURE 9.44 Types of adsorption isotherms. (Perry's Chemical Engineer's Handbook)

- **Langmuir equation**

  The Langmuir equation treats the interaction between the adsorbent and the adsorbate as a linear, reversible, monolayer chemical reaction. Developed by I. Langmuir in 1915, this equation is a relatively straightforward model that assumes that the adsorbent surface is completely homogeneous, that each adsorbent "site" can bind a maximum of one adsorbate molecule, and that there are no interactions between molecules of the adsorbate. The Langmuir adsorption model can be expressed in the following form:

$$q_e = \frac{Q_o K_L C_e}{1 + K_L C_e} \tag{9.69}$$

where q_e = equilibrium loading on the adsorbent (mg/g)
Q_o = ultimate adsorption capacity of the adsorbent (mg/g)
K_L = relative energy of adsorption, also known as the equilibrium constant—typically empirically determined (L/mg), also referred to as b
C_e = equilibrium concentration in the waste stream (mg/L)

Advantages of the Langmuir equation include simplicity and the applicability to a wide range of data. Limitations to the model include the monolayer assumption, the reversibility of bonding, and the constant uptake rates.

- **Freundlich equation**

The Freundlich equation is an empirically derived logarithmic model that attempts to factor in the effects of various adsorption energy levels. The model assumes that the number of sites associated with a particular free energy of adsorption decrease exponentially as the free energy level increases. The Freundlich adsorption model can be expressed in the following form:

$$q_e = K_F C_e^{1/n} \qquad (9.70)$$

where q_e = equilibrium loading on the adsorbent (mg/g)
K_F = adsorption capacity at unit concentration $(mg/g)(L/mg)^{1/n}$
$1/n$ = adsorption intensity β typically empirically determined, and sometimes noted as — (dimensionless)
C_e = equilibrium concentration in the waste stream (mg/L)

The equation fits a straight line when plotted on a log-log basis. The Freundlich equation can then be written in the following form:

$$\log q_e = \log K_F + (1/n)\log C_e \qquad (9.71)$$

and used to experimentally determine K_F and $1/n$. The Freundlich equation is commonly used by environmental engineers for empirical data, and is helpful in quickly providing some general information about the tendency of a compound to be adsorbed. For irreversibly adsorbed chemicals, $1/n$ (the slope of the line) is zero. For favorably adsorbed chemicals, $1/n$ is between 0 and 1. For unfavorably adsorbed chemicals, $1/n$ is greater than 1. For chemicals that are not adsorbed, $1/n$ approaches infinity (or is very large). In many cases, intermediate conditions between the Freundlich and Langmuir isotherms are observed, requiring the designer to use either a different isotherm equation or determine which of the two equations best fits the applicable data range.

9.8.3.2   Isotherm Prediction.   Because isotherm data is not always readily available, several predictive models have been developed to allow estimation of adsorptive capacities without experimental data. The most common approaches utilize quantitative structure-activity relationship (QSAR) techniques, such as the Dubinin-Polyani Adsorption Potential theory, and the Polyani-Reducskevich correlation theory. The QSAR techniques are based upon two general assumptions: knowing the isotherm of one chemical on an adsorbent, the isotherm of any other (similar) chemical can be calculated; and knowing the isotherm of a chemical at a given temperature, the isotherm at any other temperature can be calculated.

When isotherm data is available, several equations should be fit to empirical data and the optimum fit determined statistically. Most computer spreadsheet programs have analytical packages that include linear and non-linear regression methods. Comparison of the sample correlation coefficient between different adsorption equations will give the designer a good indication of the model that most accurately predicts the adsorption isotherm. As mentioned above, other equations may account better for data observed over various ranges of concentrations. The designer should be sure that the isotherm equation used accurately models the adsorption over the range of conditions applicable.

9.8.3.3   Isotherm Generation.   As mentioned above, isotherms are a measure of the capacity of an adsorbent as a function of the concentration of the adsorbate in a feed stream. The key to determining the capacity and life of an adsorbent is the development of adsorption isotherms. Treatability studies provide the most accurate means of determining the specific adsorption properties of a waste stream for a given adsorbent. Experimental data can be determined on a volumetric basis, gravimetrically, or chromatographically. According to the volumetric method, an adsorbent is exposed to several adsorbate concentrations, and a mass balance is used to determine the adsorbed concentration. The volumetric method is the most common method used because of its low cost and straightforward procedures. A typical volumetric experimental procedure based on information presented by Speitel, is described below.

1. Determine preliminary isotherm parameters for design of the experiment (may be an estimate).

2. Using the following equation, select a range of 10 to 15 values of C_e per isotherm:

$$M_i = \frac{V(C_O - C_e)}{K_F C_e^{1/n}}$$ (9.72)

where M_i = mass of activated carbon
$\quad\quad V$ = volume of bottle
$\quad\quad C_o$ = initial concentration
$\quad\quad C_e$ = final (equilibrium) concentration
$\quad\quad K_F$ = Freundlich constant
$\quad\quad n$ = Freundlich constant

3. Vary C_o and V to determine the best range of values of M_i.

4. Set up 10 to 15 air tight bottles of a constant volume (V) with the different masses of activated carbon (M_i) determined in Step 3.

5. Add the same initial chemical concentration (C_o) to each bottle.

6. Gently shake the bottles and wait for equilibrium to be established. This will typically take 4 to 8 hours, although it could take longer.

7. Measure equilibrium concentration in each bottle (C_e).

8. Calculate activated carbon loading (q_e) in each bottle using the following equation:

$$q_e = \frac{V(C_O - C_e)}{M_i}$$ (9.73)

9. Plot q_e versus C_e and determine the isotherm constants. Analysis of data for the Freundlich isotherm will need to be either plotted on log-log scale (preferable) or determined using nonlinear regression techniques.

In preparing and conducting isotherm studies, several cautions should be noted: in performing the isotherm experiments, one should avoid activated carbon dosages of less than 50 mg due to statistical errors in calculating q_e. Measurements near the analytical detection limits may also contribute to errors in calculating q_e. Also, make sure that the isotherm is developed for the concentration range of the particular application, as the isotherm parameters may exhibit a concentration dependence.

9.8.3.4 Multi-Component Isotherms.   It is always important to consider that a waste stream with multiple components may exhibit the preferential adsorption of one compound over another, even to the point of displacing previously adsorbed compounds. Individual chemicals do in fact compete with each other for adsorption sites. A common model for competitive adsorption is the Ideal Adsorbed Solution (IAS) Theory. The IAS model uses the isotherms of each of the single components to predict the competition, and therefore the equilibrium concentrations, among the chemicals. The general form of the IAS model is as follows:

$$C_{ei} = \frac{q_i}{\sum\limits_{j=1}^{N} n_j} \left[\frac{\sum\limits_{J=1}^{N} n_j q_j}{n_i K_{Fi}} \right]^{ni}$$

(9.74)

where C_{ei} = liquid phase equilibrium concentration of the ith chemical
q_i = equilibrium loading on the adsorbent for the ith chemical
K_{Fi} = Freundlich adsorption capacity constant for the ith chemical
n_i = reciprocal of the Freundlich adsorption intensity of the ith chemical
N = number of components

For a multiple component system composed of chemicals that individually follow Langmuir isotherms, the following equation can be used to predict the equilibrium relationship:

$$q_{ei} = \frac{q_{oi} K_{Li} C_{ei}}{1 + \sum\limits_{j}^{N-1} K_{Lj} C_{ej}}$$

(9.75)

where q_{ei} = equilibrium loading on the adsorbent
q_{oi} = ultimate adsorption capacity of the adsorbent for the ith chemical
K_{Lj} = relative energy of adsorption of the ith chemical, also known as the equilibrium constant—typically empirically determined
C_{ei} = equilibrium concentration of the ith chemical in the waste stream

Multiple component adsorption calculations are typically performed using computer models due to the complexity of the calculations.

9.8.3.5 Other Sources of Isotherm Data.   When an isotherm study is not practicable, or when preliminary information is all that is required, standard "pure" isotherms are available. Sources available, besides a review of current scientific literature, include Adsorption Equilibrium Data Handbook, Dobbs' and Cohen's *Carbon Adsorption Isotherms for Toxic Organics* (using the Freundlich isotherm), "Adsorption-Capacity Data for 283 Organic Compounds," (Yaws and Nijhawan) and manufacturers' literature. The designer should always remember that when literature isotherms are used, the isotherm is based upon ideal, non-competitive conditions, and that actual field conditions may vary from the predicted results.

Sample isotherm data for 143 selected organic chemicals is presented in Appendix G. In addition to being available for reference, the reader should note that the isotherms from different sources may vary slightly.

9.8.4 Treatability Studies

It is rare that a wastewater stream performs "ideally," primarily because most wastewaters do not contain pure concentrations of a compound. Frequently, a wastewater will contain multiple contaminants that will interact in complex relationships. Other waste stream characteristics such as flow rate, pH, solids content, and even temperature can affect adsorption. It is therefore strongly recommended that prior to designing a carbon adsorption treatment system, the designer should conduct a bench-scale treatability study in the lab. There are two types of treatability study: isotherm generation studies, and flow through studies. Both studies are typically conducted as part of the design process. The goal of both the studies is to generate adsorption isotherms that are then used to determine the amount of carbon needed, the loading rate, and to get a general idea of the life of the carbon bed before it needs to be replaced. Adsorption isotherm generation is described above in Sec. 9.8.3.3. After the unique adsorption isotherms have been generated, the next step is the column flow through study.

9.8.4.1   Column Flow Through Study.   As part of the design process, it is always helpful to conduct a small-scale (bench-scale) column study to evaluate the effectiveness of carbon adsorption, and in particular to evaluate the effect of loading rates on break-through. In general, up to three columns are placed in series, with sampling ports located between each column. The columns can be constructed from a variety of materials, including glass, stainless steel, or clear PVC. Columns should be sized for a contact time (Empty Bed Contact Time, "EBCT") of 10 to 20 minutes. EBCT is calculated according to the following equation:

$$EBCT = V_{column}/Q \qquad (9.76)$$

where Q = flow rate

A typical column flow through study is shown in Figs. 9.45 and 9.46. The influent waste stream can be either gravity flow or pumped. The flow rate is metered to between 0.5 and 5.0 gallons per day (gpd). Collect samples from each of the sampling ports as necessary to determine the removal efficiency, and to use in calculating the breakthrough curve (plot effluent concentration vs. time).

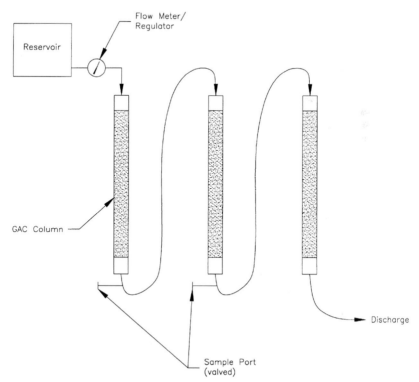

FIGURE 9.45   Typical column flow through schematic.

FIGURE 9.46 Typical column flow through treatablity study.

9.8.5 Design Process

The general design procedure for adsorption equipment involves two steps. The steps include developing adsorption isotherms by collecting experimental data and applying various equations to the data, and applying the isotherm information to be anticipated adsorbate concentrations, flow rates, and other relevant properties in order to determine the optimal design. The treatability study phase (isotherm generation, and column flow through studies) of the process is described above in Secs. 9.8.3 and 9.8.4. The isotherm data in Appendix G or from another source can be used for preliminary cost estimates and/or design or treatability study parameters.

9.8.5.1 Column Design. A primary concern is whether an adsorption unit should be designed and built by the user or whether it should be purchased as a unit from a manufacturer. The steps used in designing an adsorption system are as follows:

1. Establish the required effluent concentration (typically available in regulations or existing facility permits). Based upon the influent concentration (measured), determine the mass per a unit time period that will need to be removed (i.e. lbs. per hour).

2. Determine chemical characteristics of the various components in the wastewater stream.

3. Select an adsorbent material, including size. This can be accomplished by consulting with vendors, literature review, or through the designer's prior experience.

4. Develop adsorption isotherm data and the best equation to fit the data (as described in Sec. 9.8.3.3).

5. Determine effluent discharge operating conditions, such as flow rate, temperature, and the concentrations of the various chemicals of concern.

6. Given the adsorption isotherm, determine the theoretical adsorption capacity, defined as the mass of adsorbate per mass of adsorbent (often lbs. adsorbate/100 lbs. adsorbent).

7. Determine the operating capacity of the carbon, as a percent of the theoretical adsorption capacity. This can be done either empirically or using a percentage of the theoretical adsorption capacity.

The remaining steps in the design process are somewhat iterative, and may require several revisions to the calculations.

8. Select hydraulic loading rate for the system. The loading rate is usually around 5 gpm/ ft², though it can range as high as 10 for upflow columns and as low as 3 for downflow columns. The maximum flow rate for an adsorbent is determined by the "crushing velocity" for a particular adsorbent material. Crushing velocity information is available from the manufacturer of the adsorbent.

9. Calculate the mass of adsorbent required for one-half of an operational cycle, the adsorption phase of the cycle

$$M = \frac{QC_i t_{ads}}{q_{oper}} \tag{9.77}$$

where M = mass of adsorbent required
$\quad Q$ = the influent flow rate
$\quad C_i$ = the adsorbate influent concentration
$\quad t_{ads}$ = adsorption phase of the cycle time

10. Calculate the volume of the adsorbent required by dividing the mass of adsorbent, M, by the adsorbent bulk density, ρ_B.

11. Calculate the surface area of the adsorbent bed, A, using the hydraulic loading rate and the volume, and the bed length, L, using the volume and area. Keep in mind that longer beds usually result in greater removal.

12. Determine the steam loading rates for regeneration.

13. Calculate the Empty Bed Contact Time (EBCT), which can range from 15 to 35 minutes, and is typically of the order of 20 minutes.

14. If the EBCT is out of the acceptable range, repeat steps 8 through 13.

Other factors to consider during the design of the system include: the waste water stream must be evenly distributed at the entrance to the adsorbent bed for maximum bed efficiency; and prefilters may required to remove particulate matter that can foul an adsorbent bed (Speitel, Swindell-Dressler Co.).

9.8.5.2 Mass Transfer Zones.   During operation, an adsorbent bed consists of three zones of activity, as shown in Fig. 9.47.

The majority of adsorption occurs in the primary adsorption zone, which is also referred to as the mass transfer zone (MTZ). Behind the mass transfer zone, the adsorbent is saturated with the adsorbate. Ahead of the MTZ, the bed is essentially free of the chemical. The thickness of the MTZ is a function of concentration, adsorbent, reaction kinetics, and contact time, and is proportional to the gas velocity. A larger MTZ is often due to higher flow rates and results in decreased adsorbent bed life. A large MTZ is typically indicative of poor bed utilization. Smaller diameter adsorbent particles and "flatter isotherms" (small Freundlich 1/n values) are characteristic of small MTZ, and indicative of lower velocities and result in good adsorbent bed utilization. The mass transfer zone can be determined using the following equation:

$$L_{MTZ} = \left(\frac{L}{[t_{st}/(t_{st} - t_B)] - X} \right) \tag{9.78}$$

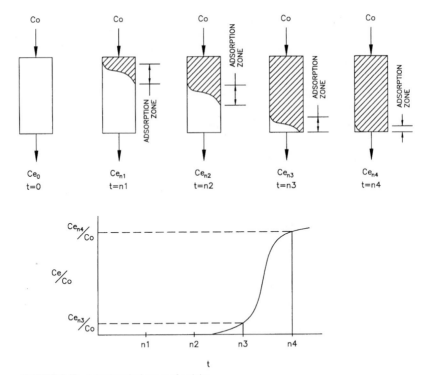

FIGURE 9.47   Adsorbent bed zones of activity.

where L_{MTZ} = length, or thickness, of the mass transfer zone
L = adsorbent bed length
t_{st} = time required for adsorbent saturation
t_B = time required for adsorbent break point
X = degree of saturation in the mass transfer zone

As the adsorbent is saturated, the MTZ moves through the bed. When a portion of the MTZ reaches the outlet of the bed (i.e. the effluent concentration is equal to the influent concentration), breakthrough is said to have occurred. Prior to breakthrough, however, the effluent concentration will equal the permitted effluent discharge standard. This point is called the break point. The times required for adsorbent break point and adsorbent saturation are most commonly empirically determined or estimated based upon experience.

Under actual operating conditions, bed adsorption capacity, known as the operating or dynamic capacity, will typically be some fraction of the theoretical adsorption capacity predicted according to the isotherms. The losses in efficiency are due to several factors, but primarily the removal of the adsorption column from the treatment system at the break point, before the adsorbent has become fully saturated. If empirical information is available, the following equation can be used to determine the operating capacity of the bed:

$$q_{oper} = q_e \left(\frac{L - L_{MTZ}}{L} \right) + 0.5 \left(\frac{L_{MTZ}}{L} \right) - HEEL \tag{9.79}$$

where q_{oper} = the operating loading on the adsorbent

$HEEL$ = residual adsorbate present after regeneration, referred to as the "heel"

Because the data required for the above equation is not often readily available, the operating capacity is assumed to be some fraction of the equilibrium loading. From a practical point-of-view, most systems operate at 25 to 50 percent of the adsorbent equilibrium capacity, with typical operation at 30 to 40 percent. In multi-component applications, the following equation can be used to calculate the operating loading of the adsorbent:

$$q_{oper} = \frac{1}{\sum_{i=1}^{N}(w_i/q_{ei})} \tag{9.80}$$

where q_{ei} = equilibrium loading on the adsorbent of the ith chemical
w_i = mass fraction of the ith chemical in N components
N = number of compounds of concern in the waste stream

If the designer is concerned that a more accurate number for column breakthrough is necessary for the application, pilot-scale tests under actual operating conditions can be used to determine the operating capacity of the adsorbate.

In a wastewater with multiple chemicals, several MTZs will be present. These zones may or may not overlap, and are also subject to competitive adsorption.

9.8.5.3 Adsorbent Materials.   It can be argued that, to some degree, all solids adsorb liquids, although the adsorption might not be measurable in significant quantities. Generally, the capacity for adsorption is directly proportional to the surface area of a given material. The most common materials used for adsorption are materials that exhibit high surface areas per unit mass, such as charcoal (activated carbon), molecular sieves, and silica gels.

- **Activated carbon**
  The most frequently used adsorbent is activated carbon. This is primarily because a wide range of organic chemicals can be adsorbed economically. It is also used for solvent recovery, odor control, and for gas stream purification. Table 9.17 lists some of the classes of organic chemicals that are readily treated with activated carbon.

  Activated carbon is produced from porous, carbonaceous materials such as wood charcoal, coal, peat, lignite, recycled tires, petroleum coke, and coconut shells. In general, the production process involves first charring the material to remove hydrocarbons by pyrolysis and then activating the carbon. The most common method for activation in the United States is achieved through high temperature (750 to 950°C) steam activation in an oxygen depleted atmosphere. The reaction between the steam and the carbon is promoted by a dehydrating agent such as zinc chloride or phosphoric acid. In Europe, it is more common to use a chemical activation process. Chemical activation occurs at a lower temperature (400 to 600°C) using the dehydrating agent without steam. The actual production processes commercially used vary and are proprietary. Materials used for activated carbon have large surface areas, typically 300 to 1500 m²/g, as shown in Table 9.16.to 9.17 However, petroleum coke-based activated carbon can have surface areas in excess of 3,000 m²/g.

  A typical carbon particle consists of a highly developed, microscopic network of pores within a crystalline matrix. The pore diameters typically range from less than 2 Å to greater than 500Å, with an average of 15 to 25Å . The carbon particles are crushed, graded, acid washed and washed with water before distribution. Activated carbon comes in four general forms: extrudates (pellets), beads, granules, and powder. For wastewater control applications, granular activated carbon is most commonly used. Granular Activated Carbon (GAC) has a particle diameter that ranges from 0.1 to 2 mm (typically 1.2 mm). Powdered Activated Carbon (PAC) has a particle diameter of less than 0.1 mm, and is typically 0.05 to 0.075 mm.

  One of the unique characteristics of the use of activated carbon is that after breakthrough is achieved, the adsorbent is not normally discarded. Activated carbon can be reactivated, restoring the carbon to approximately its original adsorptive capacity. As a rule of thumb, more than 90 percent of the activated carbon is recovered during reactivation. Losses are primarily due to spillage and overburning. Reactivation can occur in both on-line applications, where the bed remains in place, and off-line applications, where the bed is physically removed from service for reactivation. Off-line reactivation is commonly used for smaller (less than 2,000 lbs.) carbon beds that are rented or leased at a site. Carbon is often reactivated off-line by passing through a high-temperature, multiple hearth furnace. Temperatures in the furnace reach 1800° F in order to thermally destroy the organic contaminants. Other regeneration techniques include use of steam or a vacuum (particularly with regenerative techniques include use of steam or a vacuum (particularly with regenerative applications), solvent extraction (hexane), and bioregeneration.

- **Activated alumina**
  While activated alumina is primarily used to remove moisture from a gas stream, as a catalyst, for solvent recovery, or in petroleum refining applications, it does see limited use in the removal of fluoride ions from water. Prepared by heating alumina trihydrate to 400° C, activated alumina has a typical surface area of 200 to 400 m^2/g. It is available in ball, powder, pellet, and granule form. Average pore diameter is 18 to 48 Å. The bulk density ranges from 38 to 42 lb./ft^3 for granules, and from 54 to 58 lb./ft^3 for pellets.

- **Molecular sieves**
  While activated carbon, silica gel, and activated alumina are materials with amorphous structures, molecular sieves are crystalline structures in which the molecules are arranged in a definite pattern. Molecular sieves are generally manufactured from aluminosilicate gels which are dehydrated. The most common molecular sieve is based on anhydrous aluminosilicate. It is also common to refer to molecular sieves as zeolites. They are used for both contaminant and odor removal. Molecular sieves are often impregnated with potassium permanganate or other proprietary mixtures of compounds to improve removals. Molecular sieves are most often found in granular form held together by a binder, having regularly spaced cavities with interconnecting pores of a known and definite size. Aluminosilicates are effective at removing sulfur compounds, mercaptans, alcohols, hydrogen sulfide, and formaldehyde. Typical surface areas are on the order of 1200 m^2/g. Bulk density for anhydrous aluminosilicate is approximately 38 lb./ft^3, with an effective pore diameter of 13 Å. Other materials used for molecular sieves include anhydrous sodium aluminosilicate and anhydrous calcium aluminosilicate. The bulk density for both these materials is 44 lb./ft^3. The average pore diameter of anhydrous sodium aluminosilicate is 4 Å and for anhydrous calcium aluminosilicate is 5 Å.

  Molecular sieves can be regenerated with either a thermal swing process or a pressure swing process. The thermal swing process is used more often, passing a hot gas through the adsorbent bed in a counter-current direction. Regeneration processes are more fully described in Sec. 9.8.7.

- **Silica gel**
  Silica gel is produced in a process that begins with the neutralization of sodium silicate with mineral acid. Subsequent steps include washing, drying, grading, and roasting. Silica gel is most often used in its granular form, although it is available as a borosilicate glass or as an aerogel. The average surface area of granular silica gel is 750 m^2/g, although it can range from 300 to 900 m^2/g, with an average pore diameter of 22 Å. The bulk density is 44 to 46 lb./ft^3. Silica gel is used for hydrocarbon separation or in gas phase operations, but does not see common usage in water or wastewater treatment.

- **Resins**
  Resins are made from many different monomeric compounds with varying degrees of polarity. Because of these polarity differences, each resin can be specifically designed to adsorb a unique compound or contaminant. For example, non-polar organic compounds adsorb effectively onto hydrophobic resins via Van der Waal's forces, while polar organics adsorb to acrylic resins with dipole-dipole interactions. Polystyrene resins are used to remove organics and recover antibiotics. Polyacrylic ester resins purify pulping wastewater, and phenolic resins are used to decolorize and deodorize waste streams Table 18.

  One of the drawbacks to resins is that they typically have a significantly smaller surface area than activated carbon (about 100 to 700 m^2/g). In addition, the bonding forces of the contaminants to the resins is generally weaker than with GAC. This tends to make the resin adsorption capacities smaller. In contrast to GAC, however, the resin adsorbents can be designed to be very selective

TABLE 9.18 General Order of Ion Selectivity for Most Resins

Cations	Anions
Fe^{3+}	CrO_4^{2-}
Al^{3+}	SO_4^{2-}
Pb^{2+}	SO_3^{2-}
Ba^{2+}	HPO_4^{2-}
Sr^{2+}	CNS^-
Cd^{2+}	CNO^-
Zn^{2+}	NO_3^-
Cu^{2+}	NO_2^-
Fe^{2+}	Br^-
Mn^{2+}	Cl^-
Ca^{2+}	CN^-
Mg^{2+}	HCO_3^-
K^+	$HSiO_3^-$
NH^{+4}	OH^-
Na^+	F^-
H^+	
Li^+	

The Industrial Wastewater Systems Handbook, Ralph L. Stephenson, James B. Blackburn, Jr. 1998, Lewis Publishers

of the compounds adsorbed. Additionally, resins have a low ash content and tend to be very resistant to bacterial growth.

The typical resin particle is spherical and approximately 0.5 mm in diameter. Resins are more expensive than GAC. Thus resin adsorbents are not generally used with waste streams involving multiple contaminants with no recovery value. If the adsorbed material is worth recovering, resin adsorbers may be an economically viable alternative form of treatment.

The spent resin adsorbents can be regenerated with considerable ease. A solvent wash followed by distillation can be used to recover the adsorbed material without the dangers that thermal regeneration presents for GAC. The resins can be regenerated in-situ using simple aqueous solutions and solvents (Dow Chemical Co.).

9.8.6 Design Process—System Design

There are two primary types of adsorption systems for wastewater pollution control: fixed bed (with granular activated carbon) and powdered activated carbon. Often used in conjunction with other phases of treatment, the activated carbon is typically used as a polishing stage prior to discharge.

9.8.6.1 Fixed Bed Systems. A fixed bed system consists of a square or cylindrical chamber containing the adsorbent. In many installations, fixed beds are operated in several stages, either in parallel or in series. A good design practice is to install a second unit in series to catch any of the contaminants if breakthrough occurs. The direction of flow in fixed bed systems is typically downward. The adsorbent is packed into drums or other containers, which are removed when breakthrough occurs. Adsorbent manufacturers can furnish complete packaged systems, and will often offer to regenerate the spent adsorbent when necessary. Liquid adsorption media are available in a wide range of packaging, from 55-gallon plastic drums to 12,000 lb. and larger deep bed filtration systems, which can be made up of multiple filter beds in series.

Most fixed bed systems will consist of two (series flow) to four (parallel flow) carbon columns, in addition to the other processes in the treatment system. An example of both series and parallel

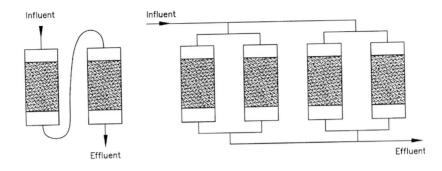

Influent

Influent

Effluent

Effluent

Series

Parallel

FIGURE 9.48   Downflow (Fixed Bed) Schematic.

systems is shown in Fig. 9.48. Figure 9.49 shows a groundwater treatment system incorporating two GAC columns in series that follow settling and air stripping. Figure 9.50 shows a photograph of this remediation system designed to remove methylene chloride and tetrachloroethene down to drinking water levels under a groundwater re-injection permit.

In this particular instance, the extracted groundwater was being reinjected at the head of the groundwater contaminant plume, and was therefore required to meet drinking water standards prior to discharge.

9.8.6.2   Powdered Activated Carbon Systems.   The addition of powdered activated carbon (PAC) is often performed in conjunction with a biological treatment system. The PAC is typically added to the effluent of a biological treatment process in a contact chamber. The contact chamber must be designed with sufficient contact time for contaminant removal, and with a PAC recovery mechanism. PAC recovery is usually accomplished through sedimentation or filtration. Because PAC is very fine, the addition of a coagulant may be required to remove all of the carbon. Advantages of PAC addition include relatively low capital costs for modifications to existing treatment systems, and lower usage rates because the PAC is applied as a "polishing" treatment, rather than as the primary treatment common with granular activated carbon applications. One of the disadvantages of PAC is the difficulty in cost effectively regenerating the carbon for reuse, which results in PAC often being used only once.

The design of a PAC treatment process is relatively straightforward. The contact chamber must be designed as a completely mixed reactor. Assuming a constant flow rate Q with an influent concentration C_i, the following steady-state equation can be used to determine the PAC dosage required:

$$W = \frac{C_i - C_0}{K_F C_O^{1/n}}$$ (9.81)

where W = PAC dosage (mg PAC/L influent)

Again, assuming that the contact chamber is properly designed (completely mixed), the PAC will equilibrate extremely rapidly with the contaminants to be removed.

9.8.7   Operation

9.8.7.1   Column Operation.   In a wastewater with multiple chemicals, several MTZs will be present. These zones may or may not overlap, and are also subject to competitive adsorption. Under competitive conditions, displacement of less readily adsorbed chemicals may occur. One of the ways that this displacement can manifest itself is through a sudden increase in the effluent concentrations (greater than the inlet concentration) of the less readily adsorbed compound. Additional controls or different operating strategies may be necessary if the displacement is a problem.

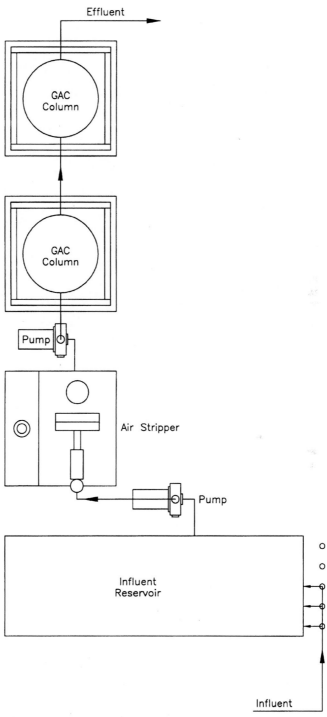

FIGURE 9.49   GAC columns following settling and air stripping.

FIGURE 9.50 Photograph of remediation system.

Factors that will influence the adsorption capacity of a carbon column include the temperature and solids (suspended or dissolved) concentration. The adsorptive capacity decreases with increasing system temperature. Solids will plug the pores in the column, reduce the sites available for adsorption, and may even plug the voids between the carbon particles.

- **Bed regeneration**
  Once the break point has been attained, the adsorbent bed must be regenerated or replaced. When working with activated carbon, on-line bed regeneration is accomplished by applying low pressure steam or a vacuum to the bed (after allowing the bed to drain). As a rule of thumb, approximately 90 percent of the granular carbon remains after regeneration. Two common regenerative techniques for on-line regeneration are thermal swing regeneration and pressure swing regeneration. Thermal swing regeneration involves heating the bed directly with either a hot inert gas or low-pressure steam, or indirectly with surface contact. The adsorbent is heated to between 300° and 600° F, and the gas is applied to purge the adsorbent. The thermal swing regeneration process requires a period in the operational cycle for cooling. Pressure swing regeneration involves applying either a low pressure or a vacuum to draw the adsorbate from the bed. Operated under essentially isothermal conditions, the pressure swing regeneration process allows for shorter cycle times and the recovery of a high purity product. Low pressure steam is the process most often used, except in solvent recovery applications where water would contaminate the recovered solvent. Applied at a rate of 1 to 4 lbs. of steam per pound of adsorbate, the saturated steam will rapidly heat the bed without polymerizing the adsorbed chemicals. During the regeneration process, condensation will occur in the adsorbent bed, leaving a portion of the adsorbate in the bed. This fraction is referred to as the heel. The most common off-line regeneration technique, particularly for activated carbon involves passing the adsorbent through a high-temperature, multiple hearth furnace.

• **Other operation considerations**

The system should be designed so that the adsorbent bed lasts for greater than a month in order to reduce maintenance costs. More frequent regeneration (bed replacement) is generally indicative of either inappropriate use of the adsorbent, or inefficient operation.

Backwashing may be necessary to flush particulate or other build-up from the bed to prevent clogging, and if no settling or filtration is provided as part of the treatment system prior to the carbon beds, should be a part of the regularly scheduled maintenance. Backwashing is often performed at 10 to 20 gpm/ft³ for 10 to 15 minutes. With the "smaller" rented units, this is typically not necessary, and it is accomplished when the bed is regenerated by the manufacturer.

9.9 Ion Exchange

9.9.1 Description

Ion exchange is a process which uses a bed of a resin which is an insoluble granular material containing cations, such as hydrogen or sodium. These nontoxic ions can be exchanged for toxic heavy metal cations such as Zn^{2+}, Ni^{2+}, Cu^{2+}, Cd^{2+}, Pb^{2+}, Ag, Au, etc. or inorganic cations such as Ca^{2+}, Mg^{2+} or NH_3^+. Likewise, anions, such as hydroxide can be exchanged for inorganic anions such as CrO_4^{3-}, SO_4^{3-}, CNS^-, or CNO^-. When all of the exchange sites have been replaced, the resin can be regenerated by passing a concentrated solution of hydroxide ions (to replace hydrogen ions), sodium ions (to replace calcium or Magnesium ions) or hydrogen ions (to replace hydroxide ions).

The ion exchange reaction is in accordance with the following general formula:

$$\text{Cation Exchange:} \quad \left.\begin{array}{l}(Cation)^+ + Na_2 \\ H_2\end{array}\right\} R \rightleftarrows \left.\begin{array}{l}(Cation)\,R + 2Na^+ \\ 2H^+\end{array}\right\} \quad (9.82)$$

$$\text{Cation Regeneration:} \quad \left.\begin{array}{l}2Na^+ \\ 2H^+\end{array}\right\} + CaR \rightleftarrows \left.\begin{array}{l}Na_2 \\ H_2\end{array}\right\} R + Ca^{2+} \quad (9.83)$$

Cation Regeneration usually accomplished with 5 to 10% brine solution of NaCl or a 2 to 10% solution of H_2SO_4.

$$\text{Anion Exchange:} \quad (Anion)^- + R(OH)_2 \rightleftarrows R(Anion) + 2OH^- \quad (9.84)$$

$$\text{Anion Regeneration:} \quad 2OH^- + R(Anion) \rightleftarrows R(OH)_2 + (Anion)^- \quad (9.85)$$

Anion regeneration usually accomplished with 5 to 10% sodium hydroxide.

When all exchange sites are filled, the contaminent appears in the effluent in a condition known as breakthrough. At this point, the resin is backwashed to remove dirt, organics and non-exchanged inorganics caught on the media. The bed is then regenerated. After regeneration, the bed should be flushed to wash out residual regnerate.

9.9.2 Treatability Studies

• A glass column with stoppered ends, approximately 1 inch in diameter and 36 inches tall should be filled with exchange resin supported by glass beads with stoppered ends.

• Rinse the column with distilled water for 10 to 15 minutes at a rate of about 50 ml/min.

• Pump 50 ml/min of known concentration wastewater through the column.

• Plot a curve similar to Figure 9.51

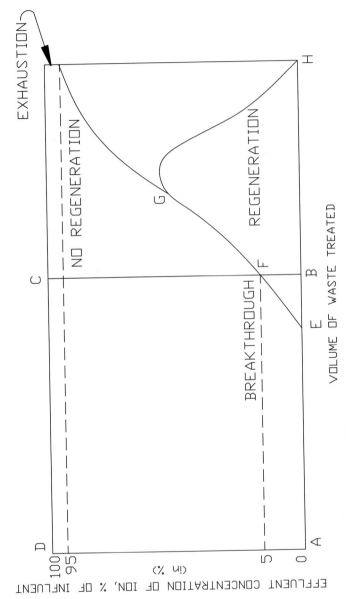

FIGURE 9.51 Ion exchange treatability plot.

9.82

In Figure 9.51, ABCD is the volume of wastewater treated, EBF is the quantity of ions breaking through, and AEFCD is the quantity of ions removed by the resin. BFGH is the quantity of ions removed during regeneration.

9.9.3 Design Hints

- Some resins can be dissolved at elevated temperature.
- Efficiency of a resin may depend on pH.
- Total suspended solids must be removed before wastewater contacts the resin.
- Oil may foul the resin.
- For hardness removal (Ca^{2+} and Mg^{2+}), approximately 8.5 pounds of salt per cubic foot will be required to regenerate the resin. About 4 pounds of salt will be removed per cubic foot.
- For ammonia removal approximately 10 bed volumes of 2% NaCl will be required at a rate of 4 to 10 bed volumes per hour to regenerate. The exchange capacity will be about 100 to 150 bed volumes at a loading rate of 7.5 to 20 bed volumes per hour using clinoptilolite media.
- For recover of chromic acid, (H_2CO_4), approximately 25 pounds/ft^3 of H_2SO_4 is required to regenerate the cation exchange resin which is exchanged at a maximum loading concentration of 14 to 16 ounces of CrO_3.
- Minimum bed depth—24"
- Exchange flow rate—2 to 5 gpm/ft^2
- Regeneration flow rate—1 to 2 gpm/ft^2
- Rinse water flow rate—1 to 1.5 gpm/ft^2
- Rinse water volume—30 to 100 gals/ft^3
- Other cations and anions can potentially be removed in accordance with Table 9.11. In this table, the higher cations or anions can be replaced by the lower ones with the more efficient replacement being between those ions fartherest apart in the table.

9.10 REFERENCES

1. Amirtharajah, Appiah, Ph.D. and P.E. and Charles R. O'Melia, Ph.D., P.E. "Coagulation Processes:
2. C.T. Butterfield, ìBacterial Properties of Chloramines and Free Chlorine in Waterî, Public Health Reports 63 (1948): 934
3. Destabilization, Mixing, and Flocculation." *Water Quality and Treatment.* American Water Works Association. McGraw-Hill, New York: 1990.
4. D. W. Green, (ed.), *Perry's Chemical Engineers Handbook,* 7th ed., McGraw-Hill, Inc., New York, 1997.
5. G. E. Speitel, Jr., University of Texas at Austin, notes, 1990–1.
6. G. L. Amy, and Cooper, William J., "Air Stripping of Volatile Organic Compounds Using Structured Media," *Journal of Environmental Engineering,* American Society of Civil Engineering, vol. 112, no. 4, August 1986, p. 729.
7. J. Haarhoff and J. L. Cleasby, "Evaluation of Air Stripping for the Removal of Organic Drinking-Water Contaminants," *Water SA,* vol. 16, no. 1, p. 13, 1990.
8. J. Michal, *An Overview: Air Stripping of Organic Compounds,* Arizona Water and Pollution Control Association, 1988.
9. J. H. Montgomery, *Groundwater Chemicals: Desk Reference,* 2nd ed., CRC Press, Inc., Boca Raton, FL, 1996.
10. Metcalf and Eddy, *Wastewater Engineering: Treatment, Disposal and Reuse,* 2nd ed., McGraw-Hill, Inc., New York, 1979.
11. R. A. Corbit, *Standard Handbook of Environmental Engineering,* McGraw-Hill, Inc., New York, 1990.

CHAPTER 10
BIOLOGICAL TREATMENT

10.1 INTRODUCTION

The purpose of this chapter is to present the theory and practice of biological treatment used for the destruction and conversion of organic chemical and biological pollutants. Inorganic chemicals cannot be destroyed by biological treatment, but they can potentially be converted in valence or oxidative state to compounds which meet permit levels. Likewise, inorganic pollutants can potentially be stripped from organic compounds biologically.

This chapter will review and recommend methods of using biological treatment to a) convert soluble organic contamination to insoluble organic and inorganic constituents, and b) convert soluble organic contamination to CO_2 and H_2O. Any insoluble organic or inorganic contamination in an untreated wastewater stream can be easily and inexpensively removed by physical treatment, usually with settling or filtration. Therefore, there is no need to use more complicated and expensive chemical or biological treatment to remove insoluble organics. Likewise, insoluble inorganics can be less expensively removed by settling or filtration, and soluble inorganics can more efficiently be removed by chemical treatment.

10.2 THE THEORY OF BIOLOGICAL TREATMENT

In biological treatment, the organic or hydrocarbon food serves as an energy source, or election donor, to the bacteria and either oxygen, nitrite and nitrate, sulfate or carbon dioxide serves as the electron acceptor. Aerobic processes use oxygen, anoxic processes use oxygen nitrite and nitrate, and anaerobic processes use sulfate or carbon dioxide as electron acceptors. In the process of metabolism, the nutrients in solution in the form of hydrocarbons, nitrogen, phosphorous and metallic micronutrients, through osmosis, catalyzed by enzymes, pass through the bacterial cell membrane and are converted into useful energy. In these oxidation chemical reactions, the electron donors in the form of organic food, primarily carbon, are removed from their molecules and accepted into oxygen molecules to form carbon dioxide in aerobic treatment, nitrogen gas in anoxic treatment and carbon dioxide, hydrogen sulfide and methane in anaerobic treatment.

A bacterium uses enzymes to gather and retrieve food in the form of hydrocarbons. In the process of converting these hydrocarbons to food for the bacteria, enzymes break the hydrocarbons down aerobically (in the presence of free oxygen) into carbon dioxide (CO_2) and water (H_2O), or anaerobically (in the presence of combined oxygen) though the Kreb's cycle into CO_2, hydrogen sulfide (H_2S), methane (CH_4) and H_2O. If these by-products of decomposition are acceptable, aerobic anoxic or anaerobic treatment is an alternative means of converting organic pollutants into non-polluting chemicals. A bacterium, whether aerobic anoxic or anaerobic, requires a certain environment to grow. Upon growth, the bacterium reproduces by splitting at a logarithmic rate. Figure 10.1 demonstrates the environment needed in an aerobic environment for bacterial cellular growth and the by-products of growth produced.

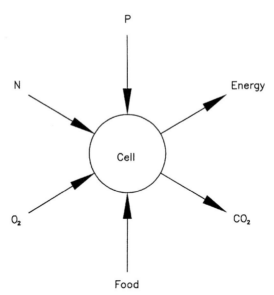

FIGURE 10.1   Bacterial cellular growth requirements.

The enzyme catalyzation rate is affected by temperature, pH and substrate concentration. Temperature affects enzyme activation generally in accordance with Fig. 10.2, although each enzyme displays maximum activity at some specific temperature. Likewise, each enzyme displays maximum activity at a certain pH range generally in accordance with Fig. 10.2.

In accordance with these requirements, an acceptable environment is required for growth, which includes an acceptable *temperature* and *pH* and an absence of toxic chemicals. In the absence of any of these requirements or in the presence of any toxic chemicals, a bacterial cell will eventually die. In the presence of these requirements and absent toxicity, a bacterial cell has a growth and death rate represented by a classic bell shaped curve as represented in Fig. 10.3.

During the initial growth phase. the bacteria split and logarithmically increase until one or more of the environmental factors is depleted. If there is no depletion, the growth phase will continue. Beginning with a given amount of food, bacterial growth will eventually reach a stationary phase upon depletion of food, followed by a death phase and finally, endogenous respiration or cannibalism. During endogenous respiration, the bacteria feed on each other in the absence of available food. As endogenous respiration declines, the remaining non-biodegradable residue (NBDR), commonly called "bug bones" by some theorists, remains solid particles. This chapter will present the theory that with sufficient time, there is no NBDR for purely organic material.

This Handbook will first present further the theory of aerobic biological treatment.

10.3   AEROBIC BIOLOGICAL TREATMENT

10.3.1   The Theory of Aerobic Suspended Growth Biological Treatment

The previously discussed theory of biological treatment can be used to develop a theory of aerobic biological treatment usable in the design of treatment system. During biological treatment, when bacteria are replicating and dying, it is possible to predict the strength of the residual food at any

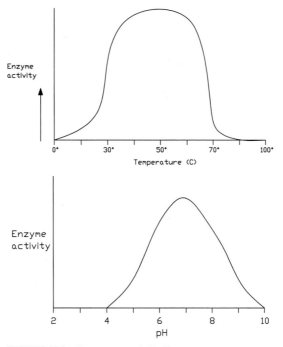

FIGURE 10.2 Temperature and pH effect.

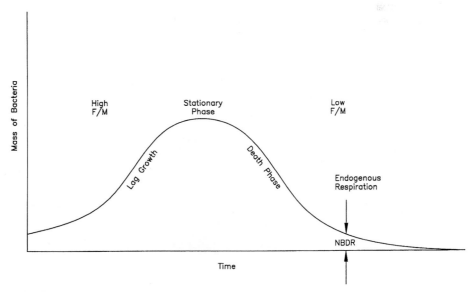

FIGURE 10.3 Cellular growth curve.

given time, called substrate (S), approximated by tests such as BOD, COD, TOC and TC, the mass of the residual bacteria at any given time, called the mixed liquor volatile suspended (MLVSS) or X_v, the concentration of dissolved substrate leaving the system, the mass of residual bacteria produced which are not needed for synthesis and oxidation, called excess suspended solids (ΔX) and the oxygen needed for synthesis and oxidation.

If in the presence of a suitable environment such as demonstrated above in Fig. 10.1, the logarithmic growth rate of bacteria will occur in accordance with Fig. 10.3. In the process of this growth, bacteria, CO_2 and H_2O are produced. The production of bacteria is known as *synthesis* and the production of CO_2 and H_2O is known as *oxidation*. These products are produced in accordance with the following equations:

Synthesis equation

$$\text{Organics (food)} + O_2 + N + P \xrightarrow{\text{cell}} \text{new cells} + CO_2 + H_2O + NBDR \qquad (10.1)$$

Oxidation equation

$$\text{Cells} + O_2 \xrightarrow{\text{organic(Gold)}} CO_2 + H_2O + N + P + NBDR \qquad (10.2)$$

The constants needed for these determinations, which can be determined experimentally or by using historical data, are the Reaction Rate of Substrate Removal, k, Synthesis Cellular Production Constant, a, the Synthesis Energy Constant, a', the Oxidation Cellular Destruction Constant, b, and the Oxidation Energy Constant, b'.

The synthesis and oxidation theory will be explained as it relates to the production of excess sludge and the use of oxygen.

10.3.1.1 Synthesis and Oxidation.   In the *Synthesis Equation*, organics in a suitable environment, in the presence of oxygen, inorganic nutrients, and bacterial cells, are converted to new cells, CO_2, H_2O, and with relatively short detention times, non-biodegradable residue, NBDR. The inorganic nutrients required for biological metabolism consist primarily of nitrogen (N), phosphorous (P) and carbon (C), but also trace amounts of iron (Fe), manganese (Mn), potassium (K) and aluminum (Al). A typical bacterial cell has the formula $C_{25}H_{35}N_5O_{10}P$ (Eckenfelder, 1970). Based on this formula a typical active cell is 11.7 percent nitrogen and 5.2 percent phosphorous as shown in Fig. 10.4.

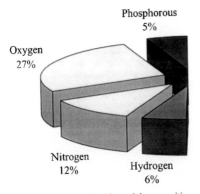
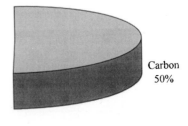

FIGURE 10.4   Typical bacterial composition.

C: 12 g/mol × 25 mol = 300 g

H: 1 × 35   = 35 g   N: 70/596 = 11.7%

N: 14 × 5   = 70 g   P: 31/596 = 5.2%

O: 16 × 10   = 160 g

+ P: 31 × 1   = 31 g

Total   596 g

A cell undergoes a conversion during oxidation as shown in Fig. 10.5.

As oxidation occurs, the nitrogen content of a cell is lowered to approximately 5 percent and the phosphorous to 1 percent in accordance with Fig. 10.5. Therefore for purposes of aerated biological treatment, the ratio of cellular material to nitrogen to phosphorous is approximately 100:5:1. Cellular material can best be approximated as shown in Chapter 3 by volatile suspended solids.

The result of the *Synthesis Equation* is a destruction of organic constituents in wastewater, and also a conversion of these soluble organics to insoluble cells which can be settled in a subsequent sedimentation basin. The result of the *Oxidation Equation* is a conversion of the remaining organics in the wastewater and the cells produced by the *Synthesis Equation* into CO_2 and H_2O. The design decision which must be made in order to apply the *Synthesis* and *Oxidation Equations* is whether to operate at short detention times to produce and settle greater excess solids, or to operate at longer detention times to minimize excess solid production and shift organic conversion from synthesis to oxidation. Pragmatically, the design question is the relative cost difference between the capital and operation cost of a wastewater treatment system and a sludge treatment and disposal system.

Aerobic biological treatment can be achieved in a *suspended growth* system or an *attached growth* system. In a suspended growth system, the bacteria are suspended in the wastewater and in an attached growth system, the bacteria are attached to a solid non-reactive media. In both cases, the requirements for growth of the bacteria are the same. In the suspended growth system, the influent wastewater (food) comes in contact with the bacteria by passing through a basin filled with suspended bacteria. In the attached growth system, the influent wastewater comes in contact with the bacteria by passing over the media on which the bacteria are growing. Oxygen is typically added to the wastewater in the suspended growth system with surface aerators or diffused air. Oxygen may also be added by algae which produce oxygen in the presence of sunlight, nitrogen and phosphorous. Oxygen is typically added to the wastewater in the attached growth system by providing a turbulent flow path allowing oxygen to diffuse into the water.

An alternative attached growth system commonly known as a *rotating biological contactor* (RBC), uses rotating plates on which the bacteria grow, which are partially submerged. The bacteria attached to the plates are exposed to oxygen when not submerged in water. Table 10.1 summarizes these three types of systems.

Figure 10.6 through 10.8 show these three systems schematically.

Wetlands, which are discussed later in this Handbook, are another type of attached growth aerobic system. The bacteria in wetlands grow on the roots of aquatic plants and provide the necessary cells for aerobic activity.

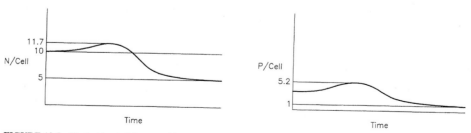

FIGURE 10.5 Typical bacterial composition.

TABLE 10.1 Suspended and Attached Growth System

	Suspended growth	Attached growth	RBC
Wastewater Flow	Horizontally through the basin	Downward over the media	Horizontally past the media
Oxygen Addition	Algae, Surface Aerators or diffused Air	Upward through the media	Plates rotate out of water through air
Bacterial Growth	Suspended in the wastewater	Attached to the Media	Attached to the Plates

This chapter will first discuss the two types of suspended growth systems; *batch and continuous plug flow and continuously mixed.*

10.3.1.2 *Suspended Growth Systems, Batch or Continuous Plug Flow.*

In a biological treatment system, a population of biological cells goes through a similar curve with time, whether the system is batch fed or plug flow. A batch system is demonstrated by a beaker or tank of wastewater injected with biological cells, O_2, N and P as shown in Fig. 10.9.

When the tank is filled in the correct proportions with wastewater, bacterial cells, nitrogen and phosphorous, and oxygen is added in the correct amount, the cells will grow in accordance with Fig. 10.5. If a sample is taken periodically (i.e. once/hr) and tested for MLVSS, the Fig. 10.3 type curve will be developed. As the bacteria continue to reproduce, the food is depleted and the bacteria begin to consume each other through endogenous respiration in the death phase. During this phase, no new cells are produced, instead they are converted to CO_2 and H_2O. This bacterial growth and death curve is also demonstrated in a continuous plug flow system, an example of which is shown in Fig. 10.10.

The ultimate continuous plug flow biological system would be a long, small diameter pipe configured so that the wastewater travels through the system in a "plug" without mixing with other previous or subsequent "plugs." If a sample is taken at various points along the length of the plug flow tank and tested for MLVSS, again a Fig. 10.3 type curve will be produced.

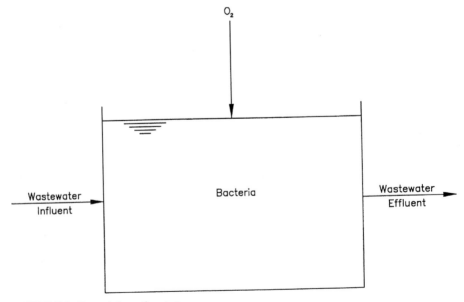

FIGURE 10.6 Suspended growth system.

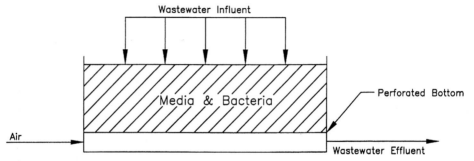

FIGURE 10.7 Attached growth system.

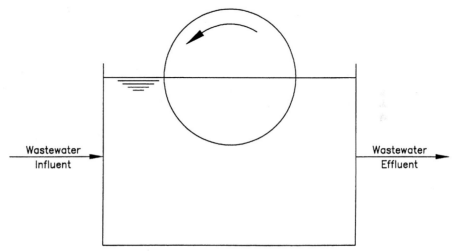

FIGURE 10.8 Rotating biological contractor (RBC).

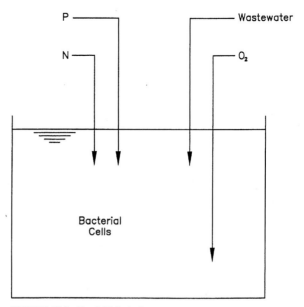

FIGURE 10.9 Batch biological system.

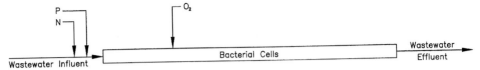

FIGURE 10.10 Continuous plug flow biological system.

At various detention times in the batch system and at various distances from the influent of the plug flow system, the cellular material produced by synthesis and subsequently destroyed by oxidation can be estimated by the Fig. 10.3 type curve developed by sampling the system.

10.3.1.3 Suspended Growth System, Continuous Feed Completely Mixed.
Another biological treatment process configuration is the continuous feed completely mixed system. In this type of system, ideally the reactor or tank is mixed with high energy and is of a shape so that each molecule of wastewater is mixed instantaneously as it enters the basin with cells, oxygen, N, and P. At every point in this ideal system, there is a mixture of cells representing every point along the Fig. 10.3 curve. With low detention times, the mixture is weighted toward the high cellular mass end of the curve (the log growth cells are maximized in the mixture), and a large amount of settleable solids are produced. With high detention times, the mixture is weighted toward the low cellular mass end of the curve (the log growth cells are minimized in the mixture), and a small amount of settleable solids is produced.

Realistically, it is impossible to construct an ideal continuously mixed basin. Theoretically, this basin should have a high energy source to mix the contents instantaneously, at which point O_2 and nutrients are added, and have all points of the basin in uniform contact with this energy source. The closest duplication of this ideal would be a sphere or less efficiently, a cube, with many points of mixing and oxygen addition, as shown in Fig. 10.11. Obviously, the larger the basin, the more energy is required for mixing.

In a continuous feed completely mixed system, the time axis shown in Fig. 10.3 is the sludge age, θ_c or total time a particle of sludge is under aeration. In a simple completely mixed system as shown in Fig. 10.11, this sludge age is equal to the hydraulic detention time, t. This type of system requires very long hydraulic detention times to lower the organic food concentration (BOD_5, COD, TOC, or TC) to an acceptable level. Oxidation ponds and aerated lagoons would be examples of this type of completely mixed system if enough energy could be added to assure that they were completely

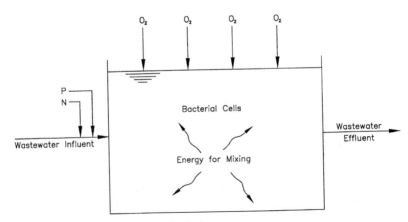

FIGURE 10.11 Continuous completely mixed system.

mixed. Realistically, this is impractical and these ponds and lagoons are a combination of plug flow and completely mixed kinetics with the oxidation pond having more characteristics of a plug flow system and the aerated lagoon characteristic of a completely mixed system.

In a completely mixed system, since fresh food is found in the effluent in equal proportions as in the influent, high levels of active bacteria are typically removed from the mixture through sedimentation. The settled bacteria are partially alive with the amount active or alive depicted by Figure 10.3 at the appropriate detention time.

A detention time which falls in the maximum mass of bacteria portion of the curve will produce more sediment of sludge than one which falls toward the tail of the curve when the bacteria have been virtually depleted through endogenous respiration.

The efficiency of a completely mixed system can be increased by adding more bacteria to the incoming food. Since active bacteria are available in the sludge, it is common practice to recirculate this sludge to the aeration basin. This type of system is known as *activated sludge* and is depicted in Fig. 10.12.

The return activated sludge (RAS) is used to provide this food and increase the efficiency of the system.

10.3.1.4 Substrate Degradation Rates in a Suspended Growth System.

Degradation rates can be computed to predict the rate of organics or substrate removal in aerobic biological treatment processes through synthesis and oxidation. In a treatment plant operation, the biological growth rate, sometimes called the kinetic rate, can be controlled to affect the degradation rate. As explained in the introduction to this chapter, inorganics and insoluble organics may be removed through less expensive methods than biological treatment.

The principal of degradation is that practically all soluble individual organic compounds are removed as a zero order reaction down to a very small concentration as shown in Eq. (10.3)

$$\frac{dS}{dt} = k \tag{10.3}$$

where S = substrate concentration remaining at time t (mg/1)
  t = time
  k = reaction constant

This equation can be plotted as shown in Fig. 10.13.

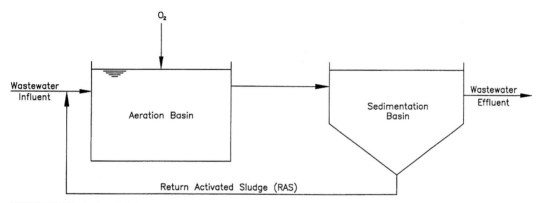

FIGURE 10.12   Activated sludge system.

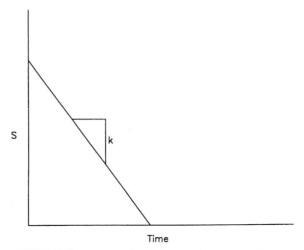

FIGURE 10.13   Zero order degradation rate, single compound.

Many mathematical models have been developed to explain substrate removal. Generally, these models recognize that at high BOD levels, the rate of BOD removal per unit of bacterial cells is constant down to a limiting BOD concentration. Below this level, the models show that the BOD removal rate will become concentration dependent and decrease. Various relationships that have been proposed in literature are described below. It is important that the relationship used for design describes the actual observed degradation rate if available.

Since most wastewaters consist of multiple soluble organic compounds, a more typical degradation rate plot than that shown in Fig. 10.13 is a first order reaction as shown in Fig. 10.15 which straightens out the curve caused by the zero order reaction rate of multiple compounds shown in Fig. 10.14.

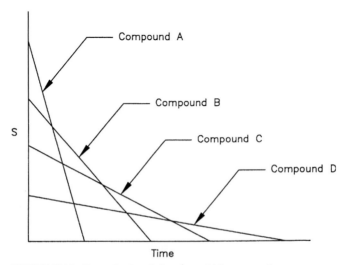

FIGURE 10.14   Zero order degradation rate, multiple compounds.

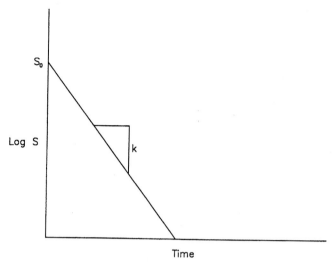

FIGURE 10.15   First order degradation rate, multiple compounds.

The first order equation from which Fig. 10.15 is plotted is shown in Eq. (10.4) which is derived from Eq. (10.5).

$$\frac{S}{S_0} = e^{-kt} \tag{10.4}$$

$$\frac{dS}{dt} = kS \tag{10.5}$$

where S_0 is the initial substrate concentration (mg/l)

The term X (in mg/l) represents total suspended solids (TSS) and the term X_v (mg/l) is used for *volatile suspended solids* (VSS). Since the test for X and X_v as explained in Chapter 3, "Pollutant Classification," can be run quickly and inexpensively compared to BOD, COD, TOC, and TC, many operators will use these tests as a surrogate for the total bacterial population under aeration. If a wastewater consists of completely soluble organics, the total suspended solids would be a reasonable indication of biological or cellular activity since the only suspended solids will be bacterial. If a wastewater consists of soluble and insoluble organics and inorganics, volatile suspended solids is a better indicator of biological activity, since all biological organics will be volatilized in the test.

In a batch or plug flow system, since bacteria build up and are depleted in accordance with Fig. 10.3, X and X_v will be variable with time. In a completely mixed system, X and X_v will be constant. In a first order reaction, Eq. (10.6) can be used to compensate for a variable X or X_v:

$$\frac{S}{S_0} = e^{-kX_v t} \tag{10.6}$$

If certain organics are removed more slowly with time than can be represented by equation (10.6) as shown in Fig. 10.16, the second order equation shown in Eq. (10.7) can be used to determine the k rate:

$$\frac{S}{S_0} = \frac{1}{1 + kt} \tag{10.7}$$

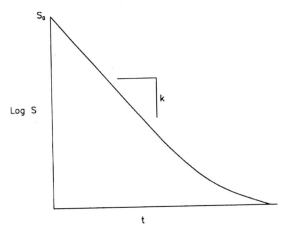

FIGURE 10.16 First order kinetics removal rate of multiple compounds with dragout.

This equation can be plotted as shown in Fig. 10.17. This equation with a variable X_v can be modelled as equation (10.8):

$$\frac{S}{S_0} = \frac{1}{1 + kX_v t} \tag{10.8}$$

In a continuous feed complete mixed process, such as shown in Fig. 10.12, a mass balance approach can be taken to compute the degradation rate as shown in Fig. 10.18.

The flow rate, $Q(\text{ft}^3/\text{s or M}^3/\text{s})$, in must equal the flow rate out. Therefore, a mass balance approach gives the following relationship:

$$QS_0 - QS_e = \frac{dS_e}{dt} V \tag{10.9}$$

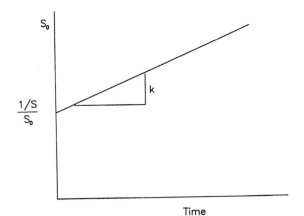

FIGURE 10.17 Second order degradation rate, multiple compounds with drag out.

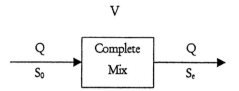

FIGURE 10.18 Continuous feed complete mixed degradation rates schematic.

where S_0 = influent substrate conc (mg/1)
S_e = effluent substrate conc (mg/l)
V = reactor volume (ft^3 or M^3)

or, the influent substrate quantity minus the effluent substrate quantity equals the substrate removal rate times the volume of the complete mixed basin. If Eq. (10.9) is divided by Q and since volume divided by flow equals hydraulic detention time:

$$S_0 - S_e = \frac{dS_e}{dt} t \qquad (10.10)$$

Assuming substrate is removed as a first order reaction as in Eq. (10.5):

$$S_0 - S_e = kS_e t \qquad (10.11)$$

If X_v is variable, Eq. (10.11) can be written as:

$$\frac{S_0 - S_e}{X_v t} = kS_e \qquad (10.12)$$

Therefore, as the concentration of substrate remaining in the basin decreases, the remaining substrate is more difficult to remove and is removed at a slower rate. If an influent varies rapidly in strength, the k rate with a standard variation must be applied.

Equation (10.12) can be plotted as shown in Fig. 10.19.

The non-biodegradable solid fraction (NBDG) will be reduced with high sludge age to zero.

10.3.1.5 Degradation in Wastewaters with Suspended and Colloidal Particles. In addition to the degradation rate equations explained above, which model the substrate removal in a soluble wastewater, many wastewaters have substrate removals which involve a two-phase reaction as shown in Fig. 10.20.

The upper curve depicts soluble solids biologically removed through synthesis and oxidation. The lower curve symbolizes suspended solids physically removed by enmeshment in the activated sludge and colloidal adsorption of colloidal solids on flocculant particles. Some soluble solids are also adsorbed.

The total influent substrate in these wastewaters consists of a combination of soluble substrate and the biodegradable portion of suspended solids as shown in Eq. (10.13):

$$S = S_0 + (f)X_0 \qquad (10.13)$$

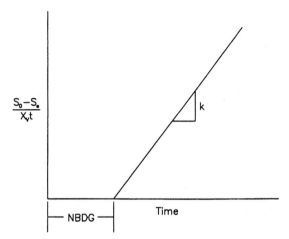

FIGURE 10.19 Complete mixed degradation rates.

where S = total influent substrate (mg/1)
$\quad S_0$ = influent soluble substrate (mg/1)
$\quad f$ = the biodegradable fraction of the influent suspended solids (dimensionless)
$\quad X_0$ = influent suspended solids (mg/1)

10.3.1.6 F/M Ratio. A useful relationship in computing total degradation rates as the influent substrate varies, is the ratio of food to microorganisms or F/M. The food can be depicted as substrate or S_0 and the microorganisms or active mass can be depicted as X_v or MLVSS for the relationship:

$$\frac{F}{M} = \frac{S_0}{X_v} \tag{10.14}$$

As explained in Chapter 3, volatile suspended solids under aeration, known as mixed liquor volatile suspended solids (MLVSS), are an indicator of active biological mass that is as accurate as has been developed. In wastewaters with a constant suspended solids X, to volatile suspended solids X_v ratio, Eq. (10.15) can be used:

$$\frac{F}{M} = \frac{S_0}{X} \tag{10.15}$$

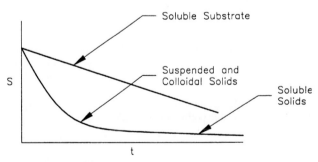

FIGURE 10.20 Two phase degradation rates.

If S_0 is expressed in lb/day and X or X_v in pounds, the units of F/M are 1/days or days^{-1}. If formula 10.14 is multiplied by 24 hours per day, and divided by the detention time in hours:

$$\frac{F}{M} = \frac{24S_0}{X_v t}$$

(10.16)

where S_0 = influent substrate conc (mg/1)
X_v = volatile suspended solids (mg/1)
t = hydraulic detention time (hours)

Bacteria, like humans, require a relatively constant diet of food to remain healthy. Therefore, in order to optimize biological activity, the F/M ratio should remain rather constant. If food (S_0) increases, active mass (X_v) must increase. Excess active mass is obtained from the sludge in the bottom of a settling basin following the activated sludge tank as shown in Fig. 10.21.

RAS is the *return activated sludge* and WAS is the *waste activated sludge* or the excess sludge (ΔX_v) from Eq. (10.18) not needed to maintain the F/M. The *F/M* ratio can be effectively used in designing and operating activated sludge systems.

10.3.1.7 Sludge Age. Another useful relationship in designing and operating activated sludge systems is sludge age, or the total time activated sludge is under aeration.

For a plug flow system which has no sludge recirculation such as a lagoon:

$$\theta_c = t$$

(10.17)

where θ_c = sludge age in days^{-1}
t = hydraulic detention time in days

For a sludge recycle system:

$$\theta_c = \frac{X_v}{\Delta X_v}$$

(10.18)

or, sludge age equals the volatile suspended solids in the tank, X_v, divided by the amount of volatile suspended solids wasted each day (ΔX_v).

In design or operation, a non-recirculation system (Eq. 10.17) can achieve a desired sludge age only by changing tank size. In a recycle system (Eq. 10.18), the desired sludge age can be achieved

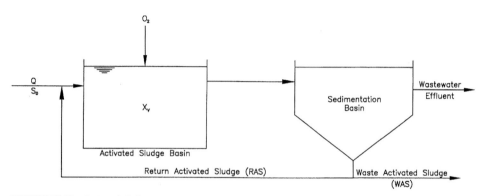

FIGURE 10.21 Activated sludge system.

by controlling the return activated sludge rate to vary the mixed liquor suspended solids or by varying the waste activated sludge rate.

If a recycle plant is operated by either F/M or sludge age, the main control mechanism is the RAS/WAS values.

Eckenfelder has shown that for high F/M and low sludge age, bacterial growth can be dispersed and/or filamentous, and for low F/M and high sludge age, bacterial growth can be dispersed as shown in Fig. 10.22. Jenkins (1986) and Richard (1989) have shown that filamentous bulking can be caused by both low and high F/M ratios. For these reasons, an operator should keep F/M between approximately 0.2 and 2.0 days^{-1} and the sludge age between 0.5 and 5 days.

10.3.1.8   Sludge Production.   The theory of synthesis and oxidation can be used to develop models for predicting sludge produced during these processes and oxygen needed for the processes.

The daily sludge produced can be predicted by determining two constants, one for synthesis and one for oxidation using Eqs. (10.1) and (10.2) as the basis for these relationships. A constant known as the Synthesis Cellular Production Constant, *a* can be determined experimentally or with historical

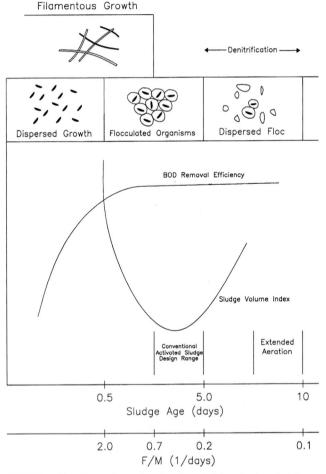

FIGURE 10.22   Flocculation and settling characteristics of activated sludge.

data to indicate the percentage of the substrate removed which is converted into cells or sludge. This constant can be considered as follows using the terms in Eq. (10.1):

$$a = \frac{\text{new cells}}{\text{organics}} \tag{10.19}$$

Likewise a constant known as the oxidation cellular destruction constant, b, can be determined to indicate the percentage of the cells present under aeration which are oxidized into CO_2 and H_2O. This constant can be determined using the terms in Eq. (10.2)

$$b = \frac{CO_2 + H_2O}{\text{Cells}} \tag{10.20}$$

For a soluble waste, these relationships can be utilized to estimate total sludge production by Eq. (10.21) as follows:

$$\Delta X_v = aS_r - bX_v \tag{10.21}$$

where ΔX_v = sludge production (lb/d or kg/d)
  aS_r = portion of substrate removed synthesized to new cells (lb/d or kg/d)
  S_r = substrate removed (lb/d or kg/d)
  bX_v = portion of volatile suspended solids oxidized (lb or kg)
  a = Synthesis Cellular Production Constant
  b = Oxidation Cellular Destruction Constant (days^{-1})

This equation can be plotted from experimentally or historically determined data as shown in Fig. 10.23.

If because of short detention times or sludge ages, there is a fraction of cells which are not biodegraded, Eq. (10.21) can be modified as follows:

$$\Delta X_v = aS_r - bX_d \tag{10.22}$$

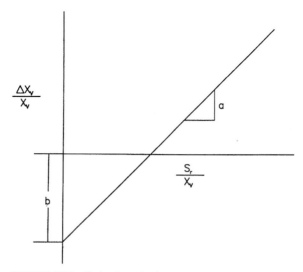

FIGURE 10.23   Sludge determination.

where X_d = degradable VSS (mg/l)

X_d can be determined by a batch type experiment by placing wastewater seeded with activated sludge in a beaker, aerating and testing the COD of the substrate daily. At the sludge age of a system of interest, the substrate removed equals X_d. This is demonstrated in Fig. 10.24.

If a wastewater contains insoluble suspended solids X_i, Eq. (10.21) can be modified as follows:

$$\Delta X_v = fX_i + aS_r - bX_v \tag{10.23}$$

where f = the fraction of influent suspended solids not degraded (NVSS)
X_i = Influent TSS (mg/l)

The value fX_i is the difference between the TSS and VSS. This formula shows that nonvolatile suspended solids in the influent will be removed as sludge without being affected by the biological process and will add to the total quantity of sludge produced.

10.3.1.9 Oxygen Requirements.   The total oxygen requirements for activated sludge can be predicted by determining two additional constants, one for synthesis and one for oxidation. Based on the Eqs. (10.1) and (10.2), it can be assumed that there is a determinable relationship between organics synthesized and the oxygen required. This constant is knows as a'. It can also be assumed that there is a determinable relationship between cells oxidized and the oxygen required. This constant is known as b'. The total oxygen required is therefore the total of the two terms. The total oxygen required for activated sludge is estimated based on Eqs. (10.1) and (10.2) as follows:

$$O_2 = a'Sr + b'X_v \tag{10.24}$$

where O_2 = oxygen required (lb/d or kg/d)
a' = Synthesis Energy Constant, the fraction of substrate oxidized for energy
S_r = substrate removed (lb/d or kg/d)
X_v = volatile suspended solids under aeration as MLVSS (lb or kg)

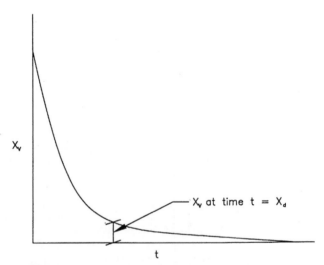

FIGURE 10.24   Determination of degradable VSS.

b' = Oxidation Energy Constant, the fraction of cells oxidized for endogenous respiration

$a'S_r$ = oxygen required per day for cellular oxidation

$b'X_v$ = oxygen required per day for endogenous respiration

The Oxidation Energy Constant, b', can be related to b the Oxidation Cellular Destruction Constant as follows:

$$b' = 1.42b \tag{10.25}$$

This relationship is determined by balancing the simplified typical cell structure $C_5H_7NO_2$ as follows:

$$C_5H_7NO_2 + 5O_2 \rightarrow 5CO_2 + 2H_2O + NH_3 \tag{10.26}$$

$C_5H_7NO_2$:

C: 5 mole × 12 g/mole = 60 g

H: 7 mole × 1 g/mole = 7 g

N: 1 mole × 14 g/mole = 14 g

O: 2 mole × 16 g/mole = 32 g

$\overline{C_5H_7NO_2 = 113 \text{ g}}$

$5O_2$:

$\underline{\text{O: } 5 \times 2 \text{ moles} \times 16 \text{ g/mole} = 160 \text{ g}}$

$5O_2 = 160$ g

$$\frac{160 \text{ g}}{113 \text{ g}} = 1.42 \text{ g O}_2/\text{g}$$

Therefore, when a cell is oxidized, it takes 1.42 g O_2 per g VSS.

The sum of the synthesis energy constant, a', and the synthesis cellular production constant, a, equals 1.0 for substrate as measured by COD, TOC, or TC:

$$a + a' = 1.0 \tag{10.27}$$

But since BOD_5 is only a fraction of COD (see Chapter 3), this is not true for substrate measured by BOD_5.

Equation (10.27) shows that all substrate is converted into either new cells or oxidized into CO_2 and H_2O.

Equation (10.24) can be plotted from experimentally or from historically determined data as shown in Fig. 10.25.

10.3.1.10 *Nitrogen Compound Removal with Biological Treatment.* Biological treatment can be used to convert ammonia to nitrate nitrogen aerobically and then to nitrogen gas under anoxic conditions. Figure 10.26 shows the nitrogen cycle.

Nitrobacter bacteria are three times as efficient as nitrosomonous bacteria, therefore the limiting growth rate is for nitrosomonous bacteria. Their growth rate is shown in Fig. 10.27 as related to sludge age.

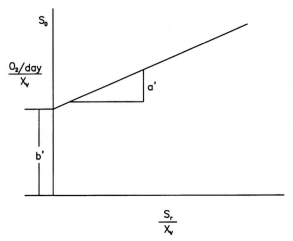

FIGURE 10.25   Oxygen determination.

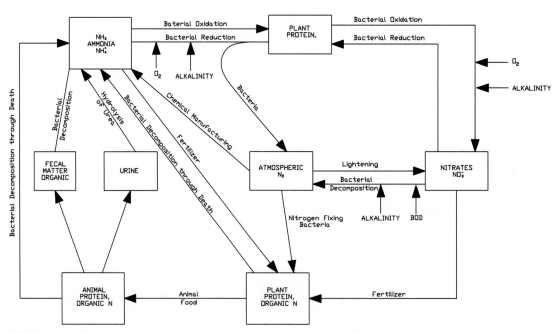

FIGURE 10.26   The nitrogen cycle.

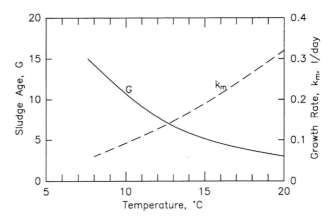

FIGURE 10.27   Growth rate of nitrosomonous bacteria.

Since the nitrogenous bacteria are less hardy than carbonaceous bacteria, they must be removed sequentially after the carbonaceous bacteria are removed, but a small amount of carbonaceous BOD_5 must remain as food for the nitrogenous bacteria. Figure 10.28 shows the relative growth of the two bacterial types.

Because of the competitive nature of the two types of bacteria, the two processes should be kept separate. This can be achieved by treating for BOD (carbonaceous bacteria) removal and ammonia NH_3 (nitrogenous bacteria) removal in series as shown in Fig. 10.29.

This system will completely separate the two types of bacteria. Since, as shown in Fig. 10.27, the reaction rate of nitrogenous bacteria is highly temperature dependent, it has been shown that in warmer climate, the system shown in Fig. 10.30 can be used to approximate the ideal system of Fig. 10.29.

The attached growth process is also used for nitrification of ammonia to nitrate. The nitrification reaction is biological and is accomplished by two genera of microorganisms: nitrosomonas and nitrobacter according to the following general reactions:

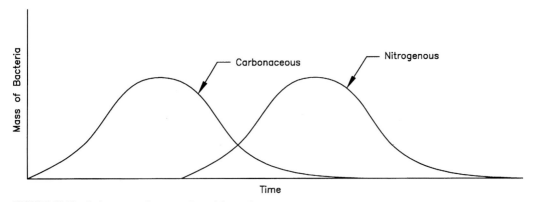

FIGURE 10.28   Carbonaceous nitrogenous bacterial growth.

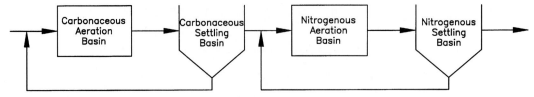

FIGURE 10.29 Biological removal of BOD_5 and NH_3.

$$NH_4^+ \ O_2 + HCO_3^- \ \underset{\longrightarrow}{\text{Nitrosomonas}} \ \text{New Cells} + NO_2^- + H_2O + H_2CO_3$$

$$NO_2 + NH_4^+ + O_2 + H_2CO_3 + HCO_3^- \ \text{Nitrobacter New Cells} + NO_3^- + H_2O$$

As can been seen from the above equations, new biological cells are generated in each type of system. In carbonaceous BOD removal, endogenous respiration of the old cells partially offsets this growth. In each system, the excess biological growth is removed from the media or the discs by shearing forces as the wastewater washes over the media or the discs rotate through the wastewater. The removed cells are passed into a clarifier where they are settled out of suspension by gravity.

10.3.1.11 Alternate Ammonia Removal Systems. As shown in the nitrogen cycle, anoxic decomposition can concert NO_3 to N_2, but the NH_3 must first be converted to NO_3 using some sort of oxidative means. Several biological treatment systems, such as oxidation ditches with brush type aerators, have been shown to successfully remove ammonia and even nitrate. These systems use plug flow with aeration at limited locations as shown in Figs. 10.31 and 10.32.

The NO_3 is converted to N_2 in the anoxic zones between the aerators. In these zones, the bacteria are starved for oxygen and when re-aerated, they uptake an excess amount of nitrogen as a nutrient. The anoxic zone must not become anaerobic or the carbonaceous removal will deteriorate. The sequencing batch reactor described in this chapter can be used to provide an anoxic zone when aeration is ceased.

Oxidation ponds and aerated lagoons have normally been unsuccessful in removing ammonia. Some of these systems approach plug flow, especially with multiple basins, but the MLVSS toward the end of the pond system is typically too low to maintain the nitrogenous bacterial growth. The system, shown in Fig. 10.33, has been developed to allow ammonia removal in an oxidation pond or aerated lagoon through the addition of a packed tower.

10.3.2 Aerobic Suspended Growth Treatability Studies

The purpose of running treatability studies is to model at bench scale a full scale treatment unit. As a minimum, two steps should be completed for the design of systems as shown in Fig. 10.34.

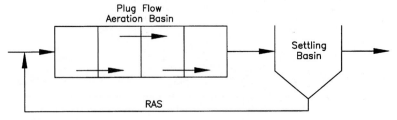

FIGURE 10.30 Warm temperature biological removal BOD_5 and NH_3.

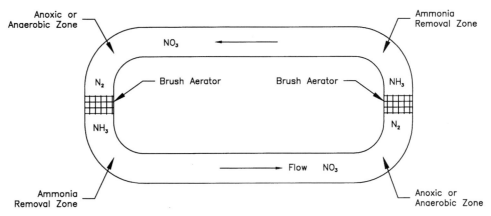

FIGURE 10.31 Oxidation ditch system.

FIGURE 10.32 Brush aerator.

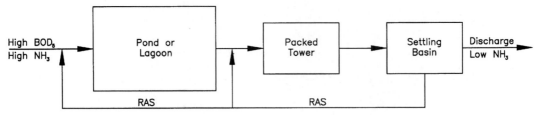

FIGURE 10.33   Lagoon ammonia removal system.

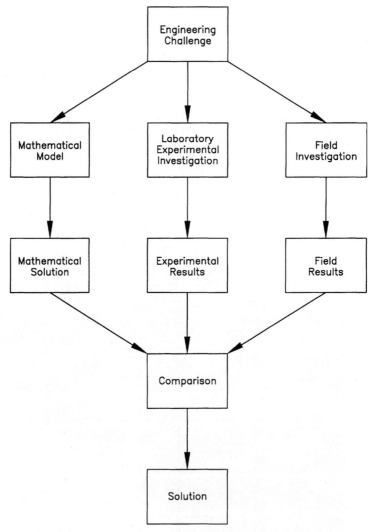

FIGURE 10.34   Optimum design procedures.

The best rule for designing a treatability study is to simulate as nearly as possible, field conditions as follows:

- Resistant and non-reactive materials
- Proportional size and shape
- Large enough to minimize wall effects
- Proportional flow rates
- Approximate vessel entrance conditions
- Appropriate vessel exit conditions
- Adequate mixing
- Adequate oxygen addition
- Proper nutrient addition
- Representative wastewater temperature

With a properly run treatability study, full scale design and operating parameters and conditions can be successfully estimated. A treatability system is normally used to develop design criteria, but it can also be run during full scale plant operation to anticipate the response of the treatment system to changes in wastewater characteristics or increased loading rate.

Treatability studies for aerobic suspended growth systems can be batch or flow-through.

The batch suspended growth aerobic system is representative of a batch or aerated lagoon system; the flow-through suspended growth aerobic system is representative of an activated sludge system.

The batch suspended growth anaerobic system is representative of a fill and draw digester; the flow-through suspended growth anaerobic system is representative of the second stage of a digester or an anaerobic treatment system; and the flow-through attached growth system is representative of an anaerobic packed column or counter current fluidized bed reactor.

Figures 10.35 to 10.38 are schematic diagrams and photographs of these various treatability systems. (See Tables 10.2 and 10.3).

10.3.2.1 Aerobic Suspended Growth Treatability Study Hints

- Reasonable basin sizes from 1 quart to 5 gallons.
- Polyethylene is easily workable if chemically resistant.
- Vessel shapes should be similar to full scale.

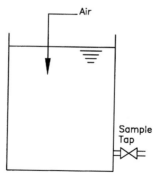

FIGURE 10.35 Schematic diagram of a batch suspended growth aerobic system.

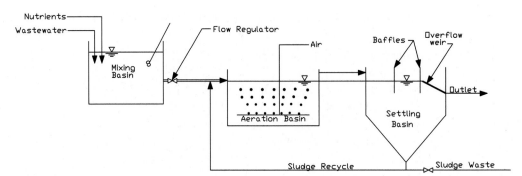

FIGURE 10.36 Diagram of a flow-through suspended growth aerobic system.

- Mixer may be needed to prevent settling.
- Flow regulator can be a pipette for close control and small flows or a plastic ball valve.
- Piping can be tubes of Tygon or other chemically resistant material. *Caution:* materials such as Tygon containing plasticizers can solublize phthalates with certain wastewater chemicals.
- An aquarium blower can be used for aeration.
- Diffusers can be aquarium diffusers for small bubble diffusers and perforated tubing for large bubble diffusers.
- Air movement in basin should be similar to full scale.
- Velocities around, over and under baffles should be minimized.
- Sludge can be returned manually periodically or with a pump such as a peristaltic pump.

10.3.3 Aerobic Suspended Growth Pilot Studies

For wastewater systems with highly variable influent and for large expensive systems, it is recommended that when an existing wastewater stream is accessible, a pilot wastewater treatment system be designed and constructed using the results of treatability studies, to further develop design criteria. The pilot system can use a slip stream off of the full scale wastewater flow. The advantages of a pilot study are as follows:

- Wastewater variability including, batch discharges, washdowns, spills stormwater flows, etc. is experienced.

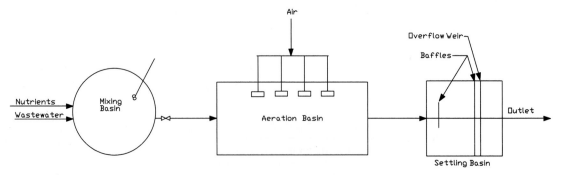

FIGURE 10.37 Plan of flow-through suspended growth aerobic system.

FIGURE 10.38   Aerobic treatability study.

TABLE 10.2   Inhibition from Inorganic Substances

Inorganic	Moderately inhibitory (mg/L)	Strongly inhibitory (mg/L)
Sodium	3,500–5,500	8,000
Potassium	2,500–4,500	12,000
Calcium	2,500–4,500	8,000
Magnesium	1,000–1,500	3,000
Ammonia–N	1,500–3,000	3,000
Sulfide	200	200
Copper		0.5 soluble
		50–70 total
Nickel		3.0 soluble
		30 total
Zinc		1.0 soluble
		3.0 soluble
Hexavalent chromium		
		200–600 total
Trivalent chromium		180–420 total

From Jones, 1990.

TABLE 10.3   Inhibition from Organic Substances

Organic	Concentration for 50% inhibition (mg/L)
Acetaldehyde	440
Acrolein	10
Chloroform	15
Creolin	1
Cyanide	1
Dinitrophenol	40
Ethylbenzene	340
Fluorinated hydrocarbons	1
Formaldehyde	70
Long chain fatty acids	500
Tannins	700

• Ambient temperature variations are experienced.

• Design parameter refinements or changes are developed.

• Operators experience wastewater plant operation on a small scale.

The pilot system expands on the theory of design development presented in Fig. 10.34 as follows.

As in the case of treatability studies, the design of a pilot system should simulate full scale field conditions as follows:

• Resistant and non-reactive materials

• Proportional size and shape

• Proportional flow rates

• Appropriate entrance conditions

• Appropriate exit conditions

• Adequate mixing

• Adequate oxygen addition

• Adequate nutrient addition

A pilot study is, in reality, a small complete wastewater treatment plant obtaining its influent from the existing wastewater treatment stream with all of its flow concentration and temperature variability.

A schematic diagram for a pilot system will be identical to the full scale system and can easily be varied in capacity, schematic layout and treatment parameters to experiment with alternative processes or design changes within a process.

Figures 10.39 and 10.40 show photographs of anaerobic pilot studies.

10.3.4   Aerobic Suspended Growth Design

Approximate design parameters will be presented for batch and flow-through suspended growth systems.

Design suggestions and options will be suggested for each of these process types. It is recommended that additional design criteria be developed using treatability studies and pilot studies.

In general, the sections of this Handbook will present the following information:

• Flow schematics

• Unit size and treatment efficiency

• Nutrient addition

FIGURE 10.39   Aerobic pilot study.

- Sludge production
- Aeration requirements
- Design

10.3.4.1   Batch Suspended Growth Aerobic Systems Design

10.3.4.1.1   Scope.   This section will describe the following types of batch suspended growth aerobic systems:

- Batch reactor
- Sequencing batch reactor
- Aerated lagoon
- Oxidation pond

10.3.4.1.2   Flow Schematics.   Please turn to section 10.3.4.1.3 for basin size and treatment efficiency recommendations once the proper schematic is selected.

The basic flow schematic in Fig. 10.41 must be followed for all batch suspended growth aerobic systems. Any system will need nutrient addition if nutrient deficiency exists. The optional systems are a fill and draw system (also known as a *Sequencing Batch Reactor* or SBR), an aerated lagoon and an oxidation pond. Figures 10.42 to 10.45 are schematic diagrams represent these systems (see also Fig. 10.46 for a photo on an aerated lagoon system).

10.3.4.1.3   Basin Size and Treatment efficiency.   Please turn to section 10.3.1.4.5 regarding nutrient additions once the basin size and treatment efficiency is selected.

The treatment efficiency of biological treatment system is directly proportional to basin size because of the typical growth curves shown in Fig. 10.47.

FIGURE 10.40   Aerobic treatability study.

10.3.4.1.3.1   Batch and Sequencing Batch Reactors.   Batch reactor systems are sized based on sludge age equaling hydraulic detention time. A zero, first or second order relation should be used for sizing the system once the best curve fit is known from treatability studies. Equations for zero, first, and second order reactions are given in Eqs. (10.3) through (10.7).

The plotted curves for these reactions are shown in Figs. 10.13 through 10.17.

Again, the best fit curve of these options should be used for sizing the basin. Please refer to previous sections of this chapter for the theoretical explanation. The choice between the formulae is summarized in Table 10.4. In all of these formulae, the units must be consistent in terms of concentration or mass loading and the k rate must be in reciprocal days or hours with time correspondingly in days or hours.

10.3.4.1.3.2   Aerated Lagoons.   Biological growth in an aerated lagoon follows the degradation rates of a batch or plug flow system as modified by the Synthesis Cellular Production Constant a, and the Oxidation Cellular Destruction Constant b, since in an aerated lagoon, the biological cells experience a once through flow pattern similar to a batch flow system with the synthesis and oxidation

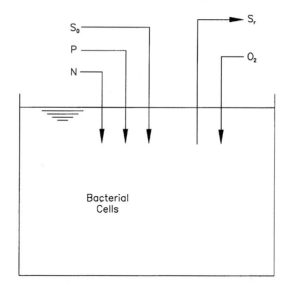

FIGURE 10.41 Batch suspended growth aerobic system schematic.

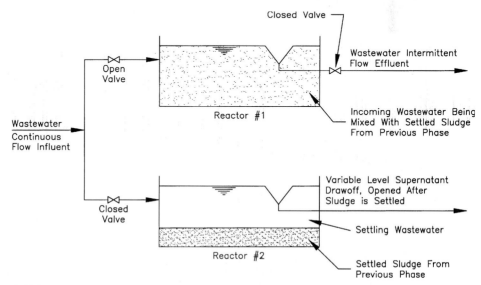

FIGURE 10.42 Batch suspended growth aerobic system sequencing batch reactor (SBR) schematic.

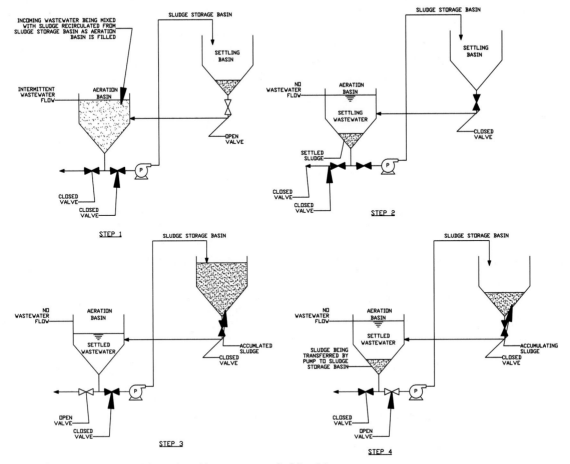

FIGURE 10.43   Batch suspended growth aerobic system stages of a full and draw system.

of a continuous mixed system. Aerated lagoons therefore have been successfully modeled by modifying Eq. (10.7):

$$\frac{S}{S_o} = \frac{1}{1 + kt}$$

to Eq. (10.37)

$$\frac{S}{S_o} = \frac{1 + bt}{akt} \tag{10.37}$$

where S = effluent soluble substrate (mg/1)
S_o = influent soluble substrate, (mg/1)
b = Oxidation Cellular Destruction Constant (1/TIME)
a = Synthesis Cellular Production Constant
k = reaction rate, (1/TIME)
t = hydraulic detention time, (TIME)

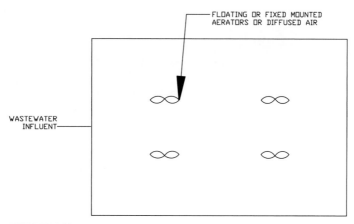

FIGURE 10.44   Aerated lagoon system schematic.

A second approach for sizing aerated lagoons as recommended by the Ten States Standard is:

$$t = \frac{E}{2.3k_1 \times (100 - E)} \tag{10.38}$$

where t = detention time, (days)

  E = percent of BOD_5 to be removed

  k_1 = reaction coefficient, 0.12/day at 68°F (20°C) and 0.06/day at 34°F (1°C) for domestic wastewater

Since algae will likely grow in the settling lagoon, contributing to TSS and BOD in the effluent, an alternate schematic is shown in Fig. 10.48. A photograph of the sand filter is shown in Fig. 10.49. Figure 10.44 shows a schematic of a normal Aerated Lagoon System.

10.3.4.1.3.3   Oxidation Pond.   An oxidation pond is essentially a plug flow system operated under the same zero, first and second order equations listed in Table 10.4. The most commonly used design is represented in Eq. (10.7). For soluble and insoluble BOD, this relationship may be modified to:

$$\frac{S}{S_o} = fX + \frac{1}{1 + kt} \tag{10.39}$$

where f = fraction of influent suspended solids not converted to soluble BOD,

  approximately $= \dfrac{NVSS}{VSS}$

  X = influent suspended solids, (mg/1)

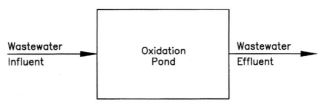

FIGURE 10.45   Oxidation pond system schematic.

FIGURE 10.46   Aerated lagoon photograph. (Ashbrook-Simon-Hartely)

Figure 10.52 shows the schematic of a typical oxidation pond. Since this system discharges algae at a high concentration which contributes to TSS and BOD, an alternative schematic is shown in Fig. 10.50.

10.3.4.1.4   Temperature Correction.   The reaction rate k, of biological activity will vary with temperature in accordance with the following equation:

$$k_T = k_{20}\theta^{(T-20)} \tag{10.40}$$

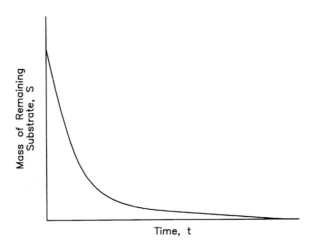

FIGURE 10.47   Biological treatment efficiency.

TABLE 10.4 Basin Size Equation Summary

	Equation	Application
$\dfrac{dS}{dt} = k$	10.3	Single soluble substrate
$\dfrac{S}{S_o} = e^{-kt}$	10.4	Multiple soluble substrates
$\dfrac{dS}{dt} = kS$	10.5	Multiple soluble substrates
$\dfrac{S}{S_o} = e^{-kX_vt}$	10.6	Multiple soluble substrates, variable X_v
$\dfrac{S}{S_o} = \dfrac{1}{1 + kt}$	10.7	Multiple soluble substrate, containing slowly removed organics

where k_T = reaction rate at temperature T
$\quad k_{20}$ = reaction rate at 20°C
$\quad \theta$ = temperature correction factor (see Table 10.5)

With colder ambient temperatures, the wastewater temperature approaches the ambient air temperatures if a large surface area is exposed to the atmosphere, such as in an aerated lagoon or an oxidation pond. The water temperature in a pond can be calculated in these conditions using the following empirical calculation:

$$\frac{T_i - T_w}{12(T_w - T_a)} = \frac{t}{d} \tag{10.41}$$

where T_i = influent temperature, °F
$\quad T_a$ = air temperature, °F

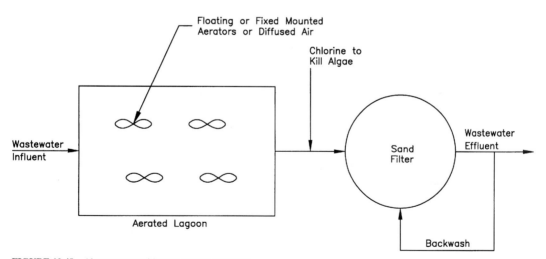

FIGURE 10.48   Alternate aerated lagoon system schematic.

FIGURE 10.49   Alternate lagoon sand filter system.

T_w = water temperature, °F
t = detention time, days
d = basin depth, feet

10.3.4.1.5 Nutrient Addition.   Please turn to section 10.3.4.1.6 for sludge production prediction recommendations once nutrient additions are selected.

As explained in section 10.3.1.1, a typical cell under oxidation has a concentration of 5 parts nitrogen and 1 part phosphorous per 100 parts cellular material. If cellular material is tested by BOD_5, the ratio of BOD_5 to nitrogen to phosphorous is therefore 100:5:1. For example, if a wastewater has a COD/BOD_5 ratio of 1.85 then, COD equals 1.85 times BOD_5, the ratio of COD to nitrogen to phosphorous should be 185:5:1.

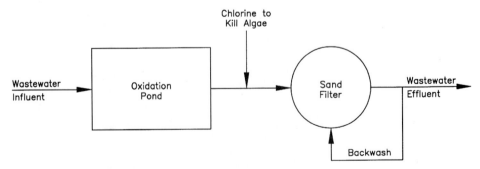

FIGURE 10.50   Alternative oxidation pond system schematic.

TABLE 10.5 Temperature Correction Factors

Parameter	θ
Aerated Lagoons	1.04
Oxidation Ponds	1.075

It should be noted that nitrogen is computed as N and phosphorous as P. Table 10.6 gives the percent of available nitrogen and phosphorous in commonly used compounds. Nutrients can be added in a liquid or solid form depending on the availability of the compound. In a solid form, they would normally be dissolved in a feed tank and fed into the wastewater stream in a liquid form for easy control.

A liquid feed system normally consists only of the nutrient mix tank, a mixer and the metering pump.

Design cautions suggest that dusty nutrients should be handled in a way as to prevent particulate air pollution or worker hazard by using dust collectors or buying chemicals in bulk so that transfers are completely covered. Another caution is that for bag feed of solid pollutants, platforms should be used for easy access.

See Figs. 10.51 and 10.52 for a schematic and a photograph of a nutrient feed system.

10.3.4.1.6 Sludge Production. Please turn to section 10.3.4.1.7 for aeration requirements once sludge production is estimated.

As shown in Fig. 10.53, sludge production increases with time for the first portion of aeration and then is reduced to a minimal or non-existent level (NBDR).

Conventional design criteria and biological treatment wisdom predicts sludge production as ΔX_v depending on sludge age, which in the case of batch treatment, equals hydraulic detention time. As explained in the theory section of this chapter, a non-biodegradable organic residue (NBDR) has been recognized as a minimum amount of sludge production with long detention times.

The author has observed that at exceptionally long detention times, the NBDR has been biologically broken down so that the sludge produced equals the influent inorganic total suspended solids. In several full scale aerated lagoons incorporating sludge recycling, after initial sludge accumulation of about 24 inches in depth, there is no longer any appreciable sludge accumulation, even after 15 years. Details of this design are given in paragraph 10.3.4.1.7.

In an aerobic batch suspended growth reactor system, the sludge produced is collected on the bottom of the tank and removed as waste sludge. In an Oxidation Pond, the sludge produced is accumulated on the bottom of the pond and removed every few months or years by dredging as needed. During this period of accumulation, depending on the depth of the pond, some of the sludge may be degraded anoxicly and/or anaerobicly so that the total volume removed will be less than calculated. In an Aerated Lagoon, the sludge will also be accumulated on the bottom usually between aerators, depending on the depth of the lagoon and the number, spacing and size of the aerators, and removed every few months or years. Again, the sludge is anoxicly and/or anaerobicly degraded during this period.

TABLE 10.6 Nutrient Percentage in Compounds

Compound	Percent N	Percent P
Urea	47	0
Ammonium nitrate	35	0
Mono Ammonium phosphate	12	27
Di Ammonium phosphate	21	23
Phosphoric acid	0	32

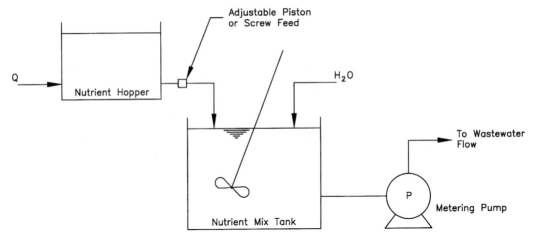

FIGURE 10.51 Solid nutrient feed system.

The following approach can be used to predict the sludge produced in a batch system but in an Oxidation Pond, the variables are too great to predict sludge production, and in an Aerated Lagoon, the sludge produced can be estimated but not necessarily the final sludge removed.

Recall Eq. (10.21) for calculating sludge production as follows:

$$\Delta X_v = aS_r - bX_v \tag{10.21}$$

FIGURE 10.52 Photograph of solid nutrient feed system.

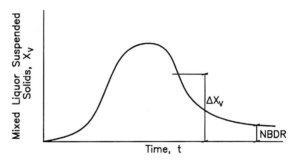

FIGURE 10.53 Sludge production vs. time.

The constants a and b can be obtained by batch treatability studies as explained in the "Theory" section of this chapter. Typical values are shown in Table 10.7.

The constant a remains relatively constant with time as X_v increases logarithmically with time, in accordance with Fig. 10.62. The constant, b is variable with time with longer detention times and must be determined for the appropriate detention time to be of use in predicting sludge production.

In a batch system, the sludge produced equals the total solids in dry pounds per day accumulated after settling, less the amount retained for subsequent cycles, such as in a sequencing batch reactor.

10.3.4.1.7 Aeration Requirements.   Please turn to section 10.3.4.1.8 for design hints once aeration requirements are selected.

In a batch system, as in any biological treatment system, the air required is the greater of the following:

• The air required to supply oxygen to optimize the bacterial metabolism
• The air required to adequately mix organics, bacteria and nutrients

The metabolism oxygen requirement should be computed using Eq. (10.24) as from the "Theory" section.

$$O_2 = a'S_r + b'X_v \tag{10.24}$$

The constants a' and b' can be obtained by batch treatability studies as explained in the "Theory" section of this chapter. Typical values are shown in Table 10.8.

Also, as explained in the "Theory" section of this chapter:

$$b' = 1.42b \tag{10.25}$$

TABLE 10.7 Synthesis Oxidation Contants

Wastewater type	a	b
Aerated Lagoon Activated Sludge	0.45	0.07
Domestic Sewage	0.75	0.075
Petroleum	0.50–0.60	0.10–0.15
Organic chemicals	0.30–0.75	0.05–0.20
Pharmaceuticals	0.70–0.80	0.15–0.20
Breweries	0.50–0.60	008–0.12
Pulp and paper	0.45–055	0.07–0.10

TABLE 10.8   Synthesis and Oxidation Energy Constants

Wastewater Type	a' (BOD basis)	b'
Domestic	0.50	0.11
Petroleum	0.40–0.80	0.14–0.21
Organic chemicals	0.30–0.75	0.07–0.28
Pharmaceuticals	0.40–0.50	0.11–0.17
Breweries	0.45–0.55	0.21–0.28
Pulp and paper	0.65–0.85	0.10–0.14

and in the case of substrate measured by COD; TOC or TC:

$$a + a' = 1.0 \tag{10.27}$$

An approximation of oxygen required in an aerated lagoon is 1.4 times the pounds of BOD_5 per day. Another more rational approach, based on the theoretical oxygen requirement is presented as follows:

$$O_2/\text{day} = 1.42S_o + X_v e \tag{10.42}$$

where O_2/day = the pounds of O_2 required per day
S_o = the influent BOD_5, (lbs/day or kg/d)
$X_v e$ = the total volatile suspended solids in the lagoon effluent, (lbs/day or kg/d)

This oxygen can be supplied either by blowing air through the liquid in the form of bubbles (diffused air) or by spraying the water through the air in conjunction with the movement of the air entrained water through the treatment tank (mechanical aerators).

Once the quantity of air needed for biological metabolism is computed as described above, the pounds per day can be converted to cubic feet per minute in order to size an air compressor as follows:

$$\text{cfm} = \frac{108 \cdot \left(\dfrac{\text{lb}}{\text{day}}\right)}{0.21} \tag{10.43}$$

where cfm = cubic feet per minute of air
lb/day = pounds of oxygen per day
0.21 = percent O_2 in air
108 = conversion factor at STP

Surface aerators are designed by their manufacturer to provide a certain efficiency in terms of $lbO_2/\text{H.P. hr}$ or $kgO_2/\text{kw hr}$ at standard conditions of 20°C and D.O. of 0.0 mg/l.

The conversion to field conditions for surface aerators is as follows:

$$N = N_o \left(\frac{B(C_w - C_L)}{9.17}\right) \alpha 1.024^{(T-20)} \tag{10.44}$$

where N = $lbO_2/\text{HP hr}$ or $kg\ O_2/\text{kw hr}$ under field conditions
K_o = $lbO_2/\text{HP hr}$ or $kgO_2/\text{kw hr}$ under standard conditions, 20°C,
D.O. = 0.0 mg/l

B = salinity/surface tension corrections factor, approximately 1.0
C_w = O_2 saturation at given T and altitude mg/l
C_L = operating O_2 concentration, mg/l, approximately 2.0 mg/l
T = temperature, °C
α = oxygen transfer correction factor for waste, approximately 0.80 to 0.85

The air requirements for mixing in an aerated lagoon using a surface aerator are:

- Approximately 0.1 HP/1000 gallons for partial mixing
- Approximately 0.5 HP/1000 gallons for complete mixing

For diffused air, the blower size is as follows:

- Approximately 4 to 6 scfm/1000 ft³ for partial mixing
- Approximately 20 to 30 scfm/1000 ft³ for complete mixing

Mixing can also be provided by mechanical mixers. In larger systems, mixing will be more efficient and will use less energy with a mechanical mixer than with aeration.

In an Oxidation Pond, the oxygen produced by algae can be calculated by:

$$O_2 = 0.25 \ FS$$

where: O_2 = Oxygen produced in pounds/day
F = Oxygenation factor
S = Solar radiation factor

The oxygenation factor, F varies between O and 4 and is the ratio of the weight of oxygen produced and the ultimate influent BOD.

The Solar Radiation Factor, S, average 150 to 270 calories/cm²—day in the summer and 0 to 270 cal/cm²—day in the winter depending on the latitude.

10.3.4.1.8   Design Hints Summary

10.3.4.1.8.1   Oxidation Ponds

- 20–50 lb BOD/acre/day (22 to 56 kg BOD/ha/day)
- 150–300 persons/acre domestic sewage (370 to 374 persons/ha)
- t = 50 days storage (200 persons/acre, 3.5 feet deep at 200 gal/capita/day) (500 persons/ha, 1.0 m deep, 750 L/capita/day)
- Depth 2.5 to 5 feet with one foot of variation for mosquito control
- Vented inlet pipe should discharge 12 inches minimum below minimum water level
- Side slopes below water—2:1
- Side slopes above water—4:1
- Minimum free board—2 feet
- Minimum width of top of dike—8 feet
- Number of ponds—two
- First pond—75 percent of total area

10.3.4.1.8.2   Aerated Lagoons

- Approximately 0.1 HP/1000 gals or 4 to 6 scfm/1000 ft³ for partial mixing
- Approximately 0.5 HP/1000 gals or 20 to 30 scfm/1000 ft³ for complete mix
- Depth 4 to 15 feet

- Side slopes below water—2:1
- Side slope above water—4:1
- Freeboard 2 to 5 feet
- Protect freeboard from erosion with synthetic liner or rip rap.
- Minimum width of top of dike—8 feet
- Shape as close to round or square as possible around each aerator (rectangular for multiple aerators)
- Corners-rounded to 20 feet radius

10.3.4.2 Flow-through Suspended Growth Aerobic Systems Design

10.3.4.2.1 Scope. This section will present design recommendations for the following types of flow-through suspended grown aerobic systems:

- Activated sludge
- Extended aeration
- Aerated lagoons with sludge recirculation
- Ammonia removal systems

10.3.4.2.2 Flow Schematics. Once a flow schematic is selected, please turn to section 10.3.4.2.3 for basin size and treatment recommendations.

The basic flow of a flow-through suspended growth aerobic biological treatment system is shown in Fig. 10.54.

The most common options for the design of an activated sludge system are as follows:

- Methods for mixing RAS
- Method for mixing influent wastewater
- Detention time
- Basin size and Treatment efficiencies
- Aeration Method
- Nutrient Addition
- Sludge Production
- Aeration Requirements
- Cation Influence.

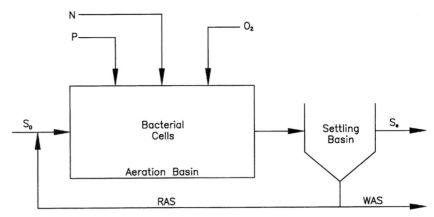

FIGURE 10.54 Typical flow-through suspended growth aerobic system.

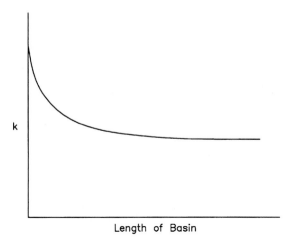

FIGURE 10.55 Contact stabilization system.

Figures 10.55, 10.56–10.72 and 10.74–10.76 are schematics which incorporate these options and have been designed by the author and successfully operated.

10.3.4.2.2.1 Methods for Mixing RAS

Step sludge: Step sludge is an activated sludge system which attempts to increase the overall efficiency of the process by feeding return activated sludge at several points along the basin length.

In biological treatment, even in a completely mixed basin, the effluent ratio F/M of substrate (S_o) and bacterial cells (MLVSS), is not equal to the ratio at the influent when the food and return activated sludge are both fed at the influent end of the basin as shown in Fig. 10.54. The k rate in such a system will decrease over the length of the basin (Fig. 10.55), similar to a batch treatment system but not as radically because of partial mixing.

To modulate this tendency, the step sludge system as shown in Fig. 10.57 has been developed. The resulting k rate variation in such a system is shown in Fig. 10.66.

Contact stabilization: Contact stabilization is an activated sludge system which attempts to reduce the total hydraulic detention time of an activated sludge system by increasing the k rate by mixing aerated return activated sludge with substrate (see Fig. 10.60).

Since the k rate can be increased by increasing X_v, the contact stabilization process aerates return activated sludge in a large contact tank at a high X_v or low F/M. This process takes advantage of the steeper part of the typical k rate curve as shown in Fig. 10.58 or the first part of Fig. 10.3, by mixing the influent substrate in an aeration tank following the contact tank.

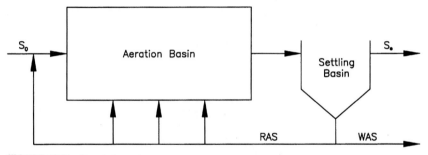

FIGURE 10.56 Step sludge system.

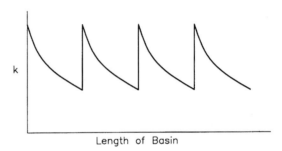

FIGURE 10.57   Step sludge k rate.

Biological contactor: A biological contactor (Fig. 10.61) is used to allow the biological degradation process to occur in the contactor and to allow efficient mixing of the return activated sludge with food.

This process, like contact stabilization, takes advantage of the very steepest part of the k rate curve as well as increasing the efficiency of mixing in a relatively small biological contactor. The process is also used to control filamentous biological growth.

Biological Contactors are also known as Selectors, when used to control sludge bulking by filamentous organisms, or to be aid in obtaining higher MLSS levels in order to meet lower effluent limits including BOD, FOG, ammonia and/or total nitrogen.

There has been much experimentation used to develop theory and design criteria for Biological Contactors, some of which is shown below under "Design Hints". These Contactors have been aerobic, anoxic and anaerobic in nature. The goal is to produce a short term, high F/M condition which favors certain floc forming bacteria and minimizers filamentous bacteria.

Most research postulates that there are two mechanisms involved in successful selector operation, high rate metabolism, which rapidly breaks down organic food and produces MLSS when the electron donor (oxygen if aerobic, oxygen, nitrite and nitrate, if anoxic, sulfate or carbon dioxide, if anaerobic) is used in biosynthesis; and substrate storage, where further substrate (organic food) is stored in the MLSS for uses in bacterial growth in the aeration basin. Figure 10.60 shows a typical Biological Contactor or Selector schematic.

Intimate mixing: The theory of intimate mixing is the complete mixing of return activated sludge with substrate sufficiently upstream from the aeration basin (Fig. 10.62). This can also be accomplished by installing a mixing basin between the RAS/Food mix point and the aeration basin as in Fig. 10.63. This system is identical to the biological contactor system except that the detention time in the intimate mix tank is typically much less than in a biological contactor.

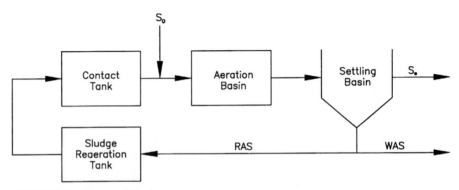

FIGURE 10.58   Contact stabilization system.

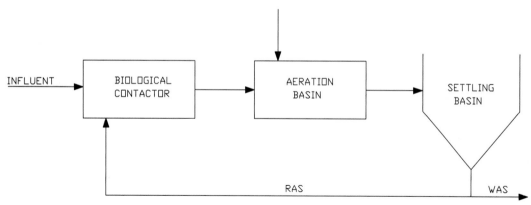

FIGURE 10.59   Contact stabilization system.

Series Activated Sludge Systems": The author has designed several activated sludge plants in series where the first plant was designed to remove highly biodegradable organics such as sugar, starch or alcohols, and the second or subsequent plants were designed to remove more refractory organics such as phthalates, lignins or tannins. The schematics for those plants are in accordance with Figure 10.57 with two or three separate systems.

In these plants, since bacteria selectively choose these easiest to biodegrade food first, the first stage system acclimates it's MLSS to the easily biodegradable food and keeps that bacteria, in the form of settled sludge, within its system. Then separately, the second stage system acclimates it's MLSS to the refractory food since there is little easily biodegradable food left, and recirculates that bacteria, in the form of settled sludge back to its aeration basin. This system allows a much lower level of refractory residual COD in the effluent that in a single stage system.

10.3.4.2.2.2   Method for Mixing Influent Wastewater

Sludge flocculation system: The sludge flocculation system (Fig. 10.64) is a method of using return activated sludge as a flocculent for the mixed liquor being transferred to the settling basin to optimize settleability. In this system, RAS is bled into the flocculation tank. Successful experiments have also been run which show that potable water treatment plant settling basin sediment can be used as a coagulant or flocculent in a sludge flocculation system.

Step aeration: This system, shown in Fig. 10.57, is a method for mixing influent wastewater at various locations along the length of the aeration basin to increase the overall efficiency of the process.

This system allows a k rate similar to that shown above in Fig. 10.59.

Hybrid Activated Sludge: Experimentation and some full scale activated sludge systems have been installed using a combination of suspended growth and fixed growth in order to increase the MLSS and therefore the amount of food that can be treated in a given volume of aeration basin. This hybrid system is usually bottom fed with the level controlled by the surface discharge weir.

This system can potentially support MLSS levels in the tens of thousands as opposed to levels in the thousands for conventional suspended growth activated sludge. Therefore a greater concentration of food (S_0) can be treated.

10.3.4.2.2.3   Detention Time

Activated sludge methods with varying detention times: The modification of Fig. 10.3 shown in Figure 10.60, along with Table 10.9, gives approximate hydraulic retention times of various activated sludge methods.

The theory of *contact stabilization* (Fig. 10.57) is to take advantage of the steepest part of the k rate curve, or the logarithmic growth section of Fig. 10.60a to minimize the hydraulic retention time (HRT) and reactor volume.

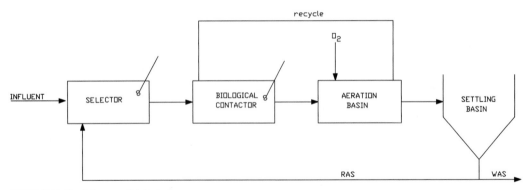

FIGURE 10.60 Selector and biological contactor.

Activated sludge (Fig. 10.54) includes all suspended growth aerobic biological systems with HRTs between contact stabilization and extended aeration.

Extended aeration (Fig. 10.54) attempts to minimize sludge production and operational difficulties by carrying HRT to the point at which the biomass produced has substantially passed the death phase and become asymptotic to the abscissa in the microbiological growth curve.

Aerated Lagoon with Returned Activated Sludge

The theory of the *aerated lagoon with return activated sludge* is that at very high HRTs, the "non-biodegradable suspended solids" are degraded and only inorganic sludge is produced. Detention times greater than 10 days have also been shown by the author to eliminate *sera daphnia* toxicity in domestic sewage.

Aerated lagoons with sludge recirculation without ammonia removal should have a schematic diagram as in Fig. 10.61. There should be no volatile WAS accumulation with this system as explained under the "Theory of Biological Treatment."

When insufficient carbonaceous bacteria exist to allow adequate settling, or if dispersed growth bacteria prevent settling, more efficient treatment can be provided with the schematic illustrated in Fig. 10.62.

A pilot study was run based on an extensive investigation by state regulatory officials of an existing wastewater treatment plant which produced, no *sera daphnia* toxicity or organic sludge. A photograph of the pilot study is shown in Figure 10.39–10.40. In this study, hydraulic retention time (HRT) was varied between one and 14 days on a flow pumped out of a municipal sewage treatment plant influent. It was found that *sera daphnia* toxicly consistently was present below 10 day HRT and absent at greater than 10 days HRT. In addition, all sludge was recycled and no accumulation

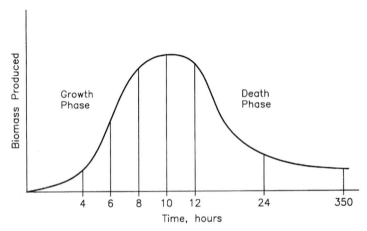

FIGURE 10.60a Biomass produced versus hydraulic retention time.

TABLE 10.9 Activated Sludge Hydraulic
Retention Times

Process	HRT, hours
Contact stabilization	6
Activated sludge	8–12
Extended aeration	24
Ammonia removal systems	24
Aerated lagoon with RAS	350

was observed during the tests after the initial accumulation. After this study, it was concluded that all toxicity and organic sludge accumulation was eliminated after 10 to 14 days HRT with an inorganic sludge accumulation in the lagoons of approximately one inch per year. The mechanism of sludge degradation appears to be anoxic conditions in the first lagooon with D.O.'s from 0.1 mg/l at the sludge interface to 0.5 mg/l on the surface, and aerobic conditions in the 2nd lagoon with D.O.'s from 1.3 mg/l at the sludge interface to 4.6 mg/l on the surface.

Further design criteria for this activated aerated lagoon treatment is shown under "Design Hints".

Based on the pilot plant results, a full scale lagoon was constructed which has experienced similar results.

When ammonia is required to be removed from an aerated lagoon system, the schematic in Fig. 10.69 is recommended.

In colder climates the schematic must be modified as in Fig. 10.70.

For the biological removal of ammonia in an activated sludge system, Figure 10.29 is recommended.

For ammonia removal systems is warm climates, the flow schematic is shown in Fig. 10.30.

Ammonia Removal Systems Ammonia removed design criteria are shown under Design Hints below. A photograph of an ammonia removal system is shown in Figure 10.31

10.3.4.2.2.4 Basin Size and Treatment Efficiencies. Once the basin size and treatment efficiencies are selected, please turn to section 10.7.2.4 for nutrient addition recommendations.

An activated sludge system of any detention time which incorporates sludge recirculation can be sized using the following equation:

$$S_0 - S_e = kS_e t \qquad (10.45)$$

If X_v is constant

where S_0 = influent substrate, (mg/l)
   S_e = effluent substrate, (mg/l)
   k = the reaction rate, (days^{-1})
   t = detention time, (days)

If X_v is variable:

$$S_o - S_e = kS_e X_v t \qquad (10.46)$$

where X_v = MLVSS in mg/l
   The k rate can be corrected for temperature by following section 10.7.1.3.4.

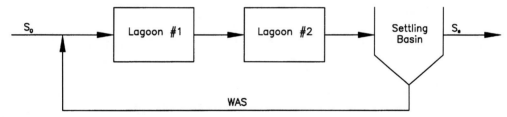

FIGURE 10.61 Aerated lagoon with sludge recirculation.

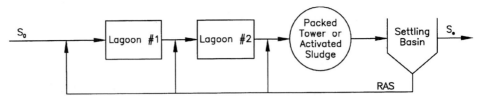

FIGURE 10.62   Alternate aerated lagoon with sludge recirculation.

10.3.4.2.3 Nutrient Addition. Once nutrient additions are selected, please turn to section 10.7.2.5 for sludge production recommendations.

The reader is referred to paragraph 10.3.4.1.5. It is recommended that nutrients (see Table 10.12) be added to optimize the biological treatment efficiency.

10.3.4.2.4 Sludge Production.   Once sludge production is estimated, pleased turn to section 10.7.2.6 for aeration requirement recommendations.

The sludge produced per day in an activated sludge system can be estimated, by Eq. (10.21.) Table 10.9 can be used for coefficient *a* and *b*.

10.3.4.2.5 Aeration Requirements.   After aeration requirements are determinded, please turn to section 10.7.2.7 for design hints.

The oxygen requirements for biological metabolism should be estimated using Eq. (10.24). Please see Table 10.10 for typical a' and b' value and Eqs. (10.25) and (10.27) relating *a* to a' and *b* to b'.

An approximation of oxygen needed in an activated sludge system is:

$$O_2/\text{day} = 1.4S_0 \qquad (10.47)$$

where O_2/day = oxygen required (lbs/day or kg/d)
$\quad\quad S_0$ = influent substrate (lbs/day or kg/d)

A second, more rational approximation is:

$$O_2/\text{day} = 1.42S_0 + X_v e \qquad (10.48)$$

Where O_2/day = oxygen required (lbs/day or kg/d)
$\quad\quad S_0$ = influent substrate, (lbs/day or kg/d)
$\quad X_v e$ = total VSS in lagoon effluent, (lb/day or kg/d)

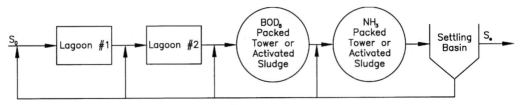

FIGURE 10.63   Warm weather aerated lagoon with sludge recirculation and NH_3 removal.

TABLE 10.10 Nutrient Addition Recommendations for a COD/BOD$_5$ Ratio of 1.85 and a TOC/BOD$_5$ Ratio of 0.69

| Basis (5) | Nutrients in lbs/day | | |
	S	N	P
BOD$_5$	100	5	1
COD	185	5	1
TOC	69	5	1

When diffused air is used and the air requirements is known in lbs/day, the conversion to cubic feet per minute can be determined using Eq. (10.43).

For surface aerators, Eq. (10.44) should be used to correct from manufacturer's standard conditions to field conditions. Air requirements for mixing can be estimated.

For ammonia removal, oxygen added must equal at least 4.6 times ammonia loading in lbs/day.

10.3.4.2.6 Cation Influence. Recent laboratory experiments and full scale trials have shown that monovalent cations interact with negatively charged biopolymers in activated sludge to change floc structure, causing settling and dewatering deterioration, whereas divalent cations tend to improve settling and dewatering properties. The author of this Handbook has had direct full scale experience with this affect in an industrial plant in which poor settleability was resolved with cation adjustment.

It is recommended that the equivalent ratio of monovalent to divalent cations be less than two to optimize flocculation and therefore settling. An overabundance of sodium (monovalent) ions is typical, especially in industrial wastewaters because sodium chloride is so commonly used in plant processes as well as in wastewater pH adjustment. Calcium (divalent) compounds can be used to balance the ratio.

10.3.4.2.7 Design Hints

10.3.4.2.7.1 Detention Time
- Activated sludge

$$t = 6 \text{ to } 20 \text{ hours}$$

- Extended aeration

$$t = 20 \text{ to } 48 \text{ hours}$$

- Aerated lagoon with sludge recirculation

$$t = 10 \text{ days (sera daphnia inhibition removal)}$$
$$t = 15 \text{ days (no organic sludge accumulation)}$$

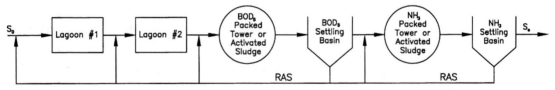

FIGURE 10.64 Cold weather aerated lagoon with sludge recirculation and NH$_3$ removal.

10.3.4.2.7.2   Oxygen Requirements

- Activated sludge
- $O_2/day = 1500$ ft^3/lb BOD$_5$ (94 M^3/kg BOD$_5$)
- Extended aeration
- $O_2/day = 2050$ ft^3/lb BOD$_5$ (128 M^3/kg BOD$_5$)
- Ammonia removal
- $O_2/day = 3000$ ft^3/ 16NH$_3$ (18 8 M$_3$/kg NH$_3$)

10.3.4.2.7.3   Mixing Requirements

- Approximately 0.1 HP/1000 gals or 4–6 scfm/1000 ft^3 for partial mixing
- Approximately 0.5 HP/1000 gals or 20–30 scfm/1000 ft^3 for complete mixing

    Air requirements for mixing are as follows:

- Diffused Air
- Approximately 4 to 6 scfm/1000 ft^3 for partial mixing
- Approximately 20 to 30 scfm/1000 ft^3 for complete mixing

    Surface Aeration:

- Approximately 0.1 HP/1000 gals for partial mixing
- Approximately 0.5 HP/1000 gals for complete mixing

    Mechanical mixing is typically more economical than mixing with air, especially for larger systems. Other general design recommendations are as follows

- Depth: 10 to 30 feet
- Diffused air width to depth ratio: 1:1 or 2:1
- Diffused air freeboard: 18 inches minimum
- Surface aeration freeboard: 36 inches minimum
- Dissolved oxygen: 2.0 mg/l minimum

    except for aerated lagoons with sludge recirculation as explained above

- The return activated sludge for domestic sewage is equal to approximately 35 percent of the BOD going into the aeration tank plus 90 percent of the sludge ash. The sludge ash equals 30 percent of the suspended solids going into the aeration tank (the other 70 percent are volatile solids).
- See tables 10.11 and 10.12 for Ten State Standards loading and RAS recommendations.

Organics Removed Biologically versus Organics Removed by Air Stripping

The removal of volatile organic compounds (VOC's) from biological treatment processes occurs through several mechanisms including biodegradation, adsorption into solids, effluent discharge and air stripping or volatilization to the atmosphere. For many industries in the U.S., fugitive emissions, resulting from volatilization of VOC's is regulated by the EPA.

The estimation of the percent of substrate removed by biodegradation versus the amount by air stripping has been historically difficult to quantify, primarily due to the inconsistency of biodegradability constants. An employee of E. Roberts Alley & Associates, Inc. has developed a procedure for modeling the percent air stripping for VOC" which can be used to estimate fugitive emissions and/or biodegradation efficiencies. (Cheng, 2006). Other methods have been proposed but with lesser accuracies.

Cheng, Joey, Modeling Air Stripping of Volatile Organic Compounds (VOCs) From Biological Treatment Processes"

May, 2006, Master of Science Thesis, Vanderbilt University.

TABLE 10.11 Ten State Standards Loading Recommendations

Process	Organic loading lbs $BOD_5/d/1000$ ft^3 or (kg/d/m^3)	F/M ratio lbs BOD_5, day/lb MLVSS	MLSS mg/l
Activated sludge	49 (0.64)	0.2–0.5	1000–3000
Contact stabilization			
Contact tank	15 (0.24)	0.2–0.6	1000–3000
Aeration tank	35 (0.56)		
Extended aeration and ammonia removal	15 (0.24)	0.05–0.1	3000–5000

10.3.5 The theory of Aerobic Attached Growth Biological Treatment

10.3.5.1 Scope. This section will describe the flow through attached growth aerobic growth systems including trickling filters and rotating biological contactors. Batch aerobic attached growth systems are not commonly used since settling occurs in a separate unit and there is therefore no reduction of the number of basins required.

10.3.5.2 The Theory of Aerobic Attached Growth Biological Treatment. It is difficult to operate a batch attached growth system since the bacteria attached to the medium require continuous food to metabolize. Attached growth systems are inherently plug flow since the bacteria are statically attached to the media and are not free to be continuously mixed. Therefore, this section of the Handbook concentrates on *continuous* plug flow attached growth systems, specifically, trickling filters and rotating biological contactors.

A trickling filter is an attached growth aerobic treatment process in which wastewater is applied in a downflow manner onto natural or synthetic filter media. Biomass attaches to the filter media and as the wastewater trickles over the biomass, organics are removed from the bulk liquid via standard respiration reactions. In trickling filters, oxygen is supplied to the biomass through natural convective currents due to the exothermic carbonaceous BOD_5 degradation reactions. In trickling filters that are primarily nitrifying towers, forced ventilation is required since the nitrification reaction does not produce sufficient heat to provide for natural convective ventilation.

The rotating biological contactor (RBC) is an attached growth aerobic treatment system in which large diameter plastic discs are partially submerged in the wastewater. The discs are closely spaced and rotated on a common center drive. During rotation, the attached biological growth is exposed to the wastewater for organic decomposition and to the ambient air for oxygen uptake.

In attached growth biological treatment, a single compound will be removed in accordance with the zero order reaction from Eq. (10.28).

$$\frac{ds}{dt} = k \tag{10.28}$$

TABLE 10.12 Ten State Standards RAS Recommendations

Process	% RAS minimum	% RAS maximum
Activated sludge	15	100
Step aeration	15	100
Contact stabilization	50	150
Extended aeration	50	150
Ammonia removal	50	200

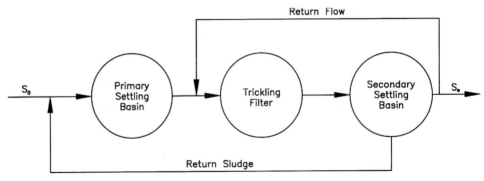

FIGURE 10.65 Single stage attached growth aerobic system.

Likewise, a multiple compound wastewater will have soluble organic compounds removed in accordance with the first order reaction time in Eq. (10.29) and Fig. (10.15):

$$\frac{S}{S_o} = e^{-kt} \qquad (10.29)$$

In an attached growth system, the sludge age is equal to the hydraulic detention time, therefore in formula 10.29:

$$t = \frac{V}{Q} \qquad (10.30)$$

where t = detention time (s)
$\qquad V$ = volume (ft^3 or M^3)
$\qquad Q$ = wastewater flow (ft^3/s or M^3/s)

and since the detention time in a plug flow system is directly to the reaction rate k, times the length or depth of the contact chamber divided by the flow in terms of ft^3/s or gpm per ft^2 of media surface area, Eq. (10.30) can be expressed as:

$$t = \frac{kD}{Q_1} \qquad (10.31)$$

where t = detention time(s)
$\qquad k$ = reaction rate (unitless)
$\qquad D$ = length or depth of media (ft or M)
$\qquad Q_1$ = hydraulic loading rate (ft^3/s, M^3/s or gpm/ft^2)

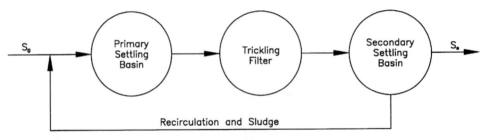

FIGURE 10.66 Alternate single stage attached growth aerobic system.

Q_1 can be converted directly to the term, million gallons per acre per day (MGAD) used historically in trickling filter design by manipulating units.

For soluble wastes, Eqs. (10.29) and (10.31) can be combined as follows:

$$\frac{S_e}{S_o} = e^{-kD/Q} \qquad (10.32)$$

When recirculation is practiced:

$$\frac{S_e}{S_o} = \frac{e^{-kD/Q}}{(1 - N) - Ne^{-kD/Q}} \qquad (10.33)$$

where N = recirculation ratio = R/Q
$\quad R$ = return flow (ft^3/s, M^3/s, or gpm/ft^2)
$\quad Q$ = flow (ft^3/s, M^3/s or gpm/ft^2)

The following empirical approach has been developed and can be adopted for more than one unit but it has generally not proven as accurate as the first order reaction. For a single stage system as shown in Figs. (10.78) and (10.79).

$$E_1 = \frac{100}{1 + 0.0561\left(\dfrac{w}{VF}\right)^{0.5}} \qquad (10.34)$$

where E = percent BOD removed
$\quad w$ = BOD load applied, (lb/day)
$\quad V$ = volume of filter media in place, (ft$^3 \times 10^{-3}$)
$\quad F$ = recirculation factor

Where

$$F = \frac{1 + N}{(1 + 0.1N)^2} \qquad (10.35)$$

and N = R/Q

where N = recirculation ratio
$\quad R$ = return flow
$\quad Q$ = flow

For the second state of two-stage system:

$$E_2 = \frac{100}{1 + \left[\dfrac{0.0561}{(1 - E_1)}\right]\left(\dfrac{w_2}{VF}\right)^{0.5}} \qquad (10.36)$$

where E_2 = percent BOD removed in the second stage
$\quad E_1$ = percent BOD removal in the first stage
$\quad w_2$ = BOD load applied to the second stage (lbs/day)

Two stage trickling filters are shown in Figures 10.80–10.82. The biological degradation of the organics in all attached growth systems including the RBC process follows the same general synthesis oxidation reactions as in all aerobic treatment systems:

$$\text{Organics} + O_2 + \text{Nutrients} \xrightarrow{\text{bacteria}} CO_2 + H_2O + \text{New Cells}$$

and

$$\text{Cells} + O_2 + \text{Nutrients} \xrightarrow{\text{bacteria}} CO_2 + H_2O + \text{energy}$$

10.3.5.3   Flow Schematics.   Trickling filters can be single or double state and can use various methods of recycling as shown in Figs. 10.67 to 10.70.

Note: A primary Setting Basin must precede the trickling filter in order to remove solids which may plug the surface distribution system.

The purpose of return flow is to prevent drying of the attached growth during periods of no flow or low flow.

Note: A two stage system can be designed to be automatically changed to a high flow schematic if flow exceeds a preset rate by converting to parallel flow. The efficiency for each side of the parallel system would then be computed as a single stage system.

The velocity of the influent flow through the distribution nozzles causes the distribution arm to rotate. The nozzles must be placed to load the surface uniformly and the timing speed must be sufficient to keep the surface wetted. At low flows, the recirculation flow explained above is not only used to increase efficiency but also to prevent surface drying during periods of low flow.

Figures 10.71, 10.72 and 10.84 show a synthetic media attached growth system and the fixed distribution system. Figure 10.71 shows a Rotating Biological Contactor.

10.3.5.4   Aerobic Attached Growth Treatability Studies.   Aerobic attached growth treatability studies can simulate a trickling filter or a rotating biological contactor (RBC). In either case, the study must be a continuous flow-through system and forward flows can vary between 0.5 and 5.0 gpd. Larger flows require a large amount of raw wastewater and can become cumbersome. If larger flows are required to be tested, an on-site pilot-scale system should be considered instead of a bench-scale system.

A trickling filter treatability study consists of a raw wastewater storage tank, a recirculation tank, transfer pumps, the filter tower, and the media.

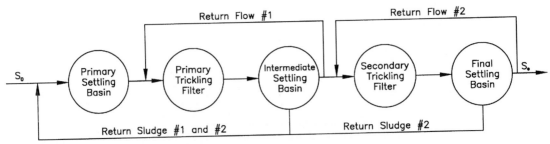

FIGURE 10.67   Double stage attached growth aerobic system.

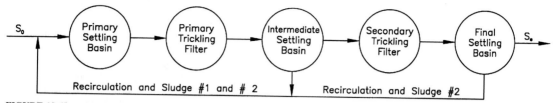

FIGURE 10.68 Alternate double stage attached growth aerobic system.

10.3.5.5 Aerobic Attached Growth Treatability Study Hints

- Distribution plate can be a perforated plate with holes large enough to prevent surface tension restrictions and small enough to get even distribution.
- Packed bed media should be inert plastic or ceramic shapes designed to maximize surface area. See laboratory supply catalogues.
- False bottom similar in design to distribution plate with holes somewhat smaller than media
- Same settling basin design as for aerobic system

An RBC treatability setup is more complicated than that of a trickling filter. RBCs are sized using an organic loading rate. To find the most effective loading rate, a bench scale system can be used prior to larger pilot-scale or full-scale systems being constructed. An RBC treatability study would use a continuous flow-through setup in which several small perforated discs were rotated through a basin on a common center drive.

To begin an RBC treatability, a biomass seed should be obtained from a biological treatment system. Preferably, the seed should come from a full-scale RBC system or a full-scale trickling filter. These seeds would reduce the time required for attachment to the bench-scale discs. The seed should be recirculated through the system with a portion of the raw wastewater until a significant biomass film has accumulated onto the discs. At that point, the RBC unit should be operated as a flow-through system.

During flow-through operation, the effluent from the RBC unit should flow to a clarifier in which biomass that has sloughed off of the discs can be settled and returned to the RBC unit. The raw wastewater should be maintained at a constant flow rate for a minimum of one to two weeks. Once steady state effluent BOD_5 or COD concentrations have been achieved at a certain flow, the flow rate should be increased and then maintained again until steady state effluent concentrations are achieved. This procedure should be followed for several flow rates in order to determine the reaction kinetics of the system.

Table 10.14 gives typical design parameters to follow during the treatability setup.

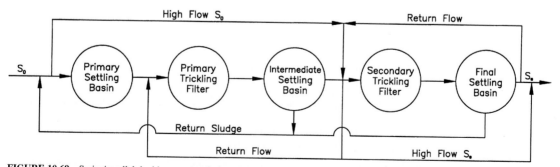

FIGURE 10.69 Series/parallel double stage attached growth aerobic system.

In a bench-scale treatability study, small flow rates of 1 to 2 gpd should be used to avoid having to store large amounts of raw wastewater in the lab. At these flow rates, the RBC is quite small which can lead to plugging of solids between the discs. Therefore, in a bench-scale application, the majority of the suspended solids must be removed prior to the RBC unit. This can be done on a batch basis with the raw wastewater or using a flow-through settling tank (Fig. 10.41)

Another factor in proper RBC design is the dissolved oxygen content of the wastewater in the RBC unit. During the treatability study, DO should be monitored on a daily basis. If the DO drops below 2 mg/L, supplemental aeration is required to provide enough oxygen to the microorganisms for proper performance.

The size of the subsequent clarifier can also be determined form settling tests conducted during the treatability study. For clarifier sizing based on laboratory settling tests, please consult section 8.5 of this Handbook.

Please refer to para 10.3.3 "Aerobic suspended growth pilot studies" for general comments also applicable to attached growth systems.

10.3.5.6   Design Hints.   As explained in the "Theory" section of this Chapter, attached growth systems can be designed rationally or empirically. Equations (10.32) through (10.36) can be used for deign.

For trickling filter design, the following are recommendations for consideration:

- Standard rate hydraulic loading: maximum—4 mgd/acre
- Standard rate organic loading: maximum—9.2 lb BOD_5/day 1000 ft³
- High rate stone media hydraulic loading: 10 to 40 MGD/acre/day
- High rate stone media organic loading: maximum—69 lb BOD_5/day/1000ft³
- High rate, synthetic media, hydraulic loading: maximum—100 MG/acre/day
- High rate, synthetic media, organic loading: maximum—120 lb BOD_5/day/1000 ft³
- Filter depth: stone media: 6 to 10 feet

   plastic media: 10 to 40 feet
- Ventilation required minimum: 1.4 lbs oxygen/lb BOD_5 (forced air required for covered filters)
- Synthetic media for carbonaceous Substrate removal—specific surface area less than 30 ft²/ft³ (100 m²/m³)
- Synthetic media for nitrogenous substrate removal—specific surface area less than 45 ft²/ft³ (150 m²/m³)
- Surface distribution: maximum deviation plus or minus 10 percent of average application rate of all points
- Clearance between distribution and surface: 12 inches minimum
- Underdrain inlet openings—minimum area of 15 percent of filter surface area
- Freeboard: stone filters: minimum—3 feet

   plastic filters: minimum—4 feet
- Recirculation rate: 0.5:1 to 4:1 (recirculation:influent flow).
- Nutrient addition at 100:5:1, BOD_5:N:P.

10.3.5.7   Rotating Biological Contactor Design Hints

The design of a RBC process can be conducted using two methods. The first method uses organic loading determination from treatability studies and the second method uses the overall removal rate coefficient. The first method involves the results from the treatability study as discussed in previous sections:

1. Determine the optimum organic loading rate from treatability results.

2. Calculate the surface area required for the given flow rate and organic concentration.

3. Determine the maximum organic loading using a peaking factor (the ratio of the highest influent organic concentration and the average influent organic concentration).

4. The maximum organic loading rate should be less than 6 pounds soluble BOD_5 per 1000 ft^2 per day.

The second method for sizing a RBC process is as follows:

1. Determine the reaction rate coefficient (k) from the treatability study results.

2. Calculate the required surface area using the following equation:

$$A = Q(S_o - S)/kS \qquad (10.50)$$

where A = surface area ft^2 (m^2)
Q = flow rate ft^3/s (m^3/s)
S_o = influent BOD_5 concentration (mg/1)
S = desired effluent BOD_5 concentration (mg/l)
k = reaction rate coefficient ft/s (m/s)

3. Determine the maximum hydraulic loading rate using the following equation:

$$Q/A = [(Q/A)S_r]_{max}/(S_o - [(Q/A)S_r]_{max.}/k) \qquad (10.51)$$

Where: Sr = BOD_5 removed

Nitrification RBC systems can be similarly sized by substituting BOD_5 parameters for NH_3.

In order to optimize the removal efficiency and size of an RBC unit, most RBS processes are designed in stages.

Pretreatment for an RBC process may involve suspended solids removal, pH adjustment, chemical addition (nutrients, alkalinity), organic load equalization, and flow equalization. For proper design of each of these treatment systems, please consult the appropriate section in this Handbook.

For both carbonaceous BOD_5 removal and nitrification, a clarifier downstream of the treatment unit is required to remove biomass from suspension prior to discharge. For complete nitrogen removal, denitrification is required. Denitrification involves the conversion of nitrate to molecular nitrogen under anoxic conditions. For denitrification design, please consult the appropriate section in this Handbook.

Correct selection of the equipment used in an RBC process plays an important role in the overall system's success in treating the wastewater.

As stated earlier, pretreatment processes may be required depending on the wastewater characteristics. Pretreatment equipment include any or all of the following:

1. Chemical conditioning

    (1) pH

        1. Tankage

        2. Sensor, transmitters, controllers

        3. Chemical feed pumps and storage

            a. Alkaline

            b. Acid

    (2) Alkalinity

        1. Tankage

        2. Sensor, transmitters, controllers

        3. Chemical feed pump and storage

   2. Equalization tank

   (1) Organic loading

    1. Tankage

    2. Mixer

      a. Mechanical mixers

      b. Diffused air mixing

      c. Hydraulic mixing

   (2) Volumetric loading

    1. Tankage

    2. Level controls

    3. Effluent pumps

   3. Primary clarification

   (1) Tank

   (2) Influent structure

   (3) Overflow weirs

   (4) Scraper mechanism and drive

   (5) Sludge pump(s)

   4. RBC reactor

   (1) Shafts

   (2) Media

   (3) Drive system

   (4) Tank

   (5) Enclosure

The installation of a rotating biological contactor system may be conducted by an RBC equipment supplier, sub-contracted by the supplier, or sub-contracted by the purchaser. Whomever performs the installation, the proper contract documents should be signed, including a legal agreement on price, completion date, insurance, and responsibilities. The degree of complexity of the installation will depend on the flow rate of wastewater to be treated, the degree of automation of the system, and the required pretreatment technologies.

10.4 ANAEROBIC BIOLOGICAL TREATMENT

10.4.1. The Theory of Anaerobic Biological Treatment

Anaerobic wastewater treatment is a biological treatment process that utilizes anaerobic microorganisms to biologically degrade organic constituents in wastewater. Anaerobic biological treatment involves a series of bacterial reactions, the slowest of which will determine the speed of the total treatment. Methane gas production is the rate limiting and therefore controlling, therefore anaerobic biological treatment design and operation should concentrate on providing the most favorable conditions for methane production. Anaerobic treatment processes are used to stabilize biological sludges prior to dewatering and/or final disposal. Anaerobic treatment processes are also used to treat a variety of industrial wastes. The organic matter present in the substrate is biologically decomposed to CO_2 and methane in the absence of molecular oxygen. Hydrogen sulfide is also produced when sulfur compounds are present in the wastewater. Anaerobic decomposition occurs in the following three steps.

10.4.1.1 Hydrolysis.   The first step comprises the hydrolytic conversion of high molecular weight organics into lower molecular weight organics suitable for further degradation.

10.4.1.2 Acidogenesis.   The next step involves the biological conversion of the lower molecular weight organics into volatile fatty acids (VFAs) such as acetic, propionic, butyric, and valeric acids.

10.4.1.3 Methanogenesis.   Finally, methane producing bacteria convert the VFFAs to methane and carbon dioxide.

Figure 10.70 graphically demonstrates the mechanisms through which organics are anaerobically degraded.

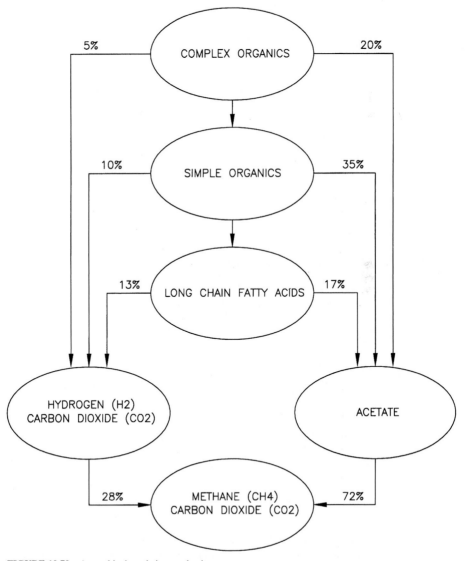

FIGURE 10.70   Anaerobic degradation mechanisms.

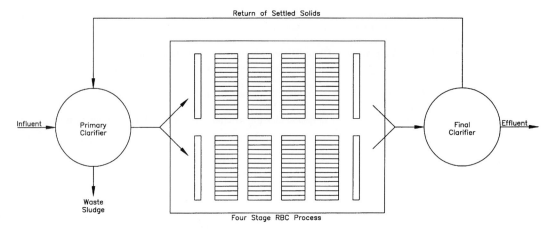

Plan View

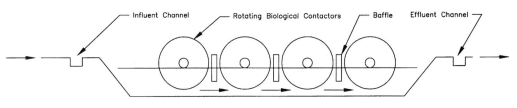

Section View

FIGURE 10.70a   Schematic diagram of a flow-through RBC system.

10.4.1.4 Reactor Configurations.   An anaerobic treatment process can be designed in several different configurations. The two varieties of anaerobic processes are attached growth and suspended growth. Suspended growth processes include anaerobic sludge digestion and the anaerobic contact process. The main attached growth processes are the anaerobic filter, the anaerobic fluidized bed, and the upflow anaerobic sludge blanket (UASB). Of the two general process configurations, the attached growth systems offer several advantages over the suspended growth systems for the treatment of industrial wastewater; 1) increased biomass immobilization; 2) higher biomass concentrations thus shorter hydraulic retention times; and 3) process stability. Typical anaerobic treatment configurations include:

1. Anaerobic digestion
2. Anaerobic contact process (Fig. 10.71)
3. Anaerobic filter (Fig. 10.72)
4. Anaerobic fluidized bed (Fig. 10.73)
5. Upflow anaerobic sludge blanket (Fig. 10.74)

Anaerobic digestion: This treatment configuration is used predominately for the stabilization of biological sludges. Anaerobic digestion can be categorized by the rate at which the anaerobic decomposition occurs: standard and high rate. A standard rate anaerobic digester is an unheated, unmixed reactor where detention times typically range from 30 to 60 days. A high rate digester is heated and completely mixed and usually requires between 10 and 15 days for complete sludge stabilization. Most current anaerobic digestion systems uses two reactors in series. The first reactor is a heated, completely mixed high rate digester and the second is an unheated, unmixed standard

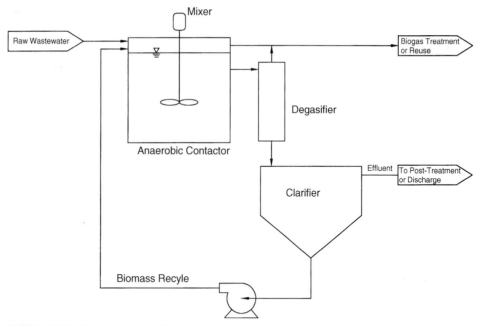

FIGURE 10.71 Schematic of anaerobic contact process.

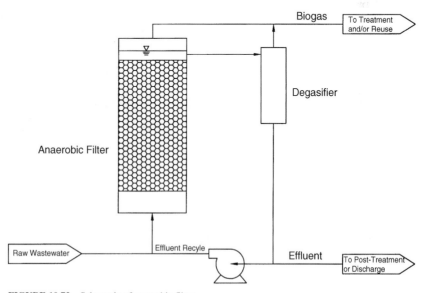

FIGURE 10.72 Schematic of anaerobic filter.

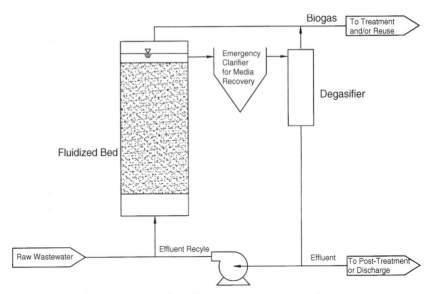

FIGURE 10.73   Schematic of anaerobic fluidized bed.

rate digester. In this system, the digestion occurs in the first digester and separation of float, sludge and supernatant occurs in the second digester. Figure 10.75 shows a two stage anaerobic digester.

Stage one typically is a constant level digester with a fixed cover to collect gas produced. Since solids and liquid are periodically drawn from stage two, the cover should float on the gas produced, or lacking gas, on the liquid.

Stage one can be mixed with gas produced and heated with energy produced by that gas as shown in Fig. 10.75.

Anaerobic contact process: The anaerobic contact process is a suspended growth treatment process that involves a reaction vessel followed by a clarifier The settled solids from the clarifier

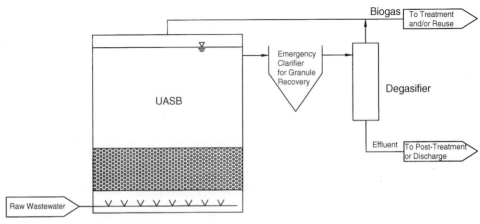

FIGURE 10.74   Schematic of UASB.

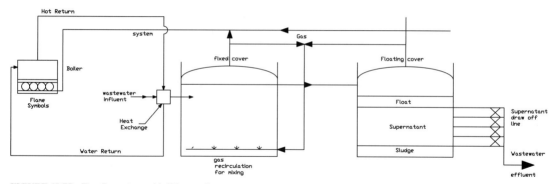

FIGURE 10.75   Two Stage Anaerobic Digester System.

are either recycled back to the reaction vessel or wasted to further sludge treatment. This configuration is very similar to the common aerobic activated sludge process and thus is a configuration with which engineers can most easily identify.

Anaerobic filter: An anaerobic filter is an attached growth biological treatment process. In an anaerobic filter, the wastewater can flow either upward or downward. As the wastewater flows through the filter, the active biomass that is attached to the surface of the filter media anaerobically degrades the organics in the wastwate. The major disadvantage of the anaerobic filter is the difficulty of the removal of accumulated biomass over time. This requires periodic dewatering of the filter media which will allow the heavy biomass to slough off the media under its own weight. For this procedure to be effective, a filter media that can 1) withstand the weight of the biomass, and 2) provide a surface and geometry that facilitates the sloughing of the biomass must be selected.

Upflow anaerobic sludge blanket: An upflow anaerobic sludge blanket (UASB) is an attached growth biological treatment process in which wastewater flows upwards through a blanket of dense granules. As the wastewater flows through the blanket, the anaerobic microorganisms attached to the granules degrade the organics producing methane and carbon dioxide. From the sludge blanket, the gases rise to the liquid surface of the reactor where they are captured for use or discharge. Once the wastewater is above the sludge blanket, the granules separate by gravity from the bulk wastewater and settle back to the blanket as the clarified effluent overflows out of the reactor.

A major issue with the UASB is the continued formation of the dense granules. If the sustained formation of the granules is possible, the UASB has distinct advantages over other anaerobic processes. The advantages include extremely high loading rates due to the high biomass concentrations (30,000 to 80,000 mg/L) and good settleability resulting in high solids retention in the reactor. One design feature included in many UASBs is a solids separation unit used to capture floating granules that may occur over time.

Anaerobic fluidized bed: An anaerobic fluidized bed is an attached growth biological treatment process in which wastewater flows upward in a column with sufficient force to fluidize a media. Upon fluidization, the active biomass attached to the fluidized media anaerobically degrades the organics in the wastewater. The fluidized bed minimizes the problems associated with the anaerobic filter namely 1) short circuiting, and 2) biomass accumulation and removal. Also, the fluidized bed has a distinct advantage over the UASB in that the supporting media can be purchased and does not need to be cultivated as is the case with the UASB granules. The most common bed media used is activated carbon due to its high surface area to volume ration and good fluidization characteristics. Other media used is sand, plastic media of any shape, or media of other non-reactive materials. A solids separation unit is typically provided directly downstream of the fluidized bed. Without the solids separation unit, rapid loss of media and biomass is possible during startups after power outages and during hydraulic surges.

Ananaerobic fixed bed: An anaerobic fixed bed system is typically upflow with a flow distribution header in a plerom under a false floor which supports the fixed media. The anaerobic bacteria

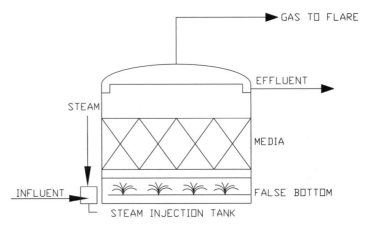

FIGURE 10.76   Anarobic fixed bed reactor.

grow as a fixed film on the surface of the media and also exist as suspended growth in the wastewater between media surfaces. Figure 10.76 shows a drawing of an anaerobic fixed bed system using PVC cross flow media at ethylene glycol removal facility. This system involved a 4 day HRT for the inflow, but used a 5:1 recirculation ratio of tank effluent back to the head of the tank. This system operates without a settling or solids capture system following the anaerobic reactor. The suspended solids above the media should settle back into the wastewater before discharge. Figure 10.77 shows a photograph of the top of the plastic cross flow media with hold down structure to prevent floating."

Anaerobic Lagoons: Any unmixed pond over about three feet deep receives a minimum of oxygen from algae and surface turbulence and will therefore become anaerobic in its lower depths. The food in this pond will be converted to CO_2, H_2S and CH_3 by the anaerobic bacteria which will grow. Typically these ponds naturally develop a floating scum layer which prevents algal growth, making the pond even more efficient anaerobically.

FIGURE 10.77   Plastic cross flow media.

10.4.1.5 Application of Anaerobic Treatment.   Anaerobic treatment has several advantages over the conventional aerobic process. These advantages include a specialized treatment process that is capable of treating high strength organic wastes not normally suitable for aerobic systems. Due to the reaction rates of an aerobic process, high strength wastes may require extremely large aeration times or more than one aeration reactor in series to provide the necessary treatment. Therefore, anaerobic processes may also be used upstream of an aerobic process to provide initial removal of high concentrations of organics in the wastewater. This combination is able to achieve very high removal rates for high strength organic wastes. The following sections give a step-by-step process to determining the applicability, design, installation, and operation of anaerobic processs.

The primary disadvantages of anaerobic treatment compared to aerobic treatment is that anaerobic treatment will not treat to low COD effluent levels, typically below 100 mg/l, and anaerobic treatment produces a gas that must be disposed of by burning or re-using as energy. The gas typically consists of H_2S which imparts a rotten odor which must be contained with air tight joints. The methane in the gas is of course flammable and safety precautions must be carefully followed in the design. The amount of methane produced is a function of TDC or COD concentration, the type of organic treated and the temperature.

Raw wastewater analysis: In order to evaluate the use of any treatment process, a detailed analysis of the wastewater to be treated should be performed. For an anaerobic process, a detailed analysis would include the following constituents:

- Chemical Oxygen Demand (COD)
- Temperature
- pH and total alkalinity
- Total and Volatile Suspended Solids (TSS and VSS)
- Dissolved anions (chlorides, fluorides, sulfates, etc.)
- Dissolved cations (sodium, potassium, magnesium, copper, zinc, iron, etc.)
- Organics

  The results of the above analysis will help determine:
- The applicability of an anaerobic process
- The appropriate pretreatment for the anaerobic process
- The necessary chemical additions
- The required pollutant removal efficiency

Chemical oxygen demand: The COD concentration of a wastewater is the first parameter used to determine if an anaerobic treatment process should be considered. Although low strength organic wastes can be treated anaerobically, in most cases, an anaerobic treatment process is not used for wastewaters with relatively dilute organics. As a general rule of thumb, an anaerobic treatment process should be considered first if the raw COD concentration of a wastewater is less than 1,000 to 2,000 mg/l (Jones, 1990). With more dilute organic wastes, anaerobic treatment may results in high residual COD, which would require post-treatment with aerobic or physical/chemical methods to meet permit limits. Therefore, economics may favor a treatment system that is 100 percent aerobic.

Temperature: The influent temperature of the wastewater plays a major role in determining the type and operation of an anaerobic process because there are two general subdivisions within anaerobic treatment: the use of thermophilic and mesophilic microorganisms. Thermophilic microorganisms thrive at temperature between 50°C and 60°C, while mesophilic microorganisms thrive at temperatures between 30°C and 40°C. These two cultures of microorganisms exhibit drastically different reaction rates as well as produce different effluent sludges. With the two cultures, heating of the wastewater and reactor play a large part in operating costs of the system; therefore, high temperature wastes are treated thermophilically, while low temperature wastewaters are treated mesophilically or heated to thermophilic range.

pH and total alkalinity: The influent pH and alkalinity of a wastewater is very important in an anaerobic treatment process. Volatile fatty acids are produced during anaerobic decomposition, resulting in an increase in pH. After acidogenesis occurs, methanogenic microorganisms degrade the VFAs into carbon dioxide and methane. The methanogenic microorganisms function over a near neutral pH range of 6.5 to 8.2 (Speece, 1996). If an influent wastewater is low in alkalinity and pH, chemicals need to be added such as sodium or magnesium bicarbonate.

Total and volatile suspended solids: Anaerobic processes will not remove inorganic suspended solids and organic suspended solids are degraded slowly under anaerobic conditions. Therefore, in cases other than anaerobic digestion of biological sludges, high concentrations of suspended solids in the raw wastewater should be removed prior to the anaerobic reactor.

Dissolved anions and cations: In high concentrations, inorganic ions in the wastewater may be inhibitory or even toxic to the anaerobic microorganisms present in the system. The following Table 10.15 shows typical inorganics reported to be inhibitory to anaerobic microorganism:

The bioavailability of trace metals in an anaerobic treatment system is crucial for the sustainability of the anaerobic biomass. The trace metals include iron, cobalt, nickel, zinc, copper, manganese, molybdenum, selenium, tungsten, and boron. In most organic wastewaters, these metals are not present or present only as non-bioavailable metals that are strongly chelated. Therefore, the routine addition of these metals into an anaerobic reactor is necessary.

Organics: Certain organics are inhibitory to methanogenic microorganisms. Table 10.16 shows several common organics inhibitory to methanogens.

10.4.2 Anaerobic Biological Treatment Treatability Studies

In order to choose an effective wastewater treatment system, a treatability study needs to be conducted. A treatability study involves 1) evaluating proper treatment technologies, and 2) determining design parameters for full-scale treatment. This section will explore the setup and operation of an anaerobic biological treatability study.

Before a treatability study is performed, screening tests are typically conducted using either the Anaerobic Toxicity Assay (ATA) or the Biochemical Methane Potential (BMP) test. These screening tests estimate the amenability of a wastewater to anaerobic treatment. For many organics, this potential is already known from literature or past experience. For those organics that are not discussed in the literature under the context of anaerobic treatment, these screening tests are relatively simple methods for determining a wastewater's degradation potential. The drawback of these tests is that they need to run for 30 to 90 days. These tests are conservative and actual anaerobic degradation in a bench-scale or full-scale flow-through system may be higher.

The most common screening method is the BMP test in which an acclimated culture of anaerobic microorganisms is added to containers with the wastewater of concern. The containers are capped with a gas-tight top and maintained at 35°C. Biogas production is measured daily and CH_4 production is measured approximately weekly. The results of this test are used to determine the wastewater's anaerobic biodegradability and subsequent COD reduction potential.

After screening has determined that an anaerobic treatment system is feasible, an anaerobic treatability study can be conducted. The first step in a treatability study of an anaerobic process involves choosing a reactor configuration. As discussed in section 10.2, there are several different anaerobic treatment configurations. The two major types of processes are those in which the anaerobic microorganisms are suspended growth and attached growth systems. Table 10.13 presents the different features and issues between suspended and attached growth systems.

The determination of the process configuration will dictate what configuration the bench-scale treatment system will be: a continuous treatment system, or a batch system. A batch or continuous system can be used when the process configuration is the anaerobic contact process, but a continuous system must be used for an attached growth system.

For most applications, an attached growth anaerobic treatment system is superior to a suspended growth system due to the reactor size. In an attached growth system, very high biomass concentrations are possible (30,000–50,000 mg/L); therefore, a smaller hydraulic detention time is required. An attached growth anaerobic system can be either a fluidized bed, anaerobic filter, anaerobic filter bed, or a UASB system. Figures 10.77a and 10.78 show a typical treatability setup using a fluidized bed reactor:

TABLE 10.13   Suspended-Attached Growth Comparison Table

Attached growth systems	Suspended growth systems
No clarifier required	Clarifier required
Able to maintain high biomass inventory (up to 60,000–70,000 mg/L)	Biomass limited to ~10,000 mg/L of suspended solids
Growth media required	No growth media required
Higher organic loading rate allowed	Mixing required
Periodic biosolids removal required	Chemical addition may be required to induce settling

To begin an anaerobic treatability study, a culture of anaerobic microorganisms needs to be obtained. Often times, the most easily accessible anaerobic culture can be acquired at a local municipal wastewater treatment facility that anaerobically digests their sludge. Once procured, the culture is added to the reactor vessel together with the wastewater, nutrients, and media (if applicable). In the case of a fluidized bed system, granular activated carbon, sand, pumice stone among others have successfully been used as media in the past. The optimal fluidization media will have a high surface area-to-volume ratio and a low specific gravity. The size of the fluidization media should be selected using the desired fluidization velocity. Using Stoke's Law, the correct media size can be determined by setting the fluidization velocity equal to the settling velocity of a particle. For activated carbon, the particle diameter for a fluidization velocity between 0.5 and 1.0 feet per second is 0.5 and 0.75 millimeters. The media is fluidized by means of a recirculation line, which may pump at a rate 100 times greater than the flow-through rate.

For an anaerobic filter, larger particles should be used for an upflow configuration on the order of 1.0 to 3.0 millimeters in diameter, while smaller particles can be used for a downflow filter (0.5 to 1.0 millimeters in diameter). The treatability study setup for an anaerobic filter is very similar to the fluidized bed setup other than the inlet and discharge locations for a downflow filter. Because the media remains in place and is not fluidized, recirculation is not required.

For an upflow anaerobic sludge blanket (UASB) reactor, granules may be obtained from a nearby industry that uses this process. UASB granules grow very slowly; therefore, obtaining already

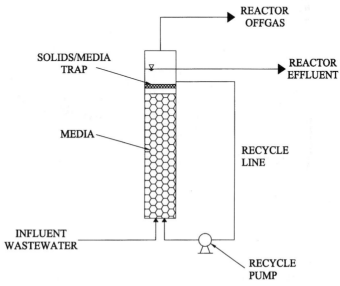

FIGURE 10.77a   Fluidized bed treatability setup PFD.

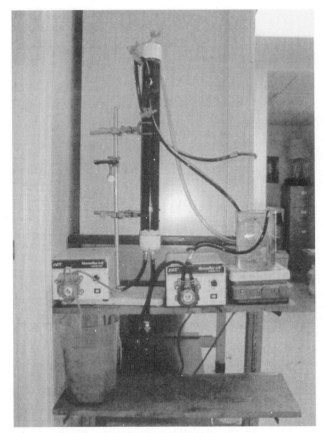

FIGURE 10.78   Fluidized bed treatability setup picture.

grown granules is a necessity in a treatability study. Caution must be taken using already grown granules because they cannot grow in certain wastewater types. If this were the case, another reactor configuration would be more suitable for anaerobic treatment.

For any attached growth anaerobic system, the COD loading rate to the reactor should be steadily increased over a period of weeks for acclimation of the biomass to the specific wastewater. Reactor pH should be measured daily to ensure adequate conditions for the bacteria. To allow for biomass attachment to media and acclimation, a treatability study should be operated for a period of 2 to 3 months. Over time, the effluent COD and gas production may be periodically measured to evaluate process performance. In order to obtain accurate data, the treatability study should be operated at steady state loading rates for 15 to 30 days.

10.4.3   Anaerobic Treatment Design

The design of anaerobic treatment processes involves the results from the treatability study as discussed in the previous section. The first step in the design involves sizing the reactor using the following method.

1. Determine optimal organic loading rate (OLR) from treatability study in grams of organics entering the reactor per day per liter of reactor volume (g/L/day).

2. Determine hydraulic detention time required for required removal using the following equation:

$$t = S_o/OLR \tag{10.49}$$

where t = hydraulic detention time, days
S_o = influent COD concentration, gram/liter
OLR = organic loading rate, grams/liter/day

3. Determine the heating required by heating the influent wastewater to 95°F (for mesophilic processes) or 130° F (for thermophilic processes) and assuming a heat loss of 1° F for every day of detention time.

4. Calculate the nutrients required for proper decompostion. This can be determined by comparing the total required nutrient loading to the actual nutrient loading from the wastewater. The required nitrogen and phosphorus concentrations in an anaerobic reactor can be calculated by the following equations:

Nitrogen Concentration Within Reactor = 40–70 mg/l

Phosphorous Concentration Within Reactor = 8–15 mg/l

Once the process design for the anaerobic reactor is complete, the overall design of the system needs to be conducted. The overall design includes pretreatment and post treatment determination and sizing, hydraulic design, and controls and instrumentation.

10.4.3.1 Pretreatment. Pretreatment of an anaerobic treatment process is very similar to an aerobic treatment process. The biological decomposition of organics occurs in a specified pH range, outside of which the microorganisms are either less productive or cannot live. The methanogenic microorganisms function over a near neutral pH range of 6.5 to 7.5. Therefore, pH adjustment may be necessary prior to the anaerobic reactor.

Since the second step in the overall anaerobic decomposition reaction involves the breakdown of low molecular weight organics into volatile fatty acids (VFAs) and the methanogenic reactions occur over a small pH range, there needs to be significant alkalinity to prevent acidic conditions from developing. Typically, a form of bicarbonate is added to wastewaters with insufficient alkalinity. This addition should occur prior to the wastewater entering the reactor.

For many wastewaters, the organic and hydraulic load fluctuations throughout the day can disturb the microbial population and prevent a consistent organic removal efficiency. Therefore, hydraulic and organic load equalization is often provided prior to the anaerobic reactor. Hydraulic equalization is provided through the storage of influent wastewater with a constant or near constant flow from the storage tank(s) to the subsequent treatment system. Organic load equalization is provided by a normally full, completely mixed basin in which the influent flowrate equals the effluent flowrate. Also, a combination of the two types of equalization can be provided together in the same tank.

10.4.3.2 Post treatment. Depending on the reactor configuration and the system's effluent limitations, treatment of the anaerobic reactor effluent may be required. This post treatment may needed for suspended solids removal, additional COD removal, refractory organics removal, or disinfection.

Typically, suspended solids are removed from the anaerobic contact process using conventional gravity separation in a clarifier. Other methods for suspended solids removal include media filtration or membrane filtration. In the case of the anaerobic fluidized bed, anaerobic filter, or UASB, it is necessary to provide some form of solids separation downstream of the reactor for periodic upsets which cause the media/granules to leave the reactor. Due to the long biomass accumulation period, loss of media/granules could cause organic removal in the anaerobic reactor to decrease significantly for a long period of time.

The only purpose of collecting solids in an Anaerobic Fixed Bed System is to catch the dead bacteria which can sluff off of the media or be carried without settling out of the reactor. These solids amounts are typically small and may be allowed to accumulate in the bottom of a downstream tank if it is cleaned periodically.

For high strength organic wastes, effluent limitations may be sufficiently low that a second stage organic removal system must be employed. Typically, this additional COD removal is achieved using an aerobic treatment process such as activated sludge, aerobic lagoons, or sequencing batch reactors. Attached growth systems can also be used such as trickling filters or rotating biological contactors.

For refractory organics removal, activated carbon adsorption can be employed. For disinfection, methods such as chlorine addition or ultraviolet radiation can be used.

For an in-depth review of each of these post treatment processes, please consult the appropriate chapters in this Handbook.

10.4.3.3 Odor Control.   The sulfate concentration in many wastewaters are significantly high that an anaerobic reactor will generate hydrogen sulfide (H_2S). Since the odor detection level of H_2S is extremely low, the production of the gas may be a nuisance problem. In order to reduce or eliminate the potential for these problems, H_2S removal from both the methane produced and the effluent wastewater stream is necessary.

All anaerobic processes produce a certain quantity of methane, which must be used or disposed of properly (flare, cogenerator, boiler, etc.). For influent wastewater that contains sulfate, the methane produced from the reactor will contain H_2S, which may need to be removed. Typically, H_2S can be effectively removed in a wet scrubber.

Also, H_2S will be dissolved in the effluent wastewater and will come out of solution upon exposure to aeration or agitation downstream of the anaerobic reactor. Therefore, H_2S removal from the effluent wastewater may also be necessary. For removal of dissolved H_2S in the reactor effluent, an air stripping tower can be employed. The tower will transfer the H_2S from solution into the stripping air. After the stripping has taken place, the air will contain H_2S, which is typically removed by a wet scrubbing system. Mixing of the air stripper gas and the methane produced from the anaerobic process could lead to a potentially dangerous situation where the methane concentration is between its upper and lower explosive limits.

10.4.3.4 Equipment Used in Anaerobic Processes.   Correct selection of the equipment used in an anaerobic treatment process plays an important role in the overall system's success in treating the wastewater.

As stated in the previous section, pretreatment processes may be required depending on the wastewater characteristics. Pretreatment equipment includes any or all of the following:

1. Chemical conditioning tank(s)

   (1) pH

      1. Sensor, transmitters, controllers

      2. Chemical feed

         a. Alkaline

         b. Acid

   (2) Alkalinity

      1. Sensor, transmitters, controllers

      2. Chemical feed

2. Equalization tank

   (1) Organic loading

      1. Tank

      2. Mixing equipment

   (2) Volumetric loading

      1. Level controls

      2. Effluent pumps

As discussed in section 10.4.1.4, the different configurations of an anaerobic reactor include: anaerobic contact process, anaerobic digestion, anaerobic filter, anaerobic fluidized bed reactor, anaerobic fixed bed, and an upflow anaerobic sludge blanket reactor. The equipment need for each configuration is listed below.

(**1**) Anaerobic contact process

1. Reactor tank
2. Heating equipment (heat exchanger, steam injection)
3. Mixer
4. Degasifier (air stripping tower)
5. Gas treatment (scrubbing, flare, cogenerator)
6. Clarifier for biomass separation
7. Recycle pumps

(**2**) Anaerobic digestion

1. Reactor tank
2. Heating equipment (heat exchanger, steam injection)
3. Mixer
4. Gas treatment (scrubbing, flare, cogenerator)

(**3**) Anaerobic filter

1. Reactor tank
2. Heating equipment (heat exchanger, steam injection)
3. Packing media
4. Effluent air stripper for H_2S removal
5. Clarifier for media capture
6. Gas treatment (scrubbing, flare, cogenerator)
7. Recycle sump and pumps

(**4**) Anaerobic fluidized bed reactor

1. Reactor tank
2. Heating equipment (heat exchanger, steam injection)
3. Fluidization media
4. Effluent air stripper for H_2S removal
5. Clarifier for media capture
6. Gas treatment (scrubbing, flare, cogenerator)
7. Recycle sump and pumps

(**5**) **Anaerobic Fixed Bed**:

1. Reactor Tank
2. Heating Equipment (heat exchanger, steam injection)
3. Fixed Bed Media
4. Gas Treatment (scrubbing, flare, cogenerator)
5. Sump and Pumps
6. Clarifier for Media Capture (optional)

(**6**) Upflow anaerobic sludge blanket

1. Reactor tank
2. Clarifier for media capture
3. Gas treatment (scrubbing, flare, cogenerator)

10.4.3.5 *Design Hints*

The following are design suggestions for anaerobic treatment:

- pH = 6.5–8.5
- Ideal T = 95 F mesophilic

  130 F thermophilic

  Note: To maintain these temperatures, it is important to minimize heat loss through insulation of walls and cover.

- Minimum Sludge Retention Time: 0.2 to 4 days
- HRT = 0.3 to 20 days
- Organic Loading Rate = 5–45 kg/m^3-d
- CH$_4$ Production = 0.25–0.40 L CH$_4$/gCOD
- Alkalinity Required = 1.0 to 1.2 g CaCO$_3$/gCOD (this can be reduced with high recycle rates.)
- Sources of Alkalinity:
  - Lime, CaO
  - Caustic Soda, NaOH
  - Soda Ash, Na$_2$CO$_3$
  - Sodium Carbonate, Na$_2$CO$_3$
  - Sodium Bicarbonate, NaHCO$_3$
  - Aqueous Ammonia, NH$_4$OH
  - Magnesium Oxide, MgO

Trace metals as follows can stimulate anaerobic treatment:

- Fe
- Co
- Ni
- Zn
- Ca
- Mn

(Fe, Co, and Ni are typically the most effective stimulators.)

10.5 ANOXIC BIOLOGICAL TREATMENT

As explained in paragraph 10.3.1.10 "Nitrogen Compound Removal with Biological Treatment," nitrate can be converted to nitrogen gas under anoxic conditions. Anoxic selectors are described in paragraph 10.3.4.2.3 under "Biological Contactors and Aerated Lagoons with Return Activated Sludge" and are also described in paragraph 10.3.4.2.3 as using anoxic conditions to more effectively degrade accumulated sludge.

Facultative ponds can be constructed with depths between 5 feet and 10 feet which employ anoxic conditions in the lower depths.

Conventional de-nitrification systems can be used following ammonia removal as described above. These systems are non aerated with a detention time based on treatability study results. The author designed a successful total nitrogen removal system for a municipal sewage treatment plant which has an anoxic hydraulic detention time of 0.6 days following an aerobic treatment for 1.89 days, is mixed, and required methanol feed as a carbon source since the preceding carbonaceous and nitrogenous removal system lowered the TOC to levels too low to provide adequate carbon as a nutrient for anoxic cellular growth.

In all of these anoxic processes, oxygen, nitrite and nitrate serve as electron acceptors in the presence of low (0.1 to 0.2 mg/l) concentrations of dissolved oxygen.

10.6 CONSTRUCTED WETLANDS FOR WASTEWATER TREATMENT

Research into the water purification potential of wetlands has been conducted since the early 1950s. This research included both industrial and municipal wastewater at different stages of treatment. The research showed that the use of natural and artificial (or constructed) wetlands had tremendous beneficial uses as a wastewater treatment technology. It wasn't until the early 1970s that the first full-scale size constructed wetlands were used for the treatment of municipal and industrial wastewater.

Since the early 1970s, much has been learned from the use of full-scale wetland systems for wastewater treatment. It has been found that wetlands have a potential for the removal of BOD_5, TSS, ammonia, nitrogen, and color. Wetlands have gained steady acceptance over the past 20 years as a viable treatment step. A common use for constructed wetlands is as a polishing step for municipal wastewater treatment systems. Wastewater from the agricultural, food processing, mining, chemical processing, and wood products industry have been treated using constructed wetlands.

10.6.1 Theory of Constructed Wetlands

Constructed wetlands can be divided into three categories: surface flow (SF), subsurface flow (SSF), and a combination of the two. SF wetlands are those systems where the primarily removal occurs above ground under flooded conditions. SSF wetlands are those systems in which the wastewater flows underground through a permeable media of gravel, sand, and/or soil. In both cases, aquatic plants are used for mineral uptake and microbial attachment sites (on stalks, leaves, and roots).

The author has developed a constructed wetland based on the established theory that wetlands function as aerobic fixed film reactors. The typical constructed wetlands receives its oxygen exclusively from algal metabolism during daylight hours plus surface turbulence of the water. The F/M ratio of this fixed film reactor is limited to the bacterial growth on attachment sites in the water. To increase the oxygen transfer, a method has been developed for varying the water level in the tanks with motor operated valves and timers to provide oxygen in the low water conditions. Likewise, media has been placed between the aquatic plants to increase the surface area on which the bacteria can grow. This system has successfully removed COD at approximately one hundredth the surface area as a parallel conventionally designed constructed wetlands when operating for over two years.

In addition, the author has experimented with but not built a constructed wetlands system followed by a clarifier with sludge recycle to the head of the wetlands tank to effectively convert the system from an aerated lagoon to activated sludge.

Further developed theory supports the use of two activated sludge wetlands in series, each with its own clarifier, designed to remove a refractory organic such as nitrogen, phthalate, tanins or lignins as explained earlier.

10.6.2 Design of Constructed Wetlands

In an SF wetland, the primary goal is to mimic the removal of pollutants in natural wetland systems. To achieve this removal, the wastewater should flow (in shallow depths) across the ground through vegetated areas. The most important design criteria for an SF wetland is the hydraulic loading through the vegetated areas. It has been shown that the organic removal efficiency of an SF wetland is not dependent on the type of vegetation used since the primary organic removal mechanism is microbiological degradation. For this mechanism, the aquatic plants provide little more than attachment sites for the microbes. In this respect, SF wetlands could be viewed as an attached growth system.

In an SSF wetland, the pollutant removal mechanisms are essentially identical other than the fact that the removal takes place underground. In an SSF wetland, the crucial design criteria again is hydraulic loading through the permeable media. The advantage of an SSF over an SF wetland is the potential for increased microbial sites due to the higher specific surface area of the plant roots and the permeable media.

The important aspects in design of either an SF or an SSF wetland are the inlet and outlet structures, the flow path of the wastewater and the choice and density of the aquatic plants. The choice of aquatic plants is based primarily on the type of plants that best grow in the climate where the wetland is located and the heartiness of the plant species to daily and seasonal weather changes.

Preliminary sizing of a constructed wetland system follows

$$A = \left(\frac{0.0365 \times Q}{k}\right) \times \ln\left(\frac{C_i - C^*}{C_e - C^*}\right) \qquad (10.52)$$

Where A = required wetland area, ha
   Q = wastewater flowrate, m³/day
   k = first order areal rate constant, m/yr
   C_i = inlet pollutant concentration, mg/L
   Ce = desired effluent pollutant concentration, mg/L
   C^* = background pollutant concentration, mg/L

The first order areal rate constant for BOD_5 removal should be determined by a bench-scale or pilot-scale treatment wetlands. For preliminary sizing, empirical data from other treatment wetlands can be used as shown in Table 10.13a

Empirical data is also available for preliminary sizing of constructed wetlands for treatment of pollutants other than BOD_5. This data is given Table 10.14–10.15.

10.6.3 Installation of Constructed Wetlands

The installation of constructed wetlands can be a difficult task for many contractors due to the infrequency of these types of systems. Because of that, it is very important that a competent civil

TABLE 10.13a BOD_5 Reduction Rate Constants

Site		k Value (m/yr)
Listowel, Ontario	System 1	13.8
	System 2	6.5
	System 3	12.4
	System 4	36.9
	System 5	42.8
Guistine, CA	Marsh 1A	18.1
	Marsh 1B	13.7
	Marsh 1C	9.4
	Marsh 1D	28.7
	Marsh 2A	22
	Marsh 2B	41.6
	Marsh 6A	33.3
	Pilot Marsh	21.6
Cobalt, Ontario	Marsh	54.2
Iron Bridge, FL	Marsh	22.5
Benton, KY	Marsh 1	93.7
	Marsh 2	59.6
Pembroke, KY	Marsh	51.4
West Jackson County, MS	Marsh	54
Lakeland, FL	Marsh 1	47.9
Average	Marsh	34.0
Standard Deviation	Marsh	22.0
Cannon Beach, OR	Forested	17.7
Bear Bay, SC	Forested	6.8
Reedy Creek, FL	Forested	34.2

TABLE 10.14   Nitrogen Removal Areal Rate Constants

Site	k Value (m/yr)
Surface-Flow Marshes	
Lakeland, FL	19.70
	8.77
	7.46
Orange Co., FL	4.70
Iron Bridge, FL	18.15
Fort Deposit, AL	4.48
West Jackson Co., MS	9.11
Leaf River, MS	14.42
	24.11
	39.20
Santa Rosa, CA	36.71
	66.36
	56.62
	4.68
Des Plaines, IL	21.49
	7.21
	20.40
Cobalt, Ontario	40.25
Benton, KY	7.97
	10.41
Listowel, Ontario	4.36
	3.55
	5.09
	5.56
	4.63
Gustine, CA	3.48
	1.43
	0.56
	2.90
	5.25
Forested Surface Flow	
Reedy Creek, FL	22.65
	50.10
Central Slough, SC	3.48
Poinciana, FL	0.78
Vereen, SC	2.92
Drummond, WI	15.91
Floating/Submergent Aquatics	
New Zealand	13.41
Richmond, NSW	14.76
Soil-based Reed Beds	
Denmark	9.73
Subsurface-flow Reed Beds	
U.K	29.76
Subsurface Flow	
Benton, KY	6.42
Hardin, K Y	15.04
	26.27
Phillips High School, AL	35.67
Mayo Peninsula, MD	36.79

TABLE 10.14 Nitrogen Removal Areal Rate Constants
(*Continued*)

Site	K Value (m/yr)
Utica, MS	7.87
Richmond, NSW	9.98
	13.19
	13.14
Baxter, TN	3.74
	3.73
	24.11
	11.11
	13.96
	2.71
	34.14
	7.85
Hamilton, New Zealand	8.36
	12.34

contractor be hired to construct the liner, berms, and inlet and outlet structures. A professional pond contractor should be hired to install the aquatic plants. Usually, these contractors can also supply the necessary plants.

10.6.4 Operation of Constructed Wetlands

One of the advantages to constructed wetlands over other more conventional treatment systems is that regular operator attention and maintenance of the system are minimal. Harvesting of the aquatic plants will be the most labor-intensive operational activity associated with the constructed wetland

TABLE 10.15 Phosphorus Removal Areal
Rate Constants

Site	k Value (m/yr)
Des Plaines, IL	23.7
Jackson Bottoms, Or	14.2
Lakeland, FL	3.4
Pembroke, KY	9.3
Great Meadows, MA	5.7
Fontanges, Quebec	11.2
Houghton Lake, MI	11.0
Cobalt, Ontario	20.9
Brookhaven, NY	8.9
Leaf River, MS	11.2
Clermont, FL	23.4
Seal Pines, SC	11.7
Benton, KY	2.4
Listowel, Ontario	8.2
Humboldt, SAS	12.8
Tarrant Country, TX	20.1
Iron Bridge, FL	13.5
Boney Marsh, FL	14.2
WCA 2A	10.2
OCESA	6.4

although, depending on the plant species chosen, lengthy harvesting may not be needed. Another operation and maintenance issue associated with constructed wetlands is regular inspection of inlet and outlet structures for clogging due to debris in the influent wastewater or debris from the wetland cell.

10.7 WASTEWATER MICROBIOLOGY

10.7.1 Introduction

Engineers have learned to harness the potential benefits of using microorganisms in a controlled environment for treating municipal wastewater and organic laden industrial wastewater. The most common wastewater treatment technology that utilizes microorganisms is the aerobic biological treatment process. These processes can be either suspended growth (activated sludge system, aerobic lagoons) or attached growth (trickling filters (s), or rotating biological contractors).

In these processes, microorganisms are either suspended in a mixture of influent wastewater or attached to a media through which the wastewater is passed. The removal mechanisms in either technology are the same, but the effort required to maintain the microorganism population in each system differs significantly. The previous sections of this chapter go into detail on the removal mechanisms and methods used to maintain the microorganism populations. This section attempts to briefly described the microorganisms found in aerobic biological treatment processes and to discuss their life cycles and uses in wastewater treatment.

Microorganisms can be divided into two main groups based on their cell structure: *eucaryotes* and *procaryotes*. Procaryotes can be further subdivided in *eubacteria* and *archaebacteria*. Microorganisms can also be classified using nutritional classification (i.e., source of energy and source of carbon required for growth). There are two sources of energy used by microorganisms: light energy and chemical energy. Microorganisms which use light as their source of energy are termed *phototrophs* and those which use chemicals as their source of energy are termed *chemotrophs*. The principal source of carbon used for growth in microorganisms is carbon dioxide (CO_2) and organic carbon. Microorganisms that use CO_2 for the principal carbon source are termed *autotrophs* and those that use organic carbon for growth are termed *heterotrophs*.

Based on these criteria, the four main categories of microorganisms are *photoautotrophs, photehererotrophs, chemoautotrophs,* and *chemoheterotrophs*. Photoautotrophic microorganisms found in wastewater treatment systems are algae and many photosynthetic bacteria. Photoheterotrophs found in wastewater treatment systems are certain purple and green bacteria. Chemautorophs found in wastewater treatment systems include only bacteria. Chemoheterotrophic microorganisms found in wastewater treatment systems include the majority of bacteria and all of the protists.

10.7.2 Eubacteria

Eubacteria are the principal microorganisms responsible for organic degradation and floc formation in aerobic biological treatment systems. Eubacteria can be subdivided into two groups: gram-positive and gram-negative bacteria. The two groups differ in the structure of their cell walls. Eubacteria take one of three different shapes: rod, sphere, or spiral and can occur individually, in pairs, packets and/or chains. Heterotrophic bacteria use organic compounds as both an energy source and carbon source for growth, and are the microorganisms responsible for the majority of the organic waste removal in biological treatment systems. Important autotrophic eubacteria in wastewater treatment are nitrifying bacteria, sulfur bacteria, and iron bacteria.

10.7.2.1 Zoogloea. The most extensively studied chemoheterotrophic eubacteria are the organisms known as aerobic pseudonomads. The pseudonomads consist of six groups: *fluorescent, pseudomallei, acidovorans, diminuta, xanthomonas,* and *zoogloea*. Zoogloea are eubacteria responsible for the formation of floc in aerobic biological reactors. The Zoogloea groups are gram-negative flagellated rod-shaped eubacteria, which are present in large numbers in polluted waters and aerobic biological reactors. The Zoogloea eubacteria form large clumps of cells held together by fibrillar extracellular polyglucose. Figure 10.79 shows a phase contrast photomicrograph of *Zoogloea ramigera* floca.

10.7.2.2   Filamentous Microorganisms.

Filamentous microorganisms are long thin macroinvertebrates that cause sludge bulking and foaming in activated sludge systems. In controlled numbers, filamentous microorganisms enhance sludge floc formation allowing for better settling. Sludge bulking occurs when the population of filamentous microorganisms begins to grow outside of the boundaries of individual floc thus bridging several floc together, which inhibits efficient settling. Filamentous microorganisms tend to outgrow floc forming bacteria under certain conditions such as low dissolved oxygen, low food to microorganism ratio (F/M), low nutrient concentrations, low pH, and/or the presence of septic wastewater/sulfides.

The control of filamentous microorganisms is an emerging science. Conventional control methods such as chemical addition of chlorine to selectively kill the filamentous microorganisms can have several disadvantages such as potential overdosing, which could kill or inhibit the growth of beneficial microorganisms in the systems, as well as resulting in high chemical costs.

For the majority of the conditions in which filamentous microorganisms thrive (low DO, low pH, low nutrients), the logical control method is to modify the treatment systems to allow for more optimal operational parameters.

Low F/M sludge bulking can be controlled using selectors. Selectors are tanks located upstream of the aeration reactor in an activated sludge system in which raw wastewater is mixed with recycled activated sludge. The mixing establishes a substrate gradient (system in which the F/M ratio is very high). The substrate gradient allows floc-forming *Zoogloea* to out compete the filamentous microorganisms due to the rapid organic uptake and storage capabilities of the floc formers. Figure 10.80 shows photomicrographs of various filamentous microorganisms.

Figure 10.81 shows foaming in an activated sludge basin which was virtually eliminated in 24 hours by converting the flow from two aeration basins in series, to two basins in parallel, thereby halfing the F/M in the two basins compared tot hat in the previous first basin.

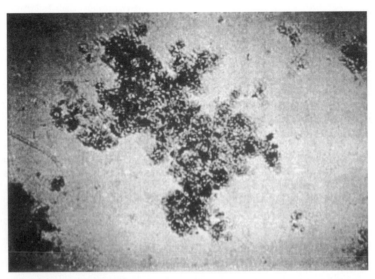

FIGURE 10.79   Zoogloea floc.

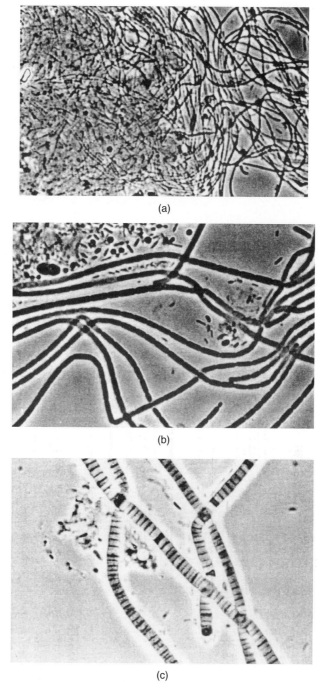

FIGURE 10.80   (*a*) *Micothrix Parvicella*, 1000×, (*b*) *Sqaerotilus natans*, 1000×, (*c*) Type 021N, 1000×, (*d*) *Thiothrix × II*, 1000×, (*e*) *No-cardia sp.*, 1000×.

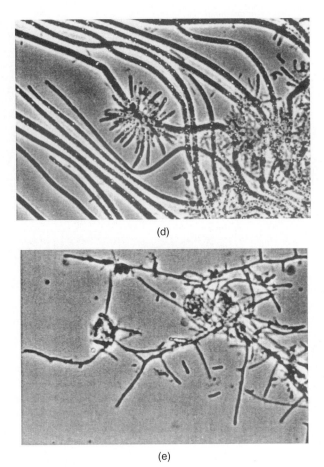

(d)

(e)

FIGURE 10.80 *(Continued).*

10.7.2.3 Coliform Bacteria. Coliform bacteria is a group of bacteria present in great quantities in human faces. Fecal coliform bacteria are coliform bacteria incubated at 40° C to duplicate temperatures in feces within warm blooded animals, and are therefore the best indicator of fecal contamination. The most notorious of all coliform bacteria is the *Escherichia coli,* or simply *E. coli. E. coli* is responsible for many illnesses and deaths when present in under-prepared meat, water supplies, and even swimming pools.

10.7.3 Eucaryotes

Eucaryotes can be subdivided into plants, animals, and protists. The eucaryotes common in wastewater treatment are nematodes (animals), rotifers (animal), protozoa (protists), algae (protists), and fungi (protists).

10.7.3.1 Protozoa. Protozoa make up the majority of the protists found in aerobic biological treatment systems. The most common protozoa are flagellates, ciliates, and amoebas. The overall community structure of the protozoan population can provide an indication of expected plant performance since the main role or protozoans in wastewater treatment systems is the predation of pathogenic bacteria. Figure 10.82 shows the three most common protozoa found in wastewater systems.

FIGURE 10.81   Foaming in an activated sludge basin.

10.7.3.2   Rotifers.   Rotifers are the smallest macroinvertebrates and are present in most aerobic biological systems. Due to their life cycle, rotifers are the dominant microorganism found in extended aeration systems. Rotifers play a role in promoting microfloral activity and decomposition, stabilizing organic wastes, enhancing oxygen penetration in attached growth systems, and harvesting of algae in lagoon systems. Rotifers also help keep the biofilm thickness in attached growth systems in check by feeding on the top layer of the biofilm. Rotifers also feed on large quantities of bacteria, keeping the bacterial population in the growth stage of their life cycle, which allows greater floc formation by young bacteria.

10.7.3.3   Nematodes.   Nematodes are the largest macroinvertebrates found in aerobic biological systems. Nematodes are present most often in attached growth systems where they help reduce excessive biofilm growth. Nematodes also help promote porous biofilm and enhance oxygen penetration due to tunneling in the biofilm.

10.7.3.4   Algae.   Algae are photoautotrophic protists commonly found in aerobic and facultative lagoon treatment systems as well as constructed wetlands. In these systems, algae supply oxygen to aerobic bacteria. Algae can also cause problems associated with total suspended solids in the effluent of lagoon and wetland systems. Algae build-up in secondary clarifiers can also attribute to high effluent suspended solids concentrations.

10.7.3.5   Fungi.   Fungi are chemoheterotrophic protists, which thrive in low moisture and pH environments. Populations of fungi are typically low in standard biological treatment systems, although certain biological treatment Systems that handle low pH industrial wastewater have a greater abundance of fungi.

10.7.4   Archaebacteria

Archaebacteria consist primarily of three groups of microorganisms: methanogens (produce methane), halophiles (live in highly saline environments), and thermophiles (live in high temperature, low pH

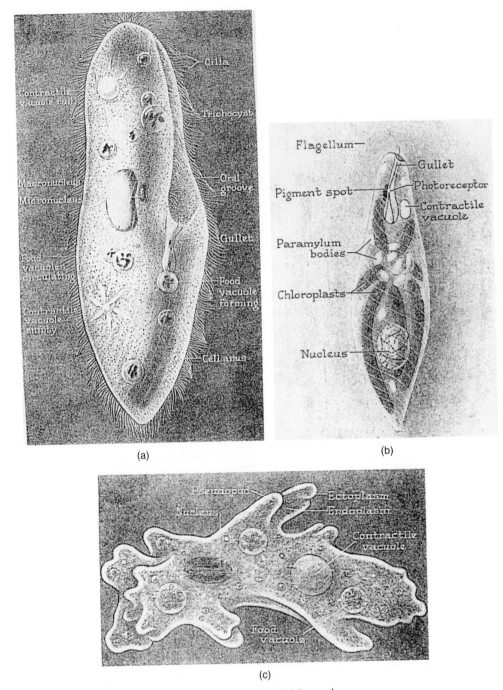

FIGURE 10.82 (a) Ciliated protozoa, (b) Flagellated protozoa, (c) An amoeba.

environments). The archaebacteria are believed to be ancient forms of bacteria, whose roots go as far back or farther than eubacteria. Of the archaebacteria, the methanogens are most important to wastewater treatment in that they represent the final step of anaerobic biological decomposition. Methanogens in wastewater treatment convert fermentation products formed by other anaerobes to methane and CO_2.

10.8   REFERENCES

Eckenfelder, W. Wesley, Jr. *Water Quality Engineering for Practicing Engineers.* Barnes and Noble, New York: 1970.

Jones, Richard M. 1990. Waste Characteristics and Treatability. Anaerobic Treatment of High Strength Waste, 3–4 December, University of Wisconsin, Milwaukee, WI.

Speece, Richard E. *Anaerobic Biotechnology for Industrial Wastewaters.* Archae Press, Nashville, TN: 1996.

Jenkins, D.J. et al. (1986) "Manual on the Causes and Control of Activated Sludge Bulking and Foaming.

Ricard, M. (1989) "Activated Sludge Microbiology", Water Pollution Control Federation, Alexandria, VA.

P · A · R · T · V

RESIDUALS

CHAPTER 11
RESIDUAL TREATMENT

Various residual solids result from wastewater treatment processes. Physical, chemical, and biological wastewater treatment all yield residuals (sludge) of various quantity and character. The primary goal of most residual treatment systems is to minimize the cost and hazard of the sludge disposal. The nature of the influent wastewater and type of wastewater treatment employed will dictate the characteristics of the resulting sludge and the means for handling, treating, and disposing these residuals. For example, sludge from a metals precipitation process will vary greatly from the sludge of an activated sludge process; therefore, the treatment and disposal options may not necessarily be the same. Figure 11.1 summarizes the various residuals from wastewater treatment processes and some of their characteristics.

11.1 THICKENING

Thickening is a process where the solids content of a wastewater sludge is increased by removing water from the sludge. Thickening is normally achieved physically by gravity settling, flotation, centrifugation, or gravity belts. Sludge thickening can have a significant effect on the cost and size of downstream sludge treatment processes, such as dewatering, digestion, and sludge hauling.

11.1.1 Gravity Thickening

The gravity thickening process operates similarly to the sedimentation process. Solids settle to the bottom of a tank by gravity while the supernatant overflows weirs at the top of the tank. The thickened sludge is pumped downstream for dewatering or digestion, while the supernatant is normally returned to the head of the treatment plant or the primary sedimentation tank. Figure 11.2 shows a typical gravity thickener.

The best success with gravity thickeners has been observed when thickening primary sludge and lime sludges. Success has also been noted when gravity thickening primary sludge in combination with trickling filter solids or anaerobically digested solids. However, gravity thickening has not been the ideal candidate for thickening waste activated sludge. Since successful gravity thickening depends on particles that settle easily, waste activated sludge with gravity of approximately 1.05 is not effectively removed by gravity thickening. Gravity thickeners are commonly circular tanks with a side-water depth of 10 to 13 ft and diameter dependent on sludge generation. The mass loading to the thickener depends on how well the sludge will settle. Mass loading rates may range from 5 lb/(ft²-d) for waste activated sludge to 20 lb/(ft²-d) for primary sludge.

Table 11.1 shows the expected underflow solids concentrations for various types of sludge and feed solids concentrations.

Please refer to Chapter 8 for design details for sedimentation which will also apply to sludge gravity thickening.

Residual	From	Properties
Screenings	Racks or screens	Various materials greater than 0.01 in, 8–20% dry solids, 80–90% volatile,
Grit	Grit chamber	Normally heavy inorganic materials with high settling velocity, 35–85% dry solids, 8–45% volatile, SG = 1.3–2.7 dry
Scum	Flotation, scum skimmers, digester scum, thickener scum	85–97% volatile, 6000–16000 BTU / # dry solids, usually contains pathogens
Primary sludge	Primary clarifiers	2–6% solids concentration, 65–93% volatile, pH 5–8, contains pathogens, SG approximately 1.4 dry
Activated sludge	Secondary clarifiers	0.7–2% solids concentration, 75–95% volatile, contains pathogens, SG approximately 1.05 dry
Trickling filter sludge	Trickling Filter	5–20 mg/L solids concentration
Digested aerobic sludge	Aerobic digester	3–6% solids concentration, 20–25 % volatile, stabilized solids, SG = 1.02–1.03 dry
Digested anaerobic sludge	Anaerobic digester	4–6% solids concentration, 20–35 % volatile, stabilized solids, SG = 1.02–1.03 dry
Chemical precipitants	Precipitation tanks	non-biologiological, wide range in properties

FIGURE 11.1   Wastewater treatment residuals.

11.1.2 Flotation

The most common sludge flotation thickening method is dissolved air flotation. A typical dissolved air flotation thickener is shown in Fig. 11.3.

In dissolved air flotation, a pressurized air/wastewater mixture is injected at the bottom of the flotation tank. As soon as the mixture is depressurized, tiny air bubbles are released from the mixture and float to the water surface. As the bubbles travel to the water surface, they collide with the sludge particles and carry them along to the water surface. When sludge particles reach the water surface, a skimmer is employed for removal. Dissolved air flotation thickening works best for sludges with a specific gravity close to 1.0, since heavier particles would be more difficult to carry to the surface. As a result, dissolved air flotation has been most successfully applied to waste sludges from suspended growth systems, such as activated sludge.

The float solids concentration that can be achieved by flotation depends on the air/solids ratio, the influent solids concentration, the settling characteristics, and solids and hydraulic loading rates. Well flocculated sludges may typically achieve a float solids concentration of 4 to 5 percent. A bulking sludge may only achieve a float solids concentration of 2 percent (Eckenfelder, 1989). The addition of polymers has shown to improve the performance of dissolved air flotation. It has been reported that high polymer doses are required for sludges with an SVI greater than 200 (WEF, 1998). Reported operating results for seven dissolved air flotation thickeners are presented in Table 11.2.

Please refer to Chapter 8 for design details for Dissolved Air Flotation which are also applicable to sludge thickening.

11.1.3 Centrifugation

Thickening by centrifugation is accomplished by inducing centrifugal forces, which impart an applied force greater than 500 times the force of gravity. Solids are driven by the centrifugal force away from the axis of rotation of the centrifuge. Centrifuges are commonly used for thickening waste activated sludge and other biological sludges. Centrifugation is not ordinarily used for primary

FIGURE 11.2 Typical gravity thickener. (Hi-Tech Environmental)

sludge since primary sludge, which settles well, can be removed more economically by gravity thickening. The most common centrifuges used for sludge thickening are the solid bowl and imperforate basket.

11.1.3.1 Solid Bowl Centrifuges for Sludge Thickening. The solid bowl centrifuge includes an imperforated cylindrical-conical bowl with an internal helical conveyor. The influent sludge is introduced to the cylindrical bowl through the conveyor discharge nozzles. Sludge is driven to the walls of the bowl by centrifugal forces and then transported to the conical section of the centrifuge by the rotating scroll, where the thickened sludge is removed. The rotational velocity of the scroll is typically 1 to 20 rpm slower than that of the bowl. The primary design parameters for the solid bowl centrifuge are bowl speed, pool volume, and scroll speed.

Figure 11.4 shows a Solid Bowl Centrifuge.

11.1.3.2 Imperforate Basket Centrifuges for Sludge Thickening. The imperforate basket centrifuge is operated in a batch sequence where sludge is introduced to the bottom of a vertically

TABLE 11.1 Typical Solids Concentration Data for Various Gravity Thickened Sludges (U.S. EPA, 1979)

Type of sludge	Feed solids, %	Thickened solids, %	Solids loading, kg/m³-d
Primary (P)	2–7	5–10	100–150
Trickling filter (TF)	1–4	3–6	40–50
Rotating Biological Contactor (RBC)	1–3.5	2–5	35–50
Waste Activated Sludge (WAS)			
WAS (air)	0.5–1.5	2–3	20–40
WAS (O$_2$)	0.5–1.5	2–3	20–40
WAS (extended aeration)	0.2–1.0	2–3	25–40
Anaerobically digested solids (primary digester)	8	12	120
Combined Sludge			
P + WAS	0.5–1.5	4–6	25–70
P + TF	2–6	5–9	40–80
P + RBC	2–6	5–8	60–100
P + Iron	2	4	50–90
P + low lime	5	7	30
P + high lime	7.5	12	100
P + (WAS + Iron)	1.5	3	120
P + (WAS + alum)	0.2–0.4	4.5–6.5	30
(P + Iron) + TF	0.4–0.6	6.5–8.5	60–80
(P + Iron) + WAS	1.8	3.6	70–100
WAS + TF	0.5–2.5	2–4	30
Anaerobically Digested			
P + WAS	4	8	70
P + (WAS + Iron)	4	6	

FIGURE 11.3 Typical dissolved air flotation thickener. (PRO-Equipment, Inc.)

TABLE 11.2   Dissolved Air Flotation Thickener Data (WEF, 1998)

Location	Activated sludge type	Feed solids concentration, mg/L	Solids loading rate, kg/m²-hr	Float concentration, %	Solids capture, %	Polymer dosage, g polymer per kg solids
Green Bay, WI	Contact Stabilization	4,000	1.5	3–4	80–85	None
San Francisco, CA	High Purity O₂	6,000	3.4	3.7	98.5	1.6
Salem, OR	High Purity O₂	14,800–20,300	19.5	5	95+	48–59
Milwaukee, WI	Conventional	5,000	4.9	3.2	90–95	1.5–2.5
Tri-Cities, OR	Conventional	11,300	–	3.9	98	1.5–2.5
Arlington, VA	Conventional	10,000	8.5	2.6	95+	1.5–2
Kenosha, WI	Conventional	8,600	5.4	4.3	95+	None

mounted spinning bowl. The solids are driven towards the walls of the bowl by centrifugal forces. When the bowl reaches its solids holding capacity, the centrate is decanted and the rotation is gradually brought to a stop. Scrapers then remove the thickened solids from the bowl. High solids removal efficiency can be achieved without the addition of polymers for this system. Figure 11.4 shows common centrifuges for sludge thickening.

The main difference between these centrifuges is the means in which solids are collected and removed from the bowl. In the same way flocculent bacteria settle poorly in the sedimentation process, flocculent bacteria also centrifuge poorly. As a result, chemical conditioning may be required to increase the specific gravity of a sludge and enhance thickening during the centrifugation process. Specific design criteria are not available, since there is great variation in centrifuge design and sludge characteristics. However, for general sizing conception, the hydraulic residence time in a centrifuge may be 30 to 60 seconds for waste activated sludge.

An empirical relationship between percent solids recovery and polymer dosage and feed rate for centrifuges has been developed by Bernard and Englande as shown in Eq. (11.1):

$$R = \frac{C_1(C_2 + P)^m}{A^n}$$

(11.1)

where: R = percent solids recovery
P = polymer dosage, lb/ton dry solids feed
Q = feed rate, gallon/(min-ft²)
C_1, C_2, m, n = constants

The constants C_1, C_2, m, and n must be developed empirically for a given centrifuge and feed sludge. Figure 11.4 shows an Imperforate Basket Centrifuge.

11.1.4   Gravity Belt Thickener

Gravity belt thickeners, which in practice are much like the upper gravity drainage zone of belt filter presses used for dewatering sludge, are becoming more popular for sludge thickening. Their popularity is due to their minimal energy and space requirements. The system consists of a gravity belt

Imperforate Basket Centrifuge (Tolhurst, Inc.)

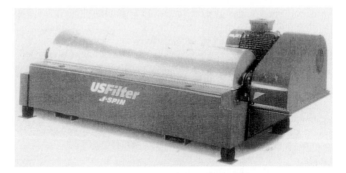

Solid Bowl Centrifuge (USFilter, JWI)

FIGURE 11.4   Common centrifuges for sludge thickening.

that passes over rollers on which sludge is distributed. After distribution on the belt, the sludge is ridged and furrowed on the belt to allow for water to pass through the belt. Polymers are usually added to the influent sludge to condition the sludge for thickening. Water contained in the sludge drains through the belt while the thickened solids are collected for further treatment or disposal. The belt is then passed through a wash cycle to clean the pores allowing for easier water passage. Figure 11.5 shows a typical gravity belt thickener.

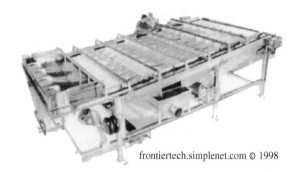

frontiertech.simplenet.com © 1998

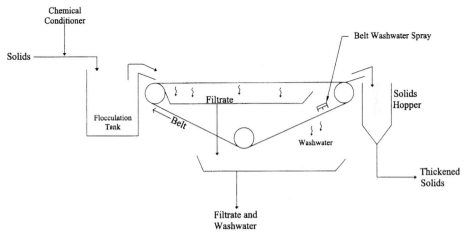

FIGURE 11.5 Gravity belt thickener (schematic and picture).

The design of a gravity thickener is based on the solids loading rate. Solids loading rates can vary from 3 to 50 lb/(ft²-day) depending on the type of sludge and influent solids concentration. The gravity belt thickener has been applied to sludge with solids concentration ranging from 0.5 to 8 percent with the addition of polymers. Polymer doses of 3 to 10 lb/ton are required to reach 4 to 8 percent solids concentration, however (WEF, 1998). Thus, the addition of polymers is essential to efficient operation of the gravity belt thickener. The polymer cost offsets the savings from low energy requirements. Sludge that contains smaller particles, such as waste activated sludge, are more costly to thicken than sludge with larger particles, such as primary sludge. Biological solids in waste activated sludge have a negative charge and therefore repel each other. Because the particles repel each other, they are difficult to flocculate. The addition of cationic (positively charged) polymer, though, reduces the particles' negative charge and allows for easier particle flocculation and sludge thickening. Because biological solids are relatively small, there is a greater negative charge barrier to overcome than with larger particles. This increased charge barrier translates to a larger polymer dose and cost. Table 11.3 shows operating results for some gravity belt thickeners.

11.1.5 Rotary Drum Thickening

The rotary drum thickener is similar to the gravity belt thickener in that sludge is placed on a moving porous media on which water may freely pass through. The fundamental component of the rotary

TABLE 11.3   Operating Results for Various Gravity Belt Thickeners (U.S. EPA, 1979 and Metcalf and Eddy, 1991)

Type of sludge	Sludge concentration (%)		Solids loading for gravity thickeners (lb/ft^2 · day)
	Unthickened	Thickened	
Separate			
Primary sludge	2–7	5–10	18–28
Trickling-filter humus sludge	1–4	3–6	7–10
Rotating biological contactor	1–3.5	2–5	7–10
Air activated sludge	0.5–1.5	2–3	2.5–7
High-purity oxygen activated sludge	0.5–1.5	2–3	2.5–7
Extended aeration activated sludge	0.2–1.0	2–3	5–7
Anaerobically digested primary sludge from primary digester	8	12	25
Combined			
Primary and trickling-filter humus sludge	2–6	4–9	12–20
Primary and rotating biological contactor	2–6	4–8	10–16
Primary and modified aeration sludge	3–4	5–10	12–20
Primary and air activated sludge	2–5	2–8	8–16
Anaerobically digested primary and waste activated sludge	4	8	14

drum thickener is an internal screw that transports thickened solids out of the rotating drum. Figure 11.6 displays a typical rotary drum thickener and schematic.

Rotary drum thickeners are typically used in small to medium size plants for waste activated sludge thickening. Their success in thickening waste activated sludge is highly dependent on the sludge characteristics. Like the gravity belt thickener, polymer addition is required and large doses may be required for sludges that do not flocculate well. Rotary drum thickeners have also been successful in thickening fibrous sludges, such as sludge from the pulp and paper industry. Typical performance data are shown in Table 11.4.

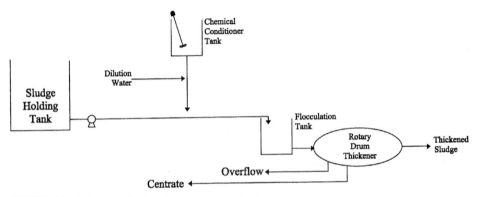

FIGURE 11.6   Rotary drum thickener (schematic and picture).

TABLE 11.4   Typical Performance Data of Rotary Drum Thickeners (WEF, 1998)

Type of solids	Feed, % TS	Water removed, %	Thickened solids, %	Solids recovery, %
Primary	3.0–6.0	40–75	7–9	93–98
WAS	0.5–1.0	70–90	4–9	93–99
Primary and WAS	2.0–4.0	50	5–9	93–98
Aerobically digested	0.8–2.0	70–80	4–6	90–98
Anaerobically digested	2.5–5.0	50	5–9	90–98
Paper fibers	4.0–8.0	50–60	9–15	97–99

11.2 STABILIZATION

Wastewater sludge can pose health hazards and be an aesthetic nuisance if not properly managed and treated. Biological sludges used to treat sanitary wastewater contain pathogens, which pose a potential health threat if improperly managed. Biological sludge also has the potential to decompose yielding undesirable odors (putrefaction). The objective of biological sludge stabilization is to reduce pathogen content and reduce the potential for putrefaction. The fulfillment of these objectives embodies processing the volatile fraction of the sludge. The means to meet these objectives is through the reduction of the sludge's volatile content by the following method:

- Chemical oxidation of the volatile content of the sludge
- Chemical addition to provide an environment in which microorganisms are unable to survive
- Disinfection through by the addition of heat
- Reduction of volatile content by anaerobic or aerobic digestion

The reasons for sludge stabilization at a given wastewater treatment plant will dictate the design of a stabilization system. For example, if a sludge is to be land applied, pathogen reduction may be the primary objective of stabilization. If the sludge is to be landfilled, stabilization may not be necessary. The most common stabilization technologies are lime stabilization, composting, anaerobic digestion, and aerobic digestion.

11.2.1 Lime Stabilization

The fundamental principle behind lime stabilization is the inactivation of microorganism by the increase of pH. Hydrated lime ($Ca(OH)_2$) or quicklime (CaO) is added to the sludge either before dewatering (called "lime pretreatment") or after dewatering (called "lime posttreatment") to raise the pH above 12. By raising the pH to a level unsuitable for biological survival for a given length of time, the sludge will not putrefy, or pose a health hazard.

An excessive amount of lime is required for lime stabilization since the pH must not only be raised above 12, but must be maintained there for several days to prevent biological reactivation, which can occur if the pH falls below 11. By adding sufficient lime, adequate residual alkalinity will be provided to buffer chemical reactions, which will lower pH. One such reaction is the sludge's gradual absorption of atmospheric carbon dioxide. The absorption of carbon dioxide slowly consumes the residual alkalinity until all residual alkalinity is removed and pH drops due to further carbon dioxide absorption. Furthermore, as pH drops low enough to reactivate biological activity, microorganisms will degrade the organics present in the sludge, produce volatile acids, and cause the pH to drop even further. Figures 11.7 and 11.8 respectively show schematics of a lime pretreatment stabilization system and lime post-treatment stabilization system.

The primary advantages of lime stabilization are its reliability, low capital costs, space efficiency, and ease of operation. For these reasons, lime stabilization is more common at small facilities than other stabilization methods. Large treatment facilities have used lime stabilization as an alternative

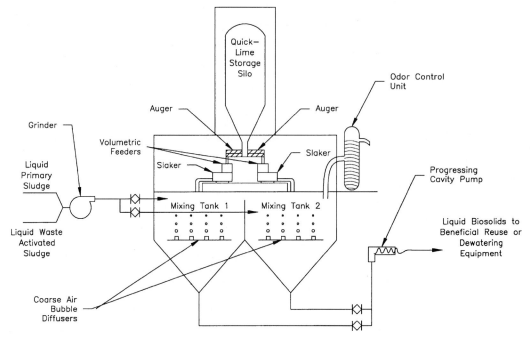

FIGURE 11.7   Lime pretreatment stabilization system.

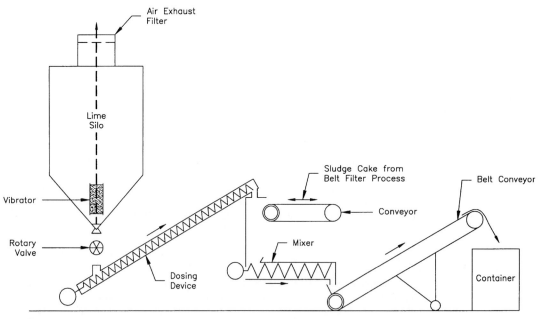

FIGURE 11.8   Lime post-treatment stabilization system.

stabilization option if their aerobic or anaerobic digestion system is off-line. The main disadvantage of lime stabilization is that there is no reduction in mass of solids. Actually, the mass of solids increases due to the chemical reactions that occur with the addition of lime. If sludge disposal fees are high in the vicinity where lime stabilization is being considered or in the case of long hauling distances, other stabilization methods should be considered since high sludge disposal costs could render lime stabilization uneconomical.

11.2.2   Anaerobic Digestion

The anaerobic digestion process is used to stabilize solids, reduce volatile solids, and reduce pathogens. A beneficial by-product of anaerobic biotransformation is methane gas, which can be used as an energy source. Anaerobic digestion has been used successfully for over 75 years stabilizing wastewater sludge. Despite its success, anaerobic digestion suffers a reputation as an unreliable process, likely due to a lack of appreciation of its process fundamentals. Process parameters such as solids retention time, hydraulic retention time, pH, alkalinity, nutrient availability, temperature, mixing, and substrate biodegradability require careful scrutiny for successful anaerobic digestion. If one of these parameters is neglected, process efficiency may suffer.

11.2.2.1   Theory of Anaerobic Digestion   The anaerobic digestion process can be divided into a sequence of three steps: (1) hydrolysis, (2) acidogenesis, and (3) methanogenesis. Figure 11.9 depicts the potential metabolic pathways of organic matter biotransformation to methane.

For organic matter to be utilized by bacteria, it must be soluble. Thus, the first step in anaerobic digestion is the hydrolysis of complex and low solubility organics. The release of enzymes by the

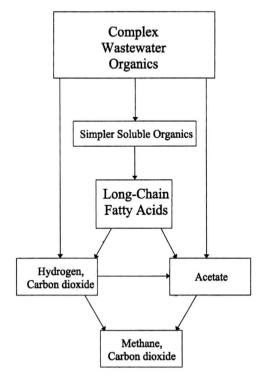

FIGURE 11.9 Potential metabolic pathways of organic waste biotransformation.

biomass population accelerates the hydrolysis of these compounds. If sufficient quantities of enzymes are unavailable to enhance hydrolysis, biotransformation will be limited by the lack of soluble substrate. Therefore, it is evident that a large biomass population is favorable to provide adequate enzyme concentration and hydrolysis. It should also be noted that not all organic matter can be hydrolyzed and made available for biotransformation, due to the hydrolysis being inhibited by their chemical structure. The portion of organic matter that cannot be hydrolyzed is called the refractory or non-biodegradable fraction. The refractory portion may be 40 to 80 percent of the sludge's volatile content depending on the nature of the sludge.

Upon hydrolysis, the soluble substrate is converted primarily to acetic acid, propionic acid, butyric acid, and valeric acid by a group of microorganisms called acidogens. In this step, organic matter is simply converted to a different form and thus, little waste stabilization occurs. Waste stabilization occurs during the final stage called methanogenesis. In the methanogenesis phase, the organic acids, primarily acetic acid, are converted to methane gas, carbon dioxide and usually hydrogen sulfide gas. The methane is quite volatile. The gases are separated from liquid easily and thus, volatile solids reduction occurs by transforming organic matter (either in the liquid or solid phase) to a gaseous end product which is separated from the sludge.

11.2.2.2 Anaerobic Digestion Processes   The three most common anaerobic digestion processes are

- Standard-rate digestion (low-rate digestion)
- High-rate digestion
- Two-stage digestion

Standard-rate digestion is normally carried out in a single stage process in which the reactors are unmixed. In this system, stabilization, sludge thickening, and supernatant formation occur simultaneously because there is no applied mixing. A schematic of a standard-rate anaerobic digester is shown in Fig. 11.10.

Because there is no mixing, four distinct layers form inside the digester. Stabilized solids settle to the bottom of the digester forming one layer. Above the solids layer is an active layer where biosolids stabilization occurs. Influent sludge is added to this layer. Just above the active layer is the supernatant layer where the effluent is removed. The top layer is a scum layer. Gas bubbles formed from microbial substrate reduction float to the surface and carry fats, oils and grease (FOG) along with them yielding a scum layer. Due to the lack of mixing, stabilization occurs only in a small portion of the digester. Heat is sometimes added to the influent to maintain a warmer temperature in the digester favorable to microbial activity. Standard-rate anaerobic digestion is the oldest of the anaerobic digestion methods and its popularity is lessening. Optimal conditions are not provided for microbial activity, and thus other anaerobic digestion methods are being favored.

High rate anaerobic digesters differ from standard-rate digesters in that sludge is mixed in a high-rate digester and heat is normally added to the system. The advent of the high-rate anaerobic digester came when designers recognized that better digester performance could be achieved if improved operating conditions were implemented. Mixing allows for a much greater volume where active stabilization occurs. Consequently, the increased active stabilization volume allows for higher solids loading rates. Like the standard rate digester, a scum layer does form at the top of the digester. Heat is added to the system to maintain a temperature in the mesophilic temperature range of 30°C to 38°C (85°F to 100°F). Figure 11.11 depicts a high-rate anaerobic digester.

Two-stage anaerobic digestion is an offspring of the high-rate anaerobic digester that utilizes two tanks in series. The first is used for stabilization and the second tank is used to separate solids from the liquid. The second reactor normally is not mixed or heated. It also provides storage capacity should the first reactor start operating inefficiently.

Thermophilic anaerobic digestion has also been practiced in the past. The thermophilic temperature range is approximately 45°C to 60°C. Advantages of thermophilic anaerobic digestion include increased pathogen destruction, increased scum digestion, greater sludge loading capacity, and improved sludge dewatering characteristics. Cited disadvantages include significant heating requirements to maintain the appropriate thermophilic operating temperature, poor process stability, and

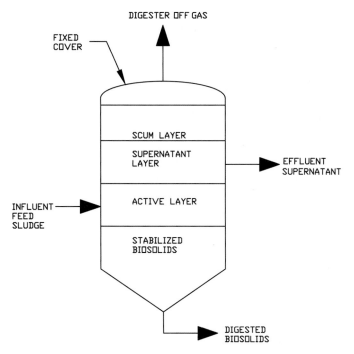

DIGESTER OFF GAS

FIXED
COVER

SCUM LAYER

SUPERNATANT
LAYER

EFFLUENT
SUPERNATANT

INFLUENT
FEED
SLUDGE

ACTIVE LAYER

STABILIZED
BIOSOLIDS

DIGESTED
BIOSOLIDS

FIGURE 11.10   Standard-rate anaerobic digester.

FIGURE 11.11   High-rate anaerobic digester. (Wes Tech Engineering, Inc.)

decreased supernatant quality containing greater dissolved solids. These disadvantages seem to outweigh the advantages since most anaerobic digesters operate in the mesophilic temperature rather than the thermophilic range.

Figure 11.12 shows a schematic diagram of a Two-Stage Digester.

11.2.2.3 Process Design Perhaps the most important system design parameter in suspended growth anaerobic digestion is retention time. Retention time is described either by *solids retention time* (SRT) or *hydraulic retention time* (HRT). SRT is defined as the mass of solids retained in the digester divided by the mass of solids wasted from the system per day. HRT is defined as the working reactor volume divided by the amount of sludge removed per day from the digester. SRT equals HRT when biomass recycle is not employed. A sufficient quantity of microorganisms must be on hand to convert organic matter to methane and carbon dioxide. Furthermore, the retention time must be long enough that the biomass have sufficient time for metabolism and do not wash out of the system. A theoretical minimum SRT is required for biomass to convert organic matter to a given effluent substrate concentration. If the operational SRT falls below this minimum SRT, biomass will not be able to metabolize the organic matter quickly enough and will wash out of the system. For a completely mixed system without biomass recycle, the minimum SRT is derived using Monod kinetics and is described in Eq. (11.2):

$$SRT_{min} = \left(\frac{YkS}{K_s + S} - b \right)^{-1} \tag{11.2}$$

where: S = effluent COD concentration (mg/L)
  Y = yield coefficient (mg/L)
  K_s = half velocity constant (mg/L)
  k = maximum substrate utilization rate (d^{-1})
  b = microorganism decay coefficient (d^{-1})

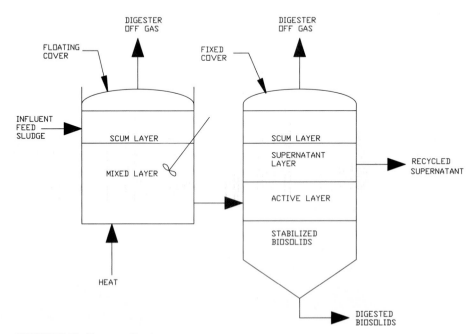

FIGURE 11.12 Two stage digester.

Typical values for Y, b, K_s, and k for municipal sludge have been established by Parkin and Owen. The coefficients K_s, b and k are temperature dependent. Temperature correction terms are included below.

$Y = 0.04$ mg VSS/mg COD removed
$K_s = 1,800$ mg COD/L X (1.112^{T-35})
$k = 6.67$ mg COD/ (mg VSS-d) X (1.035^{T-35})
$b = 0.04$ mg VSS/mg COD removed X (1.035^{T-35})
$T =$ temperature (°C)

If the effluent COD is plotted as a function of SRT_{min} as in Fig. 11.13, it can be seen that lower effluent substrate concentration can be achieved at higher retention times.

The actual SRT used in design (SRT_d) incorporates a safety factor to account for inefficient mixing, toxicity slugs, low temperatures, and changes in influent substrate characteristics. Safety factors typically range from 3 to 10. After SRT_d has been determined, expected process efficiency, sludge production, and methane production can be calculated. Process efficiency is simply calculated as the pounds of COD stabilized per pound of COD entering the digester. Process efficiency can also be determined in terms of volatile solids removed. At standard temperature and pressure (STP), 5.62 ft³ of methane are produced for every pound of COD stabilized (Parkin and Owen, 1986). Example 11.1 shows how to size a high rate anaerobic digester.

Example 11.1 Sizing of a High-Rate Anaerobic Digester   Determine the design SRT for a 10,000 gpd high rate egg shaped anaerobic digester that will be operated at 30°C. The influent COD concentration is 20,000 mg/L and the effluent COD concentration is 2,000 mg/L.

Solution: Assume the coefficients developed by Parkin and Owen for municipal sludge will apply here. K_s, b and k must be corrected for temperature, however.

$$K_s = 1,800 \times (1.112)^{T-35} = 1,800 \times (1.112)^{30-35} = 1,059 \text{ mg COD/L}$$

$$k = 6.67 \times (1.035)^{T-35} = 6.67 \times (1.035)^{30-35} = 5.62 \text{ mg COD/ (mg VSS} - d)$$

$$b = 0.04 \times (1.035)^{T-35} = 0.04 \times (1.035)^{30-35} = 0.034 \text{ mg VSS/mg COD}$$

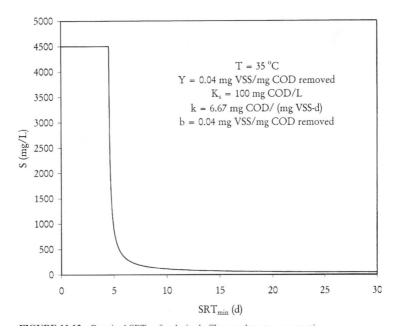

FIGURE 11.13   Required SRT_{min} for desired effluent substrate concentration.

The minimum SRT can be calculated using Eq. (11.3).

$$SRT_{min} = \left[\frac{YkS}{K_s + S} - b \right]^{-1} = \left[\frac{0.04 \times 5.62 \times 2000}{1059 + 2000} - 0.034 \right] = 8.9 \text{ days} \qquad (11.3)$$

Rather than operate at the minimum allowable SRT to avoid bacterial washout, a safety factor of 5 will be applied to allow for imperfect mixing, temperature fluctuations, pH changes, toxicity slugs, and other factors. The design SRT can then be designed.

$$SRT_{design} = SRT_{min} \times \text{safety factor} = 8.9 \text{ days} \times 5 = 44.5 \text{ or } 45 \text{ days}$$

For a completely mixed high-rate anaerobic digester without recycle, the pounds of biological solids synthesized per day can be calculated using Eq. (11.4).

$$P_x = \frac{8.34QYS_r}{1 + b(SRT)} \qquad (11.4)$$

where: P_x = daily biological solids synthesized, lb/d
Q = flowrate, mgd
Y = biological yield, lb/d
S_r = substrate removed as COD, mg/L
b = microorganism decay coefficient, d^{-1}
SRT = solid retention time, d

Another means for designing digester volume is through the use of a volatile solids loading rate, which is defined as the mass of solids added to the digester per day divided by the working volume of the digester. Design volatile solids loading rates vary between 0.12 to 0.16 lb volatile solids/day/ft^3 for high-rate digesters and 0.03 to 0.10 lb volatile solids/day/ft^3 for standard rate digesters (Metcalf and Eddy, 1991). These loading rates are based on the expected peak weekly or monthly solids production. Operating at a lower volatile solids loading rate yields a larger digester volume. Therefore, a trade-off exists between the process stability achieved with a lower loading rate and the increased cost of a larger digester volume.

The amount of volatile solids removed by anaerobic digestion normally ranges from 40 to 60 percent. Temperature and SRT affect the amount of volatile solids destroyed. Table 11.5 illustrates the differences in solids destruction for a low-rate digester (operated at ambient temperature) and a high-rate digester (operated in the mesophilic temperature range) for various SRTs.

11.2.2.4 Operating Parameters   One of the reasons a safety factor is built into the design SRT is to accommodate inefficient mixing. Efficient mixing accomplishes the following:

TABLE 11.5   Estimated Volatile Solids Removal (WEF, 1998)

	Digestion time, days	Volatile solids removal, %
High rate (mesophilic range)	30	65.5
	20	60.0
	15	56.0
Low rate	40	50.0
	30	45.0
	20	40.0

- Reduces thermal stratification
- Provides uniform contact between substrate and active biomass
- Dilutes influent toxicity and undesirable pH slugs
- Increases the effective stabilization reactor volume
- Allows for easier separation of gas and liquid

If the contents of a high-rate digester are not uniformly mixed, the actual SRT may also decrease due to short-circuiting of the reactor. Therefore, if money is not spent to provide for efficient mixing in the reactor, process failure could potentially occur if the actual SRT falls below SRT_{min}. A large SRT safety factor, however, could prevent process failure.

The most common means of digester mixing are mechanical stirring, biogas recirculation, and mechanical pumping. A list of the advantages and disadvantages of these various systems is presented in Table 11.6.

Biogas recirculation mixing is referred to as confined or unconfined. Confined biogas recirculation systems collect biogas at the top of the digester, compress the gas, then reinject the gas in the reactor through confined draft tubes. Unconfined biogas recirculation systems remove gas from the top of the digester, compress the gas, then reinject the gas through diffusors at the bottom of the reactor. Mechanical stirring is carried out by low-speed mixers with rotating impellers. Mechanical pumping utilizes a large internal recycle flowrate to provide turbulence and mixing. Schematics of the various mixing methods are presented in Fig. 11.14.

11.2.2.5 Tank Design Tank designs used in the past are cylindrical, rectangular, and egg-shaped. The common tank configuration in the United States is the cylindrical reinforced concrete tank with a conical bottom. Sidewall depths range from 20 to 50 feet, while tank diameters typically vary from 25 to 125 feet. The conical bottom slopes range from 1:3 to 1:6. The conical bottom makes for easier digester cleaning, however, the steeper slopes increase the difficulty of construction.

Egg-shaped tanks have been the most popular tank used in Europe and are becoming more popular in the United States due to their superior mixing characteristics. Figure 11.15 shows an egg-shaped anaerobic digester.

There are over 57 egg-shaped digesters now in use in the United States (WEF, 1998). The new Deer Island Wastewater Treatment Plant in Boston, MA contains eight 3 million-gallon egg-shaped digesters with 2 additional digesters used for storage. The advantages of an egg-shaped anaerobic digester are:

- Less energy is required.
- Less scum build-up occurs (which translates to less cleaning).
- Short-circuiting is minimized.

The main disadvantage to egg-shaped digesters is their higher cost. Reported construction costs range from $3.50 to $8.00 per gallon of reactor. Conventional cylindrical concrete digesters can be constructed for $1.50 to $4.00 per gallon of reactor. Egg-shaped digesters are constructed of steel or concrete. Mixing is provided by unconfined gas mixing, mechanical mixing with the use of an impeller and a draft tube, or recirculation by mechanical pumping.

11.2.2.6 Gas Production and Collection One unique feature of anaerobic digestion is that it yields methane gas, which can be recovered and used as an energy source. This facet of anaerobic digestion promotes a favorable "environmental" image because it is often self-sufficient in its energy use and does not require the use of other non-renewable resources to meet its energy needs.

Digester gas is collected under a cover that is either fixed to the top of the digester or floats at the top of the water surface. Fixed covers require additional gas storage because the headspace below the cover must be kept at a relatively constant volume for two reasons:

1. So that gas will not be lost by displacement

2. So that gas (instead of air) will be drawn into the digester in case of liquid volume falls in the digester

TABLE 11.6   Advantages and Disadvantages of Various Anaerobic Digester Mixing Methods (Metcalf and Eddy, 1991)

Type mixer	Advantages	Disadvantages
All systems	Increased sludge stabilization	Corrosion and wear of ferrous metal piping and supports. Equipment wear by grit. Equipment plugging and operational interference by rags.
Gas Injection		
Unconfined		
Cover-mounted lances	Lower maintenance and less hinderance to cleaning than bottom-mounted diffusers. Effective against scum buildup.	Corrosion of gas piping and equipment. High maintenance for compressor. Potential gas seal problem. Compressor problems if foam gets inside. Solids deposition. Plugging of gas lances.
Bottom-mounted diffusers	Better movement of bottom deposits than cover-mounted lances.	Corrosion of gas piping and equipment. High maintenance for compressor. Potential gas seal problem. Foam problem. Does not completely mix digester contents. Scum formation. Plugging of diffusers. Bottom deposits can alter mixing patterns. Breakage of bottom-mounted gas piping. Requires digester dewatering for maintenance.
Confined		
Gas lifters	Better mixing and gas production, and better movement of bottom deposits than cover-mounted lances. Lower power requirements.	Corrosion of gas piping and equipment. High maintenance for compressor. Potential gas seal problem. Corrosion of gas lifter. Lifter interferes with digester cleaning. Scum build-up. Does not provide good top mixing. Variable pumping rates. Requires digester dewatering for maintenance if bottom-mounted. Plugging of lances.
Gas Pistons	Good mixing efficiency.	Corrosion of gas piping and equipment. High maintenance for compressor. Potential gas seal problem. Equipment internally mounted. Breakage of bottom-mounted gas piping. Plugging of piston and piping. Requires digester dewatering for maintenance. Piston interferes with digester cleaning.
Mechanical stirring		
Low-speed turbines	Good mixing efficiency.	Wear of impellers and shafts. Bearing failures. Long overhung loads. Interference of impellers by rags. Requires oversized gearboxes. Gas leaks at shaft seal.
Low-speed mixers	Break up scum layers.	Not designed to mix entire tank contents. Bearing and gearbox failures. Wear of impellers. Interference of impellers by rags.
Mechanical pumping		
Internal draft tubes	Good top to bottom mixing. Minimal scum buildup.	Sensitive to liquid level. Corrosion and wear of impeller. Bearing and gear box failures. Requires oversized gearbox. Plugging of draft tube by rags.
External draft tubes	Same as internal draft tube.	Same as internal draft tube.
Pumps	Positive known quantity of mixing. Scum layer recirculated. Sludge deposits can be recirculated. Pumps easier to maintain than compressor.	Nozzle maintenance requires dewatering. Wear of impellers. Plugging of impellers and volutes by rags. Bearing failures.

If air is drawn into the digester headspace, the methane concentration can be reduced to levels that fall within its upper explosive limit (UEL) and its lower explosive limit (LEL). In this range, the biogas can explode with any outside spark source. The UEL and LEL for methane is 15 percent and 5 percent, respectively.

One disadvantage to anaerobic digestion is the production of undesirable odors. Most odor results from the reduction of sulfates in the sludge to hydrogen sulfide gas. Hydrogen sulfide is characterized by a "rotten egg" odor and has an odor detection limit of less than 0.00021 ppmv (Metcalf and Eddy, 1991). The H_2S at high enough concentrations can be corrosive to boilers and engine parts if the

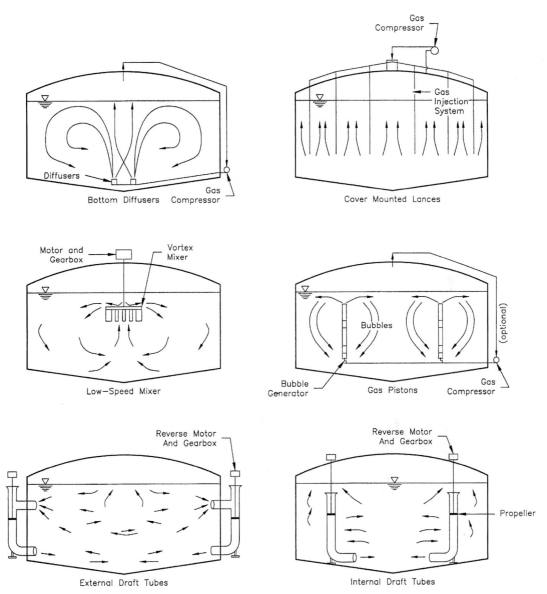

FIGURE 11.14 Various methods for anaerobic digester mixing.

biogas is used for cogeneration. Furthermore, the combustion of hydrogen sulfide results in the production of SO_x gases, which are toxic. As a result, hydrogen sulfide is normally removed prior to cogeneration if present in sufficient quantities. Hydrogen sulfide can be removed with the use of dry or wet scrubbers, adsorption systems, biofiltration, or iron salt addition.

11.2.2.7 Digester Heating   A variety of internal and external heating methods have been used to maintain the temperature in the digester in the mesophilic and thermophilic ranges. Older digesters utilized internal heating coils, which heated the contents within the digester. The popularity of this

FIGURE 11.15 Egg-shaped anaerobic digester. (CBI Walker)

method faded though due to fouling of the coils. An alternative internal heating method is the use of a heat jacketed draft tube. Internal heating methods have a major disadvantage, however, in that the digester normally must be drained to perform maintenance on heating equipment. For this reason, external heating equipment is preferred.

The common external heating equipment used in heating anaerobic digesters are

- Water baths
- Steam injection units
- Tube-in-tube heat exchangers
- Spiral-plate heat exchangers

The water bath heat exchanger is configured in one of two ways. The feed line can be passed through a hot water reservoir where heat is allowed to transfer to the feed line. Alternatively, if the digester contents are recirculated, the recirculation line can be passed through a hot water bath for heating. A steam injection system introduces steam into the feed line, thereby raising the influent feed to the desired temperature. The tube-in-tube heat exchanger contains two concentric pipes. One pipe contains the sludge and the second pipe contains hot water. The flows run counter-currently while heat is transferred from the hot water line to the sludge line. Similar to the tube-in-tube heat exchanger, the spiral-plate heat exchanger utilizes counter-current flow through two spirally wound separate streams. Boilers are often used to heat the circulating water for all heat exchanger systems. The digester gas often provides sufficient energy to heat the recirculating water.

11.2.3 Aerobic Digestion

Aerobic digestion is similar to anaerobic digestion in that the stabilization of sludge is made possible by biological reactions carried out in a suspended growth reactor. The major difference between aerobic and anaerobic digestion is that microorganisms are supplied oxygen in aerobic digestion. A typical aerobic digester is shown in Fig. 11.16.

The major advantages of aerobic digestion over anaerobic digestion are:

- Volatile solids reduction meets or exceeds that of anaerobic digestion.
- The stabilized sludge is free of offensive odor and an excellent fertilizer.
- Supernatant BOD concentrations are lower than that of anaerobic digestion.
- Operation is relatively easy.

The major disadvantage of aerobic digestion is the cost of supplying oxygen to the microorganisms. Power cost for supplying oxygen, however, may not be as significant at smaller plants. Other disadvantages are that the stabilized sludge sometimes has poor dewatering characteristics and process performance depends strongly on temperature. Aerobic digestion is more commonly used at smaller municipal wastewater treatment plant with wastewater flows less than 5 mgd.

Aerobic digestion fundamentally follows the principles of the biological mechanism of endogenous respiration. When little food is available to the biomass population, microorganisms consume their own protoplasm as an energy source, thereby reducing the volatile solids concentration of the sludge. The end products of aerobic digestion are carbon dioxide, water, and ammonia or nitrates. However, the entire cell cannot be biologically transformed to these end products. Approximately 25 percent of the cell is typically non-biodegradable.

FIGURE 11.16 Typical aerobic digester. (US Filter, Jet Tech Products)

Two major mechanisms comprise the aerobic digestion process: oxidation of biodegradable organic matter, and further oxidation of cellular material. The sum of these reactions can be expressed as:

$$C_5H_7O_2N + 5O_2 \rightarrow 5CO_2 + NH_3 + 2H_2O + Energy \tag{11.5}$$

Note that here cell mass is given the chemical formula $C_5H_7O_2N$ and also that nitrification does not occur. This equation may also be written in a form that allows for nitrification as shown below.

$$C_5H_7O_2N + 7O_2 \rightarrow 5CO_2 + NO_3 + 3H_2O + H^+ + Energy \tag{11.6}$$

If nitrification occurs, the digestion process will produce acidity as shown on the right hand side of the above reaction. If inadequate buffering capacity is available due to nitrification, the pH may drop to levels undesirable to the microorganisms. In this case, additional alkalinity may be required to maintain the pH at a sufficient level. From the above stoichiometric equations, approximately 1.5 lb of oxygen is required to convert 1 lb of active cell mass to end products when nitrification does not occur. Approximately 2 lb of oxygen is required when nitrification does occur (WEF, 1998).

11.2.3.1 Process Design   The most common aerobic digestion practice used is conventional aerobic digestion in which air is used for supplying oxygen to microorganisms and systems are operated between 20 and 30°C. Pure oxygen systems and thermophilic systems have also been used, however, the conventional method is much more popular and will be discussed here.

The most common factors in designing an aerobic digestion system include temperature, volatile solids reduction, mixing requirements, detention time, oxygen requirements, and operation method. Table 11.7 summarizes typical design criteria for aerobic digestion. Recent federal regulations (EPA 40 CFR Part 503) regarding the use or disposal of sewage sludge have focused the design objectives of aerobic digestion toward coliform reduction and controlling vector-attraction since vectors have the potential for transmitting pathogens.

Temperature: As with most biological systems, as temperature increases, biological activity increases. This is certainly true for the aerobic digestion process since the system's performance may be seriously hindered if temperature drops too low. Because most aerobic digesters are typically open tanks, the temperature of the sludge in the reactor can be strongly affected by ambient weather conditions. Aerobic digestion operation temperature typically falls in the range of

TABLE 11.7   Typical Aerobic Digestion Design Parameters (Water Pollution Control Federation, 1985)

Parameter	Value
Solids loading (lb volatile solids/ft³ · day)	0.1–0.3
Hydraulic retention time, (days) (T = 20°C)	
Waste activated sludge	10–15
Activated sludge from plant without primary settling	12–18
Primary plus waste activated sludge or trickling filter sludge	15–20
Energy requirements for mixing	
Mechanical aerators (hp/10³ ft³ · min)	20–40
Oxygen requirements (lb O_2/lb solids destroyed)	
Cell tissue	~2.3
BOD_5 in primary sludge	1.6–1.9
Reduction in volatile suspended solids (%)	40–50
Dissolved oxygen residual in liquid (mg/L)	1–2

10 to 40°C. In contrast to anaerobic digestion, aerobic digestion processes typically do not provide for sludge heating to maintain the reactor temperature in the mesophilic temperature range. Perhaps the reason for this is that upward traveling air bubbles from aeration diffusers would quickly remove the heat supplied to the system. As a result, the energy required maintaining the reactor temperature in the mesophilic range would be extremely high. Constructing with concrete rather than steel, providing insulation, roofing or constructing below grade can minimize heat losses from the reactor. Figure 11.17 shows the relationship between biological activity and operating temperature.

Solids reduction: One primary objective of the aerobic digestion process is to reduce the volatile solids concentration in order to minimize the handling cost of the residual sludge. While achieving volatile solids reduction, pathogen and vector-attraction content is reduced. Volatile solids reduction typically falls within the range of 35 to 50 percent. Recent federal regulations require 38% volatile solids reduction to meet vector-attraction standards.

Oxygen and mixing requirements: Oxygen requirements for converting cell tissue and primary sludge BOD to end products can be determined stoichiometrically using Eqs. (11.5) and (11.6). These equations show that 1.40 to 1.98 pounds of oxygen are required per pounds of cells depending on whether or not nitrification occurs. These oxygen requirements may vary based on the amount of primary sludge contained in the influent sludge. The oxidation of primary sludge requires less oxygen than the oxidation of cell tissue.

The contents of the digester should be thoroughly mixed to ensure uniform distribution of the reactor contents. Because a high concentration of biomass is present in the digester, a large amount of air is required to satisfy oxygen requirements. This amount of air normally provides adequate mixing. However, if aeration does not provide adequate mixing, provision should be made for additional mixing (either diffused air or mechanical mixing).

Reactor sizing: Sizing an aerobic digester is typically based on the desired reduction in volatile solids. Because of more stringent federal regulations regarding vector-attraction and pathogen reduction, retention times of 35 to 50 days are now typically used. Prior to these regulations,

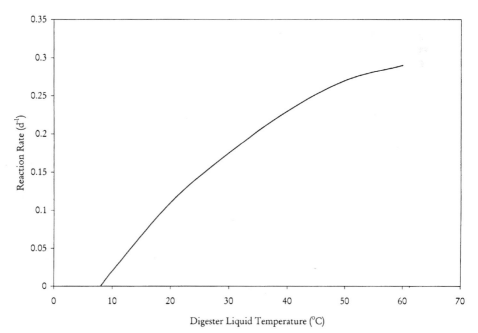

FIGURE 11.17 Biological reaction rate as a function of reactor temperature.

TSTC West Texas
Breckenridge Library

aerobic digesters were sized to reduce pathogen content based on rule of thumb detention times of 40 days at 20 °C and 60 days at 15°C (WEF, 1998).

The rate of volatile solids reduction in aerobic digestion has been well established as a first order reaction.

$$dM/dt = K_d M \qquad (11.7)$$

dM/dt = time rate of change of biodegradable volatile solids (mass/time)
$\quad K_d$ = reaction rate constant (time^{-1})
$\quad M$ = concentration of biodegradable volatile solids at time t (mass/volume)

As mentioned earlier the rate of reduction of volatile solids is highly dependent on temperature. Volatile solids reduction is also dependent on type of residuals being digested, detention time (or, solids retention time), and solids concentration. Figure 11.18 shows actual data from full-scale and pilot-scale aerobic digesters for reaction rate constants at various temperatures.

A popular method for determining the volume of an aerobic digester is shown below in Eq. (11.8).

$$V = \frac{Q_i(X_i + YS_i)}{X(K_d P_v + 1/SRT)} \qquad (11.8)$$

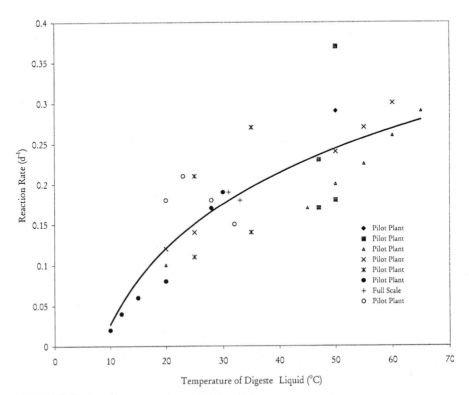

FIGURE 11.18   Reaction rate constants as a function of temperature.

where: V = volume of aerobic digester (ft^3)
Q_i = influent average flowrate (ft^3/d)
X_i = influent suspended solids (mg/L)
Y = fraction of influent BOD$_5$ consisting of raw primary sludge
S_i = influent BOD$_5$ (mg/L)
X = digester suspended solids (mg/L)
K_d = reaction rate constant (d^{-1})
P_v = volatile fraction of digester suspended solids
SRT = solids retention time (d)

Equation (11.8) is not applicable to digesters where significant nitrification will occur. It can also be noted from Eq. (11.8) that reactor volume requirements are calculated as a function of solids retention time, solids concentration, and fraction of primary solids in addition to flowrate. Volume requirements increase with shorter solids retention time, larger fractions of primary solids, increased influent solids concentration, and slower reaction rate.

Process operation: An advantage of aerobic digestion as compared to anaerobic digestion is its relative ease of operation. As long as proper pH, temperature, and toxicity control are sustained, the process does not require much maintenance.

Example 11.2 Aerobic Digester Design   Size a single stage aerobic digester for the following sludge characteristics and determine the daily oxygen requirement. The sludge contains no primary solids.

Q_i = 1,000 ft^3/d
X_i = 15,000 mg/L
X = 2,000 mg/L
K_d = 0.20 d^{-1}
P_v = 0.80
SRT = 40 days

$$V = \frac{Q_i(X_i + YS_i)}{X(K_dP_v + 1/SRT)} = \frac{1,000(15,000 + 0)}{2,000(0.20 \times 0.80 + 40^{-1})} = 40,540 \text{ or } 41,000 \text{ ft}^3$$

Assuming nitrification does not occur, approximately 1.4 lb O$_2$ will be consumed per lb of volatile solids destroyed. The amount of volatile solids destroyed is the difference between the influent suspended solids and the effluent suspended solids (which is equal to the suspended solids concentration in the reactor).

$$\text{volatile solids destroyed} = 15,000 - 2,000 = 13,000 \text{ mg/L}$$

$$\text{volatile solids destroyed per day} = (13,000 \text{ mg/L})(0.001 \text{ mgd})(8.34) = 108 \text{ lb/d}$$

$$\text{O}_2 \text{ required} = \frac{1.4 \text{ lb O}_2}{\text{lb VS destroyed}} = 1.4 \times 108 \text{ lb/d} = 151 \text{ lb/d}$$

11.2.3.2   *Thermophilic Aerobic Digestion*   Thermophilic aerobic digestion operates in the temperature range of 40 to 80°C. Large-scale pilot studies have shown that thermophilic aerobic digestion can remove up to 70 percent of degradable organics in shorter detention times (approximately 4 days) than used in mesophilic aerobic digestion. A unique feature of thermophilic aerobic digestion is that it is often self-sufficient with regard to its heating requirements. Since detention times are smaller for thermophilic reactors than mesophilic reactors for the same quantity of influent volatile solids, a smaller volume of sludge requires heating. Thus, thermophilic temperatures can be maintained in the reactor due to the smaller volume of sludge requiring heating. To minimize heat loss from the reactor, reactors should be insulated. Thermophilic aerobic digestion is regarded as a stable process that responds well to process upsets and fluctuations in ambient weather conditions.

11.2.4   Composting

Composting is an aerobic biological stabilization process where sludge is converted into a humus-like material. The process is exothermic, and the heat released by biological reactions destroys most pathogenic bacteria. The composted material contains a high organic content and can be utilized for a variety of applications. Composted biosolids have been used in landscaping applications, topsoil blending, and general gardening. The composted biosolids also enhance the physical properties of soil by improving its tilling properties, water holding capacity, and erosion control. Composted biosolids however have limited nutrient availability. Typically, composted biosolids require supplemental nutrient addition if they are to be used as a fertilizer.

The composting process consists of mixing sludge with a dry, porous media that allows for air circulation through the mixture. A wide variety of media (also called bulking agents) are used, including wood chips, various wood wastes, and yard waste. During the composting process, aerobic microorganism convert complex organics and oxygen to less complex organics, water, carbon dioxide, nitrate and sulfate. During composting, the temperature rises to about 70°C due to exothermic biological reactions. Early in the composting process, mesophilic bacteria predominate. After approximately one week of composting, the temperature increases into the thermophilic range and thermophilic bacteria, actinomycetes, and thermophilic fungi predominate. Above 70°C, spore forming bacteria predominate. As biodegradation slows, the heat released by biological reactions correspondingly decreases; the temperature drops to the mesophilic range and mesophilc bacteria once again prevail. The fundamental stabilization mechanism is the rise in temperature into the thermophilic range, since most pathogenic bacteria do not survive at this elevated temperature.

11.2.4.1 Process Description   Composting follows these general steps:

1. Mixing sludge with a bulking agent and/or amendment
2. Aeration/agitation
3. Recovery of bulking agent (if possible) by screening
4. Curing and storage
5. Disposal

An amendment is an organic agent added to increase the air voids for improved aeration and to reduce bulk density. Amendments are similar to bulking agents, but are typically smaller in size. Due to the high rate of microbial activity, bacteria require a large supply of nitrogen. For efficient composting, an ideal carbon to nitrogen ratio (C:N) exists for a given sludge/bulking agent mixture. Typically, nitrogen is limiting when the C:N ratio rises above 30:1. The C:N ratio for waste activated sludge is typically about 6:1 and 15:1 for aerobically or anaerobically digested sludge. The C:N ratio of the bulking agent may range from 50:1 to 500:1 depending on the material (Cornell Composting Homepage). Therefore, a carbon and nitrogen mass balance must be performed to ensure adequate nitrogen availability. Microorganisms also require moisture. A mass balance must also be performed on the amount of water in the influent sludge and bulking agent to ensure sufficient moisture for biological activity. At the same time, excess moisture can lead to poor air distribution.

There are three common methods of composting used in practice: aerated static pile, windrow process, and in-vessel processes. In aerated static pile composting, dewatered sludge cake is mixed with a coarse bulking agent (often wood chips). The mixture is piled over a porous bed on top of an aeration or exhaust piping network. The pile is typically 6 to 8 feet high. The material is typically

FIGURE 11.19   Aerated static pile composting.

composted for 3 to 4 weeks and then cured for at least another month. The piles are covered with a layer of composted sludge to provide insulation. After composting, the material can be screened to try recovering the bulking agent. Figure 11.19 shows a typical aerated static composting pile.

Similar to the aerated static pile composting method is the use of windrow. Windrows are long narrow piles as shown in Fig. 11.20.

The main difference between the two is that windrows do not utilize exhaust and aeration piping below the pile. Instead, air is supplied to the compost piles by mechanically turning the pile at least five times while the temperature of the pile is greater than 55°C. However, supplemental aeration is used under some circumstances. The windrows are arranged as long, narrow piles to provide extra surface area to allow for natural convection and diffusion of air through the pile. The composting period for windrow application is also approximately three to four weeks.

FIGURE 11.20   Windrow composting system.

   In-vessel composting is a somewhat advanced composting system in that composting is carried out in a closed container allowing for better control of environmental conditions, such as temperature, air flow, and oxygen concentration throughout the system. The enclosed system also minimizes odor. Since environmental conditions are optimized, required treatment time is shorter than other composting methods. As a result, less space and labor is required for composting the same amount of sludge. The two common types of in-vessel composting systems are the plug flow and dynamic, or agitated bed. In both systems, the sludge/bulking agent mixture is fed through one end of a tunnel, silo, or open channel and continuously moves to the outlet. Various reactor designs exist for both in-vessel composting treatment systems. Typical examples are shown in Fig. 11.21.

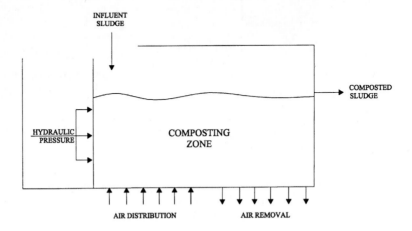

Plug Flow Reactor

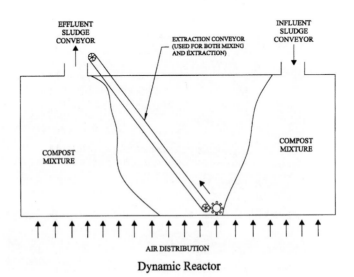

Dynamic Reactor

FIGURE 11.21   Various in-vessel composting systems.

11.2.4.2   Process Design   The major design parameters for composting processes are

- Selection of bulking agents
- Mixture ratio, moisture content
- Aeration requirements
- Temperature
- The C:N ratio

Aeration requirements are determined stoichiometrically, while the exhaust gas in enclosed systems is measured for oxygen and carbon dioxide. The oxygen concentration within the compost pile can be estimated based on the oxygen concentration of the exhaust gas. Furthermore, the carbon dioxide concentration of the exhaust gas may indicate process efficiency, since carbon dioxide is an end-product of the composting process. Temperature is also monitored using temperature probes in several locations.

A variety of bulking agents are available that each have their own advantages and disadvantages. Table 11.8 describes the advantages and disadvantages of different bulking agents. Bulking agents with high porosities are required for static pile composting to provide sufficient passage of air.

The initial mixture for composting should have a solids concentration of approximately 40 to 45 percent. Since sludge and bulking agent solids concentrations differ, the two components must be mixed at a ratio to meet the initial desired solids concentration. The sludge and bulking agent must also be mixed to achieve an initial C:N ratio of approximately 30:1. Therefore mass balances must be performed to determine the correct mixing ratio of sludge and bulking agent. Example 11.3 illustrates the process to follow to determine the proper mixing ratio.

Example 11.3 Compost Mixing Design   A secondary wastewater treatment facility treating an average flow of 10 mgd is considering developing a composting program for the beneficial use of biosolids. You have been asked to design a static pile composting process using wood chips as the bulking agent. The information below is known:

Total primary and secondary sludge = 10 dry tons per day

Sludge moisture content after dewatering = 80 percent

Sludge nitrogen content = 5 percent (dry weight basis)

TABLE 11.8   Characteristics of Various Bulking Agents (WEF, 1998)

Bulking agent	Comments
Wood chips	• Must be purchased • High recovery rate by screening • Provides supplemental carbon
Chipped brush	• Possibly available as waste material • Low recovery rate by screening • Provides supplemental carbon • Longer curing time
Leaves	• Must be ground • Wide range in moisture content • Readily available source of carbon • Relatively low porosity • Non-recoverable
Shredded tires	• Often mixed with other bulking agents • No available supplemental carbon • Nearly 100% recoverable • May contain metals
Ground waste lumber	• Possibly available as waste material • Often poor source of supplemental carbon

Sludge C:N ratio $= 11:1$ (dry weight basis)

Wood chip moisture content $= 20$ percent

Wood chip nitrogen content $= 0.2$ percent (dry weight basis)

Wood chip C:N ratio $= 200:1$ (dry weight basis)

It is assumed the initial moisture content of the mixture (MC_o) will be 60 percent and the final moisture content of the mixture (MC_f) will 40 percent at the end of the composting process. Determine the amount of wood chips needed to achieve a desirable C:N ratio between 20 and 30 for the mixture.

Solution: Since the moisture content of the dewatered sludge is 80 percent (i.e. solids concentration $= 20$ percent), the wet tonnage of solids flow can be determined.

$$\text{Total Sludge Flow} = \frac{10 \text{ dry tons per day}}{0.20} = 50 \text{ wet tons per day}$$

Therefore, of the 50 total tons per day of sludge, 40 tons must be water. The initial moisture content (MC_o) can be defined as follows,

$$MC_o = \frac{\text{Mass of water in sludge} + \text{mass of water in woodchips}}{\text{Total mass of sludge mixture}} = \frac{40 + 0.2W}{50 + W} = 0.60$$

where: $W =$ dry tons of wood chips required per day

Solving for W, we find $W = 25$ dry tons per day. Now the C:N ratio must be checked

$$\left(\frac{C}{N}\right)_{\text{mixture}} = \frac{C_{\text{wood chips}} + C_{\text{sludge}}}{N_{\text{wood chips}} + N_{\text{sludge}}} = \frac{(N_{\text{wood chips}})\left(\frac{C}{N}\right)_{\text{wood chips}} + (N_{\text{sludge}})\left(\frac{C}{N}\right)_{\text{sludge}}}{N_{\text{wood chips}} + N_{\text{sludge}}}$$

$$= \frac{[0.002(25)]\frac{200}{1} + [0.05(10)]\frac{11}{1}}{[0.002(25)] + [0.05(10)]} = 28.1$$

The C:N ratio of the mixture of approximately 28:1 is adequate and therefore adding 25 dry tons (31.3 wet tons) of wood chips per day will satisfy mixture design requirements.

Other considerations in composting processes include pH, which should be maintained between 6 and 9, site constraints, such as area availability and climatic conditions.

11.3   *SLUDGE CONDITIONING*

Prior to dewatering, chemicals are often added to the influent sludge to improve its dewatering characteristics. It is generally accepted that dewatering is not practical without conditioning. Chemical conditioning of sludge can increase the solids concentration of the sludge prior to dewatering. The increase in solids concentration depends on the chemical used and various characteristics of the sludge. Not only does the addition of chemicals increase solids concentration, but the physical properties of the sludge are altered to aid in dewatering efficiency. Typical chemical conditioners used include lime, alum, ferric chloride, and organic polymers. Inorganic conditioners such as lime and iron salts increase solids concentration in the form of precipitation, while polymers do not increase solids concentration but enhance coagulation. Another method of sludge conditioning is heat treatment, which concurrently can be used to stabilize sludge. Of the two stabilization methods, chemical conditioning is more widely applied. Diatomaceous earth can also be used to condition sludge for dewatering.

11.3.1 Lime

Lime is available in two forms: powdered hydrated lime and granular quicklime. Lime is often used at wastewater treatment plants for pH control. Lime is inexpensive (approximately 5 to 7 cents per pound (Chemical Market Reporter, 3/8/ 99)) but easily precipitates making pumping and handling troublesome. Lime is often used in combination with the use of ferric chloride as a chemical conditioner. Lime is used for two main reasons:

1. To act as a bulking agent, which enhances dewatering
2. To withdraw water out of the cellular mass for enhanced dewatering

After lime is added to water, calcium carbonate and calcium hydroxide precipitate. This precipitation increases the relative solids concentration of the sludge mixture.

11.3.2 Ferric Chloride

Ferric chloride and lime are often used in combination to precipitate ferric hydroxide. The addition of ferric chloride to wastewater may drop the pH by two units forming ferric hydroxide complexes. Lime is then added to raise the pH often above 8.5, whereupon greater amounts of ferric hydroxide may precipitate.

11.3.3 Organic Polymers

Organic polymers used for dewatering are typically long-chained, water soluble and can vary greatly in charge, density and molecular weight. A key advantage of using organic polymers in dewatering is that the volume of solids for disposal is not greatly increased compared to other chemical conditioners such as lime. The main disadvantage in the use of organic polymers is cost. They are normally more expensive than other dewatering agents.

The fundamental mechanisms behind the use of organic polymers as dewatering conditioners are charge neutralization and adsorption. These mechanisms occur through the use of cationic polymers or long chained polymers. The solids particles in sludge typically have a negative surface charge. The addition of a positively charged polymer neutralizes the solid particles surface charge, thereby reducing the repulsive forces between sludge particles and allowing for the agglomeration of sludge particles. Due to the size and shape of long-chain polymers relative to more spherical sludge particles, polymer molecules easily adsorb sludge particles allowing for the enmeshment of sludge particles in the polymer chains.

11.3.4 Dosage

The chemical dosage required for dewatering is a function of the specific gravities of the solids being dewatered and the physical nature of the sludge (primary, waste activated, or chemical sludges). Sludges that contain dense sludge particles, such as primary sludge, are easier to separate from water and therefore require less chemical conditioner. On the other hand, sludges containing particle specific gravities near that of water are more difficult to dewater and require larger chemical dosages. Dewatering waste activated sludge with a polymer may require from 10 to 30 lb dry polymer per ton of dry solids, whereas dewatering primary sludge may only require up to 10 lb dry polymer per ton of dry solids. Chemical conditioning of chemical sludges requires laboratory testing to determine the optimal conditioning agent and its dose.

11.4   *DEWATERING*

Dewatering is a physical means for reducing the water content of sludge. Dewatered sludge has several advantages as noted below:

- Dewatered sludge is easy to handle since its consistency is more like a cake than a slurry.
- Hauling costs of dewatered sludge are reduced since sludge mass and volume have been reduced.
- Sludge dewatering may be required for subsequent sludge treatment including incineration and composting.
- Dewatering may be required to meet landfill solids concentration criteria.

The common methods used for sludge dewatering are centrifugation, belt filter press, vacuum filter, pressure filter press, and sludge drying beds. The selection of the appropriate dewatering method depends on sludge characteristics, space availability, and residual sludge disposal options. For example, treatment plants with limited space would be forced to use a technology that may have high capital costs and operating costs. On the other hand, a rural treatment plant with excess space may be able to operate drying beds, which are not as cost intensive.

11.4.1   Centrifugal Dewatering

Centrifugal dewatering is similar to centrifugation for thickening, as explained earlier in this Chapter, in that the solids are separated from water with the aid of centrifugal force. The three common centrifuges used for sludge thickening (solid bowl centrifuge, imperforate basket centrifuge, and the disk nozzle) can also be used for sludge dewatering. However, the solid bowl centrifuges are the most common, and imperforate basket centrifuges the second most common for dewatering.

11.4.1.1   Solid Bowl Centrifuge for Sludge Dewatering   For the solid bowl centrifuge, influent sludge is fed into the rotating bowl where liquid and low density solids (called the *centrate*) are separated from the heavier solids (called the *sludge cake*). The centrate is returned to the head of the wastewater treatment system. However, since the solids in the centrate are too fine to be dewatered, the solids in the centrate normally will pass through the treatment plant, thereby decreasing the effluent quality. The solids bowl centrifuge may be used for dewatering a variety of sludges. The design of a solid bowl centrifuge can be based on a modified version of Stoke's Law:

$$V = \frac{(\rho_p - \rho_w)Gd^2}{1800\mu} \qquad (11.9)$$

where: V = sludge particle separation velocity in water (m/s)
ρ_p = sludge particle density (kg/m^3)
ρ_w = water density (kg/m^3)
G = acceleration due to centrifugal forces (m/s^2)
d = mean particle diameter (m)
μ = fluid viscosity (kg/m-s)

As can be seen from Eq. (11.9), the separation velocity is a function of the density of the particle being settled and an even stronger function of the diameter of the particle. Thus, large, heavy solids, such as primary solids should be more easily dewatered than lighter, less dense solids. The centrifugal acceleration applied to a particle in a centrifuge is given below.

$$G = R(0.105N)^2 \qquad (11.10)$$

where: G = centrifugal acceleration (m/s^2)
$\quad\quad\quad$ R = radius of rotating body of liquid (m)
$\quad\quad\quad$ N = centrifuge rotational velocity (rev/s)

Influent solids concentrations may range from 2 to 12 percent. Sludge cake concentrations normally range from 10 to 35 percent. The performance of the centrifuge is based on the feed flowrate, depth of the settling zone, rotational velocity of the centrifuge, sludge characteristics, and the use of chemical conditioners. Typical dewatering performance is summarized in Table 11.9.

11.4.1.2 Imperforate Basket Centrifuge for Sludge Dewatering Dewatering using the imperforate basket centrifuge is similar to thickening applications using the imperforate basket with one exception. After the basket is filled with solids and the basket decelerates, soft sludge (sludge with lower solids concentration) is removed from the inner wall of sludge in the basket. Typically, about 10 percent of the sludge volume in the bowl is removed in this additional step. Imperforate basket dewatering is common at smaller plants. Characteristic performance data for the imperforate basket centrifuge is given in Table 11.9.

11.4.2 Belt Filter Press

The belt filter press is a continuously run dewatering unit that has become one of the predominant dewatering devices used in the United States. Chemical conditioners are also typically used to aid in dewatering. The sequence of steps in belt filter press dewatering is chemical conditioning, gravity drainage, and mechanical compression to concentrate solids. A typical belt filter press is shown in Fig. 11.22.

Sludge is typically conditioned with a polymer, which can be added to a flocculation tank upstream of the belt filter press or directly injected to the influent sludge in the piping. The polymer must be thoroughly mixed with the feed sludge for adequate dewatering performance. Next, the sludge enters the gravity drainage section where a large portion of the water is removed. Conditioned

TABLE 11.9 Performance Data for the Solid Bowl and Imperforate Basket Centrifuges (Metcalf and Eddy, 1991 and U.S. EPA, 1979)

	Solid bowl			Imperforate basket		
		Solids capture, %			Solids capture, %	
	Cake solids, %	Without chemicals	With chemicals	Cake solids, %	Without chemicals	With chemicals
Untreated						
Primary (P)	25–35	75–90	90+	25–30	90–95	95+
P + Trickling Filter (TF)	20–25	60–80	90+	7–11	90+	90
P + Air Activated (AA)	12–20	55–65	90+	12–14	85–90	90+
P + Rotating				17–24	90+	95+
Biological Contactor						
Waste Sludge						
TF	10–20	60–80	90+	9–12	90+	95+
AA	5–15	60–80	90+	8–14	85–90	90+
Oxygen Activated	10–20	60–80	90+			
Anaerobically Digested						
P	25–35	65–80	85+			
P + TF	18–25	60–75	85+			
P + AA	15–20	50–65	85+	8–14	75–80	85+
Aerobically Digested	8–10	60–75	90+			
Waste Activated						

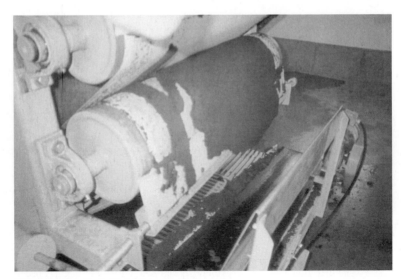

FIGURE 11.22   Typical belt filter press for dewatering.

sludge is spread uniformly on a porous, continuously moving belt with the aid of a distribution system. The centrate that passes through the porous belt is collected and returned to the headworks of the plant. After gravity drainage, the thickened sludge enters the mechanical compression stage where it is compressed between two porous belts. In this stage, water is forced through the belt by compression and shearing forces. The application of pressure to the sludge increases as it passes through this stage.

The typical variables that affect the performance of the belt filter press include sludge characteristics, method of chemical conditioning, process configuration, belt porosity, belt speed, and belt width. One drawback of the belt filter press is that it does not respond well to fluctuating sludge characteristics. If a facility's sludge exhibits variability, a sludge blending basin should be designed to ensure proper belt filter press performance. Typical belt widths range from 2 ft to 10 ft (0.5 to 3 m). Sludge loading rates vary from 100 to 500 lb/ft-h and hydraulic loading rates vary from 7 to 35 gal/ft-min. Belt filter presses are typically capable of yielding a sludge with a solids concentration of 12 to 30 percent depending on the sludge and it's conditioning. Table 11.10 presents operating data for various types of sludge.

11.4.3   Pressure Filter Press

Dewatering with the aid of the pressure filter press is performed by injecting sludge under pressure against a porous membrane, thereby effectively forcing water (the centrate) through the membrane. Pressure filter presses are capable of achieving higher dewatered sludge concentrations than most other dewatering options. For this reason when high solids concentrations are required (greater than 35 percent), pressure filter presses may be a viable option. Other advantages of the pressure filter press are:

- A high filtrate clarity can be achieved
- A variety of sludges can be treated
- Mechanical reliability

TABLE 11.10 Performance Data for the Belt Filter Press for Various Types of Sludge (WEF, 1998)

Type of sludge	Dry feed solids, %	Loading per meter of belt width (L/min)	Dry polymer (g/kg dry solids)	Cake solids, %
Raw primary sludge (P)	3–7	110–190	1–4	28
Waste activated sludge (WAS)	1–4	40–150	3–10	15
P + WAS (50:50)	3–6	80–190	2–8	23
P + Trickling filter (TF)	3–6	80–190	2–10	25
Anarobically digested:				
P	3–7	80–190	2–5	
WAS	3–4	40–150	4–10	28
P + WAS	3–6	80–190	3–8	15
Aerobically digested:				22
P + WAS (unthickened)	1–3	40–190	2–8	16
P + WAS (50:50), thickened	4–8	40–190	2–8	18
Oxygen activated WAS	1–3	40–150	4–10	18

The disadvantages of the pressure filter press are:

• Its high capital cost
• High manual labor requirements, unless automated
• Required periodic replacement of the filter cloth

Pressure filter presses are available as fixed-volume or variable-volume units. The most common are the fixed-volume, and those will be covered here. Pressure filter presses are operated as a batch process. Sludge is pumped under pressure into chambers that are separated by membranes (or filter cloth) where water is forced through the filter leaving the solids behind in the recessed chamber. The centrate is collected by internal piping and discharged out of the press. The filter cloths are held in place by a series of rectangular plates. When the chambers have reached their solids capacity, the sludge feed is shut off and the plates can be separated so that the trapped solids can manually be removed. A typical pressure filter press is shown in Fig. 11.23.

630 mm

470 mm

Worldwide patents granted and applied for:

FIGURE 11.23 Typical pressure filter press. (US Filter)

Influent sludge may require conditioning for efficient operation of the filter press. Typical chemical conditioners include polymer, lime, ferric chloride, or a combination of these. If lime alone is used, the lime acts as a bulking agent, reducing water content, but also increasing the mass of solids in the sludge cake. Lime and ferric chloride can be used in combination. Lime is added to raise the pH and then ferric chloride is added to reduce the negative charge barrier surrounding sludge particles, allowing easier agglomeration of sludge particles and more efficient dewatering. Polymers are used as bulking agents for dewatering with the pressure filter press. Diatomaceous earth may also be used a filter aid.

Various considerations must be taken into account when designing a pressure filter press. The appropriate chemical conditioner and chemical conditioner dose must be considered to ensure adequate operation. Bench scale studies can be performed to determine the optimal conditioner and dose. Pressure filter press membranes require frequent cleaning and this may result in excessive downtime of the unit. Typical loading rates to pressure filter presses based on bulk density are approximately 60 to 80 lb/ft^3. The solids concentration of the sludge cake normally ranges from 30 to 45 percent, where approximately 10 to 20 percent of the sludge cake is comprised of the chemical conditioner, depending on the conditioner used. The design of a pressure filter press is based on influent solids concentration, sludge flow rate, and the operating pressure of the unit.

11.4.4 Rotary Vacuum Filter

As its name implies, the Rotary Vacuum Filter normally consists of a horizontal cylinder or drum which rotates through the sludge with a vacuum applied to the inside of the cylinder to cause the solids to accumulate on the outside surface of the drum. The process is continuous and consists of the following phases:

- Immersion phase—sludge forms on the inside of the drum filter media under the liquid surface
- Drying phase—air continues to be pulled through the filter and its accumulated cake above the liquid level to cause drying
- Scraping phase—the dried cake is scraped from the filter surface into a conveyor or dumster

The filtrate flows out through an internal piping system. The cake thickness can vary from $1/_8$ inch to over one inch (3–30 mm).

Depending on chemical treatment and pre-coating methods, a rotary drum vacuum filter can concentrate a sludge from $1/_2$ % to 40% solids.

Figure 11.24 shows a typical Rotary Vacuum Filter.

11.4.5 Sludge Drying Beds

Sludge drying beds are a popular alternative for dewatering sludge given the appropriate climatic conditions. The most popular type of sludge drying beds used in the United States is the conventional sand drying bed. Drying beds are used to dewater digested sludge and are typically used for small treatment facilities. They are best suited to small facilities in locations with hot and dry climates, however, they may be used under any conditions for any size treatment facility. For this reason, sand drying beds are popular in the Southwestern United States. Compared to other dewatering methods, sand drying bed technology is more labor intensive and requires more land. However, the capital cost of drying beds is typically the lowest of all dewatering methods.

The primary mechanism behind sand bed drying is evaporation and drainage. Sludge is placed on a bed of sand where free water is allowed to drain through the sand bed to a water collection system. A gravel layer is located under the sand layer and the perforated plastic pipe or vitrified clay tile is used for underdrains. Sidewalls are also designed along the perimeter of the system. In many cases, the bed may need to be lined with a protective membrane to prevent groundwater contamination. This initial drainage process lasts approximately 3 to 4 days. After the drainage period, further drying occurs as a result of evaporation. The evaporation process may last an additional 10 to 15 days.

FIGURE 11.24   Rotary vacuum filter.

The typical sludge and sand bed layers are both approximately 8 to 16 inches. The design of the sand layer should be carefully performed so that adequate drainage is allowed. If the sand layer is too deep, drainage will be slower. Furthermore, if the sand contains too many fines, the sand bed may clog easily. It is recommended that the sand layer have a uniformity coefficient less than 3.5 and the mean particle diameter fall between 0.3 and 0.75 mm. A typical sand drying bed is given in Fig. 11.25.

The typical bed area is sized so that the beds can be efficiently loaded with liquid sludge and cleaned of dried sludge. The width of the bed may be approximately 20 ft, while the length of the bed will vary according to the amount of sludge to be dried. Odors will occasionally result from sludge that is not completely digested, therefore sludge drying beds may need to be located away from housing. The individual beds can be separated by concrete blocks, concrete or earthen embankments.

After the sludge has adequately dried, the sludge may be removed from the bed either manually by shovel or by a mechanical scraper system. The solids concentration of the dried sludge may reach 60 percent. This solids concentration is higher than other dewatering methods and so the cost of ultimate disposal is quite low due to the decreased volume. In non-rural areas, the drying beds may be covered in greenhouse-like structures to avoid odor complaints. Covers may also be used when sludge drying occurs year round. With a cover, the sludge will be protected from rain and snow.

11.4.6   Other Sludge Drying Beds

Other sludge drying beds include paved drying beds, wedge-wire drying beds, and vacuum-assisted drying beds. The drainage bed operates similar to the conventional sand bed where the drained leachate is collected, however since most of the bed is paved, dried sludge may be collected with a front end loader making collection easier and more efficient. The beds are sloped at approximately 1.5 percent to a drainage collection area in the center of the bed. Using paved beds allows for larger bed areas since collection of dried sludge is faster. Since less sand area is used for these beds per area of sludge applied, a larger mean diameter sand particle size will be required.

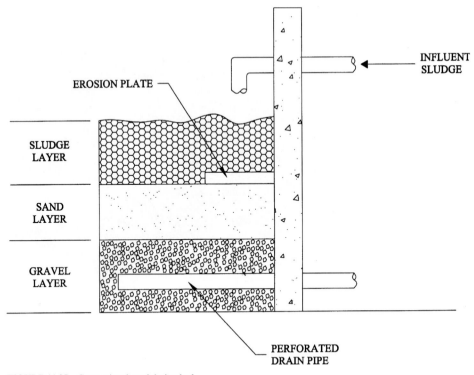

FIGURE 11.25 Conventional sand drying bed.

The wedge-wire drainage bed utilizes a false floor with openings of approximately one hundredth of an inch. Water is removed by a combination of drainage and evaporation. This type of bed is used infrequently, due to the capital cost and potential for plugging of the false bottom perforation.

Vacuum-assisted drying is used to accelerate the time required to dry sludge. Chemical conditioners are first mixed with the influent sludge as the sludge is spread about the bed by gravity. After the sludge is spread, free water is allowed to drain for approximately one hour. Next, a vacuum pump is turned on and a vacuum is created below the sludge layer enhancing drainage. The enhanced drainage cuts treatment time to about 1 to 2 days while achieving approximately 10 to 30 percent dried solids. Thus, in non-rural areas where space may be limited, less space is required to dry a given volume of solids using the vacuum-assisted drying bed compared to other methods. A schematic of a vacuum-assisted drying bed is given in Fig. 11.26.

11.5 SLUDGE DISPOSAL

All treatment processes discussed to this point in this chapter are intended to either facilitate easier and safer sludge handling for final disposal or to minimize the cost of final sludge disposal. Thickening and dewatering are practiced to reduce the amount of water in sludge to allow for easier handling and minimize the cost of hauling the sludge. Stabilization is performed to minimize potential health threats caused by components of the sludge and reduce the potential odor of the sludge. The two most common options for sludge disposal are land application and landfilling. The decision

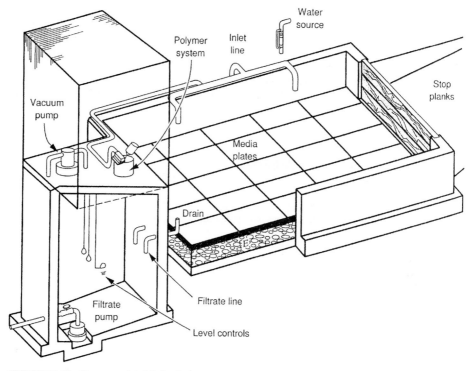

FIGURE 11.26 Vacuum-assisted drying bed.

to land apply, landfill, or ultimately dispose of sludge in some other way (including incineration, solidification, or other beneficial use) will depend on:

- Landfilling costs
- Sludge hauling costs
- Availablility of land for land application
- Sludge characteristics
- Regional regulations for sludge landfilling and land applying

11.5.1 Land Application

Limited landfill space and associated increased landfilling costs have motivated wastewater treatment facility management to explore other sludge disposal options. In light of these changes, management has explored options for utilizing the nutrient benefits of wastewater sludge. Biological wastewater sludge is high in nitrogen and phosphorous, which are required nutrients for plant growth. As a result, efforts have been made to dispose of sludge in a way that these nutrients may be used as a fertilizer through land application.

Land application of sludge refers to the spreading or injecting of sludge on top of or just below the soil surface. The sludge thereby acts as "topsoil" and can support vegetative growth. While practicing land application of sludge, the sludge must be fully characterized to ensure that other constituents in the sludge, such as toxic compounds and pathogens, do not pose a potential health threat after land application. However, after land application, trace metals are essentially rendered harmless since they are typically trapped in the soil matrix or used to a small degree by vegetation. Pathogen destruction occurs due to sunlight and desiccation. Also, if toxic organics are present in the sludge, further biodegradation of the organics may occur after land application.

11.5.1.1   Site Selection and Suitability   Approximately half of the municipal wastewater treatment facilities in the United States practice land application (WEF, 1998). Sludge can be land applied to:

- Agricultural land
- Pasture land
- Sites dedicated to sludge disposal
- Forest land

The suitability of a site for land application of sludge depends on the site's capacity to retain components of the sludge that may potentially cause a health or environmental hazard. Variables in the selection of a land application site include:

- Site geology
- Drainage characteristics
- Topography, and proximity of receiving waters
- Local residences
- Public roads
- Drinking water wells

State regulations normally outline a minimum distance that a landfill must be set back from these locales. Suggested minimum setback distances for various locales are given in Table 11.11 Slope limitations for land application of sludge are given in Table 11.12.

After a site has passed the initial screening of the above parameters, soil characteristics are examined. The soil is examined to determine pH, background metal concentrations, cation exchange capacity, and nutrient availability. A neutral pH is desired for vegetative growth. A soil's cation exchange capacity indicates the soil's ability to retain and hinder metals from leaching into the groundwater. Soil with a higher cation exchange capacity will allow greater sludge application.

11.5.1.2   Sludge Suitability   Sludge suitability for land application is determined on a case-by-case basis. Sludge characteristics will determine the rate at which sludge may be applied or whether the sludge is appropriate for land application in the first place. The presence of metals, nitrogen, phosphorous, and potassium are the primary parameters examined. Other parameters should be examined depending on the type of sludge. The amount of nitrogen, phosphorous, and potassium will dictate the nutrient value of the sludge to vegetative growth. The application rate of sludge is based on the annual nutrient requirements of the crop to be grown. Annual nutrient requirements for various crops are given in Table 11.13.

TABLE 11.11 Typical Setback Distances for Land Application

Locale	Setback distance, ft
House	400
Well	400
Public road	30
Surface water	200

TABLE 11.12   Slope Limitations for Land Application (U.S. EPA, 1995)

Slope	Comment
0–3%	Ideal; no concern for runoff or erosion of liquid or dewatered sludge
3–6%	Acceptable for surface application of liquid or dewatered sludge; slight risk of erosion
6–12%	Injection of liquid sludge required in most cases, except in closed drainage basin and/or areas with extensive runoff control. Surface application of dewatered sludge is usually acceptable.
12–15%	No liquid sludge application without effective runoff control; surface application of dewatered sludge is acceptable, but immediate incorporation is recommended.
Over 15%	Slopes greater than 15% are only suitable for sites with good permeability (e.g. forests), where the steep slope length is short (e.g. mine sites with a buffer zone downslope), and/or the steep slope is a minor part of the total application area.

Annual sludge application beyond the suggested application rate that will provide vegetative growth is discouraged. Excess application of sludge with nitrogen can lead to potential health and environmental hazards if the nitrogen is allowed to leach into the ground water or nearby surface water. Typically, the amount of nitrogen in the sludge limits the application rate of the sludge as opposed to other nutrients.

The metals content of a sludge will dictate the life of a land application site. Since the metals uptake by plants is minimal compared to nutrient uptake, metals will accumulate in the soil at the ground surface. Excess concentration of metals in soil may lead to excessive metals uptake by plants, thereby posing a health threat if the crops are used for human consumption. Excessive metals uptake by plants may also lead to plant toxicity. Criteria for maximum metals accumulation per acre are outlined by the state regulatory agency and typical values are given in Table 11.14.

Sludge may be annually re-applied to a given area so long as the area has not reached its maximum metals capacity.

11.5.2   Landfilling

Landfill fees, availability of land application sites, sludge characteristics, and sludge hauling costs all affect the decision to landfill sludge. Under many situations, landfilling may be the cheapest alternative

TABLE 11.13   Annual Nitrogen, Phosphorous and Potassium Utilization for Various Crops (in lb/acre) (U.S. EPA, 1984)

Crop	Nitrogen	Phosphorous	Potassium
Corn	269	43	235
Corn silage	251	44	380
Soybeans	370	30	138
Grain sorghum	314	50	208
Wheat	202	29	141
Oats	188	30	157
Barley	188	30	157
Alfalfa	560	44	500
Orchard grass	376	55	390
Brome fescue	208	36	264
Tall fescue	169	36	193
Bluegrass	251	30	187

TABLE 11.14 Maximum Allowable Metals Accumulation in Soil for Land Application (U.S. EPA, 1995)

Pollutant	Cumulative accumulation (lb/acre)
Arsenic	46
Cadmium	44
Chromium	3360
Copper	1680
Lead	336
Mercury	19
Nickel	470
Selenium	112
Zinc	3136

for disposing of sludge. Many industrial sludges contain unusual compounds that make land application infeasible. In areas where much landfill space is available, the cost of disposal may be small. If sludge must be hauled long distances for landfilling, this option may not be cost effective.

For example, the cost of landfilling sludge in the areas surrounding New York City is extremely high due to the lack of available landfill space. As a result, the cheapest alternative for disposing of New York City wastewater treatment plant sludge is to haul the sludge to West Texas for land application. The reality of hauling sludge nearly 2,000 miles for land application reveals for dilemma many wastewater treatment utilities confront.

11.5.3 Other Beneficial Solids Reuse Options

In addition to land application of sludge, other programs have been established for a long time that utilize the beneficial characteristics of sludge. Sludge pellets are typically composed and heat-dried biosolids and are used as a fertilizer by farmers, golf course superintendents, and landscapers. The solids concentration of sludge pellets is much higher than solids concentration of sludge that is typically land applied. As a result, they are much easier to handle and distribute.

Perhaps the greatest challenge to the marketing and distribution of sludge pellets is over-coming the stigma of sludge pellets originating from wastewater. Growing crops with the aid of a fertilizer that was once a waste yields an unfavorable image. These concerns are valid with regard to the crops' metals uptake of metals containing sludges, since the consumption of food crops with sufficiently high metal concentrations can be harmful to the consumer's health. At the same time, there is concern for the presence of coliform bacteria in sludge pellets when used in growing food chain crops. In 1993, the EPA announced the final national standards (40 FFR Part 503) for beneficial use of biosolids. These standards allowed for the continued beneficial reuse of biosolids, but at the same time limited metals loading and pathogen content of biosolids that were to be applied to cropland, golf courses, gardens, etc. As a result, these limits dictated the "safe" levels at which biosolids could be applied as a fertilizer without posing a threat to human health or the environment.

Another means to beneficially reuse sludge is through product/raw material reclamation. In several industrial waste treatment system, portions of the raw material and product escape the production process and combine with residual solids. In some cases it is economical to remove the raw material and/or product from the residual solids. The concept of reclamation is analogous to the mining industry. In the mining industry, tremendous amounts of rock may be mined to extract relatively small amounts of minerals. It is the value of such minerals that make the extraction economical. Likewise, value of some metals in a chemical industry waste residual, for example, may justify their reclamation.

The author has added sludge from a paperboard plant to the landfill cap as a fertilizer. In this case, the sludge was non-toxic and the sludge Sodium Adsorption Ratio (SAR) is monitored.

11.6 REFERENCES

1. Data on Material Composition, Appendix B. 1990. In htto://www.cals.cornell.edu [Cornell Composting Homepage] Internet. Available from /cals/dept/compost/feas.study.app.b.html.

2. W. W. Eckenfelder, and J. L. Musterman, *Activated Sludge Treatment of Industrial Wastewater* (Lancaster, PA: Technomic Publishing Co., Inc., 1995).

3. W. W. Eckenfelder, *Industrial Water Pollution Control*, 2d ed. (New York: McGraw-Hill, 1989).

4. R. W. James, *Sewage Sludge Treatment and Disposal* (Park Ridge, NJ: Noyes Data Corporation, 1976).

5. Metcalf and Eddy, Inc., *Wastewater Engineering: Treatment, Disposal, Reuse*, 3d ed. (New York: McGraw-Hill, 1991).

6. G. F. Parkin, and William F. Owen, "Fundamentals of Anaerobic Digestion of Wastewater Sludges," *ASCE Journal of Environmental Engineering* 112(5): 867–920, 1986.

7. R. E. Speece, "A Survey of Municipal Anaerobic Sludge Digesters and Diagnostic Activity Assays." *Water Research* 22(3): 365–372, 1988.

8. R. E. Speece, *Anaerobic Biotechnology for Industrial Wastewaters.* (Nashville, TN: Archae Press, 1996).

9. U.S. Environmental Protection Agency, *Process Design Manual for Sludge Treatment and Disposal*, 625179011, 1979.

10. U.S. Environmental Protection Agency, *Handbook: Septage Treatment and Disposal*, 625684009, 1984.

11. U.S. Environmental Protection Agency, "Process Design Manual: Land Application of Sewage Sludge and Domestic Septage, 625R95001, 1995.

12. P. A. Vesilind, Gerald C. Hartman and Elizabeth T. Skene, *Sludge Management and Disposal—for the Practicing Engineer* (Chelsea, MI: Lewis Publishers, 1986).

13. P. A. Vesilind, *Treatment and Disposal of Wastewater Sludges* (Ann Arbor, MI: Ann Arbor Science Publishers, 1979).

14. Water Environment Federation/American Society of Civil Engineers, *Design of Municipal Wastewater Treatment Plants*, WEF Manual of Practice 8/ASCE Manual Report on Engineering Practice No. 76, 1998.

15. "Chemical Prices," *Chemical Market Reporter,* 3/8/99.

TREATMENT SUMMARY

CHAPTER 12
WASTEWATER TREATMENT SOLUTIONS

12.1 INTRODUCTION

As explained previously in this handbook, pollution control is a factor of the characterization of the pollutants rather than the source of pollutants. The purpose of this chapter is to list some proven technologies and flow diagrams for specific pollutants and types of industries.

12.2 TREATMENT SOLUTIONS

Tables 12.1 and 12.2 list some common pollutants and the treatment technologies shown to remove these pollutants including the references in this Handbook describing appropriate design and flow schematics. Table 12.3 lists recommended wastewater treatment for various industries including the pages of this Handbook describing the appropriate design and schematic drawings.

12.3 FLOW DIAGRAMS

Figures 12.1 through 12.17 show flow diagrams of both municipal and industrial plants which have been designed by the author as examples of some solutions to practical problems which have been encountered. The reader is cautioned that treatability studies and perhaps pilot studies are needed to optimize the flow diagram. Pollutants will vary from plant to plant, even within an industry requiring variations in treatment systems.

- Municipal wastewater treatment plants
  - Activated sludge/aerobic digestion
  - Activated sludge/anaerobic digestion
  - Extended aeration
  - Extended aeration with sludge recirculation
  - Trickling filter
  - Aerated lagoon
  - Oxidation pond

TABLE 12.1 Recommended Wastewater Treatment for Various Pollutants

Pollutant	Treatment Technology	*Achievable Concentration mg/l	Achievable Percent Reduction	Handbook Reference Pages
Ammonia	Aerobic Biological	0.05–0.5		10.22
	Air Stripping	0.3–0.5		8.106
	Steam Stripping	0.1–0.3		8.106
Arsenic	Carbon Adsorption	0.6		8.125
	Charcoal Precipitation	0.06		
	Chemical Oxidation	0.2		9.7
	Ferric Chloride Precipitation	0.05–0.13		9.14
	Ferric Sulfate Precipitation	0.003–5		9.14
	Lime/Alum Precipitation	0.003–0.2		9.14
	Lime/Ferric Hydroxide Precipitation	0.005–0.2		9.14
	Lime Softening	0.03		
	Sulfide Precipitation with Filtration	0.05		9.14
Biodegradable Organics				
	Aerobic Fixed Growth	0.03	99%	10.70
	Aerobic Suspended Growth	0.03	99%	10.8
	Anaerobic Fixed Growth	1.0	95%	10.29
	Anaerobic Suspended Growth	1.0	95%	10.25
	Activated Carbon	0.03	99.9%	8.125
	Chemical Oxidation	0.03	99%	9.7
	Land Treatment	1.0	90%	
	Membrane Filtration	0.001		8.89
Boron	Ion Exchange	0.001	90%	
	Reverse Osmosis	0.14	60%	8.89
Cadmium	Hydrogen Precipitation	0.003 at pH 11.0		9.14
	Lime/Hydroxide Precipitation	0.04 at pH 10.0		9.14
	Sulfide Precipitation	0.0001 at pH 11.0		9.14
Carbon Monoxide	Air Stripping		99%	8.106
Chloride	Electrodialysis		30%	
	Reverse Osmosis	15	90%	8.89
Chromium^{+6}	Activated Carbon Adsorption	0.02		8.125
	Bisulfite	0.05		9.7
	Ion Exchange	0.02		
	Metabisulfite	0.001		9.7
	Sulfur Dioxide	0.01		9.7
Chromium^{+3}	Hydroxide Precipitation	0.02 at pH 8.5		9.14
Copper	Hydroxide Precipitation	0.02 at pH 9.0		9.14
	Sulfide Precipitation	0.01 at pH 9.0		9.14
	Ion Exchange	0.03		
Cyanide	2 Stage Alkaline Oxidation	0–0.4		9.7
	Electrodialysis	0.1		
	Ozonation	0	80%	9.27
Dissolved Solids	Electrodialysis		90%	
	Ion Exchange		80%	
	Reverse Osmosis		95%	8.89
Fluoride	Alum Precipitation	0.6		9.14
	Lime Precipitation	6		9.14
Iron^{+2}	Aeration	0.1		9.7
	Chlorination	0.05		9.7
Iron^{+3}	Hydroxide Precipitation	1.0 at pH 4.0		9.14
	Lime Precipitation	1.0		9.14
Lead	Hydroxide Precipitation	0.015 at pH 10		9.14
	Sulfide Precipitation	0.3		9.14

TABLE 12.1 Recommended Wastewater Treatment for Various Pollutants (*Continued*)

Pollutant	Treatment Technology	*Achievable Concentration mg/l	Achievable Percent Reduction	Handbook Reference Pages
Manganese^{+2}	Chlorination	0.05		9.7
	Permanganate Oxidation	0.05		9.7
Manganese^{+3}	Hydroxide Precipitation	0.05 at pH 9.5		9.14
Mercury	Activated Carbon	0.001		8.125
	Alum Precipitation	0.01 at pH 7		9.14
	Ion Exchange	0.001		
	Lime/Hydroxide Precipitation	0.001		9.14
	Sulfide Precipitation	0.01 at pH 10		9.14
Nickel	Hydroxide Precipitation	0.01 at pH 10.5		9.14
	Lime Precipitation	0.01 at pH 10		9.14
	Reverse Osmosis		99.5%	8.89
	Sulfide Precipitation	0.01 at pH 8		9.14
Nitrogen Total	Anoxic Biological Denitrification	0.5		
Oil, including Fat & Grease Emulsion Breaking	Acid Cracking			8.71
	Coagulation	0.1		8.71
	Polymer Flocculation	0.1		8.71
Oil, including Fat & Grease Removal	API Separator	15		8.72
	Dissolved Air Flotation	10		8.76
	Diatomaceous Earth Filtration	5		
	Membrane Filtration	5		8.89
	Sand Filtration	10		8.78
Pesticides	Activated Carbon Adsorption	0.01		8.125
	Aerobic Biological Treatment		99%	10.8
	Anaerobic Biological Treatment		95%	10.25
	Resin Adsorption	0.001		8.140
pH	Base Addition			9.1
	CO$_2$ Addition			9.1
	Strong Acid Addition			9.1
	Weak Acid Addition			9.1
Phenol	Activated Carbon Adsorption		99.5%	8.125
	Aerobic Biological Treatment		99.5%	10.8
	Chemical Oxidation with Lime Precipitation		99%	9.7, 9.14
	Ozonation		99.5%	9.27
Phosphorous	Alum Coagulation		99%	9.14
Selenium	Ferric Sulfate Coagulation	0.05	80%	9.14
	Ion Exchange		99.9%	
Settleable Solids	Comminution			8.13
	Flotation			8.72
	Screening			8.10
	Sedimentation (Grit Chambers)			8.15
Sulfides	Activated Carbon Adsorption		99.5%	8.125
	Air Stripping		99.5%	8.106
	Aerobic Biological Treatment		99.5%	10.8
	Anaerobic Biological Treatment		99.5%	10.25
Zinc	Carbonate Precipitation	0.3 at pH 9.0		9.14
	Hydroxide Precipitation	0.07 at pH 10.0		9.14
	Sulfide Precipitation	0.001 at pH 9.0		9.14

*Achievable concentrations taken from experience and literature and may be depended on influent concentration.

TABLE 12.2 Treatment Technologies Summary

Technology	Description of process or equipment	Example applications	Status	Considerations	Relative cost
		Physical treatment			
Magnetic processes	Magnetic-separation devices	Debris presort	Commercial	Limited applications	Low
Screening and classification	Standard manufactured units	Separation of oversize materials	Commercial	Reprocessing or disposal of miscellaneous material	Low
Crushing and grinding	Standard manufactured units	Size reduction of solid material for further processing	Commercial	Fugitive emissions	Low
Liquid/solid separation Sedimentation (with or without flocculation) Filtration, Centrifugation, Flotation, Belt presses, Filter presses	Standard manufactured units	Remove particles from liquids. Remove excess moisture from solids or sludges.	Commercial	Solid still contains some liquid	Low
Drying	Standard manufactured units	Sludge drying	Some experimental Some commercial	Mechanical problems air emissions	Expensive
Distillation	Multitray or packed column with heating and condensing device	Solvent purification for reuse	Commercial	Scaling and/or fouling Flammability hazard with some solvents	Medium
Evaporation	Single-stage, multistage or vapor-compression evaporators that may include crystallization step	Nuclear wastes Electroplating wastes	Commercial	Scaling and/or fouling Condensate is sometimes contaminated Disposal of concentrate	Moderately High
Stripping Steam, Air, Other gas	Multitray or packed column with gas injection	Sulfide stripping Trichloroethylene stripping	Commercial	Limited to volatile components Air emissions	Low to Medium
Absorption	Multitray or packed column with appropriate solvent	Usually for emission control	Commercial	Disposal of scrubbing liquor	Low
Solvent extraction Liquid-liquid, Solid-liquid, Supercritical fluid	Standard process (Supercritical fluid under development)	Extracting contaminants from soil Extracting metals from liquid	Commercial (Supercritical fluid under development)	Contaminated solvent requires further processing for disposal	Moderately High

Process	Description	Application	Status	Limitations	Cost
Adsorption Carbon, Resin (ion exchange, others); proprietary systems	Batch or continuous adsorption beds, usually with regeneration	Organic adsorption onto carbon Heavy-metal adsorption onto resins	Commercial	Limited to low concentrations Disposal of regenerate	Medium
Membrane processes Ultrafiltration, Reverse osmosis, Dialysis, Electrodialysis	Standard manufactured units with appropriate pretreatment facilities to prevent membrane fouling and/or deterioration	Removal of heavy metals or some organics from groundwater	Recently commercial	Separations are imperfect Pretreatment is complex	Medium
Freezing Crystallization, Freeze drying, Suspension freezing	Many types of units	Suspension-freezing ponds for hydrous metal hydroxides	Experimental other than drying/freezing beds	Not commercially developed	Low for drying beds, high for others

Chemical treatment

Process	Description	Application	Status	Limitations	Cost
Neutralization	Chemical addition and mix tanks	Neutralization of acid and alkaline wastes	Commercial	Heat release is concentrated Control complex	Low
Precipitation	Chemical addition, to produce an insoluble solid	Heavy-metals removal	Commercial	Solubility laws interfering substances	Low
Electrochemical processes	D.C. power and plating apparatus	Copper removal	Some commercial Some experimental	Impurities can upset process	Medium
Oxidation Chlorine-containing reagents, Ozone, Permanganate, Peroxide, Others	Chemical addition and contacting tanks	Trace-organic destruction	Some commercial Some experimental	Side-reactions may generate other hazardous constituents	Medium to Low
Reduction Dechlorination, Sulfonation, Other	Chemical addition and contacting tanks	Reduction of hexavalent chrome. Dechlorination of dioxin	Some commercial Some experimental	Side-reactions may generate other hazardous constituents	Medium to Low
Photolysis Ultraviolet Light, Natural Light	Photolamps and contacting devices	Dioxin destruction Cyanide destruction	Semi-commercial	Fouling of photo-chemical devices Kinetics	Low for natural, High for UV

(Continued)

12.7

TABLE 12.2 Treatment Technologies Summary (*Continued*)

Technology	Description of process or equipment	Example applications	Status	Considerations	Relative cost
Gamma irradiation	Shielded irradiator	Pesticide destruction	Experimental	Sophisticated irradiator design	High
Miscellaneous chemical treatments Catalysis, Hydrolysis, Others	Chemical additions and contacting tanks	Pesticide destruction	Experimental	Side-reactions may generate other hazardous constituents	Varies
Biological treatment					
Activated sludge lagoons: Aerated Facultative Anaerobic	Common commercial system designs	Removal of organic materials from water	Commercial	Only effective on biodegradable or bioadsorbable constituents Subject to toxic inhibition	Low
Anaerobic digestion Composting, Trickling filters, Aerobic biofilters, Fermentation, Waste-stabilization ponds	Common commercial system designs	Removal of organic materials from water	Commercial	Only effective on biodegradable or bioadsorbable constituents Subject to toxic inhibition	Low
New diotechnologies Enzyme, cultured bacteria, Gene splicing	Biochemical addition system	—	Experimental	Field is new, so considerations are not well understood	Low to Medium
Thermal treatment					
Established incineration processes Fluidized bed, multiple hearth Rotary kiln, Liquid injection, Shipboard	Standard commercially marketed units	Industrial incinerators, Contract hazardous-waste incinerators	Commercial	Fuel valve Destruction efficiency Disposal of ash and scrubber blowdown	Medium to High

Process	Equipment	Application	Status	Comments	Cost
Evolving incineration processes Morten salt, Microwave plasma, Plasma, arc	Developmental unit	Dioxin destruction	Experimental	Technology not well developed	High
Codisposal incineration processes Industrial boiler, Cement kiln, Lime kiln	Standard units	Waste-solvent burning	Commercial	Fuel value Effects on emissions-control equipment	Low to Medium
Pyrolysis Conventional temperature Ultra high temperature	Proprietary units	Organics destruction	Mostly experimental	Byproducts generated may be hazardous	Medium
Wet-air oxidation Autoclave, U-tube reactor Vertical-tube reactor	Proprietary units	Organics destruction	Many commercial, but mostly in non-hazardous waste applications	Process is only 85–95% efficient	Medium
Fixation/Encapsulation Sorption Flyash, Kiln, dust, Lime, Limestone, Clavs, Verricuite, Zeotites, Alumina, Carbon, Imobider beads, Proprietary agents	Stabilizing materials and contacting methods	Solidifying hazardous wastes	Commercial	Long-term effectiveness	Medium
Possolianic reaction Lime-flyash, Portland Cement	Mechanical equipment for mixing and reaction	Solidifying Hazardous wastes	Commercial	Organic agents sometimes interfere with reaction	Medium
Encapsulation Organic plymers, Asonart, Glassification, Proprietary agents	Stabilizing materials and mechanical equipment for encapsulation	Solidifying hazardous wastes	Some Experimental	Long-term effectiveness	Medium to High

TABLE 12.3 Recommended Wastewater Treatment for Various Industries

Industry	Types of Treatment	Design Reference Page	Schematic Reference Page
Baking	Biological	10.8	12.28
	DAF	8.76	
Breweries	pH Adjustment	9.1	10.76
	Biological	10.8	10.77
Cellulose	Biological	10.8	12.29
Chemical, Inorganic	Stripping	8.106	12.30
	Membrane Filtration	8.89	
Chemical, Organic	Biological	10.8	12.16–12.18
	GAC	10.125	12.13–12.15
	Stripping	8.106	12.12, 12.15
Dairies	Biological	10.8	12.31–12.32
	DAF	8.76	
Distilleries	Biological	10.8	10.76
			10.77
Edible Oils	Biological	10.8	12.21
	DAF	8.76	12.21
	O/W Separation	8.72	12.21
Electroplating	Metals Precipitation	9.14	12.12 – add new pH section dwg.
Food Processing	Biological	10.8	12.22, 12.33
Glycol Plant	Biological	10.8	12.40
Leather Tanning	Biological	10.8	12.19–12.20
Lumber Industry	Wetlands	10.71	12.35
Metals Fabrication	O/W Separation	8.72	12.12, 12.24
	Precipitation	9.14	
Metals Foundries	Emulsion Breaking	8.71	12.12, 12.24,
	O/W Separation	8.72	12.25
	Precipitation	9.14	
Metals Machinery	Emulsion Breaking	8.71	12.12
	O/W Separation	8.72	12.34
	Membrane Filtration	8.89	
	Precipitation	9.14	
Metals Working	Emulsion Breaking	8.71	12.24
	O/W Separation	8.72	
	Membrane Filtration	8.89	
	Precipitation	9.14	
Mine Drainage	Precipitation	9.14	12.22, 12.44
Municipal, Ammonia Removal	Aerobic Biological	10.8	10.23, 10.29, 10.30
Municipal, BOD Removal	Aerobic Biological	10.8	12.16, 12.23, 12.25, 12.26
Municipal, Nitrogen Removal	Anoxic Biological	?	? new dwg.
Municipal, Phosphorous Removal	Precipitation	9.14	12.36
Personal Hygiene Products	Emulsion Breaking	8.71	10.21
	DAF	8.76	

TABLE 12.3   Recommended Wastewater Treatment for Various Industries (*Continued*)

Industry	Types of Treatment	Design Reference Page	Schematic Reference Page
Petroleum Refineries	Biological	10.8	10.12, 12.38
	DAF	8.76	10.29, 12.37, 12.38
	GAC	10.125	10.30
	Stripping	8.106	12.27
	O/W Separation	8.72	12.28, 12.37
Pharmaceuticals	Biological	10.8	10.12
	DAF	8.76	10.29
	Stripping	8.106	10.30
Pipe Foundry	Sedimentation	8.19	12.35
	Emulsion Breaking	8.71	12.35
	CN⁻ Reduction	9.7	12.35
	Metals Precipitation	210	12.35
Polymer/Monomer	Emulsion Breaking	8.71	12.39
	DAF	8.76	
Pthalate	Biological	10.8	New dwg.
Pulp & Paper	Primary Sedimentation	8.19	12.47
	Aerated Lagoons	10.62	12.47
Soap	Biological	10.8	12.47
Sweet's Goods	Emulsion Breaking	8.71	12.41
	DAF	8.76	12.42
	Biological	10.8	12.42
Timber Runoff	Artificial wetlands	10.71	12.45
Tire Stem Manufacturing	Emulsion Breaking	8.71	12.46
	Precipitation	8.106	
Urea	Stripping	8.106	12.43
Vegetable Oil Refineries	Biological	10.8	12.11
	DAF	8.76	
	O/W Separation	10.71	
Vegetable Processing	Biological	10.8	12.11
	DAF	8.76	
Viscose, Rayon	Biological	10.8	10.12

Note: *Achievable concentrations taken from experience and literature may be dependent on influent concentration. Basic Treatment systems such as equalization, screening, pH control, coagulation, flocculation, settling, are not listed in this table since they are appropriate for most types of industrial wastewater treatment systems.

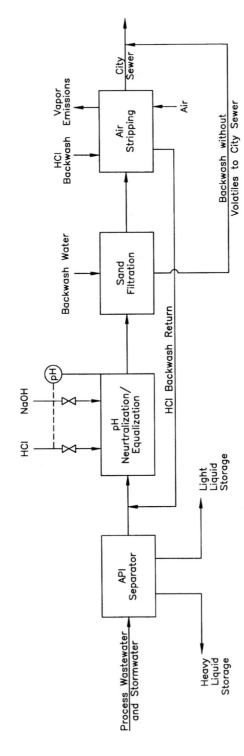

FIGURE 12.1 Schematic diagram of organic chemicals wastewater treatment plant.

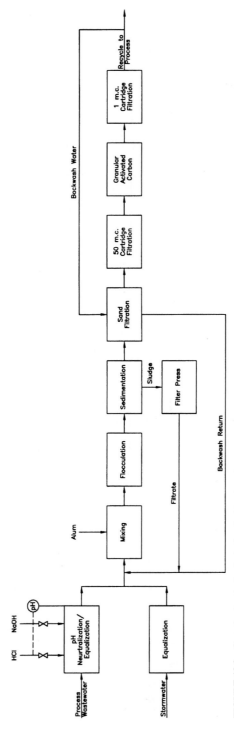

FIGURE 12.2 Organic chemicals plant, chemical treatment.

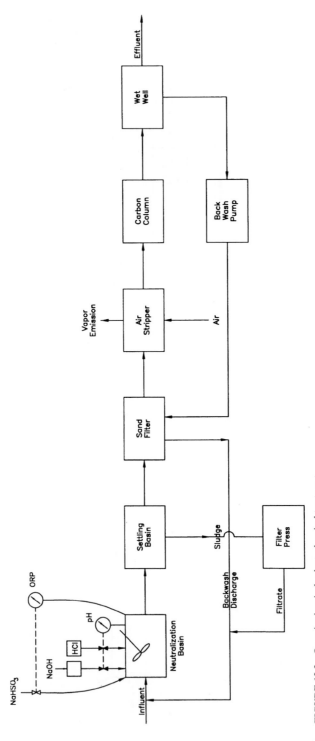

FIGURE 12.3 Organic chemicals plant, chemical treatment.

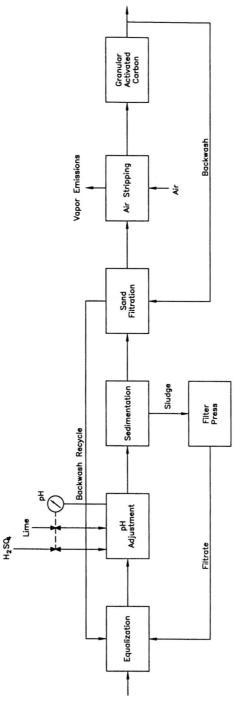

FIGURE 12.4 Organic chemicals plant, chemical treatment.

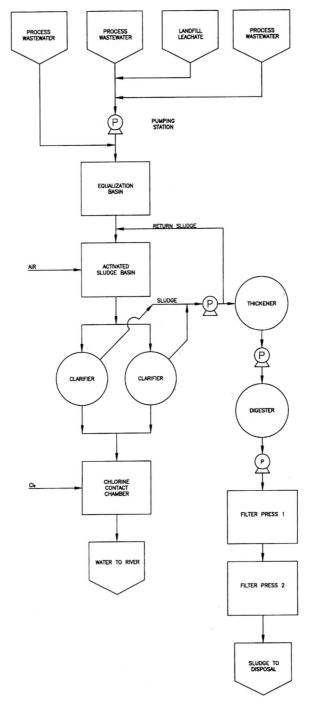

FIGURE 12.5   Organic chemicals plant, biological treatment.

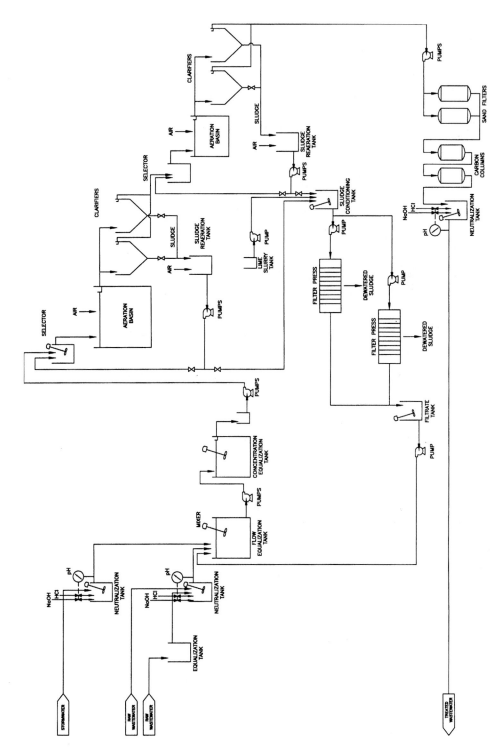

FIGURE 12.6 Organic chemical plant, aerobic biological pretretment, aerobic biological treatment.

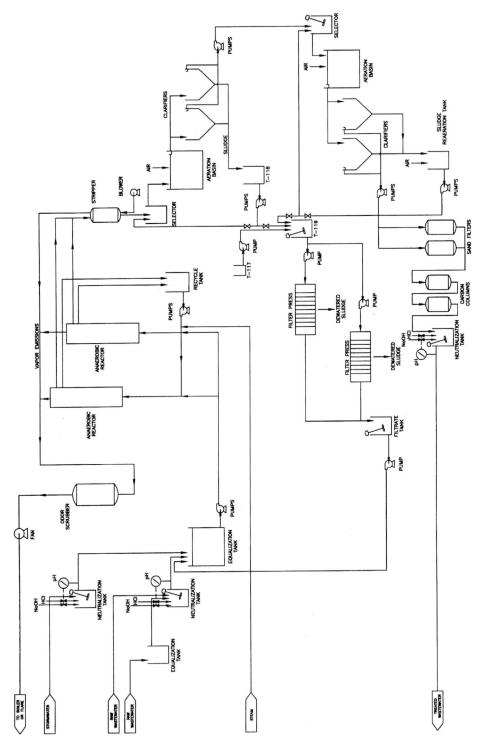

FIGURE 12.7 Organic chemical plant, anaerobic biological pretreatment, aerobic biological treatment.

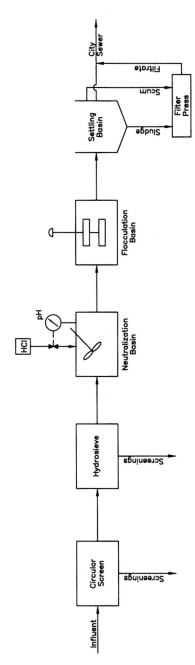

FIGURE 12.8 Leather tannery, chemical treatment.

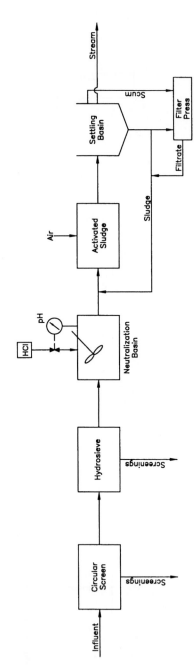

FIGURE 12.9 Leather tannery, biological treatment.

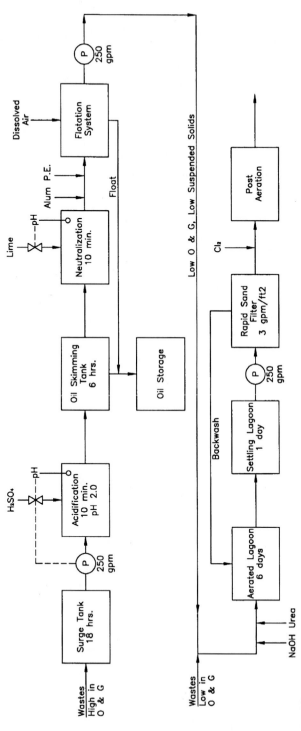

FIGURE 12.10 Typical edible oil refinery.

12.21

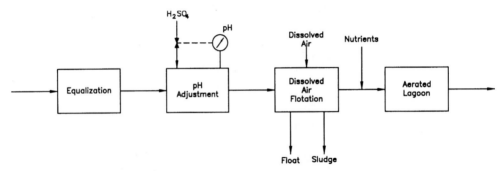

FIGURE 12.11   Typical food processing wastewater treatment plant.

- Industrial wastewater treatment plants

SIC Major Group	Industry
20	Food and kindred products
26	Paper and allied products
28	Chemical and allied products
29	Petroleum refining and related industry
31	Leather and leather products
33	Primary metal industry
34	Fabricated metal products

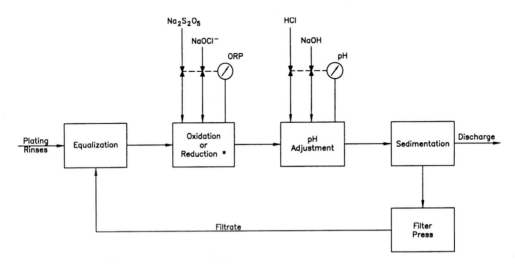

* Note
The use of an oxidation or reduction step is
dependent on the metal being removed.

FIGURE 12.12   Typical metals removal wastewater treatment plant.

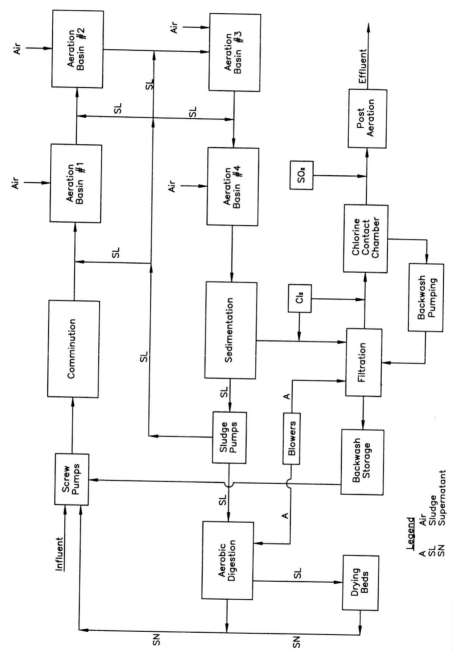

FIGURE 12.13 Typical industrial park treatment plant.

Legend
A   Air
SL Sludge
SN Supernatant

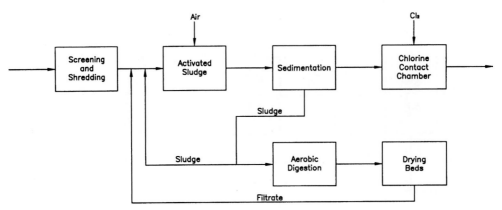

FIGURE 12.14 Typical municipal activated sludge plant.

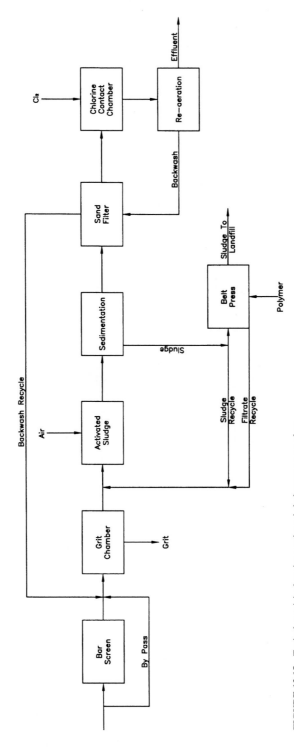

FIGURE 12.15 Typical municipal tertiary activated sludge treatment plant.

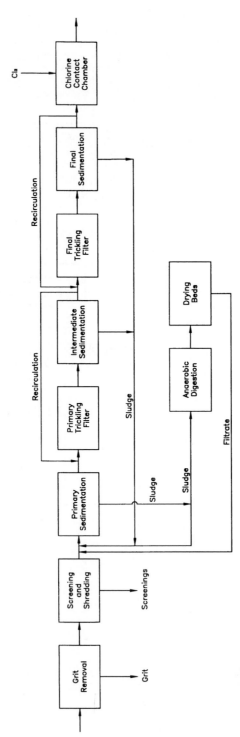

FIGURE 12.16 Typical municipal trickling filter plant.

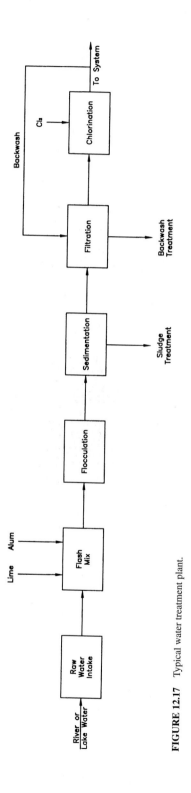

FIGURE 12.17 Typical water treatment plant.

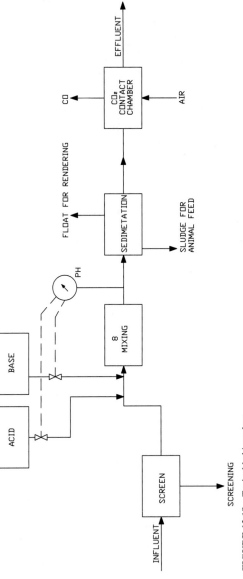

FIGURE 12.18 Typical baking plant.

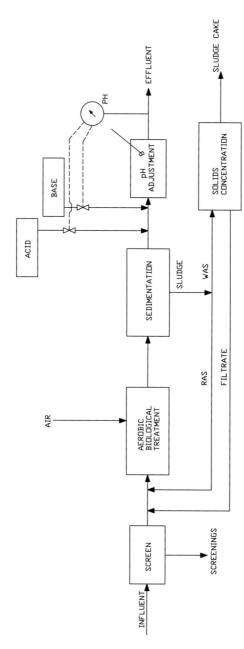

FIGURE 12.19 Typical cellulose plant.

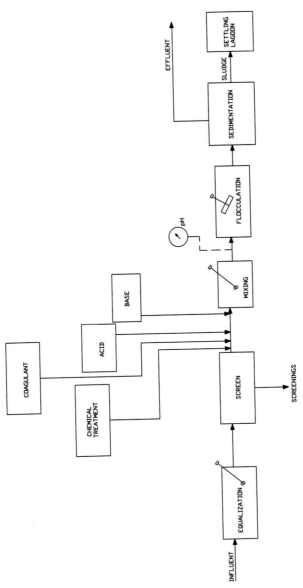

FIGURE 12.20 Typical inorganic chemical plant.

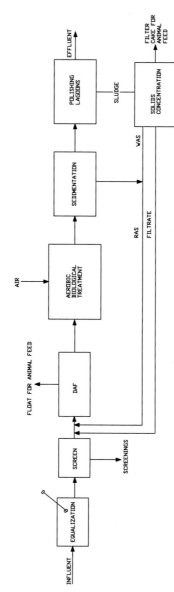

FIGURE 12.21 Typical dairy treatment.

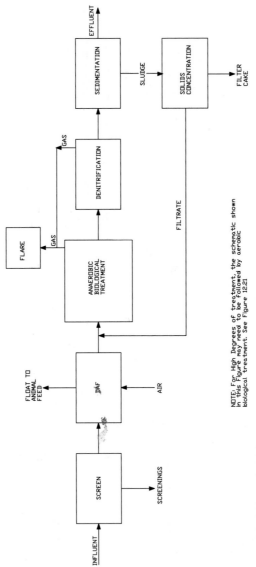

FIGURE 12.22 Typical dairy treatment.

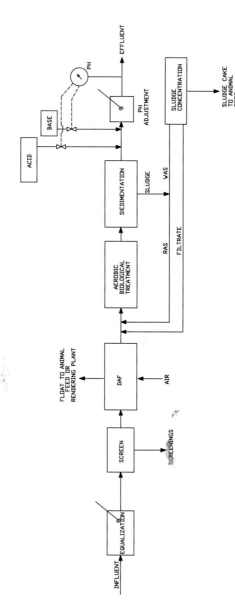

FIGURE 12.23 Typical meat processing plant.

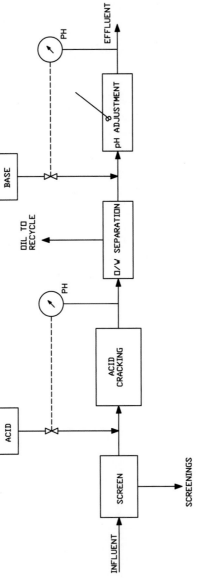

FIGURE 12.24 Typical metals oil removal plant.

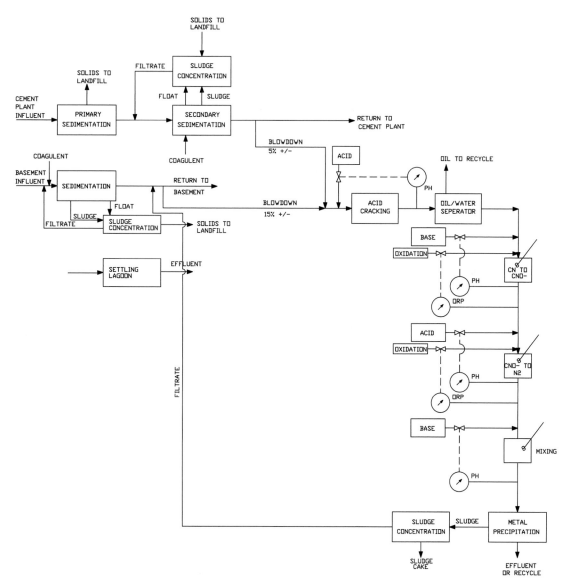

FIGURE 12.25   Typical pipe foundry recycle system.

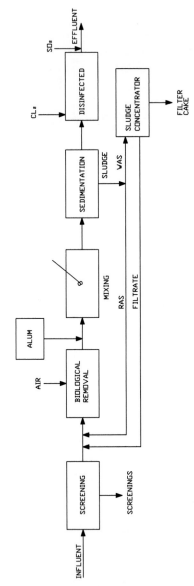

FIGURE 12.26 Typical municipal phosphorous removal.

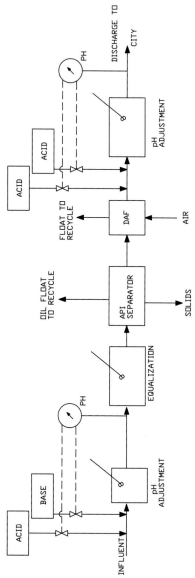

FIGURE 12.27 Typical refinery.

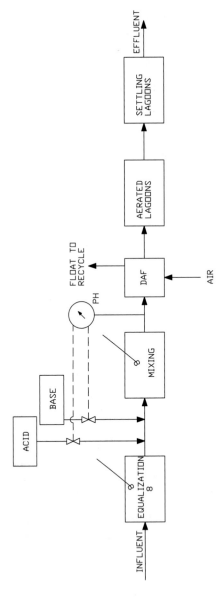

FIGURE 12.28 Typical refinery.

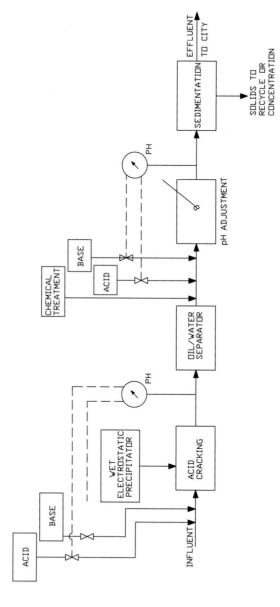

FIGURE 12.29 Typical polymer/monomer plant.

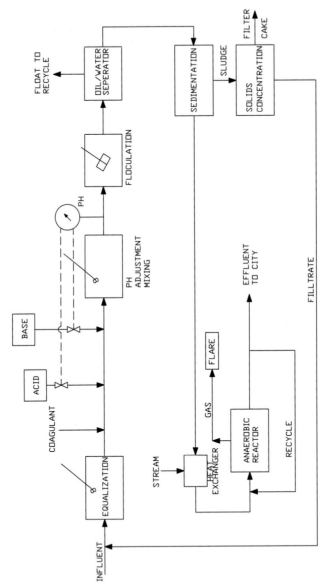

FIGURE 12.30 Typical glycol plant.

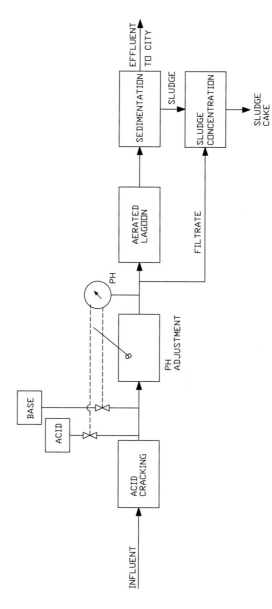

FIGURE 12.31 Typical soap plant.

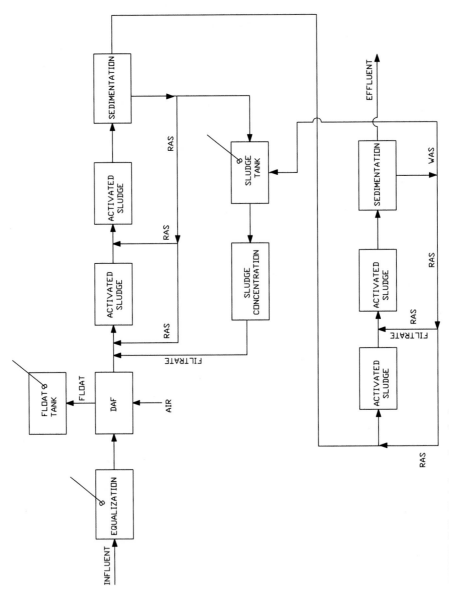

FIGURE 12.32 Typical sweet's goods plant.

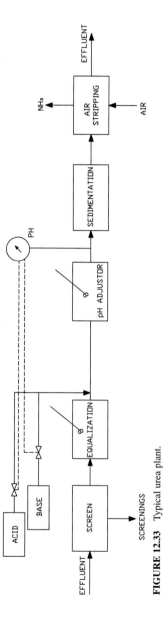

FIGURE 12.33 Typical urea plant.

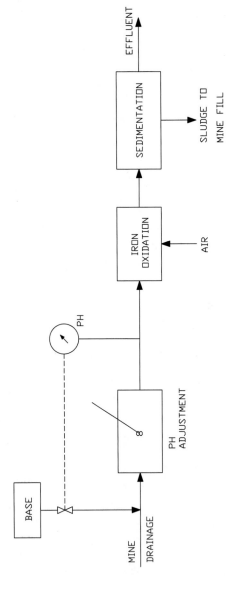

FIGURE 12.34 Typical mine drainage plant.

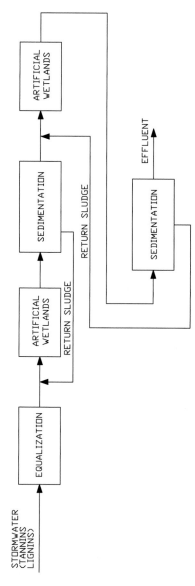

FIGURE 12.35 Typical timber runoff.

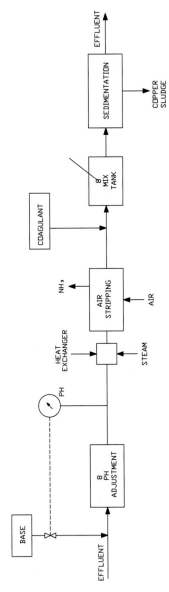

FIGURE 12.36 Tire stem manufacturing.

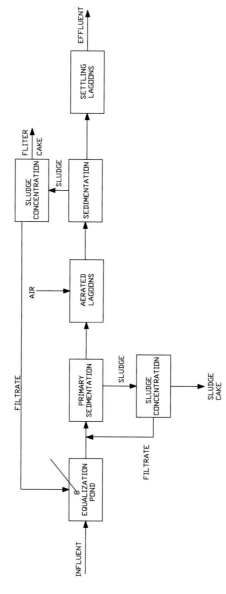

FIGURE 12.37 Typical pulp & paper plant.

CHAPTER 13
POLLUTANT INFORMATION

13.1 INTRODUCTION

In this chapter, the pollutants listed in the EPA publication "Quality Criteria for Water" are described with recommended criteria for discharge pollutant characteristics, and when appropriate, certain wastewater treatment and pollutant removal parameters and regulatory limits and recommendations. The sources of information presented are as follows:

1. Discharge Criteria: "Quality Criteria for Water," Environmental Protection Agency (EPA), July 1976, Publication No. PB-263943.
2. Carbon Adsorption: "Carbon Adsorption Isotherms for Toxic Organics," EPA, 1980, Publication No. EPA-60018-80-023.
3. Values in Natural Soils: "Hazardous Waste Land Treatment," EPA, Publication No. SW-874, 1983, and "Reclamation and Redevelopment of Contaminated Land," EPA, Publication No. EPA/600/2-86/066, 1986.

13.2 POLLUTANT INFORMATION, INORGANIC CHEMICALS

13.2.1 Alkalinity

- Discharge criteria–Greater than 20 mg/L as $CaCO_3$ for freshwater aquatic life except where natural concentrations are less.

13.2.2 Ammonia, NH_3

- Formula weight = 17.03 g/mol
- Specific gravity = 0.77
- Solubility at 0°C = 89.9 g/100 mL
- Solubility at 98°C = 7.4 g/100 mL
- Discharge criteria: 0.02 mg/L as unionized ammonia for freshwater aquatic life.
- Concentrations found under various conditions: Table 13.1 shows concentrations of Total Ammonia ($NH_3 + NH_4^+$) which contain a unionized ammonia concentration of 0.02 mg/L.

Table 13.2 shows the percent of unionized ammonia (NH^3) in aqueous solution.

TABLE 13.1 Total Ammonia ($NH_3 + NH_4^+$) Concentration at Various pH and Temperature for a Unionized Ammonia Concentration of 0.02 mg/L

T, °C	pH								
	6.0	6.5	7.0	7.5	8.0	8.5	9.0	9.5	10.0
5	160	51	16	5.1	1.6	0.53	0.18	0.071	0.036
10	110	34	11	3.4	1.1	0.36	0.13	0.054	0.031
15	73	12	7.3	2.3	0.75	0.25	0.093	0.043	0.027
20	50	16	5.1	1.6	0.52	0.18	0.070	0.036	0.025
25	35	11	3.5	1.1	0.37	0.13	0.055	0.031	0.024
30	25	7.9	2.5	0.81	0.27	0.099	0.045	0.028	0.022

13.2.3 Arsenic, As

- Formula weight = 299.64 g/mol
- Specific gravity = 5.727 @ 14°C
- Insoluble in water
- Discharge criteria: 50 μg/L for domestic water supplies and 100 μg/L for crop irrigation
- TCLP regulatory limit: 5.0 mg/L
- RFI clean closure concentration: 50.0 μg/L
- Common concentration in natural soils: 0.1 to 40 mg/L
- Average concentration in natural soils: 5 mg/L

13.2.4 Barium, Ba

- Formula weight = 137.36 g/mol
- Specific gravity = 3.5
- Insoluble in water (decomposes)
- Discharge criteria: 1 mg/L for domestic water supply
- TCLP regulatory limit: 100 mg/L
- Common concentration in natural soils: 100 to 3500 mg/L
- Average concentration in natural soils: 430 mg/L

TABLE 13.2 Percent of Unionized Ammonia (NH^3) in Aqueous Solutions

T, °C	pH								
	6.0	6.5	7.0	7.5	8.0	8.5	9.0	9.5	10.0
5	0.013	0.040	0.12	0.39	1.2	3.8	11	28	56
10	0.019	0.059	0.19	0.59	1.8	5.6	16	37	65
15	0.027	0.087	0.21	0.86	2.7	8	211	46	73
20	0.040	0.13	0.40	1.2	3.8	11	28	56	80
25	0.057	0.18	0.57	1.8	5.4	15	36	64	85
30	0.080	0.25	0.80	2.5	7.5	20	45	72	89

13.2.5 Beryllium, Be

- Formula weight = 9.01 g/mol
- Specific gravity = 1.816
- Insoluble in cold water
- Soluble but decomposes in 100 parts of hot water
- Discharge criteria:
  - 11 μg/L for the protection of aquatic life in soft fresh water
  - 1100 μg/L for the protection of aquatic life in hard fresh water
  - 100 μg/L for continuous irrigation on all soils, except 500 μg/L for irrigation on neutral to alkaline fine-textured soils
- RFI clean closure concentration: 0.008 μg/L
- Common concentration in natural soils: 0.1 to 40 mg/L
- Average concentration in natural soils: 6 mg/L

13.2.6 Boron, B

- Formula weight = 10.82 g/mol
- Specific gravity = 2.32
- Insoluble in water
- Discharge criteria: 750 mg/L for long-term irrigation on sensitive crops
- Common concentration in natural soils: 2 to 130 mg/L
- Average concentration in natural soils: 10 mg/L

13.2.7 Cadmium, Cd

- Formula weight = 12.41 g/mol
- Specific gravity = 8.65 @ 20°C
- Insoluble in water
- Discharge criteria:
  - 10 μg/L for domestic water supply
  - 0.4 μg/L for cladocerans and salmonid fishes in soft fresh water, and 4.0 μg/L for other less sensitive aquatic life in soft fresh water.
  - 1.2 μg/L for caldocerans and salmonid fishes in hard fresh water, and 12.0 μg/L for other, less sensitive aquatic life in hard fresh water.
- TCLP regulatory limit: 1.0 mg/L
- RFI clean closure concentrations: 5.0 μg/L
- Common concentration in natural soils: 0.01 to 7 mg/L
- Average concentration in natural soils: 0.06 mg/L

Figure 13.1 shows the solubility curves for $Cd(OH)_2$ and CdS as a function. (Note that this figure also presents the solubility curves for other metal hydroxides and sulfides found in this section.)

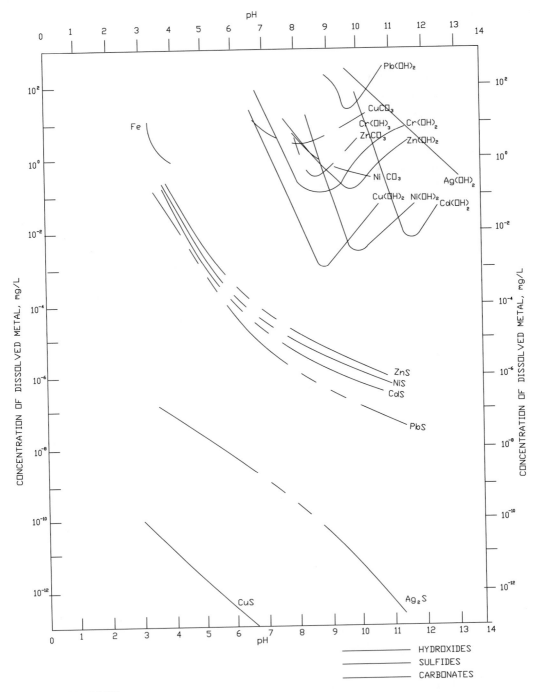

FIGURE 13.1 Solubility curve.

13.2.8   Chlorine, Cl$_2$

- Formula weight = 70.91 g/mol
- Specific gravity = 1.56 @ −33.6°C as liquid and 2.49 @ 0°C with reference to air = 1.0
- Solubility at 0°C = 1.46 g/100mL
- Solubility at 30°C = 0.57 g/100mL
- Discharge criteria: 2.0 μg/L for salmonid fish and 10.0 μg/L for other freshwater and marine organisms

13.2.9   Chromium, Cr

- Formula weight = 52.01 g/mol
- Specific gravity = 7.1
- Elemental chromium is insoluble in water
- Cr^{+6} soluble in water at all pH values
- Cr^{+3} solubility is dependent on pH
- Discharge criteria: 50 μg/L for domestic water supply and 100 μg/L for freshwater aquatic life
- TCLP regulatory limit: 5 mg/L
- RFI clean closure concentration: 40 mg/L for Cr^{+3} and 0.10 mg/L for Cr^{+6}
- Common concentration in natural soils: 5 to 3000 mg/L
- Average concentration in natural soils: 100 mg/L

13.2.10   Color

- Discharge criteria:
  - 75 color units on the platinum-cobalt scale for domestic water supplies
  - Increase in color (in combination with turbidity) must not reduce the depth of the compensation point for the photosynthetic activity by more than 10 percent of the seasonably established norm for aquatic life.

13.2.11   Copper, Cu

- Formula weight = 63.57 g/mol
- Specific gravity = 8.92 @ 20°C
- Elemental copper is insoluble in water
- Copper ion solubility is dependent on pH
- Discharge criteria: 1.0 mg/L for domestic water supply and 0.1 times 96 hour LC$_{50}$ using a non-aerated bioassay and a sensitive aquatic species for freshwater and marine aquatic life
- RFI clean closure concentration: 1300 μg/L
- Common concentration in natural soils: 2 to 100 mg/L
- Average concentration in natural soils: 30 mg/L

13.2.12   Cyanide Ion, CN$^-$

- Formula weight = 40.02 g/mol
- Formula weight of potassium cyanide = 65.11 g/mol

- Specific gravity of potassium cyanide = 1.52 @ 16°C
- Solubility of potassium cyanide in cold water: complete
- Solubility of potassium cyanide in water @ 103.5°C: 122.2 g/100 m/L
- Formula weight of sodium cyanide = 49.02 g/mol
- Solubility of sodium cyanide in water @ 10°C: 48 g/100 mL
- Solubility of sodium cyanide in water @ 35°C: 82 g/100 mL
- Discharge criteria: 5.0 μg/L for freshwater and marine aquatic life and wildlife

13.2.13   Dissolved Solids and Salinity

- Discharge criteria:
  - 250 mg/L for chlorides and sulfates in domestic water supplies

13.2.14   Fecal Coliform Bacteria

- Discharge criteria:
  - 200 per 100 mL for log mean of any 5 samples over a 30 day period and 400 per 100 mL for 10% of total samples over a 30-day period for bathing waters
  - 14 MPN per 100 mL median concentration and 43 MPN per 100 mL for 10 percent for shellfish harvesting

13.2.15   Hydrogen Sulfide, H$_2$S

- Formula weight = 34.08 g/mol
- Specific gravity = 1.1895 with reference to air = 1.0
- Solubility in water @ 0°C = 437 mL/100 mL
- Solubility in water @ 40°C = 186 mL/100 mL
- Discharge criteria: 2 μg/L for fish and aquatic life

13.2.16   Iron, Fe

- Formula weight = 55.85 g/mol
- Specific gravity = 7.03
- Elemental iron is insoluble in water
- Ferrous iron (Fe^{+2}) is soluble in water at all temperatures
- Ferric iron (Fe^{+3}) solubility in water is pH dependent
- Discharge criteria: 0.3 mg/L for domestic water supply and 1.0 μg/L for freshwater aquatic life

13.2.17   Lead, Pb

- Formula weight = 207.21 g/mol
- Specific gravity = 11.337

- Elemental lead is insoluble in water
- Discharge criteria: 50 μg/L for domestic water supply and 0.01 times 96 hour LC_{50} for sensitive freshwater residence species
- TCLP regulatory limit: 5.0 mg/L
- RFI clean closure action level: 15μg/L
- Common concentration in natural soils: 2 to 200 mg/L
- Average concentration in natural soils: 10 mg/L

13.2.18 Manganese, Mn

- Formula weight = 54.93 g/mol
- Specific gravity = 7.2 @ 20°C
- Decomposes in water
- Discharge criteria: 50 μg/L for domestic water supply and 100 μg/L for protection of consumers of marine mollusks
- Common concentration in natural soils: 100 to 4000 mg/L
- Average concentration in natural soils: 600 mg/L

13.2.19 Mercury, Hg

- Formula weight = 200.61 g/mol
- Specific gravity = 13.546 @ 20°C
- Elemental mercury is insoluble in water
- Discharge criteria:
  - 2.0 μg/L for domestic water supply
  - 0.05 μg/L for freshwater aquatic life and wildlife
  - 0.10 μg/L for marine aquatic life
- TCLP regulatory limit: 0.2 mg/L
- RFI clean closure concentration: 2.00 μg/L
- Common concentration in natural soils: 0.01 to 0.8 mg/L
- Average concentration in natural soils: 0.11 mg/L

13.2.20 Nickel, Ni

- Formula weight = 58.69 g/mol
- Specific gravity = 8.90 @ 20°C
- Elemental mercury is insoluble in water
- Ionic nickel water solubility is pH dependent
- Discharge criteria: 0.01 times the 96 hour LC_{50} for freshwater and marine aquatic life
- RFI clean closure concentration: 100 μg/L
- Common concentration in natural soils: 5 to 1000 mg/L
- Average concentration in natural soils: 40 mg/L

13.2.21   Nitrate, NO^3

- Formula weight = 62.02 g/mol
- Discharge criteria: 10 μg/L for domestic water supply

13.2.22   Oil and Grease

- Discharge criteria:
  - Virtually free for domestic water supply
  - 0.01 times the lowest 96 hour LC_{50} for selected freshwater and marine species

13.2.23   Phosphorous, P

- Formula weight = 123.92 g/mol
- Specific gravity = 1.82 @ 20°C
- Elemental phosphorous is slightly soluble in water (0.0003 g/100 mL)
- Discharge criteria: 0.10 μg/L for marine or estuarine waters

13.2.24   Selenium, Se

- Formula weight = 78.96 g/mol
- Specific gravity = 4.26 @ 25°C
- Elemental selenium is insoluble in water
- Discharge criteria:
  - 10 μg/L for domestic water supply
  - 0.01 times 96 hour LC_{50} using a sensitive resident species for marine and freshwater aquatic life
- TCLP regulatory limit: 1.0 mg/L
- RFI clean closure action level: 50 μg/L
- Common concentration in natural soils: 0.1 to 2.0 mg/L
- Average concentration in natural soils: 0.3 mg/L

13.2.25   Silver, Ag

- Formula weight = 107.88 g/mol
- Specific gravity = 10.5 @ 20°C
- Elemental silver is insoluble in water
- Discharge criteria:
  - 50 μg/L for domestic water supply
  - 0.1 times the 96 hour LC_{50} using a sensitive resident species to marine and freshwater aquatic life

13.2.26   Zinc, Zn

- Formula weight = 65.38 g/mol
- Specific gravity = 7.140 @ 25°C

- Elemental zinc is insoluble in water
- Ionic zinc water solubility is dependent on pH
- Discharge criteria:
  - 5.0 mg/L for domestic water supply
  - 0.1 times 96 hour LC_{50} using a sensitive resident species for freshwater aquatic life
- RFI clean closure action level: 7 μg/L
- Common concentration in natural soils: 10 to 300 mg/L
- Average concentration in natural soils: 50 mg/L

13.3   CARBON ADSORPTION ISOTHERMS FOR TOXIC ORGANICS

13.3.1   Introduction

The EPA has published a study (Publication No. EPA-600/8-80-023) of the adsorptive capacity of various toxic organics on activated carbon using batch equilibrium carbon adsorption isotherms with the results plotted according to the Freundlich adsorption equation. The adsorbent used for the tests was Filtrasorb 300 granular activated carbon (GAC), a product of the Calgon Corporation. The GAC was pulverized and the quantity between a 200-and 400-mesh screen was used. The protocol for the tests can be found in the above referenced EPA report.

13.3.2   Isotherms for Individual Compounds

Table 13.3 as follows summarizes the isotherm tests in terms of adsorption capacity in milligram of compound per gram of GAC. Appendix G gives an alphabetical list of compounds and a series of test results giving detailed information on each compound tested.

TABLE 13.3 Carbon Adsorption Capacities (US EPA, Publication No. EPA-600/8-80-023)

Compound	Adsorption (a) capacity, mg/g	Compound	Adsorption (a) capacity, mg/g
bis (2-Ethylhexyl) phthalate	11,300	2,4-Dinitrotoluene	146
Butylbenzyl phthalate	1,520	2,6-Dinitrotoluene	145
Heptachlor	1,220	4-Bromophenyl phenyl ether	144
Heptachlor epoxide	1,038	p-Nitroaniline*	140
Endosulfan sulfate	686	1,1-Diphenylhydrazine	135
Endrin	666	Naphthalene	132
Fluoranthene	664	1-Chloro-2-nitrobenzene	130
Aldrin	651	1,2-Dichlorobenzene	219
PCB-1232	630	p-Chlorometacresol	124
beta-Endosulfan	615	1,4-Dichlorobenzene	121
Dieldrin	606	Benzothiazole*	120
Hexachlorobenzene	450	Diphenylamine	120
Anthracene	376	Guanine*	120
4-Nitrobiphenyl	370	Styrene	120
Fluorene	330	1,3-Dichlorobenzene	118
DDT	322	Acenaphthylene	115
2-Acetylaminofluorene	318	4-Chlorophenyl phenyl ether	111
alpha-BHC	303	Diethyl phthalate	110
Anethole*	300	2-Nitrophenol	99
3,3-Dichlorobenzidine	300	Dimethyl phthalate	97
2-Chloronaphthalene	280	Hexachloroethane	97
Phenylmercuric Acetate	270	Chlorobenzene	91
Hexachlorobutadiene	258	p-Xylene	85
gamma-BHC (lindane)	256	2,4-Dimethylphenol	78
p-Nonylphenol	250	4-Nitrophenol	76
4-Dimethylaminoazobenzene	249	Acetophenone	74
Chlordane	245	1,2,3,4-Tetrahydronaphthalene	74
PCB-1221	242	Adenine*	71
DDE	232	Dibenzo(a,h)anthracene	69
Acridine yellow*	230	Nitrobenzene	68
Benzidine dihydrochloride	220	3,4-Benzofluoranthene	57
beta-BHC	220	1,2-Dibromo-3-chloropropane	53
N-Butylphthalate	220	Ethylbenzene	53
N-Nitrosodiphenylamine	220	2-Chlorophenol	51
Phenanthrene	215	Tetrachloroethene	51
Dimethylphenylcarbinol*	210	o-Anisidine*	50
4-Aminobiphenyl	200	5 Bromouracil	44
beta-Naphthol*	200	Benzo(a)pyrene	34
alpha-Endosulfan	194	2,4-Dinitrophenol	33
Acenaphthene	190	Isophorone	32
4,4′Methylene-bis-(2-chloroaniline)	190	Trichloroethene	28
		Thymine*	27
Benzo(k)fluoranthene	181	Toluene	26
Acridine orange*	180	5-Chlorouracil*	25
alpha-Naphthol	180	N-Nitrosodi-n-propylamine	24
4,6-Dinitro-o-cresol	169	bis(2-Chloroisopropyl)ether	24
alpha-Naphthylamine	160	Phenol	21
2,4-Dichlorophenol	157	Bromoform	20
1,2,4-Trichlorobenzene	157	Carbon tetrachloride	11
2,4,6-Trichlorophenol	155	bis(2-Chloroethoxy)methane	11
beta-Naphthylamine	150	Uracil*	11

TABLE 13.3   Carbon Adsorption Capacities (US EPA, Publication No. EPA-600/8-80-023) (*Continued*)

Compound	Adsorption (a) capacity, mg/g	Compound	Adsorption (a) capacity, mg/g
pentachlorophenol	150	Benzo(ghi)perylene	11
1,1,2,2-Tetrachloroethane	11	Chloroform	2.6
1,2-Dichloropropene	8.2	1,1,1-Trichloroethane	2.5
Dichlorobromomethane	7.9	1,1-Dichloroethane	1.8
Cyclohexanone*	6.2	Acrylonitrile	1.4
1,2-Dichloropropane	5.9	Methylene chloride	1.3
1,1,2-Trichloroethane	5.8	Acrolein	1.2
Trichlorofluoromethane	5.6	Cytosine*	1.1
5-Fluorouracil*	5.5	Benzene	1.0
1,1-Dichloroethylene	4.9	Ethylenediaminetetra-acetic acid	0.86
Dibromochloromethane	4.8	Benzoic acid	0.76
2-Chlorethyl vinyl ether	3.9	Chloroethane	0.59
1,2-Dichloroethane	3.6	N-Dimethylnitrosamine	6.8×10^{-5}
1,2-trans-Dichlorethane	3.1		

Not adsorbed

Acetone cyanohydrin	Adipic acid
Butylamine	Choline chloride
Cyclohexylamine	Diethylene glycol
Ethanol	Hexamethylenediamine
Hydroquinone	Morpholine
	Triethanolamine

*Compounds prepared in "mineralized" distilled water containing the following composition:

Ion	Conc., mg/L	Ion	Conc., mg/L
Na^+	92	$PO_4^=$	10
K^+	12.6	$SO_4^=$	100
Ca^{++}	100	Cl^-	177
Mg^{++}	25.3	Alkalinity	200

(a) Adsorption capacities are calculated for an equilibrium concentration of 1.0 mg/l at neutral pH.

P · A · R · T · VII

APPENDICES

APPENDIX A
HYDRAULIC INFORMATION

EXPLANATION OF APPENDICES

Appendix A-1, Entrance Losses for Pipe: When water flows from a tank with zero velocity to a pipe flowing full with a certain velocity, the entrance loss can be determined from Appendix A-1 for various pipe flow velocities and for various types of entrances. i.e. The loss of head for water to reach a velocity of 2 ft/s for an inward projecting pipe is 0.05 feet.

Appendix A-2, Surface Roughness Constants: Appendix A-2 lists ranges, average values and design recommendations for values of C for various types of pipes for use in the Hazen Williams Formula.

Appendix A-3, Friction Head for Pipe, Table for Hazen Williams Formula: Appendix A-3 lists velocity in ft/s and friction head in feet per 100 feet of pipe for various flows (gpm) for pipes flowing full with a C = 100. For other C values, the numbers in Appendix A-3 should be multiplied by the table correction values at the bottom of Appendix A-3. This Appendix lists pipes from 3/8″ to 42″ diameter.

Appendix A-4, Friction Head for Pipes, Nomograph for Hazen Williams Formula: These pages in Appendix A-4 allow the use of a nomograph for computing head loss in ft/100 ft for pipes flowing full from 2″ to 16″ diameter with C values of 100, 120, 130 and 140.

Appendix A-5, Friction Head for Pipes, Nomograph for Kutter's Formula: These nomographs are used for open channel flow, computed from Kutter's Formula for pipes from 6″ to 96″ diameter flowing full. One nomograph is for an N of 0.013 and one for an N of 0.015. Appendix A-6 can be used to convert these values to values for partially full pipes.

Appendix A-6, Hydraulic Properties of Circular Sewers: This graph is used to convert the values in Appendix A-5 for velocities and discharges in pipes flowing partially full.

Appendix A-7, Friction Head for Pipes, Nomograph for Manning's Equation: This nomograph uses Manning's Equation, a simplified version of Kutter's Formula, to compute flows, velocities and slopes for open channel flows in pipes flowing full from 8″ to 144″ in diameter. Again, Appendix A-6 can be used to correct these flows and velocities to pipes flowing partially full.

Appendix A-8, Equivalent Length of Pipe: When calculating total friction loss in a piping system containing various pipe sizes, it is convenient to convert all pipe to feet of a selected equivalent pipe size.

Appendix A-9, Friction Loss in Pipe Fittings: This Appendix is used to obtain an equivalent length of straight pipe for various fittings in order to determine total friction loss in a piping system. If various sizes of fittings are encountered, Appendix A-8 can be used to convert these equivalent pipe lengths to a standard size.

Appendix A-10, Free Discharge of Pipes: This Appendix lists free discharges in gpm under pressure from 10 to 100 psi for various lengths of pipes. Manning's Equation can be used to calculate additional pipe sizes, pressures or pipe lengths.

Appendix A-11, Parshall Flume Dimensions and Flows: The Parshall Flume has been developed and standardized to allow the ratio between flow and water depth to be modeled. The Parshall Flume is considered the most accurate method of measuring freely flowing water. The Appendix shows the standard dimensions and capacities on one page and the flows versus head for 3″, 6″, 9″, 12″, 18″ and 24″ flumes on subsequent pages.

Appendix A-12, Recommended Flow Rates for Palmer Bowlus Flume: The Palmer Bowlus Flume is a simplified version of the Parshall Flume which can possibly be inserted into a manhole or junction box without the loss of head required by the Parshall Flume. This Appendix lists the maximum and minimum recommended flows for 4″ through 30″ flumes.

Appendix A-13, Discharge from Rectangular Sharp Crested Weirs: A sharp crested weir is a discharge shape which has the most narrow width upstream with the thickness of the weir plate beveled outward downstream as shown on the drawings in Appendices A-13 and A-14. Appendix A-13 shows the discharges in gpm from 1, 3 and 5 foot wide rectangular weirs with limited end contractions and additional discharges above 5 feet for heads from 1 to 19.5 inches. The model for this weir is shown as Francis Formula in the Appendix.

Appendix A-14, Discharge from Triangular Sharp Crested Weirs: See the description for A-13 for the definition of a sharp crested weir. Appendix A-14 shows the discharges from 60 and 90 degree V-notch or triangular weirs with limited end contractions for heads from 1 to 25 inches. The model for the flows in gpm is shown in the Appendix.

Appendix A-15, Orifice Flow: This Appendix shows the formula for water discharging through an orifice at various heads. The discharge coefficient k, for weir entrance conditions can be found in the discharge coefficient chart.

Appendix A-16, Discharge Curves for Slip Pipes or Telescopic Valves: This Appendix gives flows in gpm for various heads over the opening of a telescopic valve or slip pipe from 3 to 10 inches in diameter.

Appendix A-17, Discharge Curves for Air Lift Pumps: Flows in gpm and velocities in fps for 3 and 4 inch air lift pumps are presented in this Appendix for various lifts.

Appendix A-18, Water Hammer Nomograph: This nomograph compares the maximum pressure increase P_m for water hammer conditions caused by an instantaneous valve closing for steel and cast iron pipes for flows from 3 to 10,000 gpm. The pressure increase formula is shown in the Appendix.

APPENDIX A-1   Entrance Losses for Pipe in Feet of Liquid

These data apply to any liquid

Type of entrance	Value of k	Velocity in feet per second (v)							
		2	3	4	5	6	7	8	9
Inward projecting pipe	.78	.05	.11	.19	.30	.44	.59	.78	.98
Sharp cornered	.50	.03	.07	.12	.19	.28	.38	.50	.63
Slightly rounded	.23	.01	.03	.06	.09	.13	.18	.23	.29
Bell mouthed	.04	.00	.01	.01	.02	.02	.03	.04	.05

Type of entrance	Value of k	Velocity in feet per second (v)							
		10	12	15	20	25	30	35	40
Inward projecting pipe	.78	1.21	1.75	2.73	4.85	7.58	10.91	14.82	19.41
Sharp cornered	.50	.78	1.12	1.75	3.11	4.86	7.00	9.50	12.44
Slightly rounded	.23	.36	.51	.81	1.43	2.24	3.22	4.37	5.72
Bell mouthed	.04	.06	.09	.14	.25	.39	.56	.76	1.00

Based on formula: $h = k \dfrac{v^2}{2g}$

In which   h = entrance loss in feet of liquid.
   k = constant depending on shape of entrance
   v = velocity of flow in feet per second.
   g = acceleration of gravity: at 45° latitude and sea level. G = 32.174 ft/sec^2

APPENDIX A-2 Surface Roughness Constants

Type of pipe	Values of C		
	Range	Average value for new pipe	Design value
Steel, welded, seamless	80–150	140	100
Steel, corrugated	50–70	60	60
Iron, cast, ductile or wrought	80–150	130	60
Iron, cast, tar coated	80–145	130	100
Iron or steel, centrifugally applied bitumastic lined	130–160	150	140
Iron or steel, centrifugally applied cement lined	130–160	150	140
Copper, brass, lead, tin, glass	120–150	140	130
Concrete	80–150	120	100
Vitrified day	80–130	110	100
PVC	130–160	160	140
FRP	130–160	150	140
HDPE	130–160	160	140

APPENDIX A-3 Friction Head for Pipe

Friction head in feet per hundred feet of pipe; velocity in feet per second; based on Hazen-Williams formula with C=100 for steel.

⅜-inch pipe

Cap. in GPM	Vel. in FPS	Frict. head ft per 100 ft
0.8	1.35	4.3
1.0	1.68	6.5
1.5	2.52	13.8
2.0	3.36	23.4
2.5	4.21	35.4
3.0	5.05	49.6
3.5	5.89	66.0
4.0	6.73	84.5
5.0	8.41	134.
6.0	10.1	179.

½-inch pipe

Cap. in GPM	Vel. in FPS	Frict. head ft per 100 ft
0.5	0.53	0.58
1.0	1.06	2.1
2.0	2.11	7.6
3.0	3.17	16.0
4.0	4.23	27.3
5.0	5.28	41.2
6.0	6.34	57.8
7.0	7.39	76.8
8.0	8.45	98.3
9.0	9.51	122.
10.0	10.6	149.

¾-inch pipe

Cap. in GPM	Vel. in FPS	Frict. head ft per 100 ft
2	1.201	1.934
3	1.81	4.08
4	2.41	6.94
5	3.01	10.5
6	3.61	14.7
7	4.21	19.6
8	4.82	25.0
9	5.42	31.1
10	6.02	37.8
11	6.62	45.1
12	7.22	53.0
13	7.82	61.5
14	8.43	70.5
16	9.63	90.2
18	10.8	112.
20	12.0	136.

1-inch pipe

Cap. in GPM	Vel. in FPS	Frict. head ft per 100 ft
2	0.74	0.60
3	1.11	1.26
4	1.49	2.14
5	1.86	3.24
6	2.23	4.54
8	2.97	7.73
10	3.71	11.7
12	4.46	16.4
14	5.20	21.8
16	5.94	27.9
18	6.68	34.7
20	7.43	42.1
25	9.30	64.0
30	11.1	89.2
35	13.0	119.
40	14.9	152.
45	16.7	189.

1¼-inch pipe

Cap. in GPM	Vel. in FPS	Frict. head ft per 100 ft
4	0.86	0.56
5	1.07	0.85
6	1.29	1.20
7	1.50	1.59
8	1.72	2.04
10	2.15	3.08
12	2.57	4.31
14	3.00	5.73
16	3.43	7.34
18	3.86	9.13
20	4.29	11.1
25	5.36	16.8
30	6.43	23.5
35	7.51	31.2
40	8.58	40.0
50	10.7	60.4
60	12.9	84.7
70	15.0	114.
80	17.2	144.

1½-inch pipe

Cap. in GPM	Vel. in FPS	Frict. head ft per 100 ft
5	0.79	0.40
6	0.95	0.57
8	1.26	0.96
10	1.58	1.45
12	1.89	2.04
14	2.21	2.71
16	2.52	3.47
18	2.84	4.31
20	3.15	5.24
25	3.94	7.9
30	4.73	11.1
40	6.30	18.9
50	7.88	28.5
60	9.46	40.0
70	11.0	53.2
80	12.6	68.1
90	14.2	84.7
100	15.8	103.
120	18.9	144.

2-inch pipe

Cap. in GPM	Vel. in FPS	Frict. head ft per 100 ft
5	0.48	0.12
10	0.96	0.43
15	1.49	0.92
20	1.91	1.55
25	2.39	2.35
30	2.87	3.29
35	3.35	4.37
40	3.82	5.60
50	4.78	8.46
60	5.74	11.9
70	6.69	15.8
80	7.65	20.2
90	8.61	25.1
100	9.56	30.5
110	10.5	36.4
120	11.5	42.7
130	12.4	49.6
140	13.4	56.9
150	14.3	64.7

2½-inch pipe

Cap. in GPM	Vel. in FPS	Frict. Head ft per 100 ft
10	0.67	0.18
15	1.01	0.39
20	1.34	0.65
25	1.68	0.99
30	2.01	1.39
35	2.35	1.84
40	2.68	2.36
50	3.35	3.56
60	4.02	4.99
70	4.69	6.64
80	5.36	8.50
90	6.03	10.6
100	6.70	12.8
120	8.04	18.0
140	9.38	23.9
160	10.7	30.7
180	12.1	38.1
200	13.4	46.3
220	14.7	55.3
240	16.1	66.4
260	17.4	75.3

APPENDIX A-3 Friction Head for Pipe (*Continued*)

3-inch pipe			4-inch pipe			5-inch pipe			6-inch pipe			8-inch pipe			10-inch pipe			12-inch pipe		
Cap. in GPM	Vel. in FPS	Frict. head ft per 100 ft	Cap. in GPM	Vel. in FPS	Frict. head ft per 100 ft	Cap. in GPM	Vel. in FPS	Frict. head ft per 100 ft	Cap. in GPM	Vel. in FPS	Frict. head ft per 100 ft	Cap. in GPM	Vel. in FPS	Frict. head ft per 100 ft	Cap. in GPM	Vel. in FPS	Frict. head ft per 100 ft	Cap. in GPM	Vel. in FPS	Frict. head ft per 100 ft
20	0.87	0.23	40	1.01	0.22	50	0.80	0.110	50	0.56	0.045	150	0.96	0.091	200	0.81	0.051	300	0.85	0.045
25	1.09	0.34	60	1.51	0.46	70	1.12	0.205	100	1.11	0.162	200	1.28	0.154	250	1.02	0.077	500	1.42	0.115
30	1.30	0.48	80	2.02	0.79	100	1.60	0.395	150	1.67	0.345	250	1.61	0.233	300	1.22	0.108	1000	2.84	0.415
35	1.52	0.64	100	2.52	1.19	150	2.40	0.848	200	2.22	0.584	300	1.92	0.325	350	1.42	0.143	1200	3.41	0.518
40	1.74	0.82	120	3.02	1.67	200	3.20	1.43	250	2.78	0.885	350	2.24	0.433	400	1.63	0.183	1400	3.98	0.773
50	2.17	1.24	140	3.53	2.22	250	4.01	2.16	300	3.33	1.24	400	2.57	0.554	450	1.83	0.228	1600	4.55	0.990
60	2.60	1.74	160	4.03	2.84	300	4.81	3.03	350	3.89	1.65	450	2.88	0.689	500	2.04	0.277	1800	5.11	1.23
70	3.04	2.31	180	4.54	3.53	350	5.61	4.03	400	4.44	2.11	500	3.20	0.838	600	2.44	0.388	2000	5.68	1.50
80	3.47	2.96	200	5.05	4.29	400	6.41	5.15	450	5.00	2.62	600	3.85	1.17	800	3.25	0.660	2200	6.25	1.78
90	3.91	3.67	250	6.30	6.49	450	7.27	6.41	500	5.56	3.19	700	4.49	1.56	1000	4.07	0.998	2400	6.81	2.10
100	4.34	4.47	300	7.57	9.09	500	8.02	7.79	550	6.11	3.80	800	5.13	1.99	1200	4.89	1.40	2600	7.38	2.43
120	5.21	6.26	350	8.83	12.1	550	8.82	9.28	600	6.66	4.46	900	5.77	2.48	1400	5.70	1.86	2800	7.95	2.78
140	6.08	8.32	400	10.1	15.5	600	9.62	10.9	650	7.22	5.17	1000	6.41	3.02	1600	6.51	2.38	3000	8.52	3.17
160	6.94	10.7	450	11.4	19.3	700	11.2	14.5	700	7.78	5.93	1200	7.69	4.23	1800	7.32	2.96	3500	9.95	4.21
180	7.81	13.2	500	12.6	23.4	800	12.8	18.6	800	8.90	7.60	1400	8.97	5.62	2000	8.14	3.60	4000	11.4	5.39
200	8.68	16.1	550	13.9	27.9	900	14.4	23.1	900	10.0	9.44	1600	10.3	7.20	2500	10.18	5.44	4500	12.8	6.70
220	9.55	19.2	600	15.1	32.8	1000	16.0	28.1	1000	11.1	11.5	1800	11.5	8.95	3000	12.2	7.61	5000	14.2	8.15
240	10.4	22.6	650	16.4	38.0	1200	19.2	39.3	1200	13.3	16.1	2000	12.8	10.9	3500	14.2	10.2	5500	15.6	9.72
260	11.3	26.2	700	17.6	43.6	1400	22.4	52.3	1400	15.6	21.4	2500	16.1	16.5	4000	16.3	13.0	6000	17.0	11.4
280	12.2	30.0	800	20.2	55.8	1600	25.6	66.9	1600	17.8	27.4	3000	19.2	23.0	4500	18.3	16.1	7000	19.9	15.2
300	13.0	34.1							1800	20.0	34.0	3500	22.4	30.6	5000	20.3	19.6	8000	22.7	19.4

3-inch pipe			4-inch pipe			5-inch pipe			6-inch pipe			8-inch pipe			10-inch pipe			12-inch pipe		
Cap. in GPM	Vel. in FPS	Frict. head ft per 100 ft	Cap. in GPM	Vel. in FPS	Frict. head ft per 100 ft	Cap. in GPM	Vel. in FPS	Frict. head ft per 100 ft	Cap. in GPM	Vel. in FPS	Frict. head ft per 100 ft	Cap. in GPM	Vel. in FPS	Frict. head ft per 100 ft	Cap. in GPM	Vel. in FPS	Frict. head ft per 100 ft	Cap. in GPM	Vel. in FPS	Frict. head ft per 100 ft
500	1.16	0.071	1000	1.76	0.129	1000	1.11	0.042	1000	0.712	0.014	2000	0.91	0.017	4000	1.26	0.026	6000	1.39	0.025
1000	2.23	0.256	1200	2.11	0.181	1500	1.67	0.090	1500	1.07	0.038	2500	1.14	0.025	6000	1.89	0.054	7000	1.62	0.033
1100	2.56	0.306	1400	2.46	0.241	2000	2.22	0.153	2000	1.42	0.051	3000	1.38	0.037	8000	2.52	0.092	8000	1.85	0.043
1200	2.79	0.359	1600	2.81	0.308	2500	2.78	0.231	3000	2.14	0.108	4000	1.83	0.070	9000	2.83	0.114	10000	2.31	0.065
1400	3.26	0.477	1800	3.16	0.383	3000	3.33	0.323	4000	2.85	0.18	5000	2.28	0.094	10000	3.14	0.139	15000	3.47	0.140
1600	3.72	0.611	2000	3.51	0.466	3500	3.89	0.430	5000	3.55	0.27	6000	2.74	0.131	12000	3.78	0.193	20000	4.62	0.238
1800	4.19	0.760	2500	4.39	0.704	4000	4.45	0.551	6000	4.25	0.38	7000	3.17	0.175	14000	4.40	0.260	25000	5.79	0.358
2000	4.65	0.924	3000	5.27	0.987	4500	5.00	0.692	7000	4.95	0.50	8000	3.65	0.224	16000	5.03	0.330	30000	6.94	0.501
2500	5.81	1.40	3500	6.15	1.31	5000	5.55	0.832	8000	5.68	0.69	9000	4.10	0.278	18000	5.56	0.412	32000	7.40	0.566
3000	6.98	1.96	4000	7.03	1.68	6000	6.67	1.17	9000	6.40	0.82	10000	4.56	0.337	19000	5.98	0.454	34000	7.86	0.632
3500	8.15	2.60	4500	7.91	2.09	7000	7.78	1.55	10000	7.11	1.00	11000	5.01	0.401	20000	6.20	0.504	36000	8.33	0.702
4000	9.31	3.32	5000	8.79	2.54	8000	8.89	1.98	11000	7.82	1.19	12000	5.47	0.473	21000	6.60	0.544	38000	8.80	0.778
4500	10.5	4.13	5500	9.65	3.05	9000	10.0	2.49	12000	8.55	1.40	13000	5.94	0.55	22000	6.92	0.590	40000	9.25	0.855
5000	11.6	5.03	6000	10.5	3.56	10000	11.1	3.00	13000	9.25	1.63	14000	6.40	0.63	23000	7.24	0.640	42000	9.72	0.936
6000	14.0	7.05	7000	12.3	4.73	12000	13.3	4.20	14000	9.95	1.86	15000	6.85	0.71	24000	7.55	0.695	44000	10.18	1.01
7000	16.3	9.38	8000	14.1	6.06	14000	15.5	5.59	15000	10.06	2.11	16000	7.30	0.81	26000	8.18	0.806	46000	10.63	1.10
8000	18.6	12.0	9000	15.8	7.53	16000	17.8	7.15	16000	11.38	2.42	18000	8.20	1.00	28000	8.80	0.935	48000	11.10	1.19
9000	20.9	14.9	10000	17.6	9.15	18000	20.0	8.90	17000	12.09	2.62	20000	9.12	1.22	30000	9.44	1.07	50000	11.58	1.28
10000	23.3	18.1	12000	21.1	12.8	20000	22.2	10.8	18000	12.80	2.98	22000	9.61	1.48	34000	10.70	1.34	52000	12.01	1.38
11000	25.6	21.6	14000	24.6	17.1	25000	27.8	16.3	19000	13.50	3.28	24000	10.09	1.73	38000	11.95	1.65	54000	12.49	1.49
12000	27.9	25.4	16000	28.1	21.8	30000	33.3	22.9	20000	14.20	3.61	28000	12.75	2.27	42000	13.20	1.99	56000	12.93	1.60

Value of C	150	140	130	120	110	100	90	80	70	60
Multiplier to Correct Table	0.47	0.54	0.62	0.71	0.84	1.00	1.22	1.50	1.93	2.57

APPENDIX A-4 Friction Head for Pipe

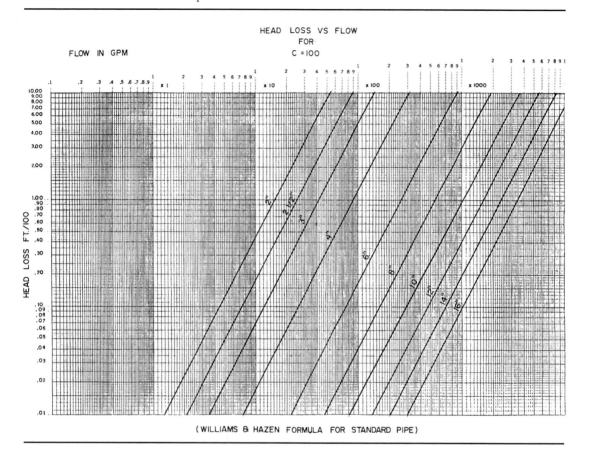

(WILLIAMS & HAZEN FORMULA FOR STANDARD PIPE)

APPENDIX A-4   Friction Head for Pipe (*Continued*)

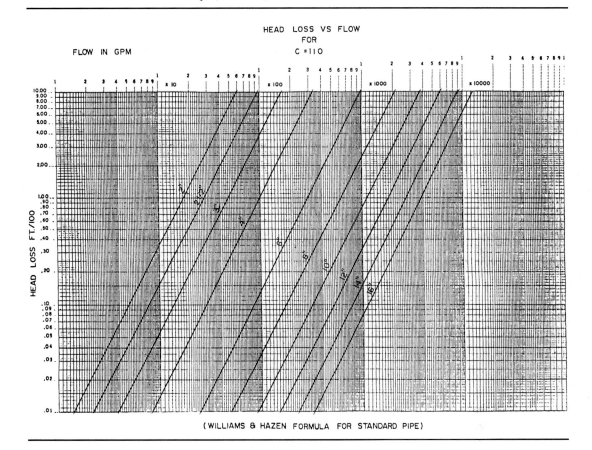

HEAD LOSS VS FLOW
FOR
C = 110

FLOW IN GPM

HEAD LOSS FT./100

(WILLIAMS & HAZEN FORMULA FOR STANDARD PIPE)

APPENDIX A-4   Friction Head for Pipe (*Continued*)

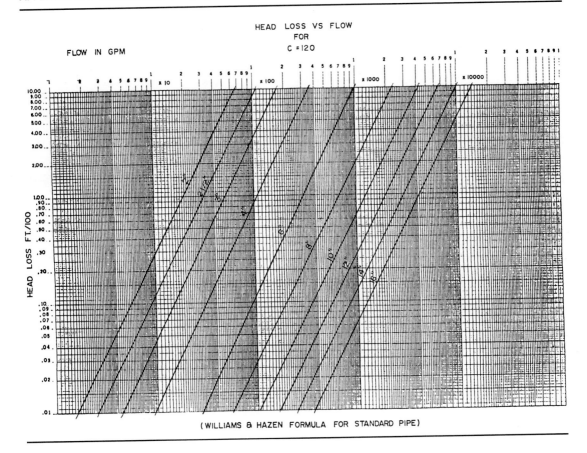

APPENDIX A-4 Friction Head for Pipe (*Continued*)

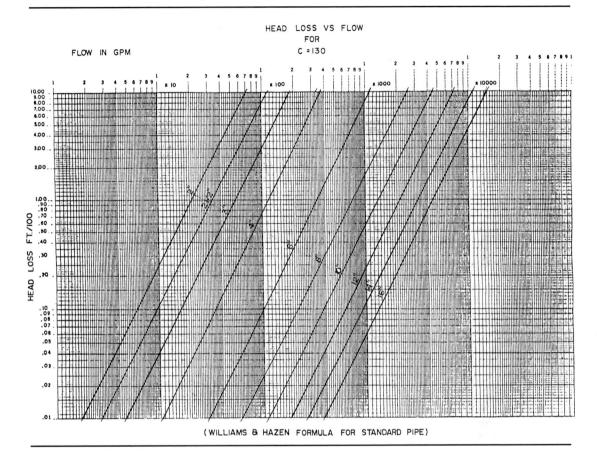

APPENDIX A-4   Friction Head for Pipe (*Continued*)

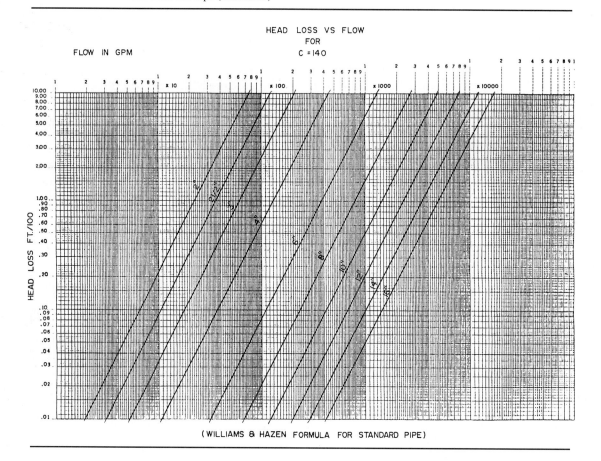

APPENDIX A-5   Friction Head for Pipe

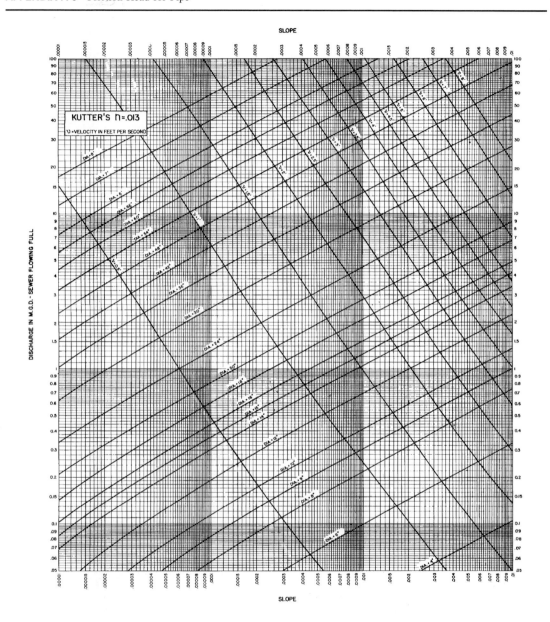

DISCHARGE OF CIRCULAR SEWERS FLOWING FULL

for

KUTTER'S N = .013

APPENDIX A-5   Friction Head for Pipe (*Continued*)

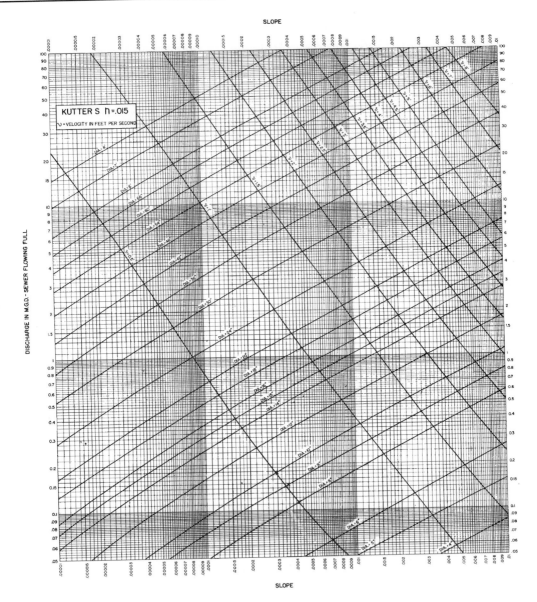

DISCHARGE OF CIRCULAR SEWERS FLOWING FULL

for

KUTTER'S N = .015

APPENDIX A-6   Hydraulic Properties of Circular Sewers

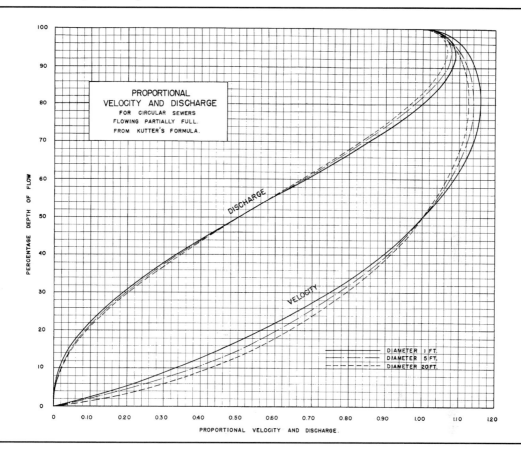

APPENDIX A-7   Friction Head for Pipe

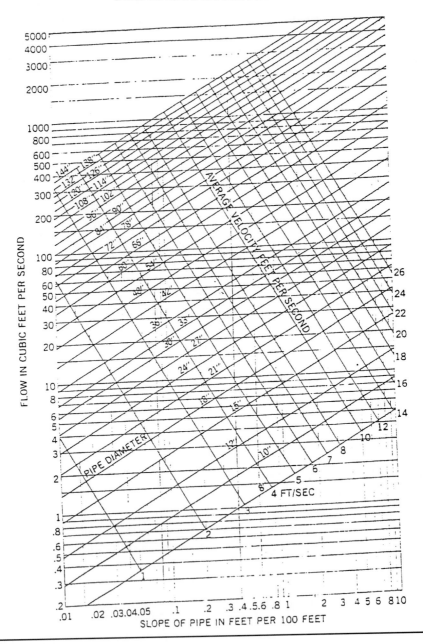

FLOW FOR CIRCULAR PIPE FLOWING FULL
BASED ON MANNING'S EQUATION n=0.010

APPENDIX A-8 Equivalent Length of Pipe

Given diametr (inches)	Division factor	Diameter of equivalent length (inches)
4	7.22	6
6	4.04	8
6	12.0	10
6	29.1	12
8	2.98	10
8	7.22	12
8	15.3	14
8	29.3	16
8	51.9	18
10	2.43	12
10	29.1	20
10	71.2	24
16	7.22	24
16	21.4	30
18	1.67	20
18	12.04	30
18	27.2	36

Example:
    Given: 10 ft of 6-inch diameter pipe
    Find: Equivalent length in 12-inch diameter pipe.

$$\text{Solution:} \quad \frac{10 \ feet \ (6'' \ dia. \ pipe)}{29.1} = 0.344 \ feet \ (12'' \ dia. \ pipe)$$

APPENDIX A-9   Friction Loss in Pipe fittings (in terms of equivalent feet of straight pipe)

Nominal pipe size std wt. steel	45° Elbow	Long-radius elbow or run of std. tee	Std. elbow or run of tee reduced ½	Std. tee thru side outlet	Close return bend	Gate valve FULL OPEN	Angle valve FULL OPEN	Globe valve FULL OPEN	Butterfly valve FULL OPEN	Swing check valve FULL OPEN
				These data may be applied to any liquid or gas						
Resistance factor	0.42	0.6	0.9	1.8	2.2	0.19	5.0	10.0		2.3
½	0.78	1.11	1.7	3.3	4.1	0.35	9.3	18.6		4.3
¾	0.97	1.4	2.1	4.2	5.1	0.44	11.5	23.1		5.3
1	1.23	1.8	2.6	5.3	6.5	0.56	14.7	29.4		6.8
1¼	1.6	2.3	3.5	7.0	8.5	0.74	19.3	38.6		8.9
1½	1.9	2.7	4.1	8.1	9.9	0.86	22.6	45.2		10.4
2	2.4	3.5	5.2	10.4	12.8	1.10	29	58		13.4
2½	2.9	4.2	6.2	12.4	15.2	1.32	35	69		15.9
3	3.6	5.2	7.7	15.5	18.9	1.6	43	86		19.8
4	4.7	6.8	10.2	20.3	24.8	2.1	57	113	15	26.0
5	5.9	8.5	12.7	25.4	31	2.7	71	142		33
6	7.1	10.2	15.3	31	37	3.2	85	170	20	39
7	8.3	11.8	17.7	35	43	3.7	98	197		45
8	9.4	13.4	20.2	40	49	4.3	112	224	20	52
10	11.8	16.9	25.3	51	62	5.3	141	281	25	65
12	14.1	20.2	30	61	74	6.4	168	336	25	77
14	16.5	23.5	35	71	86	7.5			35	90
16	18.8	26.9	40	81	99	8.5			35	104
18	21.2	30	45	91	111	9.6			50	116
20	23.5	34	50	101	123	10.7			50	129
24	28.2	40	61	121	148	12.8			60	155
30	35.3	50	76	151	185	16.0				193
36	42.4	61	91	181	222	19.2				232
42	49.4	71	106	212	259	22.4				271
48	57.6	81	121	242	296	25.6				310

APPENDIX A-10   Free Discharge of Pipes
Discharge of Pipes in gpm: ¾″ to 6″
Free discharge form horizontal straight pipes against no backpressure

C = 100

Lbs./ Sq. In.	Size	Length in Feet					Lbs./ Sq. In.	Size	Length in Feet					
		25	50	100	200	400			25	50	100	200	400	1000
10		16	11	8	5	4	10		290	200	137	95	65	
20		23	16	11	8	5	20		420	290	200	137	95	
40	¾″	34	24	16	11	8	40	2½″	620	420	290	200	137	
60		43	29	20	14	10	60		770	525	360	250	172	
80		50	34	24	16	11	80		890	620	420	290	200	
100		57	39	28	18	13	100		1160	695	480	330	225	
10		31	21	14	10	7	10		575	395	270	185	130	
20		44	31	21	14	10	20		835	575	395	270	185	
40	1″	65	44	31	21	14	40		1215	835	575	395	270	
60		81	55	38	26	18	60	3″	1515	1040	715	490	340	
80		94	65	44	31	21	80		1765	1215	835	575	395	
100		107	73	50	35	24	100		2010	1370	945	650	445	
10		84	58	39	27	19	10		1070	725	495	340	235	144
20		121	84	58	39	27	20		1535	1070	725	495	340	208
40	1¼″	177	121	84	58	39	40	4″	2225	1535	1070	725	495	305
60		220	151	104	72	50	60		2780	1910	1315	900	625	380
80		257	177	121	84	58	80		3250	2225	1535	1070	725	445
100		292	200	137	95	65	100		3680	2520	1735	1190	820	500
10		94	65	44	31	21	10		1915	1315	900	620	425	260
20		137	94	65	45	31	20		2780	1915	1315	900	620	370
40	1½″	200	137	94	65	45	40	5″	4040	2780	1915	1315	1000	550
60		250	170	117	81	56	60		5040	3460	2380	1635	1135	685
80		290	200	137	94	65	80		5880	4040	2780	1915	1315	800
100		330	225	155	106	73	100		6680	4560	3140	2150	1490	910
10		183	126	86	59	41	10		3100	2125	1450	1000	690	425
20		265	183	126	86	59	20		4480	3100	2125	1450	1000	600
40	2″	385	265	183	126	85	40	6″	6520	4480	3100	2125	1450	890
60		480	330	227	156	108	60		8140	5580	3880	2640	1630	1110
80		560	385	265	183	126	80		9500	6520	4480	3100	2125	1300
100		640	435	300	205	142	100		10800	7360	5075	3480	2410	1460

APPENDIX A-11 Parshall Flume Dimensions and Flow

Dimensions of Parshall Flume

| Throat width | | | | | | | | | | | | | Free flow capacity | | | |
| | | | | | | | | | | | | | Maximum | | Minimum | |
W	A	2/3 A	B	C	D	E	F	G	K	N	X	Y	Head	Discharge	Discharge	Head
3"	1'-63/8"	1'-0¼"	1'-6"	7"	103/16"	1'-3"	6"	1'	1"	2¼"	1"	1½"	1.125	½ Sec. Ft. 0.777 mgd	0.03 Sec. Ft. 19400 gpd	0
6"	2'-07/16"	1'-45/16"	2'-0"	1'-3½"	1'-3½"	1'-6"	12"	2'	3"	4½"	2"	3"	1.24	2.9 Sec. Ft. 1.87 mgd	0.05 Sec. Ft. 32300 gpd	0.
9"	2'-105/8"	1'-111/8"	2'-10"	1'-3"	1'-105/8"	2'-0"	12"	1'-6"	3"	4½"	2"	3"	1.5	5.7 Sec. Ft. 3.69 mgd	0.10 Sec. Ft. 64700 gpd	0.
1'-0"	4'-6"	3'-0"	4'-4⅞"	2"	2'-9¼"	3'-0"	2'	3'	3"	9"	2"	3"	2.5	16.1 Sec. Ft. 10.4 mgd	0.35 Sec. Ft. 0.226 mgd	0.2'
1'-6"	4'-9"	3'-2"	4'-7⅞"	2'-6"	3'-4⅜"	3'-0"	2'	3'	3"	9"	2"	3"	2.5	24.6 Sec. Ft. 15.9 mgd		
2'-0"	5'-0"	3'-4"	4'-10⅞"	3'	3'-11½"	3'-0"	2'	3'	3"	9"	2"	3"	2.5	33.1 Sec. Ft. 21.45 mgd	0.66 Sec. Ft. 0.427 mgd	0.2'
3'-0"	5'-6"	3'-8"	5'-4¾"	4'	5'-1⅞"	3'-0"	2'	3'	3"	9"	2"	3"	2.5	50.4 Sec. Ft. 32.6 mgd	0.97 Sec. Ft. 0.627 mgd	0.2'
4'-0"	6'-0"	4'-0"	5'-10⅝"	5'	6'-4¼"	3'-0"	2'	3'	3"	9"	2"	3"	2.5	67.9 Sec. Ft. 44.0 mgd	1.26 Sec. Ft. 0.815 mgd	0.2'
6'-0"	7'-0"	4'-8"	6'-103/8"	7'	8'-9"	3'-0"	2'	3'	3"	9"	2"	3"	2.5	103.5 Sec. Ft. 66.9 mgd	2.63 Sec. Ft. 1.70 mgd	0.25'
8'-0"	8'-0"	5'-4"	7'-101/8"	9'	11'-1¾"	3'-0"	2'	3'	3"	9"	2"	2"	2.5	139.5 Sec. Ft. 90.2 mgd	4.62 Sec. Ft. 2.99 mgd	0.3'

A.23

APPENDIX A-11   Parshall Flume Dimensions and Flow (*Continued*)

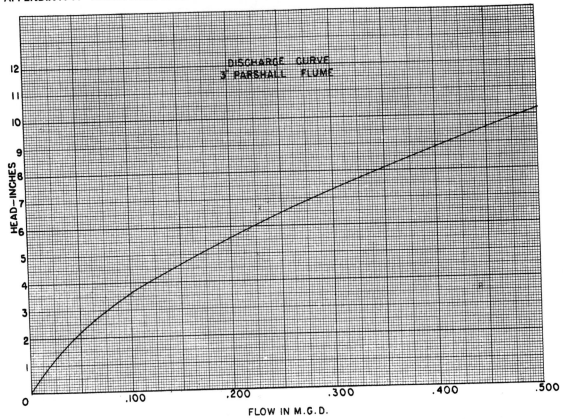

APPENDIX A-11   Parshall Flume Dimensions and Flow (*Continued*)

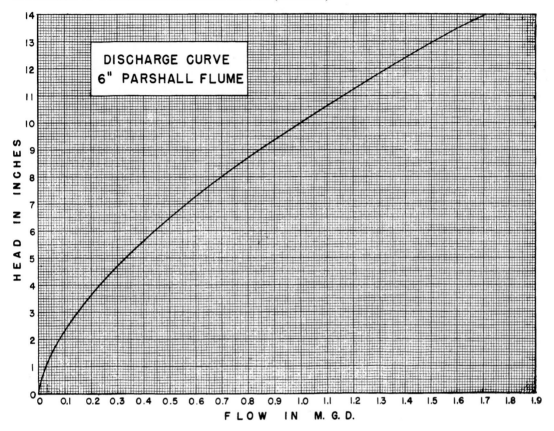

APPENDIX A-11 Parshall Flume Dimensions and Flow (*Continued*)

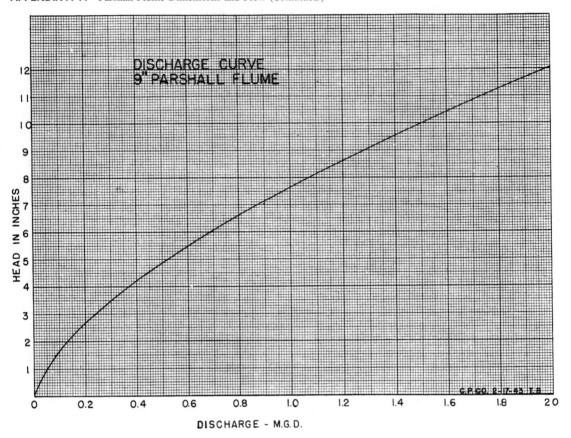

APPENDIX A-11   Parshall Flume Dimensions and Flow (*Continued*)

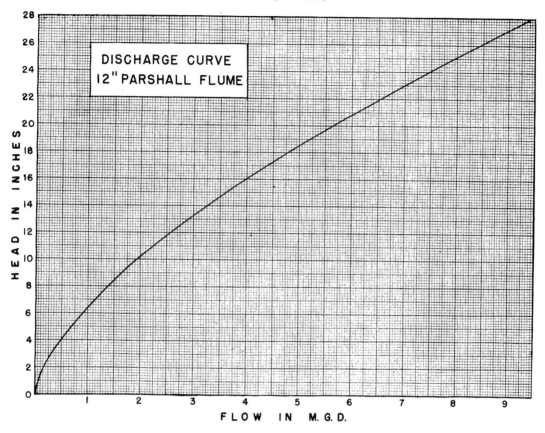

APPENDIX A-11   Parshall Flume Dimensions and Flow (*Continued*)

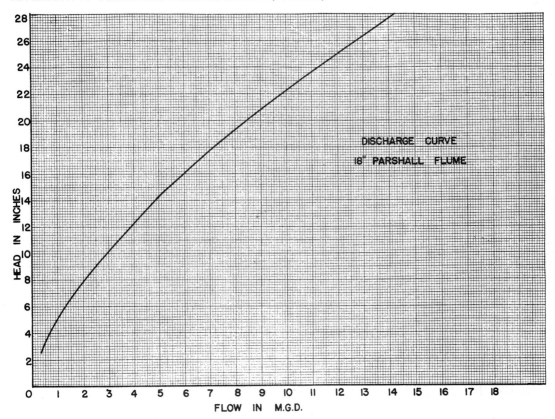

APPENDIX A-11   Parshall Flume Dimensions and Flow (*Continued*)

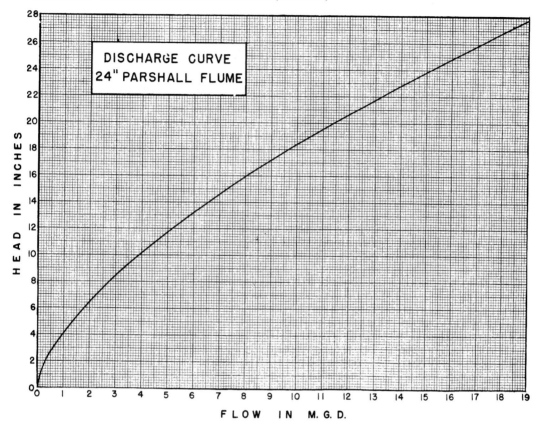

APPENDIX A-12 Recommended Flow Rates for Palmer Bowlus Flume

Flume size (in.)	Maximum upstream slope (%)	Min head (ft)	Minimum flow rate			Max head (ft)	Maximum flow rate		
			CFS	GPM	MGD		CFS	GPM	MGD
4	2.2	0.06	0.009	3.87	0.006	0.25	0.121	54.4	0.078
6	2.2	0.07	0.016	7.02	0.010	0.35	0.295	132	0.191
8	2.0	0.09	0.031	13.7	0.020	0.50	0.690	310	0.446
10	1.8	0.11	0.052	23.3	0.034	0.60	1.12	502	0.723
12	1.6	0.12	0.069	31.2	0.045	0.70	1.68	752	1.08
15	1.5	0.14	0.109	48.8	0.070	0.90	3.09	1380	1.99
18	1.4	0.16	0.156	69.9	0.101	1.05	4.61	2070	2.98
21	1.4	0.18	0.222	99.8	0.144	1.25	7.04	3160	4.55
24	1.3	0.20	0.294	132	0.190	1.40	9.47	4250	6.12
27	1.3	0.22	0.377	169	0.244	1.60	13.1	5870	8.46
30	1.3	0.24	0.482	216	0.311	1.75	16.5	7410	10.7

Note: These figures represent minimum and maximum recommended flow rates for free flow through Plasti-Fab Palmer Bowlus Flumes.

APPENDIX A-13   Discharge from Rectangular Sharp Crested Weirs

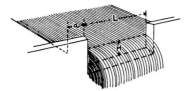

Figures in Table are In Gallons Per Minute

Head (H) in inches	Length (L) of weir in feet				Head (H) in inches	Length (L) of weir in feet		
	1	3	5	Additional g.p.m. for each ft. over 5 ft.		3	5	Additional g.p.m. for each ft. over 5 ft.
1	35.4	107.5	179.8	36.05	8	2338	3956	814
1¼	49.5	150.4	250.4	50.4	8¼	2442	4140	850
1½	64.9	197	329.5	66.2	8½	2540	4312	890
1¾	81	248	415	83.5	8¾	2556	4511	929
2	98.5	302	508	102	9	2765	4699	970
2¼	117	361	605	122	9¼	2876	4899	1011
2½	136.2	422	706	143	9½	2985	5098	1051
2¾	157	485	815	165	9¾	3101	5288	1091
3	177.8	552	926	187	10	3216	5490	1136
3¼	199.8	624	1047	211	10½	3480	5940	1230
3½	222	695	1167	236	11	3716	6355	1320
3¾	245	769	1292	261	11½	3960	6780	1410
4	269	846	1424	288	12	4185	7165	1495
4¼	293.6	925	1559	316	12½	4430	7595	1575
4½	318	1006	1696	345	13	4660	8010	1660
4¾	344	1091	1835	374	13½	4950	8510	1780
5	370	1175	1985	405	14	5215	8980	1885
5¼	395.6	1262	2130	434	14½	5475	9440	1985
5½	421.6	1352	2282	465	15	5740	9920	2090
5¾	449	1442	2440	495	15½	6015	10400	2165
6	476.5	1535	2600	528	16	6290	10900	2300
6¼		1632	2760	560	16½	6585	11380	2410
6½		1742	2920	596	17	6925	11970	2620
6¾		1826	3094	630	17½	7140	12410	2640
7		1928	3260	668	18	7410	12900	2745
7¼		2029	3436	701.5	18½	7695	13410	2855
7½		2130	3609	736	19⅓	7980	13940	2970
7¾		2238	3785	774	19½	8280	14460	3090

This table is based on Francis formula:
$$Q = 3.33 (L - 0.2H) H^{1.1}$$
which
Q = cu. ft. of water flowing per second.
L = length of weir opening in feet. (should be 4 to 8 times H).
H = head on weir in feet (to be measured at least 8 ft. back of weir opening).
a = should be at least 3 H.

APPENDIX A-14 Discharge from Triangular Sharp Crested Weirs

Head (h) in inches	Flow in gallons per min.		Head (h) in inches	Flow in gallons per min.		Head (h) in inches	Flow in gallons per min.	
	90° Notch	60° Notch		90° Notch	60° Notch		90° Notch	60° Notch
1	2.19	1.27	6¾	260	150	15	1912	1104
1¼	3.83	2.21	7	284	164	15½	2073	1197
1½	6.05	3.49	7¼	310	179	16	2246	1297
1¾	8.89	5.13	7½	338	195	16½	2426	1401
2	12.4	7.16	7¾	367	212	17	2614	1509
2¼	16.7	9.62	8	397	229	17½	2810	1623
2½	21.7	12.5	8¼	429	248	18	3016	1741
2¾	27.5	15.9	8½	462	267	18½	3229	1864
3	34.2	19.7	8¾	498	287	19	3452	1993
3¼	41.8	24.1	9	533	308	19½	3684	2127
3½	50.3	29.0	9¼	571	330	20	3924	2266
3¾	59.7	34.5	9½	610	352	20½	4174	2410
4	70.2	40.5	9¾	651	376	21	4433	2560
4¼	81.7	47.2	10	694	401	21½	4702	2715
4½	94.2	54.4	10½	784	452	22	4980	2875
4¾	108	62.3	11	880	508	22½	5268	3041
5	123	70.8	11½	984	568	23	4565	3213
5¼	139	80.0	12	1094	632	23½	5873	3391
5½	156	89.9	12½	1212	700	24	6190	3574
5¾	174	100	13	1337	772	24½	6518	3762
6	193	112	13½	1469	848	25	6855	3958
6¼	214	124	14	1609	929			
6½	236	136	14½	1756	1014			

Based on formula:
$$Q = (C)\,(4/15)\,(L)\,(H)\,\sqrt{2gH}$$
in which Q = flow of water in cu. ft. per sec.
       L = width of notch in ft. at H distance above apex.
       H = head of water above apex of notch in ft.
       C = constant varying with conditions, .57 being used for this table.
       a = should be not less than ¾ L.
For 90° notch the formula becomes
$$Q = 2.4381\ H^{1/2}$$
For 60° notch the formula becomes
$$Q = 1.4076\ H^{1/2}$$

APPENDIX A-15   Orifice Flow

(For Sharp or Square Edged Orifice)

$$Q = kA(2gh)^{1/2}$$

where:

Q	=	flow in cfs
A	=	cross sectional area in ft^2
g	=	acceleration due to gravity = 32.2 ft/s^2
h	=	head in feet over centerline of orifice
k	=	discharge coefficient

Re−entrant Tube	Sharp Edged	Square Edged	Re−entrant Tube	Square Edged	Well Rounded
K = 0.52	K = 0.61	K = 0.61	K = 0.73	K = 0.82	K = 0.98

Discharge Coefficients

APPENDIX A-16   Discharge Curves for Slip Pipes or Telescopic Valves

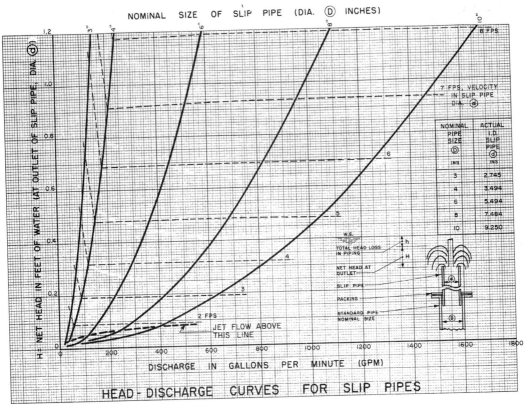

HEAD - DISCHARGE CURVES FOR SLIP PIPES

REFERENCE:
A.H. GIBSON, "HYDRAULICS & ITS APPLICATION",
CONSTABLE & CO. LTD., 4TH ED., P 156

APPENDIX A-17   Discharge Curves for Airlift Pumps

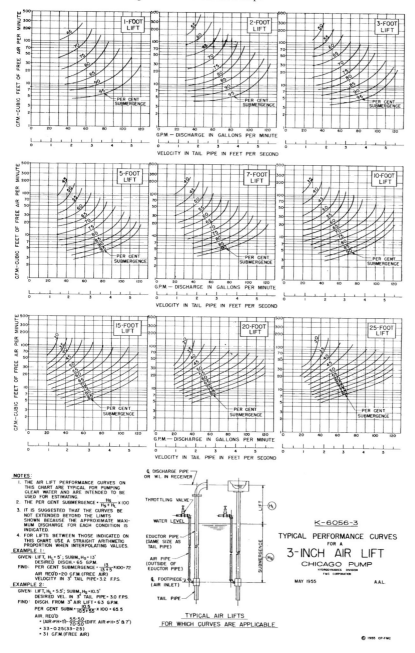

K-6056-3

TYPICAL PERFORMANCE CURVES

FOR A

3-INCH AIR LIFT

CHICAGO PUMP

HYDRODYNAMICS DIVISION
FMC CORPORATION

MAY 1955   A.A.L.

© 1955 CP-FMC

APPENDIX A-17   Discharge Curves for Airlift Pumps (*Continued*)

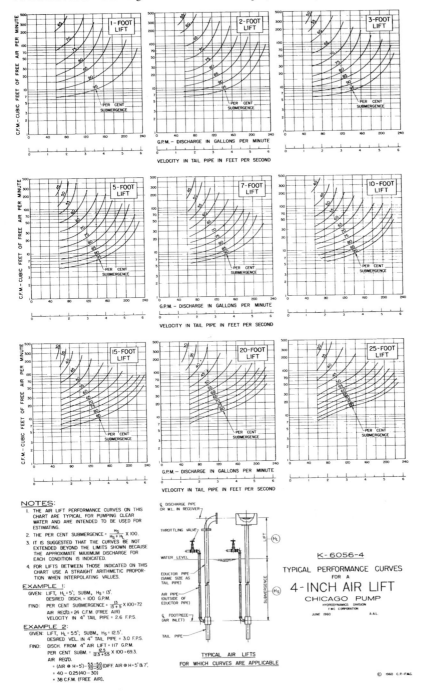

NOTES:
1. THE AIR LIFT PERFORMANCE CURVES ON THIS CHART ARE TYPICAL FOR PUMPING CLEAR WATER AND ARE INTENDED TO BE USED FOR ESTIMATING.
2. THE PER CENT SUBMERGENCE = $\frac{H_S}{H_S + H_L}$ X 100.
3. IT IS SUGGESTED THAT THE CURVES BE NOT EXTENDED BEYOND THE LIMITS SHOWN BECAUSE THE APPROXIMATE MAXIMUM DISCHARGE FOR EACH CONDITION IS INDICATED.
4. FOR LIFTS BETWEEN THOSE INDICATED ON THIS CHART USE A STRAIGHT ARITHMETIC PROPORTION WHEN INTERPOLATING VALUES.

EXAMPLE 1:
GIVEN: LIFT, H_L = 5'; SUBM., H_S = 13'.
     DESIRED DISCH. = 100 G.P.M.
FIND: PER CENT SUBMERGENCE = $\frac{13}{13 + 5}$ X 100. = 72
      AIR REQ'D. = 24 C.F.M. (FREE AIR)
      VELOCITY IN 4" TAIL PIPE = 2.6 F.P.S.

EXAMPLE 2:
GIVEN: LIFT, H_L = 5.5'; SUBM., H_S = 12.5'.
     DESIRED VEL. IN 4" TAIL PIPE = 3.0 F.P.S.
FIND: DISCH. FROM 4" AIR LIFT = 117 G.P.M.
      PER CENT SUBM. = $\frac{12.5}{12.5 + 5.5}$ X 100 = 69.3.
      AIR REQ'D.
      = (AIR @ H = 5') $\frac{5.5 - 5.0}{7.0 - 5.0}$ (DIFF. AIR @ H = 5' B 7'.
      = 40 - 0.25(40 - 30)
      = 38 C.F.M. (FREE AIR).

K-6056-4

TYPICAL PERFORMANCE CURVES
FOR A
4-INCH AIR LIFT
CHICAGO PUMP
HYDRODYNAMICS DIVISION
FMC CORPORATION
JUNE 1960   A.A.L.

TYPICAL AIR LIFTS
FOR WHICH CURVES ARE APPLICABLE

© 1960 C.P.-F.M.C.

APPENDIX A-18   Water Hammer Nomograph

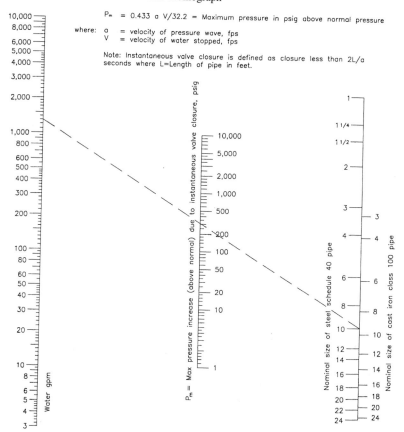

$P_m = 0.433\ a\ V/32.2 =$ Maximum pressure in psig above normal pressure

where: a = velocity of pressure wave, fps
V = velocity of water stopped, fps

Note: Instantaneous valve closure is defined as closure less than $2L/a$ seconds where L=Length of pipe in feet.

APPENDIX B
CONVERSION DATA

Appendix Number	Description
B-1	SI to English Length Conversion
B-2	SI Units and Conversion Factors
B-3	Conversion Factors for Commonly Used Wastewater Treatment Plant Design Parameters
B-4	Equivalent Temperature Readings for Fahrenheit and Celsius Scales
B-5	Conversion Table—Cu. Ft/sec to gpm to gal/24 hr

EXPLANATION OF APPENDICES

Appendix B-1, SI to English Length Conversion: This Appendix is a quick reference for the conversion of length in centimeters and meters to inches and feet.

Appendix B-2, SI Units and Conversion Factors: This Appendix lists English units in the first column which can be converted to SI units as listed in the second column and abbreviated in the third column by multiplying the English units by the factors in the fourth column.

Appendix B-3, Conversion Factors for Commonly Used Wastewater Treatment Plant Design Parameters: In this Appendix, the first column lists English units and the fourth column, SI units. To convert from English to SI, multiply the English units by the factors in the second column. To convert from SI units to English, multiply the SI units by the factors in the third column.

Appendix B-4, Equivalent Temperature Readings in Fahrenheit and Celsius Scales: This Appendix lists the conversion from degrees Fahrenheit to Celsius from $-273°C$ to $+121°C$. The conversion formula is listed in the Appendix.

Appendix B-5, Conversion Table—Cu. ft/sec to gpm to gal/24 hr: The first half of this Table lists Cu. Ft per second in uniform steps from 0.2 to 150 with equivalent gpm and gals/24 hr values. The second half of the table converts from gals/24 hrs in uniform steps between 100,000 and 1,000,000,000 to gpm and Cu. Ft/sec values.

APPENDIX B-1   SI to English Length
Conversion

Centimeters	Inches
2.54	1
5	2
15	6

Feet	
30.5	1
60	2

Meters	Feet
1	3
2	6
3	11
4	13
5	16
6	20

APPENDIX B-2 SI Units and Conversion Factors
Units underlined are those selected for common use

To convert from	to		Multiply by
LENGTH			
Feet	meter	m	0.3048
Inch	millimeter	mm	25.4
Microinches	micrometer	μm	0.0254
Statute miles	kilometer	km	1.609
AREA			
Square inches	square milimeter	mm^2	645.2
Square inches	square centimeter	cm^2	6.452
Square inches	square meter	m^2	0.000645
Square feet	square meter	m^2	0.09290
Acres	hectare	ha	0.4047
VOLUME			
Cubic inches	cubic millimeter	mm^3	16387.
Cubic inches	cubic centimeter	cm^3	16.387
Cubic inches	cubic meter	m^3	.00001639
Cubic feet	cubic meter	m^3	0.02832
Fluid ounce	milliliter	mL	29.57
Quarts (U.S.)	liter	L	0.9464
Gallons (U.S.)	liter	L	3.785
MASS			
Pounds	kilogram	kg	0.45359
Ton (short)	metric ton (†)	t	0.9072
Ton (long)	metric ton (†)	t	1.016
FORCE			
Pound force	newton	N	4.448
Kilogram force	newton	N	9.807

†Note: The unit tonne is used in place of metric ton in other countries using ISO symbols.

APPENDIX B-2   SI Units Conversion Factors (*Continued*)
Units underlined are those selected for common use

To convert from	to		Multiply by
PRESSURE, STRESS			
Pounds/square inch	pascal	Pa	6895.
Pounds/square inch	kilopascal	kPa	6.895
Pounds/square inch	megapascal	MPa	0.006895
Kilogram/square meter	pascal	Pa	9.807
Bar*	kilopascal	kPa	100.
Millibar*	pascal	Pa	100.
SPEED, VELOCITY			
Feet/second	meter per second	m/s	0.3048
Feet/minute	meter per second	m/s	0.00508
Miles/hour	kilometer per hour	km/h	1.609
ENERGY, WORK			
British Thermal Units BTU	joule	J	1055.
Foot pound force	joule	J	1.356
Calorie	joule	J	4.1868
POWER			
BTU/hour	watt	W	0.2931
BTU/second	watt	W	1055.
Horsepower	kilowatt	kW	0.746
TORQUE: BENDING MOMENT			
Pound feet	newton meter	N · m	1.356
Kilogram meter	newton meter	N · m	9.807
DENSITY, MASS/VOLUME			
Pound-mass/cubic foot	kilogram per cubic meter	kg/m³	16.018
FLOW RATE VOLUME			
Cubic feet/minute	cubic meter per minute	m³/min	0.02832
Gallons (U.S.)/minute	liter per minute	L/min	3.785
FLOW RATE MASS			
Pounds/minute	kilogram per minute	kg/min	0.4536

*Note: Some European countries have adopted the Bar for pressure units for its practical value and/or use in specialized fields.

APPENDIX B-2 SI Units and Conversion Factors (*Continued*)
Units underlined are those selected for common use

To convert from	to		Multiply by
WATER HARDNESS			
Grains/gallon (U.S.) (GPG)	<u>grams per liter</u>	g/L	0.01712
ACCELERATION			
Feet/second^2	meter per second^2	m/s^2	0.3048
Free fall, standard	meter per second^2	m/s^2	9.8067
ENERGY/AREA TIME			
BTU/feet^2 second	watt per meter^2	W/m^2	11348
BTU/feet^2 hour	watt per meter^2	W/m^2	3.1525
THERMAL CONDUCTIVITY			
BTU · inch/hour · feet^2 · deg F	watt per meter—kelvin	W/(m · K)	0.1442
THERMAL CONDUCTANCE			
BTU/hour · feet^2 · deg F	watt per meter^2—kelvin	W/(m^2 · K)	5.678
CAPACITY, DISPLACEMENT			
inches^3/revolution	liter per revolution	L/r	0.01639
inches^3/revolution	milliliter per revolution	mL/r	16.39
SPECIFIC ENERGY, LATENT HEAT			
BTU/pound	joule/kilogram	J/kg	2326
ENERGY DENSITY			
BTU/cubic foot	kilojoule/meter^3	kJ/m^3	37.25
SPECIFIC HEAT, SPECIFIC ENTROPY			
BTU/pound-deg F	joules/kilogram—kelvin	J/(Kg · K)	4184

APPENDIX B-3 Conversion Factors for Commonly Used Wastewater Treatment Plant
Design Parameters

U.S. units	To convert, multiply in direction shown by arrows $\longrightarrow$	$\longleftarrow$	SI units
acre/(Mgal/d)	0.1069	9.3536	ha/(10^3m^3/d)
Btu	1.0551	0.9478	kJ
Btu/lb	2.3241	0.4303	kJ/kg
Btu/ft$^2 \cdot$ F $\cdot$ h	5.6735	0.1763	W/m$^2 \cdot$ C
bu/acre $\cdot$ yr	2.4711	0.4047	bu/ha $\cdot$ yr
ft/h	0.3048	3.2808	m/h
ft/min	18.2880	0.0547	m/h
ft^2/capita	0.0929	10.7639	m^2/capita
ft^3/capita	0.0283	35.3147	m^3/capita
ft^3/gal	7.4805	0.1337	m^3/m^3
ft^2/ft $\cdot$ min	0.0929	10.7639	m^3/m $\cdot$ min
ft^3/lb	0.0624	16.0185	m^3/kg
ft^3/Mgal	7.04805×10^{-3}	133.6805	m^3/10^3m^3
ft^2/Mgal $\cdot$ d	407.4611	0.0025	m^2/10^3m$^3 \cdot$ d
ft^3/ft$^2 \cdot$ h	0.3048	3.2808	m^3/m$^2 \cdot$ h
ft^3/10^3 gal $\cdot$ min	7.04805×10^{-3}	133.6805	m^3/m$^3 \cdot$ min
ft^3/min	1.6990	0.5886	m^3/h
ft^3/s	2.8317×10^{-2}	35.3145	m^3/s
ft^3/10^3 ft$^3 \cdot$ min	0.001	1,000.0	m^3/m$^2 \cdot$ min
gal	3.7854	0.2642	L
gal/acre $\cdot$ d	0.0094	106.9064	m^3/ha $\cdot$ d
gal/ft $\cdot$ d	0.0124	80.5196	m^2/m $\cdot$ d
gal/ft$^2 \cdot$ d	0.0407	24.5424	m^3/m$^2 \cdot$ d
gal/ft$^2 \cdot$ d	0.0017	589.0173	m^3/m$^2 \cdot$ h
gal/ft$^2 \cdot$ d	0.0283	35.3420	L/m$^2 \cdot$ min
gal/ft$^2 \cdot$ d	40.7458	$2,4542 \times 10^{-2}$	L/m$^2 \cdot$ d
gal/ft$^2 \cdot$ min	2.4448	0.4090	m/h
gal/ft$^2 \cdot$ min	40.7458	0.0245	L/m$^2 \cdot$ min
gal/ft$^2 \cdot$ min	58.6740	0.0170	m^3/m$^2 \cdot$ d
gal/min $\cdot$ ft	12.4193	8.052×10^{-2}	L/min $\cdot$ m
hp/10^3 gal	0.1970	5.0763	kW/m^3
hp/10^3 ft^3	26.3342	0.0380	kW/10^3 m^3
in	25.4	3.9370×10^{-2}	mm
in Hg (60 F)	3.3768	0.2961	kPa Hg (60°F)
lb	0.4536	2.2046	kg
lb/acre	1.1209	0.8922	kg/ha
lb/10^3 gal	0.1198	8.3452	kg/m^3
lb/hp $\cdot$ h	0.6083	1.6440	kg/kW $\cdot$ h
lb/Mgal	0.1198	8.3454	g/m^3
lb/Mgal	1.1983×10^{-3}	8345.4	kg/m^3
lb/ft^2	4.8824	0.2048	kg/m^2
lb/in^2 (gage)	6.8948	0.1450	kPa (gage)
lb/ft$^2 \cdot$ h	16.0185	0.0624	kg/m$^3 \cdot$ h
lb/10^3 ft$^3 \cdot$ d	0.0160	62.4280	kg/m$^3 \cdot$ d
lb/ton	0.5000	2.0000	kg/tonne
Mgal/acre $\cdot$ d	0.9354	1.0691	m^3/m$^2 \cdot$ d
Mgal/d	3.7854×10^3	0.264×10^{-3}	m^3/d
Mgal/d	4.3813×10^{-2}	22.8245	m^3/s
min/in	3.9370	0.2540	min/10^2 mm
tons/acre	2.2417	0.4461	Mg/ha
yd^3	0.7646	1.3079	m^3

APPENDIX B-4 Equivalent Temperature Readings for Fahrenheit and Celsius Scales
$°F = 9/5 \ °C + 32°$

$°C = 5/9 \ (°F − 32°)$

°Fahrenheit	°Celsius	°Fahrenheit	°Celsius	°Fahrenheit	°Celsius	°Fahrenheit	°Celsius
−459.4	−273.	−21.	−29.4	17.6	−8.	56.	13.3
−436.	−260.	−20.2	−29.	18.	−7.8	57.	13.9
−418.	−250.	−20.	−28.9	19.	−7.2	57.2	14.
−400.	−240.	−19.	−28.3	19.4	−7.	58.	14.4
−382.	−230.	−18.4	−28.	20.	−6.7	59.	15.
−364.	−220.	−18.	−27.8	21.	−6.1	60.	15.6
−346.	−210.	−17.	−27.2	21.2	−6.	60.8	16.
−328.	−200.	−16.6	−27.	22.	−5.6	61.	16.1
−310.	−190.	−16.	−26.7	23.	−5.	62.	16.7
−292.	−180.	−15.	−26.1	24.	−4.4	62.6	17.
−274.	−170.	−14.8	−26.	24.8	−5.	63.	17.2
−256.	−160.	−14.	−25.6	25.	−3.9	64.	17.8
−238.	−150.	−13.	−25.	26.	−3.3	64.4	18.
−220.	−140.	−12.0	−24.4	26.6	−3.	65.	18.3
−202.	−130.	−11.2	−24.	27.	−2.8	66.	18.9
−184.	−120.	−11.	−23.9	28.	−2.2	66.2	19.
−166.	−110.	−10.0	−23.3	28.4	−2.	67.	19.4
−148.	−100.	−9.4	−23.	29.	−1.7	68.	20.
−139.	−95.	−9.	−23.8	30.	−1.1	69.	20.6
−130.	−90.	−8.	−22.2	30.2	−1.	69.8	21.
−121.	−85.	−7.6	−22.	31.°	−0.6	70.	21.1
−112.	−80.	−7.	−21.7	32.	0.	71.	21.7
−103.	−75.	−6.	−21.1	33.	+0.6	71.6	22.
−94.	−70.	−5.8	−21.	33.8	1.	72.	22.2
−85.	−65.	−5.	−20.6	34.	1.1	73.	22.8
−76.	−60.	−4.	−20.	35.	1.7	73.4	23.
−67.	−55.	−3.	−19.4	35.6	2.	74.	23.3
−58.	−50.	−2.2	−19.	36.	2.2	75.	23.9
−49.	−45.	−2.	−18.9	37.	2.8	75.2	24.
−40.	−40.	−1.	−18.3	37.4	3.	76.	24.4
−39.	−39.4	−0.4	−18.	38.	3.3	77.	25.
−38.2	−39.	0.	−17.8	39.	3.9	78.	25.6
−38.	−38.9	+1.	−17.2	39.2	4.	78.8	26.
−37.	−38.3	1.4	−17.	40.	4.4	79.	26.1
−36.4	−38.	2.	−16.7	41.	5.	80.	26.7
−36.	−37.8	−3.	−16.1	42.	5.6	80.6	27.
−35.	−37.2	3.2	−16.	42.8	6.	81.	27.2
−34.6	−37.	4.	−15.6	43.	6.1	82.	27.8
−34.	−36.7	5.	−15.	44.	6.7	82.4	28.
−33.	−36.1	6.	−14.4	44.6	7.	83.	28.3

APPENDIX B-4 Equivalent Temperature Readings for Fahrenheit and Celsius Scales (*Continued*)
°F = 9/5 °C + 32°   °C = 5/9 (°F − 32°)

°Fahrenheit	°Celsius	°Fahrenheit	°Celsius	°Fahrenheit	°Celsius	°Fahrenheit	°Celsius
−32.8	−36.	−6.8	−14.	45.	7.2	84.	28.9
−32.	−35.6	7.	−13.9	46.	7.8	84.2	29.
−31.	−35.	8.	−13.3	46.4	8.	85.	29.4
−30.	−34.4	8.6	−13.	47.	8.3	86.	30.
−29.2	−34.	9.	−12.8	48.	8.9	87.	30.6
−29.	−33.9	10.	−12.2	48.2	9.	87.8	31.
−28.	−33.3	10.4	−12.	49.	9.4	88.0	31.1
−27.4	−33.	11.	−11.7	50.	10.0	89.	31.7
−27.	−32.8	12.	−11.1	51.	10.6	89.6	32.
−26.	−32.2	12.2	−11.	51.8	11.	90.	32.2
−25.6	−32.	13.	−10.6	52.	11.1	91.	32.8
−25.	−32.9	14.	−10.	53.	11.7	91.4	33.
−24.	−31.7	15.	−9.4	53.6	12.	92.	33.3
−23.8	−31.1	15.8	9.	54.	12.2	93.	33.9
−23.	−30.5	16.	−8.9	55.	12.8	93.2	34.
−22.	−30.	17.	−8.3	55.4	13.	94.	34.4
95.	35.	134.	56.7	172.4	78.	211.	99.4
96.	35.6	134.6	57.0	173.	78.3	212.	100.
96.8	36.	135.	57.2	174.	78.9	213.	100.6
97.	36.1	136.	57.8	174.2	79.	213.8	101.
98.	36.7	136.4	58.	175.	79.4	214.	101.1
98.6	37.	137.	58.3	176.	80.	215.	101.7
99.	37.2	138.	58.9	177.	80.6	215.6	102.
100.	37.8	138.2	59.0	177.8	81.	216.	102.2
100.4	38.	139.	59.4	178.	81.1	217.	102.8
101.	38.3	140.	60.	179.	81.7	217.4	103.
102.	38.9	141.	60.6	179.6	82.	218.	103.3
102.2	39.	141.8	61.	180.	82.2	219.	103.9
103.	39.4	142.	61.1	181.	82.8	219.2	104.
104.	40.	143.	61.7	181.4	83.	220.	104.4
105.	40.6	143.6	62.	182.	83.3	221.	105.
105.8	41.	144.	62.2	183.	83.9	222.	105.6
106.	41.1	145.	62.8	183.2	84.	222.8	106.
107.	41.7	145.4	63.	184.	84.4	223.	106.1
107.6	42.	146.	63.3	185.	85.	224.	106.7
108.	42.2	147.	63.9	186.	85.6	224.6	107.
109.	42.8	147.2	64.	186.8	86.	225.	107.2
109.4	43.	148.	64.4	187.	86.1	226.	107.8
110.	43.3	149.	65.	188.	86.7	226.4	108.
111.	43.9	150.	65.6	188.6	87.	227.	108.3
111.2	44.	150.8	66.	189.	87.2	228.	108.9

APPENDIX B-4 Equivalent Temperature Readings for Fahrenheit and Celsius Scales (*Continued*)
°F = 9/5 °C + 32°

°C = 5/9 (°F − 32°)

°Fahrenheit	°Celsius	°Fahrenheit	°Celsius	°Fahrenheit	°Celsius	°Fahrenheit	°Celsius
112.	44.4	151.	66.1	190.	87.8	228.2	109.
113.	45.	152.	66.7	190.4	88.	229.	109.4
114.	45.6	152.6	67.	191.	88.3	230.	110.
114.8	46.	153.	67.2	192.	88.9	231.	110.6
115.	46.1	154.	67.8	192.2	89.	231.8	111.
116.	46.7	154.4	68.	193.	89.4	232.	111.1
116.6	47.	155.	68.3	194.	90.	233.	111.7
117.	47.2	156.	68.9	195.	90.6	233.6	112.
118.	47.8	156.2	69.	195.8	91.	234.	112.3
118.4	48.	157.	69.4	196.	91.1	235.	112.8
119.	48.3	158.	70.	197.	91.7	235.4	113.
120.	48.9	159.	70.6	197.6	92.	236.	113.3
120.2	49.	159.8	71.	198.	92.2	237.	113.9
121.	49.4	160.	71.1	199.	92.8	237.2	114.
122.	50.	161.	71.7	199.4	93.	238.	114.4
123.	50.6	161.6	72.	200.	93.3	239.	115.
123.8	51.	162.	72.2	201.	93.9	240.	115.6
124.	51.1	163.	72.8	201.2	94.	240.8	116.
125.	51.7	163.4	73.	202.	94.4	241.	116.1
125.6	52.	164.	73.3	203.	95.	242.	116.7
126.	52.2	165.	73.9	204.	95.6	242.6	117.
127.	52.8	165.2	74.	204.8	96.	243.	117.2
127.4	53.	166.	74.4	205.	96.1	244.	117.8
128.	53.3	167.	75.	206.	96.7	244.4	118.
129.	53.9	168.	75.6	206.6	97.	245.	118.3
129.2	54.	168.8	76.	207.	97.2	246.	118.9
130.	54.4	169.	76.1	208.	97.8	246.2	119.
131.	55.	170.	76.7	208.4	98.	247.	119.4
132.	55.6	170.6	77.	209.	98.3	248.	120.
132.8	56.	171.	77.2	210.	98.9	249.	120.6
133.	56.1	172.	77.8	210.2	99.	249.8	121.

APPENDIX B-5   Conversion Table—Cu. Ft/sec to gpm to gal/24 hr

Cu. ft. per sec.	to	Gallons per minute	to	Gallons per 24 hrs.	Gallons per 24 hrs.	to	Gallons per minute	to	Cu. ft. per sec.
0.2		90		129,254	100,000		69		0.15
0.4		180		258,508	125,000		87		0.19
0.6		269		387,763	200,000		139		0.31
0.8		359		517,017	400,000		278		0.62
1.0		449		646,272	500,000		347		0.77
1.2		539		775,526	600,000		417		0.98
1.4		628		904,780	700,000		486		1.08
1.6		718		1,034,036	800,000		556		1.24
1.8		808		1,163,290	900,000		625		1.89
2.0		898		1,292,544	1,000,000		694		1.55
2.2		987		1,421,798	2,000,000		1,389		3.09
2.4		1,077		1,551,053	3,000,000		2,088		4.64
2.6		1,167		1,680,307	4,000,000		2,778		6.19
2.8		1,257		1,809,562	5,000,000		3,472		7.74
3.0		1,346		1,938,816	6,000,000		4,167		9.28
3.2		1,436		2,068,070	7,000,000		4,861		10.83
3.4		1,526		2,197,325	8,000,000		5,556		12.88
3.6		1,616		2,826,579	9,000,000		6,250		18.92
3.8		1,705		2,455,834	10,000,000		6,944		18.47
4.0		1,795		2,585,088	12,000,000		8,888		18.56
4.2		1,885		2,714,342	12,500,000		8,680		19.34
4.4		1,975		1,843,597	14,000,000		9,722		21.65
4.6		2,068		2,972,851	15,000,000		10,417		23.20
4.8		2,164		3,102,106	16,000,000		11,111		24.75
5.0		2,244		3,231,360	18,000,000		12,500		26.85
10.0		4,488		6,462,720	20,000,000		13,889		30.94
20.0		8,987		12,925,440	24,000,000		17,861		38.68
30.0		13,464		19,388,160	30,000,000		20,833		46.41
40.0		17,952		25,850,880	40,000,000		27,778		61.88
50.0		22,440		32,313,600	50,000,000		34,722		77.35
60.0		26,928		38,776,320	60,000,000		41,667		92.82
70.0		31,416		45,239,040	70,000,000		48,611		108.29
75.0		33,660		48,470,000	75,000,000		52,083		116.04
80.0		35,904		51,701,760	80,000,000		55,556		123.76
90.0		40,392		58,164,480	90,000,000		62,500		129.23
100.0		44,880		64,627,200	100,000,000		69,444		154.72
101.0		45,329		65,273,472	125,000,000		86,805		193.40
102.0		45,778		65,919,744	150,000,000		104,167		232.08
103.0		46,226		66,866,016	175,000,000		121,528		270.76
104.0		46,678		67,212,288	200,000,000		138,889		309.44

APPENDIX B-5 Conversion Table—Cu. Ft/sec to gpm to gal/24 hr (*Continued*)

Cu. ft. per sec.	to	Gallons per minute	to	Gallons per 24 hrs.	Gallons per 24 hrs.	to	Gallons per minute	to	Cu. ft. per sec.
105.0		47,124		67,858,560	225,000,000		156,250		348.12
106.0		47,572		68,504,832	250,000,000		178,611		386.80
107.0		48,022		69,181,104	300,000,000		208,333		464.16
108.0		48,470		69,797,876	400,000,000		277,778		618.88
109.0		48,919		70,443,648	500,000,000		347,220		772.60
110.0		49,368		71,089,920	600,000,000		416,664		928.82
120.0		53,856		77,552,640	700,000,000		486,108		1,088.04
125.0		56,100		80,784,000	750,000,000		520,328		1,160.40
130.0		58,344		84,015,360	800,000,000		555,552		1,237.76
140.0		62,832		90,478,080	900,000,000		624,996		1,392.48
150.0		67,820		96,940,800	1,000,000,000		694,440		1,547.20

NOTE: gpm and gal per 24 hr given to the nearest whole number. The value 7.48 gallons equals 1 cu ft is used in calculating above table.

$$\text{cfs} \times 646{,}272 = \text{gpd}$$

APPENDIX C
WATER CHARACTERISTICS

Appendix Number	Description
C-1	Pressure, Lb per Sq. In. to Feet (head) of Water
C-2	Pressure, Feet (head) of Water to Lb per Sq. In.
C-3	Physical Properties of Water (English Units)
C-4	Physical Properties of Water (SI Units)
C-5	Dissolved Oxygen Concentration in Water as a Function of Temperature and Salinity
C-6	Dissolved Oxygen Concentration in Water as a Function of Temperature and Barometric Pressure

EXPLANATION OF APPENDICES

Appendix C-1, Pressure, Lb per Sq. In. to Feet (head) of Water: This Appendix converts water pressures in psi from 1 to 3000 to feet of head.

Appendix C-2, Pressure, Feet (head) of Water to Lb per Sq. In.: This Appendix converts water pressures in feet of head from 1 to 3000, to psi.

Appendix C-3, Physical Properties of Water (English Units): This Appendix lists weights, densities, elasticities, viscosities, surface tensions and vapor pressures of water from 32° to 212° F.

Appendix C-4, Physical Properties of Water (SI Units): This Appendix lists weights, densities, elasticities, viscosities, surface tensions and vapor pressures of water from 0° to 100° C.

Appendix C-5, Dissolved Oxygen Concentration in Water as a Function of Temperature and Salinity: This Appendix lists D.O. concentrations in mg/L for various salinities for water temperatures from 0° to 40° C.

Appendix C-6, Dissolved Oxygen Concentration in Water as a Function of Temperature and Barometric Pressure: This Appendix lists D.O. concentrations in mg/L for various barometric pressures for water temperatures from 0° to 40° C.

APPENDIX C-1 Pressure, Lb per Sq. In. to Feet (head) of Water

Based on water at its greatest density (39.2°F)

Pressure pounds per square inch	Feet head	Pressure pounds per square inch	Feet head	Pressure pounds per square inch	Feet head	Pressure pounds per square inch	Feet head	Pressure pounds per square inch	Feet head	Pressure pounds per square inch	Feet head	Pressure pounds per square inch	Feet head
1	2.31	53	122.43	105	242.55	157	362.67	209	482.79	261	602.91	365	843.15
2	4.62	54	124.74	106	244.86	158	364.98	210	485.10	262	605.22	370	854.70
3	6.93	55	127.05	107	247.17	159	367.29	211	487.41	263	607.53	375	866.25
4	9.23	56	129.36	108	249.48	160	369.60	212	489.72	264	609.84	380	877.80
5	11.55	57	131.67	109	251.79	161	371.91	213	492.03	265	612.15	385	899.35
6	13.86	58	133.98	110	254.10	162	374.22	214	494.34	266	614.46	390	900.90
7	16.17	59	136.29	111	256.41	163	375.53	215	496.65	267	616.77	395	912.45
8	18.48	60	138.60	112	258.72	164	378.84	216	498.96	268	619.08	400	924.00
9	20.79	61	140.91	113	261.03	165	381.15	217	501.27	269	621.39	405	931.55
10	23.10	62	143.22	114	263.34	166	383.46	218	503.58	270	623.70	410	947.10
11	25.41	63	145.53	115	265.65	167	385.77	219	505.89	271	626.01	415	958.65
12	27.72	64	147.84	116	267.96	168	388.08	220	508.20	272	628.32	420	970.20
13	30.03	65	150.15	117	270.27	169	390.39	221	510.51	273	630.03	425	981.75
14	32.34	66	152.46	118	272.58	170	392.70	222	512.82	274	632.94	430	993.30
15	34.65	67	154.77	119	274.89	171	395.01	223	515.13	275	635.25	435	1004.85
16	36.96	68	157.08	120	277.20	172	397.32	224	517.44	276	637.56	440	1016.40
17	39.27	69	159.39	121	279.51	173	399.63	225	519.75	277	639.87	445	1027.95
18	41.58	70	161.70	122	281.82	174	410.94	226	522.06	278	642.18	450	1039.50
19	43.89	71	164.01	123	284.13	175	404.25	227	524.37	279	644.49	455	1051.05
20	46.20	72	166.32	124	286.44	176	406.56	228	526.68	280	646.80	460	1062.60
21	48.51	73	168.63	125	288.75	177	408.87	229	528.99	281	649.11	465	1074.15
22	50.82	74	170.94	126	291.06	178	411.18	230	531.30	282	651.42	470	1085.70
23	53.13	75	173.25	127	293.37	179	413.49	231	533.61	283	653.73	475	1097.25
24	55.44	76	175.56	128	295.68	180	415.80	232	535.92	284	656.04	450	1108.60
25	57.75	77	177.87	129	297.99	181	418.11	233	538.23	285	658.35	485	1120.35
26	60.06	78	180.18	130	300.30	182	420.42	234	540.54	286	660.66	490	1131.90
27	62.37	79	182.49	131	302.61	183	422.73	235	542.85	287	662.97	495	1143.45
28	64.68	80	184.80	132	304.92	184	425.04	236	545.16	288	665.28	500	1155.00
29	66.99	81	187.11	133	307.23	185	427.35	237	547.47	289	667.59	525	1212.75
30	69.30	82	189.42	134	309.54	186	429.66	238	549.78	290	669.90	550	1270.50
31	71.71	83	191.73	135	311.85	187	431.97	239	552.09	291	672.21	575	1328.25
32	73.92	84	194.04	136	314.16	188	434.28	240	554.40	292	674.52	600	1386.00
33	76.23	85	196.35	137	316.47	189	436.59	241	556.71	293	676.83	625	1443.75
34	78.54	86	198.66	138	318.78	190	438.90	242	559.02	294	679.14	650	1501.50
35	80.85	87	200.97	139	321.09	191	441.21	243	561.33	295	681.45	675	1559.25
36	83.16	88	203.28	140	323.40	192	443.52	244	563.64	296	683.76	700	1617.00
37	85.47	89	205.59	141	325.71	193	445.83	245	565.95	297	686.07	725	1674.75
38	87.78	90	207.90	142	328.02	194	448.14	246	568.62	298	688.38	750	1732.50
39	90.09	91	210.21	143	330.33	195	450.45	247	570.57	299	690.69	775	1740.25
40	92.40	92	212.52	144	332.64	196	452.76	248	572.88	300	693.00	800	1848.00
41	94.71	93	214.83	145	334.95	197	445.07	249	575.19	305	704.55	825	1905.75
42	97.02	94	217.14	146	337.26	198	457.38	250	577.50	310	716.10	850	1963.50
43	99.33	95	219.45	147	339.57	199	459.69	251	579.81	315	727.65	875	2021.25
44	101.64	96	221.76	148	341.68	200	462.00	252	582.12	320	739.20	900	2079.00
45	103.95	97	224.07	149	344.19	201	464.31	253	584.43	325	750.75	925	2136.75
45	106.26	98	226.38	150	346.50	202	466.62	254	586.74	330	762.30	950	2194.50
47	108.57	99	228.69	151	348.81	203	468.93	255	589.05	335	773.85	975	2252.25
48	110.88	100	231.00	152	351.12	204	471.24	256	591.36	340	785.40	1000	2310.00
49	113.19	101	233.31	153	353.43	305	473.55	257	593.67	345	796.95	1500	3465
50	115.50	102	235.62	154	355.74	206	475.86	258	595.98	350	808.50	2000	4620
51	117.81	103	237.93	155	358.05	207	478.17	259	598.29	355	820.15	3000	6930
52	120.12	104	240.24	156	360.36	208	480.48	260	600.60	360	831.60		

APPENDIX C-2 Pressure, Feet (head) of Water to Lb per Sq. In.

Based on water at its greatest density (39.2°F)

Feet head	Pressure pounds per square inch	Feet head	Pressure pounds per square inch	Feet head	Pressure pounds per square inch	Feet head	Pressure pounds per square inch	Feet head	Pressure pounds per square inch	Feet head	Pressure pounds per square inch
1	0.43	54	23.39	107	46.34	160	69.31	213	92.20	285	123.45
2	0.86	55	23.82	108	46.78	161	69.74	214	92.69	290	125.62
3	1.30	56	24.26	109	47.21	162	70.17	215	93.13	295	127.78
4	1.73	57	24.69	110	47.64	163	70.61	216	93.56	300	129.95
5	2.16	58	25.12	111	48.08	164	71.04	217	93.99	305	132.12
6	2.59	59	25.55	112	48.51	165	71.47	218	94.43	310	134.28
7	3.03	60	25.99	113	48.94	166	71.91	219	94.86	315	136.46
8	3.46	61	26.42	114	49.38	167	72.34	220	95.30	320	138.62
9	3.89	62	26.85	115	49.81	168	72.77	221	95.73	325	140.79
10	4.33	63	27.29	116	50.24	169	73.20	222	96.16	330	142.95
11	4.76	64	27.72	117	50.68	170	73.64	223	96.60	335	145.12
12	5.20	65	28.15	118	51.11	171	74.07	224	97.03	340	147.28
13	5.63	66	28.58	119	51.54	172	74.50	225	97.46	345	149.45
14	6.06	67	29.02	120	51.98	173	74.94	226	97.90	350	151.61
15	6.49	68	29.45	121	52.41	174	75.37	227	98.33	355	153.78
16	6.93	69	29.88	122	52.84	175	75.80	228	98.76	360	155.94
17	7.36	70	30.32	123	53.28	176	76.23	229	99.20	365	158.10
18	7.79	71	30.75	124	53.71	177	76.67	230	99.63	370	160.27
19	8.22	72	31.18	125	54.15	178	77.10	231	100.0	375	162.45
20	8.66	73	31.62	126	54.58	179	77.53	232	100.49	380	164.61
21	9.09	74	32.05	127	55.01	180	77.97	233	100.93	385	166.78
22	9.53	75	32.48	128	55.44	181	78.40	234	101.36	390	168.94
23	9.96	76	32.92	129	55.88	182	78.84	235	101.70	395	171.11
24	10.39	77	33.35	130	56.31	183	79.27	236	102.23	400	173.27
25	10.82	78	33.78	131	56.74	184	79.70	237	102.66	425	134.10
26	11.26	79	34.21	132	57.18	185	80.14	238	103.09	450	195.0
27	11.69	80	34.65	133	57.61	186	80.57	239	103.53	475	205.77
28	12.12	81	35.08	134	58.04	187	81.0	240	103.96	500	216.58
29	12.55	82	35.52	135	58.48	188	81.43	241	104.39	525	227.42
30	12.99	83	35.95	136	58.91	189	81.87	242	104.83	550	238.25
31	13.42	84	36.39	137	59.34	190	82.30	243	105.26	575	249.09
32	13.86	85	36.82	138	59.77	191	82.73	244	105.69	600	259.90
33	14.29	86	37.25	139	60.21	192	83.17	245	106.13	625	270.73
34	14.72	87	37.68	140	60.64	193	83.60	246	106.56	650	281.56
35	15.16	88	38.12	141	61.07	194	84.03	247	106.99	675	292.40
36	15.59	89	38.55	142	61.51	195	84.47	248	107.43	700	303.22
37	16.02	90	38.98	143	61.94	196	84.90	249	107.86	725	314.05
38	16.45	91	39.42	144	62.37	197	85.33	250	108.29	750	324.88
39	16.89	92	39.85	145	62.81	198	85.76	251	108.73	775	335.72
40	17.32	93	40.28	146	63.24	199	86.20	252	109.16	800	346.54
41	17.75	94	40.72	147	63.67	200	86.63	253	109.59	825	357.37
42	18.19	95	41.15	148	64.10	201	87.07	254	110.03	850	368.20
43	18.62	96	41.58	149	64.54	202	87.60	255	110.46	875	379.03
44	19.05	97	42.01	150	64.97	203	87.93	256	110.89	900	389.86
45	19.49	98	42.45	151	65.40	204	88.36	257	111.32	925	400.70

APPENDIX C-2 Pressure, Feet (head) of Water to Lb per Sq. In. (*Continued*)

Based on water at its greatest density (39.2°F)

Feet head	Pressure pounds per square inch	Feet head	Pressure pounds per square inch	Feet head	Pressure pounds per square inch	Feet head	Pressure pounds per square inch	Feet head	Pressure pounds per square inch	Feet head	Pressure pounds per square inch
46	19.92	99	42.88	152	65.84	205	88.80	258	111.76	960	411.54
47	20.35	100	43.31	153	66.27	206	89.21	259	112.19	975	422.35
48	20.79	101	43.76	154	66.70	207	89.66	260	112.62	1000	433.18
49	21.22	102	44.18	155	67.14	208	90.10	261	113.06	1500	649.7
50	21.65	103	44.61	156	67.57	209	90.53	262	113.49	2000	866.3
51	22.09	104	45.05	157	68.0	210	90.96	270	116.96	3000	1,299.5
52	22.52	105	45.48	158	68.43	211	91.39	275	119.12		
53	22.95	106	45.91	159	68.87	212	91.83	280	121.29		

APPENDIX C-3 Physical Properties of Water (English Units)

Temperature, °F	Specific weight, γ, lb/ft^3	Density,[b] ρ, slug/ft^3	Modulus of elasticity,[b] $E/10^3$, lb$_f$/in^2	Dynamic viscosity, $\mu \times 10^5$, lb $\cdot$ s/ft^2	Kinematic viscosity, $\nu \times 10^5$, ft^2/s	Surface tension,[c] σ, lb/ft	Vapor pressure, ρ_v, lb$_f$/in^2
32	62.42	1.940	287	3.746	1.931	0.00518	0.09
40	62.43	1.940	296	3.229	1.664	0.00614	0.12
50	62.41	1.940	305	2.735	1.410	0.00509	0.18
60	62.37	1.938	313	2.359	1.271	0.00504	0.26
70	62.30	1.936	319	2.050	1.059	0.00498	0.36
80	62.22	1.934	324	1.799	0.930	0.00492	0.51
90	62.11	1.931	328	1.595	0.826	0.00486	0.70
100	62.00	1.927	331	1.424	0.739	0.00480	0.95
110	61.86	1.923	332	1.284	0.667	0.00473	1.27
120	61.71	1.918	332	1.168	0.609	0.00467	1.69
130	61.55	1.913	331	1.069	0.558	0.00460	2.22
140	61.38	1.908	330	0.981	0.514	0.00454	2.89
150	61.20	1.902	328	0.905	0.476	0.00447	3.72
160	61.00	1.896	326	0.838	0.442	0.00441	4.74
170	60.80	1.890	322	0.780	0.413	0.00434	5.99
180	60.58	1.883	318	0.726	0.385	0.00427	7.51
190	60.36	1.876	313	0.678	0.362	0.00420	9.34
200	60.12	1.868	308	0.637	0.341	0.00413	11.52
212	59.83	1.860	300	0.593	0.319	0.00404	14.70

[a] Adapted from "Hydraulic Modes," Manual of Engineering Practice No. 25, American Society of Civil Engineers.
[b] At atmospheric pressure.
[c] In contact with air.

APPENDIX C-4   Physical Properties of Water (SI Units)

Temperature, °F	Specific weight, γ, kN/ m^3	Density,[b] g/cm^3	Modulus of elasticity,[b] $E/10^6$, kN/m^2	Viscosity poises of g/cm.s	Kinematic viscosity, $v \times 10^6$, m^2/s	Surface tension,[c] σ, N/m	Vapor pressure, ρ_v, kN/m^2
0	9.805	0.99987	1.98	179.21	1.785	0.0765	0.61
5	9.807	0.99999	2.05	151.88	1.519	0.0749	0.87
10	9.804	0.99973	2.10	130.77	1.306	0.0742	1.23
15	9.798	0.99913	2.15	114.04	1.139	0.0735	1.70
20	9.789	0.99823	2.17	100.50	1.003	0.0728	2.34
25	9.777	0.99707	2.22	89.37	0.893	0.0720	3.17
30	9.764	0.99567	2.25	80.07	0.800	0.0712	4.24
40	9.730	0.99224	2.28	65.60	0.658	0.0696	7.38
50	9.689	0.98807	2.29	54.94	0.553	0.0679	12.33
60	9.642	0.98324	2.28	46.88	0.474	0.0662	19.92
70	9.589	0.97781	2.25	40.61	0.413	0.0644	31.16
80	9.530	0.97183	2.20	35.65	0.364	0.0626	47.34
90	9.466	0.96534	2.14	31.65	0.326	0.0608	70.10
100	9.399	0.95838	2.07	28.38	0.294	0.0589	101.33

[a]Adapted from "Hydraulic Modes," Manual of Engineering Practice No. 25, American Society of Civil Engineers.
[b]At atmospheric pressure.
[c]In contact with air.

APPENDIX C-5 Dissolved Oxygen Concentration in Water as a Function of Temperature and Salinity

Dissolved-oxygen concentration in water as a function of temperature and salinity (barometric pressure = 760 mm Hg)[a]

Temp, °C	Dissolved-oxygen concentration, mg/L									
	Salinity, parts per thousand									
	0	5	10	15	20	25	30	35	40	45
0	14.60	14.11	13.64	13.18	12.74	12.31	11.90	11.50	11.11	10.74
1	14.20	13.73	13.27	12.83	12.40	11.98	11.58	11.20	10.83	10.46
2	13.81	13.36	12.91	12.49	12.07	11.67	11.29	10.91	10.55	10.20
3	13.45	13.00	12.58	12.16	11.76	11.38	11.00	10.64	10.29	9.95
4	13.09	12.67	12.25	11.85	11.47	11.09	10.73	10.38	10.04	9.71
5	12.76	12.34	11.94	11.56	11.18	10.82	10.47	10.13	9.80	9.48
6	12.44	12.04	11.65	11.27	10.91	10.56	10.22	9.89	9.57	9.27
7	12.13	11.74	11.37	11.00	10.65	10.31	9.98	9.66	9.35	9.06
8	11.83	11.46	11.09	10.74	10.40	10.07	9.75	9.44	9.14	8.85
9	11.55	11.19	10.83	10.49	10.16	9.84	9.53	9.23	8.94	8.66
10	11.28	10.92	10.58	10.25	9.93	9.62	9.32	9.03	8.75	8.47
11	11.02	10.67	10.34	10.02	9.71	9.41	9.12	8.83	8.56	8.30
12	10.77	10.43	10.11	9.80	9.50	9.21	8.92	8.65	8.38	8.12
13	10.53	10.20	9.89	9.59	9.30	9.01	8.74	8.47	8.21	7.96
14	10.29	9.98	9.68	9.38	9.10	8.82	8.55	8.30	8.04	7.80
15	10.07	9.77	9.47	9.19	8.91	8.64	8.38	8.13	7.88	7.65
16	9.86	9.56	9.28	9.00	8.73	8.47	8.21	7.97	7.73	7.50
17	9.65	9.36	9.09	8.82	8.55	8.30	8.05	7.81	7.58	7.36
18	9.45	9.17	8.90	8.64	8.39	8.14	7.90	7.66	7.44	7.22
19	9.26	8.99	8.73	8.47	8.22	7.98	7.75	7.52	7.30	7.09
20	9.08	8.81	8.56	8.31	8.07	7.83	7.60	7.38	7.17	6.96
21	8.90	8.64	8.39	8.15	7.91	7.69	7.46	7.25	7.04	6.84
22	8.73	8.48	8.23	8.00	7.77	7.54	7.33	7.12	6.91	6.72
23	8.56	8.32	8.08	7.85	7.63	7.41	7.20	6.99	6.79	6.60
24	8.40	8.16	7.93	7.71	7.49	7.28	7.07	6.87	6.68	6.49
25	8.24	8.01	7.79	7.57	7.36	7.15	6.95	6.75	6.56	6.38
26	8.09	7.87	7.65	7.44	7.23	7.03	6.83	6.64	6.46	6.28
27	7.95	7.73	7.51	7.31	7.10	6.91	6.72	6.53	6.35	6.17
28	7.81	7.59	7.38	7.18	6.98	6.79	6.61	6.42	6.25	6.08
29	7.67	7.46	7.26	7.06	6.87	6.68	6.50	6.32	6.15	5.98
30	7.54	7.33	7.14	6.94	6.75	6.57	6.39	6.22	6.05	5.89
31	7.41	7.21	7.02	6.83	6.65	6.47	6.29	6.12	5.96	5.80
32	7.29	7.09	6.90	6.72	6.54	6.36	6.19	6.03	5.87	5.71
33	7.17	6.98	6.79	6.61	6.44	6.26	6.10	5.94	5.78	5.63
34	7.05	6.86	6.68	6.51	6.33	6.17	6.01	5.85	5.69	5.54
35	6.93	6.75	6.58	6.40	6.24	6.07	5.92	5.76	5.61	5.46
36	6.82	6.65	6.47	6.31	6.14	5.98	5.83	5.68	5.53	5.39
37	6.72	6.54	6.37	6.21	6.05	5.89	5.74	5.59	5.45	5.31
38	6.61	6.44	6.28	6.12	5.96	5.81	5.66	5.51	5.37	5.24
39	6.51	6.34	6.18	6.03	5.87	5.72	5.58	5.44	5.30	5.16
40	6.41	6.25	6.09	5.94	5.79	5.64	5.50	5.36	5.22	5.09

[a] From Colt, J.: "Computation of Dissolved Gas Concentrations in Water as Functions of Temperature, Salinity, and Pressure," *American Fisheries Society Special Publication 14*, Bethesda, MD, 1984.

APPENDIX C-6 Dissolved Oxygen Concentration in Water as a Function of Temperature and Barometric Pressure
Dissolved-oxygen concentration in water as a function of temperature and barometric pressure (salinity = 0 ppt)[a]

Temp, °C	Dissolved-oxygen concentration, mg/L									
	Barometric pressure, millimeters of mercury									
	735	740	745	750	755	760	765	770	775	780
0	14.12	14.22	14.31	14.41	14.51	14.60	14.70	14.80	14.89	14.99
1	13.73	13.82	13.92	14.01	14.10	14.20	14.29	14.39	14.48	14.57
2	13.36	13.45	13.54	13.63	13.72	13.81	13.90	14.00	14.09	14.18
3	13.00	13.09	13.18	13.27	13.36	13.45	13.53	13.62	13.71	13.80
4	12.66	12.75	12.83	12.92	13.01	13.09	13.18	13.27	13.35	13.44
5	12.33	12.42	12.50	12.59	12.67	12.76	12.84	12.93	13.01	13.10
6	12.02	12.11	12.19	12.27	12.35	12.44	12.52	12.60	12.68	12.77
7	11.72	11.80	11.89	11.97	12.05	12.13	12.21	12.29	12.37	12.45
8	11.44	11.52	11.60	11.67	11.75	11.83	11.91	11.99	12.07	12.15
9	11.16	11.24	11.32	11.40	11.47	11.55	11.63	11.70	11.78	11.86
10	10.90	10.98	11.05	11.13	11.20	11.28	11.35	11.43	11.50	11.58
11	10.65	10.72	10.80	10.87	10.94	11.02	11.09	11.16	11.24	11.31
12	10.41	10.48	10.55	10.62	10.69	10.77	10.84	10.91	10.98	11.05
13	10.17	10.24	10.31	10.38	10.46	10.53	10.60	10.67	10.74	10.81
14	9.95	10.02	10.09	10.16	10.23	10.29	10.36	10.43	10.50	10.57
15	9.73	9.80	9.87	9.94	10.00	10.07	10.14	10.21	10.27	10.34
16	9.53	9.59	9.66	9.73	9.79	9.86	9.92	9.99	10.06	10.12
17	9.33	9.39	9.46	9.52	9.59	9.65	9.72	9.78	9.85	9.91
18	9.14	9.20	9.26	9.33	9.39	9.45	9.52	9.58	9.64	9.71
19	8.95	9.01	9.07	9.14	9.20	9.26	9.32	9.39	9.45	9.51
20	8.77	8.83	8.89	8.95	9.02	9.08	9.14	9.20	9.26	9.32
21	8.60	8.66	8.72	8.78	8.84	8.90	8.96	9.02	9.08	9.14
22	8.43	8.49	8.55	8.61	8.67	8.73	8.79	8.84	8.90	8.96
23	8.27	8.33	8.39	8.44	8.50	8.56	8.62	8.68	8.73	8.79
24	8.11	8.17	8.23	8.29	8.34	8.40	8.46	8.51	8.57	8.63
25	7.96	8.02	8.08	8.13	8.19	8.24	8.30	8.36	8.41	8.47
26	7.82	7.87	7.93	7.98	8.04	8.09	8.15	8.20	8.26	8.31
27	7.68	7.73	7.79	7.84	7.89	7.95	8.00	8.06	8.11	8.17
28	7.54	7.59	7.65	7.70	7.75	7.81	7.86	7.91	7.97	8.02
29	7.41	7.46	7.51	7.57	7.62	7.67	7.72	7.78	7.83	7.88
30	7.28	7.33	7.38	7.44	7.49	7.54	7.59	7.64	7.69	7.75
31	7.16	7.21	7.26	7.31	7.36	7.41	7.46	7.51	7.46	7.62
32	7.04	7.09	7.14	7.19	7.24	7.29	7.34	7.39	7.44	7.49
33	6.92	6.97	7.02	7.07	7.12	7.17	7.22	7.27	7.31	7.36
34	6.80	6.85	6.90	6.95	7.00	7.05	7.10	7.15	7.20	7.24
35	6.69	6.74	6.79	6.84	6.89	6.93	6.98	7.03	7.08	7.13
36	6.59	6.63	6.68	6.73	6.78	6.82	6.87	6.92	6.97	7.01
37	6.48	6.53	6.57	6.62	6.67	6.72	6.76	6.81	6.86	6.90
38	6.38	6.43	6.47	6.52	6.56	6.61	6.66	6.70	6.75	6.80
39	6.28	6.33	6.37	6.42	6.46	6.51	6.56	6.60	6.65	6.69
40	6.18	6.23	6.27	6.32	6.36	6.41	6.46	6.50	6.55	6.59

[a] From Colt, J.: "Computation of Dissolved Gas Concentrations in Water as Functions of Temperature, Salinity, and Pressure," *American Fisheries Society Special Publication 14,* Bethesda, MD, 1984.
Note: ppt = parts per thousand

APPENDIX D
CHEMICAL CHARACTERISTICS

Appendix Number	Description
D-1	Characteristics of the Elements
D-2	Physical Constants of Various Fluids
D-3	Physical Constants of Hydrocarbons
D-4	Weights of Chemicals for Structural Design
D-5	Commonly Used Water Treatment Information
D-6	Typical Chemical Feed Rates
D-7	Maximum Concentration of Contaminants for Toxicity Characteristics
D-8	Electromotive Series with Respect to Hydrogen
D-9	Corrosion Data
D-10	Thermal Expansion Coefficients
D-11	Chemicals Used in Coagulation Process
D-12	Chemicals Used in Stabilization and Corrosion Control
D-13	Chemicals Used in Softening Process
D-14	Chemicals Used in Taste and Odor Control
D-15	Various Properties of Anhydrous Ammonia, Chlorine, and Sulfur Dioxide

EXPLANATION OF APPENDICES

Appendix D-1, Characteristics of the Elements: This Appendix lists the elements from the Periodic Table with their Symbol, Atomic Number, stable or longest lived isotope Mass Number, Melting Point and Boiling Point.

Appendix D-2, Physical Constants of Various Fluids: Appendix D-2 lists various fluids and their Formula, Molecular Weight, Boiling Point, Vapor Pressure, Critical Temperature, Critical Pressure and Specific Gravity as a liquid and as a gas if appropriate.

Appendix D-3, Physical Constants of Hydrocarbons: This Appendix lists some of the common hydrocarbons, their Formula, Molecular Weight, Boiling Point, Vapor Pressure, Freezing Point, Critical Temperature, Critical Pressure and Specific Gravity as a liquid and as a gas.

Appendix D-4, Weights of Chemicals for Structural Design: Appendix D-4 lists chemicals used in wastewater treatment, the normal packaging and the weight. The purpose of this Appendix is to give weights for structural design purposes for storage of chemicals.

Appendix D-5, Commonly Used Water Treatment Information: This Appendix lists commonly used water and wastewater treatment chemicals with their Molecular Weight, Density, Solubility in Water, pH of 1% solution and typical composition of commercial product.

Appendix D-6, Typical Chemical Feed Rates: Appendix D-6 lists some typical chemical feed rates for alum, chlorine and lime for various treatment processes and conditions. The reader is cautioned to use these rates only as a guide and to establish more accurate rates from treatability studies.

Appendix D-7, Maximum Concentration of Contaminants for Toxicity Characteristics: This table lists the Regulatory Level for the maximum concentration for all of the contaminants required under the TCLP test for the RCRA determination of a hazardous toxic waste. THe EPA Hazardous Waste Number (EPA HW No.) and the Chemical Abstracts Service Number (CAS No.) are listed for each contaminant.

Appendix D-8, Electromotive Series with Respect to Hydogen: This table lists elements in order of Standard Oxidation Potential. When two metals come in contact, the one with the higher oxidation potential becomes the cathode and the lower potential becomes the anode. The electric potential caused by this contact (or when they are connected in an electrical pathway in the presence of a conducting solution or electrolyte) causes the more active metal (the one closest to the anode) to corrode or to go into solution or be dissolved. The farther apart the metals are in the Table, the greater the tendency to corrode, all other factors being equal.

Appendix D-9, Corrosion Data: Appendix D-9 lists corrosion resistance ratings from *Excellent* down to *Not Recommended* for several metals and coatings for various chemicals. Caution is urged in the use of this table with extreme conditions of temperature or with chemical mixtures.

Appendix D-10, Thermal Expansion Coefficients: This table lists various chemicals. The table is based on the formula:

$$\beta_{liq} = a \left(1 - T/T_c\right)^m$$

where

β_{liq} = thermal expansion coefficient of a chemical compound in liquid phase 1/°C
a, m = regression coefficients for the compound
T = temperature, K
T_c = critical temperature, K

T_{min}, K and T_{max}, K are the temperature ranges fro which the Equation can be used. The table lists the thermal expansion coefficients and the liquid densities oof the chemicals at 25°C.

Appendix D-11, Chemicals Used in Coagulation Process: This Appendix lists several chemicals commonly used for coagulation along with their common names, shipping, containers, handling, storage or transporting materials, available forms, weight, solubility, commercial strength and characteristics.

Appendix D-12, Chemicals Used in Stabilization and Corrosion Control: This Appendix lists chemicals which may be present in wastewater which have been used for pH or alkalinity stabilization and corrosion control. The table lists common names, shipping containers, handling materials, available forms, weight, solubility, commercial strength and characteristics.

Appendix D-13, Chemicals Used in Softening Process: This Appendix lists chemicals which may be present in wastewater which have been used to soften process water. The table lists common names, shipping containers, suitable handling materials, available forms, weight, solubility, commercial strength and characteristics.

Appendix D-14, Chemicals Used in Taste and Odor Control: This Appendix lists chemicals which could be present in wastewater which have been used in taste and odor control of a process water. The table lists common names, shipping containers, suitable handling materials, available forms, weight, solubility, commercial strength and characteristics.

Appendix D-15, Various Properties of Anhydrous Ammonia, Chlorine and Sulfur Dioxide: This Appendix lists properties of these three chemicals commonly used or found in wastewater treatment in both the liquid and gaseous forms.

APPENDIX D-1 Characteristics of the Elements

Element	Symbol	Atomic number	Mass number*	Melting point (°C)	Boiling point (°C)
actnium	Ac	89	(227)	1600†	
aluminum	Al	13	27	659.7	2057
americium	Am	95	(243)		
antimony (stibium)	Sb	51	121	630.5	1380
argon	Ar	18	40	−189.2	−185.7
arsenic	As	33	75	sublimes at 615	
astatine	At	85	(210)		
barium	Ba	56	138	850	1140
berkelium	Bk	97	(247)		
berylium	Be	4	9	1278 ± 5	2970
bismuth	Bi	83	209	271.3	1550 ± 5
boron	B	5	11	2300	2550
bromine	Br	35	79	−7.2	58.78
cadmium	Cd	48	114	320.9	767 ± 2
calcium	Ca	20	40	842 ± 8	1240
californium	Cf	98	(249)		
carbon	C	6	12	>3550	4200
cenom	Ce	58	140	840	1400
cesium	Cs	55	133	28.5	670
chlorine	Cl	17	35	−103 ± 5	−34.6
chromium	Cr	24	52	1890	2480
cobalt	Co	27	59	1495	2900
copper	Cu	29	63	1083	2336
curium	Cm	96	(248)		
dysprosium	Dy	66	164		
einsteinium	Es	99	(254)		
erbium	Er	68	166		
europium	Eu	63	1153	1150 ± 50	
fermium	Fm	100	(252)		
fluorine	F	9	19	−223	−188
francium	Fr	87	(223)		
gadolinium	Gd	64	158		
gallium	Ga	31	69	29.78	1983
germanium	Ge	32	74	958.5	2700
gold	Au	79	197	1063	2600
hafnium	Hf	72	180	1700†	>3200
helium	He	2	4	−272	−268.9
holmium	Ho	67	165		
hyrogen	H	1	1	−259.14	−252.8
indium	In	49	115	156.4	2000 ± 10

APPENDIX D-1   Characteristics of the Elements (*Continued*)

Element	Symbol	Atomic number	Mass number*	Melting point (°C)	Boiling point (°C)
iodine	I	53	127	113.7	184.35
iridium	Ir	77	193	2454	>4800
iron	Fe	26	56	1535	3000
krypton	Kr	36	84	−156.6	−152.9
lanthanum	La	57	139	826	
lawrencium	Lw	103	(257)		
lead	Pb	82	208	327.43	1620
lithium	Li	3	7	186	1336 ± 5
lutetium	Lu	71	175		
magnesium	Mg	12	24	651	1107
manganese	Mn	25	55	1260	1900
mendelevium	Mv	101	(256)		
mercury	Hg	80	202	−38.87	356.58
molybdenum	Mo	42	98	2620 ± 20	4800
neodymium	Nd	60	142	840	
neon	Ne	10	20	−248.67	−245.9
neptunium	Np	93	(237)		
nickel	Ni	28	58	1455	2900
niobium	Nb	41	93	2500 ± 50	3700
nitrogen	N	7	14	−209.86	−195.8
nobelium	No	102	(253)		
osmium	Os	76	192	2700	>5300
oxygen	O	8	16	−218.4	−182.86
palladium	Pd	46	106	1549.4	2000
phosphorous	P	15	31		
platinum	Pt	78	195	1773.5	4300
plutonium	Pu	94	(242)		
polonium	Po	84	(209)		
potassium	K	19	39	63.3	760
praseodymium	Pr	59	141	940	
promethium	Pm	61	(145)		
protactinium	Pa	91	(231)		
radium	Ra	88	(226)	700	1140
radon	Rn	86	(222)	−71	−61.8
rhenium	Re	75	187	3167 ± 60	
rhodium	Rh	45	103	1966 ± 3	>2500
rubidium	Rb	37	85	38.5	700
ruthenium	Ru	44	102	2450	2700
samarium	Sm	62	152	>1300	
scandium	Sc	21	45	1200	2400

APPENDIX D-1   Characteristics of the Elements (*Continued*)

Element	Symbol	Atomic number	Mass number*	Melting point (°C)	Boiling point (°C)
selenium	Se	34	80	217	688
silicon	Si	14	28	1420	2355
silver	Ag	47	107	960.8	1950
sodium	Na	11	23	97.5	880
strontium	Sr	38	88	800	1150
sulfur	S	16	32		
tantalum	Ta	73	180	2996 ± 50	c.4100
technetium	Tc	43	(99)		
tellurium	Te	52	130	452	1390
terbium	Tb	65	159	327 ± 5	
thallium	Tl	81	205	302	1457 ± 10
thorium	Th	90	232	1845	4500
thulium	Tm	69	169		
tin	Sn	50	120	231.89	2270
titanium	Ti	22	48	1800	>3000
tungsten (wolfram)	W	74	184	3370	5900
uranium	U	92	238	1133	
vanadium	V	23	51	1710	3000
xenon	Xe	54	132	−112	−107.1
ytterbium	Yb	70	174	1800	
yttrium	Y	39	89	1490	2500
zinc	Zn	30	64	419.47	907
zirconium	Zr	40	90	1857	>2900

APPENDIX D-2   Physical Constants of Various Fluids

Fluid	Formula	Molecular weight	Boiling point (°F at 15.696 PSIA)	Vapor pressure @ 70°F (PSIG)	Critical temp. (°F)	Critical pressure (PSIA)	Specific gravity Liquid 60/60°F	Specific gravity Gas
Acetic Acid	$HC_2H_3O_2$	60.05	245				1.05	
Acetone	C_3H_6O	58.08	133		455	691	0.79	2.01
Air	N_2O_2	28.97	−317		−221	547	0.86‡	1.0
Alcohol, Ethyl	C_2H_6O	46.07	173	2.3†	470	925	0.794	1.59
Alcohol, Methyl	CH_4O	32.04	148	4.63†	463	1174	0.796	1.11
Ammonia	NH_3	17.03	−28	114	270	1636	0.62	0.59
Ammonium Chloride*	NH_4Cl						1.07	
Ammonium Hydroxide*	NH_4OH						0.91	
Ammonium Sulfate*	$(NH_4)_2SO_4$						1.15	
Aniline	C_6H_7N	93.12	365		798	770	1.02	
Argon	A	39.94	−302		−188	705	1.65	1.38
Beer							1.01	
Bromine	Br_2	159.84	138		575		2.93	5.52
Calcium Chloride*	$CaCl_2$						1.23	
Carbon Dioxide	CO_2	44.01	−109	839	88	1072	0.801‡	1.52
Carbon Disulfide	CS_2	76.1	115				1.29	2.63
Carbon Monoxide	CO	28.01	−314		−220	507	0.80	0.97
Carbon Tetrachloride	CCl_4	153.84	170		542	661	1.59	5.31
Chlorine	Cl_2	70.91	−30	85	291	1119	1.42	2.45
Chromic Acid	H_2CrO_4	118.03					1.21	
Citric Acid	$C_6H_8O_7$	192.12					1.54	
Copper Sulfate*	$CuSO_4$						1.17	
Ether	$(C_2H_5)_2O$	174.12	34				0.74	2.55
Ferric Chloride*	$FeCl_2$						1.23	
Fluorine	F_2	38.00	−305	300	−200	809	1.11	1.31
Formaldehyde	H_2CO	30.03	−6				0.82	1.08
Formic Acid	HCO_2H	46.03	214				1.23	
Furfural	$C_5H_4O_2$	96.08	324				1.16	
Glycerine	$C_3H_6O_3$	92.09	554				1.26	
Glycol	$C_2H_6O_2$	62.07	387				1.11	
Helium	He	4.003	−454		−450	33	0.18	0.14
Hydrochloric Acid	HCl	36.47	−115				1.64	
Hydrofluoric Acid	HF	20.01	66	0.9	446		0.92	
Hydrogen	H_2	2.016	−422		−400	188	0.07‡	0.07
Hyrogen Chloride	HCl	36.47	−115	613	125	1198	0.86	1.26
Hydrogen Sulfide	H_2S	34.07	−76	252	213	1307	0.79	1.17
Isopropyl Alcohol	C_3H_8O	60.09	180				0.78	2.08
Linseed Oil			538				0.93	
Magnesium Chloride*	$MgCl_2$						1.22	
Mercury	Hg	200.61	670				13.6	6.93
Methyl Bromide	CH_3Br	94.95	38	13	376		1.73	3.27
Methyl Chloride	CH_3Cl	50.49	−11	59	290	969	0.99	1.74
Naphthalene	$C_{10}H_8$	128.16	424				1.14	4.43
Nitric Acid	HNO_3	63.02	187				1.5	
Nitrogen	N_2	28.02	−320		−233	493	0.81‡	0.97
Oil, Vegetable							0.91–0.94	
Oxygen	O_2	32	−297		−181	737	1.14‡	1.105
Phosgene	$COCl_2$	98.92	47	10.7	360	823	1.39	3.42
Phosphoric Acid	H_3PO_4	98.00	415				1.83	
Potassium Carbonate*	K_2CO_3						1.24	

APPENDIX D-2   Physical Constants of Various Fluids (*Continued*)

Fluid	Formula	Molecular weight	Boiling point (°F at 15.696 PSIA)	Vapor pressure @ 70°F (PSIG)	Critical temp. (°F)	Critical pressure (PSIA)	Specific gravity Liquid 60/60°F	Gas
Potassium Chloride*	KCl						1.16	
Potassium Hydroxide*	KOH						1.24	
Refrigerant 11	CCl_3F	137.38	75	13.4	388	635		5.04
Refrigerant 12	CCl_2F_2	120.93	−22	70.2	234	597		4.2
Refrigerant 13	$CClF_3$	104.47	−115	458.7	84	561		
Refrigerant 21	$CHCl_2F$	102.93	48	8.4	353	750		3.82
Refrigerant 22	$CHClF_2$	86.48	−41	122.5	205	716		
Refrigerant 23	CHF_3	70.02	−119	635	91	691		
Sodium Chloride*	NaCl						1.19	
Sodium Hydroxide*	NaOH						1.27	
Sodium Sulfate*	Na_2SO_4						1.24	
Sodium Thiosulfate*	$Na_2S_2O_3$						1.23	
Starch	$(C_6H_{10}O_5)x$						1.50	
Sugar Solutions*	$C_{12}H_{22}O_{11}$						1.10	
Sulfuric Acid	H_2SO_4	98.08	626				1.83	
Sulfur Dioxide	SO_2	64.6	14	34.4	316	1145	1.39	2.21
Turpentine			320				0.87	
Water	H_2O	18.016	212	0.9492†	706	3208	1.00	0.62
Zinc Chloride*	$ZnCl_2$						1.24	
Zinc Sulfate*	$ZnSo_4$						1.31	

* Aqueous Solution—25% by weight of compound.
† Vapor pressure in psis at 100°F.
‡ Density of liquid, gm/ml at normal cooling point.

APPENDIX D-3 Physical Constants of Hydrocarbons

No.	Compound	Formula	Molecular weight	Boiling point at 14.696 psia (°F)	Vapor pressure at 100°F (psia)	Freezing point at 14.696 psia (°F)	Critical constants Critical temperature (°F)	Critical pressure (psia)	Specific gravity at 14.696 psia Liquid [3,4] 60°F/60°F	Gas at 60°F (air = 1)'
1	Methane	CH_4	16.043	−258.69	(5000)[2]	−296.46[5]	−116.64	667.8	0.3[8]	0.5539
2	Ethane	C_2H_6	30.070	−127.48	(800)[2]	−297.89[5]	90.09	707.8	0.3564[7]	1.0382
3	Propane	C_3H_8	44.097	−43.67	190.	−305.84[5]	206.01	616.3	0.5077[7]	1.5225
4	n-Butane	C_4H_{10}	58.124	31.10	51.6	−217.05	305.65	550.7	0.5844[7]	2.0068
5	Isobutane	C_4H_{10}	58.124	10.90	72.2	−255.29	274.98	529.1	0.5631[7]	2.0068
6	n-Pentane	C_5H_{12}	72.151	96.92	15.570	−201.51	385.7	488.6	0.6310	2.4911
7	Isopentane	C_5H_{12}	72.151	82.12	20.44	−255.83	369.10	490.4	0.6247	2.4911
8	Neopentane	C_5H_{12}	72.151	49.10	35.9	2.17	321.13	464.0	0.5967[7]	2.4911
9	n-Hexane	C_6H_{14}	86.178	155.72	4.956	−139.58	453.7	436.9	0.6640	2.9753
10	2-Methylpentane	C_6H_{14}	86.178	140.47	6.767	−244.63	435.83	436.6	0.6579	2.9753
11	3-Methylpentane	C_6H_{14}	86.178	145.89	6.098	...	448.3	453.1	0.6689	2.9753
12	Neohexane	C_6H_{14}	86.178	121.52	9.856	−147.72	420.13	446.8	0.6540	2.9753
13	2,3-Dimethylbutane	C_6H_{14}	86.178	136.36	7.404	−199.38	440.29	453.5	0.6664	2.9753
14	n-Heptane	C_7H_{16}	100.205	209.17	1.620	−131.05	512.8	396.8	0.6882	3.4596
15	2-Methylhexane	C_7H_{16}	100.205	194.09	2.271	−180.89	495.00	396.5	0.6830	3.4596
16	3-Methylhexane	C_7H_{16}	100.205	197.32	2.130	...	503.78	408.1	0.6917	3.4596
17	3-Ethylpentane	C_7H_{16}	100.205	200.25	2.012	−181.48	513.48	419.3	0.7028	3.4596
18	2,2-Dimethylpentane	C_7H_{16}	100.205	174.54	3.492	−190.86	477.23	402.2	0.6782	3.4596
19	2,4-Dimethylpentane	C_7H_{16}	100.205	176.89	3.292	−182.63	475.95	396.9	0.6773	3.4596
20	3,3-Dimethylpentane	C_7H_{16}	100.205	186.91	2.773	−210.01	505.85	427.2	0.6976	3.4596
21	Triptane	C_7H_{16}	100.205	177.58	3.374	−12.82	496.44	428.4	0.6946	3.4596
22	n-Octane	C_8H_{18}	114.232	258.22	0.537	−70.18	564.22	360.6	0.7068	3.9439
23	Diisobutyl	C_8H_{18}	114.232	228.39	1.101	−132.07	530.44	360.6	0.6979	3.9439
24	Isooctane	C_8H_{18}	114.232	210.63	1.708	−161.27	519.46	372.4	0.6962	3.9439
25	n-Nonane	C_9H_{20}	128.259	303.47	0.179	−64.28	610.68	332.	0.7217	4.4282
26	n-Decane	$C_{10}H_{22}$	142.286	345.48	0.0597	−21.36	652.1	304.	0.7342	4.9125
27	Cyclopentane	C_5H_{10}	70.135	120.65	9.914	−136.91	:61.5	653.8	0.7504	2.4215
28	Methylcyclopentane	C_6H_{12}	84.162	161.25	4.503	−224.44	499.35	548.9	0.7536	2.9057
29	Cyclohexane	C_6H_{12}	84.162	177.29	3.264	43.77	536.7	591.	0.7834	2.9057
30	Methylcyclohexane	C_7H_{14}	98.189	213.68	1.609	−195.87	570.27	503.5	0.7740	3.3900

APPENDIX D-3 Physical Constants of Hydrocarbons (*Continued*)

No.	Compound	Formula	Molecular weight	Boiling point at 14.696 psia (°F)	Vapor pressure at 100°F (psia)	Freezing point at 14.696 psia (°F)	Critical constants		Specific gravity at 14.696 psia	
							Critical temperature (°F)	Critical pressure (psia)	Liquid, [3,4] 60°F/60°F	Gas at 60°F (air = 1)
31	Ethylene	C_2H_4	28.054	−154.62	...	−272.45[5]	48.58	729.8	...	0.9686
32	Propene	C_3H_6	42.081	−53.90	226.4	−301.45[5]	196.9	669.	0.5220[7]	1.4529
33	1-Butene	C_4H_8	56.108	20.75	63.05	−301.63[5]	295.6	583.	0.6013[7]	1.9372
34	Cis-2-Butene	C_4H_8	56.108	38.69	45.54	−218.06	324.37	610.	0.6271[7]	1.9372
35	Trans-2-Butene	C_4H_8	56.108	33.58	49.80	−157.96	311.86	595.	0.6100[7]	1.9372
36	Isobutene	C_4H_8	56.108	19.59	63.40	−220.61	292.55	590.	0.6004[7]	2.4215
37	1-Pentene	C_5H_{10}	70.135	85.93	19.115	−265.39	376.93	590.	0.6457	1.8676
38	1,2-Butadiene	C_4H_6	54.092	51.53	(20.)[2]	−213.16	(339.)[2]	(653.)[2]	0.658[7]	1.8676
39	1,3-Butadiene	C_4H_6	54.092	24.06	(60.)[2]	−164.02	306.	628.	0.6272[7]	2.3519
40	Isoprene	C_5H_8	68.119	93.30	16.672	−230.74	(412.)[2]	(558.4)[2]	0.6861	0.8990
41	Acetylene	C_2H_2	26.038	−119.[6]	...	−114.[5]	95.31	890.4	0.615[9]	2.6969
42	Benzene	C_6H_6	78.114	176.17	3.224	41.96	552.22	710.4	0.8844	3.1812
43	Toluene	C_7H_8	92.141	231.13	1.032	−138.94	605.55	595.9	0.8718	3.6655
44	Ethylbenzene	C_8H_{10}	106.168	277.16	0.371	−138.91	651.24	523.5	0.8718	3.6655
45	o-Xylene	C_8H_{10}	106.168	291.97	0.264	−13.30	675.0	541.4	0.8848	3.6655
46	m-Xylene	C_8H_{10}	106.168	282.41	0.326	−54.12	651.02	513.6	0.8687	3.6655
47	p-Xylene	C_8H_{10}	106.168	281.05	0.342	55.86	649.6	509.2	0.8657	3.5959
48	Styrene	C_8H_8	104.152	293.29	(0.24)[2]	−23.10	706.0	580.	0.9110	4.1498
49	Isopropylbenzene	C_9H_{12}	120.195	306.34	0.188	−140.82	676.4	465.4	0.8663	

[1] Calculated values.
[2] ( )-Estimated values.
[3] Air saturated hydrocarbons.
[4] Absolute values from weights in vacuum.
[5] At saturation pressure (triple point).
[6] Sublimation point.
[7] Saturation pressure and 60°F.
[8] Apparent value for methane at 60°F.
[9] Specific gravity, 119°F/60°F (sublimation point).

APPENDIX D-4   Weights of Chemicals for Structural Steel

Chemical	Shipping and storage data	Weight, lb per cu ft except as stated
Activated carbon*	Bags 35 lb (3×21×39 in.); drums 5, 25 lb; bulk	Powder 8 to 28, average 12
Activated silica*	—	About 11.6 lb per gal.
Alumium ammonium sulfate*	Fiber drums, kegs 100–400 lb; bags 100 lb; bulk	60 to 70
Alumium chloride solution*	Carboys; rubber lined tank trucks	Specific gravity 1.15 to 1.16
Aluminum potassium sulfate*	Bags 100 lb; bbl 250, 350 lb; drums 100, 350 lb; bulk	60 to 70
Aluminum sulfate*	Bags 100, 200 lb; bbl 325, 400 lb; drums 25, 100, 250 lb; bulk	60 to 75 (powder is lighter); to calculate hooper capacities use 60
Alum liquid*	Manufactured near site because high freight cost precludes distant shipment; 6000–8000 gal. steel tank cars; 2000–4000 gal. rubber lined steel tank trucks	At 60 F 32.2* Be specific gravity 1.285; 10.7 lb per gal.
Ammonia, anhydrous NH_3	Steel cylinders 50, 100, 150 lb; tank car, 50,000 lb	Specific gravity of liquid is 0.68 at −28 F
Ammonia, aqua	Carboys 5, 10 gal.; drums 375, 750 lb; 8000 gal. tank car	At 60 F 26* Be specific gravity 0.8974
Ammonia silicofluoride	Pkgs 4 to 26.5 lb; kegs 100 lb; bbl or drums 400 lb	70–80
Ammonium sulfate*	Boxes 25 lb; kegs 100 lb; bags 100 lb; bbl 300, 400 lb; bulk	60
Barium carbonate	Boxes 25 lb; kegs 100 lb; bags 200 lb	52–78
Bentonite	Bags 50, 100 lb; bulk	Powder 45 to 60; granules 65 to 75
Bromine	Glass bottles 6.5 lb; earthenware bottles	Specific gravity 3.119 at 20 per 15 C 26 lb per gal.
Calcium carbonate	Bags 50 lb; drums, bulk	Powder 35 to 60; granules 100 to 115
Calcium hydroxide	Bags 50 lb; bbl 100 lb; bulk; store in dry place	20 to 50; to calculate hopper capacity use 40
Calcium hypochlorite	Bbl 415 lb; cans 5, 15, 100, 300, lb; drums 800 lb; store dry and cool; avoid contact with organic matter	Granules 68 to 80; powder 32 to 50
Calcium oxide	Moistureproof bags 100 lb; wood bbl, bulk; C/L; store dry, max 60 days; keep container closed	55 to 70; to calculate hopper capacity use 60
Carbon dioxide	Steel cylinders for compressed gas, 150 lb; dry ice delivered as required and evaporated on site in large steel cylinders	—
Chlorinated copperas	See chlorine and ferrous sulfate as reagents needed	—
Chlorinated lime	Druums 100, 300, 800 lb; store cool and dry	45 to 50
Chlorine	Steel cylinders 100, 150 lb; ton containers, 15 ton containers; 16, 30, 55 ton tank cars	Specific gravity with respect to air 2.49
Copper sulfate*	Bags 100 lb; bbl 450 lb; drums	Crystal 75 to 90, powder 60–68
Diatomaceous earth	Bags 50 lb, bulk	Natural 5 to 18; calcined 6 to 3; flux calcined 10 to 25
Disodium phosphate	Bags 100 lb; kegs 125 lb; drums 25, 100, 125 lb; bbl 325, 350 lb	Crystal hydrate 80–90; anhydrous 53 to 62

APPENDIX D-4 Weights of Chemicals for Structural Steel

Chemical	Shipping and storage data	Weight, lb per cu ft except as stated
Dolomitic hydrated lime	Bags 50 lb; bbl. bulk	30 to 50; to calculate hopper capacity use 40
Ferric chloride	Solution—carboys 5, 13 gal. truck, tank car; Crystal—keg 100, 400, 450 lb; drums 150, 350, 630 lb	Solution 11.2 to 12.4 lb; crystal 60 to 64; anhydrous 45 to 60
Ferric sulfate*	Bags 100 lb; drums 400, 425 lb; bulk	70 to 72
Ferrous sulfate	Bags 100 lb; bbl 400 lb; bulk	63 to 66
Fluosilicic acid	Kegs 5 gal.; pitch lined drums, 50 gal.; rubber lined bbl 420 lb	30 percent is 10.5 lb per gal.
Hydrofluoric acid	Drums steel 20, 30, 100 gal.; bulk	—
Potassium permanganate	USP 25, 110, 125 lb; steel keg; tech 25 110, 600 lb steel drum	86–102
Sodium aluminate	Ground bags 50, 100 lb; liquid in drums	High purity 50, standard 60
Sodium bicarbonate	Bags 100 lb and drums	59–62
Sodium bisulfate	Bags 100 lb; drums 100, 400 lb	70 to 80
Sodium carbonate	Bags 100 lb; bbl 100 lb; drums 25, 100 lb; bulk	Dense 65, medium 40; light 30
Sodium chloride	Bags 100 lbs; bbl, drums 25 lb; bulk	Rock 50 to 60; crystal 58 to 70
Sodium chlorite	Drums 100 lb	65 to 75
Sodium fluoride	Bags 100 lb; drums 25, 125, 375 lb	Powder 65 to 100, granules; crystal 90 to 106
Sodium Hexametaphosphate	Bag 100 lb; drums 100, 300, 320 lb	Glass 64–100; powder and granular 44–60
Sodium hydroxide	Drums 25, 50, 350, 400, 700 lb; bulk-solution	Pellets 60–70; flakes 46–62
Sodium hypochlorite	Carboys 5, 13 gal.; drums 30 gal.; bulk 1300, 1800, 2000 gal.; percent per ton	15 percent 10.2 lb per gal.
Sodium silicate	Drums 1, 5, 55 gal.; bulk	41*Be, 11.6 lb per gal.; 42.2*Be, 11.73 lb per gal.
Sodium silicofluoride	Bags 100 lb; drums 25, 125, 375 lb	Granular 85–105; powder-granular 60–96
Sodium sulfate	Bags bbl, drums, kegs	70 to 100
Sodium thiosulfate	Bags, bbl, drums, kegs	53 to 60
Sulfur dioxide	Steel cylinders 100, 150, 200 lb	—
Sulfuric acid	Bottle, carboys 5, 13 gal.; drums 55, 110 gal.; bulk	66*Be 15.1 lb per gal.
Tetrasodium pyrophosphate	Bags 100, 200 lb; bbl 350 lb; drums 25, 100, 300, 350 lb; kegs 125 lb	Crystal 50–70; powder 46–66
Trisodium phosphate	Bags 100, 200 lb; bbl 325, 400 lb; kegs 125 lb	Crystal 55 to 60, monohydrate 65

* Aggressive action on concrete.

APPENDIX D-5 Commonly Used Water Treatment Information

			Properties of water treating chemicals		
Chemical	Molecular weight	Apparent density	Approx. max. solubility in water at 75°F. % by weight	pH of 1% solution	Typical analysis or composition of commercial product
Aluminum Sulfate	594	38–71	50%	3.4	17% Al_2O_3 $Al_2(SO_4)_3 \cdot 14H_2O$
Sodium Aluminate		50–66	45%	11.5	45% Al_2O_3
Soda Ash	106	30–60	22%	11.0	58% Na_2O or 99% Na_2CO_3
Calcium Carbonate (200 mesh)		45–70	Not Soluble		$CaCO_3$ or $CaCo_3$ + $MgCo_3$
Activated Carbon		10–25	Not Soluble		Carbon plus variable % of Ash
Hydrated Lime	74	30–50	0.15%	12.2	90% available hydrate
Pulverized Quicklime	56	50–70	0.1%	12.2	90% available CaO
Diatomite		8–10	Not Soluble		Essentially SiO_2
Copper Sulfate	250	87	18%	4.5	$CuSo_4 \cdot 5H_2O$
Disodium Phosphate	142	50–82	11%	8.9	Na_2HPO_4 49% P_2O_5
Disodium Phosphate	178	73	13%	8.9	$Na_2HPO_4 \cdot 2H_2O$, 40% P_2O_3
Trisodium Phosphate	164	58–66	12%	11.4	Na_4PO_4, 42% P_2O_5
Trisodium Phosphate	182	45	13%	11.4	$Na_4PO_4 \cdot H_2O$, 39% P_2O_3
Trisodium Phosphate	380	55–60	27%	11.4	$Na_3PO_4 \cdot 12H_2O$, 19% P_2O_5
Sodium Tripolyphosphate	368	49–70	14%	9.7	$Na_2P_4O_{10}$, 58% P_2O_5
Tetrasodium Pyrophosphate	266	45–65	5%	10.2	$Na_4P_2O_7$, 53% P_2O_5
Hemisodium Phosphate	218	50	80%	2.2	$NaH_2PO_3 \cdot H_2PO_3$, 64% P_2O_5
Monosodium Phosphate	120	50–60	48%	4.5	NaH_2PO_4, 59% P_2O_5
Monosodium Phosphate	138	62	48%	4.5	$NaH_2PO_4 \cdot H_2O$, 53% P_2O_5
Sodium Bicarbonate	84	71	9%	8.1	$NaHCO_3$
Clay		65	Not Soluble		Aluminum Silicates
Magnesium Oxide	40	5–60	Not Soluble		97% MgO
Ferric Sulfate	400	70–80	>30%	2.0	$Fe_2(SO_4)_2$
Ferrous Sulfate	278	65–70	23%	3.8	$FeSO_4 \cdot 7H_2O$
Sodium Fluoride	42	50–86	4%	9.5	NaF
Sodium Silicofluoride	188	70–95	0.7%		Na_2SiF_8
Ammonium Sulfate	132	47	40	5.6	$(NH_4)_2SO_4$
Ammonium Chloride	53.4	50	28%	5.4	NH_4Cl

APPENDIX D-6 Typical Chemical Feed Rates

Chlorination	Range (mg/L)	Chlorinator feed capacity (mg/L)
Raw Sewage	6–12	25
Septic Raw Sewage	12–25	50
Primary Treatment Effluent	5–10	20
Trickling Filter Effluent	3–10	15
Activated Sludge Effluent	2–8	10
Sand Filter Effluent	1–5	7

Chlorine dioxide	Range (mg/L)
Raw Sewage	2–5
Septic Raw Sewage	4–10
Primary Treatment Effluent	2–5
Trickling Filter Effluent	1–4
Activated Sludge Effluent	1–3
Sand Filter Effluent	0.4–2

Oxidation of Sulfides

Oxidation with	mg/L per mg/L of sulfide
Chlorine	10–15
Hydrogen Peroxide	1–1.5
Sodium Nitrate	10–30

Control of Filamentous Growth

Control with	Range (mg/L)
Chlorine	10–20
Hydrogen Peroxide	>100

Coagulant Feed

Coagulant	Range (mg/L)	Optimum pH
Aluminum Sulfate (Alum)	75–250	4.0–7.0
Ferric Chloride	45–90	3.5–6.5 or >8.5
Lime	200–400	
Ferrous Sulfate		>8.5
Ferric Sulfate		4.0–7.0 or >9.0

APPENDIX D-6   Typical Chemical Feed Rates (*Continued*)

Theoretical Auxilliary Chemicals Required for Coagulation with

Alum	mg/L
Natural alkalinity as $CaCO_3$	7.7
Lime as CaO	4.3
Lime as $Ca(OH)_2$	5.7
Soda Ash as Na_2CO_2	8.2

Ferrous Sulfate	mg/L
Lime as CaO	3.4
Lime as $Ca(OH)_2$	4.6
Chlorine	2.1

Ammonia Removal

Removal with	Dosage
Chlorine	10 mg/L per mg/L NH_3-N

Phosphorus Removal

Removal with	Dosage
Alum	75–250 mg/L

pH Increase to 11 with Lime (CaO)

Wastewater alkalinity (mg/L as $CaCO_3$)	Lime dosage (mg/L)
100	200
200	320
300	400
400	450
500	500

APPENDIX D-7 Maximum Concentration of Contaminants for Toxicity Characteristics

EPA HW No.[1]	Contaminant	CAS No.[2]	Regulatory level (mg/L)
D004	Arsenic	7440-38-2	5.0
D005	Barium	7440-39-3	100.0
D018	Benzene	71-43-2	0.5
D006	Cadmium	7440-43-9	1.0
D019	Carbon tetrachloride	56-23-5	0.5
D020	Chlordane	57-74-9	0.03
D021	Chlorobenzene	108-90-7	100.0
D022	Chloroform	67-66-3	6.0
D007	Chromium	7440-47-3	5.0
D023	o-Cresol	95-48-7	[4]200.0
D024	m-Cresol	108-39-4	[4]200.0
D025	p-Cresol	106-44-5	[4]200.0
D026	Cresol		[4]200.0
D016	2,4-D	94-75-7	10.0
D027	1,4-Dichlorobenzene	106-46-7	7.5
D028	1,2-Dichloroethane	107-06-2	0.5
D029	1,1-Dichloroethylene	75-35-4	0.7
D030	2,4-Dinitrotoluene	121-14-2	[3]0.13
D012	Endrin	72-20-8	0.02
D031	Heptachlor (and its epoxide)	76-44-8	0.008
D032	Hexachlorobenzene	118-74-1	[3]0.13
D033	Hexachlorobutadiene	87-68-3	0.5
D034	Hexachloroethane	67-72-1	3.0
D008	Lead	7439-92-1	5.0
D013	Lindane	58-89-9	0.4
D009	Mercury	7439-97-6	0.2
D014	Methoxychlor	72-43-5	10.0
D035	Methyl ethyl ketone	78-93-3	200.0
D036	Nitrobenzene	98-95-3	2.0
D037	Pentrachlorophenol	87-86-5	100.0
D038	Pyridine	110-86-1	[3]5.0
D010	Selenium	7782-49-2	1.0
D011	Silver	7440-22-4	5.0
D039	Tetrachloroethylene	127-18-4	0.7
D015	Toxaphene	8001-35-2	0.5
D040	Trichloroethylene	79-01-6	0.5
D041	2,4,5-Trichlorophenol	95-95-4	400.0
D042	2,4,6-Trichlorophenol	88-06-2	2.0
D017	2,4,5-TP (Silvex)	93-72-1	1.0
D043	Vinyl chloride	75-01-4	0.2

[1] Hazardous waste number.

[2] Chemical abstracts service number.

[3] Quantitation limit is greater than the calculated regulatory level. The quantitation limit therefore becomes the regulatory level.

[4] If o-, m-, and p-Cresol concentrations cannot be differentiated, the total cresol (D026) concentration is used. The regulatory level of total cresol is 200 mg/l.

APPENDIX D-8 Electromotive Series with Respect to Hydrogen

Electrode reaction	Standard oxidation potential, $E°$, V, 25°C*
Li $= Li^+ + e^-$	3.05
K $= K^+ + e^-$	2.93
Ca $= Ca^{++} + 2e^-$	2.87
Na $= Na^+ + e^-$	2.71
Mg $= Mg^{++} + 2e^-$	2.37
Be $= Be^{++} + 2e^-$	1.85
U $= U^{+3} + 3e^-$	1.80
Hf $= Hf^{+4} + 4e^-$	1.70
Al $= Al^{+3} + 3e^-$	1.66
Ti $= Ti^{++} + 2e^-$	1.63
Zr $= Zr^{+4} + 4e^-$	1.53
Mn $= Mn^{++} + 2e^-$	1.18
Nb $= Nb^{+3} + 3e^-$	ca. 1.1
Zn $= Zn^{++} + 2e^-$	0.763
Cr $= Cr^{+3} + 3e^-$	0.74
Ga $= Ga^{+3} + 3e^-$	0.53
Fe $= Fe^{++} + 2e^-$	0.440
Cd $= Cd^{++} + 2e^-$	0.403
In $= In^{+3} + 3e^-$	0.342
Tl $= Tl^+ + e^-$	0.336
Co $= Co^{++} + 2e^-$	0.277
Ni $= Ni^{++} + 2e^-$	0.250
Mo $= Mo^{+3} + 3e^-$	ca. 0.2
Sn $= Sn^{++} + 2e^-$	0.136
Pb $= Pb^{++} + 2e^-$	0.126
H_2 $= 2H^+ + 2e^-$	0.000
Cu $= Cu^{++} + 2e^-$	−0.337
Cu $= Cu^+ + e^-$	−0.521
2Hg $= Hg_2^{++} + 2e^-$	−0.789
Ag $= Ag^+ + e^-$	−0.800
Pd $= Pd^{++} + 2e^-$	−0.987
Hg $= Hg^{++} + 2e^-$	−0.854
Pt $= Pt^{++} + 2e^-$	−1.2
Au $= Au^{+3} + 3e^-$	−1.50

*Standard reduction potentials, ○, have the opposite sign.

From Uhig, H.H., *"Corrosion and Corrosion Control,"* 2nd ed., p. 29.

APPENDIX D-9 Corrosion Data

Chemicals	Aluminum	Ductile iron	Bronze	Carbon steel	S. S. 304	S. S. 316	Alloy 20	Monel	Nickel	Hastelloy	TFE/FEP
Acetaldehyde	B	C	D	C	A	A	A	A	A	A	A
Acetate Solvents	A	B	A	A	A	A	A	C	A		A
Acetic Acid (Aerated 0 to 50%)	C	D	D	D	B	A	A	C	D	A	A
Acetic Acid (Air Free 0 to 50%)	A	D	B	D	C	A	A	C	D	A	A
Acetic Acid (Aerated 55 to 100%)	B	D	D	D	A	B	A	D	D	A	A
Acetic Acid (Air Free 55 to 100%)	A	D	B	D	B	B	A	B	B	A	A
Acetic Anhydride	B	D	C	D	B	B	B	B	B	A	A
Acetone	B	B	B	B	B	B	B	A	A	A	A
Acetylene (Dry Only)	A	A	B	A	A	A	A	A		A	A
Acrylonitrile	B	C	A	A	A	A	A	A			A
Alcohols—Methyl, Ethyl	B	B	B	B	A	A	A	A	A	A	A
Alcohol—Amyl	A	B	B	B	A	A	A	A	A	A	A
Alcohol—Butyl	A	B	B	B	A	A	A	A	C	A[b]	A
Aluminum Chloride (Dry)	D	D	D	D	C	C	B	C	B	B	A
Aluminum Sulfate (Alums)	B	C	C	C	B	A	A	A	B	B	A
Alums	B	C	C	C	B	A	A	A	A	B	A
Amines	A	A	A	A	A	A	A	A	A	A	A
Ammonia, Anhydrous	B	B	D	A	A	A	A	B	A	B	A
Ammonia (Aqueous)	B	A	D	A	A	A	A	B	A	B	A
Ammonia Solutions	D	B	D	B	A	A	A	B			A
Ammonium Bicarbonate	B	B	B	C	B	B	B	B	B	B	A
Ammonium Carbonate	B	B	D	B	B	B	B	B	B	A	A
Ammonium Chloride	C	D	D	D	B	B	B	B	D	A	A
Ammonium Hydroxide (28%)	C	C	D	C	B	B	B	D	D	A	A
Ammonium Hydroxide (Conc.)	C	C	D	C	B	B	B	D	D	A	A
Ammonium Monophosphate	B	D	D	D	B	B	B	C		A[c]	A
Ammonium Nitrate	B	D	D	D	A	A	A	D	D	B[c]	A
Ammonium Phosphate (Dibasic)	B	D	C	D	B	B	B	C		B	A
Ammonium Phosphate (Tribasic)	B	D	C	B	B	B	B	C	B	B	A
Ammonium Sulfate	C	C	B	C	B	B	B	C		B[c]	A

D.18

Substance													
Amyl Acetate	B	C	B	C	B	B	A	A	A	A	A	A	A
Aniline	C	C	C	C	C	B	B	B	B	B	B	B	A
Aniline Dyes	C	C	C	C	C	A	A	A	A	A			A
Antimony Trichloride	D	D	D	D	D	D	B	D	B	B	B	B	A
Apple Juice	B	C	C	D	B	B	A	B	A	A			A
Arsenic Acid	D	D	D	B	D	B	B	A	D	A	A		A
Asphalt Emulsion	C	A	A	A	A	A	A	A	A	A	A		A
Asphalt Liquid	C	A	A	A	A	A	A	A	A	A	A		A
Barium Carbonate	B	B	B	B	B	A	B	B	B	B	B	B	A
Barium Chloride	D	C	B	C	A	C	C	B	B	B	A	A	A
Barium Hydroxide	D	B	B	B	B	B	A	A	B	B	A	B	A
Barium Sulfate	D	C	C	C	B	B	B	B	B	B	B		A
Barium Sulfide	D	C	C	C	B	B	B	B	C	A	A	A	A
Beer (Alcohol Industry)	A	C	B	B	A	A	A	A	A	A	A	A	A
Beet Sugar Liquors	A	B	A	B	A	A	A	A	A	A	A	A	A
Benzene (Benzol)	B	B	B	B	B	B	A	A	A	A	B	B	A
Benzaldehyde	B	D	B	D	B	B	B	B	B	B	B	B	A
Benzoic Acid	B	D	B	D	B	B	B	B	B	B	A	A	A
Borax Liquors	C	C	B	C	A	A	A	A	A	A	A	A	A
Boric Acid	B	D	B	D	A	B	A	A	A	A	A	A	A
Brines	B	C	B	C	B	B	B	B	A	A		A	A
Bromine (Dry)	C	D	D	D	D	D	D	D	A	A	A	A(c)	A
Bromine (Wet)	D	D	D	D	D	D	D	D	A	C	C	A(c)	A
Bunker Oils (Fuel Oils)	A	B	B	B	B	A	A	A	D	A			A
Butadiene	A	B	C	B	A	A	A	A	C	C			A
Butane	A	B	A	B	A	A	A	A	A	A	A	A	A
Butylene	A	A	A	A	A	A	A	A	A	A	A	A	A
Buttermilk	A	D	D	D	D	A	A	A	D	D			A
Butyric Acid	B	D	C	D	B	B	B	B	C	B	A	A	A
Calcium Bisulfite	C	D	B	D	C	B	C	B	D	B	B	B	A
Calcium Carbonate	C	D	C	D	B	B	B	B	D	B	B	B	A
Calcium Chloride	C	B	B	B	C	B	B	B	C	A	A	A	A
Calcium Hydroxide	C	B	B	B	B	B	B	B	B	B	B	A(c)	A
Calcium Hypochlorite	D	D	D	D	C	C	C	C	D	C	D	B	A
Calcium Sulfate	B	C	C	C	B	B	B	B	B	B	B	B	A

D.19

APPENDIX D-9 Corrosion Data (*Continued*)

Chemicals	Aluminum	Ductile iron	Bronze	Carbon steel	S. S. 304	S. S. 316	Alloy 20	Monel	Nickel	Hastelloy	TFE/FEP
Carbolic Acid	A	D	A	D	B	B	B	B	A	A	A
Carbon Bisulfide	A	B	C	B	B	B	B	B	B	B	A
Carbon Dioxide (Dry)	A	B	A	A	A	A	A	A	A		A
Carbonic Acid	B	D	D	D	B	B	B	B	B	A	A
Carbon Tetrachloride (Dry)	C	C	C	C	B	B	B	A	A	B	A
Carbon Tetrachloride (Wet)	C	D	D	D	B	B	B	B	A	A(c)	A
China Wood Oil (Tung)	A	C	C	C	A	A	A	A	A		A
Chlorinated Solvents (Dry)	D	C	C	C	B	B	B	B	B	B(c)	A
Chlorine Gas (Dry)	D	D	C	D	B	B	B	B			A
Chlorine (Wet)	D	D	D	D	D	D	D	B			A
Chloroacetic Acid	C	D	C	D	C	C	C	B	A	A(c)	A
Chlorobenzene (Dry)	B	B	B	B	B	B	B	A	B	B	A
Chloroform (Dry)	D	B	B	B	A	A	A	B	A	A	A
Chlorosulphonic Acid (Dry)	B	B	B	B	B	B	B	B	B	A	A
Chlorosulphonic Acid (Wet)	D	D	D	D	D	D	D	C			A
Chrome Alum	C	B	C	B	A	A	A	B	D	B	A
Chromic Acid	C	D	D	D	C	C	B	D	B	A	A
Citrus Juices	C	D	B	D	B	B	A	B			A
Coconut Oil	B	C	B	C	B	B	B	B	A		A
Coffee Extracts (Hot)	A	C	B	C	A	A	A	A			A
Coke Oven Gas	A	B	C	B	A	A	A	B	B	B	A
Copper Acetate	C	D	D	D	B	B	B	C	D	B(c)	A
Copper Chloride	D	D	D	D	D	D	D	C	D	B	A
Copper Nitrate	D	D	D	D	B	D	B	D	B	A(c)	A
Copper Sulfate	C	D	D	D	B	B	A	B	B		A
Corn Oil	B	C	B	C	B	B	B	B	B	B	A
Creosote Oil	B	B	B	B	B	B	A	B	B	B	A
Cresylic Acid	C	D	C	C	B	B	B	B		B	A
Crude Oil, Sweet	A	B	B	B	A	A	A	A			A
Crude Oil, Sour	A	C	C	B	A	A	A	A			A

Chemical	1	2	3	4	5	6	7	8	9	10	11
Cutting Oils, Water Emulsions	A	A		A	A	A	A	A	B	A	A
Cyclohexane	A	A		A	A	A	A	A	B	A	A
Diacetone Alcohol	A	A		A	D	A	A	A		A	A
Diethylamine	A	A	A	A	A	A	A	A		A	A
Dowtherms	A	A		A	B	A	A	A		A	A
Drilling Mud	B	B		B	B	B	B	B		B	A
Drip Cocks, Gas	B	B		B	B	B	B	B		B	A
Dry Cleaning Fluids	A	B		C	C	B	B	C		B	A
Drying Oil	C	B	B	C	D	B	B	C	B	B	A
Epsom Salt	A	C		B	D	B	C	C		B	A
Ethane	A	A		A	A	A	A	A		A	A
Ethers	B	B	B	B	B	A	A	B	B	B	A
Ethyl Acetate	A	C		C	B	B	B	B	B	B	A
Ethyl Acrylate	A			A	A	A	A	A	A	A	A
Ethyl Chloride (Dry)	B	B		B	B	A	B	B	B	B	A
Ethyl Chloride & Ethyl Fluoride (wet)	D	D		C	D	C	C	B	B	B	A
Ethylene Glycol	A	B	B	B	B	A	B	B	B	B	A
Ethylene Oxide	A	B		D	B	B	B	B	B	A	A
Fatty Acids	B	D	B	D	D	A	B	B	A	A	A
Ferric Chloride	D	D	B	D	D	D	D	D	D	B	A
Ferric Nitrate	D	D		D	D	B	B	B	D	B	A
Ferric Sulfate	D	D		D	D	A	B	C	D	B	A
Ferrous Chloride	D	D		B	D	C	D	C	D	B	A
Ferrous Sulfate	D	D		B	D	A	B	D	D	B	A
Fertilizer Solutions	B	C		C	B	B	B	B	B		A
Fluorine (Dry)	B	B		B	B	B	B	B	B	A	A
Fluorosilicic Acid	D	B		A	D	A	B	A	D	B	A
Food Fluids and Pastes	A	C		B	C	C	A	B	A	A	A
Formaldehyde (Cold)	A	B		A	B	A	B	A	B	B	A
Formaldehyde (Hot)	B	D		A	D	A	B	B	B	B	A
Formic Acid (Cold)	D	D	B	B	D	A	A	C	B	B	A
Formic Acid (Hot)	D	D	B	B	D	B	B	C	B	A	A
Freon (Dry)	B	B	A	B	B	A	A	A	A	A	A
Fruit Juices	B	D	A	D	D	A	A	A	A		A
Furfural	B	B	B	B	B	A	A	B	B	B	A

APPENDIX D-9 Corrosion Data (*Continued*)

Chemicals	Aluminum	Ductile iron	Bronze	Carbon steel	S. S. 304	S. S. 316	Alloy 20	Monet	Nickel	Hastelloy	TFE/FEP
Gallic Acid	B	D	C	D	B	B	B	B	B	B	A
Gas, Manufactured	B	B	B	B	B	B	B	A	A		A
Gas, Natural	B	B	B	B	A	A	A	A	A		A
Gas Odorizers	A	B	A	B	B	B	B	B			A
Gasoline	A	B	A	A	A	A	A	A	A	A	A
Gasoline (Sour)	A	B	B	B	A	A	A	A	D		
Gelatin	A	D	A	D	A	A	A	A	A		
Glucose	A	B	A	B	A	A	A	A			
Glue	A	A	B	A	B	B	B	A		A	
Glycerine or Glycerol	A	B	B	B	A	A	A	B	A	A	
Glycols	B	B	B	B	B	B	A	B			
Grease	A	A	B	A	A	A	A	B			
Heptane	A	B	A	B	A	A	B	B		A	
Hexane	A	B	B	B	B	B	B	B			
Hexanol, Tertiary	A	A	A	A	A	A	A	A			
Hydraulic Oil, Petroleum Base	A	B	B	A	A	A	A	A			
Hydrobromic Acid	D	D	D	D	D	D	D	D	D	B	
Hydrochloric Acid (Air Free)	D	D	D	D	D	D	D	B	B	A[b]	
Hydrocyanic Acid	B	C	D	C	B	B	B	B	B	B	
Hydrofluoric Acid	D	D	D	D	D	D	C	B	B	B[c]	
Hydrogen Gas (Cold)	A	B	B	B	A	A	A	A	A		
Hydrogen Peroxide (Dilute)	A	D	B	D	B	B	B	B	B	B	
Hydrogen Peroxide (Conc.)	A	D	D	D	B	B	B	B	B	B	
Hydrogen Sulfide (Dry)	B	B	C	B	B	A	A	B	B	B	
Hydrogen Sulfide (Wet)	C	D	D	C	B	B	A	B	B	B	
Hydrofluosilicic Acid	D	D	A	D	C	C	B	B			
Hypo (Sodium Thiosulfate)	A	C	C	D	B	B	B	B	B		
Hypochlorites, Sodium	C	D	D	D	C	C	C	B	D	A[c]	
Ink	B	B	C	D	B	A	A	B	A	B	
Iodine (Wet)	D	D	D	D	D	D	D	D	C	B[c]	

Iodoform	B	C	C	B	A	A	A	C	B	C	
Iso-octane	A	B	A	A	A	A	A	A		A	
Isopropyl Alcohol	B	B	B	B	B	B	B	B		B	
Isopropyl Ether	A	B	A	A	A	A	A	A			
Jet Fuel	A	A	A	A	A	A	A	A		A	
Kerosene	A	B	A	B	A	A	A	A	A	A	
Ketchup	D	D	D	D	A	A	A	B		B	
Ketones	A	A	A	A	A	A	A	A		A	
Lacquers (and Solvents)	A	C	A	C	A	A	A	A		A	
Lactic Acid (Dilute Cold)	A	D	D	D	B	A	A	C	B	A	
Lactic Acid (Dilute Hot)	B	D	D	D	B	A	A	D	B	A	
Lactic Acid (Conc. Cold)	C	D	D	D	B	A	A	D	B	A	
Lactic Acid (Conc. Hot)	C	D	D	D	B	B	A	D	B	A	
Lard Oil	A	C	A	C	A	A	A	B		A	
Lead Acetate	D	D	C	D	B	B	B	B	B	B	
Linoleic Acid	A	B	B	B	A	A	A	B			
Linseed Oil	A	A	B	A	A	A	A	B	A	A	
Liquefied Pet. Gas (LPG)	A	B	A	B	B	B	B	B			
Magnesium Bisulfate	B	B	B	B	A	A	A	B			
Magnesium Chloride	D	D	B	C	B	B	B	B	B	B	
Magnesium Hydroxide	D	B	B	B	B	B	B	B	B	B	
Magnesium Hydroxide (Hot)	D	B	D	B	B	B	B	B	B	B	
Magnesium Sulfate	B	B	B	B	B	A	A	B	B	A	
Maleic Acid	B	B	B	B	B	B	B	B	B	B	A
Malic Acid	B	D	B	D	A	A	A	B	A		
Mayonnaise	D	D	D	D	A	A	A	B	A	B	A
Mercuric Chloride	D	D	D	D	B	B	B	C	C	B(c)	A
Mercuric Cyanide	D	D	D	D	B	B	B	B	A	B	A
Mercury	C	A	D	A	B	B	B	B	B	B	A
Methane	A	B	A	A	A	A	A	A	A	A	A
Methyl Acetate							B			A	A
Methyl Acetone	A	A	A	A	A	A	A	A			A
Methylamine	B	B	D	B	B	B	B	C			A
Methyl Cellosolve	A	B	A	B	A	A	A	B			A
Methyl Chloride (Dry)	D	B	A	B	B	A	A	B	B	B(c)	A

Chemicals	Aluminum	Ductile iron	Bronze	Carbon steel	S. S. 304	S. S. 316	Alloy 20	Monel	Nickel	Hastelloy	TFE/FEP
Methyl Ethyl Ketone	A	A	A	A	A	A	A	A		A	A
Methyl Formate	A	C	A	C	A	A	A	A	B	B	A
Methylene Chloride (Dry)	C	B	B	B	B	B	A	B	A	A	A
Milk	A	D	A	D	A	A	A	B	A	D	A
Mine Waters (Acid)	D	D	C	D	B	B	B	B	A	D	A
Mineral Spirits	A	B	B	B	B	B	B	B	A		A
Mixed Acids (Cold)	D	C	D	C	B	B	A	D	D		A
Molasses, Edible	A	A	A	A	A	A	A	A	A	A	A
Molasses, Crude	A	A	A	A	A	A	A	A	A	A	A
Muriatic Acid	D	D	D	D	D	D	D	B			A
Mustard	B	B	A	B	A	A	A	A	B	B	A
Naphtha	A	B	B	B	B	B	B	B		B	A
Naphthalene	B	B	B	A	B	B	B	B			A
Nickel Ammonium Sulfate	D	D	D	D				B	C		A
Nickel Chloride	D	D	D	D	B	B	B	B	C	A[b]	A
Nickel Nitrate	C	D	D	D	B	B	A	B	D	B	A
Nickel Sulfate	D	D	D	D	B	B	A	B	B	B	A
Nicotinic Acid									C	B[c]	A
Nitric Acid (10%)	D	D	D	D	A	A	A	D	C	B[c]	A
Nitric Acid (30%)	D	D	D	D	A	A	A	D	C	D	A
Nitric Acid (80%)	B	D	D	D	A	A	A	D	C	B[c]	A
Nitric Acid (100%)	A	D	D	A	A	A	A	D	C	B[c]	A
Nitric Acid Anhydrous	B	A	D	A	A	A	A	D	D	D	A
Nitrobenzene	C	B	D	B	B	B	B	B	B	B	A
Nitrogen	A	A	A	A	A	A	A	A	A	A	A
Nitrous Acid (10%)	D	D	D	D	B	B	B	D	D	D	A
Nitrous Gases	B	C	D	B	A	A	A	D			A
Nitrous Oxide	C	C	D	B	B	B	B	D	D	D	A
Oils, Animal	A	A	A	A	A	A	A	A			A
Oil, Cottonseed	B	C	B	C	B	B	B	B			A

D.24

Material	1	2	3	4	5	6	7	8	9
Oils, Fish	A			A	A	A	B	B	B
Oils, Fuel	A			A	A	A	B	B	A
Oils, Lube	A		A	B	A	A	A	B	A
Oils, Mineral	A			A	A	A	B	B	B
Oil, Petroleum (Refined)	A			A	A	A	A	B	A
Oil, Petroleum (Sour)	A			A	A	A	B	C	A
Oil-Water Mixtures	A			A	A	A	B	B	A
Oleic Acid	A		A	B	B	B	B	B	A
Oleum	A	B	A	D	B	B	B	D	B
Olive Oil	A	B		A	A	A	B	B	A
Oxalic Acid	A	B	C	B	B	B	D	B	C
Oxygen	A			A	A	A	B	A	A
Ozone (Wet)	A			A	A	A	C	B	B
Ozone (Dry)	A			A	A	A	A	A	A
Paints and Solvents	A			A	A	A	A	A	A
Palmitic Acid	A	B	B	B	B	B	C	B	B
Palm Oil	A			A	A	A	C	A	A
Paraffin	A	A	B	A	A	A	B	A	A
Paraformaldehyde	A	B	B	B	B	B	B	B	B
Pentane	A	A	B	B	B	B	B	B	A
Perchlorethylene (Dry)	A		A	A	B	B	B	B	B
Petrolatum	A			A	B	B	C	B	B
Phenol (100%)	A	A	A	B	B	B	B	B	A
Phosphoric Acid (10%) Cold	A	A[b]	B	B	B	B	D	D	D
Phosphoric Acid (10%) Hot	A	A[b]	C	B	B	B	D	D	D
Phosphoric Acid (50%) Cold	A	A[b]	C	B	B	B	D	D	D
Phosphoric Acid (50%) Hot	A	A[b]	C	B	B	B	D	D	D
Phosphoric Acid (85%) Cold	A	B[b]	B	A	B		D	D	D
Phosphoric Acid (85%) Hot	A	B[b]	D	B	B		C	D	C
Phthalic Acid		A	B	B	B	A	A	B	B
Phthalic Anhydride	A	A	A	A	B	B	C	B	B
Picric Acid	A	B[c]	D	C	B	B	D	D	B
Pine Oil	A			A	A	A	B	B	A
Pineapple Juice	A			A	A	A	C	C	A
Potassium Bisulfite	A	D		D	B	B	D	C	C

Chemicals	Aluminum	Ductile iron	Bronze	Carbon steel	S. S. 304	S. S. 316	Alloy 20	Monel	Nickel	Hastelloy	TFE/FEP
Potassium Bromide	C	D	C	D	B	B	B	B	B	A	A
Potassium Carbonate	C	B	B	B	B	B	B	B	A	B	A
Potassium Chlorate	C	B	B	B	B	B	B	C	C	B	A
Potassium Chloride	B	B	B	C	A	A	A	B	B	B	A
Potassium Cyanide	D	B	D	B	B	B	B	B	B	B	A
Potassium Dichromate	A	B	B	B	A	A	A	B	B	B	A
Potassium Diphosphate	B	A	B	A	A	A	A	B			A
Potassium Ferricyanide	B	B	C	B	B	B	B	B	B	B	A
Potassium Ferrocyanide	A	B	B	B	B	B	B	B	A	B	A
Potassium Hydroxide (Dilute Cold)	D	B	D	B	B	B	B	A	A	B	A
Potassium Hydroxide (Dilute Hot)	D	B	D	B	B	B	B	A	A	B	A
Potassium Hydroxide (to 70% Cold)	D	B	D	B	B	B	B	A	A	B	A
Potassium Hydroxide (to 70% Hot)	D	B	D	B	B	B	B	A	A	B	A
Potassium Iodide	C	C	B	C	B	B	B	B	B	B	A
Potassium Nitrate	A	B	B	B	B	B	B	A	B	B	A
Potassium Permanganate	A	B	B	B	B	B	B	B	B	A	A
Potassium Sulfate	B	C	B	B	B	B	B	B	B	B	A
Potassium Sulfide	D	D	D	D	B	B	B	D	B	B	A
Potassium Sulfite	B	D	D	D	B	B	B	D	D	B	A
Producer Gas	B	B	B	B	B	B	B	A	A	B	A
Propane	A	A	B	A	A	A	A	A	A	A	A
Propyl Alcohol	A	B	A	B	A	A	A	A			A
Propylene Glycol	A	B	B	B	B	B	B	B	B		A
Pyrogallic Acid	B	B	B	B	B	B	B	B	B		A
Quench Oil	A	B	B	B	A	A	A	A	A	A	A
Resins and Rosins	B	C	B	C	B	B	B	A	A	A	A
Road Tar	A	A	A	A	A	A	A	A			A
Roof Pitch		A	A	A	A	A	A	A			A
Rubber Latex Emulsions	A	B	A	B	A	A	A		B	A	A
Rubber Solvent	A	A	A	A	A	A	A	A			A

Chemical											
Salad Oil	B	C	B	B	C	B	B	B	B		A	A
Salicylic Acid	C	D	C	C	D	A	A	A	A	A	A	A
Salt	B	C	B	B	B	B	B	B	A			A
Sea Water	C	D	C	C	D	B	B	B	B		A(c)	A
Shellac (Bleached)	A	B	A	A	A	A	A	A	A		A(c)	A
Shellac (Orange)	A	B	A	A	B	A	A	A	A		A	A
Silver Nitrate	D	D	D	D	B	B	B	B	D		A	A
Soap Solutions (Stearates)	C	A	A	A	A	A	A	A	A		B	A
Sodium Acetate	A	B	B	B	B	B	B	B	A		B	A
Sodium Aluminate	C	C	B	B	C	B	B	B	B		B	A
Sodium Bicarbonate	B	C	B	B	B	B	B	B	B		B	A
Sodium Bisulfate (10%)	D	D	B	B	B	A	A	B	B		B(b)	A
Sodium Bisulfite (10%)	D	D	B	B	B	B	B	B	B		B(c)	A
Sodium Borate	B	C	B	B	B	B	B	B	B		B	A
Sodium Bromide (10%)	B	D	B	B	B	B	B	A	B	B	A	A
Sodium Carbonate	D	B	B	B	B	B	B	B	B		B	A
Sodium Chlorate	B	C	B	B	B	B	B	B	B		B	A
Sodium Chloride	B	C	B	B	B	B	B	A	A		B	A
Sodium Chromate	D	B	C	B	B	B	B	B	B		B	A
Sodium Cyanide	D	B	D	A	A	A	A	A	C		B	A
Sodium Fluoride	C	D	B	C	B	B	B	B	D		B	A
Sodium Hydroxide (Cold) 20%	D	A	A	A	A	A	A	A	A		A(b)	A
Sodium Hydroxide (Hot) 20%	D	B	B	B	A	A	A	A	A		A(b)	A
Sodium Hydroxide (Cold) 50%	D	A	A	A	A	A	A	A	A		A(b)	A
Sodium Hydroxide (Hot) 50%	D	B	A	A	B	A	A	A	A		A(b)	A
Sodium Hydroxide (Cold) 70%	D	D	D	D	B	B	B	B	A		A(b)	A
Sodium Hydroxide (Hot) 70%	D	D	D	D	B	B	B	B	A		A(b)	A
Sodium Hypochlorite		D	D	D	D	D	D	D	D		A	A
Sodium Metaphosphate	D	B	C	B	B	B	B	B	B		B	A
Sodium Metasilicate (Cold)	B	C	B	B	C	A	A	A	A		A	A
Sodium Metasilicate (Hot)	B	D	B	B	A	A	A	A	A		A	A
Sodium Nitrate	A	B	B	B	B	B	A	A	B		A(c)	A
Sodium Perborate	B	B	B	B	B	B	B	B	B		B	A
Sodium Peroxide	C	D	D	C	B	B	B	C	B		B	A
Sodium Phosphate (Dibasic)	D	B	B	B	B	B	B	B	B		B	A

APPENDIX D-9 Corrosion Data (*Continued*)

Chemicals	Aluminum	Ductile iron	Bronze	Carbon steel	S. S. 304	S. S. 316	Alloy 20	Monet	Nickel	Hastelloy	TFE/FEP
Sodium Phosphate (Tribasic)	D	B	B	B	B	B	B	B	B	B	A
Sodium Silicate	B	B	B	B	B	B	B	B	B	B	A
Sodium Silicate (Hot)	C	B	B	B	B	B	B	B	B	B	A
Sodium Sulfate	A	B	B	B	B	A	A	B	B	B	A
Sodium Sulfide	C	B	D	B	B	B	B	B	B	B	A
Sodium Sulfide (Hot)	D	C	D	C	B	B	B	B	B	B	A
Sodium Thiosulfate	A	B	B	B	B	B	B	B			A
Soybean Oil	B	C	B	C	A	A	A	A	D	A	A
Stannic Chloride	D	D	C	D	D	D	D	D	D	B	A
Stannous Chloride	D	D	D	D	D	A	A	C	B	B	A
Starch	A	A	B	A	A	A	A	A			A
Stearic Acid	B	C	C	C	B	B	B	B	B	A	A
Stoddard Solvent	A	B	B	B	B	B	B	B			A
Styrene	A	B	A	A	A	A	A	A			A
Sugar Liquids	A	B	A	B	A	A	A	A	A	A	A
Sulfate, Black Liquor	D	B	D	B	B	A	B	B	D	A	A
Sulfate, Green Liquor	D	B	D	B	B	A	B	C	B	B	A
Sulfate, White Liquor	B	C	C	C	B	A	B	C	A	B	A
Sulphur	A	A	D	A	A	A	A	A	A	A[c]	A
Sulphur Dioxide (Dry)	B	B	B	B	B	B	B	B	B	A[c]	A
Sulphur Trioxide (Dry)	C	B	B	B	B	B	B	B	B	B	A
Sulfuric Acid (0–7%)	B	D	C	D	C	B	A	A	D	A[b]	A
Sulfuric Acid (20%)	D	D	C	D	D	D	A	B	D	A[b]	A
Sulfuric Acid (50%)	D	D	C	D	D	D	A	B	D	B	A
Sulfuric Acid (100%)	D	B	C	B	B	B	A	B	D	B[b]	A
Sulfurous Acid	C	D	C	D	B	B	B	D	C	A[c]	A
Synthesis Gas	B	B	B	B	B	B	B	A	A		A
Tall Oil	D	B	B	B	B	B	B	B	B	A	A
Tannic Acid	C	C	B	C	B	B	B	B	B	B	A
Tartaric Acid	B	D	C	D	B	A	B	B	B	B	A

	1	2	3	4	5	6	7	8	9	10	11
Tetraethyl Lead	A		A	A	B	B	B	C	B	C	B
Toluene or Toluol	A	A	A	A	A	A	A	A	A	A	A
Tomato Juice	A	A	A	A	A	A	A	C	C	C	A
Transformer Oil	A	A	A	A	A	A	A	A	B	B	A
Tributyl Phosphate	A	A	A	A	A	A	A	A	A	A	A
Trichloroethylene	A	A	A	A	B	B	B	B	B	C	A
Tung Oil	A			C	A	A	A	B	B	B	B
Turpentine	A	A	A	B	A	A	A	B	B	B	A
Urea	A	B	B	B	B	B	B	C	B	C	B
Varnish	A	A	A	A	A	A	A	C	A	C	A
Vegetable Oi, Edible	A	B	A	B	A	A	A	B	B	B	A
Vegetable Oil, Non-edible	A	B	A	B	A	A	A	B	B	B	A
Vinegar	A	A	A	A	A	A	A	D	B	D	C
Water, Distilled (Aerated)	A	A	A	A	A	A	A	D	A	D	A
Water, Fresh	A	A	A	A	A	A	A	C	A	C	A
Water, Sea	A	A	A	A	A	A	A	D	B	D	B
Wax Emulsions	A	A	A	A	A	A	A	A	A	B	A
Whiskey and Wine	A	A	A	A	A	A	A	D	A	D	D
Xylene (Dry)	A	A	A	A	A	A	A	A	A	A	A
Zinc Chloride	A	B[b]	B	B	A	D	D	D	D	C	D
Zinc Hydrosulfite	A	B	B	B	A	A	A	B	C	B	D
Zinc Sulfate	A	B	B	B	A	A	A	D	B	D	D

Ratings: A—Excellent, B—Good, C—Fair, D—Not Recommended, Blank—Insufficient Data

APPENDIX D-10 Thermal Expansion Coefficients

Formula	Name	a	T_c, K	m	T_{min}, K	T_{max}, K	β_{liq} @25°C	ρ_{liq}, g/cm³ @25°C
CBrClF₂	Bromochlorodifluoromethane	8.448E-04	426.15	-0.7356	113.65	404.84	2.046E-03	1.810
CBrCl₃	Bromotrichloromethane	6.001E-04	606.00	-0.7143	252.15	575.70	9.735E-04	1.994
CBrF₃	Bromotrifluoromethane	1.078E-03	340.15	-0.7199	105.15	323.14	4.861E-03	1.536
CBr₂F₂	Dibromodifluoromethane	7.047E-04	478.00	-0.7374	163.05	454.10	1.449E-03	2.274
CClF₃	Chlorotrifluoromethane	1.229E-03	301.96	-0.7093	92.15	286.86	NA	0.841
CCIN	Cyanogen chloride	7.816E-04	449.00	-0.7614	266.65	426.55	1.793E-03	1.172
CCl₂F₂	Dichlorodifluoromethane	9.838E-04	384.95	-0.7035	115.15	365.70	2.805E-03	1.307
CCl₂O	Phosgene	7.801E-04	455.00	-0.7280	145.37	432.25	1.694E-03	1.363
CCl₃F	Trichlorofluoromethane	7.815E-04	471.20	-0.7143	162.04	447.64	1.598E-03	1.477
CCl₄	Carbon tetrachloride	6.699E-04	556.35	-0.7100	250.33	528.53	1.155E-03	1.583
CF₂O	Carbonyl fluoride	1.069E-03	297.00	-0.7143	161.89	282.15	NA	NA
CF₄	Carbon tetrafluoride	1.610E-03	227.50	-0.7091	89.56	216.13	NA	NA
CHBr₃	Tribromomethane	6.030E-04	696.00	-0.6994	281.20	661.20	8.916E-04	2.876
CHClF₂	Chlorodifluoromethane	1.008E-03	369.30	-0.7188	115.73	350.84	3.294E-03	1.193
CHCl₂F	Dichlorofluoromethane	8.258E-04	451.58	-0.7143	138.15	429.00	1.786E-03	1.367
CHCl₃	Chloroform	7.376E-04	536.40	-0.7123	209.63	509.58	1.315E-03	1.480
CHF₃	Trifluoromethane	1.322E-03	298.89	-0.7109	117.97	283.95	NA	0.667
CHI₃	Triiodomethane	4.565E-04	794.55	-0.7143	396.16	754.82	NA	NA
CHN	Hydrogen cyanide	1.039E-03	456.65	-0.7179	259.91	433.82	2.222E-03	0.680
CHNS	Isothiocyaic acid	NA	NA	NA	NA	NA	NA	NA
CH₂BrCl	Bromochloromethane	6.585E-04	557.00	-0.7143	185.20	529.15	1.138E-03	1.926
CH₂Br₂	Dibromomethane	6.288E-04	611.00	-0.7242	220.60	580.45	1.021E-03	2.482
CH₂ClF	Chlorofluoromethane	9.882E-04	424.91	-0.7143	140.16	403.66	2.345E-03	1.256
CH₂Cl₂	Dicloromethane	7.736E-04	510.00	-0.7098	178.01	484.50	1.443E-03	1.318
CH₂F₂	Difluoromethane	1.101E-03	351.60	-0.7190	137.00	334.2	4.266E-03	0.957
CH₂I₂	Diiodomethane	4.926E-04	747.00	-0.7346	279.25	709.65	7.161E-04	3.306
CH₂O	Formaldehyde	1.053E-03	408.00	-0.7143	181.15	387.60	2.687E-03	0.736
CH₂O₂	Formic acid	5.772E-04	580.00	-0.7634	281.55	551.00	1.001E-03	1.214
CH₃Br	Methylbromide	8.018E-04	467.00	-0.7197	179.55	443.65	1.667E-03	1.662
C₂H₃N	Acetonitrile	7.659E-04	545.50	-0.7187	229.32	518.23	1.352E-03	0.779
C₂H₃NO	Methylisocyanate	8.207E-04	505.00	-0.7143	256.15	479.75	1.553E-03	0.926
C₂H₄	Ethylene	1.286E-03	282.36	-0.7143	104.01	268.24	NA	NA
C₂H₄Br₂	1,1-Dibromoethane	6.331E-04	628.00	-0.7018	210.15	596.60	9.948E-04	2.045
C₂H₄Br₂	1,2-Dibromoethane	5.814E-04	650.15	-0.7143	282.94	617.64	9.012E-04	2.169
C₂H₄Cl₂	1,1-Dichloroethane	7.281E-04	523.30	-0.7130	176.19	496.55	1.329E-03	1.168
C₂H₄Cl₂	1,2-Dichloroethane	6.899E-04	561.00	-0.6896	237.49	532.95	1.164E-03	1.246
C₂H₄Cl₂O	Bis(chloromethyl)ether	6.760E-04	579.00	-0.7143	231.65	550.05	1.133E-03	1.312

Formula	Name							
$C_2H_4F_2$	1,1-Difluoroethane	9.847E-04	386.60	-0.7203	156.15	367.27	2.849E-03	0.898
$C_2H_4F_2$	1,2-Difluoroethane	9.031E-04	476.00	-0.7143	215.00	452.20	1.824E-03	1.016
$C_2H_4I_2$	1,2-Diiodoethane	5.359E-04	749.91	-0.7143	356.16	712.41	NA	NA
C_2H_4O	Acetaldehyde	8.111E-04	461.00	-0.7224	150.15	437.95	1.720E-03	0.774
C_2H_4O	Ethylene oxide	8.092E-04	469.15	-0.7175	161.45	445.69	1.669E-03	0.862
C_2H_4OS	Thioacetic acid	5.685E-04	577.34	-0.7143	150.16	548.47	9.553E-04	1.059
$C_2H_4O_2$	Acetic acid	5.939E-04	592.71	-0.7316	289.81	563.07	9.903E-04	1.043
$C_2H_4O_2$	Methyl formate	7.689E-04	487.20	-0.7232	174.15	462.84	1.525E-03	0.967
C_2H_4S	Thiacyclopropane	7.937E-04	555.00	-0.7143	165.37	527.25	1.376E-03	1.007
C_2H_5Br	Bromoethane	7.499E-04	503.80	-0.7202	154.55	478.61	1.430E-03	1.450
C_2H_5Cl	Ethyl chloride	6.496E-04	460.35	-0.7686	136.75	437.33	1.448E-03	0.890
C_2H_5ClO	2-Chloroethanol	5.018E-04	585.00	-0.7811	205.65	555.75	8.756E-04	1.196
C_2H_5F	Ethylfluoride	8.496E-04	375.31	-0.7558	129.95	356.54	2.808E-03	0.712
C_2H_5I	Ethyliodide	7.186E-04	561.00	-0.7015	162.05	532.95	1.223E-03	1.920
C_2H_5N	Ethyleneimine	6.338E-04	537.00	-0.7664	195.20	510.15	1.179E-03	0.831
C_2H_5NO	Acetamide	5.701E-04	761.00	-0.7143	354.15	722.95	NA	NA
C_2H_5NO	N-Methylformamide	5.696E-04	721.00	-0.7253	269.35	684.95	8.387E-04	0.999
$C_2H_5NO_2$	Nitroethane	6.758E-04	593.00	-0.7220	183.63	563.35	1.119E-03	1.043
$C_2H_5NO_3$	Ethyl nitrate	NA	NA	NA	NA	NA	NA	NA
C_2H_6	Ethane	1.203E-03	305.42	-0.7167	90.35	290.15	NA	0.315
C_2H_5AlCl	Dimethylaluminum chloride	6.878E-04	619.00	-0.7143	252.15	488.05	1.100E-03	0.988
C_2H_6O	Dimethyl ether	9.360E-04	400.10	-0.7194	131.66	380.10	2.503E-03	0.655
C_2H_6O	Ethanol	6.107E-04	516.25	-0.7633	159.05	490.44	1.179E-03	0.787
C_2H_6OS	Dimethyl sulfoxide	6.087E-04	726.00	-0.6780	291.67	689.70	8.712E-04	1.095
$C_2H_6O_2$	Ethylene glycol	3.644E-04	645.00	-0.8280	260.15	612.75	6.091E-04	1.110
$C_2H_6O_4S$	Dimethyl sulfate	6.593E-04	758.00	-0.6298	241.35	720.10	9.032E-04	1.322
C_2H_6S	Dimethyl sulfide	7.483E-04	503.04	-0.7143	174.88	477.89	1.421E-03	0.850
C_2H_6S	Ethyl mercaptan	7.322E-04	499.15	-0.7213	125.26	474.19	1.411E-03	0.833
$C_2H_6S_2$	Dimethyl disulfide	6.596E-04	606.00	-0.6886	188.44	575.70	1.052E-03	1.057
C_2H_7N	Dimethylamine	7.465E-04	437.65	-0.7520	180.96	415.77	1.764E-03	0.650
C_2H_7N	Ethylamine	8.527E-04	456.15	-0.7141	192.15	433.34	1.818E.03	0.677
C_2H_7NO	Monoethanolamine	4.724E-04	638.00	-0.7985	283.65	606.10	7.811E-04	1.014
$C_2H_8N_2$	Ethylenediamine	5.354E-04	593.00	-0.7983	284.29	563.35	9.352E-04	0.893
C_2H_8Si	Dimethyl silane	9.420E-04	402.00	-0.7158	122.93	381.90	2.484E-03	0.578
C_2N_2	Cyanogen	8.052E-04	400.15	-0.7937	245.25	380.14	2.382E-03	0.866
C_3F_6	Hexafluoropropylene	1.200E-03	368.00	-0.6887	116.65	349.60	3.770E-03	1.304
C_3F_6O	Hexafluoroacetone	9.315E-04	357.14	-0.7710	151.15	339.28	3.734E-03	1.321
C_3F_8	Octafluoropropane	1.062E-03	345.05	-0.7183	125.46	327.80	4.452E-03	1.317
$C_3H_2N_2$	Malononitrile	5.790E-04	715.00	-0.7391	304.90	679.25	NA	NA
C_3H_3Cl	Propargyl chloride	7.920E-04	541.00	-0.6868	293.00	513.95	1.373E-03	1.024
C_3H_3N	Acrylonitrile	7.966E-04	535.00	-0.7106	189.63	508.25	1.421E-03	0.801
C_3H_3NO	Oxazole	5.796E-04	554.00	-0.7143	189.15	526.30	1.007E-03	0.718
C_3H_4	Methylacetylene	9.222E-04	402.39	-0.7210	170.45	382.27	2.442E-03	0.607
C_3H_4	Propadiene	1.009E-03	393.15	-0.6970	136.87	373.49	2.714E-03	0.579

APPENDIX D-10 Thermal Expansion Coefficients (*Continued*)

Formula	Name	T_c, K	a	m	T_{min}, K	T_{max}, K	β_{liq} @25°C	ρ_{liq} g/cm³ @25°C
$C_3H_4Cl_2$	2,3-Dichloropropene	577.00	7.282E-04	-0.6937	191.50	548.15	1.206E-03	1.201
C_3H_4O	Acrolein	506.00	6.603E-04	-0.7511	185.45	480.70	1.288E-03	0.834
C_3H_4O	Propargyl alcohol	580.00	6.848E-04	-0.7143	221.35	551.00	1.147E-03	0.945
$C_3H_4O_2$	Acrylic acid	615.00	6.759E-04	-0.6930	286.65	584.25	1.070E-03	1.046
$C_3H_4O_2$	Beta-propilactone	686.00	6.014E-04	-0.7143	239.75	651.70	9.037E-04	1.262
$C_3H_4O_2$	Vinyl formate	498.00	8.164E-04	-0.7055	200.00	473.10	1.555E-03	0.954
$C_3H_4O_3$	Ethylene carbonate	790.00	5.213E-04	-0.7140	309.55	750.50	NA	NA
$C_3H_4O_3$	Pyruvic acid	634.52	5.277E-04	-0.7660	286.75	602.79	8.580E-04	1.265
C_3H_5Br	3-Bromo-1-propene	540.20	6.695E-04	-0.7143	153.76	513.19	1.188E-03	1.389
C_3H_5Cl	2-Chloropropene	478.00	7.790E-04	-0.7143	135.75	454.10	1.566E-03	0.895
C_3H_5Cl	3-Chloropropene	514.15	7.495E-04	-0.7143	138.65	488.44	1.392E-03	0.931
C_3H_5ClO	Alpha-epichlorohydrin	610.00	6.603E-04	-0.6969	215.95	579.50	1.054E-03	1.174
$C_3H_5ClO_2$	Methyl chloroacetate	600.00	7.129E-04	-0.6905	241.03	570.00	1.146E-03	1.229
$C_3H_5ClO_2$	Ethyl chloroformate	508.15	7.483E-04	-0.7170	192.00	482.74	1.410E-03	1.127
$C_3H_5Cl_3$	1,2,3-Trichloropropane	652.00	6.136E-04	-0.7143	258.45	619.40	9.495E-04	1.384
C_3H_5I	3-Iodo-1-propene	595.81	6.667E-04	-0.7143	173.86	566.02	1.094E-03	1.839
C_3H_5N	Propionitrile	564.40	7.205E-04	-0.7196	180.26	536.18	1.237E-03	0.777
C_3H_5NO	Acrylamide	710.00	5.547E-04	-0.7143	357.65	674.50	NA	NA
C_3H_5NO	Hydracrylonitrile	690.00	4.688E-04	-0.7755	227.15	655.50	2.270E-04	1.040
C_3H_5NO	Lactonitrile	643.00	5.090E-04	-0.7666	233.00	610.85	8.206E-04	0.983
$C_3H_5N_3O_9$	Nitroglycerine	680.00	5.633E-04	-0.7001	286.15	646.00	8.437E-04	1.586
C_3H_6	Cyclopropane	397.91	9.296E-04	-0.7143	145.73	378.01	2.497E-03	0.619
C_3H_6	Propylene	364.76	1.070E-03	-0.6975	87.90	346.52	3.503E-03	0.504
CH_3Cl	Methyl chloride	416.25	9.256E-04	-0.7131	175.45	395.44	2.273E-03	0.913
CH_3Cl_3Si	Methyl trichlorosilane	517.00	7.217E-04	-0.7221	195.35	491.15	1.343E-03	1.266
CH_3F	Methyl fluoride	317.70	1.276E-03	-0.7146	131.35	301.82	9.359E-03	0.566
CH_3I	Methyl iodide	528.00	6.863E-04	-0.7321	206.70	501.60	1.262E-03	2.265
CH_3NO	Formamide	771.00	5.199E-04	-0.7482	275.70	732.45	7.495E-04	1.129
CH_3NO_2	Nitromethane	588.15	7.061E-04	-0.7097	244.60	558.74	1.166E-03	1129
CH_3NO_2	Methyl nitrite	NA	NA	NA	NA	NA	NA	NA
CH_3NO_3	Methyl nitrate	NA	NA	NA	NA	NA	NA	NA
CH_4	Methane	190.58	1.809E-03	-0.7230	90.67	181.05	NA	NA
CH_4Cl_2Si	Methyl dichlorosilane	483.00	5.903E-04	-0.7747	182.55	458.85	1.242E-03	1.103
CH_4O	Methanol	512.58	5.922E-04	-0.7669	175.47	486.95	1.155E-03	0.787
CH_4O_3S	Methanesulfonic acid	NA	NA	NA	NA	NA	NA	1.477
CH_4S	Methyl mercaptan	469.95	7.722E-04	-0.7148	150.18	446.45	1.588E-03	0.862
CH_5ClSi	Methyl chlorosilane	442.00	8.038E-04	-0.7343	139.05	419.90	1.833E-03	0.884
CH_5N	Methylamine	430.05	8.155E-04	-0.7725	179.69	408.55	2.032E-03	0.655
CH_6Si	Methyl silane	352.50	1.077E-03	-0.7120	116.34	334.88	4.077E-03	0.486

CN_4O_8	Tetranitromethane	9.030E-04	540.00	−0.7143	287.05	513.00	1.603E-03	1.626
CO	Carbon monoxide	2.809E-03	132.92	−0.7095	68.15	126.27	NA	NA
COS	Carbonyl sulfide	8.883E-04	378.80	−0.7286	134.35	359.86	2.742E-03	1.005
CO_2	Carbon dioxide	1.280E-03	304.19	−0.7097	216.58	288.98	NA	0.713
CS_2	Carbon disulfide	7.285E-04	552.00	−0.6774	161.58	524.40	1.233E-03	1.256
C_2BrF_3	Bromotrifluoroethylene	9.911E-04	432.00	−0.7000	173.00	410.40	2.251E-03	1.830
$C_2Br_2F_4$	1,2-Dibromotetrafluoroethane	7.295E-04	487.50	−0.7345	162.65	463.51	1.460E-03	2.162
C_2ClF_3	Chlorotrifluoroethylene	1.094E-03	379.15	−0.6926	115.00	360.19	3.187E-03	1.275
C_2ClF_5	Chloropentafluoroethane	1.015E-03	353.15	−0.7147	173.71	335.49	3.836E-03	1.287
$C_2Cl_2F_4$	1,2-Dichlorotetrafluoroethane	8.600E-04	418.85	−0.7255	179.15	397.91	2.121E-03	1.455
$C_2Cl_3F_3$	1,1,2-Trichlorotrifluoroethane	7.497E-04	487.25	−0.7196	238.15	462.89	1.481E-03	1.564
C_2Cl_4	Tetrachloroethylene	6.414E-04	620.00	−0.6437	250.80	589.00	9.781E-04	1.613
$C_2Cl_4F_2$	1,1,2,2-tetrachlorodifluoroethane	6.839E-04	551.00	−0.7143	299.15	523.45	NA	NA
C_2Cl_4O	Trichloroacetyl chloride	6.292E-04	590.00	−0.7143	273.15	560.50	1.040E-03	1.613
C_2Cl_6	Hexachloroethane	5.893E-04	698.00	−0.7143	459.95	663.10	NA	NA
C_2F_4	Tetrafluoroethylene	1.225E-03	306.45	−0.7143	142.00	291.13	NA	0.920
C_2F_6	Hexafluoroethane	1.344E-03	292.80	−0.7021	172.45	278.16	NA	NA
$C_2HBrClF_3$	Halothane	7.215E-04	521.00	−0.7143	223.15	494.95	1.323E-03	1.869
C_2HClF_2	2-Chloro-1,1-difluoroethylene	9.484E-04	400.55	−0.7118	134.65	380.52	2.504E-03	1.217
C_2HCl_3	Trichloroethylene	6.560E-04	571.00	−0.7143	188.40	542.45	1.112E-03	1.458
C_2HCl_3O	Dichloroacetyl chloride	6.466E-04	579.00	−0.7143	298.15	550.05	NA	1.519
C_2HCl_3O	Trichloroacetaldehyde	6.526E-04	565.00	−0.7216	216.00	536.75	1.121E-03	1.499
C_2HCl_5	Pentachloroethane	5.669E-04	665.00	−0.7151	244.15	631.75	8.675E-04	1.675
C_2HF_3	Trifluoroethene	1.030E-03	347.22	−0.7143	94.53	329.86	4.167E-03	0.919
$C_2HF_3O_2$	Trifluoroacetic acid	7.966E-04	491.25	−0.6971	257.90	466.69	1.527E-03	1.480
C_2HF_5	Pentafluoroethane	1.122E-03	342.00	−0.7143	170.15	324.90	4.867E-03	1.174
C_2H_2	Acetylene	1.210E-03	308.32	−0.7143	192.40	292.90	NA	0.377
$C_2H_2Br_4$	1,1,2,2-Tetrabromoethane	3.495E-04	824.00	−0.7860	273.15	782.80	4.975E-04	2.927
$C_2H_2Cl_2$	1,1-Dichloroethylene	7.338E-04	482.00	−0.7143	150.65	457.90	1.461E-03	1.117
$C_2H_2Cl_2$	Cis-1,2-dichloroethylene	7.276E-04	527.00	−0.7143	193.15	500.65	1.320E-03	1.265
$C_2H_2Cl_2$	Trans-1,2-dichloroethylene	7.427E-04	508.00	−0.7143	223.35	482.60	1.397E-03	1.244
$C_2H_2Cl_2O$	Chloroacetyl chloride	6.735E-04	581.00	−0.7143	251.15	551.95	1.126E-03	1.434
$C_2H_2Cl_2O$	Dichloroacetaldehyde	7.056E-04	555.00	−0.7143	223.00	527.25	1.223E-03	1.433
$C_2H_2Cl_2O_2$	Dichloroacetic acid	6.631E-04	686.00	−0.6745	286.55	651.70	9.742E-04	1.553
$C_2H_2Cl_3F$	1,1,-Trichlorofluoroethane	7.010E-04	565.00	−0.7143	173.00	536.75	1.198E-03	1.575
$C_2H_2Cl_4$	1,1,1,2-Tetrachloroethane	6.173E-04	624.00	−0.7143	202.94	592.80	9.818E-04	1.535
$C_2H_2Cl_4$	1,1,2,2-Tetrachloroethane	6.189E-04	645.00	−0.7041	229.35	612.75	9.578E-04	1.587
$C_2H_2F_2$	1,1-Difluoroethylene	1.466E-03	302.80	−0.6906	129.15	287.66	NA	0.594
$C_2H_2F_2$	Cis-1,2,difluoroethene	1.039E-03	394.67	−0.7143	107.90	374.94	2.842E-03	1.023
$C_2H_2F_2$	Trans-1,2-difluoroethene	1.039E-03	394.67	−0.7143	107.90	374.90	2.842E-03	1.023
$C_2H_2F_4$	1,1,1,2-Tetrafluoroethane	1.173E-03	380.00	−0.6863	172.15	361.00	3.365E-03	1.199
C_2H_2O	Ketene	1.005E-03	370.00	−0.7143	122.00	351.50	3.241E-03	0.660
$C_2H_2O_4$	Oxalic acid	5.462E-04	804.00	−0.7143	462.65	763.80	NA	NA
C_2H_3Br	Vinyl bromide	8.392E-04	473.00	−0.7110	135.35	449.35	1.703E-03	1.499

Formula	Name	a	T_c, K	m	T_{min}, K	T_{max}, K	β_{liq} @25°C	ρ_{liq}, g/cm³ @25°C
C_2H_3Cl	Vinyl chloride	8.216E-04	432.00	-0.7284	119.36	410.40	1.929E-03	0.903
$C_2H_3ClF_2$	1-Chloro-1,1-difluoroethane	1.021E-03	410.20	-0.7020	142.35	389.69	2.538E-03	1.107
C_2H_3ClO	Acetyl chloride	7.391E-04	508.00	-0.7143	160.30	482.60	1.390E-03	1.102
C_2H_3ClO	Chloroacetaldehyde	7.355E-04	555.00	-0.7143	293.00	527.25	1.275E-03	1.200
$C_2H_3ClO_2$	Chloroacetic acid	6.076E-04	686.00	-0.7143	333.15	651.70	NA	NA
$C_2H_3ClO_2$	Methyl chloroformate	6.978E-04	525.00	-0.7219	192.00	498.75	1.279E-03	1.213
$C_2H_3Cl_3$	1,1,1-Trichloroethane	6.996E-04	545.00	-0.7067	242.75	517.75	1.224E-03	1.330
$C_2H_3Cl_3$	1,1,2-Trichloroethane	7.042E-04	602.00	-0.6900	236.50	571.90	1.129E-03	1.435
C_2H_3F	Vinyl fluoride	1.129E-03	327.80	-0.7143	112.65	311.41	6.281E-03	0.620
$C_2H_3F_3$	1,1,1-Trifluoroethane	1.015E-03	346.25	-0.7375	161.85	328.94	4.353E-03	0.953
$C_3H_6Br_2$	1,2-Dibromopropane	4.995E-04	634.11	-0.7143	217.96	602.40	7.864E-04	1.925
$C_3H_6Cl_2$	1,1-Dichloropropane	6.764E-04	560.00	-0.7143	200.00	532.00	1.164E-03	1.126
$C_3H_6Cl_2$	1,2-Dichloropropane	6.718E-04	572.00	-0.7143	172.71	543.40	1.137E-03	1.150
$C_3H_6Cl_2$	1,3-Dichloropropane	6.015E-04	603.00	-0.7290	173.65	572.85	9.891E-04	1.181
$C_3H_6Cl_2$	2,2-Dichloropropane	7.014E-04	539.46	-0.7143	239.36	512.49	1.246E-03	1.106
$C_3H_6I_2$	1,2-Diiodopropane	5.194E-04	780.49	-0.7143	253.16	741.47	7.325E-04	2.566
C_3H_6O	Acetone	7.981E-04	508.20	-0.7010	178.45	482.79	1.483E-03	0.786
C_3H_6O	Allyl alcohol	7.182E-04	545.05	-0.7143	144.15	517.80	1.264E-03	0.845
C_3H_6O	Methyl vinyl ether	7.856E-04	437.00	-0.7420	151.15	415.15	1.839E-03	0.744
C_3H_6O	n-Propionaldehyde	8.136E-04	496.00	-0.7140	193.15	471.20	1.568E-03	0.796
C_3H_6O	1,2-Propytene oxide	7.828E-04	482.25	-0.7143	161.22	458.14	1.546E-03	0.823
C_3H_6O	1,3-Propylene oxide	7.541E-04	520.00	-0.7143	255.00	494.00	1.386E-03	0.894
$C_3H_6O_2$	Ethyl formate	7.573E-04	508.40	-0.7065	193.55	482.98	1.413E-03	0.917
$C_3H_6O_2$	Methyl acetate	7.326E-04	506.80	-0.7255	175.15	481.46	1.395E-03	0.927
$C_3H_6O_2$	Propionic acid	6.180E-04	604.00	-0.7236	252.45	573.80	1.011E-03	0.988
$C_3H_6O_2S$	3-Mercaptopropionic acid	8.010E-04	729.00	-0.5953	290.65	692.55	1.096E-03	1.213
$C_3H_6O_3$	Lactic acid	6.186E-04	616.00	-0.7143	291.15	585.20	9.923E-04	1.201
$C_3H_6O_3$	Methoxyacetic acid	5.483E-04	691.00	-0.7255	281.00	656.45	8.259E-04	1.170
$C_3H_6O_3$	Trioxane	6.397E-04	604.00	-0.7143	334.65	573.80	NA	NA
C_3H_6S	Thiacyclobutane	7.005E-04	603.00	-0.7143	199.96	572.85	1.140E-03	1.014
C_3H_7Br	1-Bromopropane	7.218E-04	544.00	-0.7084	163.15	516.80	1.267E-03	1.345
C_3H_7Br	2-Bromopropane	9.959E-04	532.00	-0.6200	184.15	505.40	1.658E-03	1.282
C_3H_7Cl	Isopropyl chloride	7.542E-04	489.00	-0.7143	155.97	464.55	1.477E-03	0.855
C_3H_7Cl	n-Propyl chloride	7.358E-04	503.15	-0.7143	150.35	477.99	1.397E-03	0.856
C_3H_7F	1-Fluoropropane	9.043E-04	422.00	-0.7143	114.16	400.90	2.171E-03	0.787
C_3H_7F	2-Fluoropropane	9.206E-04	415.68	-0.7143	139.80	394.90	2.270E-03	0.733
C_3H_7I	Isopropyl iodide	6.914E-04	578.00	-0.6977	183.15	549.10	1.147E-03	1.695
C_3H_7I	n-Propyl iodide	6.817E-04	593.00	-0.6988	171.85	563.35	1.111E-03	1.739
C_3H_7N	Allylamine	5.873E-04	505.00	-0.7924	184.95	479.75	1.191E-03	0.757

Formula	Name							
C_3H_7N	Propyleneimine	7.354E-04	529.00	-0.7140	229.00	502.55	1.329E-03	0.802
C_3H_7NO	N,N-Dimethylformamide	6.274E-04	647.00	-0.7237	212.72	614.65	9.810E-04	0.945
C_3H_7NO	N-Methylacetamide	5.511E-04	718.00	-0.7262	301.15	682.10	NA	NA
$C_3H_7NO_2$	1-Nitropropane	6.367E-04	605.00	-0.7264	169.16	574.75	1.043E-03	0.996
$C_3H_7NO_2$	2-Nitropropane	6.449E-04	594.00	-0.7263	181.83	564.30	1.070E-03	0.983
C_37NO_3	Propyl-nitrate	NA	NA	NA	NA	NA	NA	NA
$C_3H_7NO_3$	Isopropyl-nitrate	NA	NA	NA	NA	NA	NA	NA
C_3H_8	Propane	9.950E-04	369.82	-0.7130	85.46	351.33	3.206E-03	0.493
C_3H_8O	Isopropanol	6.353E-04	508.31	-0.7570	185.28	482.89	1.240E-03	0.783
C_3H_8O	Methyl ethyl ether	8.204E-04	437.80	-0.7105	160.00	415.91	1.848E-03	0.692
C_3H_8O	n-Propanol	6.050E-04	536.71	-0.7506	146.95	509.87	1.112E-03	0.802
$C_3H_8O_2$	2-Methoxyethanol	6.921E-04	564.00	-0.7143	188.05	535.80	1.184E-03	0.960
$C,3H_8O_2$	Methylal	7.829E-04	480.60	-0.6825	168.35	456.57	1.516E-03	0.854
$C_3H_8O_2$	1,2-Propylene glycol	4.389E-04	626.00	-0.7954	213.15	594.70	7.342E-04	1.033
$C_3H_8O_2$	1,3-Propylene glycol	6.005E-04	658.00	-0.7143	246.45	625.10	9.242E-04	1.052
$C_3H_8O_3$	Glycerol	2.963E-04	723.00	-0.8459	291.33	686.85	4.646E-04	1.257
C_3H_8S	n-Propylmercaptan	6.411E-04	536.00	-0.7308	159.95	509.20	1.161E-03	0.836
C_3H_8S	Isopropyl mercaptan	7.016E-04	517.00	-0.7143	142.61	491.15	1.297E-03	0.809
C_3H_8S	Ethyl-methyl-sulfide	7.183E-04	532.80	-0.7143	167.20	506.16	1.290E-03	0.832
C_3H_9N	n-Propylaine	7.093E-04	496.95	-0.7539	190.15	472.10	1.415E-03	0.714
C_3H_9N	Isopropylamine	7.955E-04	471.85	-0.7028	177.95	448.26	1.606E-03	0.684
C_3H_9N	Trimethylamine	8.426E-04	433.25	-0.7313	156.08	411.59	1.976E-03	0.629
C_3H_9NO	1-Amino-2-propanol	5.281E-04	614.00	-0.7787	274.89	583.30	8.862E-04	0.957
C_3H_9NO	3-Amino1-propanol	6.083E-04	649.00	-0.7143	284.15	616.55	9.440E-04	0.972
C_3H_9NO	Methylethanolamine	6.245E-04	630.00	-0.7143	268.65	598.50	9.871E-04	0.934
$C_3H_9O_4P$	Trimethyl phosphate	NA	NA	NA	NA	NA	NA	1.202
$C_3H_{10}N_2$	1,2-Propanediamine	6.773E-04	587.00	-0.7438	236.53	557.65	1.148E-03	0.856
$C_3H_{10}Si$	Trimethyl silane	9.004E-04	432.00	-0.7077	137.26	410.40	2.063E-03	0.614
C_4Cl_4S	Tetrachlorothiophene	5.245E-04	753.00	-0.7143	301.97	715.35	NA	NA
C_4Cl_6	Hexachloro-1,3-butadiene	5.299E-04	741.00	-0.7143	252.15	703.95	7.654E-04	1.556
C_4F_8	Octafluoro-2-butene	9.830E-04	392.00	-0.7187	138.15	372.40	2.747E-03	1.442
C_4F_8	Octafluorocyclobutane	9.628E-04	388.37	-0.7223	232.96	368.95	2.763E-03	1.495
C_4F_{10}	Decafluorobutane	9.417E-04	386.35	-0.7330	144.95	367.03	2.780E-03	1.497
C_4H_2	Butadiyne (Blacetylene)	7.804E-04	478.02	-0.7143	237.16	454.12	1.569E-03	0.709
$C_4H_2O_3$	Maleic anhydride	6.622E-04	721.00	-0.6442	326.00	684.95	NA	NA
C_4H_4	Vinylacetylene	8.394E-04	454.00	-0.7143	179.95	431.30	1.801E-03	0.680
$C_4H_4N_2$	Succinonitrile	5.347E-04	770.00	-0.7298	331.30	731.50	NA	NA
C_4H_4O	Furan	7.427E-04	490.15	-0.7395	187.55	465.64	1.485E-03	0.935
$C_4H_4O_2$	Diketene	6.022E-04	616.00	-0.7143	266.65	585.20	9.660E-04	1.050
$C_4H_4O_3$	Succinic anhydride	5.286E-04	811.00	-0.7143	393.00	770.45	NA	NA
$C_4H_4O_4$	Furmaric acid	5.430E-04	771.00	-0.7143	560.15	732.45	NA	NA
$C_4H_4O_4$	Maleic acid	5.497E-04	773.00	-0.7100	403.45	734.35	NA	NA
C_4H_4S	Thiophene	6.724E-04	579.35	-0.6923	234.94	550.38	1.109E-03	1.059
C_4H_5Cl	Chloroprene	7.062E-04	525.00	-0.7214	143.15	498.75	1.294E-03	0.950

APPENDIX D-10 Thermal Expansion Coefficients (*Continued*)

Formula	Name	a	T_c, K	m	T_{min}, K	T_{max}, K	β_{liq} @25°C	ρ_{liq}, g/cm³ @25°C
C_5H_5N	Trans-crotonitrile	7.288E-04	586.00	−0.7143	222.00	556.70	1.211E-03	0.807
C_5H_5N	Cis-crotonitrile	7.197E-04	568.00	−0.7178	200.55	539.60	1.228E-03	0.819
C_5H_5N	Methacrylonitrile	7.694E-04	554.00	−0.7037	237.35	526.30	1.325E-03	0.795
C_4H_5N	Pyrrole	5.419E-04	639.75	−0.7521	249.74	607.76	8.686E-04	0.965
C_4H_5N	Vinylacetonitrile	7.102E-04	584.00	−0.7101	186.15	554.80	1.180E-03	0.829
$C_5H_5NO_2$	Methyl cyanoacetate	6.896E-04	687.00	−0.6805	260.08	652.65	1.016E-03	1.119
C_4H_6	Cyclobutene	8.195E-04	446.33	−0.7143	153.76	424.01	1.801E-03	0.704
C_4H_6	1,2-Butadiene	8.498E-04	444.00	−0.7143	136.95	421.80	1.882E-03	0.646
C_4H_6	1,3-Butadiene	8.892E-04	425.37	−0.7093	164.25	404.10	2.093E-03	0.615
C_4H_6	Dimethylacetylene	7.680E-04	488.15	−0.7143	240.91	463.74	1.507E-03	0.686
C_4H_6	Ethylacetylene	8.569E-04	443.20	−0.7169	147.43	421.04	1.908E-03	0.648
$C_4H_6Cl_2$	1,3-Dichloro-trans-2-butene	5.952E-04	618.00	−0.7212	276.00	587.10	9.571E-04	1.153
$C_4H_6Cl_2$	1,4-Dichloro-cis-2-butene	6.307E-04	640.00	−0.7143	225.15	608.00	9.870E-04	1.188
$C_4H_6Cl_2$	1,4-Dichloro-trans-2-butene	6.262E-04	646.00	−0.7143	274.15	613.70	9.744E-04	1.187
$C_4H_6Cl_2$	3,4-Dichloro-1-butene	6.560E-04	589.00	−0.7143	212.00	559.55	1.086E-03	1.148
C_4H_6O	Trans-crotonaldehyde	7.951E-04	571.00	−0.6765	196.65	542.45	1.310E-03	0.847
C_4H_6O	2,5-Dihydrofuran	7.031E-04	542.00	−0.7143	273.00	514.90	1.244E-03	0.939
C_4H_6O	Divinyl ether	7.944E-04	463.00	−0.7143	172.05	439.85	1.661E-03	0.731
C_4H_6O	Methacrolein	7.591E-04	530.00	−0.7143	192.15	503.50	1.370E-03	0.840
$C_4H_6O_2$	2-Butyne-1,4-diol	5.538E-04	695.00	−0.7143	331.00	660.25	NA	NA
$C_4H_6O_2$	Gamma-butyrolactone	5.107E-04	739.00	−0.7350	229.78	702.05	7.466E-04	1.125
$C_4H_6O_2$	Cis-crotonic acid	6.270E-04	647.00	−0.7143	288.65	614.65	9.747E-04	1.023
$C_4H_6O_2$	Trans-crotonic acid	6.124E-04	666.00	−0.7143	344.55	632.70	NA	NA
$C_4H_6O_2$	Methyacrylic acid	6.271E-04	643.00	−0.7143	288.15	610.85	9.786E-04	1.012
$C_4H_6O_2$	Methyl acrylate	7.277E-04	536.00	−0.7143	196.32	509.20	1.300E-03	0.949
$C_4H_6O_2$	Vinyl acetate	7.309E-04	524.00	−0.7173	180.35	497.80	1.337E-03	0.926
$C_4H_6O_3$	Acetic anhydride	6.752E-04	569.15	−0.7301	200.15	540.69	1.161E-03	1.077
$C_4H_6O_4$	Succinic acid	5.517E-04	806.00	−0.7143	461.15	765.50	NA	NA
$C_4H_6O_5$	Diglycolic acid	5.356E-04	820.00	−0.7143	421.15	779.00	NA	NA
$C_4H_6O_5$	Malic acid	4.956E-04	781.00	−0.7143	403.15	741.95	NA	NA
$C_4H_6O_6$	Tartaric acid	5.071E-04	828.00	−0.7143	479.15	786.60	NA	NA
C_4H_7N	n-Butyronitrile	6.939E-04	582.25	−0.7141	161.25	553.14	1.158E-03	0.786
C_4H_7N	Isobutyronitrile	7.475E-04	565.00	−0.7002	201.70	536.75	1.264E-03	0.766
C_4H_7NO	Acetone cyanohydrin	6.305E-04	647.00	−0.7143	253.15	614.65	9.801E-04	0.928
C_4H_7NO	2-Methyacrylamide	5.212E-04	741.00	−0.7100	383.65	703.95	NA	NA
C_4H_7NO	3-Methoxypropionitrile	6.744E-04	638.00	−0.7143	210.12	606.10	1.058E-03	0.924
C_4H_7NO	2-Pyrrolidone	5.314E-04	792.00	−0.7036	298.15	752.40	NA	1.108
C_4H_8	1-Butene	8.997E-04	419.59	−0.7147	87.80	398.61	2.182E-03	0.588
C_4H_8	Cis-2-butene	8.575E-04	435.58	−0.7143	134.26	413.80	1.955E-03	0.617

	Name							
C_4H_8	Trans-2-butene	8.674E-04	428.63	−0.7143	167.62	407.20	2.029E-03	0.599
C_4H_8	Cyclobutane	6.297E-04	459.93	−0.7619	182.48	436.93	1.396E-03	0.689
C_4H_8	Isobutene	8.846E-04	417.90	−0.7204	132.81	397.00	2.177E-03	0.589
$C_4H_8Br_2$	1,2-Dibromobutane	4.871E-04	659.28	−0.7143	207.76	626.32	7.487E-04	1.785
$C_4H_8Br_2$	2,3-Dibromobutane	4.901E-04	656.96	−0.7143	238.66	624.11	7.550E-04	1.774
$C_4H_8Cl_2$	1,4-Dichlorobutane	6.258E-04	641.00	−0.7027	235.85	608.95	9.715E-04	1.135
$C_4H_8I_2$	1,2-Diiodobutane	5.223E-04	726.41	−0.7143	279.06	690.09	7.618E-04	2.280
C_4H_8O	n-Butyraldehyde	7.583E-04	525.00	−0.7143	176.75	498.75	1.381E-03	0.797
C_4H_8O	Isobutyraldehyde	7.658E-04	507.00	−0.7143	208.15	481.65	1.443E-03	0.784
C_4H_8O	1,2-Epoxybutane	7.390E-04	526.00	−0.7143	123.15	499.70	1.343E-03	0.824
C_4H_8O	Methyl ethyl ketone	7.366E-04	535.50	−0.7143	186.48	508.73	1.317E-03	0.799
C_4H_8O	Ethyl vinyl ether	7.919E-04	475.15	−0.7143	157.35	451.39	1.603E-03	0.749
C_4H_8O	Tetrahydrofuran	6.847E-04	540.15	−0.7088	164.65	513.14	1.210E-03	0.880
$C_4H_8O_2$	Ci-2-butene-1,4-diol	5.880E-04	677.88	−0.7143	284.15	643.99	8.895E-04	1.070
$C_4H_8O_2$	Trans-2-butene-1,4-diol	5.856E-04	681.00	−0.7143	300.45	646.95	NA	NA
$C_4H_8O_2$	Isobutyric acid	6.027E-04	609.15	−0.7314	227.15	578.69	9.855E-04	0.946
$C_4H_8O_2$	n-Butyric acid	5.973E-04	628.00	−0.7200	267.95	596.60	9.495E-04	0.953
$C_4H_8O_2$	1,4-Dioxane	6.584E-04	587.00	−0.6953	284.95	557.65	1.078E-03	1.029
$C_4H_8O_2$	Ethyl acetate	7.186E-04	523.30	−0.7220	189.60	497.13	1.321E-03	0.894
$C_4H_8O_2$	Methyl propionate	7.060E-04	530.60	−0.7230	185.65	504.07	1.282E-03	0.909
$C_4H_8O_2$	n-Propyl formate	6.984E-04	538.00	−0.7200	180.25	511.10	1.249E-03	0.900
$C_4H_8O_2S$	Sulfolane	4.691E-04	849.00	−0.6960	300.75	806.55	NA	NA
C_4H_8S	Tetrahydrothiophene	4.777E-04	631.95	−0.7512	176.99	600.35	7.715E-04	0.997
C_4H_9Br	1-Bromobutane	6.696E-04	577.00	−0.7109	160.75	548.15	1.123E-03	1.269
C_4H_9Br	2-Bromobutane	5.435E-04	567.00	−0.7600	161.25	538.65	9.582E-04	1.253
C_4H_9Cl	n-Butyl chloride	7.320E-04	537.00	−0.7046	150.55	510.15	1.295E-03	0.880
C_4H_9Cl	Sec-butyl chloride	7.031E-04	520.60	−0.7209	141.85	494.57	1.298E-03	0.868
C_4H_9Cl	Tert-butyl chloride	7.229E-04	507.00	−0.7143	247.75	481.65	1.362E-03	0.836
C_4H_9I	2-Iodo-2-methylpropane	6.418E-04	587.90	−0.7143	234.96	558.51	1.064E-03	1.536
C_4H_9N	Pyrrolidine	6.197E-04	568.55	−0.7367	215.31	540.12	1.071E-03	0.860
C_4H_9NO	N,N-Dimethylacetamide	5.987E-04	658.00	−0.7298	253.15	625.10	9.300E-04	0.937
C_4H_9NO	Morpholine	5.644E-04	618.00	−0.7436	270.05	587.10	9.210E-04	0.996
$C_4H_9NO_2$	1-Nitrobutane	6.168E-04	624.00	−0.7143	191.83	592.80	9.810E-04	0.968
$C_4H_9NO_2$	2-Nitrobutane	6.187E-04	615.00	−0.7143	141.16	584.25	9.936E-04	0.978
C_4H_{10}	n-Butane	8.757E-04	425.18	−0.7137	134.86	403.92	2.074E-03	0.573
C_4H_{10}	Isobutane	8.686E-04	408.14	−0.7270	113.54	387.73	2.253E-03	0.552
$C_4H_{10}N_2$	Piperazine	5.061E-04	638.00	−0.7143	379.15	606.10	NA	NA
$C_4H_{10}O$	n-Butanol	5.768E-04	562.93	−0.7543	183.85	534.78	1.019E-03	0.806
$C_4H_{10}O$	Sec-butanol	6.479E-04	536.01	−0.7396	158.45	509.21	1.182E-03	0.805
$C_4H_{10}O$	Tert-butanol	7.357E-04	506.20	−0.7263	298.97	480.89	NA	NA
$C_4H_{10}O$	Diethyl ether	8.096E-04	466.70	−0.7064	156.85	443.37	1.662E-03	0.708
$C_4H_{10}O$	Methyl-propyl-ether	8.013E-04	476.20	−0.7143	156.87	452.39	1.618E-03	0.723
$C_4H_{10}O$	Methyl isopropyl ether	6.609E-04	464.50	−0.7556	127.93	441.28	1.436E-03	0.714
$C_4H_{10}O$	Isobutanol	5.570E-04	547.73	−0.7657	165.15	520.34	1.017E-03	0.797

APPENDIX D-10 Thermal Expansion Coefficients (*Continued*)

Formula	Name	a	T_c, K	m	T_{min}, K	T_{max}, K	β_{liq} @25°C	ρ_{liq}, g/cm³ @25°C
$C_4H_{10}O_2$	1,3-Butanediol	5.989E-04	643.00	-0.7143	196.15	610.85	9.346E-04	1.002
$C_4H_{10}O_2$	1,4-Butanediol	5.916E-04	667.00	-0.7143	293.05	633.65	9.032E-04	1.013
$C_4H_{10}O_2$	2,3-Butanediol	6.131E-04	611.00	-0.7140	280.75	580.45	9.887E-04	0.994
$C_5H_{10}O_2$	t-Butyl hydroperoxide	6.527E-04	576.00	-0.7143	277.45	547.20	1.099E-03	0.886
$C_4H_{10}O_2$	1,2-Dimethoxyethane	7.141E-04	536.15	-0.7143	215.15	509.34	1.276E-03	0.865
$C_4H_{10}O_2$	2-Ethoxyethanol	6.767E-04	569.00	-0.7143	183.00	540.55	1.150E-03	0.925
$C_4H_{10}O_3$	Diethylene glycol	4.368E-04	744.60	-0.7578	262.70	707.37	6.436E-04	1.114
$C_4H_{10}O_4S$	Diethylsulfate	6.920E-04	792.00	-0.5916	248.00	752.40	9.151E-04	1.172
$C_4H_{10}S$	n-Butyl mercaptan	6.880E-04	569.00	-0.7006	157.46	540.55	1.157E-03	0.837
$C_4H_{10}S$	Isobutyl mercaptan	5.873E-04	559.00	-0.7405	128.31	531.05	1.033E-03	0.830
$C_4H_{10}S$	Sec-butyl mercaptan	7.247E-04	554.00	-0.6932	133.02	526.30	1.238E-03	0.825
$C_4H_{10}S$	Tert-butyl mercaptan	7.828E-04	530.00	-0.6800	274.26	503.50	1.373E-03	0.795
$C_4H_{10}S$	Diethyl sulfide	6.573E-04	557.15	-0.7256	169.20	529.29	1.146E-03	0.832
$C_4H_{10}S$	Isopropyl-methyl-sulfide	6.809E-04	551.00	-0.7143	171.65	523.45	1.188E-03	0.825
$C_4H_{10}S$	Methyl-propyl-sulfide	6.720E-04	563.00	-0.7143	160.19	534.85	1.152E-03	0.837
$C_4H_{10}S_2$	Diethyl disulfide	6.000E-04	642.00	-0.6994	171.63	609.90	9.285E-04	0.988
$C_4H_{11}N$	n-Butylamine	6.425E-04	531.90	-0.7572	224.05	505.31	1.198E-03	0.741
$C_4H_{11}N$	Isobutylamine	6.421E-04	513.73	-0.7635	188.55	488.04	1.246E-03	0.730
$C_4H_{11}N$	Sec-Butylamine	6.642E-04	514.30	-0.7529	168.65	488.58	1.276E-03	0.720
$C_4H_{11}N$	Tert-Butylamine	6.788E-04	483.90	-0.7460	206.19	459.71	1.387E-03	0.688
$C_4H_{11}N$	Diethylamine	7.536E-04	496.60	-0.7272	223.35	471.77	1.468E-03	0.702
$C_4H_{11}NO$	Dimethyl ethanomiamine	6.708E-04	571.82	-0.7143	214.15	543.23	1.135E-03	0.882
$C_4H_{11}NO_2$	Diethanolamine	3.780E-04	715.00	-0.8108	301.15	679.25	NA	NA
$C_4H_{11}NO_2$	2-Aminoethoxyethanol	5.713E-04	699.00	-0.7143	293.15	664.05	8.497E-04	1.051
$C_4H_{12}N_2O$	N-Aminoethyl ethanolamine	5.625E-04	698.00	-0.7143	273.15	663.10	8.375E-04	1.022
$C_4H_{12}Si$	Tetramethylsilane	8.546E-04	450.40	-0.7062	174.07	427.88	1.838E-03	0.641
$C_4H_{13}N_3$	Diethylene traimine	5.747E-04	676.00	-0.7143	234.15	642.20	8.708E-04	0.954

APPENDIX D-11 Chemicals Used in Coagulation Process

Chemical name and formula	Common or trade name	Shipping containers	Suitable handling materials	Available forms	Weight lb/cu ft	Solubility lb/gal	Commercial strength per cent	Characteristics
Aluminum sulfate $Al_2(SO_4)_3$ 14 H_2O	alum. filter alum sulfate of alumina	100–200-lb bags 300–400-lb bbls. bulk (carloads) tank truck tank car	dry-iron, steel, solution lead-lined rubber, silicon asphalt, 316 stainless steel	ivory-colored powder granule lump liquid	38–45 60–63 62–67 10 (lb/g)	4.2 (60°F)	15–22 (Al_2O_3) 8 (Al_2O_3)	pH of 1 per cent solution 3.4
Ammonium aluminum sulfate $Al_2(SO_4)_3$ $(NH_4)_2 \cdot SO_4 \cdot 24$ H_2O	ammonia alum crystal alum	bags, bbls. bulk	duriron lead rubber silicon iron stoneware	lump nut pea powdered	64–68 62 65 60	0.3 (32°F) 8.3 (212°F)	11 (Al_2O_3)	pH of 1 per cent solution 3.5
Bentonite	colloidal clay volclay wilkinite	100-lb bags bulk	iron, steel	powder pellet mixed sizes	60	insoluble (colloidal sol used)		
Ferric chloride $FeCl_3$ (35–45 per cent solution)	"ferrichlor" chloride of iron	5–13 · gal carboys, trucks tank cars	glass, rubber, stoneware, synthetic resins	dark brown syrapy liquid		complete	37–47 $(FeCl_3)$ 2–21 (Fe)	hygroscopic (store lumps and powder in tight container) no dry feed; optimum pH, 4.0–11.0
$FeCl_3$–6 H_2O	crystal ferric chloride	300-bl bbls.		yellow-brown lump			59–61 $(FeCl_3)$ 20–21(Fe)	
$FeCl_3$	anhydrous ferric chloride	500-lb casks; 100–300, 400-lb kegs		green-black powder			98 $(FeCl_3)$ 34 (Fe)	
Ferric sulfate $Fe_2(SO_4)_3$ 9 H_2O	"ferrifloc" ferrisul	100–175-lb bags 400–425-lb drums	ceramics, lead plastic rubber 18-8 stainless steel	red-brown powder 70- or granule 72	63–66	soluble in 2–4 parts cold water	90–94 $(Fe)(SO_4)_3$ 25–26 (Fe)	mildly hygroscopic coagulant at pH 3.5–11.0
Ferrous sulfate $FeSo_4$ 7 H_2O	copperos, green vitriol	bags, bbls. bulk	asphalt, concrete lead, tin, wood	green-crystal granule, lump			55 $(FeSo_4)$ 20 (Fe)	hygroscopic; cakes in storage; optimum pH 8.5–11
Potassium aluminum sulfate $K_2SO_4 \cdot Al_2(SO_4)_3$ 24 H_2O	potash alum	bags, lead-lined bulk (carloads)	lead, lead-lined rubber, stoneware	lump granule powder	62–67 60–65 60	0.5 (32°F) 1.0 (68°F) 1.4 (86°F)	10–11 (Al_2O_3)	low, even solubility; pH of 1 per cent solution, 3.5

APPENDIX D-11 Chemicals Used in Coagulation Process

Chemical name and formula	Common or trade name	Shipping containers	Suitable handling materials	Available forms	Weight $lb/cu\,ft$	Solubility lb/gal	Commercial strength per cent	Characteristics
Sodium aluminate $Na_2O\ Al_2O_3$	soda alum	100–150-lb bags 250–440 lbs drums, solution	iron, plastics, rubber, steel	brown powder liquid (27°Be)	50–60	3.0 (68°F) 3.3 (86°F)	70–80 (Na_2) Al_2O_4 min. 32 Na_2 Al_2O_4	hopper agitation required for dry feed
Sodium silicate $Na_2O\ SiO_2$	water glass	drums, bulk (tank trucks, tank cars)	cast iron, rubber, steel	opaque, viscous liquid		complete	38–42°Be	variable ratio of Na_2O to SiO_2; pH of 1 per cent solution, 12.3

APPENDIX D-12 Chemicals Used in Stabilization and Corrosion Control

Chemical name and formula	Common or trade name	Shipping containers	Suitable handling materials	Available forms	Weight lb/cu ft	Solubility lb/gal	Commercial strength per cent	Characteristics
Disodium phosphate $Na_2HPO_4 \cdot 12H_2O$	basic sodium phosphate, DSP, secondary sodium phosphate	125-lb kegs, 200-lb bags, 325-lb bbls.	cast iron, steel	crystal	60–64	0.4 (32°F) 6.4 (86°F)	19,19 5 (P_2O_5)	precipitates ca, Mg, pH of 1 per cent solution, 9.1
Sodium hexametolphosphate $Na(PO_3)_6$	"Calgon" glassy phosphate vitreous phosphate	100-lb bags	hard rubber, plastics, stainless steel	crystal flake powder	47	1–4.2	66 (P_2O_5 unadjusted)	pH of 0.25 per cent solution 6.0–8.3
Sodium hydroxide NaOH	caustic soda, soda lye	100–700-lb drums; bulk (trucks, tank cars)	cast iron, rubber, steel	flake, lump liquid		2.4 (32°F) 4.4 (68°F) 4.8 (104°F)	98.9 (NaOH) 74–76 (NaO_2)	solid hygroscopic pH of 1 per cent solution, 12.9
Sulfuric acid H_2SO_4	oil of vitriol, vitriol	bottles, carboys, drums, trucks, tank cars	concentrated iron, steel; dilute glass, lead, porcelain, rubber	solution	(60-66°) Be	complete	60°Be 77.7 (H_2SO_4) 66°Be 93.2 (H_2SO_4)	approx. pH of 0.5 per cent solution, 1.2
Tetrasodium pyro-phosphate $Na_4P_2O_7 \cdot 10\,H_2O$	alkaline sodium pyrophosphate TSPP	125-lb kegs, 200-lb bags, 300-lb bbls.	cast iron, steel	white powder	68	0.6 (80°F) 3.3 (212°F)	53 (P_2O_5)	pH of 1 per cent solution, 10.8
Trisodium phosphate $Na_3\,PO_4 \cdot 12\,H_2O$	normal sodium phosphate, teriary sodium phosphate TSP	125-lb kegs, 200-lb bags, 325-lb bbls.	cast iron, steel	crystal— course medium standard	56 58 61	0.1 (32°F) 13.0 (158°F)	19 (P_2O_5)	pH of 1 per cent solution, 11.9

APPENDIX D-13 Chemicals Used in Softening Process

Chemical name and formula	Common or trade name	Shipping containers	Suitable handling materials	Available forms	Weight lb/cu ft	Solubility lb/gal	Commercial strength per cent	Characteristics
Calcium oxide CaO	burnt lime, chemical lime, quicklime, unslaked lime	50-lb bags, 100-lb bbls. bulk (carloads)	asphalt, cement, iron, rubber steel	lump pebble granule		slaked to form hydrated lime	75–99 (CaO)	pH of saturated solution, on detention time temp. amount of water critical for efficient slaking
Sodium carbonate Na_2CO_3	soda ash	bags, bbls., bulk (carloads), trucks	iron, rubber, steel	white powder extra light light dense	23 35 65	1.5 (68°F) 2.3 (86°F)	99.4 (Na_2CO_3) 38 (Na_2O)	hopper agitation required for dry feed of light and extra light forms pH of 1 per cent solution, 11.3
Sodium chloride $NaCl$	common salt salt	bags, bbls., bulk (carloads)	bronze, cement, rubber	rock rine		2.9 (32°F) 3.0 (68.F) 86°F	98 ($NaCl$)	
Calcium hydroxide $Ca(OH)_2$	hydrated lime, slaked lime	50-lb bags, 100-lb bbls. bulk (carloads) bulk trucks	asphalt, cement, iron, rubber, steel	white powder light dense		0.014 (68°F) 0.012 (90°F)	85–99 ($Ca(OH)_2$) 63–73 (CaO)	hopper agitation required for dry feed of light form

APPENDIX D-14 Chemicals Used in Taste and Odor Control

Chemical name and formula	Common or trade name	Shipping containers	Suitable handling materials	Available forms	Weight lb/cu ft	Solubility lb/gal	Commercial strength per cent	Characteristics
Activated carbon C	"Aqua Nuchor" "Hydrodarco" "Herite"	bags, bulk	dry iron, steel; wet rubber, silicon, iron, stainless steel	black granules powder	15	insoluble (suspension used)		
Chlorine Cl_3	chlorine gas, liquid chlorine	100-, 150-lb cylinders; 1-ton tanks; 16, 30-, 55-ton tank cars	dry black iron, copper, steel; wet gas glass, hard rubber, silver	liquified gas under pressure	91.7	0.07 (60°F) 0.04 (100°F)	99.8 (Cl_2)	
Chlorine dioxide ClO_2	chlorine dioxide	generated as used	plastics, soft rubber (avoid hard rubber)	yellow-red gas		0.02 (30 mm)	26.3 (available Cl_2)	
Copper sulfate $CuSO_4 \cdot$ 5 H_2O	blue vitriol, blue stone	100-lb bags, 450-lb bbls. drums	asphalt, silicon, iron, stainless steel	crystal lump powder	73–90 73–80 60–64	1.6 (32°F) 2.2 (68°F) 2.6 (86°F)	99 ($CuSO_4$)	
Ozone O_3	ozone	generated at site of application	aluminum, ceramics, glass	colorless gas				
Potassium permanganate $KMnO_4$	purple salt	bulk, bbls., drums	iron, steel wool	purple crystals		infinite	100	danger of explosion in contact organic matters

APPENDIX D-15 Various Properties of Anhydrous Ammonia, Chlorine and Sulfur Dioxide

Property	Anhydrous Ammonia		Chlorine		Sulfur Dioxide	
	Liquid	Gas	Liquid	Gas	Liquid	Gas
Formula	NH_3	NH_3	Cl_2	Cl_2	SO_2	SO_2
Molecular weight	17.032	17.032	70.906	70.906	64.06	64.06
Strength—*per cent*	99.50 NH_3	99.50	99.50 Cl_2	99.50 Cl_2	about 99.90 SO_2	99.90 SO_2
Affinity for water	considerable	considerable	slight	slight	moderate	moderate
Boiling point (at 1 atmos.)—°C	−33.35 (−28°F)		−34.05 (−29.29°F)		−10°C (14°F)	
Color	colorless	colorless	clear amber	greenish-yellow	water white	colorless
Density—*lb/cu ft*	38.50 (at 60°F)	0.04813 (at 32°F and 1 atmos.)	88.79 (at 60°F)	0.2003 (at 32°F and 1 atmos.)	89.58 (at 32°F)	0.1827 (at 32°F and 1 atmos.)
Explosive limits in air—*per cent by volume*		16–25	non-explosive	non-explosive	non-explosive	non-explosive
Flammability	slight	slight	non-flammable	non-flammable	non-flammable	non-flammable
Melting (freezing point) (at 1 atmos.)—°C	−77.7 (−107.9°F)		−100.98 (−149.76°F)		−75.5 (−103.9°F)	
Odor	suffocating; pungent	same as liquid	penetrating; irritating	same as liquid	suffocating; pungent	same as liquid
Solubility (g/100 g H_2O at 20°C and 1 atmos.)		53.1		0.7293		11.28
Specific gravity (compared to 4°C H_2O) at 32°F	0.639		1.468		1.436	
Relative vapor density (air=1) at 32°F and 1 atmos.)		0.597		2.482		2.2636
Viscosity—centipoises	0.255 (at −33.5°C)	0.00918 (at 0°C) 0.01279 (at 100°C)	0.3885 (at 0°C) 0.729 (at −76.5°C)	0.01679 (at 100°C)	0.4904 (at −4°F) 0.3923 (at 32°F) 0.2785 (at 68°F)	0.0117 (at 32°F) 0.0145 (at 68°F) 0.0177 (at 104°F)

APPENDIX E
DESIGN INFORMATION

Appendix Number	Description
E-1	Standard Weight Welded and Seamless Steel Pipe
E-2	American Standard Steel Pipe Flanges
E-3	General Dimensions, Standard Ductile Iron Flanged Fittings
E-4	Water Distribution Reaction at Bends
E-5	Tank Drainage Times
E-6	Flow of Air in Pipes
E-7	Horsepower Calculations
E-8	Stair Design
E-9	Discrete Compounding

EXPLANATION OF APPENDICES

Appendix E-1, Standard Weight Welded and Seamless Steel Pipe: This Table lists for $\frac{1}{8}$ inch through 20 inch diameter steel pipe, the External and Internal Diameters, the Wall Thickness, the External and Internal Circumference, the External, Internal Metal Transverse Area, the External and Internal Length of Pipe per Square Foot of Surface Area, the Weight per Foot of Length, and the Length of Pipe Containing One Cubic Foot.

Appendix E-2, American Standard Steel Pipe Flanges: This Table lists for steel pipe from $\frac{1}{2}$ inch to 24 inches in diameter, the Flange Diameter, Thickness, Bolt Circle Diameter, Number of Bolts and Bolt Size for 150 to 600 psi ratings.

Appendix E-3, General Dimensions, Standard Ductile Iron Flanged Fittings: This Appendix shows the standard dimensions for Ductile Iron Flanged Fittings from 1 inch to 96 inches in diameter.

Appendix E-4, Water Distribution Reaction at Bends: This Appendix gives the reaction on the pipe in pounds for various bends and heads for pipes from 2 to 72 inches in diameter.

Appendix E-5, Tank Drainage Times: This Nomograph gives the time in minutes to drain a tank of various shapes. Begin with an initial fluid volume, V_3 in the right column and an initial head, in feet, H in the middle column. Connect V_3 through the appropriate shape point to Pivot Line 1 (example Point A) and connect H through the Orifice Diameter, d_0, to Pivot Line 2 (example Point B). Connect points A and B. Read the drain time where the A-B line crosses the Drain Time, T line.

Appendix E-6, Flow of Air in Pipes: This Nomograph enables the friction head and velocity head of air flow to be computed. Connect the CFM (14,000 in the example) vertically upward to the pipe size (24″ in the example). Connect this point horizontally to the left to the temperature (120°F in the example). Connect this point vertically downward to the pressure line (8 psi in the example). Connect this point horizontally to the left abscissa to read the Drop in Pressure Due to Friction (0.67 in the example). This horizontal line can be extended to the left to intersect the pipe size (24″ in the example) and downward vertically to the ordinate to read the velocity head (0.9 inches in the example). The friction loss and velocity head are computed similarly to that explained for water in this Handbook.

Appendix E-7, Horsepower Calculations: This Appendix lists the formula for computing horsepower, kilowatts and power cost.

Appendix E-8, Stair Design: This chart and table recommends riser and tread relationships for the preferred stair design.

Appendix E-9, Discrete Compounding: This Appendix lists factors for discrete interest compounding for interest rates from 3% to 15%. To compute Present Worth, P, multiply the Equal Annual Payment, A by the Present Worth Factor for the appropriate Interest Rate, i and Year Term, N. To compute the Equal Annual Payment, A, multiply the Present Worth, A, by the Capital Recovery Factor for the appropriate Interest Rate, i and Term, N.

APPENDIX E-1 Standard Weight Welded and Seamless Pipe

Size	Diameter External Inches	Diameter Internal Inches	Thickness Inches	Circumference External Inches	Circumference Internal Inches	Transverse area External Sq. in.	Transverse area Internal Sq. in.	Transverse area Metal Sq. in.	Length of pipe per square foot External surface Feet	Length of pipe per square foot Internal surface Feet	Weight per foot of length lbs.	Contents in gallons per foot of length	Length of pipe containing one cubic foot
1/8	.405	.269	.068	1.272	.845	.129	.057	.072	9.481	14.199	.244	.002961	2533.775
1/4	.540	.364	.088	1.696	1.144	.229	.104	.125	7.073	10.493	.424	.005403	1383.789
3/8	.675	.493	.091	2.121	1.549	.358	.191	.167	5.658	7.748	.567	.009922	754.360
1/2	.840	.622	.109	2.639	1.954	.554	.304	.250	4.547	6.141	.850	.01579	473.906
3/4	1.050	.824	.113	3.299	2.589	.866	.533	.333	3.637	4.635	1.130	.02769	270.034
1	1.315	1.049	.133	4.131	3.296	1.358	.864	.494	2.904	3.641	1.673	.04488	166.618
1 1/4	1.660	1.380	.140	5.215	4.335	2.164	1.495	.669	2.301	2.768	2.272	.07766	96.275
1 1/2	1.900	1.610	.145	5.969	5.058	2.835	2.036	.799	2.010	2.372	2.717	.1058	70.733
2	2.375	2.067	.154	7.461	6.494	4.430	3.355	1.075	1.608	1.847	3.652	.1743	42.913
2 1/2	2.875	2.469	.203	9.032	7.757	6.492	4.788	1.704	1.328	1.547	5.793	.2487	30.077
3	3.500	3.068	.216	10.996	9.638	9.621	7.393	2.228	1.091	1.245	7.575	.3841	19.479
3 1/2	4.000	3.548	.226	12.566	11.146	12.566	9.886	2.680	.954	1.076	9.109	.5136	14.565
4	4.500	4.026	.237	14.137	12.648	15.904	12.730	3.174	.848	.948	10.790	.6613	11.312
5	5.563	5.047	.258	17.477	15.856	24.306	20.006	4.300	.686	.756	14.617	1.0393	7.198
6	6.625	6.065	.280	20.813	19.054	34.472	28.891	5.581	.576	.629	18.974	1.501	4.984
8	8.625	7.981	.322	27.096	25.073	58.426	50.027	8.399	.443	.478	24.696	2.5989	2.878
10	10.750	10.020	.365	33.772	31.479	90.763	78.855	11.908	.355	.381	40.483	4.0965	1.826
12	12.750	12.000	.375	40.055	37.699	127.676	113.097	14.579	.299	.318	49.562	5.8754	1.273
*14 O.D.	14.000	13.250	.375	43.982	42.626	153.938	137.886	16.052	.272	.288		7.1632	1.044
*15 O.D.	15.000	14.250	.375	47.124	44.768	176.715	159.485	17.230	.254	.268		8.2852	.903
*16 O.D.	16.000	15.250	.375	50.265	47.909	201.062	182.654	18.408	.238	.250		9.4889	.788
*17 O.D.	17.000	16.214	.393	53.407	50.938	226.980	226.476	20.504	.224	.235		10.7264	.697
*18 O.D.	18.000	17.182	.409	56.549	53.979	245.469	231.866	22.603	.212	.222		12.0454	.621
*20 O.D.	20.000	19.182	.409	62.832	60.262	314.159	288.986	25.173	.191	.199		15.0128	.498

*There is no standard thickness for pipes above 14" O.D. These are as published by National Tube Company. Wrought Iron Welded pipe have same outside diameters, but wall thickness .002 to .007 thicker. Based on ASTM Standard Specs "A53-33."

E.3

APPENDIX E-2 American Standard Steel Pipe Flanges

Nominal pipe diameter	Diameter of flange			Flange thickness*				Diameter bolt circle			Number of bolts			Size of bolts			
	150 lbs.	300 400 lbs.	600 lbs.	150 lbs.	300 lbs.	400 lbs.	600 lbs.	150 lbs.	300 400 lbs.	600 lbs.	150 lbs.	300 400 lbs.	600 lbs.	150 lbs.	300 lbs.	400 lbs.	600 lbs.
$\frac{1}{2}$	$3\frac{1}{2}$	$3\frac{3}{4}$	$3\frac{3}{4}$	$\frac{7}{16}$	$\frac{9}{16}$	$\frac{9}{16}$	$\frac{9}{16}$	$2\frac{3}{8}$	$2\frac{5}{8}$	$2\frac{5}{8}$	4	4	4	$\frac{1}{2}$	$\frac{1}{2}$	$\frac{1}{2}$	$\frac{1}{2}$
$\frac{3}{4}$	$3\frac{7}{8}$	$4\frac{5}{8}$	$4\frac{5}{8}$	$\frac{1}{2}$	$\frac{5}{8}$	$\frac{5}{8}$	$\frac{5}{8}$	$2\frac{3}{4}$	$3\frac{1}{4}$	$3\frac{1}{4}$	4	4	4	$\frac{1}{2}$	$\frac{5}{8}$	$\frac{5}{8}$	$\frac{5}{8}$
1	$4\frac{1}{4}$	$4\frac{7}{8}$	$4\frac{7}{8}$	$\frac{9}{16}$	$\frac{11}{16}$	$\frac{11}{16}$	$\frac{11}{16}$	$3\frac{1}{8}$	$3\frac{1}{2}$	$3\frac{1}{2}$	4	4	4	$\frac{1}{2}$	$\frac{5}{8}$	$\frac{5}{8}$	$\frac{5}{8}$
$1\frac{1}{4}$	$4\frac{5}{8}$	$5\frac{1}{4}$	$5\frac{1}{4}$	$\frac{5}{8}$	$\frac{3}{4}$	$\frac{3}{4}$	$\frac{3}{4}$	$3\frac{1}{2}$	$3\frac{7}{8}$	$3\frac{7}{8}$	4	4	4	$\frac{1}{2}$	$\frac{5}{8}$	$\frac{5}{8}$	$\frac{5}{8}$
$1\frac{1}{2}$	5	$6\frac{1}{8}$	$6\frac{1}{8}$	$\frac{11}{16}$	$\frac{13}{16}$	$\frac{13}{16}$	$\frac{7}{8}$	$3\frac{7}{8}$	$4\frac{1}{2}$	$4\frac{1}{2}$	4	4	4	$\frac{1}{2}$	$\frac{3}{4}$	$\frac{3}{4}$	$\frac{3}{4}$
2	6	$6\frac{1}{2}$	$6\frac{1}{2}$	$\frac{3}{4}$	$\frac{7}{8}$	$\frac{7}{8}$	1	$4\frac{3}{4}$	5	5	4	8	8	$\frac{5}{8}$	$\frac{5}{8}$	$\frac{5}{8}$	$\frac{5}{8}$
$2\frac{1}{2}$	7	$7\frac{1}{2}$	$7\frac{1}{2}$	$\frac{7}{8}$	1	1	$1\frac{1}{8}$	$5\frac{1}{2}$	$5\frac{7}{8}$	$5\frac{7}{8}$	4	8	8	$\frac{5}{8}$	$\frac{3}{4}$	$\frac{3}{4}$	$\frac{3}{4}$
3	$7\frac{1}{2}$	$8\frac{1}{4}$	$8\frac{1}{4}$	$\frac{15}{16}$	$1\frac{1}{8}$	$1\frac{1}{8}$	$1\frac{1}{4}$	6	$6\frac{5}{8}$	$6\frac{5}{8}$	4	8	8	$\frac{5}{8}$	$\frac{3}{4}$	$\frac{3}{4}$	$\frac{3}{4}$
$3\frac{1}{2}$	$8\frac{1}{2}$	9	9	$\frac{15}{16}$	$1\frac{3}{16}$	$1\frac{3}{16}$	$1\frac{3}{8}$	7	$7\frac{1}{4}$	$7\frac{1}{4}$	8	8	8	$\frac{5}{8}$	$\frac{3}{4}$	$\frac{7}{8}$	$\frac{7}{8}$
4	9	10	$10\frac{3}{4}$	$\frac{15}{16}$	$1\frac{1}{4}$	$1\frac{3}{8}$	$1\frac{1}{2}$	$7\frac{1}{2}$	$7\frac{7}{8}$	$8\frac{1}{2}$	8	8	8	$\frac{5}{8}$	$\frac{3}{4}$	$\frac{7}{8}$	1
5	10	11	13	$\frac{15}{16}$	$1\frac{3}{8}$	$1\frac{1}{2}$	$1\frac{3}{4}$	$8\frac{1}{2}$	$9\frac{1}{4}$	$10\frac{1}{2}$	8	8	8	$\frac{3}{4}$	$\frac{3}{4}$	$\frac{7}{8}$	1
6	11	$12\frac{1}{2}$	14	1	$1\frac{7}{16}$	$1\frac{5}{8}$	$1\frac{7}{8}$	$9\frac{1}{2}$	$10\frac{5}{8}$	$11\frac{1}{2}$	8	12	12	$\frac{3}{4}$	$\frac{3}{4}$	1	$1\frac{1}{8}$
8	$13\frac{1}{2}$	15	$16\frac{1}{2}$	$1\frac{1}{8}$	$1\frac{5}{8}$	$1\frac{7}{8}$	$2\frac{3}{16}$	$11\frac{3}{4}$	13	$13\frac{3}{4}$	8	12	12	$\frac{3}{4}$	$\frac{7}{8}$	$1\frac{1}{8}$	$1\frac{1}{4}$
10	16	$17\frac{1}{2}$	20	$1\frac{3}{16}$	$1\frac{7}{8}$	$2\frac{1}{8}$	$2\frac{1}{2}$	$14\frac{1}{4}$	$15\frac{1}{4}$	17	12	16	16	$\frac{7}{8}$	1	$1\frac{1}{4}$	$1\frac{1}{4}$
12	19	$20\frac{1}{2}$	22	$1\frac{1}{4}$	2	$2\frac{1}{4}$	$2\frac{5}{8}$	17	$17\frac{3}{4}$	$19\frac{1}{4}$	12	16	20	$\frac{7}{8}$	$1\frac{1}{8}$	$1\frac{1}{4}$	$1\frac{3}{8}$
14 O.D.	21	23	$23\frac{1}{4}$	$1\frac{3}{8}$	$2\frac{1}{8}$	$2\frac{3}{8}$	$2\frac{3}{4}$	$18\frac{3}{4}$	$20\frac{1}{4}$	$20\frac{3}{4}$	12	20	20	1	$1\frac{1}{8}$	$1\frac{3}{8}$	$1\frac{1}{2}$
16 O.D.	$23\frac{1}{2}$	$25\frac{1}{2}$	27	$1\frac{7}{16}$	$2\frac{1}{4}$	$2\frac{1}{2}$	3	$21\frac{1}{4}$	$22\frac{1}{2}$	$23\frac{3}{4}$	16	20	20	1	$1\frac{1}{4}$	$1\frac{3}{8}$	$1\frac{5}{8}$
18 O.D.	25	28	$29\frac{1}{4}$	$1\frac{9}{16}$	$2\frac{3}{8}$	$2\frac{5}{8}$	$3\frac{1}{4}$	$22\frac{3}{4}$	$24\frac{3}{4}$	$25\frac{3}{4}$	16	24	20	$1\frac{1}{8}$	$1\frac{1}{4}$	$1\frac{3}{8}$	$1\frac{5}{8}$
20 O.D.	$27\frac{1}{2}$	$30\frac{1}{2}$	32	$1\frac{11}{16}$	$2\frac{1}{2}$	$2\frac{3}{4}$	$3\frac{1}{2}$	25	27	$28\frac{1}{2}$	20	24	24	$1\frac{1}{8}$	$1\frac{1}{4}$	$1\frac{1}{2}$	$1\frac{5}{8}$
24 O.D.	32	35	37	$1\frac{7}{8}$	$2\frac{3}{4}$	3	4	$29\frac{1}{2}$	32	33	20	24	24	$1\frac{1}{4}$	$1\frac{1}{2}$	$1\frac{1}{4}$	$1\frac{7}{8}$

APPENDIX E-3   General Dimensions, Standard Ductile Iron Flanged Fittings

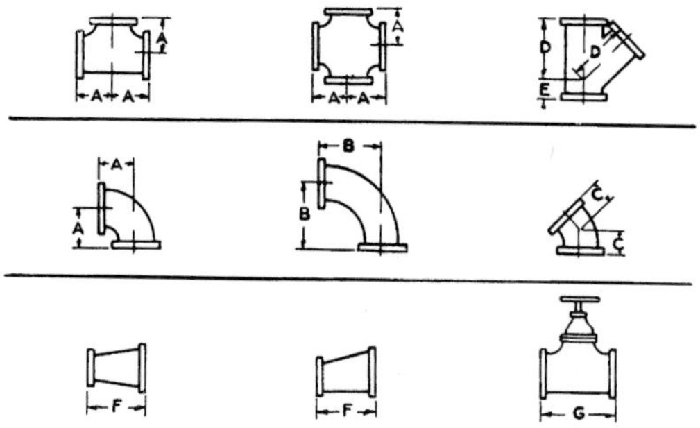

Size	1	1¼	1½	2	2½	3	3½	4	5	6	7	8	9	10	12
A Elbows, Tees & Crosses	3½	3¾	4	4½	5	5½	6	6½	7½	8	8½	9	10	11	12
B Long Radius Elbows	5	5½	6	6½	7	7¾	8½	9	10¼	11½	12¾	14	15¼	16½	19
C C to F 45° Elbows	1¾	2	2¼	2½	3	3	3½	4	4½	5	5½	5½	6	6½	7½
D C to F 45° Laterals	5¾	6¼	7	8	9½	10	11½	12	13½	14½	16½	17½	19½	20½	24½
E C to F 45° Laterals	1¾	1¾	2	2½	2½	3	3	3	3½	3½	4	4½	4½	5	5½
F F to F Reducers				5	5½	6	6½	7	8	9	10	11	11½	12	14
G F to F Gate Valves Std Chapman No. 571				7	7½	8	8½	9	10	10½	11	11½	12	13	14

Size	14	16	18	20	24	30	36	42	48	54	60	72	84	96
A Elbows, Tees & Crosses	14	15	16½	18	22	25	28	31	34	39	44	53	62	71
B Long Radius Elbows	21½	24	26½	29	34	41½	49	56½	64	71½	79	94	109	124
C C to F 45° Elbows	7½	8	8½	9½	12	15	18	21	24	27	30	36	42	48
D C to F 45° Laterals	27	30	32	35	40½	49								
E C to F 45° Laterals	6	6½	7	8	9	10								
F F to F Reducers	16	18	19	20	24	30	36	42	48	54	60	72	84	96
G Gate Valves, Low Press Chapman No. 250 B M	11½	12	12½	13	13½	15	16	17½	19½	21	25	28½		

All dimensions in inches.
American Standards Association "Cast Iron Pipe Flanges and Fittings" except dimension "G".
The National Bureau of Standards has recommended the elimination of some sizes on certain of these fittings.

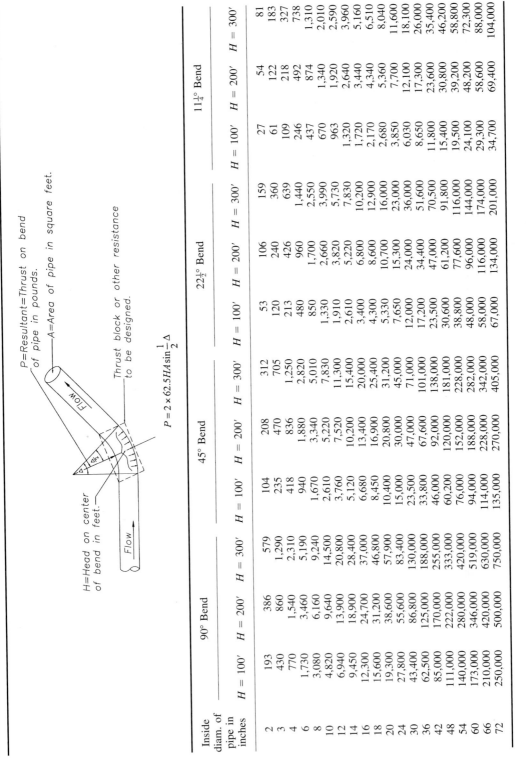

P = Resultant = Thrust on bend of pipe in pounds.

A = Area of pipe in square feet.

H = Head on center of bend in feet.

Thrust block or other resistance to be designed.

$$P = 2 \times 62.5\,HA\sin\tfrac{1}{2}\Delta$$

Inside diam. of pipe in inches	90° Bend			45° Bend			22½° Bend			11¼° Bend		
	$H=100'$	$H=200'$	$H=300'$	$H=100'$	$H=200'$	$H=300'$	$H=100'$	$H=200'$	$H=300'$	$H=100'$	$H=200'$	$H=300'$
2	193	386	579	104	208	312	53	106	159	27	54	81
3	430	860	1,290	235	470	705	120	240	360	61	122	183
4	770	1,540	2,310	418	836	1,250	213	426	639	109	218	327
6	1,730	3,460	5,190	940	1,880	2,820	480	960	1,440	246	492	738
8	3,080	6,160	9,240	1,670	3,340	5,010	850	1,700	2,550	437	874	1,310
10	4,820	9,640	14,500	2,610	5,220	7,830	1,330	2,660	3,990	670	1,340	2,010
12	6,940	13,900	20,800	3,760	7,520	11,300	1,910	3,820	5,730	963	1,920	2,590
14	9,450	18,900	28,400	5,120	10,200	15,400	2,610	5,220	7,830	1,320	2,640	3,960
16	12,300	24,700	37,000	6,680	13,400	20,000	3,400	6,800	10,200	1,720	3,440	5,160
18	15,600	31,200	46,800	8,450	16,900	25,400	4,300	8,600	12,900	2,170	4,340	6,510
20	19,300	38,600	57,900	10,400	20,800	31,200	5,330	10,700	16,000	2,680	5,360	8,040
24	27,800	55,600	83,400	15,000	30,000	45,000	7,650	15,300	23,000	3,850	7,700	11,600
30	43,400	86,800	130,000	23,500	47,000	71,000	12,000	24,000	36,000	6,030	12,100	18,100
36	62,500	125,000	188,000	33,800	67,600	101,000	17,200	34,400	51,600	8,650	17,300	26,000
42	85,000	170,000	255,000	46,000	92,000	138,000	23,500	47,000	70,500	11,800	23,600	35,400
48	111,000	222,000	333,000	60,200	120,000	181,000	30,600	61,200	91,800	15,400	30,800	46,200
54	140,000	280,000	420,000	76,000	152,000	228,000	38,800	77,600	116,000	19,500	39,200	58,800
60	173,000	346,000	519,000	94,000	188,000	282,000	48,000	96,000	144,000	24,100	48,200	72,300
66	210,000	420,000	630,000	114,000	228,000	342,000	58,000	116,000	174,000	29,300	58,600	88,000
72	250,000	500,000	750,000	135,000	270,000	405,000	67,000	134,000	201,000	34,700	69,400	104,000

APPENDIX E-5 Tank Drainage Times

VESSEL CONFIGURATION	VOLUME	VESSEL CONFIGURATION	VOLUME
	$V = \pi R^2 H$		$V = \dfrac{\pi R^2 H}{3}$
	$V = \dfrac{4}{3}\pi R^3$		$V = \dfrac{HLW}{3}$
	$V = \dfrac{\pi}{3}H^2(3R - H)$		$V = \dfrac{HLW}{2}$

APPENDIX E-6 Flow of Air in Pipes

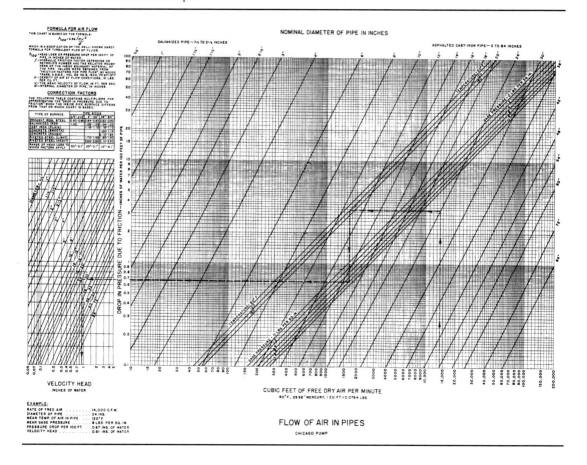

FLOW OF AIR IN PIPES
CHICAGO PUMP

APPENDIX E-7 Horsepower Calculations

$$HP = \frac{gpm \times TDH}{3960 \times eff}$$

$$KW = HP \times 0.746$$

Power Cost

$$= \frac{Power\ Rate/KWH \times gpm \times TDH \times 0.746 \times HR}{3960 \times eff}$$

where:

$$HP = \text{Horsepower in Foot Pounds per Hour}$$
$$TDH = \text{Total Dynamic Head in Feet}$$
$$Eff = \text{Pump Efficiency in Percent}$$
$$Power\ Rate = \text{Cost of Power in Cents per Kilowat Hour}$$
$$HR = \text{Hours}$$

APPENDIX E-8 Stair Design

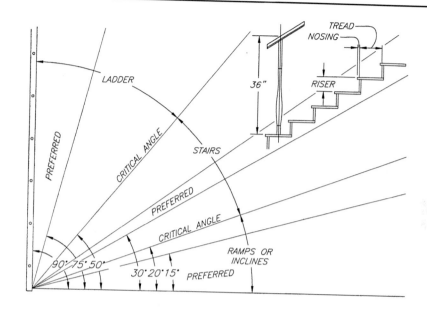

Table of Riser and Treads for Stairs

(tread + riser = 17 1/2)

angle with horizontal	riser in in.	tread in in.
22°–00'	5	12 1/2
23°–14'	5 1/4	12 1/4
24°–38'	5 1/2	12
26°–00'	5 3/4	11 3/4
27°–33'	6	11 1/2
29°–03'	6 1/4	11 1/4
30°–35' *	6 1/2	11
32°–08' *	6 3/4	10 3/4
33°–41' *	7	10 1/2
35°–16' *	7 1/4	10 1/4

* Preferred

angle with horizontal	riser in in.	tread in in.
36°–52'	7 1/2	10
38°–29'	7 3/4	9 3/4
40°–08'	8	9 1/2
41°–44'	8 1/4	9 1/4
43°–22'	8 1/2	9
45°–00'	8 3/4	8 3/4
46°–38'	9	8 1/2
48°–16'	9 1/4	8 1/4
49°–54'	9 1/2	8

APPENDIX E-9   Discrete Compounding

Discrete compounding; $i = 3\%$

	Single payment		Uniform series				
	Compound amount factor	Present worth factor	Compound amount factor	Present worth factor	Sinking fund factor	Capital recovery factor	
N	To find F given P F/P	To find P given F P/F	To find F given A F/A	To find P given A P/A	To find A given F A/F	To find A given P A/P	N
1	1.0300	0.9709	1.0000	0.9709	1.0000	1.0300	1
2	1.0609	0.9426	2.0300	1.9135	0.4926	0.5226	2
3	1.0927	0.9151	3.0909	2.8286	0.3235	0.3535	3
4	1.1255	0.8885	4.1836	3.7171	0.2390	0.2690	4
5	1.1593	0.8626	5.3091	4.5797	0.1884	0.2184	5
6	1.1941	0.8375	6.4684	5.4172	0.1546	0.1846	6
7	1.2299	0.8131	7.6625	6.2303	0.1305	0.1605	7
8	1.2668	0.7894	8.8923	7.0197	0.1125	0.1425	8
9	1.3048	0.7664	10.1591	7.7861	0.0984	0.1284	9
10	1.3439	0.7441	11.4639	8.5302	0.0872	0.1172	10
11	1.3842	0.7224	12.8078	9.2526	0.0781	0.1081	11
12	1.4258	0.7014	14.1920	9.9540	0.0705	0.1005	12
13	1.4685	0.6810	15.6178	10.6349	0.0640	0.0940	13
14	1.5126	0.6611	17.0863	11.2961	0.0585	0.0885	14
15	1.5580	0.6419	18.5989	11.9379	0.0538	0.0838	15
16	1.6047	0.6232	20.1569	12.5611	0.0496	0.0796	16
17	1.6528	0.6050	21.7616	13.1661	0.0460	0.0760	17
18	1.7024	0.5874	23.4144	13.7535	0.0427	0.0727	18
19	1.7535	0.5703	25.1168	14.3238	0.0398	0.0698	19
20	1.8061	0.5537	26.8703	14.8775	0.0372	0.0672	20
21	1.8603	0.5375	28.6765	15.4150	0.0349	0.0649	21
22	1.9161	0.5219	30.5367	15.9369	0.0327	0.0627	22
23	1.9736	0.5067	32.4528	16.436	0.0308	0.0608	23
24	2.0328	0.4919	34.4264	16.9455	0.0290	0.0590	24
25	2.0938	0.4776	36.4592	17.4131	0.0274	0.0574	25
26	2.1566	0.4637	38.5530	17.8768	0.0259	0.0559	26
27	2.2213	0.4502	40.7096	18.3270	0.0246	0.0546	27
28	2.2879	0.4371	42.9309	18.7641	0.0233	0.0533	28
29	2.3566	0.4243	45.2188	19.1884	0.0221	0.0521	29
30	2.4273	0.4120	47.5754	19.6004	0.0210	0.0510	30
35	2.8139	0.3554	60.4620	21.4872	0.0165	0.0465	35
40	3.2620	0.3066	75.4012	23.1148	0.0133	0.0433	40
45	3.7816	0.2644	92.7197	24.5187	0.0108	0.0408	45
50	4.3839	0.2281	112.797	25.7298	0.0089	0.0389	50
55	5.0821	0.1968	136.071	26.7744	0.0073	0.0373	55
60	5.8916	0.1697	163.053	27.6756	0.0061	0.0361	60
65	6.8300	0.1464	194.332	28.4529	0.0051	0.0351	65
70	7.9178	0.1263	230.594	29.1234	0.0043	0.0343	70
75	9.1789	0.1089	272.630	29.7018	0.0037	0.0337	75
80	10.6409	0.0940	321.362	30.2008	0.0031	0.0331	80
85	12.3357	0.0811	377.856	30.6311	0.0026	0.0326	85
90	14.3004	0.0699	443.35	31.0024	0.0023	0.0323	90
95	16.5781	0.0603	519.27	31.3227	0.0019	0.0319	95
100	19.2186	0.0520	607.29	31.5989	0.0016	0.0316	100
∞				33.3333		0.0300	∞

APPENDIX E-9 Discrete Compounding (*Continued*)

Discrete compounding; $i = 5\%$

	Single payment		Uniform series				
	Compound amount factor	Present worth factor	Compound amount factor	Present worth factor	Sinking fund factor	Capital recovery factor	
N	To find F given P F/P	To find P given F P/F	To find F given A F/A	To find P given A P/A	To find A given F A/F	To find A given P A/P	N
1	1.0500	0.9524	1.0000	0.9524	1.0000	1.0500	1
2	1.1025	0.9070	2.0500	1.8594	0.4878	0.5378	2
3	1.1576	0.8638	3.1525	2.7232	0.3172	0.3672	3
4	1.2155	0.8227	4.3101	3.5460	0.2320	0.2820	4
5	1.2763	0.7835	5.5256	4.3295	0.1810	0.2310	5
6	1.3401	0.7462	6.8019	5.0757	0.1470	0.1970	6
7	1.4071	0.7107	8.1420	5.7864	0.1228	0.1728	7
8	1.4775	0.6768	9.5491	6.4632	0.1047	0.1547	8
9	1.5513	0.6446	11.0266	7.1078	0.0907	0.1407	9
10	1.6289	0.6139	12.5779	7.7217	0.0795	0.1295	10
11	1.7103	0.5847	14.2068	8.3064	0.0704	0.1204	11
12	1.7959	0.5568	15.9171	8.8633	0.0628	0.1128	12
13	1.8856	0.5303	17.7130	9.3936	0.0565	0.1065	13
14	1.9799	0.5051	19.5986	9.8986	0.0510	0.1010	14
15	2.0789	0.4810	21.5786	10.3797	0.0463	0.0963	15
16	2.1829	0.4581	23.6575	10.8378	0.0423	0.0923	16
17	2.2920	0.4363	25.8404	11.2741	0.0387	0.0887	17
18	2.4066	0.4155	28.1324	11.6896	0.0355	0.0855	18
19	2.5269	0.3957	30.5390	12.0853	0.0327	0.0827	19
20	2.6533	0.3769	33.0659	12.4622	0.0302	0.0802	20
21	2.7860	0.3589	35.7192	12.8212	0.0280	0.0780	21
22	2.9253	0.3418	38.5052	13.1630	0.0260	0.0760	22
23	3.0715	0.3256	41.4305	13.4886	0.0241	0.0741	23
24	3.2251	0.3101	44.5020	13.7986	0.0225	0.0725	24
25	3.3864	0.2953	47.7271	14.0939	0.0210	0.0710	25
26	3.5557	0.2812	51.1134	14.3752	0.0196	0.0696	26
27	3.7335	0.2678	54.6691	14.6430	0.0183	0.0683	27
28	3.9201	0.2551	58.4026	14.8981	0.0171	0.0671	28
29	4.1161	0.2429	62.3227	15.1411	0.0160	0.0660	29
30	4.3219	0.2314	66.4388	15.3725	0.0151	0.0651	30
35	5.5160	0.1813	90.3203	16.3742	0.0111	0.0611	35
40	7.0400	0.1420	120.800	17.1591	0.0083	0.0583	40
45	8.9850	0.1113	159.700	17.7741	0.0063	0.0563	45
50	11.4674	0.0872	209.348	18.2559	0.0048	0.0548	50
55	14.6356	0.0683	272.713	18.6335	0.0037	0.0537	55
60	18.6792	0.0535	353.584	18.9293	0.0028	0.0528	60
65	23.8399	0.0419	456.798	19.1611	0.0022	0.0522	65
70	30.4264	0.0329	588.528	19.3427	0.0017	0.0517	70
75	38.8327	0.0258	756.653	19.4850	0.0013	0.0513	75
80	49.5614	0.0202	971.228	19.5965	0.0010	0.0510	80
85	63.2543	0.0158	1245.09	19.6838	0.0008	0.0508	85
90	80.7303	0.0124	1594.61	19.7523	0.0006	0.0506	90
95	103.035	0.0097	2040.69	19.8059	0.0005	0.0505	95
100	131.501	0.0076	2610.22	19.8479	0.0004	0.0504	100
∞				20.0000		0.0500	∞

APPENDIX E-9 Discrete Compounding (*Continued*)

Discrete compounding; $i = 8\%$

	Single payment		Uniform series				
	Compound amount factor	Present worth factor	Compound amount factor	Present worth factor	Sinking fund factor	Capital recovery factor	
N	To find F given P F/P	To find P given F P/F	To find F given A F/A	To find P given A P/A	To find A given F A/F	To find A given P A/P	N
1	1.0800	0.9259	1.0000	0.9259	1.0000	1.0800	1
2	1.1664	0.8573	2.0800	1.7833	0.4808	0.5608	2
3	1.2597	0.7938	3.2464	2.5771	0.3080	0.3880	3
4	1.3605	0.7350	4.5061	3.3121	0.2219	0.3019	4
5	1.4693	0.6806	5.8666	3.9927	0.1705	0.2505	5
6	1.5869	0.6302	7.3359	4.6229	0.1363	0.2163	6
7	1.7138	0.5835	8.9228	5.2064	0.1121	0.1921	7
8	1.8509	0.5403	10.6366	5.7466	0.0940	0.1740	8
9	1.9990	0.5002	12.4876	6.2469	0.0801	0.1601	9
10	2.1589	0.4632	14.4866	6.7101	0.0690	0.1490	10
11	2.3316	0.4289	16.6455	7.1390	0.0601	0.1401	11
12	2.5182	0.3971	18.9771	7.5361	0.0527	0.1327	12
13	2.7196	0.3677	21.4953	7.9038	0.0465	0.1265	13
14	2.9372	0.3405	24.2149	8.2442	0.0413	0.1213	14
15	3.1722	0.3152	27.1521	8.5595	0.0368	0.1168	15
16	3.4259	0.2919	30.3243	8.8514	0.0330	0.1130	16
17	3.7000	0.2703	33.7502	9.1216	0.0296	0.1096	17
18	3.9960	0.2502	37.4502	9.3719	0.0267	0.1067	18
19	4.3157	0.2317	41.4463	9.6036	0.0241	0.1041	19
20	4.6610	0.2145	45.7620	9.8181	0.0219	0.1019	20
21	5.0338	0.1987	50.4229	10.0168	0.0198	0.0998	21
22	5.4365	0.1839	55.4567	10.2007	0.0180	0.0980	22
23	5.8715	0.1703	60.8933	10.3711	0.0164	0.0964	23
24	6.3412	0.1577	66.7647	10.5288	0.0150	0.0950	24
25	6.8485	0.1460	73.1059	10.6748	0.0137	0.0937	25
26	7.3964	0.1352	79.9544	10.8100	0.0125	0.0925	26
27	7.9881	0.1252	87.3507	10.9352	0.0114	0.0914	27
28	8.6271	0.1159	95.3388	11.0511	0.0105	0.0905	28
29	9.3173	0.1073	103.966	11.1584	0.0096	0.0896	29
30	10.0627	0.0994	113.283	11.2578	0.0088	0.0888	30
35	14.7853	0.0676	172.317	11.6546	0.0058	0.0858	35
40	21.7245	0.0460	259.056	11.9246	0.0039	0.0839	40
45	31.9204	0.0313	386.506	12.1084	0.0026	0.0826	45
50	46.9016	0.0213	573.770	12.2335	0.0017	0.0817	50
55	68.9138	0.0145	848.923	12.3186	0.0012	0.0812	55
60	101.257	0.0099	1253.21	12.3766	0.0008	0.0808	60
65	148.780	0.0067	1847.25	12.4160	0.0005	0.0805	65
70	218.606	0.0046	2720.08	12.4428	0.0004	0.0804	70
75	321.204	0.0031	4002.55	12.4611	0.0002	0.0802	75
80	471.955	0.0021	5886.93	12.4735	0.0002	0.0802	80
85	693.456	0.0014	8655.71	12.4820	0.0001	0.0801	85
90	1018.92	0.0010	12723.9	12.4877	a	0.0801	90
95	1497.12	0.0007	18071.5	12.4917	a	0.0801	95
100	2199.76	0.0005	27484.5	12.4943	a	0.0800	100
∞				12.5000		0.0800	∞

APPENDIX E-9   Discrete Compounding (*Continued*)

Discrete compounding; $i = 10\%$

	Single payment		Uniform series				
	Compound amount factor	Present worth factor	Compound amount factor	Present worth factor	Sinking fund factor	Capital recovery factor	
N	To find F given P F/P	To find P given F P/F	To find F given A F/A	To find P given A P/A	To find A given F A/F	To find A given P A/P	N
1	1.1000	0.9091	1.0000	0.9091	1.0000	1.1000	1
2	1.2100	0.8264	2.1000	1.7355	0.4762	0.5762	2
3	1.3310	0.7513	3.3100	2.4869	0.3021	0.4021	3
4	1.4641	0.6830	4.6410	3.1699	0.2155	0.3155	4
5	1.6105	0.6209	6.1051	3.7908	0.1638	0.2638	5
6	1.7716	0.5645	7.7156	4.3553	0.1296	0.2296	6
7	1.9487	0.5132	9.4872	4.8684	0.1054	0.2054	7
8	2.1436	0.4665	11.4359	5.3349	0.0874	0.1874	8
9	2.3579	0.4241	13.5795	5.7590	0.0736	0.1736	9
10	2.5937	0.3855	15.9374	6.1446	0.0627	0.1627	10
11	2.8531	0.3505	18.5312	6.4951	0.0540	0.1540	11
12	3.1384	0.3186	21.3843	6.8137	0.0468	0.1468	12
13	3.4523	0.2897	24.5227	7.1034	0.0408	0.1408	13
14	3.7975	0.2633	27.9750	7.3667	0.0357	0.1357	14
15	4.1772	0.2394	31.7725	7.6061	0.0315	0.1315	15
16	4.5950	0.2176	35.9497	7.8237	0.0278	0.1278	16
17	5.0545	0.1978	40.5447	8.0216	0.0247	0.1247	17
18	5.5599	0.1799	45.5992	8.2014	0.0219	0.1219	18
19	6.1159	0.1635	51.1591	8.3649	0.0195	0.1195	19
20	6.7275	0.1486	57.2750	8.5136	0.0175	0.1175	20
21	7.4002	0.1351	64.0025	8.6487	0.0156	0.1156	21
22	8.1403	0.1228	71.4027	8.7715	0.0140	0.1140	22
23	8.9543	0.1117	79.5430	8.8832	0.0126	0.1126	23
24	9.8497	0.1015	88.4973	8.9847	0.0113	0.1113	24
25	10.8347	0.0923	98.3470	9.0770	0.0102	0.1102	25
26	11.9182	0.0839	109.182	9.1609	0.0092	0.1092	26
27	13.1100	0.0763	121.100	9.2372	0.0083	0.1083	27
28	14.4210	0.0693	134.210	9.3066	0.0075	0.1075	28
29	15.9631	0.0630	148.631	9.3696	0.0067	0.1067	29
30	17.4494	0.0573	164.494	9.4269	0.0061	0.1061	30
35	28.1024	0.0356	271.024	9.6442	0.0037	0.1037	35
40	45.2592	0.0221	442.592	9.7791	0.0023	0.1023	40
45	72.8904	0.0137	718.905	9.8628	0.0014	0.1014	45
50	117.391	0.0085	1163.91	9.9148	0.0009	0.1009	50
55	189.059	0.0053	1880.59	9.9471	0.0005	0.1005	55
60	304.481	0.0033	3034.81	9.9672	0.0003	0.1003	60
65	490.370	0.0020	4893.71	9.9796	0.0002	0.1002	65
70	789.746	0.0013	7887.47	9.9873	0.0001	0.1001	70
75	1271.89	0.0008	12708.9	9.9921	*a*	0.1001	75
80	2048.40	0.0005	20474.0	9.9951	*a*	0.1000	80
85	3298.97	0.0003	32979.7	9.9970	*a*	0.1000	85
90	5313.02	0.0002	53120.2	9.9981	*a*	0.1000	90
95	8556.67	0.0001	85556.7	9.9988	*a*	0.1000	95
100	13780.6	*a*	137796	9.9993	*a*	0.1000	100
∞				10.0000		0.1000	∞

APPENDIX E-9   Discrete Compounding (*Continued*)

Discrete compounding; $i = 12\%$

	Single payment		Uniform series				
	Compound amount factor	Present worth factor	Compound amount factor	Present worth factor	Sinking fund factor	Capital recovery factor	
N	To find F given P F/P	To find P given F P/F	To find F given A F/A	To find P given A P/A	To find A given F A/F	To find A given P A/P	N
1	1.1200	0.8929	1.0000	0.8929	1.0000	1.1200	1
2	1.2544	0.7972	2.1200	1.6901	0.4717	0.5917	2
3	1.4049	0.7118	3.3744	2.4018	0.2963	0.4163	3
4	1.5735	0.6355	4.7793	3.0373	0.2092	0.3292	4
5	1.7623	0.5674	6.3528	3.6048	0.1574	0.2774	5
6	1.9738	0.5066	8.1152	4.1114	0.1232	0.2432	6
7	2.2107	0.4523	10.0890	4.5638	0.0991	0.2191	7
8	2.4760	0.4039	12.2997	4.9676	0.0813	0.2013	8
9	2.7731	0.3606	14.7757	5.3282	0.0677	0.1877	9
10	3.1058	0.3220	17.5487	5.6502	0.0570	0.1770	10
11	3.4785	0.2875	20.6546	5.9377	0.0484	0.1684	11
12	3.8960	0.2567	24.1331	6.1944	0.0414	0.1614	12
13	4.3635	0.2292	28.0291	6.4235	0.0357	0.1557	13
14	4.8871	0.2046	32.3926	6.6282	0.0309	0.1509	14
15	5.4736	0.1827	37.2797	6.8109	0.0268	0.1468	15
16	6.1304	0.1631	42.7533	6.9740	0.0234	0.1434	16
17	6.8660	0.1456	48.8837	7.1196	0.0205	0.1405	17
18	7.6900	0.1300	55.7497	7.2497	0.0179	0.1379	18
19	8.6128	0.1161	63.4397	7.3658	0.0158	1.1358	19
20	9.6463	0.1037	72.0524	7.4694	0.0139	0.1339	20
21	10.8038	0.0926	81.6987	7.5620	0.0122	0.1322	21
22	12.1003	0.0826	92.5026	7.6446	0.0108	0.1308	22
23	13.5523	0.0738	104.603	7.7184	0.0096	0.1296	23
24	15.1786	0.0659	118.155	7.7843	0.0085	0.1285	24
25	17.0001	0.0588	133.334	7.8431	0.0075	0.1275	25
26	19.0401	0.0525	150.334	7.8957	0.0067	0.1267	26
27	21.3249	0.0469	169.374	7.9426	0.0059	0.1259	27
28	23.8839	0.0419	190.699	7.9844	0.0052	0.1252	28
29	26.7499	0.0374	214.583	8.0218	0.0047	0.1247	29
30	29.9599	0.0334	241.333	8.0552	0.0041	0.1241	30
35	52.7996	0.0189	431.663	8.1755	0.0023	0.1223	35
40	93.0509	0.0107	767.091	8.2438	0.0013	0.1213	40
45	163.988	0.0061	1358.23	8.2825	0.0007	0.1207	45
50	289.002	0.0035	2400.02	8.3045	0.0004	0.1204	50
55	509.320	0.0020	4236.00	8.3170	0.0002	0.1202	55
60	897.596	0.0011	7471.63	8.3240	0.0001	0.1201	60
65	1581.87	0.0006	13173.9	8.3281	a	0.1201	65
70	2787.80	0.0004	23223.3	8.3303	a	0.1200	70
75	4913.05	0.0002	40933.8	8.3316	a	0.1200	75
80	8658.47	0.0001	72145.6	8.3324	a	0.1200	80
∞				8.333		0.1200	∞

APPENDIX E-9   Discrete Compounding (*Continued*)

Discrete compounding; $i = 15\%$

	Single payment		Uniform series				
	Compound amount factor	Present worth factor	Compound amount factor	Present worth factor	Sinking fund factor	Capital recovery factor	
N	To find F given P F/P	To find P given F P/F	To find F given A F/A	To find P given A P/A	To find A given F A/F	To find A given P A/P	N
1	1.1500	0.8696	1.0000	0.8696	1.0000	1.1500	1
2	1.3225	0.7561	2.1500	1.6257	0.4651	0.6151	2
3	1.5209	0.6575	3.4725	2.2832	0.2880	0.4380	3
4	1.7490	0.5718	4.9934	2.8550	0.2003	0.3503	4
5	2.0114	0.4972	6.7424	3.3522	0.1483	0.2983	5
6	2.3131	0.4323	8.7537	3.7845	0.1142	0.2642	6
7	2.6600	0.3759	11.0668	4.1604	0.0904	0.2404	7
8	3.0590	0.3269	13.7268	4.4873	0.0729	0.2229	8
9	3.5179	0.2843	16.7858	4.7716	0.0596	0.2096	9
10	4.0456	0.2472	20.3037	5.0188	0.0493	0.1993	10
11	4.6524	0.2149	24.3493	5.2337	0.0411	0.1911	11
12	5.3502	0.1869	29.0017	5.4206	0.0345	0.1845	12
13	6.1528	0.1625	34.3519	5.5831	0.0291	0.1791	13
14	7.0757	0.1413	40.5047	5.7245	0.0247	0.1747	14
15	8.1371	0.1229	47.5804	5.8474	0.0210	0.1710	15
16	9.3576	0.1069	55.7175	5.9542	0.0179	0.1679	16
17	10.7613	0.0929	65.0751	6.0472	0.0154	0.1654	17
18	12.3755	0.0808	75.8363	6.1280	0.0132	0.1632	18
19	14.2318	0.0703	88.2118	6.1982	0.0113	0.1613	19
20	16.3665	0.0611	102.444	6.2593	0.0098	0.1598	20
21	18.8215	0.0531	118.810	6.3125	0.0084	0.1584	21
22	21.6447	0.0462	137.632	6.3587	0.0073	0.1573	22
23	24.8915	0.0402	159.276	6.3988	0.0063	0.1563	23
24	28.6252	0.0349	184.168	6.4338	0.0054	0.1554	24
25	32.9189	0.0304	212.793	6.4641	0.0047	0.1547	25
26	37.8568	0.0264	245.712	6.4906	0.0041	0.1541	26
27	43.5353	0.0230	283.569	6.5135	0.0035	0.1535	27
28	50.0656	0.0200	327.104	6.5335	0.0031	0.1531	28
29	57.5754	0.0174	377.170	6.5509	0.0027	0.1527	29
30	66.2118	0.0151	434.745	6.5660	0.0023	0.1523	30
35	133.176	0.0075	881.170	6.6166	0.0011	0.1511	35
40	267.863	0.0037	1779.09	6.6418	0.0006	0.1506	40
45	538.769	0.0019	3585.13	6.6543	0.0003	0.1503	45
50	1083.66	0.0009	7217.71	6.6605	0.0001	0.1501	50
55	2179.62	0.0005	14524.1	6.6636	a	0.1501	55
60	4384.00	0.0002	29220.0	6.6651	a	0.1500	60
65	8817.78	0.0001	58778.5	6.6659	a	0.1500	65
70	17735.7	a	118231	6.6663	a	0.1500	70
75	35672.8	a	237812	6.6665	a	0.1500	75
80	71750.8	a	478332	6.6666	a	0.1500	80
∞				6.667		0.1500	∞

a Less than 0.0001.

APPENDIX F
SAMPLING AND ANALYTICAL METHODS

Appendix Number	Description
F-1	Physical Polutants
F-2	Chemical Pollutants
F-3	Biological Pollutants

EXPLANATION OF APPENDICES

Appendix F-1, Physical Pollutants: This Appendix is a reference for the sampling and analysis of physical pollutants. Source—EPA 600/4-79-020.

Appendix F-2, Chemical Pollutants: This Appendix is a reference for the sampling and analysis of chemical pollutants. Sources—EPA 600/4-79-020 and 40 CFR 136 Appendix A.

Appendix F-3, Biological Pollutants: This Appendix is a reference for the sampling and analysis of biological pollutants. Sources—EPA 600/4-79-020 and EPA 600/8-78-017.

Pollutant Type	Parameter	Water/Wastewater Method EPA 600/4-79-020	Sample Volume Required (mL)	Preservative	Holding Time
Physical	Residue, Total	160.3	100	Cool 4° C	7 Days
	Residue, Filterable	160.1	100	Cool 4° C	7 Days
	Residue, Non-Filterable	160.2	100	Cool 4° C	7 Days
	Residue, Settleable	160.5	1000	Cool 4° C	48 Hours
	Residue, Volatile	160.4	100	Cool 4° C	7 Days
	Color	110.1	50	Cool 4° C	48 Hours
	Odor	140.1	200	Cool 4° C	24 Hours
	Specific Conductance	120.1	100	Cool 4° C	28 Days
	Temperature	170.1	1000	None	Analyze Immediately
	Turbidity	180.1	100	Cool 4° C	48 Hours

Pollutant Type	Parameter	Water/Wastewater Method EPA 600/4-79-020 or 40CFR136	Sample Volume Required (mL)	Container Type	Preservative	Holding Time
Chemical	Acidity	305.1	100	HDPE	Cool 4° C	14 Days
	Alkalinity	310.1	100	HDPE	Cool 4° C	14 Days
	Ammonia	350.2	500	HDPE	Cool 4° C, H2SO4 to pH < 2	28 Days
	Bromide	320.1	100	HDPE	None	28 Days
	Chloride	325.3	50	HDPE	None	28 Days
	Chlorine (Total Residual)	330.1	200	HDPE	None	Analyze Immediately
	Cyanide (Total and Amenable)	335.2, 335.1	500	HDPE	NaOH to pH > 12, 0.6g Ascorbic Acid	14 Days
	Fluoride	340.2	300	HDPE	None	28 Days
	Hardness	130.1	100	HDPE	HNO3, or H2SO4 to pH < 2	6 Months
	Hydrogen Ion (pH)	150.1	25	HDPE	None	Analyze Immediately
	Chromium VI	218.4	100	HDPE	Cool 4° C	24 Hours
	Mercury	245.1	100	HDPE	HNO3 to pH < 2	28 Days
	Metals (Except Chromium VI and Hg)	200.7 Trace Amounts or 200 Series	200	HDPE	HNO3 to pH < 2	6 Months
	Kjeldahl & Organic Nitrogen	351.3	500	HDPE	Cool 4° C, H2SO4 to pH < 2	28 Days
	Nitrate	352.1 or 300	100	HDPE	Cool 4° C	48 Hours
	Nitrate-Nitrite	353.3 or 301	100	HDPE	Cool 4° C, H2SO4 to pH < 2	28 Days
	Nitrite	354.1 or 302	50	HDPE	Cool 4° C	48 Hours
	Oil & Grease/TPH	413.1	1000	Clear Glass Wide Mouth	Cool 4° C, H2SO4 to pH < 2	28 Days
	Orthophosphate	365.2	50	HDPE	Filter Immediately, Cool 4° C	48 Hours
	Oxygen, Dissolved (Probe)	360.1	300	Glass BOD Bottle	None	Analyze Immediately
	Winkler D.O.	360.2	300	Glass BOD Bottle	Fix on Site, Store in Dark	8 Hours
	Phenols	420.1	500	Clear Glass Bottle	Cool 4° C, H2SO4 to pH < 2	28 Days
	Phosphorus (Elemental)	365.2	50	Clear Glass Bottle	Cool 4° C	48 Hours
	Phosphorus (Total)	365.4	50	HDPE	Cool 4° C, H2SO4 to pH < 2	28 Days
	Silica	370.1	50	HDPE	Cool 4° C	28 Days
	Sulfate	375.3	50	HDPE	Cool 4° C	28 Days
	Sulfide	376.2	500	HDPE	ZnAc and NaOH to pH > 9	7 Days
	Sulfite	377.1	50	HDPE	None	Analyze Immediately
	Surfactants (MBAs)	425.1	250	HDPE	Cool 4° C	48 Hours
	Purgeable Halocarbons	601, 624	40	40 mL Glass Vial	0.008% Na2S2O3 and Cool 4° C	14 Days
	Purgeable Aromatic Hydrocarbons	602	40	40 mL Glass Vial	0.008% Na2S2O3, Cool 4° C and HCl to pH < 2	14 Days
	Acrolein & Acrylonitrile	603	40	40 mL Glass Vial	0.008% Na2S2O3 and Cool 4° C and for Acrolein adjust pH to 4 - 5	14 Days
	Phenols	604	1000	Amber Glass Bottle	0.008% Na2S2O3 and Cool 4° C	7 Days until Extraction, 40 days after Extraction
	Benzidine	605	1000	Amber Glass Bottle	0.008% Na2S2O3 and Cool 4° C	7 Days until Extraction, 40 days after Extraction
	Phthalate Esters	606	1000	Amber Glass Bottle	Cool 4° C	7 Days until Extraction, 40 days after Extraction
	Nitrosamines	607	1000	Amber Glass Bottle	0.008% Na2S2O3, Cool 4° C, and Store in Dark	7 Days until Extraction, 40 days after Extraction
	PCB's	608	1000	Amber Glass Bottle	Cool 4° C	7 Days until Extraction, 40 days after Extraction
	Nitroaromatics & Isophorone	609	1000	Amber Glass Bottle	0.008% Na2S2O3, Cool 4° C, and Store in Dark	7 Days until Extraction, 40 days after Extraction
	Polynuclear Aromatic Hydrocarbons	610	1000	Amber Glass Bottle	0.008% Na2S2O3, Cool 4° C, and Store in Dark	7 Days until Extraction, 40 days after Extraction
	Haloethers	611	1000	Amber Glass Bottle	0.008% Na2S2O3 and Cool 4° C	7 Days until Extraction, 40 days after Extraction
	Chlorinated Hydrocarbons	612	1000	Amber Glass Bottle	Cool 4° C	7 Days until Extraction, 40 days after Extraction
	Chlorinated Herbicides	5096	1000	Amber Glass Bottle	0.008% Na2S2O3, Cool 4° C, and Store in Dark	7 Days until Extraction, 40 days after Extraction
	Dioxins & Furans (T.C.D.D.)	613, 1613	1000	Amber Glass Bottle	Cool 4° C and adjust pH 5 - 9 with NaOH or H2SO4	7 Days until Extraction, 40 days after Extraction
	Pesticides, Chlorinated	608	1000	Amber Glass Bottle	Cool 4° C and adjust pH 5 - 9 with NaOH or H2SO4	7 Days until Extraction, 40 days after Extraction
	Radiological Test: Gross Alpha	Not Included	1000	HDPE	HNO3 to pH < 2	6 Months
	Beta	Not Included	1000	HDPE	HNO3 to pH < 2	6 Months
	Radium - Total	Not Included	1000	HDPE	HNO3 to pH < 2	6 Months
	Volatile Organics	624, 1624	40	40 mL Glass Vial	0.008% Na2S2O3 and Cool 4° C	14 Days
	Semi Volatile Organics	625, 1625	1000	Clear Glass Bottle	Cool 4° C	7 Days until Extraction, 40 days after Extraction

Pollutant Type	Parameter	Water/Wastewater Method EPA 600/4-79-020 and EPA 600/8-78-017	Sample Volume Required (mL)	Container Type	Preservative	Holding Time
Biological	BOD	405.1	1000	HDPE	Cool 4° C	48 Hours
	CBOD	405.1	1000	HDPE	Cool 4° C	48 Hours
	COD	410.1	50	HDPE	Cool 4° C, H2SO4 to pH < 2	28 Days
	Organic Carbon	415.1	25	HDPE	Cool 4° C, HCl or H2SO4 to pH < 2	28 Days
	Coliform, Fecal & Total	Section B, Section C		HDPE	0.008% Na2S2O3 and Cool 4° C	6 Hours
	Fecal Streptococci	Section D	100	HDPE	0.008% Na2S2O3 and Cool 4° C	6 Hours

Appendix F-1

Pollutant Type	Parameter	Water/Wastewater Method EPA 600/4-79-020	Sample Volume Required (mL)	Container Type	Preservative	Holding Time
Physical	Residue, Total	160.3	100	HDPE	Cool 4°C	7 Days
	Residue, Filterable	160.1	100	HDPE	Cool 4°C	7 Days
	Residue, Non-Filterable	160.2	100	HDPE	Cool 4°C	7 Days
	Residue, Settleable	160.5	1000	Imhoff Cone	Cool 4°C	48 Hours
	Residue, Volatile	160.4	100	HDPE	Cool 4°C	7 Days
	Color	110.1	50	HDPE	Cool 4°C	48 Hours
	Odor	140.1	200	Clear Glass Bottle	Cool 4°C	24 Hours
	Specific Conductance	120.1	100	HDPE	Cool 4°C	28 Days
	Temperature	170.1	1000	HDPE	None	Analyze Immediately
	Turbidity	180.1	100	HDPE	Cool 4°C	48 Hours

Appendix F-2

Pollutant Type	Parameter	Water/Wastewater Method EPA 600/4-79-020 or 40CFR136	Sample Volume Required (mL)	Container Type	Preservative	Holding Time
Chemical	Acidity	305.1	100	HDPE	Cool 4°C	14 Days
	Alkalinity	310.1	100	HDPE	Cool 4°C	14 Days
	Ammonia	350.2	500	HDPE	Cool 4°C, H_2SO_4 to pH < 2	28 Days
	Bromide	320.1	100	HDPE	None	28 Days
	Chloride	325.3	50	HDPE	None	28 Days
	Chlorine (Total Residual)	330.1	200	HDPE	None	Analyze Immediately
	Cyanide (Total and Amenable)	335.2, 335.1	500	HDPE	NaOH to pH > 12, 0.6g Ascorbic Acid	14 Days
	Fluoride	340.2	300	HDPE	None	28 Days
	Hardness	130.1	100	HDPE	HNO_3 or H_2SO_4 to pH < 2	6 Months
	Hydrogen Ion (pH)	150.1	25	HDPE	None	Analyze Immediately
	Chromium VI	218.4	100	HDPE	Cool 4°C	24 Hours
	Mercury	245.1	100	HDPE	HNO_3 to pH < 2	28 Days
	Metals (Except Chromium VI and Hg)	200.7 Trace Amounts or 200 Series	200	HDPE	HNO_3 to pH < 2	6 Months
	Kjeldahl & Organic Nitrogen	351.3	500	HDPE	Cool 4°C, H_2SO_4 to pH < 2	28 Days
	Nitrate	352.1 or 300	100	HDPE	Cool 4°C	48 Hours
	Nitrate-Nitrite	353.3 or 301	100	HDPE	Cool 4°C, H_2SO_4 to pH < 2	28 Days
	Nitrite	354.1 or 302	50	HDPE	Cool 4°C	48 Hours
	Oil & Grease/TPH	413.1	1000	Clear Glass Wide Mouth	Cool 4°C, H_2SO_4 to pH < 2	28 Days
	Orthophosphate	365.2	50	HDPE	Filter Immediately, Cool 4°C	48 Hours
	Oxygen, Dissolved (Probe)	360.1	300	Glass BOD Bottle	None	Analyze Immediately
	Winkler D.O.	360.2	300	Glass BOD Bottle	Fix on Site, Store in Dark	8 Hours
	Phenols	420.1	500	Clear Glass Bottle	Cool 4°C, H_2SO_4 to pH < 2	28 Days
	Phosphorous (Elemental)	365.2	50	Clear Glass Bottle	Cool 4°C	48 Hours
	Phosphorous (Total)	365.2	50	HDPE	Cool 4°C, H_2SO_4 to pH < 2	28 Days
	Silica	370.1	50	HDPE	Cool 4°C	28 Days
	Sulfate	375.3	50	HDPE	Cool 4°C	28 Days
	Sulfide	376.2	500	HDPE	Cool 4°C	7 Days
	Sulfite	377.1	50	HDPE	Zinc Acetate and NaOH to pH ≥ 9	Analyze Immediately
	Surfactants (MBAs)	425.1	250	HDPE	Cool 4°C	48 Hours
	Purgeable Halocarbons	601, 624	40	40 mL Glass Vials	0.008% $Na_2S_2O_3$ and Cool 4°C	14 Days
	Purgeable Aromatic Hydrocarbons	602	40	40 mL Glass Vials	0.008% $Na_2S_2O_3$, Cool 4°C, and HCl to pH < 2	14 Days
	Acrolein & Acrylonitrile	603	40	40 mL Glass Vials	0.008% $Na_2S_2O_3$ and Cool 4°C and for Acrolein adjust pH to 4 - 5	14 Days
	Phenols	604	1000	Amber Glass Bottle	0.008% $Na_2S_2O_3$ and Cool 4°C	7 Days until Extraction, 40 days after Extraction
	Benzidines	605	1000	Amber Glass Bottle	0.008% $Na_2S_2O_3$ and Cool 4°C	7 Days until Extraction, 40 days after Extraction
	Phthalate Esters	606	1000	Amber Glass Bottle	0.008% $Na_2S_2O_3$ and Cool 4°C	7 Days until Extraction, 40 days after Extraction
	Nitrosamines	607	1000	Amber Glass Bottle	0.008% $Na_2S_2O_3$, Cool 4°C, and Store in Dark	7 Days until Extraction, 40 days after Extraction
	PCB's	608	1000	Amber Glass Bottle	0.008% $Na_2S_2O_3$ and Cool 4°C	7 Days until Extraction, 40 days after Extraction
	Nitroaromatics & Isophorone	609	1000	Amber Glass Bottle	0.008% $Na_2S_2O_3$, Cool 4°C, and Store in Dark	7 Days until Extraction, 40 days after Extraction
	Polynuclear Aromatic Hydrocarbons	610	1000	Amber Glass Bottle	0.008% $Na_2S_2O_3$, Cool 4°C, and Store in Dark	7 Days until Extraction, 40 days after Extraction
	Haloethers	611	1000	Amber Glass Bottle	0.008% $Na_2S_2O_3$ and Cool 4°C	7 Days until Extraction, 40 days after Extraction
	Chlorinated Hydrocarbons	612	1000	Amber Glass Bottle	0.008% $Na_2S_2O_3$ and Cool 4°C	7 Days until Extraction, 40 days after Extraction
	Chlorinated Herbicides	5090	1000	Amber Glass Bottle	Cool 4°C	7 Days until Extraction, 40 days after Extraction
	Dioxins & Furans (T C D D)	613, 1613	1000	Amber Glass Bottle	0.008% $Na_2S_2O_3$ and Cool 4°C	7 Days until Extraction, 40 days after Extraction
	Pesticides, Chlorinated	608	1000	Amber Glass Bottle	Cool 4°C and adjust pH to 5 - 9 with NaOH or H_2SO_4	7 Days until Extraction, 40 days after Extraction
	Radiological Test - Gross Alpha	Not Included	1000	HDPE	HNO_3 to pH < 2	6 Months
	Beta	Not Included	1000	HDPE	HNO_3 to pH < 2	6 Months
	Radium, Total	Not Included	1000	HDPE	HNO_3 to pH < 2	6 Months
	Volatile Organics	624, 1624	40	40 mL Glass Vials	0.008% $Na_2S_2O_3$ to pH < 2 and Cool 4°C	14 Days
	Semi Volatile Organics	625, 1625	1000	Clear Glass Bottles	Cool 4°C	7 Days until Extraction, 40 days after Extraction

Appendix F-3

Pollutant Type	Parameter	Water/Wastewater Method EPA 600/4-79-020 and EPA 600/4-78-017	Sample Volume Required (mL)	Container Type	Preservative	Holding Time
Biological	BOD	405.1	1000	HDPE	Cool 4°C	48 Hours
	CBOD	405.1	1000	HDPE	Cool 4°C	48 Hours
	COD	410.1	50	HDPE	Cool 4°C, H_2SO_4 to pH < 2	28 Days
	Organic Carbon	415.1	25	HDPE	Cool 4°C, HCl or H_2SO_4 to pH < 2	28 Days
	Coliform, Fecal & Total	Section B, Section C	500	HDPE	0.008% $Na_2S_2O_3$ and Cool 4°C	6 Hours
	Fecal Streptococci	Section D	100	HDPE	0.008% $Na_2S_2O_3$ and Cool 4°C	6 Hours

Appendix F-1 Physical Pollutants

Parameter	Water/Wastewater Method EPA 600/4-79-020	Sample Volume Required (mL)	Container Type	Preservative	Holding Time
Residue, Total	160.3	100	HDPE	Cool 4° C	7 Days
Residue, Filterable	160.1	100	HDPE	Cool 4° C	7 Days
Residue, Non-Filterable	160.2	100	HDPE	Cool 4° C	48 Hours
Residue, Settlable	160.5	1000	Imhoff Cone	Cool 4° C	7 Days
Residue, Volatile	160.4	100	HDPE	Cool 4° C	48 Hours
Color	110.1	50	HDPE	Cool 4° C	24 Hours
Odor	140.1	200	Clear Glass Bottle	Cool 4° C	28 Days
Specific Conductance	120.1	100	HDPE	Cool 4° C	Analyze Immediately
Temperature	170.1	1000	HDPE	None	
Turbidity	180.1	100	HDPE	Cool 4° C	48 Hours

RESIDUE, TOTAL

Method 160.3 (Gravimetric, Dried at 103–105°C)

STORET NO. 00500

1. Scope and Application
   1.1 This method is applicable to drinking, surface, and saline waters, domestic and industrial wastes.
   1.2 The practical range of the determination is from 10 mg/1 to 20,000 mg/1.
2. Summary of Method
   2.1 A well mixed aliquot of the sample is quantitatively transferred to a pre-weighed evaporating dish and evaporated to dryness at 103–105°C.
3. Definitions
   3.1 Total Residue is defined as the sum of the homogenous suspended and dissolved materials in a sample.
4. Sample Handling and Preservation
   4.1 Preservation of the sample is not practical; analysis should begin as soon as possible. Refrigeration or icing to 4°C, to minimize microbiological decomposition of solids, is recommended.
5. Interferences
   5.1 Non-representative particulates such as leaves, sticks, fish and lumps of fecal matter should be excluded from the sample if it is determined that their inclusion is not desired in the final result.
   5.2 Floating oil and grease, if present, should be included in the sample and dispersed by a blender device before aliquoting.
6. Apparatus
   6.1 Evaporating dishes, porcelain, 90 mm, 100 ml capacity. (Vycor or platinum dishes may be substituted and smaller size dishes may be used if required.)
7. Procedure
   7.1 Heat the clean evaporating dish to 103–105°C for one hour, if Volatile Residue is to be measured, heat at 550 ±50°C for one hour in a muffle furnace. Cool, desiccate, weigh and store in desiccator until ready for use.
   7.2 Transfer a measured aliquot of sample to the pre-weighed dish and evaporate to dryness on a steam bath or in a drying oven.
      7.2.1 Choose an aliquot of sample sufficient to contain a residue of at least 25 mg. To obtain a weighable residue, successive aliquots of sample may be added to the same dish.
      7.2.2 If evaporation is performed in a drying oven, the temperature should be lowered to approximately 98°C to prevent boiling and splattering of the sample.

Approved for NPDES
Issued 1971

7.3   Dry the evaporated sample for at least 1 hour at 103–105°C. Cool in a desiccator and weigh. Repeat the cycle of drying at 103–105°C, cooling, desiccating and weighing until a constant weight is obtained or until loss of weight is less than 4% of the previous weight, or 0.5 mg, whichever is less.

8.   Calculation

8.1   Calculate total residue as follows:

$$\text{Total residue, mg/l} = \frac{(A - B) \times 1,000}{C}$$

where:

A = weight of sample + dish in mg
B = weight of dish in mg
C = volume of sample in ml

9.   Precision and Accuracy

9.1   Precision and accuracy data are not available at this time.

Bibliography

1.   Standard Methods for the Examination of Water and Wastewater, 14th Edition, p 91, Method 208A, (1975).

RESIDUE, FILTERABLE

Method 160.1 (Gravimetric, Dried at 180°C)

STORET NO. 70300

1. Scope and Application
   1.1 This method is applicable to drinking, surface, and saline waters, domestic and industrial wastes.
   1.2 The practical range of the determination is 10 mg/1 to 20,000 mg/1.
2. Summary of Method
   2.1 A well-mixed sample is filtered through a standard glass fiber filter. The filtrate is evaporated and dried to constant weight at 180°C.
   2.2 If Residue, Non-Filterable is being determined, the filtrate from that method may be used for Residue, Filterable.
3. Definitions
   3.1 Filterable residue is defined as those solids capable of passing through a glass fiber filter and dried to constant weight at 180°C.
4. Sample Handling and Preservation
   4.1 Preservation of the sample is not practical; analysis should begin as soon as possible. Refrigeration or icing to 4°C, to minimize microbiological decomposition of solids, is recommended.
5. Interferences
   5.1 Highly mineralized waters containing significant concentrations of calcium, magnesium, chloride and/or sulfate may be hygroscopic and will require prolonged drying, desiccation and rapid weighing.
   5.2 Samples containing high concentrations of bicarbonate will require careful and possibly prolonged drying at 180°C to insure that all the bicarbonate is converted to carbonate.
   5.3 Too much residue in the evaporating dish will crust over and entrap water that will not be driven off during drying. Total residue should be limited to about 200 mg.
6. Apparatus
   6.1 Glass fiber filter discs, 4.7 cm or 2.1 cm, without organic binder, Reeve Angel type 934-AH, Gelman type A/E, or equivalent.
   6.2 Filter holder, membrane filter funnel or Gooch crucible adapter.
   6.3 Suction flask, 500 ml.
   6.4 Gooch crucibles, 25 ml (if 2.1 cm filter is used).
   6.5 Evaporating dishes, porcelain, 100 ml volume. (Vycor or platinum dishes may be substituted).
   6.6 Steam bath.
   6.7 Drying oven, 180°C ±2°C.
   6.8 Desiccator.

Approved for NPDES
Issued 1971

6.9   Analytical balance, capable of weighing to 0.1 mg.

7.   Procedure

7.1   Preparation of glass fiber filter disc: Place the disc on the membrane filter apparatus or insert into bottom of a suitable Gooch crucible. While vacuum is applied, wash the disc with three successive 20 ml volumes of distilled water. Remove all traces of water by continuing to apply vacuum after water has passed through. Discard washings.

7.2   Preparation of evaporating dishes: If Volatile Residue is also to be measured heat the clean dish to 550 ±50°C for one hour in a muffle furnace. If only Filterable Residue is to be measured heat the clean dish to 180 ±2°C for one hour. Cool in desiccator and store until needed. Weigh immediately before use.

7.3   Assemble the filtering apparatus and begin suction. Shake the sample vigorously and rapidly transfer 100 ml to the funnel by means of a 100 ml graduated cylinder. If total filterable residue is low, a larger volume may be filtered.

7.4   Filter the sample through the glass fiber filter, rinse with three 10 ml portions of distilled water and continue to apply vacuum for about 3 minutes after filtration is complete to remove as much water as possible.

7.5   Transfer 100 ml (or a larger volume) of the filtrate to a weighed evaporating dish and evaporate to dryness on a steam bath.

7.6   Dry the evaporated sample for at least one hour at 180 ±2°C. Cool in a desiccator and weigh. Repeat the drying cycle until a constant weight is obtained or until weight loss is less than 0.5 mg.

8.   Calculation

8.1   Calculate filterable residue as follows:

$$\text{Filterable residue, mg/l} = \frac{(A - B) \times 1,000}{C}$$

where:

A = weight of dried residue + dish in mg

B = weight of dish in mg

C = volume of sample used in ml

9.   Precision and Accuracy

9.1   Precision and accuracy are not available at this time.

Bibliography

1.   Standard Methods for the Examination of Water and Wastewater, 14th Edition, p 92, Method 208B, (1975).

RESIDUE, NON-FILTERABLE

Method 160.2 (Gravimetric, Dried at 103–105°C)

STORET NO. 00530

1. Scope and Application
   1.1 This method is applicable to drinking, surface, and saline waters, domestic and industrial wastes.
   1.2 The practical range of the determination is 4 mg/1 to 20,000 mg/1.
2. Summary of Method
   2.1 A well-mixed sample is filtered through a glass fiber filter, and the residue retained on the filter is dried to constant weight at 103–105°C.
   2.2 The filtrate from this method may be used for Residue, Filterable.
3. Definitions
   3.1 Residue, non-filterable, is defined as those solids which are retained by a glass fiber filter and dried to constant weight at 103–105°C.
4. Sample Handling and Preservation
   4.1 Non-representative particulates such as leaves, sticks, fish, and lumps of fecal matter should be excluded from the sample if it is determined that their inclusion is not desired in the final result.
   4.2 Preservation of the sample is not practical; analysis should begin as soon as possible. Refrigeration or icing to 4°C, to minimize microbiological decomposition of solids, is recommended.
5. Interferences
   5.1 Filtration apparatus, filter material, pre-washing, post-washing, and drying temperature are specified because these variables have been shown to affect the results.
   5.2 Samples high in Filterable Residue (dissolved solids), such as saline waters, brines and some wastes, may be subject to a positive interference. Care must be taken in selecting the filtering apparatus so that washing of the filter and any dissolved solids in the filter (7.5) minimizes this potential interference.
6. Apparatus
   6.1 Glass fiber filter discs, without organic binder, such as Millipore AP-40, Reeves Angel 934-AH, Gelman type A/E, or equivalent.
   NOTE: Because of the physical nature of glass fiber filters, the absolute pore size cannot be controlled or measured. Terms such as "pore size", collection efficiencies and effective retention are used to define this property in glass fiber filters. Values for these parameters vary for the filters listed above.
   6.2 Filter support: filtering apparatus with reservoir and a coarse (40–60 microns) fritted disc as a filter support.

Approved for NPDES
Issued 1971

NOTE: Many funnel designs are available in glass or porcelain. Some of the most common are Hirsch or Buchner funnels, membrane filter holders and Gooch crucibles. All are available with coarse fritted disc.

6.3   Suction flask.

6.4   Drying oven, 103–105°C.

6.5   Desiccator.

6.6   Analytical balance, capable of weighing to 0.1 mg.

7.   Procedure

7.1   Preparation of glass fiber filter disc: Place the glass fiber filter on the membrane filter apparatus or insert into bottom of a suitable Gooch crucible with wrinkled surface up. While vacuum is applied, wash the disc with three successive 20 ml volumes of distilled water. Remove all traces of water by continuing to apply vacuum after water has passed through. Remove filter from membrane filter apparatus or both crucible and filter if Gooch crucible is used, and dry in an oven at 103–105°C for one hour. Remove to desiccator and store until needed. Repeat the drying cycle until a constant weight is obtained (weight loss is less than 0.5 mg). Weigh immediately before use. After weighing, handle the filter or crucible/filter with forceps or tongs only.

7.2   Selection of Sample Volume
For a 4.7 cm diameter filter, filter 100 ml of sample. If weight of captured residue is less than 1.0 mg, the sample volume must be increased to provide at least 1.0 mg of residue. If other filter diameters are used, start with a sample volume equal to 7 ml/cm^2 of filter area and collect at least a weight of residue proportional to the 1.0 mg stated above.
NOTE: If during filtration of this initial volume the filtration rate drops rapidly, or if filtration time exceeds 5 to 10 minutes, the following scheme is recommended: Use an unweighed glass fiber filter of choice affixed in the filter assembly. Add a known volume of sample to the filter funnel and record the time elapsed after selected volumes have passed through the filter. Twenty-five ml increments for timing are suggested. Continue to record the time and volume increments until fitration rate drops rapidly. Add additional sample if the filter funnel volume is inadequate to reach a reduced rate. Plot the observed time versus volume filtered. Select the proper filtration volume as that just short of the time a significant change in filtration rate occurred.

7.3   Assemble the filtering apparatus and begin suction. Wet the filter with a small volume of distilled water to seat it against the fritted support.

7.4   Shake the sample vigorously and quantitatively transfer the predetermined sample volume selected in 7.2 to the filter using a graduated cylinder. Remove all traces of water by continuing to apply vacuum after sample has passed through.

7.5   With suction on, wash the graduated cylinder, filter, non-filterable residue and filter funnel wall with three portions of distilled water allowing complete drainage between washing. Remove all traces of water by continuing to apply vacuum after water has passed through.
NOTE: Total volume of wash water used should equal approximately 2 ml per cm^2. For a 4.7 cm filter the total volume is 30 ml.

7.6   Carefully remove the filter from the filter support. Alternatively, remove crucible and filter from crucible adapter. Dry at least one hour at 103–105°C. Cool in a desiccator and weigh. Repeat the drying cycle until a constant weight is obtained (weight loss is less than 0.5 mg).

8.   Calculations

8.1   Calculate non-filterable residue as follows:

$$\text{Non-filterable residue, mg/l} = \frac{(A - B) \times 1{,}000}{C}$$

where:

A = weight of filter (or filter and crucible) + residue in mg
B = weight of filter (or filter and crucible) in mg
C = ml of sample filtered

9.   Precision and Accuracy

9.1   Precision data are not available at this time.

9.2   Accuracy data on actual samples cannot be obtained.

Bibliography

1.   NCASI Technical Bulletin No. 291, March 1977. National Council of the Paper Industry for Air and Stream Improvement, Inc., 260 Madison Ave., NY.

SETTLEABLE MATTER

Method 160.5 (Volumetric, Imhoff Cone)

STORET NO. 50086

1. Scope and Application
    1.1   This method is applicable to surface and saline waters, domestic and industrial wastes.
    1.2   The practical lower limit of the determination is about 0.2 ml/l/hr.

2. Summary of Method
    2.1   Settleable matter is measured volumetrically with an Imhoff cone.

3. Comments
    3.1   For some samples, a separation of settleable and floating materials will occur; in such cases the floating materials are not measured.
    3.2   Many treatment plants, especially plants equipped to perform gravimetric measurements, determine residue non-filterable (suspended solids), in preference to settleable matter, to insure that floating matter is included in the analysis.

4. Precision and Accuracy
    4.1   Data on this determination are not available at this time.

5. References
    5.1   The procedure to be used for this determination is found in:
    Standard Methods for the Examination of Water and Wastewater, 14th Edition, p 95, Method 208F, Procedure 3a (1975).

Approved for NPDES
Issued 1974

RESIDUE, VOLATILE

Method 160.4 (Gravimetric, Ignition at 550°C)

STORET NO. Total 00505
Non-Filterable 00535
Filterable 00520

1. Scope and Application
   1.1 This method determines the weight of solid material combustible at 550°C.
   1.2 The test is useful in obtaining a rough approximation of the amount of organic matter present in the solid fraction of sewage, activated sludge, industrial wastes, or bottom sediments.

2. Summary of Method
   2.1 The residue obtained from the determination of total, filterable or non-filterable residue is ignited at 550°C in a muffle furnace. The loss of weight on ignition is reported as mg/1 volatile residue.

3. Comments
   3.1 The test is subject to many errors due to loss of water of crystallization, loss of volatile organic matter prior to combustion, incomplete oxidation of certain complex organics, and decomposition of mineral salts during combustion.
   3.2 The results should not be considered an accurate measure of organic carbon in the sample, but may be useful in the control of plant operations.
   3.3 The principal source of error in the determination is failure to obtain a representative sample.

4. Sample Handling and Preservation
   4.1 Preservation of the sample is not practical; analysis should begin as soon as possible. Refrigeration or icing to 4°C, to minimize microbiological decompostion of solids is recommended.

5. Precision and Accuracy
   5.1 A collaborative study involving three laboratories examining four samples by means of ten replicates showed a standard deviation of ±11 mg/1 at 170 mg/1 volatile residue concentration.

6. Reference
   6.1 The procedure to be used for this determination is found in:
   Standard Methods for the Examination of Water and Wastewater, 14th Edition, p 95, Method 208E, (1975).

Approved for NPDES
Issued 1971

COLOR

Method 110.1 (Colorimetric, ADMI)

STORET NOS.
00082 at pH 7.6
00083 at ORIGINAL SAMPLE pH

1. Scope and Application
    1.1 This method is applicable to colored waters and waste that have color characteristics significantly different from the yellow platinum-cobalt standard.
    1.2 A working range of 25 to 250 color units is recommended. Sample values above 250 units may be determined by quantitative dilution.

2. Summary of Method
    2.1 This method is an extension of the Tristimulus Filter Method[1]. Tristimulus values are converted to an ADMI single number color difference, of the same magnitude assigned to platinum-cobalt standards, using the Adams Nickerson Color Difference (DE).
    2.2 Tristimulus values obtained by Spectrophotometric Method 204B[1] may be used to calculate ADMI values as outlined in this procedure under Calculation 9.2.

3. Interferences
    3.1 Since very slight amounts of turbidity interfere with the determination, turbid samples must be filtered prior to analysis. The optimum filter media to remove turbidity without removing color has not been found. Membrane and glass fiber filters with functional pore sizes of approximately 0.45 u are convenient to use. Other techniques such as centrifuging and/or filter aids may be used.

4. Sample Handling and Preservation
    4.1 Since biological activity may change the color characteristics of a sample, the determination should be made as soon as possible. Refrigeration at 4°C is recommended.

5. Calibration
    5.1 Standard curves must be established (as outlined in Procedure 8.3) for each photometer used, and are not interchangeable. For color values less than 250, a 5 cm cell path is recommended. Less than 5 cm cell paths may be used if calibration is performed with the shorter cell.

6. Apparatus
    6.1 Spectrophotometer or filter photometer capable of transmission measurements using tristimulus filters listed below:

Approved for NPDES
Issued 1978

Filter number	Wavelength of maximum transmittance in nm	Corning designation*
1	590	CS 3–107
2	540	CS 4–98
3	438	CS 5–70

*Available from Corning Glass Works, Optical Products Department, Corning, NY 14830.

7. Reagents

7.1 Standard chloroplatinate solution: Dissolve 1.246 g potassium chloroplatinate, K_2PtCl_6, (equivalent to 0.500 g metallic Pt) and 1 g crystalline cobaltous chloride, $CoCl_2 \cdot 6H_2O$, in distilled water containing 100 ml of conc. HCl. Dilute to 1000 ml with distilled water. This standard solution is defined as 500 ADMI color units.

7.2 Sulfuric acid, concentrated.

7.3 Sodium hydroxide, 10 N: Dissolve 40 g of sodium hydroxide in 80 ml of distilled water. Cool to room temperature and dilute to 100 ml with distilled water.

8. Procedure

8.1 Prepare two 100 ml volumes of sample by maintaining the original pH of one aliquot and adjusting the second aliquot as necessary to pH 7.6 with sulfuric acid (7.2) or sodium hydroxide (7.3).

8.2 Filter samples to remove turbidity through a 0.45 u membrane filter, glass fiber filter or other suitable media (see interferences 3.1).

8.3 Use distilled water to set the transmittance at 100% and then determine the transmittance of the clarified sample or standard with each of the three tristimulus filters. Calibration standards from 25 to 250 units are recommended.

9. Calculations

9.1 Calculate intermediate tristimulus values for samples and standards from the transmittance data in 8.3 using the following equations:

$$X_s = (T_3 \times 0.1899) + (T_1 \times 0.791)$$

$$Y_s = T_2$$

$$Z_s = T_3 \times 1.1835$$

where:

T_1 = transmittance value in % using filter number 1

T_2 = transmittance value in % using filter number 2

T_3 = transmittance value in % using filter number 3

9.2 Convert tristimulus values to the corresponding Munsell values V_x, V_y and V_z by the use of published tables[2,3,4], or the equation suggested by Bridgeman[5].

9.3 Calculate DE values for samples and standards, construct a calibration curve by plotting DE against ADMI units of standards and determine ADMI color units of samples from the calibration curve.

$$ DE = \sqrt{[0.23 \times (V_{yc} - V_{ys})]^2 + [(V_{xc} - V_{yc}) - (V_{xs} - V_{ys})]^2 + [0.4[(V_{yc} - V_{zc}) - (V_{ys} - V_{zs})]]^2} $$

where V_{xs}, V_{ys} and V_{zs} = the Munsell values for X_s, Y_s and Z_s respectively and V_{xc}, V_{yc} and V_{zc} = the Munsell values for a blank solution whose tristimulus values are X_c, Y_c and Z_c.

NOTE 1: If the photometer used is set at 100% transmittance with distilled water, the tristimulus values for the blank are 98.09, 100.00 and 118.35 for X_c, Y_c and Z_c respectively. If necessary, tristimulus values for the blank are determined as in calculation (9.1).

9.4 Report ADMI color values at pH 7.6 and at the original pH.

NOTE 2: The intermediate tristimulus values calculated under 9.1, using the three tristimulus filters, are used only to calculate the ADMI color value. They should not be reported as tristimulus values or used to determine dominant wavelength, luminance and purity.

10. Precision and Accuracy

10.1 Accuracy data on actual samples cannot be obtained.

10.2 Precision data are not available at this time.

Bibliography

1. Standard Methods for the Examination of Water and Wastewater, 14th Edition (1975), p 64.
2. J. Soc. Dyers and Colorists, 86, No. 8, 354 (1970).
3. Wyszecki and Stiles, Color Science, Wiley, N.Y., 1967, Tables 6.4 A, B and C.
4. Judd and Wyszecki, Color in Business, Science and Industry, 2nd Edition, Wiley, N.Y. (1963) Tables A, B and C in Appendix.
5. J. Opt. Soc. Am., Volume 53, page 499, April 1963.
6. Dyes and the Environment–Report on Selected Dyes and Their Effects, Volume 1, Sept. 1973 Appendix; American Dye Manufacturers Institute, Inc.

ODOR

Method 140.1 (Threshold Odor, Consistent Series)

STORET NO. 60°C: 00086
Room Temp: 00085

1. Scope and Application
    1.1   This method is applicable to the determination of threshold odor of drinking, surface, and saline waters, domestic and industrial wastes.
    1.2   Highly odorous samples are reduced in concentration proportionately before being tested. Thus, the method is applicable to samples ranging from nearly odorless natural waters to industrial wastes with threshold odor numbers in the thousands.
2. Summary of Method[1]
    2.1   The sample of water is diluted with odor-free water until a dilution that is of the least definitely perceptible odor to each tester is found. The resulting ratio by which the sample has been diluted is called the "threshold odor number" (T.O.N.).
    2.2   People vary widely as to odor sensitivity, and even the same person will not be consistent in the concentrations they can detect from day to day. Therefore, panels of not less than five persons, and preferably 10 or more, are recommended to overcome the variability of using one observer.[2] As an absolute minimum, two persons are necessary: One to make the sample dilutions and one to determine the threshold odor.
3. Sample Handling and Preservation
    3.1   Water samples must be collected in glass bottles with glass or Teflon-lined closures. Plastic containers are not reliable for odor samples and must not be used.
    3.2   Odor tests should be completed as soon as possible after collection of the sample. If storage is necessary, collect at least 1000 ml of sample in a bottle filled to the top. Refrigerate, making sure no extraneous odors can be drawn into the sample as the water cools.
4. Interferences
    4.1   Most tap waters and some waste waters are chlorinated. It is often desirable to determine the odor of the chlorinated sample as well as of the same sample after removal of chlorine. Dechlorination is achieved using sodium thiosulfate in exact stoichiometric quantity.
        4.1.1 It is important to check a blank to which a similar amount of dechlorinating agent has been added to determine if any odor has been imparted. Such odor usually disappears upon standing if excess reagent has not been added.
5. Apparatus
    5.1   Odor-free glassware: Glassware must be freshly cleaned shortly before use, with non-odorous soap and acid cleaning solution followed by rinsing with odor-free water (6.1).

Issued 1971

Glassware used in odor testing should be reserved for that purpose only. Rubber, cork, and plastic stoppers must not be used.

5.2   Constant temperature bath: A water bath or electric hotplate capable of maintaining a temperature control of $\pm 1°C$ for performing the odor test at 60°C. The temperature bath must not contribute any odor to the odor flasks.

5.3   Odor flasks: Glass stoppered 500 ml ($\bar{\text{T}}$ 32) Erlenmeyer flasks, or wide-mouthed 500 ml Erlenmeyer flasks equipped with Petri dishes as cover plates.
       NOTE: Narrow-mouth vessels are not suitable for running odor tests. Potential positive bias due to color and/or turbidity of water sample under observation can be eliminated by wrapping odor flasks in aluminum foil, painting flasks with non-odorous paint, or by using red actinic Erlenmeyer flasks.

5.4   Sample bottles: Glass bottles with glass or Teflon-lined closures.

5.5   Pipets, measuring: 10.0 and 1.0 ml graduated in tenths.

5.6   Graduated cylinders: 250, 200, 100, 50, and 25 ml.

5.7   Thermometer: 0–110°C ($\pm 1°C$), chemical or metal stem dial type.

5.8   Odor-free water generator: See Figure 1.

6.   Reagents

6.1   Odor-free water: Odor-free dilution water must be prepared as needed by filtration through a bed of activated carbon. Most tap waters are suitable for preparation of odor-free waters, except that it is necessary to check the filtered water for chlorine residual, unusual salt concentrations, or unusually high or low pH. All these may affect some odorous samples.
       Where supplies are adequate, distilled water avoids these problems as a source of odor-free water. A convenient odor-free water generator may be made as shown in Figure 1. Pass tap or distilled water through the odor-free water generator at a rate of 0.1 liter/minute. When the generator is first started, it should be flushed to remove carbon fines before the odor-free water is used.

       6.1.1 The quality of water obtained from the odor-free water generator should be checked daily at the temperature tests are to be conducted (room temperature and/or 60°C). The life of the carbon will vary with the condition and amount of water filtered. Subtle odors of biological origin are often found if moist carbon filters are permitted to stand idle between test periods. Detection of odor in the water coming through the carbon indicates a change of carbon is needed.

7.   Procedure

7.1   <u>Precaution</u>: Selection of persons to make odor tests should be carefully made. Extreme sensitivity is not required, but insensitive persons should not be used. A good observer has a sincere interest in the test. Extraneous odor stimuli such as those caused by smoking and eating prior to the test or through the use of scented soaps, perfumes, and shaving lotions must be avoided. The tester should be free from colds or allergies that affect odor-response. Frequent rests in an odor-free atmosphere are recommended. The room in which the tests are to be conducted should be free from distractions, drafts, and other odor. In certain industrial atmospheres, a special odor-free room may be required, ventilated by air filtered through activated carbon and maintained at a constant comfortable temperature and humidity. For precise work a panel of five or more testers should be used. The persons making the odor measurements should not prepare the samples and should not know the dilution concentrations being evaluated. These persons should have been made familiar with the procedure before participating in a panel test. Always start with the most dilute sample to avoid tiring the senses with the concentrated sample. The temperature of the samples during testing should be kept within 1 degree of the specified temperature for the test.

7.2   Threshold measurement: The ratio by which the odor-bearing sample has to be diluted with odor-free water for the odor to be just detectable by the odor test is the "threshold odor number" (T.Ó.N.). The total volume of sample and odor-free water used in each test is 200 ml. The proper volume of odor-free water is put into the flask first; the sample is then added to the water. Table 1 gives the dilutions and corresponding threshold numbers.

Table 1

Threshold Odor Number
Corresponding to Various Dilutions

Sample Volume (ml) Diluted to 200 ml	Threshold Odor Number
200	1
100	2
50	4
25	8
12.5	16
6.3	32
3.1	64
1.6	128
0.8	256

7.3   Determine the approximate range of the threshold odor by:

    7.3.1 Adding 200 ml, 50 ml, 12.5 ml, and 3.1 ml of the sample to separate 500 ml glass-stoppered Erlenmeyer flasks containing odor-free water to make a total volume of 200 ml. A separate flask containing only odor-free water serves as the reference for comparison. If run at 60°C, heat the dilutions and the reference in the constant temperature bath at 60°C (±1°C).

    7.3.2 Shake the flask containing the odor-free water, remove the stopper, and sniff the vapors. Test the sample containing the least amount of odor-bearing water in the same way. If odor can be detected in this dilution, more dilute samples must be prepared as described in (7.3.3). If odor cannot be detected in the first dilution, repeat the above procedure using the sample containing the next higher concentration of the odor-bearing water, and continue this process until odor is clearly detected.

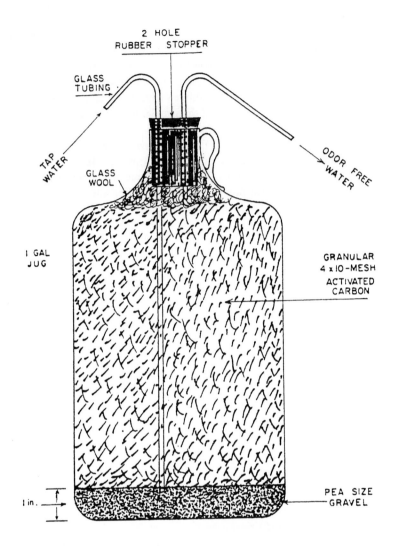

FIGURE 1. ODOR-FREE WATER GENERATOR

7.3.3 If the sample being tested requires more extensive dilution than is provided by Table 1, an intermediate dilution is prepared from 20 ml of sample diluted to 200 ml with odor-free water. Use this dilution for the threshold determination. Multiply the T.O.N. obtained by ten to correct for the intermediate dilution. In rare cases more than one tenfold intermediate dilution step may be required.

7.4 Based on the results obtained in the preliminary test, prepare a set of dilutions using Table 2 as a guide. One or more blanks are inserted in the series, in the vicinity of the expected threshold, but avoiding any repeated pattern. The observer does not know which dilutions are odorous and which are blanks. He smells each flask in sequence, beginning with the least concentrated sample and comparing with a known flask of odor-free water, until odor is detected with utmost certainty.

Table 2
Dilutions for Various Odor Intensities

\	Sample Volume in Which Odor First Noted		
200 ml	50 ml	12.5 ml	3.1 ml

Volume (ml) of Sample to be Diluted to 200 ml

200	100	50	(intermediate
100	50	25	Dilution
50	25	12.5	See 7.3.3)
25	12.5	6.3	
12.5	6.3	3.1	

7.5 Record the observations of each tester by indicating whether odor is noted (+ sign) in each test flask.

For example:

ml sample diluted to 200 ml	12.5	0	25	0	50	100	200
Response	−	−	+	−	+	+	+

8. Calculations

8.1 The threshold odor number is the dilution ratio at which odor is just detectable. In the example above (7.5), the first detectable odor occurred when 25 ml sample was diluted to 200 ml. Thus, the threshold is 200 divided by 25, equals 8. Table 1 lists the threshold odor numbers that correspond to common dilutions.

8.2 Anomalous responses sometimes occur; a low concentration may be called positive and a higher concentration in the series may be called negative. In such a case, the threshold is designated as that point of detection after which no further anomalies occur.

For instance:

ml sample diluted to 200 ml	6.3	12.5	0	25	50	100
Response threshold	+	–	–	+	+	+

Threshold (arrow pointing to 25 column)

8.3   Calculations of panel results to find the most probable average threshold are best accomplished by appropriate statistical methods. For most purposes, the threshold of a group can be expressed as the geometric mean of the individual thresholds. The geometric mean is calculated in the following manner:

8.3.1 Obtain odor response as outlined in Procedure and record results.

For example:

Table 3
Sample Odor Series

ml of Odor-free water	ml of Sample	Observer Response* 1	2	3	4	5
188	12.5	–	–	–	+	–
175	25	–	⊕	–	+	⊕
200	0	–	–	–	–	–
150	50	⊕	+	–	–	+
200	0	–	–	–	–	–
100	100	+	+	⊕	⊕	+
0	200	+	+	+	+	+

*Circled plus equals threshold level.

8.3.2 Obtain individual threshold odor numbers from Table 1.

Observer	T.O.N.
1	4
2	8
3	2
4	2
5	8

8.3.3 The geometric mean is equal to the nth root of the product of n numbers. Therefore:

$4 \times 8 \times 2 \times 2 \times 8 = 1,024$

and $\sqrt[5]{1,024} = \dfrac{\log 1,024}{5} = \dfrac{3.0103}{5} = 0.6021$

and anti-log of $0.6021 = 4 =$ T.O.N.

9.   Precision and Accuracy

   9.1   Precision and accuracy data are not available at this time.

   9.2   A threshold number is not a precise value. In the case of the single observer, it represents
         a judgment at the time of testing. Panel results are more meaningful because individual
         differences have less influence on the result. One or two observers can develop useful data
         if comparison with larger panels has been made to check their sensitivity. Comparisons
         of data from time to time or place to place should not be attempted unless all test
         conditions have been carefully standardized and some basis for comparison of observer
         intensities exists.

Bibliography

1.   Standard Methods for the Examination of Water and Wastewater, 14th Edition, p 75, Method
     206, (1975).

2.   ASTM, Comm E-18, STP 433, "Basic Principles of Sensory Evaluation"; STP 434, Manual on
     Sensory Testing Methods; STP 440, "Correlation of Subjective-Objective Methods in the Study
     of Odors and Taste"; Phil., Pennsylvania (1968).

3.   Baker, R. A., "Critical Evaluation of Olfactory Measurement". Jour. WPCF, 34, 582 (1962).

CONDUCTANCE

Method 120.1 (Specific Conductance, umhos at 25°C)

STORET NO. 00095

1. Scope and Application
    1.1 This method is applicable to drinking, surface, and saline wates, domestic and industrial wastes and acid rain (atmospheric deposition).

2. Summary of Method
    2.1 The specific conductance of a sample is measured by use of a self-contained conductivity meter, Wheatstone bridge-type, or equivalent.
    2.2 Samples are preferable analyzed at 25°C. If not, temprature corrections aremade and results reported at 25°C.

3. Comments
    3.1 Instrument must be standardized with KCl solution before daily use.
    3.2 Conductivity cell must be kept clean.
    3.3 Field measurements with comparable instruments are reliable.
    3.4 Temperature variations and corrections represent the largest source of potential error.

4. Sample Handling and Preservation
    4.1 Analyses can be performed either in the field or laboratory.
    4.2 If analysis is not completed within 24 hours of sample collection, sample should be filtered through a 0.45 micron filter and stored at 4°C. Filter and apparatus must be washed with high quality distilled water and pre-rinsed with sample before use.

5. Apparatus
    5.1 Conductivity bridge, range 1 to 1000 μmho per centimeter.
    5.2 Conductivity cell, cell constant 1.0 or micro dipping type cell with 1.0 constant. YSI #3403 or equivalent.
    5.4 Thermometer

6. Reagents
    6.1 Standard potassium chloride solutions, 0.01 M: Dissolve 0.7456 gm of pre-dried (2 hour at 105°C) KCl in distilled water and dilute to 1 liter at 25°C.

7. Cell Calibration
    7.1 The analyst should use the standard potassium chloride solution (6.1) and the table below to check the accuracy of the cell constant and conductivity bridge.

Approved for NPDES
Issued 1971.
Editorial revision, 1982

Conductivity 0.01 m KCl

°C	Micromhos/cm
21	1305
22	1332
23	1359
24	1386
25	1413
26	1441
27	1468
28	1496

8. Procedure

   8.1 Follow the direction of the manufacturer for the operation of the instrument.

   8.2 Allow samples to come to room temperature (23 to 27°C), if possible.

   8.3 Determine the temperature of samples within 0.5°C. If the temperature of the samples is not 25°C, make temperature correction in accordance with the instruction in Section 9 to convert reading to 25°.

9. Calculation

   9.1 These temperature corrections are based on the standard KCl solution.

      9.1.1 If the temperature of the sample is below 25°C, add 2% of the reading per degree.

      9.1.2 If the temperature is above 25°C, subtract 2% of the reading per degree.

   9.2 Report results as Specific Conductance, μmhos/cm at 25°.

10. Precision and Accuracy

    10.1 Forty-one analysts in 17 laboratories analyzed six synthetic water samples containing increments of inorganic salts, with the following results:

Increment as Specific Conductance	Precision as Standard Deviation	Accuracy as	
		Bias, %	Bias, μmhos/cm
100	7.55	−2.02	−2.0
106	8.14	−0.76	−0.8
808	66.1	−3.63	−29.3
848	79.6	−4.54	−38.5
1640	106	−5.36	−87.9
1710	119	−5.08	−86.9

(FWPCA Method Study 1, Mineral and Physical Analyses.)

   10.2 In a single laboratory (EMSL) using surface water samples with an average conductivity of 536 μmhos/cm at 25°C, the standard deviation was ±6.

Bibliography

1.   The procedure to be used for this determination is found in:
     Annual Book of ASTM Standards Part 31, "Water," Standard D1125-64, p. 120 (1976).

2.   Standard Methods for the Examination of Water and Wastewater, 14th Edition, p. 71, Method 205 (1975).

3.   Instruction Manual for YSI Model 31 Conductivity Bridge.

4.   Peden, M. E., and Skowron. "Ionic Stability of Precipitation Samples," Atmospheric Environment, Vol. 12, p. 2343-2344, 1978.

TEMPERATURE

Method 170.1 (Thermometric)

STORET NO. 00010

1. Scope and Application
   1.1 This method is applicable to drinking, surface, and saline waters, domestic and industrial wastes.

2. Summary of Method
   2.1 Temperature measurements may be made with any good grade of mercury-filled or dial type centigrade thermometer, or a thermistor.

3. Comments
   3.1 Measurement device should be routinely checked against a precision thermometer certified by the National Bureau of Standards.

4. Precision and Accuracy
   4.1 Precision and accuracy for this method have not been determined.

5. Reference
   5.1 The procedure to be used for this determination is found in:
   Standard Methods for the Examination of Water and Wastewater, 14th Edition, p 125, Method 212 (1975).

TURBIDITY
Method 180.1 (Nephelometric)

STORET NO. 00076

1. Scope and Application

   1.1 This method is applicable to drinking, surface, and saline waters in the range of turbidity from 0 to 40 nephelometric turbidity units (NTU). Higher values may be obtained with dilution of the sample.

   NOTE 1: NTU's are considered comparable to the previously reported Formazin Turbidity Units (FTU) and Jackson Turbidity Units (JTU).

2. Summary of Method

   2.1 The method is based upon a comparison of the intensity of light scattered by the sample under defined conditions with the intensity of light scattered by a standard reference suspension. The higher the intensity of scattered light, the higher the turbidity. Readings, in NTU's, are made in a nephelometer designed according to specifications outlined in Apparatus. A standard suspension of Formazin, prepared under closely defined conditions, is used to calibrate the instrument.

      2.1.1 Formazin polymer is used as the turbidity reference suspension for water because it is more reproducible than other types of standards previously used for turbidity standards.

      2.1.2 A commercially available polymer standard is also approved for use for the National Interim Primary Drinking Water Regulations. This standard is identified as AMCO-AEPA-1 available from Amco Standard International, Inc.

3. Sample Handling and Preservation

   3.1 Preservation of the sample is not practical; analysis should begin as soon as possible. Refrigeration or icing to 4°C, to minimize microbiological decomposition of solids, is recommended.

4. Interferences

   4.1 The presence of floating debris and coarse sediments which settle out rapidly will give low readings. Finely divided air bubbles will affect the results in a positive manner.

   4.2 The presence of true color, that is the color of water which is due to dissolved substances which absorb light, will cause turbidities to be low, although this effect is generally not significant with finished waters.

5. Apparatus

   5.1 The turbidimeter shall consist of a nephelometer with light source for illuminating the sample and one or more photo-electric detectors with a readout device to indicate the intensity of light scattered at right angles to the path of the incident light. The turbidimeter should be so designed that little stray light reaches the detector in the

Approved for NPDES and SDWA
Issued 1971
Editorial revision 1974
Editorial revision 1978

absence of turbidity and should be free from significant drift after a short warm-up period.

5.2 The sensitivity of the instrument should permit detection of a turbidity difference of 0.02 unit or less in waters having turbidities less than 1 unit. The instrument should measure from 0 to 40 units turbidity. Several ranges will be necessary to obtain both adequate coverage and sufficient sensitivity for low turbidities.

5.3 The sample tubes to be used with the available instrument must be of clear, colorless glass. They should be kept scrupulously clean, both inside and out, and discarded when they become scratched or etched. They must not be handled at all where the light strikes them, but should be provided with sufficient extra length, or with a protective case, so that they may be handled.

5.4 Differences in physical design of turbidimeters will cause differences in measured values for turbidity even though the same suspension is used for calibration. To minimize such differences, the following design criteria should be observed:

5.4.1 Light source: Tungsten lamp operated at a color temperature between 2200–3000°K.

5.4.2 Distance traversed by incident light and scattered light within the sample tube: Total not to exceed 10 cm.

5.4.3 Detector: Centered at 90° to the incident light path and not to exceed ±30° from 90°. The Detector, and filter system if used, shall have a spectral peak response between 400 and 600nm.

5.5 The Hach Turbidimeter, Model 2100 and 2100 A, is in wide use and has been found to be reliable; however, other instruments meeting the above design criteria are acceptable.

Reagents

6.1 Turbidity-free water: Pass distilled water through a 0.45u pore size membrane filter if such filtered water shows a lower turbidity than the distilled water.

6.2 Stock formazin turbidity suspension:

Solution 1: Dissolve 1.00 g hydrazine sulfate, $(NH_2)_2 \cdot H_2SO_4$, in distilled water and dilute to 100 ml in a volumetric flask.

Solution 2: Dissolve 10.00 g hexamethylenetetramine in distilled water and dilute to 100 ml in a volumetric flask.

In a 100 ml volumetric flask, mix 5.0 ml Solution 1 with 5.0 ml Solution 2. Allow to stand 24 hours at 25 ±3°C, then dilute to the mark and mix.

6.3 Standard formazin turbidity suspension: Dilute 10.00 ml stock turbidity suspension to 100 ml with turbidity-free water. The turbidity of this suspension is defined as 40 units. Dilute portions of the standard turbidity suspension with turbidity-free water as required.

6.3.1 A new stock turbidity suspension should be prepared each month. The standard turbidity suspension and dilute turbidity standards should be prepared weekly by dilution of the stock turbidity suspension.

6.4 The AMCO-AEPA-1 standard as supplied requires no preparation or dilution prior to use.

7.   Procedure

7.1   Turbidimeter calibration: The manufacturer's operating instructions should be followed. Measure standards on the turbidimeter covering the range of interest. If the instrument is already calibrated in standard turbidity units, this procedure will check the accuracy of the calibration scales. At least one standard should be run in each instrument range to be used. Some instruments permit adjustments of sensitivity so that scale values will correspond to turbidities. Reliance on a manufacturer's solid scattering standard for setting overall instrument sensitivity for all ranges is not an acceptable practice unless the turbidimeter has been shown to be free of drift on all ranges. If a pre-calibrated scale is not supplied, then calibration curves should be prepared for each range of the instrument.

7.2   Turbidities less than 40 units: Shake the sample to thoroughly disperse the solids. Wait until air bubbles disappear then pour the sample into the turbidimeter tube. Read the turbidity directly from the instrument scale or from the appropriate calibration curve.

7.3   Turbidities exceeding 40 units: Dilute the sample with one or more volumes of turbidity-free water until the turbidity falls below 40 units. The turbidity of the original sample is then computed from the turbidity of the diluted sample and the dilution factor. For example, if 5 volumes of turbidity-free water were added to 1 volume of sample, and the diluted sample showed a turbidity of 30 units, then the turbidity of the original sample was 180 units.

7.3.1   The Hach Turbidimeters, Models 2100 and 2100A, are equipped with 5 separate scales: 0–0.2, 0–1.0, 0–100, and 0–1000 NTU. The upper scales are to be used only as indicators of required dilution volumes to reduce readings to less than 40 NTU. **NOTE 2:** Comparative work performed in the MDQAR Laboratory indicates a progressive error on sample turbidities in excess of 40 units.

8.   Calculation

8.1   Multiply sample readings by appropriate dilution to obtain final reading.

8.2   Report results as follows:

NTU	Record to Nearest:
0.0 – 1.0	0.05
1 – 10	0.1
10 – 40	1
40 – 100	5
100 – 400	10
400 – 1000	50
> 1000	100

9.   Precision and Accuracy

9.1   In a single laboratory (EMSL), using surface water samples at levels of 26, 41, 75 and 180 NTU, the standard deviations were ±0.60, ±0.94, ±1.2 and ±4.7 units, respectively.

9.2   Accuracy data are not available at this time.

Bibliography

1. Annual Book of ASTM Standards, Part 31, "Water", Standard D1889–71, p 223 (1976).
2. Standard Methods for the Examination of Water and Wastewater, 14th Edition, p 132, Method 214A, (1975).

ACIDITY

Method 305.1 (Titrimetric)

STORET NO. 70508

1.   Scope and Application
   1.1   This method is applicable to surface waters, sewages and industrial wastes, particularly mine drainage and receiving streams, and other waters containing ferrous iron or other polyvalent cations in a reduced state.
   1.2   The method covers the range from approximately 10 mg/1 acidity to approximately 1000 mg/1 as $CaCO_3$, using a 50 ml sample.

2.   Summary of Method
   2.1   The pH of the sample is determined and a measured amount of standard acid is added, as needed, to lower the pH to 4 or less. Hydrogen peroxide is added, the solution boiled for several minutes, cooled, and titrated electrometrically with standard alkali to pH 8.2.

3.   Definitions
   3.1   This method measures the mineral acidity of a sample plus the acidity resulting from oxidation and hydrolysis of polyvalent cations, including salts of iron and aluminum.

4.   Interferences
   4.1   Suspended matter present in the sample, or precipitates formed during the titration may cause a sluggish electrode response. This may be offset by allowing a 15–20 second pause between additions of titrant or by slow dropwise addition of titrant as the endpoint pH is approached.

5.   Apparatus
   5.1   pH meter, suitable for electrometric titrations.

6.   Reagents
   6.1   Hydrogen peroxide (H_2O_2, 30% solution).
   6.2   Standard sodium hydroxide, 0.02 N.
   6.3   Standard sulfuric acid, 0.02 N.

7.   Procedure
   7.1   Pipet 50 ml of the sample into a 250 ml beaker.
   7.2   Measure the pH of the sample. If the pH is above 4.0, add standard sulfuric acid (6.3) in 5.0 ml increments to lower the pH to 4.0 or less. If the initial pH of the sample is less than 4.0, the incremental addition of sulfuric acid is not required.
   7.3   Add 5 drops of hydrogen peroxide (6.1).
   7.4   Heat the sample to boiling and continue boiling for 2 to 4 minutes. In some instances, the concentration of ferrous iron in a sample is such that an additional amount of hydrogen peroxide and a slightly longer boiling time may be required.

Approved for NPDES
Issued 1971
Technical revision 1974

7.5   Cool the sample to room temperature and titrate electrometrically with standard sodium hydroxide (6.2) to pH 8.2.

8.   Calculations

8.1   Acidity, as mg/l $CaCO_3 = \dfrac{[(A \times B) - (C \times D)] \times 50,000}{ml\ of\ sample}$

where:

A = vol. of standard sodium hydroxide used in titration
B = normality of standard sodium hydroxide
C = volume of standard sulfuric acid used to reduce pH to 4 or less
D = normality of standard sulfuric acid

8.2   If it is desired to report acidity in millequivalents per liter, the reported values as $CaCO_3$ are divided by 50, as follows:

$$Acidity\ as\ meq/l = \frac{mg/l\ CaCO_3}{50}$$

9.   Precision

9.1   On a round robin conducted by ASTM on 4 acid mine waters, including concentrations up to 2000 mg/l, the precision was found to be ±10 mg/l.

Bibliography

1.   Annual Book of ASTM Standards, Part 31, "Water", p 116, D 1067, Method E(1976).
2.   Standard Methods for the Examination of Water and Wastewater, 14th Edition, p 277, Method 402(4d) (1975).

ALKALINITY

Method 310.1 (Titrimetric, pH 4.5)

STORET NO. 00410

1. Scope and Application
   1.1 This method is applicable to drinking, surface, and saline waters, domestic and industrial wastes.
   1.2 The method is suitable for all concentration ranges of alkalinity; however, appropriate aliquots should be used to avoid a titration volume greater than 50 ml.
   1.3 Automated titrimetric analysis is equivalent.
2. Summary of Method
   2.1 An unaltered sample is titrated to an electrometrically determined end point of pH 4.5. The sample must not be filtered, diluted, concentrated, or altered in any way.
3. Comments
   3.1 The sample should be refrigerated at 4°C and run as soon as practical. Do not open sample bottle before analysis.
   3.2 Substances, such as salts of weak organic and inorganic acids present in large amounts, may cause interference in the electrometric pH measurements.
   3.3 For samples having high concentrations of mineral acids, such as mine wastes and associated receiving waters, titrate to an electrometric endpoint of pH 3.9, using the procedure in:
   Annual Book of ASTM Standards, Part 31, "Water", p 115, D-1067, Method D, (1976).
   3.4 Oil and grease, by coating the pH electrode, may also interfere, causing sluggish response.
4. Apparatus
   4.1 pH meter or electrically operated titrator that uses a glass electrode and can be read to 0.05 pH units. Standardize and calibrate according to manufacturer's instructions. If automatic temperature compensation is not provided, make titration at 25 ±2° C.
   4.2 Use an appropriate sized vessel to keep the air space above the solution at a minimum. Use a rubber stopper fitted with holes for the glass electrode, reference electrode (or combination electrode) and buret.
   4.3 Magnetic stirrer, pipets, flasks and other standard laboratory equipment.
   4.4 Burets, Pyrex 50, 25 and 10 ml.
5. Reagents
   5.1 Sodium carbonate solution, approximately 0.05 N: Place 2.5 ±0.2 g (to nearest mg) Na_2CO_3 (dried at 250°C for 4 hours and cooled in desiccator) into a 1 liter volumetric flask and dilute to the mark.

Approved for NPDES
Issued 1971
Editorial revision 1978

5.2 Standard acid (sulfuric or hydrochloric), 0.1 N: Dilute 3.0 ml conc H_2SO_4 or 8.3 ml conc HCl to 1 liter with distilled water. Standardize versus 40.0 ml of 0.05 N Na_2CO_3 solution with about 60 ml distilled water by titrating potentiometrically to pH of about 5. Lift electrode and rinse into beaker. Boil solution gently for 3–5 minutes under a watch glass cover. Cool to room temperature. Rinse cover glass into beaker. Continue titration to the pH inflection point. Calculate normality using:

$$N = \frac{A \times B}{53.00 \times C}$$

where:

A = g Na_2CO_3 weighed into 1 liter
B = ml Na_2CO_3 solution
C = ml acid used to inflection point

5.3 Standard acid (sulfuric or hydrochloric), 0.02 N: Dilute 200.0 ml of 0.1000 N standard acid to 1 liter with distilled water. Standardize by potentiometric titration of 15.0 ml 0.05 N Na_2CO_3 solution as above.

6. Procedure

6.1 Sample size

6.1.1 Use a sufficiently large volume of titrant (> 20 ml in a 50 ml buret) to obtain good precision while keeping volume low enough to permit sharp end point.

6.1.2 For < 1000 mg $CaCO_3$/l use 0.02 N titrant

6.1.3 For > 1000 mg $CaCO_3$/l use 0.1 N titrant

6.1.4 A preliminary titration is helpful.

6.2 Potentiometric titration

6.2.1 Place sample in flask by pipetting with pipet tip near bottom of flask

6.2.2 Measure pH of sample

6.2.3 Add standard acid (5.2 or 5.3), being careful to stir thoroughly but gently to allow needle to obtain equilibrium.

6.2.4 Titrate to pH 4.5. Record volume of titrant.

6.3 Potentiometric titration of low alkalinity

6.3.1 For alkalinity of < 20 mg/l titrate 100–200 ml as above (6.2) using a 10 ml microburet and 0.02 N acid solution (5.3).

6.3.2 Stop titration at pH in range of 4.3–4.7, record volume and exact pH. Very carefully add titrant to lower pH exactly 0.3 pH units and record volume.

7. Calculations

7.1 Potentiometric titration to pH 4.5

$$\text{Alkalinity, mg/l } CaCO_3 = \frac{A \times N \times 50,000}{\text{ml of sample}}$$

where:

A = ml standard acid

N = normality standard acid

7.2   Potentiometric titration of low alkalinity:

$$\text{Total alkalinity, mg/1 } CaCO_3 = \frac{(2B - C) \times N \times 50,000}{\text{ml of sample}}$$

where:

B = ml titrant to first recorded pH

C = total ml titrant to reach pH 0.3 units lower

N = normality of acid

8.   Precision and Accuracy

8.1   Forty analysts in seventeen laboratories analyzed synthetic water samples containing increments of bicarbonate, with the following results:

Increment as Alkalinity mg/liter, $CaCO_3$	Precision as Standard Deviation mg/liter, $CaCO_3$	Bias, %	Accuracy as	Bias, mg/l, $CaCO_3$
8	1.27	+10.61		+0.85
9	1.14	+22.29		+2.0
113	5.28	- 8.19		-9.3
119	5.36	- 7.42		-8.8

(FWPCA Method Study 1, Mineral and Physical Analyses)

8.2   In a single laboratory (EMSL) using surface water samples at an average concentration of 122 mg $CaCO_3$/1, the standard deviation was ±3.

Bibliography

1.   Standard Methods for the Examination of Water and Wastewater, 14th Edition, p 278, Method 403, (1975).

2.   Annual Book of ASTM Standards, Part 31, "Water", p 113, D-1067, Method B, (1976).

NITROGEN, AMMONIA

Method 350.2 (Colorimetric; Titrimetric; Potentiometric – Distillation Procedure)

STORET NO. Total 00610
Dissolved 00608

1. Scope and Application
   1.1 This distillation method covers the determination of ammonia-nitrogen exclusive of total Kjeldahl nitrogen, in drinking, surface and saline waters, domestic and industrial wastes. It is the method of choice where economics and sample load do not warrant the use of automated equipment.
   1.2 The method covers the range from about 0.05 to 1.0 mg NH_3–N/1 for the colorimetric procedure, from 1.0 to 25 mg/1 for the titrimetric procedure, and from 0.05 to 1400 mg/1 for the electrode method.
   1.3 This method is described for macro glassware; however, micro distillation equipment may also be used.
2. Summary of Method
   2.1 The sample is buffered at a pH of 9.5 with a borate buffer in order to decrease hydrolysis of cyanates and organic nitrogen compounds, and is then distilled into a solution of boric acid. The ammonia in the distillate can be determined colorimetrically by nesslerization, titrimetrically with standard sulfuric acid with the use of a mixed indicator, or potentiometrically by the ammonia electrode. The choice between the first two procedures depends on the concentration of the ammonia.
3. Sample Handling and Preservation
   3.1 Samples may be preserved with 2 ml of conc. H_2SO_4 per liter and stored at 4°C.
4. Interferences
   4.1 A number of aromatic and aliphatic amines, as well as other compounds, both organic and inorganic, will cause turbidity upon the addition of Nessler reagent, so direct nesslerization (i.e., without distillation), has been discarded as an official method.
   4.2 Cyanate, which may be encountered in certain industrial effluents, will hydrolyze to some extent even at the pH of 9.5 at which distillation is carried out. Volatile alkaline compounds, such as certain ketones, aldehydes, and alcohols, may cause an off-color upon nesslerization in the distillation method. Some of these, such as formaldehyde, may be eliminated by boiling off at a low pH (approximately 2 to 3) prior to distillation and nesslerization.
   4.3 Residual chlorine must also be removed by pretreatment of the sample with sodium thiosulfate before distillation.

Approved for NPDES
Issued 1971
Editorial revision 1974

5.   Apparatus
    5.1   An all-glass distilling apparatus with an 800–1000 ml flask.
    5.2   Spectrophotometer or filter photometer for use at 425 nm and providing a light path of 1 cm or more.
    5.3   Nessler tubes: Matched Nessler tubes (APHA Standard) about 300 mm long, 17 mm inside diameter, and marked at 225 mm ± 1.5 mm inside measurement from bottom.
    5.4   Erlenmeyer flasks: The distillate is collected in 500 ml glass-stoppered flasks. These flasks should be marked at the 350 and the 500 ml volumes. With such marking, it is not necessary to transfer the distillate to volumetric flasks.

6.   Reagents
    6.1   Distilled water should be free of ammonia. Such water is best prepared by passage through an ion exchange column containing a strongly acidic cation exchange resin mixed with a strongly basic anion exchange resin. Regeneration of the column should be carried out according to the manufacturer's instructions.
        NOTE 1: All solutions must be made with ammonia-free water.
    6.2   Ammonium chloride, stock solution: 1.0 ml = 1.0 mg NH_3–N. Dissolve 3.819 g NH_4Cl in distilled water and bring to volume in a 1 liter volumetric flask.
    6.3   Ammonium chloride, standard solution: 1.0 ml = 0.01 mg. Dilute 10.0 ml of stock solution (6.2) to 1 liter in a volumetric flask.
    6.4   Boric acid solution (20 g/1): Dissolve 20 g H_3BO_3 in distilled water and dilute to 1 liter.
    6.5   Mixed indicator: Mix 2 volumes of 0.2% methyl red in 95% ethyl alcohol with 1 volume of 0.2% methylene blue in 95% ethyl alcohol. This solution should be prepared fresh every 30 days.
        NOTE 2: Specially denatured ethyl alcohol conforming to Formula 3A or 30 of the U.S. Bureau of Internal Revenue may be substituted for 95% ethanol.
    6.6   Nessler reagent: Dissolve 100 g of mercuric iodide and 70 g of potassium iodide in a small amount of water. Add this mixture slowly, with stirring, to a cooled solution of 160 g of NaOH in 500 ml of water. Dilute the mixture to 1 liter. If this reagent is stored in a Pyrex bottle out of direct sunlight, it will remain stable for a period of up to 1 year.
        NOTE 3: This reagent should give the characteristic color with ammonia within 10 minutes after addition, and should not produce a precipitate with small amounts of ammonia (0.04 mg in a 50 ml volume).
    6.7   Borate buffer: Add 88 ml of 0.1 N NaOH solution to 500 ml of 0.025 M sodium tetraborate solution (5.0 g anhydrous $Na_2B_4O_7$ or 9.5 g $Na_2B_4O_7 \cdot 10H_2O$ per liter) and dilute to 1 liter.
    6.8   Sulfuric acid, standard solution: (0.02 N, 1 ml = 0.28 mg NH_3–N). Prepare a stock solution of approximately 0.1 N acid by diluting 3 ml of conc. H_2SO_4 (sp. gr. 1.84) to 1 liter with CO_2-free distilled water. Dilute 200 ml of this solution to 1 liter with CO_2-free distilled water.
        NOTE 4: An alternate and perhaps preferable method is to standardize the approximately 0.1 N H_2SO_4 solution against a 0.100 N Na_2CO_3 solution. By proper dilution the 0.02 N acid can then be prepared.

6.8.1 Standardize the approximately 0.02 N acid against 0.0200 N Na_2CO_3 solution. This last solution is prepared by dissolving 1.060 g anhydrous Na_2CO_3, oven-dried at 140°C, and diluting to 1000 ml with CO_2-free distilled water.

6.9   Sodium hydroxide, 1 N: Dissolve 40 g NaOH in ammonia-free water and dilute to 1 liter.

6.10 Dechlorinating reagents: A number of dechlorinating reagents may be used to remove residual chlorine prior to distillation. These include:

    a.   Sodium thiosulfate (1/70 N): Dissolve 3.5 g $Na_2S_2O_3 \cdot 5H_2O$ in distilled water and dilute to 1 liter. One ml of this solution will remove 1 mg/1 of residual chlorine in 500 ml of sample.

    b.   Sodium arsenite (1/70 N): Dissolve 1.0 g $NaAsO_2$ in distilled water and dilute to 1 liter.

7.   Procedure

7.1   Preparation of equipment: Add 500 mL of distilled water to an 800 ml Kjeldahl flask. The addition of boiling chips which have been previously treated with dilute NaOH will prevent bumping. Steam out the distillation apparatus until the distillate shows no trace of ammonia with Nessler reagent.

7.2   Sample preparation: Remove the residual chlorine in the sample by adding dechlorinating agent equivalent to the chlorine residual. To 400 ml of sample add 1 N NaOH (6.9), until the pH is 9.5, checking the pH during addition with a pH meter or by use of a short range pH paper.

7.3   Distillation: Transfer the sample, the pH of which has been adjusted to 9.5, to an 800 ml Kjeldahl flask and add 25 ml of the borate buffer (6.7). Distill 300 ml at the rate of 6–10 ml/min. into 50 ml of 2% boric acid (6.4) contained in a 500 ml Erlenmeyer flask.

NOTE 5: The condenser tip or an extension of the condenser tip must extend below the level of the boric acid solution.

Dilute the distillate to 500 ml with distilled water and nesslerize an aliquot to obtain an approximate value of the ammonia-nitrogen concentration. For concentrations above 1 mg/1 the ammonia should be determined titrimetrically. For concentrations below this value it is determined colorimetrically. The electrode method may also be used.

7.4   Determination of ammonia in distillate: Determine the ammonia content of the distillate titrimetrically, colorimetrically or potentiometrically as described below.

7.4.1 Titrimetric determination: Add 3 drops of the mixed indicator to the distillate and titrate the ammonia with the 0.02 N H_2SO_4, matching the end point against a blank containing the same volume of distilled water and H_3BO_3 solution.

7.4.2 Colorimetric determination: Prepare a series of Nessler tube standards as follows:

ml of Standard 1.0 ml = 0.01 mg NH₃–N	mg NH₃–N/50.0 ml
0.0	0.0
0.5	0.005
1.0	0.01
2.0	0.02
3.0	0.03
4.0	0.04
5.0	0.05
8.0	0.08
10.0	0.10

Dilute each tube to 50 ml with distilled water, add 2.0 ml of Nessler reagent (6.6) and mix. After 20 minutes read the absorbance at 425 nm against the blank. From the values obtained plot absorbance vs. mg NH_3–N for the standard curve. Determine the ammonia in the distillate by nesslerizing 50 ml or an aliquot diluted to 50 ml and reading the absorbance at 425 nm as described above for the standards. Ammonia-nitrogen content is read from the standard curve.

7.4.3 Potentiometric determination: Consult the method entitled Nitrogen, Ammonia: Selective Ion Electrode Method (Method 350.3) in this manual.

7.5   It is not imperative that all standards be distilled in the same manner as the samples. It is recommended that at least two standards (a high and low) be distilled and compared to similar values on the curve to insure that the distillation technique is reliable. If distilled standards do not agree with undistilled standards the operator should find the cause of the apparent error before proceeding.

8.   Calculations

8.1   Titrimetric

$$\text{mg/l NH}_3 - \text{N} = \frac{A \times 0.28 \times 1{,}000}{S}$$

where:
A = ml 0.02 N H_2SO_4 used.
S = ml sample.

8.2   Spectrophotometric

$$\text{mg/l NH}_3 - \text{N} = \frac{A \times 1{,}000}{D} \times \frac{B}{C}$$

where:
A = mg NH_3–N read from standard curve.
B = ml total distillate collected, including boric acid and dilution.
C = ml distillate taken for nesslerization.
D = ml of original sample taken.

8.3   Potentiometric

$$mg/l\ NH_3 - N = \frac{500}{D} \times A$$

where:

A = mg NH_3–N/1 from electrode method standard curve.

D = ml of original sample taken.

9.   Precision and Accuracy

9.1   Twenty-four analysts in sixteen laboratories analyzed natural water samples containing exact increments of an ammonium salt, with the following results:

Increment as Nitrogen, Ammonia mg N/liter	Precision as Standard Deviation mgN/liter	Accuracy as	
		Bias, %	Bias, mg N/liter
0.21	0.122	−5.54	−0.01
0.26	0.070	−18.12	−0.05
1.71	0.244	+0.46	+0.01
1.92	0.279	−2.01	−0.04

(FWPCA Method Study 2, Nutrient Analyses)

Bibliography

1.   Standard Methods for the Examination of Water and Wastewater, 14th Edition, p 410, Method 418A and 418B (1975).

2.   Annual Book of ASTM Standards, Part 31, "Water", Standard D1426–74, Method A, p 237 (1976).

CHLORIDE

Method 325.3 (Titrimetric, Mercuric Nitrate)

STORET NO. 00940

1. Scope and Application
   1.1 This method is applicable to drinking, surface, and saline waters, domestic and industrial wastes.
   1.2 The method is suitable for all concentration ranges of chloride content; however, in order to avoid large titration volume, a sample aliquot containing not more than 10 to 20 mg Cl per 50 ml is used.
   1.3 Automated titration may be used.
2. Summary of Method
   2.1 An acidified sample is titrated with mercuric nitrate in the presence of mixed diphenylcarbazone-bromophenol blue indicator. The end point of the titration is the formation of the blue-violet mercury diphenylcarbazone complex.
3. Comments
   3.1 Anions and cations at concentrations normally found in surface waters do not interfere.
   3.2 Sulfite interference can be eliminated by oxidizing the 50 ml of sample solution with 0.5 to 1 ml of H_2O_2.
4. Apparatus
   4.1 Standard laboratory titrimetric equipment including a 1 ml or 5 ml microburet with 0.01 ml graduations.
5. Reagents
   5.1 Standard sodium chloride, 0.025 N: Dissolve 1.4613 g $\pm$ 0.0002 g sodium chloride (dried at 600 °C for 1 hour) in chloride-free water in a 1 liter volumetric flask and dilute to the mark 1 ml = 886.5 μg Cl.
   5.2 Nitric acid, HNO_3 solution (3 + 997)
   5.3 Sodium hydroxide solution, NaOH, (10 g/1)
   5.4 Hydrogen peroxide (30%), H_2O_2
   5.5 Hydroquinone solution (10 g/liter): Dissolve 1 g of purified hydroquinone in water in a 100 ml volumetric and dilute to the mark.
   5.6 Mercuric nitrate titrant (0.141 N): Dissolve 25 g $Hg(NO_3)_2 \cdot H_2O$ in 900 ml of distilled water acidified with 5.0 ml conc. HNO_3 in a 1 liter volumetric flask and dilute to the mark with distilled water. Filter if necessary. Standardize against standard sodium chloride solution (5.1) using procedure 6. Adjust to exactly 0.141 N and check. Store in a dark bottle. A 1.00 ml aliquot is equivalent to 5.00 mg of chloride.
   5.7 Mercuric nitrate titrant (0.025 N): Dissolve 4.2830 g $Hg(NO_3)_2 \cdot H_2O$ in 50 ml of distilled water acidified with 0.5 ml conc. HNO_3 (sp. gr. 1.42) in a 1 liter volumetric flask and dilute to the mark with distilled water. Filter if necessary. Standardize against standard

Approved for NPDES
Issued 1971
Editorial revision 1978 and 1982

sodium chloride solution (5.1) using procedure 6. Adjust to exactly 0.025 N and check. Store in a dark bottle.

5.8   Mercuric nitrate titrant (0.0141 N): Dissolve 2.4200 g $Hg(NO_3)_2 \cdot H_2O$ in 25 ml of distilled water acidified with 0.25 ml of conc. HNO_3 (sp. gr. 1.42) in a 1 liter volumetric flask and dilute to the mark with distilled water. Filter if necessary. Standardize against standard sodium chloride solution (5.1) using procedure 6. Adjust to exactly 0.0141 N and check. Store in a dark bottle. A 1 ml aliquot is equivalent to 500 ug of chloride.

5.9   Mixed indicator reagent: Dissolve 0.5 g crystalline diphenylcarbazone and 0.05 g bromophenol blue powder in 75 ml 95% ethanol in a 100 ml volumetric flask and dilute to the mark with 95% ethanol. Store in brown bottle and discard after 6 months.

5.10 Xylene cyanole FF solution: Dissolve 0.005 g of xylene cyanole FF dye in 95% ethanol or isopropanol in a 100 ml volumetric and dilute to the mark with 95% ethanol or isopropanol.

6.   Procedure

6.1   Use 50 ml of sample or an aliquot of sample diluted to 50 ml with distilled water, so that the concentration of chloride does not exceed 20 mg aliquot. If the sample or aliquot contains more than 2.5 mg of chloride, use 0.025N mercuric nitrate titrant (5.7) in step 6.6. If the sample or aliquot contains less than 2.5 mg of chloride, use 0.0141N mercuric nitrate titrant (5.8) in step 6.6. Determine an indicator blank on 50 ml chloride-free water using step 6.6. If the sample contains less than 0.1 mg l of chloride concentrate an appropriate volume to 50 ml.

6.2   Add 5 drops of mixed indicator reagent (5.9), shake or swirl solution.

6.3   If a blue-violet or red color appears add HNO_3 solution (5.2) dropwise until the color changes to yellow.

6.4   If a yellow or orange color forms immediately on addition of the mixed indicator, add NaOH solution (5.3) dropwise until the color changes to blue-violet; then add HNO_3 solution (5.2) dropwise until the color changes to yellow.

6.5   Add 1 ml excess HNO_3 solution (5.2).

6.6   Titrate with 0.025 N mercuric nitrate titrant (5.7) until a blue-violet color persists throughout the solution.   See 6.1 for choice of titrant normality.   Xylene cyanol FF solution (5.10) may be added with the indicator to sharpen the end point. This will change color shades. Practice runs should be made.

6.7   Additional steps to eliminate particular interferences:

6.7.1   If chromate is present and iron is not present the end point may be difficult to detect.

be an olive-purple color.

6.7.2   If chromate is present at > 100 mg/1 and iron is not present, add 2 ml of fresh hydroquinone solution (5.5).

6.7.3   If ferric ion is present use volume containing no more than 2.5 mg of ferric ion or ferric ion plus chromate ion. Add 2 ml fresh hydroquinone solution (5.5).

6.7.4   If sulfite ion is present, add 0.5 ml of H_2O_2 solution (5.4) to 50 ml sample and mix for 1 minute.

7.   Calculation

$$\text{mg chloride/l} = \frac{(A - B)N \times 35,450}{\text{ml of sample}}$$

where:

A = ml titrant for sample
B = ml titrant for blank
N = normality mercuric nitrate titrant

mg NaCl/l = mg chloride/l x 1.65

8.   Precision and Accuracy

8.1   Forty two analysts in eighteen laboratories analyzed synthetic water samples containing exact increments of chloride, with the following results:

Increment as Chloride mg/liter	Precision as Standard Deviation mg/liter	Bias, %	Accuracy as Bias, mg/liter
17	1.54	+2.16	+0.4
18	1.32	+3.50	+0.6
91	2.92	+0.11	+0.1
97	3.16	−0.51	−0.5
382	11.70	−0.61	−2.3
398	11.80	−1.19	−4.7

(FWPCA Method Study 1, Mineral and Physical Analyses)

8.2   In a single laboratory (EMSL), using surface water samples at an average concentration of 34 mg Cl/l, the standard deviation was ±1.0.

8.3   A synthetic unknown sample containing 241 mg/l chloride, 108 mg/l Ca, 82 mg/l Mg, 3.1 mg/l K, 19.9 mg/l Na, 1.1 mg/l nitrate N, 0.25 mg/l nitrite N, 259 mg/l sulfate and 42.5 mg/l total alkalinity (contributed by $NaHCO_3$) in distilled water was analyzed in 10 laboratories by the mercurimetric method, with a relative standard deviation of 3.3% and a relative error of 2.9%.

Bibliography

1.   Annual Book of ASTM Standards, Part 31, "Water", Standard D512–67, Method A, p 270 (1976).

CHLORINE, TOTAL RESIDUAL

Method 330.1 (Titrimetric, Amperometric)

STORET NO. 50060

1. Scope and Application
   1.1 The amperometric titration method is applicable to all types of waters and wastes that do not contain a substantial amount of organic matter.
2. Summary of Method
   2.1 Chlorine (hypochlorite ion, hypochlorous acid) and chloramines stoichiometrically liberate iodine from potassium iodide at pH 4 or less.
   2.2 The iodine is titrated with standard reducing agent such as sodium thiosulfate or phenylarsine oxide using an amperometer to determine the end point.
   2.3 The results are calculated as mg/1 Cl even though the actual measurement is of total oxidizing power because chlorine is the dominant oxidizing agent present.
3. Interferences
   3.1 Manganese, nitrite and iron do not interfere.
   3.2 Stirring can lower chlorine values by volatilization.
   3.3 If dilution is necessary, it must be done with distilled water which is free of chlorine, chlorine-demand and ammonia.
   3.4 Copper and silver poison the electrode.
4. Apparatus
   4.1 An amperometer consisting of a microammeter with necessary electrical accessories, a cell unit with a salt bridge, reference electrode and an agitator: Commercially available. If the entire system (including titrant delivery system) is to be used, make sure that the volume read off the pipet or buret is really being delivered to the sample cell. Reservoir-type system sometimes back up, producing false readings.
   4.2 A microburet, 0–2 ml or 0–10 ml, depending on required precision, accuracy and range.
5. Reagents
   5.1 Phenylarsine oxide solution (0.00564N), commercially available, Wallace and Tiernan or equivalent. Standardize with potassium biiodate (5.8, 5.9).
   5.2 Potassium Iodide, KI, crystals.
   5.3 Potassium Iodide Solution: Dissolve 50 g KI in freshly boiled and cooled distilled water and dilute to 1 liter. Store in colored, glass-stoppered bottle in refrigerator. Discard when yellow color develops.
   5.4 Commercially available starch indicators such as thyodene or equivalent may be used.
   5.5 Acetate buffer solution (pH 4): Dissolve 146g anhydrous $NaC_2H_3O_2$ or 243g $NaC_2H_3O_2 \cdot 3H_2O$ in 400 ml distilled water, add 480g conc acetic acid and dilute to 1 liter with distilled water.

Approved for NPDES
Issued 1974
Editorial revision 1978

5.6   Sulfuric Acid (1:4): Slowly add 200 ml H_2SO_4 (sp. gr. 1.84) to 800 ml of distilled water.

5.7   Potassium biiodate (0.1N): Dissolve 3.249g potassium biiodate, previously dried 2 hours at 103°C, in distilled water and dilute to 1.0 liters. Store in a glass stoppered bottle.

5.8   Potassium biiodate (0.005N): Dilute 50 ml of 0.1N potassium biiodate (5.7) to 1-liter in a volumetric flask. Store in a glass stoppered bottle.

5.9   Standardization of 0.00564N phenylarsine oxide: Dissolve approximately 2g (±1g) KI (5.2) in 100 to 150 ml distilled water; add 10 ml H_2SO_4 solution (5.6) followed by 20 ml 0.005N potassium biiodate solution (5.8). Place in dark for 5 minutes; dilute to 300 ml and titrate with 0.00564N phenylarsine oxide solution (5.1) to a pale straw color. Add a small scoop of indicator (5.4). Wait until homogeneous blue color develops and continue the titration drop by drop until the color disappears. Run in duplicate. Duplicate determinations should agree within ±0.05 ml.

$$N\,PAO = \frac{20 \times 0.005}{ml\,PAO}$$

Adjust PAO solution if necessary and recheck.

6.   Procedure

6.1   Place 200 ml of sample in the sample container. This volume is convenient because the buret reading in milliliters is equivalent to mg/1 Cl. Up to 2 mg/1 is reliably titrated this way. Smaller sample aliquots diluted to 200 ml are used for concentrations greater than 2 mg/1. The construction of the cell and electrode component usually require 200 ml of sample.

6.2   Place on electrode assembly.

6.3   Add 1.0 ml KI solution (5.3).

6.4   Add 1 ml acetate buffer (5.5).

6.5   Titrate with 0.00564N PAO (5.1).

6.6   As each increment is added the needle deflects toward rest. When the needle no longer deflects subtract the last drop added from the buret reading to obtain the mg/1 Cl. Less and/or slower deflection signals that the end point is near.

7.   Calculations

7.1   For 0.00564N PAO and a 200 ml sample there are no calculations. The buret reading is in mg/1. The last increment, when the needle does not deflect toward rest, must be subtracted.

8.   Precision and Accuracy

8.1   More than 20 laboratories analyzed prepared samples of 0.64 and 1.83 mg/1 total Cl. The relative standard deviations were 24.8% and 12.5% respectively and the relative errors were 8.5% and 8.8% respectively.

In a single operator, single laboratory situation the following results were obtained.

Sample Matrix	Average mg/l	Stand Dev. mg/l	Rel. Stand. Dev. %
Distilled Water	0.38	0.02	6.1
	3.50	0.01	0.2
Drinking Water	0.97	0.03	2.6
River Water	0.57	0.02	3.0
Domestic Sewage	0.41	0.03	6.9

For these samples the results were compared to the iodometric titration as a means of obtaining a relative accuracy.

Sample Matrix	Iodometric Titration mg/l	Amperometric Titration. mg/l	% Recovery
Drinking Water	0.94	0.97	103.2
River Water	0.56	0.57	101.8
Domestic Sewage	0.50	0.41	82.0

Bibliography

1. Standard Methods for the Examination of Water and Wastewater, 14th Edition, p 322, Method 409C (1975).

2. Annual Book of ASTM Standards, Part 31, "Water", Standard D 1253–76, p 277, Method A (1976).

3. Bender, D. F., "Comparison of Methods for the Determination of Total Available Residual Chlorine in Various Sample Matrices", EPA Report-600/4–78–019.

CYANIDE, TOTAL

Method 335.2 (Titrimetric; Spectrophotometric)

STORET NO. 00720

1. Scope and Application
   1.1 This method is applicable to the determination of cyanide in drinking, surface and saline waters, domestic and industrial wastes.
   1.2 The titration procedure using silver nitrate with p-dimethylamino-benzal-rhodanine indicator is used for measuring concentrations of cyanide exceeding 1 mg/l (0.25 mg/250 ml of absorbing liquid).
   1.3 The colorimetric procedure is used for concentrations below 1 mg/l of cyanide and is sensitive to about 0.02 mg/l.
2. Summary of Method
   2.1 The cyanide as hydrocyanic acid (HCN) is released from cyanide complexes by means of a reflux-distillation operation and absorbed in a scrubber containing sodium hydroxide solution. The cyanide ion in the absorbing solution is then determined by volumetric titration or colorimetrically.
   2.2 In the colorimetric measurement the cyanide is converted to cyanogen chloride, CNCl, by reaction with chloramine-T at a pH less than 8 without hydrolyzing to the cyanate. After the reaction is complete, color is formed on the addition of pyridine-pyrazolone or pyridine-barbituric acid reagent. The absorbance is read at 620 nm when using pyridine-pyrazolone or 578 nm for pyridine-barbituric acid. To obtain colors of comparable intensity, it is essential to have the same salt content in both the sample and the standards.
   2.3 The titrimetric measurement uses a standard solution of silver nitrate to titrate cyanide in the presence of a silver sensitive indicator.
3. Definitions
   3.1 Cyanide is defined as cyanide ion and complex cyanides converted to hydrocyanic acid (HCN) by reaction in a reflux system of a mineral acid in the presence of magnesium ion.
4. Sample Handling and Preservation
   4.1 The sample should be collected in plastic or glass bottles of 1 liter or larger size. All bottles must be thoroughly cleansed and thoroughly rinsed to remove soluble material from containers.
   4.2 Oxidizing agents such as chlorine decompose most of the cyanides. Test a drop of the sample with potassium iodide-starch test paper (KI-starch paper); a blue color indicates the need for treatment. Add ascorbic acid, a few crystals at a time, until a drop of sample produces no color on the indicator paper. Then add an additional 0.06 g of ascorbic acid for each liter of sample volume.

Approved for NPDES
Issued 1974
Editorial revision 1974 and 1978
Technical Revision 1980

4.3 Samples must be preserved with 2 ml of 10 N sodium hydroxide per liter of sample (pH ≥ 12) at the time of collection.

4.4 Samples should be analyzed as rapidly as possible after collection. If storage is required, the samples should be stored in a refrigerator or in an ice chest filled with water and ice to maintain temperature at 4°C.

5. Interferences

5.1 Interferences are eliminated or reduced by using the distillation procedure described in Procedure 8.1, 8.2 and 8.3.

5.2 Sulfides adversely affect the colorimetric and titration procedures. Samples that contain hydrogen sulfide, metal sulfides or other compounds that may produce hydrogen sulfide during the distillation should be distilled by the optional procedure described in Procedure 8.2. The apparatus for this procedure is shown in Figure 3.

5.3 Fatty acids will distill and form soaps under the alkaline titration conditions, making the end point almost impossible to detect.

5.3.1 Acidify the sample with acetic acid (1+9) to pH 6.0 to 7.0.

<u>Caution</u>: This operation must be performed in the hood and the sample left there until it can be made alkaline again after the extraction has been performed.

5.3.2 Extract with iso-octane, hexane, or chloroform (preference in order named) with a solvent volume equal to 20% of the sample volume. One extraction is usually adequate to reduce the fatty acids below the interference level. Avoid multiple extractions or a long contact time at low pH in order to keep the loss of HCN at a minimum. When the extraction is completed, immediately raise the pH of the sample to above 12 with NaOH solution.

5.4 High results may be obtained for samples that contain nitrate and/or nitrite. During the distillation nitrate and nitrite will form nitrous acid which will react with some organic compounds to form oximes. These compounds formed will decompose under test conditions to generate HCN. The interference of nitrate and nitrite is eliminated by pretreatment with sulfamic acid.

6. Apparatus

6.1 Reflux distillation apparatus such as shown in Figure 1 or Figure 2. The boiling flask should be of 1 liter size with inlet tube and provision for condenser. The gas absorber may be a Fisher-Milligan scrubber.

6.2 Microburet, 5.0 ml (for titration).

6.3 Spectrophotometer suitable for measurements at 578 nm or 620 nm with a 1.0 cm cell or larger.

6.4 Reflux distillation apparatus for sulfide removal as shown in Figure 3. The boiling flask same as 6.1. The sulfide scrubber may be a Wheaton Bubber #709682 with 29/42 joints, size 100 ml. The air inlet tube should not be fritted. The cyanide absorption vessel should be the same as the sulfide scrubber. The air inlet tube should be fritted.

6.5 Flow meter, such as Lab Crest with stainless steel float (Fisher 11-164-50).

7. Reagents

7.1 Sodium hydroxide solution, 1.25N: Dissolve 50 g of NaOH in distilled water, and dilute to 1 liter with distilled water.

7.2 Lead acetate: Dissolve 30 g of $Pb(C_2H_3O_2) \cdot 3H_2O$ in 950 ml of distilled water. Adjust the pH to 4.5 with acetic acid. Dilute to 1 liter.

7.5 Sulfuric acid; 18N: Slowly add 500 ml of concentrated H_2SO_4 to 500 ml of distilled water.

7.6 Sodium dihydrogenphosphate, 1 M: Dissolve 138 g of $NaH_2PO_4 \cdot H_2O$ in 1 liter of distilled water. Refrigerate this solution.

7.7 Stock cyanide solution: Dissolve 2.51 g of KCN and 2 g KOH in 900 ml of distilled water. Standardize with 0.0192 N $AgNO_3$. Dilute to appropriate concentration so that 1 ml = 1 mg CN.

7.8 Standard cyanide solution, intermediate: Dilute 100.0 ml of stock (1 ml = 1 mg CN) to 1000 ml with distilled water (1 ml = 100.0 ug).

7.9 Working standard cyanide solution: Prepare fresh daily by diluting 100.0 ml of intermediate cyanide solution to 1000 ml with distilled water and store in a glass stoppered bottle. 1 ml = 10.0 ug CN.

7.10 Standard silver nitrate solution, 0.0192 N: Prepare by crushing approximately 5 g $AgNO_3$ crystals and drying to constant weight at 40°C. Weigh out 3.2647 g of dried $AgNO_3$, dissolve in distilled water, and dilute to 1000 ml (1 ml = 1mg CN).

7.11 Rhodanine indicator: Dissolve 20 mg of p-dimethyl-amino-benzalrhodanine in 100 ml of acetone.

7.12 Chloramine T solution: Dissolve 1.0 g of white, water soluble Chloramine T in 100 ml of distilled water and refrigerate until ready to use. Prepare fresh daily.

7.13 Color Reagent — One of the following may be used:

    7.13.1 Pyridine-Barbituric Acid Reagent: Place 15 g of barbituric acid in a 250 ml volumetric flask and add just enough distilled water to wash the sides of the flask and wet the barbituric acid. Add 75 ml of pyridine and mix. Add 15 ml of conc. HCl, mix, and cool to room temperature. Dilute to 250 ml with distilled water and mix. This reagent is stable for approximately six months if stored in a cool, dark place.

    7.13.2 Pyridine-pyrazolone solution:

        7.13.2.1 3-Methyl-1-phenyl-2-pyrazolin-5-one reagent, saturated solution: Add 0.25 g of 3-methyl-1-phenyl-2-pyrazolin-5-one to 50 ml of distilled water, heat to 60°C with stirring. Cool to room temperature.

        7.13.2.2 3,3'Dimethyl-1, 1'-diphenyl-[4,4'-bi-2 pyrazoline]-5,5'dione (bispyrazolone): Dissolve 0.01 g of bispyrazolone in 10 ml of pyridine.

        7.13.2.3 Pour solution (7.13.2.1) through non-acid-washed filter paper. Collect the filtrate. Through the same filter paper pour solution (7.13.2.2) collecting the filtrate in the same container as filtrate from (7.13.2.1). Mix until the filtrates are homogeneous. The mixed reagent develops a pink color but this does not affect the color production with cyanide if used within 24 hours of preparation.

7.14 Magnesium chloride solution: Weight 510 g of $MgCl_2 \cdot 6H_2O$ into a 1000 ml flask, dissolve and dilute to 1 liter with distilled water.

7.15 Sulfamic acid.

8. Procedure
   - 8.1 For samples without sulfide.
     - 8.1.1 Place 500 ml of sample, or an aliquot diluted to 500 ml in the 1 liter boiling flask. Pipet 50 ml of sodium hydroxide (7.1) into the absorbing tube. If the apparatus in Figure 1 is used, add distilled water until the spiral is covered. Connect the boiling flask, condenser, absorber and trap in the train. (Figure 1 or 2)
     - 8.1.2 Start a slow stream of air entering the boiling flask by adjusting the vacuum source. Adjust the vacuum so that approximately two bubbles of air per second enters the boiling flask through the air inlet tube. Proceed to 8.4.
   - 8.2 For samples that contain sulfide.
     - 8.2.1 Place 500 ml of sample, or an aliquot diluted to 500 ml in the 1 liter boiling flask. Pipet 50 ml of sodium hydroxide (7.1) to the absorbing tube. Add 25 ml of lead acetate (7.2) to the sulfide scrubber. Connect the boiling flask, condenser, scrubber and absorber in the train. (Figure 3) The flow meter is connected to the outlet tube of the cyanide absorber.
     - 8.2.2 Start a stream of air entering the boiling flask by adjusting the vacuum source. Adjust the vacuum so that approximately 1.5 liters per minute enters the boiling flask through the air inlet tube. The bubble rate may not remain constant while heat is being applied to the flask. It may be necessary to readjust the air rate occasionally. Proceed to 8.4.
   - 8.3 If samples contain NO_3 and or NO_2 add 2 g of sulfamic acid solution (7.15) after the air rate is set through the air inlet tube. Mix for 3 minutes prior to addition of H_2SO_4.
   - 8.4 Slowly add 50 ml 18N sulfuric acid (7.5) through the air inlet tube. Rinse the tube with distilled water and allow the airflow to mix the flask contents for 3 min. Pour 20 ml of magnesium chloride (7.14) into the air inlet and wash down with a stream of water.
   - 8.5 Heat the solution to boiling. Reflux for one hour. Turn off heat and continue the airflow for at least 15 minutes. After cooling the boiling flask, disconnect absorber and close off the vacuum source.
   - 8.6 Drain the solution from the absorber into a 250 ml volumetric flask. Wash the absorber with distilled water and add the washings to the flask. Dilute to the mark with distilled water.
   - 8.7 Withdraw 50 ml or less of the solution from the flask and transfer to a 100 ml volumetric flask. If less than 50 ml is taken, dilute to 50 ml with 0.25N sodium hydroxide solution (7.4). Add 15.0 ml of sodium phosphate solution (7.6) and mix.
     - 8.7.1 Pyridine-barbituric acid method: Add 2 ml of chloramine T (7.12) and mix. See Note 1. After 1 to 2 minutes, add 5 ml of pyridine-barbituric acid solution (7.13.1) and mix. Dilute to mark with distilled water and mix again. Allow 8 minutes for color development then read absorbance at 578 nm in a 1 cm cell within 15 minutes.
     - 8.7.2 Pyridine-pyrazolene method: Add 0.5 ml of chloramine T (7.12) and mix. See Note 1 and 2. After 1 to 2 minutes add 5 ml of pyridine-pyrazolone solution

(7.13.1) and mix. Dilute to mark with distilled water and mix again. After 40 minutes read absorbance at 620 nm in a 1 cm cell.

NOTE 1: Some distillates may contain compounds that have a chlorine demand. One minute after the addition of chloramine T, test for residual chlorine with KI-starch paper. If the test is negative, add an additional 0.5 ml of chlorine T. After one minute, recheck the sample.

NOTE 2: More than 05. ml of chloramine T will prevent the color from developing with pyridine-pyrazolone.

8.8   Standard curve for samples without sulfide.

8.8.1   Prepare a series of standards by pipeting suitable volumes of standard solution (7.9) into 250 ml volumetric flasks. To each standard add 50 ml of 1.25 N sodium hydroxide and dilute to 250 ml with distilled water. Prepare as follows:

ML of Working Standard Solution (1 ml = 10 μg CN)	Conc. μg CN per 250 ml
0	BLANK
1.0	10
2.0	20
5.0	50
10.0	100
15.0	150
20.0	200

8.8.2   It is not imperative that all standards be distilled in the same manner as the samples. It is recommended that at least two standards (a high and low) be distilled and compared to similar values on the curve to insure that the distillation technique is reliable. If distilled standards do not agree within ±10% of the undistilled standards the analyst should find the cause of the apparent error before proceeding.

8.8.3   Prepare a standard curve by plotting absorbance of standard vs. cyanide concentrations.

8.8.4   To check the efficiency of the sample distillation, add an increment of cyanide from either the intermediate standard (7.8) or the working standard (7.9) to 500 ml of sample to insure a level of 20 μg/l. Proceed with the analysis as in Procedure (8.1.1).

8.9   Standard curve for samples with sulfide.

8.9.1   It is imperative that all standards be distilled in the same manner as the samples. Standards distilled by this method will give a linear curve, but as the concentration increases, the recovery decreases. It is recommended that at least 3 standards be distilled.

8.9.2   Prepare a standard curve by plotting absorbance of standard vs. cyanide concentrations.

8.10 Titrimetric method.

    8.10.1 If the sample contains more than 1 mg/l of CN, transfer the distillate or a suitable aliquot diluted to 250 ml, to a 500 ml Erlenmeyer flask. Add 10-12 drops of the benzalrhodanine indicator.

    8.10.2 Titrate with standard silver nitrate to the first change in color from yellow to brownish-pink. Titrate a distilled water blank using the same amount of sodium hydroxide and indicator as in the sample.

    8.10.3 The analyst should familiarize himself with the end point of the titration and the amount of indicator to be used before actually titrating the samples.

9. Calculation

    9.1 If the colorimetric procedure is used, calculate the cyanide, in ug/l, in the original sample as follows:

$$CN, ug/l = \frac{A \times 1,000}{B} \times \frac{50}{C}$$

where:

A = ug CN read from standard curve
B = ml of original sample for distillation
C = ml taken for colorimetric analysis

9.2   Using the titrimetric procedure, calculate concentration of CN as follows:

$$CN, \text{mg/l} = \frac{(A - B)1,000}{\text{ml orig. sample}} \times \frac{250}{\text{ml of aliquot titrated}}$$

where:

A = volume of $AgNO_3$ for titration of sample.
B = volume of $AgNO_3$ for titration of blank.

10. Precision and Accuracy
    10.1 In a single laboratory (EMSL), using mixed industrial and domestic waste samples at concentrations of 0.06, 0.13, 0.28 and 0.62 mg/l CN, the standard deviations were ±0.005, ±0.007, ±0.031 and ±0.094, respectively.
    10.2 In a single laboratory (EMSL), using mixed industrial and domestic waste samples at concentrations of 0.28 and 0.62 mg/l CN, recoveries were 85% and 102%, respectively.

Bibliography

1.   Bark, L. S., and Higson, H. G. "Investigation of Reagents for the Colorimetric Determination of Small Amounts of Cyanide", Talanta, 2:471–479 (1964).
2.   Elly, C. T. "Recovery of Cyanides by Modified Serfass Distillation". Journal Water Pollution Control Federation 40:848–856 (1968).
3.   Annual Book of ASTM Standards, Part 31, "Water", Standard D2036–75, Method A, p 503 (1976).
4.   Standard Methods for the Examination of Water and Wastewater, 14th Edition, p 367 and 370, Method 413B and D (1975).
5.   Egekeze, J. O., and Oehne, F. W., "Direct Potentiometric Determination of Cyanide in Biological Materials," J. Analytical Toxicology, Vol. 3, p. 119, May/June 1979.
6.   Casey, J. P., Bright, J. W., and Helms, B. D., "Nitrosation Interference in Distillation Tests for Cyanide," Gulf Coast Waste Disposal Authority, Houston, Texas.

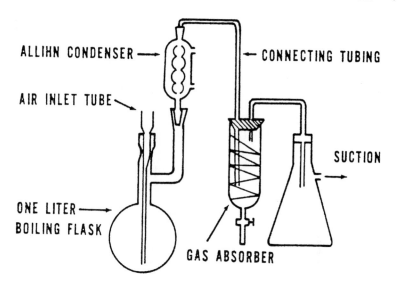

FIGURE 1
CYANIDE DISTILLATION APPARATUS

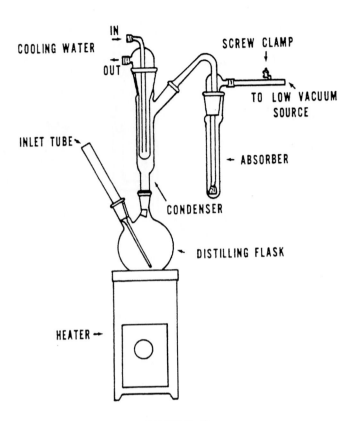

COOLING WATER

IN

OUT

SCREW CLAMP

TO LOW VACUUM
SOURCE

INLET TUBE

ABSORBER

CONDENSER

DISTILLING FLASK

HEATER

FIGURE 2
CYANIDE DISTILLATION APPARATUS

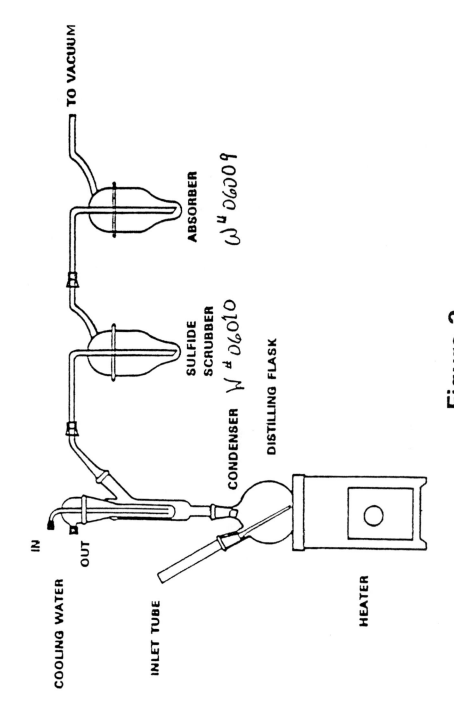

Figure 3.
Cyanide Distillation Apparatus

CYANIDES, AMENABLE TO CHLORINATION

Method 335.1 (Titrimetric; Spectrophotometric)

STORET NO. 00722

1. Scope and Application
   1.1 This method is applicable to the determination of cyanides amenable to chlorination in drinking, surface and saline waters, domestic and industrial wastes.
   1.2 The titration procedure is used for measuring concentrations of cyanide exceeding 1 mg/1 after removal of the cyanides amenable to chlorination. Below this level the colorimetric determination is used.

2. Summary of Method
   2.1 A portion of the sample is chlorinated at a pH > 11 to decompose the cyanide. Cyanide levels in the chlorinated sample are then determined by the method for Cyanide, Total, in this manual. Cyanides amenable to chlorination are then calculated by difference.

3. Reagents
   3.1 Calcium Hypochlorite solution: Dissolve 5 g of calcium hypochlorite $(Ca(OCl)_2)$ in 100 ml of distilled water.
   3.2 Sodium Hydroxide solution: Dissolve 50 g of sodium hydroxide (NaOH) in distilled water and dilute to 1 liter.
   3.3 Ascorbic acid: crystals.
   3.4 Potassium Iodide–starch test paper.

4. Procedure
   4.1 Two sample aliquots are required to determine cyanides amenable to chlorination. To one 500 ml aliquot or a volume diluted to 500 ml, add calcium hypochlorite solution (3.1) dropwise while agitating and maintaining the pH between 11 and 12 with sodium hydroxide (3.2).
   Caution: The initial reaction product of alkaline chlorination is the very toxic gas cyanogen chloride; therefore, it is recommended that this reaction be performed in a hood. For convenience, the sample may be agitated in a 1 liter beaker by means of a magnetic stirring device.
   4.2 Test for residual chlorine with KI-starch paper (3.4) and maintain this excess for one hour, continuing agitation. A distinct blue color on the test paper indicates a sufficient chlorine level. If necessary, add additional hypochlorite solution.
   4.3 After one hour, add 0.5 g portions of ascorbic acid (3.3) until KI-starch paper shows no residual chlorine. Add an additional 0.5 g of ascorbic acid to insure the presence of excess reducing agent.
   4.4 Test for total cyanide in both the chlorinated and unchlorinated aliquots as in the method Cyanide, Total, in this manual.

Approved for NPDES
Issued 1974

5. Calculation

 5.1 Calculate the cyanide amendable to chlorination as follows:

 $$CN, mg/1 = A - B$$

 where:

 A = mg/1 total cyanide in unchlorinated aliquot

 B = mg/1 total in chlorinated aliquot

Bibliography

1. Annual Book of ASTM Standards, Part 31, "Water", Standard D 2036–75, Method B, p 505 (1976).

2. Standard Methods for the Examination of Water and Wastewater, 14th Edition, p 376 and 370, Method 413F and D (1975).

FLUORIDE

Method 340.2 (Potentiometric, Ion Selective Electrode)

STORET NO: Total 00951
Dissolved 00950

1. Scope and Application
    1.1 This method is applicable to the measurement of fluoride in drinking, surface and saline waters, domestic and industrial wastes.
    1.2 Concentration of fluoride from 0.1 up to 1000 mg/liter may be measured.
    1.3 For Total or Total Dissolved Fluoride, the Bellack distillation is required for NPDES monitoring but is not required for SDWA monitoring.
2. Summary of Method
    2.1 The fluoride is determined potentiometrically using a fluoride electrode in conjunction with a standard single junction sleeve-type reference electrode and a pH meter having an expanded millivolt scale or a selective ion meter having a direct concentration scale for fluoride.
    2.2 The fluoride electrode consists of a lanthanum fluoride crystal across which a potential is developed by fluoride ions. The cell may be represented by Ag/Ag Cl, $Cl^-(0.3)$, $F^-(0.001)$ LaF/test solution/SCE/.
3. Interferences
    3.1 Extremes of pH interfere; sample pH should be between 5 and 9. Polyvalent cations of Si^{+4}, Fe^{+3} and Al^{+3} interfere by forming complexes with fluoride. The degree of interference depends upon the concentration of the complexing cations, the concentration of fluoride and the pH of the sample. The addition of a pH 5.0 buffer (described below) containing a strong chelating agent preferentially complexes aluminum (the most common interference), silicon and iron and eliminates the pH problem.
4. Sampling Handling and Preservation
    4.1 No special requirements.
5. Apparatus
    5.1 Electrometer (pH meter), with expanded mv scale, or a selective ion meter such as the Orion 400 Series.
    5.2 Fluoride Ion Activity Electrode, such as Orion No. 94–09[1].
    5.3 Reference electrode, single junction, sleeve-type, such as Orion No. 90–01, Beckman No. 40454, or Corning No. 476010.
    5.4 Magnetic Mixer, Teflon-coated stirring bar.

Approved for NPDES and SDWA
Issued 1971
Editorial revision 1974

6. Reagents

    6.1 Buffer solution, pH 5.0–5.5: To approximately 500 ml of distilled water in a 1 liter beaker add 57 ml of glacial acetic acid, 58 g of sodium chloride and 4 g of CDTA[2]. Stir to dissolve and cool to room temperature. Adjust pH of solution to between 5.0 and 5.5 with 5 N sodium hydroxide (about 150 ml will be required). Transfer solution to a 1 liter volumetric flask and dilute to the mark with distilled water. For work with brines, additional NaCl should be added to raise the chloride level to twice the highest expected level of chloride in the sample.

    6.2 Sodium fluoride, stock solution: 1.0 ml = 0.1 mg F. Dissolve 0.2210 g of sodium fluoride in distilled water and dilute to 1 liter in a volumetric flask. Store in chemical-resistant glass or polyethylene.

    6.3 Sodium fluoride, standard solution: 1.0 ml = 0.01 mg F. Dilute 100.0 ml of sodium fluoride stock solution (6.2) to 1000 ml with distilled water.

    6.4 Sodium hydroxide, 5N: Dissolve 200 g sodium hydroxide in distilled water, cool and dilute to 1 liter.

7. Calibration

    7.1 Prepare a series of standards using the fluoride standard solution (6.3) in the range of 0 to 2.00 mg/1 by diluting appropriate volumes to 50.0 ml. The following series may be used:

Millimeters of Standard (1.0 ml = 0.01 mg/F)	Concentration when Diluted to 50 ml, mg F/liter
0.00	0.00
1.00	0.20
2.00	0.40
3.00	0.60
4.00	0.80
5.00	1.00
6.00	1.20
8.00	1.60
10.00	2.00

    7.2 Calibration of Electrometer: Proceed as described in (8.1). Using semilogarithmic graph paper, plot the concentration of fluoride in mg/liter on the log axis vs. the electrode potential developed in the standard on the linear axis, starting with the lowest concentration at the bottom of the scale. Calibration of a selective ion meter: Follow the directions of the manufacturer for the operation of the instrument.

8. Procedure

    8.1 Place 50.0 ml of sample or standard solution and 50.0 ml of buffer (See Note) in a 150 ml beaker. Place on a magnetic stirrer and mix at medium speed. Immerse the electrodes in the solution and observe the meter reading while mixing. The electrodes must remain in the solution for at least three minutes or until the reading has stabilized. At concentrations under 0.5 mg/liter F, it may require as long as five minutes to reach a stable meter reading; high concentrations stabilize more quickly. If a pH meter is used, record the potential measurement for each unknown sample and convert the potential reading to the fluoride ion concentration of the unknown using the standard curve. If a selective ion meter is used, read the fluoride level in the unknown sample directly in mg/l on the fluoride scale.

NOTE: For industrial waste samples, this amount of buffer may not be adequate. Analyst should check pH first. If highly basic (> 9), add 1 N HCl to adjust pH to 8.3.

9. Precision and Accuracy
    9.1 A synthetic sample prepared by the Analytical Reference Service, PHS, containing 0.85 mg/l fluoride and no interferences was analyzed by 111 analysts; a mean of 0.84 mg/l with a standard deviation of ±0.03 was obtained.
    9.2 On the same study, a synthetic sample containing 0.75 mg/l fluoride, 2.5 mg/l polyphosphate and 300 mg/l alkalinity, was analyzed by the same 111 analysts; a mean of 0.75 mg/l fluoride with a standard deviation of ±0.036 was obtained.

Bibliography

1. Patent No. 3,431,182 (March 4, 1969).
2. CDTA is the abbreviated designation of 1,2-cyclohexylene dinitrilo tetraacetic acid. (The monohydrate form may also be used.) Eastman Kodak 15411, Mallinckrodt 2357, Sigma D 1383, Tridom-Fluka 32869-32870 or equivalent.
3. Standard Methods for the Examination of Water and Wastewaters, p 389, Method No. 414A, Preliminary Distillation Step (Bellack), and p 391, Method No. 414B, Electrode Method, 14th Edition (1975).
4. Annual Book of ASTM Standards, Part 31, "Water", Standard D1179–72, Method B, p 312 (1976).

HARDNESS, Total (mg/1 as CaCO₃)

Method 130.1 (Colorimetric, Automated EDTA)

STORET NO. 00900

1. Scope and Application
   1.1 This automated method is applicable to drinking, surface, and saline waters. The applicable range is 10 to 400 mg/1 as $CaCO_3$. Approximately 12 samples per hour can be analyzed.

2. Summary of Method
   2.1 The magnesium EDTA exchanges magnesium on an equivalent basis for any calcium and/or other cations to form a more stable EDTA chelate than magnesium. The free magnesium reacts with calmagite at a pH of 10 to give a red-violet complex. Thus, by measuring only magnesium concentration in the final reaction stream, an accurate measurement of total hardness is possible.

3. Sample Handling and Preservation
   3.1 Cool to 4°C, HNO_3 to pH < 2.

4. Interferences
   4.1 No significant interferences.

5. Apparatus
   5.1 Technicon AutoAnalyzer consisting of:
       5.1.1 Sampler I.
       5.1.2 Continuous Filter.
       5.1.3 Manifold.
       5.1.4 Proportioning Pump.
       5.1.5 Colorimeter equipped with 15 mm tubular flow cell and 520 nm filters.
       5.1.6 Recorder equipped with range expander.

6. Reagents
   6.1 Buffer: Dissolve 67.6 g NH_4Cl in 572 ml of NH_4OH and dilute to 1 liter with distilled water.
   6.2 Calmagite Indicator: Dissolve 0.25 g in 500 ml of distilled water by stirring approximately 30 minutes on a magnetic stirrer. Filter.
   6.3 Monomagnesium ethylenediamine-tetraacetate (MgEDTA): Dissolve 0.2 g of MgEDTA in 1 liter of distilled water.
   6.4 Stock Solution: Weigh 1.000 g of calcium carbonate (pre-dried at 105°C) into 500 ml Erlenmeyer flask; add 1:1 HCl until all $CaCO_3$ has dissolved. Add 200 ml of distilled water and boil for a few minutes. Cool, add a few drops of methyl red indicator, and adjust to the orange color with 3N NH_4OH and dilute to 1000 ml with distilled water. 1.0 ml = 1.0 mg $CaCO_3$.

Approved for NPDES
Issued 1971

6.4.1 Dilute each of the following volumes of stock solutions to 250 ml in a volumetric flask for appropriate standards:

Stock Solution, ml	$CaCO_3$, mg/l
2.5	10.0
5.0	20.0
10.0	40.0
15.0	60.0
25.0	100.0
35.0	140.0
50.0	200.0
75.0	300.0
100.0	400.0

6.5 Ammonium Hydroxide, 1N: Dilute 70 ml of conc. NH_4OH to 1 liter with distilled water.

7. Procedure
   7.1 Pretreatment
      7.1.1 For drinking waters, surface waters, saline waters, and dilutions thereof, no pretreatment steps are necessary. Proceed to 7.2.
      7.1.2 For most wastewaters, and highly polluted waters, the sample must be digested as given in the Atomic Absorption Methods section of this manual, paragraphs 4.1.3 and 4.1.4. Following this digestion, proceed to 7.2.
   7.2 Neutralize 50.0 ml of sample with 1N ammonium hydroxide (6.5) and note volume of NH_4OH used.
   7.3 Set up manifold as shown in Figure 1.
   7.4 Allow both colorimeter and recorder to warm up for 30 minutes. Run a baseline with all reagents, feeding distilled water through the sample line. Adjust dark current and operative opening on colorimeter to obtain stable baseline.
   7.5 Place distilled water wash tubes in alternate openings in Sampler and set sample timing at 2.5 minutes.
   7.6 Arrange working standards in Sampler in order of decreasing concentrations. Complete loading of Sampler tray with unknown samples.
   7.7 Switch sample line from distilled water to Sampler and begin analysis.
8. Calculation
   8.1 Prepare standard curve by plotting peak heights of processed standards against concentration values. Compute concentration of samples by comparing sample peak heights with standard curve. Correct for amount of NH_4OH used in 7.2 as follows:

   $$mg/l = \frac{A}{50} \times B$$

   where:

   A = Vol. of sample plus volume of NH_4OH
   B = Concentration from standard curve

9. Precision and Accuracy

    9.1   In a single laboratory (EMSL), using surface water samples at concentrations of 19, 120, 385, and 366 mg/1 as $CaCO_3$, the standard deviations were ±1.5, ±1.5, ±4.5, and ±5.0, respectively.

    9.2   In a single laboratory (EMSL), using surface water samples at concentrations of 39 and 296 mg/1 as $CaCO_3$, recoveries were 89% and 93%, respectively.

Bibliography

1. Technicon AutoAnalyzer Methodology, Bulletin No. 2, Technicon Controls, Inc., Chauncey, New York (July 1960).

2. Standard Methods for the Examination of Water and Wastewater, 14th Edition, p 202, Method 309B (1975).

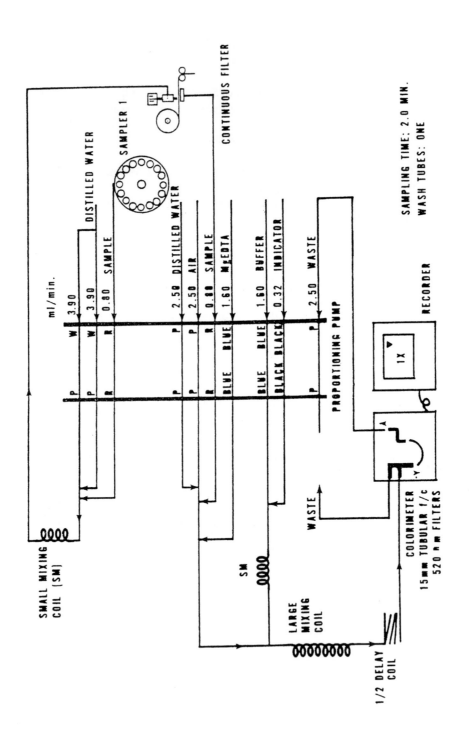

FIGURE 1. HARDNESS MANIFOLD AA-I

pH

Method 150.1 (Electrometric)

STORET NO.

Determined on site   00400

Laboratory   00403

1. Scope and Application
   1.1 This method is applicable to drinking, surface, and saline waters, domestic and industrial wastes and acid rain (atmospheric deposition).
2. Summary of Method
   2.1 The pH of a sample is determined electrometrically using either a glass electrode in combination with a reference potential or a combination electrode.
3. Sample Handling and Preservation
   3.1 Samples should be analyzed as soon as possible preferably in the field at the time of sampling.
   3.2 High-purity waters and waters not at equilibrium with the atmosphere are subject to changes when exposed to the atmosphere, therefore the sample containers should be filled completely and kept sealed prior to analysis.
4. Interferences
   4.1 The glass electrode, in general, is not subject to solution interferences from color, turbidity, colloidal matter, oxidants, reductants or high salinity.
   4.2 Sodium error at pH levels greater than 10 can be reduced or eliminated by using a "low sodium error" electrode.
   4.3 Coatings of oily material or particulate matter can impair electrode response. These coatings can usually be removed by gentle wiping or detergent washing, followed by distilled water rinsing. An additional treatment with hydrochloric acid (1 + 9) may be necessary to remove any remaining film.
   4.4 Temperature effects on the electrometric measurement of pH arise from two sources. The first is caused by the change in electrode output at various temperatures. This interference can be controlled with instruments having temperature compensation or by calibrating the electrode-instrument system at the temperature of the samples. The second source is the change of pH inherent in the sample at various temperatures. This error is sample dependent and cannot be controlled, it should therefore be noted by reporting both the pH and temperature at the time of analysis.
5. Apparatus
   5.1 pH Meter-laboratory or field model. A wide variety of instruments are commercially available with various specifications and optional equipment.

Approved for NPDES

Issued 1971

Editorial revision 1978 and 1982

5.2   Glass electrode.

5.3   Reference electrode–a calomel, silver-silver chloride or other reference electrode of constant potential may be used.

NOTE 1:   Combination electrodes incorporating both measuring and reference functions are convenient to use and are available with solid, gel type filling materials that require minimal maintenance.

5.4   Magnetic stirrer and Teflon-coated stirring bar.

5.5   Thermometer or temperature sensor for automatic compensation.

6.   Reagents

6.1   Primary standard buffer salts are available from the National Bureau of Standards and should be used in situations where extreme accuracy is necessary.

6.1.1   Preparation of reference solutions from these salts require some special precautions and handling[1] such as low conductivity dilution water, drying ovens, and carbon dioxide free purge gas. These solutions should be replaced at least once each month.

6.2   Secondary standard buffers may be prepared from NBS salts or purchased as a solution from commercial vendors. Use of these commercially available solutions, that have been validated by comparison to NBS standards, are recommended for routine use.

7.   Calibration

7.1   Because of the wide variety of pH meters and accessories, detailed operating procedures cannot be incorporated into this method. Each analyst must be acquainted with the operation of each system and familiar with all instrument functions. Special attention to care of the electrodes is recommended.

7.2   Each instrument/electrode system must be calibrated at a minimum of two points that bracket the expected pH of the samples and are approximately three pH units or more apart.

7.2.1   Various instrument designs may involve use of a "balance" or "standardize" dial and/or a slope adjustment as outlined in the manufacturer's instructions. Repeat adjustments on successive portions of the two buffer solutions as outlined in procedure 8.2 until readings are within 0.05 pH units of the buffer solution value.

8.   Procedure

8.1   Standardize the meter and electrode system as outlined in Section 7.

8.2   Place the sample or buffer solution in a clean glass beaker using a sufficient volume to cover the sensing elements of the electrodes and to give adequate clearance for the magnetic stirring bar.

8.2.1   If field measurements are being made the electrodes may be immersed directly in the sample stream to an adequate depth and moved in a manner to insure sufficient sample movement across the electrode sensing element as indicated by drift free (< 0.1 pH) readings.

8.3   If the sample temperature differs by more than 2°C from the buffer solution the measured pH values must be corrected. Instruments are equipped with automatic or manual

[1]National Bureau of Standards Special Publication 260.

compensators that electronically adjust for temperature differences. Refer to manufacturer's instructions.

8.4 After rinsing and gently wiping the electrodes, if necessary, immerse them into the sample beaker or sample stream and stir at a constant rate to provide homogeneity and suspension of solids. Rate of stirring should minimize the air transfer rate at the air water interface of the sample. Note and record sample pH and temperature. Repeat measurement on successive volumes of sample until values differ by less than 0.1 pH units. Two or three volume changes are usually sufficient.

8.5 For acid rain samples it is most important that the magnetic stirrer is not used. Instead, swirl the sample gently for a few seconds after the introduction of the electrode(s). Allow the electrode(s) to equilibrate. The air-water interface should not be disturbed while measurement is being made. If the sample is not in equilibrium with the atmosphere, pH values will change as the dissolved gases are either absorbed or desorbed. Record sample pH and temperature.

9. Calculation

9.1 pH meters read directly in pH units. Report pH to the nearest 0.1 unit and temperature to the nearest °C.

10. Precision and Accuracy

10.1 Forty-four analysts in twenty laboratories analyzed six synthetic water samples containing exact increments of hydrogen-hydroxyl ions, with the following results:

pH Units	Standard Deviation pH Units	Bias, %	Accuracy as Bias, pH Units
3.5	0.10	−0.29	−0.01
3.5	0.11	−0.00	
7.1	0.20	+1.01	+0.07
7.2	0.18	−0.03	−0.002
8.0	0.13	−0.12	−0.01
8.0	0.12	+0.16	+0.01

(FWPCA Method Study 1, Mineral and Physical Analyses)

10.2 In a single laboratory (EMSL), using surface water samples at an average pH of 7.7, the standard deviation was ±0.1.

Bibliography

1. Standard Methods for the Examination of Water and Wastewater, 14th Edition, p 460, (1975).
2. Annual Book of ASTM Standards, Part 31, "Water", Standard D1293–65, p 178 (1976).
3. Peden, M. E. and Skowron, L. M., Ionic Stability of Precipitation Samples, Atmospheric Environment, Vol. 12, pp. 2343-2349, 1978.

NITROGEN, KJELDAHL, TOTAL

Method 351.3 (Colorimetric; Titrimetric; Potentiometric)

STORET NO. 00625

1. Scope and Application
   1.1 This method covers the determination of total Kjeldahl nitrogen in drinking, surface and saline waters, domestic and industrial wastes. The procedure converts nitrogen components of biological origin such as amino acids, proteins and peptides to ammonia, but may not convert the nitrogenous compounds of some industrial wastes such as amines, nitro compounds, hydrazones, oximes, semicarbazones and some refractory tertiary amines.
   1.2 Three alternatives are listed for the determination of ammonia after distillation: the titrimetric method which is applicable to concentrations above 1 mg N/liter; the Nesslerization method which is applicable to concentrations below 1 mg N/liter; and the potentiometric method applicable to the range 0.05 to 1400 mg/1.
   1.3 This method is described for macro and micro glassware systems.
2. Definitions
   2.1 Total Kjeldahl nitrogen is defined as the sum of free-ammonia and organic nitrogen compounds which are converted to ammonium sulfate $(NH_4)_2SO_4$, under the conditions of digestion described below.
   2.2 Organic Kjeldahl nitrogen is defined as the difference obtained by subtracting the free-ammonia value (Method 350.2, Nitrogen, Ammonia, this manual) from the total Kjeldahl nitrogen value. This may be determined directly by removal of ammonia before digestion.
3. Summary of Method
   3.1 The sample is heated in the presence of conc. sulfuric acid, K_2SO_4 and $HgSO_4$ and evaporated until SO_3 fumes are obtained and the solution becomes colorless or pale yellow. The residue is cooled, diluted, and is treated and made alkaline with a hydroxide-thiosulfate solution. The ammonia is distilled and determined after distillation by Nesslerization, titration or potentiometry.
4. Sample Handling and Preservation
   4.1 Samples may be preserved by addition of 2 ml of conc. H_2SO_4 per liter and stored at 4°C. Even when preserved in this manner, conversion of organic nitrogen to ammonia may occur. Preserved samples should be analyzed as soon as possible.
5. Interference
   5.1 High nitrate concentrations (10X or more than the TKN level) result in low TKN values. The reaction between nitrate and ammonia can be prevented by the use of an anion exchange resin (chloride form) to remove the nitrate prior to the TKN analysis.

Approved for NPDES
Issued 1971
Editorial revision 1974 and 1978

6. Apparatus

    6.1   Digestion apparatus: A Kjeldahl digestion apparatus with 800 or 100 ml flasks and suction takeoff to remove SO_3 fumes and water.

    6.2   Distillation apparatus: The macro Kjeldahl flask is connected to a condenser and an adaptor so that the distillate can be collected. Micro Kjeldahl steam distillation apparatus is commercially available.

    6.3   Spectrophotometer for use at 400 to 425 nm with a light path of 1 cm or longer.

7. Reagents

    7.1   Distilled water should be free of ammonia. Such water is best prepared by the passage of distilled water through an ion exchange column containing a strongly acidic cation exchange resin mixed with a strongly basic anion exchange resin. Regeneration of the column should be carried out according to the manufacturer's instructions.

           NOTE 1: All solutions must be made with ammonia-free water.

    7.2   Mercuric sulfate solution: Dissolve 8 g red mercuric oxide (HgO) in 50 ml of 1:4 sulfuric acid (10.0 ml conc. H_2SO_4 : 40 ml distilled water) and dilute to 100 ml with distilled water.

    7.3   Sulfuric acid-mercuric sulfate-potassium sulfate solution: Dissolve 267 g K_2SO_4 in 1300 ml distilled water and 400 ml conc. H_2SO_4. Add 50 ml mercuric sulfate solution (7.2) and dilute to 2 liters with distilled water.

    7.4   Sodium hydroxide-sodium thiosulfate solution: Dissolve 500 g NaOH and 25 g $Na_2S_2O_3 \cdot 5H_2O$ in distilled water and dilute to 1 liter.

    7.5   Mixed indicator: Mix 2 volumes of 0.2% methyl red in 95% ethanol with 1 volume of 0.2% methylene blue in ethanol. Prepare fresh every 30 days.

    7.6   Boric acid solution: Dissolve 20 g boric acid, H_3BO_3, in water and dilute to 1 liter with distilled water.

    7.7   Sulfuric acid, standard solution: (0.02 N) 1 ml = 0.28 mg NH_3–N. Prepare a stock solution of approximately 0.1 N acid by diluting 3 ml of conc. H_2SO_4 (sp. gr. 1.84) to 1 liter with CO_2-free distilled water. Dilute 200 ml of this solution to 1 liter with CO_2-free distilled water. Standardize the approximately 0.02 N acid so prepared against 0.0200 N Na_2CO_3 solution. This last solution is prepared by dissolving 1.060 g anhydrous Na_2CO_3, oven-dried at 140°C, and diluting to 1 liter with CO_2-free distilled water.

           NOTE 2: An alternate and perhaps preferable method is to standardize the approximately 0.1 N H_2SO_4 solution against a 0.100 N Na_2CO_3 solution. By proper dilution the 0.02 N acid can the be prepared.

    7.8   Ammonium chloride, stock solution: 1.0 ml = 1.0 mg NH_3–N. Dissolve 3.819 g NH_4Cl in water and make up to 1 liter in a volumetric flask with distilled water.

    7.9   Ammonium chloride, standard solution: 1.0 ml = 0.01 mg NH_3–N. Dilute 10.0 ml of the stock solution (7.8) with distilled water to 1 liter in a volumetric flask.

    7.10   Nessler reagent: Dissolve 100 g of mercuric iodide and 70 g potassium iodide in a small volume of distilled water. Add this mixture slowly, with stirring, to a cooled solution of 160 g of NaOH in 500 ml of distilled water. Dilute the mixture to 1 liter. The solution is stable for at least one year if stored in a pyrex bottle out of direct sunlight.

NOTE 3: Reagents 7.7, 7.8, 7.9, and 7.10 are identical to reagents 6.8, 6.2, 6.3, and 6.6 described under Nitrogen, Ammonia (Colorimetric; Titrimetric; Potentiometric-Distillation Procedure, Method 350.2).

8. Procedure

8.1 The distillation apparatus should be pre-steamed before use by distilling a 1:1 mixture of distilled water and sodium hydroxide-sodium thiosulfate solution (7.4) until the distillate is ammonia-free. This operation should be repeated each time the apparatus is out of service long enough to accumulate ammonia (usually 4 hours or more).

8.2 Macro Kjeldahl system

8.2.1 Place a measured sample or the residue from the distillation in the ammonia determination (for Organic Kjeldahl only) into an 800 ml Kjeldahl flask. The sample size can be determined from the following table:

Kjeldahl Nitrogen in Sample, mg/l	Sample Size ml
0–5	500
5–10	250
10–20	100
20–50	50.0
50–500	25.0

Dilute the sample, if required, to 500 ml with distilled water, and add 100 ml sulfuric acid-mercuric sulfate-potassium sulfate solution (7.3). Evaporate the mixture in the Kjeldahl apparatus until SO_3 fumes are given off and the solution turns colorless or pale yellow. Continue heating for 30 additional minutes. Cool the residue and add 300 ml distilled water.

8.2.2 Make the digestate alkaline by careful addition of 100 ml of sodium hydroxide – thiosulfate solution (7.4) without mixing.

NOTE 5: Slow addition of the heavy caustic solution down the tilted neck of the digestion flask will cause heavier solution to underlay the aqueous sulfuric acid solution without loss of free-ammonia. Do not mix until the digestion flask has been connected to the distillation apparatus.

8.2.3 Connect the Kjeldahl flask to the condenser with the tip of condenser or an extension of the condenser tip below the level of the boric acid solution (7.6) in the receiving flask.

8.2.4 Distill 300 ml at the rate of 6–10 ml/min., into 50 ml of 2% boric acid (7.6) contained in a 500 ml Erlenmeyer flask.

8.2.5 Dilute the distillate to 500 ml in the flask. These flasks should be marked at the 350 and the 500 ml volumes. With such marking, it is not necessary to transfer the distillate to volumetric flasks. For concentrations above 1 mg/l, the ammonia can be determined titrimetrically. For concentrations below this value, it is determined colorimetrically. The potentiometric method is applicable to the range 0.05 to 1400 mg/l.

8.3 Micro Kjeldahl system

    8.3.1 Place 50.0 ml of sample or an aliquot diluted to 50 ml in a 100 ml Kjeldahl flask and add 10 ml sulfuric acid-mercuric sulfate-potassium sulfate solution (7.3). Evaporate the mixture in the Kjeldahl apparatus until SO_3 fumes are given off and the solution turns colorless or pale yellow. Then digest for an additional 30 minutes. Cool the residue and add 30 ml distilled water.

    8.3.2 Make the digestate alkaline by careful addition of 10 ml of sodium hydroxide-thiosulfate solution (7.4) without mixing. Do not mix until the digestion flask has been connected to the distillation apparatus.

    8.3.3 Connect the Kjeldahl flask to the condenser with the tip of condenser or an extension of the condenser tip below the level of the boric acid solution (7.6) in the receiving flask or 50 ml short-form Nessler tube.

    8.3.4 Steam distill 30 ml at the rate of 6–10 ml/min., into 5 ml of 2% boric acid (7.6).

    8.3.5 Dilute the distillate to 50 ml. For concentrations above 1 mg/l the ammonia can be determined titrimetrically. For concentrations below this value, it is determined colorimetrically. The potentiometric method is applicable to the range 0.05 to 1400 mg/l.

8.4 Determination of ammonia in distillate: Determine the ammonia content of the distillate titrimetrically, colorimetrically, or potentiometrically, as described below.

    8.4.1 Titrimetric determination: Add 3 drops of the mixed indicator (7.5) to the distillate and titrate the ammonia with the 0.02 N H_2SO_4 (7.7), matching the endpoint against a blank containing the same volume of distilled water and H_3BO_3 (7.6) solution.

    8.4.2 Colorimetric determination: Prepare a series of Nessler tube standards as follows:

ml of Standard 1.0 ml = 0.01 mg NH_3-N	mg NH_3-N/50.0 ml
0.0	0.0
0.5	0.005
1.0	0.010
2.0	0.020
4.0	0.040
5.0	0.050
8.0	0.080
10.0	0.10

Dilute each tube to 50 ml with ammonia free water, add 1 ml of Nessler Reagent (7.10) and mix. After 20 minutes read the absorbance at 425 nm against the blank. From the values obtained for the standards plot absorbance vs. mg NH_3-N for the standard curve. Develop color in the 50 ml diluted distillate in exactly the same manner and read mg NH_3-N from the standard curve.

    8.4.3 Potentiometric determination: Consult the method entitled Nitrogen, Ammonia: Potentiometric, Ion Selective Electrode Method, (Method 350.3) in this manual.

    8.4.4 It is not imperative that all standards be treated in the same manner as the samples. It is recommended that at least 2 standards (a high and low) be digested, distilled, and compared to similar values on the curve to insure that the digestion-distillation technique is reliable. If treated standards do not agree with untreated standards the operator should find the cause of the apparent error before proceeding.

9.   Calculation

9.1   If the titrimetric procedure is used, calculate Total Kjeldahl Nitrogen, in mg/1, in the original sample as follows:

$$\text{TKN, mg/l} = \frac{(A - B)N \times F \times 1{,}000}{S}$$

where:

A = milliliters of standard 0.020 N H_2SO_4 solution used in titrating sample.
B = milliliters of standard 0.020 N H_2SO_4 solution used in titrating blank.
N = normality of sulfuric acid solution.
F = milliequivalent weight of nitrogen (14 mg).
S = milliliters of sample digested.

If the sulfuric acid is exactly 0.02 N the formula is shortened to:

$$\text{TKN, mg/l} = \frac{(A - B) \times 280}{S}$$

9.2   If the Nessler procedure is used, calculate the Total Kjeldahl Nitrogen, in mg/1, in the original sample as follows:

$$\text{TKN, mg/l} = \frac{A \times 1{,}000}{D} \times \frac{B}{C}$$

where:

A = mg NH_3–N read from curve.
B = ml total distillate collected including the H_3BO_3.
C = ml distillate taken for Nesslerization.
D = ml of original sample taken.

9.3   Calculate Organic Kjeldahl Nitrogen in mg/1, as follows:
Organic Kjeldahl Nitrogen = TKN –(NH_3–N.)

9.4   Potentiometric determination: Calculate Total Kjeldahl Nitrogen, in mg/l, in the original sample as follows:

$$TKN, mg/l = \frac{B}{D} \times A$$

where:

A = mg NH_3–N/l from electrode method standard curve.

B = volume of diluted distillate in ml.

D = ml of original sample taken.

10.   Precision

10.1   Thirty-one analysts in twenty laboratories analyzed natural water samples containing exact increments of organic nitrogen, with the following results:

Increment as Nitrogen, Kjeldahl mg N/liter	Precision as Standard Deviation mg N/liter	Accuracy as	
		Bias, %	Bias, mg N/liter
0.20	0.197	+ 15.54	+0.03
0.31	0.247	+ 5.45	+0.02
4.10	1.056	+ 1.03	+0.04
4.61	1.191	− 1.67	–0.08

(FWPCA Method Study 2, Nutrient Analyses)

Bibliography

1.   Standard Methods for the Examination of Water and Wastewater, 14th Edition, p 437, Method 421 (1975).

2.   Schlueter, Albert, "Nitrate Interference In Total Kjeldahl Nitrogen Determinations and Its Removal by Anion Exchange Resins", EPA Report 600/7–77–017.

NITROGEN, NITRATE

Method 352.1 (Colorimetric, Brucine)

STORET NO. Total 00620

1. Scope and Application
    1.1 This method is applicable to the analysis of drinking, surface and saline waters, domestic and industrial wastes. Modification can be made to remove or correct for turbidity, color, salinity, or dissolved organic compounds in the sample.
    1.2 The applicable range of concentrations is 0.1 to 2 mg NO_3–N/liter.
2. Summary of Method
    2.1 This method is based upon the reaction of the nitrate ion with brucine sulfate in a 13 N H_2SO_4 solution at a temperature of 100°C. The color of the resulting complex is measured at 410 nm. Temperature control of the color reaction is extremely critical.
3. Sample Handling and Preservation
    3.1 Analysis should be made as soon as possible. If analysis can be made within 24 hours, the sample should be preserved by refrigeration at 4°C. When samples must be stored for more than 24 hours, they should be preserved with sulfuric acid (2 ml conc. H_2SO_4 per liter) and refrigeration.
4. Interferences
    4.1 Dissolved organic matter will cause an off color in 13 N H_2SO_4 and must be compensated for by additions of all reagents except the brucine-sulfanilic acid reagent. This also applies to natural color present not due to dissolved organics.
    4.2 The effect of salinity is eliminated by addition of sodium chloride to the blanks, standards and samples.
    4.3 All strong oxidizing or reducing agents interfere. The presence of oxidizing agents may be determined with a total residual chlorine test kit.
    4.4 Residual chlorine interference is eliminated by the addition of sodium arsenite.
    4.5 Ferrous and ferric iron and quadrivalent manganese give slight positive interferences, but in concentrations less than 1 mg/1 these are negligible.
    4.6 Uneven heating of the samples and standards during the reaction time will result in erratic values. The necessity for absolute control of temperature during the critical color development period cannot be too strongly emphasized.
5. Apparatus
    5.1 Spectrophotometer or filter photometer suitable for measuring absorbance at 410 nm.
    5.2 Sufficient number of 40–50 ml glass sample tubes for reagent blanks, standards and samples.
    5.3 Neoprene coated wire racks to hold sample tubes.
    5.4 Water bath suitable for use at 100°C. This bath should contain a stirring mechanism so that all tubes are at the same temperature and should be of sufficient capacity to accept

Approved for NPDES and SDWA
Issued 1971

the required number of tubes without significant drop in temperature when the tubes are immersed.

5.5 Water bath suitable for use at 10–15°C.

6. Reagents

6.1 Distilled water free of nitrite and nitrate is to be used in preparation of all reagents and standards.

6.2 Sodium chloride solution (30%): Dissolve 300 g NaCl in distilled water and dilute to 1 liter.

6.3 Sulfuric acid solution: Carefully add 500 ml conc. H_2SO_4 to 125 ml distilled water. Cool and keep tightly stoppered to prevent absorption of atmospheric moisture.

6.4 Brucine-sulfanilic acid reagent: Dissolve 1 g brucine sulfate $[(C_{23}H_{26}N_2O_4)_2 \cdot H_2SO_4 \cdot 7H_2O]$ and 0.1 g sulfanilic acid $(NH_2C_6H_4SO_3H \cdot H_2O)$ in 70 ml hot distilled water. Add 3 ml conc. HCl, cool, mix and dilute to 100 ml with distilled water. Store in a dark bottle at 5°C. This solution is stable for several months; the pink color that develops slowly does not effect its usefulness. <u>Mark bottle with warning CAUTION: Brucine Sulfate is toxic; take care to avoid ingestion.</u>

6.5 Potassium nitrate stock solution: 1.0 ml = 0.1 mg NO_3–N. Dissolve 0.7218 g anhydrous potassium nitrate (KNO_3) in distilled water and dilute to 1 liter in a volumetric flask. Preserve with 2 ml chloroform per liter. This solution is stable for at least 6 months.

6.6 Potassium nitrate standard solution: 1.0 ml = 0.001 mg NO_3–N. Dilute 10.0 ml of the stock solution (6.5) to 1 liter in a volumetric flask. This standard solution should be prepared fresh weekly.

6.7 Acetic acid (1 + 3): Dilute 1 volume glacial acetic acid (CH_3COOH) with 3 volumes of distilled water.

6.8 Sodium hydroxide (1N): Dissolve 40 g of NaOH in distilled water. Cool and dilute to 1 liter.

7. Procedure

7.1 Adjust the pH of the samples to approximately 7 with acetic acid (6.7) or sodium hydroxide (6.8). If necessary, filter to remove turbidity.

7.2 Set up the required number of sample tubes in the rack to handle reagent blank, standards and samples. Space tubes evenly throughout the rack to allow for even flow of bath water between the tubes. This should assist in achieving uniform heating of all tubes.

7.3 If it is necessary to correct for color or dissolved organic matter which will cause color on heating, a set of duplicate samples must be run to which all reagents except the brucine-sulfanilic acid have been added.

7.4 Pipette 10.0 ml of standards and samples or an aliquot of the samples diluted to 10.0 ml into the sample tubes.

7.5 If the samples are saline, add 2 ml of the 30% sodium chloride solution (6.2) to the reagent blank, standards and samples. For fresh water samples, sodium chloride solution may be omitted. Mix contents of tubes by swirling and place rack in cold water bath (0–10°C).

7.6 Pipette 10.0 ml of sulfuric acid solution (6.3) into each tube and mix by swirling. Allow tubes to come to thermal equilibrium in the cold bath. Be sure that temperatures have equilibrated in all tubes before continuing.

7.7 Add 0.5 ml brucine-sulfanilic acid reagent (6.4) to each tube (except the interference control tubes, 7.3) and carefully mix by swirling, then place the rack of tubes in the 100°C water bath for exactly 25 minutes.

Caution: Immersion of the tube rack into the bath should not decrease the temperature of the bath more than 1 to 2°C. In order to keep this temperature decrease to an absolute minimum, flow of bath water between the tubes should not be restricted by crowding too many tubes into the rack. If color development in the standards reveals discrepancies in the procedure, the operator should repeat the procedure after reviewing the temperature control steps.

7.8 Remove rack of tubes from the hot water bath and immerse in the cold water bath and allow to reach thermal equilibrium (20–25°C).

7.9 Read absorbance against the reagent blank at 410 nm using a 1 cm or longer cell.

8. Calculation

8.1 Obtain a standard curve by plotting the absorbance of standards run by the above procedure against mg NO_3–N/1. (The color reaction does not always follow Beer's law).

8.2 Subtract the absorbance of the sample without the brucine-sulfanilic reagent from the absorbance of the sample containing brucine-sulfanilic acid and determine mg NO_3–N/1. Multiply by an appropriate dilution factor if less than 10 ml of sample is taken.

9. Precision and Accuracy

9.1 Twenty-seven analysts in fifteen laboratories analyzed natural water samples containing exact increments of inorganic nitrate, with the following results:

Increment as Nitrogen, Nitrate mg N/liter	Precision as Standard Deviation mg N/liter	Accuracy as	
		Bias, %	Bias, mg N/liter
0.16	0.092	−6.79	−0.01
0.19	0.083	+8.30	+0.02
1.08	0.245	+4.12	+0.04
1.24	0.214	+2.82	+0.04

(FWPCA Method Study 2, Nutrient Analyses).

Bibliography

1. Standard Methods for the Examination of Water and Wastewater, 14th Edition, p 427, Method 419D (1975).

2. Annual Book of ASTM Standards, Part 31, "Water", Standard D 992–71, p 363 (1976).

3. Jenkins, D., and Medsken, L., "A Brucine Method for the Determination of Nitrate in Ocean, Estuarine, and Fresh Waters", Anal Chem., 36, p 610, (1964).

NITROGEN, NITRATE-NITRITE

Method 353.3 (Spectrophotometric, Cadmium Reduction)

STORET NO. Total 00630

1. Scope and Application
    1.1 This method is applicable to the determination of nitrite singly, or nitrite and nitrate combined in drinking, surface and saline waters, domestic and industrial wastes. The applicable range of this method is 0.01 to 1.0 mg/1 nitrate-nitrite nitrogen. The range may be extended with sample dilution.

2. Summary of Method
    2.1 A filtered sample is passed through a column containing granulated copper-cadmium to reduce nitrate to nitrite. The nitrite (that originally present plus reduced nitrate) is determined by diazotizing with sulfanilamide and coupling with N–(1-naphthyl)–ethylenediamine dihydrochloride to form a highly colored azo dye which is measured spectrophotometrically. Şeparate, rather than combined nitrate-nitrite, values are readily obtained by carrying out the procedure first with, and then without, the Cu-Cd reduction step.

3. Sample Handling and Preservation
    3.1 Analysis should be made as soon as possible. If analysis can be made within 24 hours, the sample should be preserved by refrigeration at 4°C. When samples must be stored for more than 24 hours, they should be preserved with sulfuric acid (2 ml H_2SO_4 per liter) and refrigeration.
    <u>Caution</u>: Samples for reduction column must not be preserved with mercuric chloride.

4. Interferences
    4.1 Build up of suspended matter in the reduction column will restrict sample flow. Since nitrate-nitrogen is found in a soluble state, the sample may be pre-filtered through a glass fiber filter or a 0.45u membrane filter. Highly turbid samples may be pretreated with zinc sulfate before filtration to remove the bulk of particulate matter present in the sample.
    4.2 Low results might be obtained for samples that contain high concentrations of iron, copper or other metals. EDTA is added to the samples to eliminate this interference.
    4.3 Samples that contain large concentrations of oil and grease will coat the surface of the cadmium. This interference is eliminated by pre-extracting the sample with an organic solvent.
    4.4 This procedure determines both nitrate and nitrite. If only nitrate is desired, a separate determination must be made for nitrite and subsequent corrections made. The nitrite may be determined by the procedure below without the reduction step.

Approved for NPDES and SDWA
Issued 1974

5. Apparatus

5.1 Reduction column: The column in Figure I was constructed from a 100 ml pipet by removing the top portion. This column may also be constructed from two pieces of tubing joined end to end. A 10 mm length of 3 cm I.D. tubing is joined to a 25 cm length of 3.5 mm I.D. tubing.

5.2 Spectrophotometer for use at 540 nm, providing a light path of 1 cm or longer.

6. Reagents

6.1 Granulated cadmium: 40-60 mesh (MCB Reagents).

6.2 Copper-Cadmium: The cadmium granules (new or used) are cleaned with dilute HCl and copperized with 2% solution of copper sulfate in the following manner:

6.2.1 Wash the cadmium with dilute HCl (6.10) and rinse with distilled water. The color of the cadmium should be silver.

6.2.2 Swirl 25 g cadmium in 100 ml portions of a 2% solution of copper sulfate (6.11) for 5 minutes or until blue color partially fades, decant and repeat with fresh copper sulfate until a brown colloidal precipitate forms.

6.2.3 Wash the copper-cadmium with distilled water (at least 10 times) to remove all the precipitated copper. The color of the cadmium so treated should be black.

6.3 Preparation of reaction column: Insert a glass wool plug into the bottom of the reduction column and fill with distilled water. Add sufficient copper-cadmium granules to produce a column 18.5 cm in length. Maintain a level of distilled water above the copper-cadmium granules to eliminate entrapment of air. Wash the column with 200 ml of dilute ammonium chloride solution (6.5). The column is then activated by passing through the column 100 ml of a solution composed of 25 ml of a 1.0 mg/1 NO_3–N standard and 75 ml of ammonium chloride – EDTA solution (6.4). Use a flow rate between 7 and 10 ml per minute.

6.4 Ammonium chloride – EDTA solution: Dissolve 13 g ammonium chloride and 1.7 g disodium ethylenediamine tetracetate in 900 ml of distilled water. Adjust the pH to 8.5 with conc. ammonium hydroxide (6.9) and dilute to 1 liter.

6.5 Dilute ammonium chloride-EDTA solution: Dilute 300 ml of ammonium chloride-EDTA solution (6.4) to 500 ml with distilled water.

6.6 Color reagent: Dissolve 10 g sulfanilamide and 1 g N(1-naphthyl)–ethylene-diamine dihydrochloride in a mixture of 100 ml conc. phosphoric acid and 800 ml of distilled water and dilute to 1 liter with distilled water.

6.7 Zinc sulfate solution: Dissolve 100 g $ZnSO_4 \cdot 7H_2O$ in distilled water and dilute to 1 liter.

6.8 Sodium hydroxide solution, 6N: Dissolve 240 g NaOH in 500 ml distilled water, cool and dilute to 1 liter.

6.9 Ammonium hydroxide, conc.

6.10 Dilute hydrochloric acid, 6N: Dilute 50 ml of conc. HCl to 100 ml with distilled water.

6.11 Copper sulfate solution, 2%: Dissolve 20 g of $CuSO_4 \cdot 5H_2O$ in 500 ml of distilled water and dilute to 1 liter.

6.12 Stock nitrate solution: Dissolve 7.218 g KNO_3 in distilled water and dilute to 1000 ml. Preserve with 2 ml of chloroform per liter. This solution is stable for at least 6 months. 1.0 ml = 1.00 mg NO_3–N.

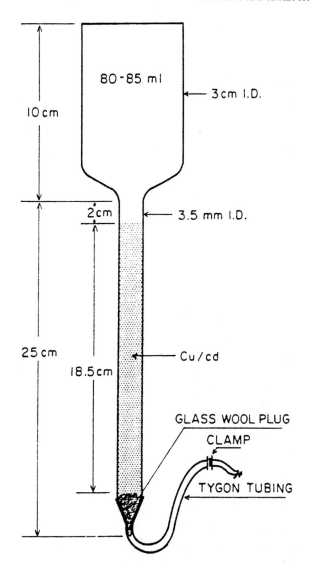

FIGURE 1. REDUCTION COLUMN

6.13 Standard nitrate solution: Dilute 10.0 ml of nitrate stock solution (6.12) to 1000 ml with distilled water. 1.0 ml = 0.01 mg NO_3-N.

6.14 Stock nitrite solution: Dissolve 6.072 g KNO_2 in 500 ml of distilled water and dilute to 1000 ml. Preserve with 2 ml of chloroform and keep under refrigeration. Stable for approximately 3 months. 1.0 ml = 1.00 mg NO_2-N.

6.15 Standard nitrite solution: Dilute 10.0 ml of stock nitrite solution (6.14) to 1000 ml with distilled water. 1.0 ml = 0.01 mg NO_2-N.

6.16 Using standard nitrate solution (6.13) prepare the following standards in 100 ml volumetric flasks:

Conc., mg-NO_3-N/l	ml of Standard Solution/100.0 ml
0.00	0.0
0.05	0.5
0.10	1.0
0.20	2.0
0.50	5.0
1.00	10.0

Procedure

7.1 Turbidity removal: One of the following methods may be used to remove suspended matter.

    7.1.1 Filter sample through a glass fiber filter or a $0.45u$ membrane filter.

    7.1.2 Add 1 ml zinc sulfate solution (6.7) to 100 ml of sample and mix thoroughly. Add 0.4–0.5 ml sodium hydroxide solution (6.8) to obtain a pH of 10.5 as determined with a pH meter. Let the treated sample stand a few minutes to allow the heavy flocculent precipitate to settle. Clarify by filtering through a glass fiber filter or a $0.45u$ membrane filter.

7.2 Oil and grease removal: Adjust the pH of 100 ml of filtered sample to 2 by addition of conc. HCl. Extract the oil and grease from the aqueous solution with two 25 ml portions of a non-polar solvent (Freon, chloroform or equivalent).

7.3 If the pH of the sample is below 5 or above 9, adjust to between 5 and 9 with either conc. HCl or conc. NH_4OH. This is done to insure a sample pH of 8.5 after step 7.4.

7.4 To 25.0 ml of sample or an aliquot diluted to 25.0 ml, add 75 ml of ammonium chloride-EDTA solution (6.4) and mix.

7.5 Pour sample into column and collect sample at a rate of 7–10 ml per minute.

7.6 Discard the first 25 ml, collect the rest of the sample (approximately 70 ml) in the original sample flask. Reduced samples should not be allowed to stand longer than 15 minutes before addition of color reagent, step 7.7.

7.7 Add 2.0 ml of color reagent (6.6) to 50.0 ml of sample. Allow 10 minutes for color development. Within 2 hours measure the absorbance at 540 nm against a reagent blank. NOTE: If the concentration of sample exceeds 1.0 mg NO_3-N/l, the remainder of the reduced sample may be used to make an appropriate dilution before proceeding with step 7.7.

7.8 Standards: Carry out the reduction of standards exactly as described for the samples. At least one nitrite standard should be compared to a reduced nitrate standard at the same concentration to verify the efficiency of the reduction column.

8. Calculation

8.1 Obtain a standard curve by plotting the absorbance of standards run by the above procedure against NO_3–N mg/l. Compute concentration of samples by comparing sample absorbance with standard curve.

8.2 If less than 25 ml of sample is used for the analysis the following equation should be used:

$$mgNO_2 + NO_3 - N/l = \frac{A \times 25}{ml\ sample\ used}$$

where:

A = Concentration of nitrate from standard curve.

9. Precision and Accuracy

9.1 In a single laboratory (EMSL), using sewage samples at concentrations of 0.04, 0.24, 0.55 and 1.04 mg $NO_3 + NO_2$-N/l, the standard deviations were ±0.005, ±0.004, ±0.005 and ±0.01, respectively.

9.2 In a single laboratory (EMSL), using sewage samples at concentrations of 0.24, 0.55, and 1.05 mg $NO_3 + NO_2$–N/l, the recoveries were 100%, 102% and 100%, respectively.

Bibliography

1. Standard Methods for the Examination of Water and Wastewater, 14th Edition, p 423, Method 419C (1975).

2. Henrikson, A., and Selmer-Olsen, "Automatic Methods for Determining Nitrate and Nitrite in Water and Soil Extracts". Analyst, May 1970, Vol. 95, p 514–518.

3. Grasshoff, K., "A Simultaneous Multiple Channel System for Nutrient Analysis in Sea Water with Analog and Digital Data Record", "Advances in Automated Analysis", Technicon International Congress, 1969, Vol. 11, p 133–145.

4. Brewer, P. G., Riley, J. P., "The Automatic Determination of Nitrate in Sea Water", Deep Sea Research, 1965, Vol. 12, p 765–772.

NITROGEN, NITRITE

Method 354.1 (Spectrophotometric)

STORET NO. Total 00615

1.   Scope and Application
   1.1   This method is applicable to the determination of nitrite in drinking, surface and saline waters, domestic and industrial wastes.
   1.2   The method is applicable in the range from 0.01 to 1.0 mg NO_2–N/1.

2.   Summary of Method
   2.1   The diazonium compound formed by diazotation of sulfanilamide by nitrite in water under acid conditions is coupled with N–(1-naphthyl)–ethylenediamine dihydrochloride to produce a reddish-purple color which is read in a spectrophotometer at 540 nm.

3.   Sample Handling and Preservation
   3.1   Samples should be analyzed as soon as possible. They may be stored for 24 to 48 hours at 4°C.

4.   Interferences
   4.1   There are very few known interferences at concentrations less than 1,000 times that of the nitrite; however, the presence of strong oxidants or reductants in the samples will readily affect the nitrite concentrations. High alkalinity (> 600 mg/1) will give low results due to a shift in pH.

5.   Apparatus
   5.1   Spectrophotometer equipped with 1 cm or larger cells for use at 540 nm.
   5.2   Nessler tubes, 50 ml or volumetric flasks, 50 ml.

6.   Reagents
   6.1   Distilled water free of nitrite and nitrate is to be used in preparation of all reagents and standards.
   6.2   Buffer-color reagent: To 250 ml of distilled water, add 105 ml conc. hydrochloric acid, 5.0 g sulfanilamide and 0.5 g N–(1-naphthyl) ethylenediamine dihydrochloride. Stir until dissolved. Add 136 g of sodium acetate ($CH_3COONa•3H_2O$) and again stir until dissolved. Dilute to 500 ml with distilled water. This solution is stable for several weeks if stored in the dark.
   6.3   Nitrite stock solution: 1.0 ml = 0.10 mg NO_2–N. Dissolve 0.1493 g of dried anhydrous sodium nitrite (24 hours in desiccator) in distilled water and dilute to 1000 ml. Preserve with 2 ml chloroform per liter.
   6.4   Nitrite standard solution: 1.0 ml = 0.001 mg NO_2–N. Dilute 10.0 ml of the stock solution (6.3) to 1000 ml.

7.   Procedure
   7.1   If the sample has a pH greater than 10 or a total alkalinity in excess of 600 mg/1, adjust to approximately pH 6 with 1:3 HCl.

Approved for NPDES
Issued 1971

7.2 If necessary, filter the sample through a 0.45 *u* pore size filter using the first portion of filtrate to rinse the filter flask.

7.3 Place 50 ml of sample, or an aliquot diluted to 50 ml, in a 50 ml Nessler tube; hold until preparation of standards is completed.

7.4 At the same time prepare a series of standards in 50 ml Nessler tubes as follows:

ml of Standard Solution 1.0 ml = 0.001 mg NO_2-N	Conc., When Diluted to 50 ml, mg/l of NO_2-N
0.0	(Blank)
0.5	0.01
1.0	0.02
1.5	0.03
2.0	0.04
3.0	0.06
4.0	0.08
5.0	0.10
10.0	0.20

7.5 Add 2 ml of buffer-color reagent (6.2) to each standard and sample, mix and allow color to develop for at least 15 minutes. The color reaction medium should be between pH 1.5 and 2.0.

7.6 Read the color in the spectrophotometer at 540 nm against the blank and plot concentration of NO_2-N against absorbance.

8. Calculation

8.1 Read the concentration of NO_2-N directly from the curve.

8.2 If less than 50.0 ml of sample is taken, calculate mg/l as follows:

$$NO_2 - N, mg/l = \frac{mg/l \text{ from std. curve} \times 50}{ml \text{ sample used}}$$

9. Precision and Accuracy

9.1 Precision and Accuracy data are not available at this time.

Bibliography

1. Standard Methods for the Examination for Water and Wastewater, 14th Edition, p 434, Method 420, (1975).

OIL AND GREASE, TOTAL, RECOVERABLE

Method 413.1 (Gravimetric, Separatory Funnel Extraction)

STORET NO. 00556

1. Scope and Application
   1.1 This method includes the measurement of fluorocarbon-113 extractable matter from surface and saline waters, industrial and domestic wastes. It is applicable to the determination of relatively non-volatile hydrocarbons, vegetable oils, animal fats, waxes, soaps, greases and related matter.
   1.2 The method is not applicable to measurement of light hydrocarbons that volatilize at temperatures below 70°C. Petroleum fuels from gasoline through #2 fuel oils are completely or partially lost in the solvent removal operation.
   1.3 Some crude oils and heavy fuel oils contain a significant percentage of residue-type materials that are not soluble in fluorocarbon-113. Accordingly, recoveries of these materials will be low.
   1.4 The method covers the range from 5 to 1000 mg/1 of extractable material.
2. Summary of Method
   2.1 The sample is acidified to a low pH (<2) and serially extracted with fluorocarbon-113 in a separatory funnel. The solvent is evaporated from the extract and the residue weighed.
3. Definitions
   3.1 The definition of oil and grease is based on the procedure used. The nature of the oil and/or grease, and the presence of extractable non-oily matter will influence the material measured and interpretation of results.
4. Sampling and Storage
   4.1 A representative sample of 1 liter volume should be collected in a glass bottle. If analysis is to be delayed for more than a few hours, the sample is preserved by the addition of 5 ml HCl (6.1) at the time of collection and refrigerated at 4°C.
   4.2 Because losses of grease will occur on sampling equipment, the collection of a composite sample is impractical. Individual portions collected at prescribed time intervals must be analyzed separately to obtain the average concentration over an extended period.
5. Apparatus
   5.1 Separatory funnel, 2000 ml, with Teflon stopcock.
   5.2 Vacuum pump, or other source of vacuum.
   5.3 Flask, boiling, 125 ml (Corning No. 4100 or equivalent).
   5.4 Distilling head, Claisen or equivalent.
   5.5 Filter paper, Whatman No. 40, 11 cm.
6. Reagents
   6.1 Hydrochloric acid, 1:1. Mix equal volumes of conc. HCl and distilled water.

Approved for NPDES
Issued 1974
Editorial revision 1978

SAMPLING AND ANALYTICAL METHODS **F.87**

6.2 Flurocarbon-113,(1,1,2-trichloro-1,2,2-trifluoroethane), b. p. 48°C.

6.3 Sodium sulfate, anhydrous crystal.

7. Procedure

7.1 Mark the sample bottle at the water meniscus for later determination of sample volume. If the sample was not acidified at time of collection, add 5 ml hydrochloric acid (6.1) to the sample bottle. After mixing the sample, check the pH by touching pH-sensitive paper to the cap to insure that the pH is 2 or lower. Add more acid if necessary.

7.2 Pour the sample into a separatory funnel.

7.3 Tare a boiling flask (pre-dried in an oven at 103°C and stored in a desiccator).

7.4 Add 30 ml fluorocarbon-113 (6.2) to the sample bottle and rotate the bottle to rinse the sides. Transfer the solvent into the separatory funnel. Extract by shaking vigorously for 2 minutes. Allow the layers to separate, and filter the solvent layer into the flask through a funnel containing solvent moistened filter paper.

NOTE: An emulsion that fails to dissipate can be broken by pouring about 1 g sodium sulfate (6.3) into the filter paper cone and slowly draining the emulsion through the salt. Additional 1 g portions can be added to the cone as required.

7.5 Repeat (7.4) twice more, with additional portions of fresh solvent, combining all solvent in the boiling flask.

7.6 Rinse the tip of the separatory funnel, the filter paper, and then the funnel with a total of 10–20 ml solvent and collect the rinsings in the flask.

7.7 Connect the boiling flask to the distilling head and evaporate the solvent by immersing the lower half of the flask in water at 70°C. Collect the solvent for reuse. A solvent blank should accompany each set of samples.

7.8 When the temperature in the distilling head reaches 50°C or the flask appears dry remove the distilling head. Sweep out the flask for 15 seconds with air to remove solvent vapor by inserting a glass tube connected to a vacuum source. Immediately remove the flask from the heat source and wipe the outside to remove excess moisture and fingerprints.

7.9 Cool the boiling flask in a desiccator for 30 minutes and weigh.

8. Calculation

8.1 mg/1 total oil and grease $= \dfrac{R - B}{V}$

where:

R = residue, gross weight of extraction flask minus the tare weight, in milligrams.

B = blank determination, residue of equivalent volume of extraction solvent, in milligrams.

V = volume of sample, determined by refilling sample bottle to calibration line and correcting for acid addition if necessary, in liters.

9.   Precision and Accuracy

    9.1   The two oil and grease methods in this manual were tested by a single laboratory (EMSL) on sewage. This method determined the oil and grease level in the sewage to be 12.6 mg/1. When 1 liter portions of the sewage were dosed with 14.0 mg of a mixture of $\neq$2 fuel oil and Wesson oil, the recovery was 93% with a standard deviation of $=$0.9 mg/1.

Bibliography

1.   Standard Methods for the Examination of Water and Wastewater, 14th Edition, p 515, Method 502A, (1975).

2.   Blum, K. A., and Taras, M. J., "Determination of Emulsifying Oil in Industrial Wastewater", JWPCF Research Suppl. 40, R404 (1968).

PHOSPHORUS, ALL FORMS

Method 365.2 (Colorimetric, Ascorbic Acid, Single Reagent)

STORET NO. See Section 4

1. Scope and Application
   1.1 These methods cover the determination of specified forms of phosphorus in drinking, surface and saline waters, domestic and industrial wastes.
   1.2 The methods are based on reactions that are specific for the orthophosphate ion. Thus, depending on the prescribed pre-treatment of the sample, the various forms of phosphorus given in Figure 1 may be determined. These forms are defined in Section 4.
      1.2.1 Except for in-depth and detailed studies, the most commonly measured forms are phosphorus and dissolved phosphorus, and orthophosphate and dissolved orthophosphate. Hydrolyzable phosphorus is normally found only in sewage-type samples and insoluble forms of phosphorus are determined by calculation.
   1.3 The methods are usable in the 0.01 to 0.5 mg P/1 range.
2. Summary of Method
   2.1 Ammonium molybdate and antimony potassium tartrate react in an acid medium with dilute solutions of phosphorus to form an antimony-phospho-molybdate complex. This complex is reduced to an intensely blue-colored complex by ascorbic acid. The color is proportional to the phosphorus concentration.
   2.2 Only orthophosphate forms a blue color in this test. Polyphosphates (and some organic phosphorus compounds) may be converted to the orthophosphate form by sulfuric acid hydrolysis. Organic phosphorus compounds may be converted to the orthophosphate form by persulfate digestion[2].
3. Sample Handling and Preservation
   3.1 If benthic deposits are present in the area being sampled, great care should be taken not to include these deposits.
   3.2 Sample containers may be of plastic material, such as cubitainers, or of Pyrex glass.
   3.3 If the analysis cannot be performed the day of collection, the sample should be preserved by the addition of 2 ml conc. H_2SO_4 per liter and refrigeration at 4°C.
4. Definitions and Storet Numbers
   4.1 Total Phosphorus (P) — all of the phosphorus present in the sample, regardless of form, as measured by the persulfate digestion procedure. (00665)
      4.1.1 Total Orthophosphate (P, ortho) — inorganic phosphorus $[(PO_4)^{-3}]$ in the sample as measured by the direct colorimetric analysis procedure. (70507)
      4.1.2 Total Hydrolyzable Phosphorus (P, hydro) - phosphorus in the sample as measured by the sulfuric acid hydrolysis procedure, and minus pre-determined orthophosphates. This hydrolyzable phosphorus includes polyphosphorus. $[(P_2O_7)^{-4}, (P_3O_{10})^{-5},$ etc.] plus some organic phosphorus. (00669)

Approved for NPDES
Issued 1971

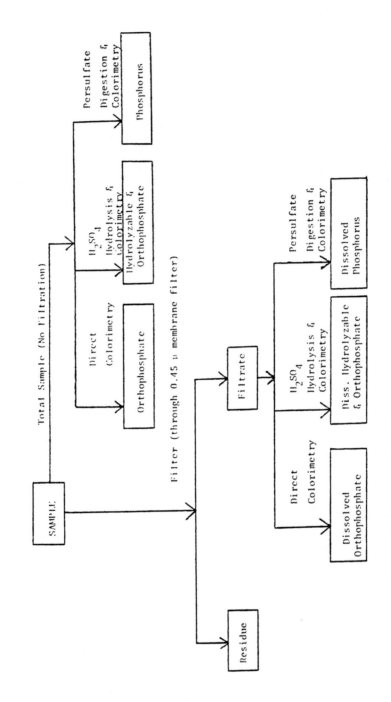

FIGURE 1. ANALYTICAL SCHEME FOR DIFFERENTIATION OF PHOSPHORUS FORMS

F.90

4.1.3 Total Organic Phosphorus (P, org) — phosphorus (inorganic plus oxidizable organic) in the sample measured by the persulfate digestion procedure, and minus hydrolyzable phosphorus and orthophosphate. (00670)

4.2 Dissolved Phosphorus (P–D) — all of the phosphorus present in the filtrate of a sample filtered through a phosphorus-free filter of 0.45 micron pore size and measured by the persulfate digestion procedure. (00666)

4.2.1 Dissolved Orthophosphate (P–D, ortho) — as measured by the direct colorimetric analysis procedure. (00671)

4.2.2 Dissolved Hydrolyzable Phosphorus (P–D, hydro) — as measured by the sulfuric acid hydrolysis procedure and minus pre-determined dissolved orthophosphates. (00672)

4.2.3 Dissolved Organic Phosphorus (P–D, org) — as measured by the persulfate digestion procedure, and minus dissolved hydrolyzable phosphorus and orthophosphate. (00673)

4.3 The following forms, when sufficient amounts of phosphorus are present in the sample to warrant such consideration, may be calculated:

4.3.1 Insoluble Phosphorus (P-I)=(P)-(P-D). (00667)

4.3.1.1 Insoluble orthophosphate (P-I, ortho)=(P, ortho)-(P-D, ortho). (00674)

4.3.1.2 Insoluble Hydrolyzable Phosphorus (P-I, hydro)=(P, hydro)-(P-D, hydro). (00675)

4.3.1.3 Insoluble Organic Phosphorus (P-I, org)=(P, org) – (P-D, org). (00676)

4.4 All phosphorus forms shall be reported as P, mg/1, to the third place.

5. Interferences

5.1 No interference is caused by copper, iron, or silicate at concentrations many times greater than their reported concentration in sea water. However, high iron concentrations can cause precipitation of and subsequent loss of phosphorus.

5.2 The salt error for samples ranging from 5 to 20% salt content was found to be less than 1%.

5.3 Arsenate is determined similarly to phosphorus and should be considered when present in concentrations higher than phosphorus. However, at concentrations found in sea water, it does not interfere.

6. Apparatus

6.1 Photometer – A spectrophotometer or filter photometer suitable for measurements at 650 or 880 nm with a light path of 1 cm or longer.

6.2 Acid-washed glassware: All glassware used should be washed with hot 1:1 HCl and rinsed with distilled water. The acid-washed glassware should be filled with distilled water and treated with all the reagents to remove the last traces of phosphorus that might be adsorbed on the glassware. Preferably, this glassware should be used only for the determination of phosphorus and after use it should be rinsed with distilled water and kept covered until needed again. If this is done, the treatment with 1:1 HCl and reagents is only required occasionally. Commercial detergents should never be used.

7.   Reagents

   7.1   Sulfuric acid solution, 5N: Dilute 70 ml of conc. H_2SO_4 with distilled water to 500 ml.

   7.2   Antimony potassium tartrate solution: Weigh 1.3715 g $K(SbO)C_4H_4O_6 \cdot 1/2H_2O$, dissolve in 400 ml distilled water in 500 ml volumetric flask, dilute to volume. Store at 4°C in a dark, glass-stoppered bottle.

   7.3   Ammonium molybdate solution: Dissolve 20 g $(NH_4)_6Mo_7O_{24} \cdot 4H_2O$ in 500 ml of distilled water. Store in a plastic bottle at 4°C.

   7.4   Ascorbic acid, 0.1M: Dissolve 1.76 g of ascorbic acid in 100 ml of distilled water. The solution is stable for about a week if stored at 4°C.

   7.5   Combined reagent: Mix the above reagents in the following proportions for 100 ml of the mixed reagent: 50 ml of 5N H_2SO_4, (7.1), 5 ml of antimony potassium tartrate solution (7.2), 15 ml of ammonium molybdate solution (7.3), and 30 ml of ascorbic acid solution (7.4). Mix after addition of each reagent. All reagents must reach room temperature before they are mixed and must be mixed in the order given. If turbidity forms in the combined reagent, shake and let stand for a few minutes until the turbidity disappears before proceeding. Since the stability of this solution is limited, it must be freshly prepared for each run.

   7.6   Sulfuric acid solution, 11 N: Slowly add 310 ml conc. H_2SO_4 to 600 ml distilled water. When cool, dilute to 1 liter.

   7.7   Ammonium persulfate.

   7.8   Stock phosphorus solution: Dissolve in distilled water 0.2197 g of potassium dihydrogen phosphate, KH_2PO_4, which has been dried in an oven at 105°C. Dilute the solution to 1000 ml; 1.0 ml = 0.05 mg P.

   7.9   Standard phosphorus solution: Dilute 10.0 ml of stock phosphorus solution (7.8) to 1000 ml with distilled water; 1.0 ml = 0.5 ug P.

     7.9.1 Using standard solution, prepare the following standards in 50.0 ml volumetric flasks:

ml of Standard Phosphorus Solution (7.9)	Conc., mg/l
0	0.00
1.0	0.01
3.0	0.03
5.0	0.05
10.0	0.10
20.0	0.20
30.0	0.30
40.0	0.40
50.0	0.50

   7.10   Sodium hydroxide, 1 N: Dissolve 40 g NaOH in 600 ml distilled water. Cool and dilute to 1 liter.

8.   Procedure
   8.1   Phosphorus
      8.1.1   Add 1 ml of H_2SO_4 solution (7.6) to a 50 ml sample in a 125 ml Erlenmeyer flask.
      8.1.2   Add 0.4 g of ammonium persulfate.
      8.1.3   Boil gently on a pre-heated hot plate for approximately 30–40 minutes or until a final volume of about 10 ml is reached. Do not allow sample to go to dryness. Alternatively, heat for 30 minutes in an autoclave at 121°C (15–20 psi).
      8.1.4   Cool and dilute the sample to about 30 ml and adjust the pH of the sample to 7.0 ±0.2 with 1 N NaOH (7.10) using a pH meter. If sample is not clear at this point, add 2–3 drops of acid (7.6) and filter. Dilute to 50 ml.
      Alternatively, if autoclaved see **NOTE 1**.
      8.1.5   Determine phosphorus as outlined in 8.3.2 Orthophosphate.
   8.2   Hydrolyzable Phosphorus
      8.2.1   Add 1 ml of H_2SO_4 solution (7.6) to a 50 ml sample in a 125 ml Erlenmeyer flask.
      8.2.2   Boil gently on a pre-heated hot plate for 30–40 minutes or until a final volume of about 10 ml is reached. Do not allow sample to go to dryness. Alternatively, heat for 30 minutes in an autoclave at 121°C (15–20 psi).
      8.2.3   Cool and dilute the sample to about 30 ml and adjust the pH of the sample to 7.0 ±0.2 with NaOH (7.10) using a pH meter. If sample is not clear at this point, add 2–3 drops of acid (7.6) and filter. Dilute to 50 ml.
      Alternatively, if autoclaved see **NOTE 1**.
      8.2.4   The sample is now ready for determination of phosphorus as outlined in 8.3.2 Orthophosphate.
   8.3   Orthophosphate
      8.3.1   The pH of the sample must be adjusted to 7±0.2 using a pH meter.
      8.3.2   Add 8.0 ml of combined reagent (7.5) to sample and mix thoroughly. After a minimum of ten minutes, but no longer than thirty minutes, measure the color absorbance of each sample at 650 or 880 nm with a spectrophotometer, using the reagent blank as the reference solution.
      NOTE 1: If the same volume of sodium hydroxide solution is not used to adjust the pH of the standards and samples, a volume correction has to be employed.
9.   Calculation
   9.1   Prepare a standard curve by plotting the absorbance values of standards versus the corresponding phosphorus concentrations.
      9.1.1   Process standards and blank exactly as the samples. Run at least a blank and two standards with each series of samples. If the standards do not agree within ±2% of the true value, prepare a new calibration curve.
   9.2   Obtain concentration value of sample directly from prepared standard curve. Report results as P, mg/1. **SEE NOTE 1**.

10. Precision and Accuracy

10.1 Thirty-three analysts in nineteen laboratories analyzed natural water samples containing exact increments of organic phosphate, with the following results:

Increment as Total Phosphorus mg P/liter	Precision as Standard Deviation mg P/liter	Accuracy as	
		Bias, %	Bias mg P/liter
0.110	0.033	+3.09	+0.003
0.132	0.051	+11.99	+0.016
0.772	0.130	+2.96	+0.023
0.882	0.128	−0.92	−0.008

(FWPCA Method Study 2, Nutrient Analyses)

10.2 Twenty-six analysts in sixteen laboratories analyzed natural water samples containing exact increments of orthophosphate, with the following results:

Increment as Orthophosphate mg P/liter	Precision as Standard Deviation mg P/liter	Accuracy as	
		Bias, %	Bias, mg P/liter
0.029	0.010	−4.95	−0.001
0.038	0.008	−6.00	−0.002
0.335	0.018	−2.75	−0.009
0.383	0.023	−1.76	−0.007

(FWPCA Method Study 2, Nutrient Analyses)

Bibliography

1. Murphy, J., and Riley, J., "A modified Single Solution for the Determination of Phosphate in Natural Waters", Anal. Chim. Acta., 27, 31 (1962).
2. Gales, M., Jr., Julian, E., and Kroner, R., "Method for Quantitative Determination of Total Phosphorus in Water", Jour. AWWA, 58, No. 10, 1363 (1966).
3. Annual Book of ASTM Standards, Part 31, "Water", Standard D515–72, Method A, p 389 (1976).
4. Standard Methods for the Examination of Water and Wastewater, 14th Edition, p 476 and 481, (1975).

OXYGEN, DISSOLVED

Method 360.1 (Membrane Electrode)

STORET NO. 00299

1. Scope and Application
   1.1 The probe method for dissolved oxygen is recommended for those samples containing materials which interfere with the modified Winkler procedure such as sulfite, thiosulfate, polythionate, mercaptans, free chlorine or hypochlorite, organic substances readily hydrolyzed in alkaline solutions, free iodine, intense color or turbidity and biological flocs.
   1.2 The probe method is recommended as a substitute for the modified Winkler procedure in monitoring of streams, lakes, outfalls, etc., where it is desired to obtain a continuous record of the dissolved oxygen content of the water under observation.
   1.3 The probe method may be used as a substitute for the modified Winkler procedure in BOD determinations where it is desired to perform nondestructive DO measurements on a sample.
   1.4 The probe method may be used under any circumstances as a substitute for the modified Winkler procedure provided that the probe itself is standardized against the Winkler method on samples free of interfering materials.
   1.5 The electronic readout meter for the output from dissolved oxygen probes is normally calibrated in convenient scale (0 to 10, 0 to 15, 0 to 20 mg/1 for example) with a sensitivity of approximately 0.05 mg/liter.
2. Summary of Method
   2.1 The most common instrumental probes for determination of dissolved oxygen in water are dependent upon electrochemical reactions. Under steady-state conditions, the current or potential can be correlated with DO concentrations. Interfacial dynamics at the probe-sample interface are a factor in probe response and a significant degree of interfacial turbulence is necessary. For precision performance, turbulence should be constant.
3. Sample Handling and Preservation
   3.1 See 4.1, 4.2, 4.3, 4.4 under Modified Winkler Method (360.2).
4. Interferences
   4.1 Dissolved organic materials are not known to interfere in the output from dissolved oxygen probes.
   4.2 Dissolved inorganic salts are a factor in the performance of dissolved oxygen probe.
       4.2.1 Probes with membranes respond to partial pressure of oxygen which in turn is a function of dissolved inorganic salts. Conversion factors for seawater and brackish waters may be calculated from dissolved oxygen saturation versus salinity data. Conversion factors for specific inorganic salts may be developed experimentally.

Approved for NPDES
Issued 1971

Broad variations in the kinds and concentrations of salts in samples can make the use of a membrane probe difficult.

4.3 Reactive compounds can interfere with the output or the performance of dissolved oxygen probes.

4.3.1 Reactive gases which pass through the membrane probes may interfere. For example, chlorine will depolarize the cathode and cause a high probe-output. Long-term exposures to chlorine will coat the anode with the chloride of the anode metal and eventually desensitize the probe. Alkaline samples in which free chlorine does not exist will not interfere. Hydrogen sulfide will interfere with membrane probes if the applied potential is greater than the half-wave potential of the sulfide ion. If the applied potential is less than the half-wave potential, an interfering reaction will not occur, but coating of the anode with the sulfide of the anode metal can take place.

4.4 Dissolved oxygen probes are temperature sensitive, and temperature compensation is normally provided by the manufacturer. Membrane probes have a temperature coefficient of 4 to 6 percent/°C dependent upon the membrane employed.

5. Apparatus

5.1 No specific probe or accessory is especially recommended as superior. However, probes which have been evaluated or are in use and found to be reliable are the Weston & Stack DO Analyzer Model 30, the Yellow Springs Instrument (YSI) Model 54, and the Beckman Fieldlab Oxygen Analyzer.

6. Calibration
Follow manufacturer instructions.

7. Procedure
Follow manufacturer instructions.

8. Calculation
Follow manufacturer instructions.

9. Precision and Accuracy
Manufacturer's specification claim 0.1 mg/1 repeatability with ±1% accuracy.

Bibliography

1. Standard Methods for the Examination of Water and Wastewater, 14th Edition, p 450, Method 422F (1975).

OXYGEN, DISSOLVED

Method 360.2 (Modified Winkler, Full-Bottle Technique)

STORET NO. 00300

1. Scope and Application
    1.1   This method is applicable for use with most wastewaters and streams that contain nitrate nitrogen and not more than 1 mg/1 of ferrous iron. Other reducing or oxidizing materials should be absent. If 1 ml of fluoride solution is added before acidifying the sample and there is no delay in titration, the method is also applicable in the presence of 100–200 mg/1 ferric iron.
    1.2   The Dissolved Oxygen (DO) Probe technique gives comparable results on all samples types.
    1.3   The azide modification is not applicable under the following conditions: (a) samples containing sulfite, thiosulfate, polythionate, appreciable quantities of free chlorine or hypochlorite; (b) samples high in suspended solids; (c) samples containing organic substances which are readily oxidized in a highly alkaline solution, or which are oxidized by free iodine in an acid solution; (d) untreated domestic sewage; (e) biological flocs; and (f) where sample color interferes with endpoint detection. In instances where the azide modification is not applicable, the DO probe should be used.

2. Summary of Method
    2.1   The sample is treated with manganous sulfate, potassium hydroxide, and potassium iodide (the latter two reagents combined in one solution) and finally sulfuric acid. The initial precipitate of manganous hydroxide, $Mn(OH)_2$, combines with the dissolved oxygen in the sample to form a brown precipitate, manganic hydroxide, $MnO(OH)_2$. Upon acidification, the manganic hydroxide forms manganic sulfate which acts as an oxidizing agent to release free iodine from the potassium iodide. The iodine, which is stoichiometrically equivalent to the dissolved oxygen in the sample is then titrated with sodium thiosulfate or phenylarsine oxide (PAO).

3. Interferences
    3.1   There are a number of interferences to the dissolved oxygen test, including oxidizing and reducing agents, nitrate ion, ferrous iron, and organic matter.
    3.2   Various modifications of the original Winkler procedure for dissolved oxygen have been developed to compensate for or eliminate interferences. The Alsterberg modification is commonly used to successfully eliminate the nitrite interference, the Rideal-Stewart modification is designed to eliminate ferrous iron interference, and the Theriault procedure is used to compensate for high concentration of organic materials.
    3.3   Most of the common interferences in the Winkler procedure may be overcome by use of the dissolved oxygen probe.

Approved for NPDES
Issued 1971

4.   Sample Handling and Preservation

4.1   Where possible, collect the sample in a **300 ml BOD** incubation bottle. Special precautions are required to avoid entrainment or solution of atmospheric oxygen or loss of dissolved oxygen.

4.2   Where samples are collected from shallow depths (less than 5 feet), use of an APHA-type sampler is recommended. Use of a Kemmerer type sampler is recommended for samples collected from depths of greater than 5 feet.

4.3   When a Kemmerer sampler is used, the **BOD** sample bottle should be filled to overflowing. (overflow for approximately 10 seconds). Outlet tube of Kemmerer should be inserted to bottom of **BOD** bottle. Care must be taken to prevent turbulence and the formation of bubbles when filling bottle.

4.4   At time of sampling, the sample temperature should be recorded as precisely as required.

4.5   Do not delay the determination of dissolved oxygen in samples having an appreciable iodine demand or containing ferrous iron. If samples must be preserved either method (4.5.1) or (4.5.2) below, may be employed.

4.5.1   Add 2 ml of manganous sulfate solution (6.1) and then 2 ml of alkaline iodide-azide solution (6.2) to the sample contained in the **BOD** bottle. Both reagents must be added well below the surface of the liquid. Stopper the bottle immediately and mix the contents thoroughly. The sample should be stored at the temperature of the collection water, or water sealed and kept at a temperature of 10 to 20°C, in the dark. Complete the procedure by adding 2 ml H_2SO_4 (see 7.1) at time of analysis.

4.5.2   Add 0.7 ml of conc. H_2SO_4 (6.3) and 1 ml sodium azide solution (2 g NaN_3 in 100 ml distilled water) to sample in the **BOD** bottle. Store sample as in (4.5.1). Complete the procedure using 2 ml of manganous sulfate solution (6.1), 3 ml alkaline iodide-azide solution (6.2), and 2 ml of conc. H_2SO_4 (6.3) at time of analysis.

4.6   If either preservation technique is employed, complete the analysis within 4–8 hours after sampling.

5.   Apparatus

5.1   Sample bottles-300 ml ±3 ml capacity **BOD** incubation bottles with tapered ground glass pointed stoppers and flared mouths.

5.2   Pipets-with elongated tips capable of delivering 2.0 ml ±0.10 ml of reagent.

6.   Reagents

6.1   Manganous sulfate solution: Dissolve 480 g manganous sulfate ($MnSO_4$•$4H_2O$ in distilled water and dilute to 1 liter.

6.1.1   Alternatively, use 400 g of $MnSO_4$•$2H_2O$ or 364 g of $MnSO_4$•H_2O per liter. When uncertainty exists regarding the water of crystallization, a solution of equivalent strength may be obtained by adjusting the specific gravity of the solution to 1.270 at 20°C.

6.2   Alkaline iodide-azide solution: Dissolve 500 g of sodium hydroxide (NaOH) or 700 g of potassium hydroxide (KOH) and 135 g of sodium iodide (NaI) or 150 g of potassium iodide (KI) in distilled water and dilute to 1 liter. To this solution add 10 g of solution azide (NaN_3) dissolved in 40 ml of distilled water.

6.3   Sulfuric acid: concentrated.

6.4   Starch solution: Prepare an emulsion of 10 g soluble starch in a mortar or beaker with a small quantity of distilled water. Pour this emulsion into 1 liter of boiling water, allow to boil a few minutes, and let settle overnight. Use the clear supernate. This solution may be preserved by the addition of 5 ml per liter of chloroform and storage in a 10°C refrigerator.

    6.4.1   Dry, powdered starch indicators such as "thyodene" may be used in place of starch solution.

6.5   Potassium fluoride solution: Dissolve 40 g KF•2H$_2$O in distilled water and dilute to 100 ml.

6.6   Sodium thiosulfate, stock solution, 0.75 N: Dissolve 186.15 g Na$_2$S$_2$O$_3$•5H$_2$O in boiled and cooled distilled water and dilute to 1 liter. Preserve by adding 5 ml chloroform.

6.7   Sodium thiosulfate standard titrant, 0.0375 N: Prepare by diluting 50.0 ml of stock solution to 1 liter. Preserve by adding 5 ml of chloroform. Standard sodium thiosulfate, exactly 0.0375 N is equivalent to 0.300 mg of DO per 1.00 ml. Standardize with 0.0375 N potassium biiodate.

6.8   Potassium biiodate standard, 0.0375 N: For stock solution, dissolve 4.873 g of potassium biiodate, previously dried 2 hours at 103°C, in 1000 ml of distilled water. To prepare working standard, dilute 250 ml to 1000 ml for 0.0375 N biiodate solution.

6.9   Standardization of 0.0375 N sodium thiosulfate: Dissolve approximately 2 g (±1.0 g) KI in 100 to 150 ml distilled water; add 10 ml of 10% H$_2$SO$_4$ followed by 20.0 ml standard potassium biiodate (6.8). Place in dark for 5 minutes, dilute to 300 ml, and titrate with the standard sodium thiosulfate (6.7) to a pale straw color. Add 1–2 ml starch solution and continue the titration drop by drop until the blue color disappears. Run in duplicate. Duplicate determinations should agree within ±0.05 ml.

6.10   As an alternative to the sodium thiosulfate, phenylarsine oxide (**PAO**) may be used. This is available, already standardized, from commercial sources.

7.   Procedure

7.1   To the sample collected in the **BOD** incubation bottle, add 2 ml of the manganous sulfate solution (6.1) followed by 2 ml of the alkaline iodide-azide solution (6.2), well below the surface of the liquid; stopper with care to exclude air bubbles, and mix well by inverting the bottle several times. When the precipitate settles, leaving a clear supernatant above the manganese hydroxide floc, shake again. When settling has produced at least 200 ml of clear supernatant, carefully remove the stopper and immediately add 2 ml of conc. H$_2$SO$_4$ (6.3) (sulfamic acid packets, 3 g may be substituted for H$_2$SO$_4$)[1] by allowing the acid to run down the neck of the bottle, re-stopper, and mix by gentle inversion until the iodine is uniformly distributed throughout the bottle. Complete the analysis within 45 minutes.

7.2   Transfer the entire bottle contents by inversion into a 500 ml wide mouth flask and titrate with 0.0375 N thiosulfate solution (6.7) (0.0375 N phenyarsine oxide (**PAO**) may be substituted as titrant) to pale straw color. Add 1–2 ml of starch solution (6.4) or 0.1 g of powdered indicator and continue to titrate to the first disappearance of the blue color.

7.3 If ferric iron is present (100 to 200 mg/1), add 1.0 ml of KF (6.5) solution before acidification.

7.4 Occasionally, a dark brown or black precipitate persists in the bottle after acidication. This precipitate will dissolve if the solution is kept for a few minutes longer than usual or, if particularly persistent, a few more drops of H_2SO_4 will effect dissolution.

8. Calculation

8.1 Each ml of 0.0375N sodium thiosulfate (or PAO) titrant is equivalent to 1 mg DO when the entire bottle contents are titrated.

8.2 If the results are desired in milliliters of oxygen gas per liter at 0°C and 760 mm pressure multiply mg/1 DO by 0.698.

8.3 To express the results as percent saturation at 760 mm atmospheric pressure, the solubility data in Table 422:1 (Whipple & Whipple, p 446–447, Standard Methods, 14th Edition) may be used. Equations for correcting the solubilities to barometric pressures other than mean sea level are given below the table.

8.4 The solubility of DO in distilled water at any barometric pressure, p (mm Hg), temperature, T°C, and saturated vapor pressure, u (mm Hg), for the given T, may be calculated between the temperature of 0° and 30°C by:

$$ml/1\ DO = \frac{(P - u) \times 0.678}{35 + T}$$

and between 30° and 50°C by:

$$ml/1\ DO = \frac{(P - u) \times 0.827}{49 + T}$$

9. Precision and Accuracy

9.1 Exact data are unavailable on the precision and accuracy of this technique; however, reproducibility is approximately 0.2 mg/1 of DO at the 7.5 mg/1 level due to equipment tolerances and uncompensated displacement errors.

Bibliography

1. Kroner, R. C., Longbottom, J. E., Gorman, R.A., "A Comparison of Various Reagents Proposed for Use in the Winkler Procedure for Dissolved Oxygen", PHS Water Pollution Surveillance System Applications and Development, Report #12, Water Quality Section, Basic Data Branch, July 1964.

2. Annual Book of ASTM Standards, Part 31, "Water", Standard D1589–60, Method A, p 373 (1976).

3. Standard Methods for the Examination of Water and Wastewater, 14th Edition, p 443, method 422 B (1975).

PHENOLICS, TOTAL RECOVERABLE

Method 420.1 (Spectrophotometric, Manual 4-AAP with Distillation)

STORET NO. 32730

1. Scope and Application
    1.1 This method is applicable to the analysis of drinking, surface and saline waters, domestic and industrial wastes.
    1.2 The method is capable of measuring phenolic materials at the 5 ug/1 level when the colored end product is extracted and concentrated in a solvent phase using phenol as a standard.
    1.3 The method is capable of measuring phenolic materials that contain more than 50 ug/1 in the aqueous phase (without solvent extraction) using phenol as a standard.
    1.4 It is not possible to use this method to differentiate between different kinds of phenols.
2. Summary of Method
    2.1 Phenolic materials react with 4-aminoantipyrine in the presence of potassium ferricyanide at a pH of 10 to form a stable reddish-brown colored antipyrine dye. The amount of color produced is a function of the concentration of phenolic material.
3. Comments
    3.1 For most samples a preliminary distillation is required to remove interfering materials.
    3.2 Color response of phenolic materials with 4-amino antipyrine is not the same for all compounds. Because phenolic type wastes usually contain a variety of phenols, it is not possible to duplicate a mixture of phenols to be used as a standard. For this reason phenol has been selected as a standard and any color produced by the reaction of other phenolic compounds is reported as phenol. This value will represent the minimum concentration of phenolic compounds present in the sample.
4. Sample Handling and Preservation
    4.1 Biological degradation is inhibited by the addition of 1 g/1 of copper sulfate to the sample and acidification to a pH of less than 4 with phosphoric acid. The sample should be kept at 4°C and analyzed within 24 hours after collection.
5. Interference
    5.1 Interferences from sulfur compounds are eliminated by acidifying the sample to a pH of less than 4 with H_3PO_4 and aerating briefly by stirring and adding $CuSO_4$.
    5.2 Oxidizing agents such as chlorine, detected by the liberation of iodine upon acidification in the presence of potassium iodide, are removed immediately after sampling by the addition of an excess of ferrous ammonium sulfate (7.10). If chlorine is not removed the phenolic compounds may be partially oxidized and the results may be low.

Approved for NPDES
Issued 1971
Editorial revision 1978

6. Apparatus

    6.1   Distillation apparatus, all glass consisting of a 1 liter pyrex distilling apparatus with Graham condenser.

    6.2   pH meter.

    6.3   Spectrophotometer, for use at 460 or 510 nm.

    6.4   Funnels.

    6.5   Filter paper.

    6.6   Membrane filters.

    6.7   Separatory funnels, 500 or 1,000 ml.

    6.8   Nessler tubes, short or long form.

7. Reagents

    7.1   Phosphoric acid solution, 1 + 9: Dilute 10 ml of 85% H_3PO_4 to 100 ml with distilled water.

    7.2   Copper sulfate solution: Dissolve 100 g $CuSO_4 \cdot 5H_2O$ in distilled water and dilute to 1 liter.

    7.3   Buffer solution: Dissolve 16.9 g NH_4Cl in 143 ml conc. NH_4OH and dilute to 250 ml with distilled water. Two ml should adjust 100 ml of distillate to pH 10.

    7.4   Aminoantipyrine solution: Dissolve 2 g of 4AAP in distilled water and dilute to 100 ml.

    7.5   Potassium ferricyanide solution: Dissolve 8 g of $K_3Fe(CN)_6$ in distilled water and dilute to 100 ml.

    7.6   Stock phenol solution: Dissolve 1.0 g phenol in freshly boiled and cooled distilled water and dilute to 1 liter. 1 ml = 1 mg phenol.

    7.7   Working solution A: Dilute 10 ml stock phenol solution to 1 liter with distilled water. 1 ml = 10 ug phenol.

    7.8   Working solution B: Dilute 100 ml of working solution A to 1000 ml with distilled water. 1 ml = 1 ug phenol.

    7.9   Chloroform

    7.10 Ferrous ammonium sulfate: Dissolve 1.1 g ferrous ammonium sulfate in 500 ml distilled water containing 1 ml conc. H_2SO_4 and dilute to 1 liter with freshly boiled and cooled distilled water.

8. Procedure

    8.1   Distillation

        8.1.1 Measure 500 ml sample into a beaker. Lower the pH to approximately 4 with 1 + 9 H_3PO_4 (7.1), add 5 ml $CuSO_4$ solution (7.2) and transfer to the distillation apparatus. Omit adding H_2PO_4 and $CuSO_4$ if sample was preserved as described in 4.1.

        8.1.2 Distill 450 ml of sample, stop the distillation, and when boiling ceases add 50 ml of warm distilled water to the flask and resume distillation until 500 ml have been collected.

        8.1.3 If the distillate is turbid, filter through a prewashed membrane filter.

    8.2   Direct photometric method

        8.2.1 Using working solution A (7.7), prepare the following standards in 100 ml volumetric flasks.

ml of working solution A	Conc. ug/l
0	0.0
0.5	50.0
1.0	100.0
2.0	200.0
5.0	500.0
8.0	800.0
10.0	1000.0

8.2.2 To 100 ml of distillate or an aliquot diluted to 100 ml and/or standards, add 2 ml of buffer solution (7.3) and mix. The pH of the sample and standards should be 10 ±0.2.

8.2.3 Add 2.0 ml aminoantipyrine solution (7.4) and mix.

8.2.4 Add 2.0 ml potassium ferricyanide solution (7.5) and mix.

8.2.5 After 15 minutes read absorbance at 510 nm.

8.3 Chloroform extraction method

8.3.1 Using working solution B (7.8), prepare the following standards. Standards may be prepared by pipetting the required volumes into the separatory funnels and diluting to 500 ml with distilled water.

ml of working solution B	Conc. ug/l
0.0	0.0
3.0	6.0
5.0	10.0
10.0	20.0
20.0	40.0
25.0	50.0

8.3.2 Place 500 ml of distillate or an aliquot diluted to 500 ml in a separatory funnel. The sample should not contain more than 25 ug phenol.

8.3.3 To sample and standards add 10 ml of buffer solution (7.3) and mix. The pH should be 10 ±0.2.

8.3.4 Add 3.0 ml aminoantipyrine solution (7.4) and mix.

8.3.5 Add 3.0 ml potassium ferricyanide solution (7.5) and mix.

8.3.6 After three minutes, extract with 25 ml of chloroform (7.9). Shake the separatory funnel at least 10 times, let CHCl3 settle, shake again 10 times and let chloroform settle again. Vent chloroform fumes into hood.

8.3.7 Filter chloroform extracts through filter paper. Do not add more chloroform. Carryout filtration in a hood. Dispose of chloroform in environmentally acceptable manner.

8.3.8 Read the absorbance of the samples and standards against the blank at 460 nm.

9. Calculation

9.1 Prepare a standard curve by plotting the absorbance value of standards versus the corresponding phenol concentrations.

9.2 Obtain concentration value of sample directly from standard curve.

10. Precision and Accuracy
    10.1 Using the extraction procedure for concentration of color, six laboratories analyzed samples at concentrations of 9.6, 48.3, and 93.5 ug/1. Standard deviations were ±0.99, ±3.1 and ±4.2 ug/1, respectively.
    10.2 Using the direct photometric procedure, six laboratories analyzed samples at concentrations of 4.7, 48.2 and 97.0 mg/1. Standard deviations were ±0.18, ±0.48 and ±1.58 mg/1, respectively.

Bibliography

1. Annual Book of ASTM Standards, Part 31, "Water", Standard D1783–70, p553 (1976).
2. Standard Methods for the Examination of Water and Wastewater, 14th Edition, p574–581, Method 510 through 510C, (1975).

SILICA, DISSOLVED

Method 370.1 (Colorimetric)

STORET NO. Dissolved 00955

1. Scope and Application
    1.1 This method is applicable to drinking, surface and saline waters, domestic and industrial wastes.
    1.2 The working range of the method is approximately 2 to 25 mg silica/1. The upper range can be extended by taking suitable aliquots; the lower range can be extended by the addition of amino-naphthol-sulfonic acid solution, as described in (6.8).
2. Summary of Method
    2.1 A well-mixed sample is filtered through a 0.45 u membrane filter. The filtrate, upon the addition of molybdate ion in acidic solution, forms a greenish-yellow color complex proportional to the dissolved silica in the sample. The color complex is then measured spectrophotometrically.
    2.2 In the low concentration modification the yellow (410 nm) molybdosilicic acid color is reduced by 1-amino-2-naphthol-4-sulfonic acid to a more intense heteropoly blue (815 nm or 650 nm).
3. Interferences
    3.1 Excessive color and/or turbidity interfere. Correct by running blanks prepared without addition of the ammonium molybdate solution. See (6.7).
    3.2 Tannin interference may be eliminated and phosphate interferences may be decreased with oxalic acid.
    3.3 Large amounts of iron and sulfide interfere.
    3.4 Contact with glass should be minimized, silica free reagents should be used as much as possible. A blank should be run.
4. Apparatus
    4.1 Platinum dishes, 100 ml.
    4.2 Colorimetric equipment—one of the following:
        4.2.1 Spectrophotometer for use at 410 nm, 650 nm and/or 815 nm with a 1 cm or longer cell.
        4.2.2 Filter photometer with a violet filter having maximum transmittance as near 410 nm as possible and a 1 cm or longer cell.
        4.2.3 Nessler tubes, matched, 50 ml, tall form.
5. Reagents
    5.1 Use chemicals low in silica and store in plastic containers.
    5.2 Sodium bicarbonate, $NaHCO_3$, powder.
    5.3 Sulfuric acid, H_2SO_4, 1 $\underline{N}$.

Approved for NPDES
Issued 1971
Editorial revision 1978

5.4   Hydrochloric acid, HCl, $1 + 1$.

5.5   Ammonium molybdate reagent: Place 10 g $(NH_4)_6Mo_7O_{24} \cdot 4H_2O$ in distilled water in a 100 ml volumetric. Dissolve by stirring and gently warming. Dilute to the mark. Filter if necessary. Adjust to pH 7 to 8 with silica free NH_4OH or NaOH. Store in plastic bottle.

5.6   Oxalic acid solution: Dissolve 10 g $H_2C_2O_4 \cdot 2H_2O$ in distilled water in a 100 ml volumetric flask, dilute to the mark. Store in plastic.

5.7   Stock silica solution: Dissolve 4.73 g sodium metasilicate nonahydrate, $Na_2SiO_3 \cdot 9H_2O$, in recently boiled and cooled distilled water. Dilute to approximately 900 ml. Analyze 100.0 ml portions by gravimetry (ref. 1, p. 484). Adjust concentration to 1.000 mg/1 SiO_2. Store in tightly stoppered plastic bottle.

5.8   Standard silica solution: Dilute 10.0 ml stock solution to 1 liter with recently boiled and cooled distilled water. This is 10 mg/1 SiO_2 (1.00 ml = 10.0 ug SiO_2). Store in a tightly stoppered plastic bottle.

5.9   Permanent color solutions

5.9.1   Potassium chromate solution: Dissolve 630 mg K_2CrO_4 in distilled water in a 1 liter volumetric flask and dilute to the mark.

5.9.2   Borax solution: Dissolve 10 g sodium borate decahydrate, $(Na_2B_4O_7 \cdot 10H_2O)$ in distilled water in a 1 liter volumetric flask and dilute to the mark.

5.10   Reducing agent: Dissolve 500 mg of 1-amino-2-naphthol-4-sulfonic acid and 1 g Na_2SO_3 in 50 ml distilled water with gentle warming if necessary. Dissolve 30 g $NaHSO_3$ in 150 ml distilled water. Mix these two solutions. Filter into a plastic bottle. Refrigerate and avoid exposure to light. Discard when it darkens. If there is incomplete solubility or immediate darkening of the aminonaphthosulfonic acid solution do not use.

6.   Procedure

6.1   Filter sample through a 0.45 u membrane filter.

6.2   Digestion: If molybdate unreactive silica is present and its inclusion in the analysis is desired, include this step, otherwise proceed to 6.3.

6.2.1   Place 50 ml, or a smaller portion diluted to 50 ml, of filtered (6.1) sample in a 100 ml platinum dish.

6.2.2   Add 200 mg silica-free $NaHCO_3$ (5.2) and digest on a steam bath for 1 hour. Cool.

6.2.3   Add slowly and with stirring 2.4 ml H_2SO_4 (5.3).

6.2.4   Immediately transfer to a 50 ml Nessler tube, dilute to the mark with distilled water and proceed to 6.3 without delay.

6.3   Color development

6.3.1   Place 50 ml sample in a Nessler tube.

6.3.2   Add rapidly 1.0 ml of $1+1$ HCl (5.4) and 2.0 ml ammonium molybdate reagent (5.5).

6.3.3   Mix by inverting at least 6 times.

6.3.4   Let stand 5 to 10 minutes.

6.3.5   Add 1.5 ml oxalic acid solution (5.6) and mix thoroughly.

6.3.6   Read color (spectrophotometrically or visually) after 2 minutes but before 15 minutes from the addition of oxalic acid.

6.4 Preparation of Standards

6.4.1 If digestion (6.2) was used add 200 mg $NaHCO_3$ (5.2) and 2.4 ml H_2SO_4 (5.3) to standards to compensate for silica introduced by these reagents and for effect of the salt on the color intensity.

6.5 Photometric measurement

6.5.1 Prepare a calibration curve using approximately six standards to span the range shown below with the selected light path.

Selection of Light Path Length for Various
Silica Concentrations

Light Path cm	Silica in 54.5 ml final volume (ug)
1	200–1300
2	100–700
5	40–250
10	20–130

6.5.2 Carry out the steps in 6.3 using distilled water as the reference. Read a blank.

6.5.3 Plot photometric reading versus ug of silica in the final solution of 54.5 ml. Run a reagent blank and at least one standard with each group of samples.

6.6 Visual Comparison

6.6.1 Prepare a set of permanent artificial color standards according to the table. Use well stoppered, properly labelled 50 ml Nessler tubes.

Silica value mg	Potassium chromate solution (5.9.1) ml	Borax solution (5.9.2) ml	Distilled water ml
0.00	0.0	25	30
0.10	1.0	25	29
0.20	2.0	25	28
0.40	4.0	25	26
0.50	5.0	25	25
0.75	7.5	25	22
1.00	10.0	25	20

6.6.2 Verify permanent standards by comparison to color developed by standard silica solutions.

6.6.3 These permanent artificial color standards are only for color comparison procedure, not for photometric procedure.

6.7 Correction for color or turbidity

6.7.1 A special blank is run using a portion of the sample and carrying out the procedure in 6.1, 6.2 if used, and 6.3 except for the addition of ammonium molybdate (6.3.2).

6.7.2 Zero the photometer with this blank before reading the samples.

6.8   Procedure for low concentration (< 1000 ug/1)

6.8.1 Perform steps 6.1 and 6.2 if needed.

6.8.2 Place 50 ml sample in a Nessler tube.

6.8.3 In rapid succession add 1.0 ml of 1 + 1 HCl (5.4).

6.8.4 Add 2.0 ml ammonium molybdate reagent (5.5).

6.8.5 Mix by inverting at least six times.

6.8.6 Let stand 5 to 10 minutes.

6.8.7 Add 1.5 ml oxalic acid solution (5.6).

6.8.8 Mix thoroughly.

6.8.9 At least 2, but not more than 15 minutes after oxalic acid addition, add 2.0 ml reducing agent (5.10).

6.8.10 Mix thoroughly.

6.8.11 Wait 5 minutes, read photometrically or visually.

6.8.12 If digestion (6.2) was used see (6.4).

6.8.13 Photometric measurement

6.8.13.1   Prepare a calibration curve using approximately 6 standards and a reagent blank to span the range shown below with the selected light path.

Selection of Light Path Length for
Various Silica Concentrations

Light Path cm	Silica in 56.5 ml Final volume, ug	
	650 nm	815 nm
1	40–300	20–100
2	20–150	10–50
5	7–50	4–20
10	4–30	2–10

6.8.13.2   Read versus distilled water.

6.8.13.3   Plot photometric reading at 650 nm or at 815 nm versus ug of silica in 56.5 ml.

6.8.13.4   For turbidity correction use 6.1, 6.2 if used and 6.8.2–6.8.11 omitting 6.8.4 and 6.8.9.

6.8.13.5   Run a reagent blank and at least one standard (to check calibration curve drift) with each group of samples.

6.8.14   Visual comparison

6.8.14.1   Prepare not less than 12 standards covering the range of 0 to 120 ug SiO_2 by placing the calculated volumes of standard silica (5.8) in 50 ml Nessler tubes, diluting to the mark and develop the color as in 6.8.2–6.8.11.

7. Calculations

    7.1   Read ug SiO_2 from calibration curve or by visual comparison

    7.2   mg/1 SiO_2 =

$$\frac{ug\,/SiO_2}{ml\ sample}$$

    7.3   Report whether $NaHCO_3$ digestion (6.2) was used

8. Precision and Accuracy

    8.1   A synthetic unknown sample containing 5.0 mg/1 SiO_2, 10 mg/1 chloride, 0.200 mg/1 ammonia N, 1.0 mg/1 nitrate N, 1.5 mg/1 organic N, and 10.0 mg/1 phosphate in distilled water was analyzed in 19 laboratories by the molybdosilicate method, with a relative standard deviation of 14.3% and a relative error of 7.8%.

    8.2   Another synthetic unknown sample containing 15.0 mg/1 SiO_2, 200 mg/1 chloride, 0.800 mg/1 ammonia N, 1.0 mg/1 nitrate N, 0.800 mg/1 organic N, and 5.0 mg/1 phosphate in distilled water was analyzed in 19 laboratories by the molybdosilicate method, with a relative standard deviation of 8.4% and a relative error of 4.2%.

    8.3   A third synthetic unknown sample containing 30.0 mg/1 SiO_2, 400 mg/1 chloride, 1.50 mg/1 ammonia N, 1.0 mg/1 nitrate N, 0.200 mg/1 organic N, and 0.500 mg/1 phosphate in distilled water was analyzed in 20 laboratories by the molybdosilicate method, with a relative standard deviation of 7.7% and a relative error of 9.8%. All results were obtained after sample digestion with $NaHCO_3$.

    8.4   Photometric evaluations by the amino-naphthol-sulfonic acid procedure have an estimated precision of ±0.10 mg/1 in the range from 0 to 2 mg/1 (ASTM).

    8.5   Photometric evaluations of the silico-molybdate color in the range from 2 to 50 mg/1 have an estimated precision of approximately 4% of the quantity of silica measured (ASTM).

Bibliography

1. Annual Book of ASTM Standards, Part 31, "Water", Standard D859–68, p 401 (1976).

2. Standard Methods for the Examination of Water and Wastewater, 14th Edition, p 487, Method 426B, (1975).

SULFATE

Method 375.3 (Gravimetric)

STORET NO. Total 00945

1. Scope and Application
    1.1 This method is applicable to drinking, surface and saline water, domestic and industrial wastes.
    1.2 This method is the most accurate method for sulfate concentrations above 10 mg/l. Therefore, it should be used whenever results of the greatest accuracy are required.
2. Summary of Method
    2.1 Sulfate is precipitated as barium sulfate in a hydrochloric acid medium by the addition of barium chloride. After a period of digestion, the precipitate is filtered, washed with hot water until free of chloride, ignited, and weighed as $BaSO_4$.
    2.2 Preserve by refrigeration at 4°C.
3. Interferences
    3.1 High results may be obtained for samples that contain suspended matter, nitrate, sulfite and silica.
    3.2 Alkali metal sulfates frequently yield low results. This is especially true of alkali hydrogen sulfates. Occlusion of alkali sulfate with barium sulfate causes the substitution of an element of lower atomic weight than barium in the precipitate. Hydrogen sulfate of alkali metal acts similarly and decomposes when heated. Heavy metals such as chromium and iron, cause low results by interfering with complete precipitation and by formation of heavy metal sulfates.
4. Apparatus
    4.1 Steam bath
    4.2 Drying oven, equipped with thermostatic control.
    4.3 Muffle furnace with heat indicator.
    4.4 Desiccator
    4.5 Analytical balance, capable of weighing to 0.1 mg.
    4.6 Filter paper, acid-washed, ashless hard-finish filter paper sufficiently retentive for fine precipitates.
5. Reagents
    5.1 Methyl red indicator solution: Dissolve 100 mg methyl red sodium salt in distilled water in a 100 ml volumetric flask and dilute to the mark with distilled water.
    5.2 Hydrochloric acid, HCl, 1 + 1
    5.3 Barium chloride solution: Dissolve 100 g $BaCl_2 \cdot 2H_2O$ in 1 liter of distilled water. Filter through a membrane filter or hard-finish filter paper. One ml of this reagent is capable of precipitating approximately 40 mg SO_4.

Approved for NPDES
Issued 1974
Editorial revision 1978

5.4   Silver nitrate-nitric acid reagent: Dissolve 8.5 g $AgNO_3$ and 0.5 ml conc. HNO_3 in 500 ml distilled water.

6.   Procedure

6.1   Removal of silica: If silica concentration is greater than 25 mg/1

6.1.1   Evaporate sample nearly to dryness in a platinum dish on a steam bath.

6.1.2   Add 1 ml HCl solution (5.2), tilt dish and rotate until acid contacts all of the residue.

6.1.3   Continue evaporation to dryness.

6.1.4   Complete drying in an oven at 180°C.

6.1.5   If organic matter present, char over a flame.

6.1.6   Moisten with 2 ml distilled water and 1 ml HCl solution (5.2).

6.1.7   Evaporate to dryness on a steam bath.

6.1.8   Add 2 ml HCl solution (5.2).

6.1.9   Take up soluble residue in hot distilled water and filter.

6.1.10   Wash the insoluble silica with several small portions of hot distilled water.

6.1.11   Combine filtrate and washings.

6.2   Precipitation of barium sulfate

6.2.1   If necessary, treat clarified sample to remove interfering agents.

6.2.2   Adjust to contain approximately 50 mg SO_4 ion in a 250 ml volume.

6.2.3   Adjust acidity with HCl solution (5.2) to pH 4.5 to 5.0, using pH meter or orange color of methyl red indicator (5.1).

6.2.4   Add an additional 1 to 2 ml HCl solution (5.2).

6.2.5   For lower concentrations of sulfate ion fix the total volume at 150 ml.

6.2.6   Heat to boiling and, while stirring gently, add warm $BaCl_2$ solution (5.3) slowly, until precipitation appears to be complete; then add approximately 2 ml in excess.

6.2.7   If amount of precipitate is small, add a total of 5 ml $BaCl_2$ solution (5.3).

6.2.8   Digest the precipitate at 80 to 90°C preferably overnight but for not less than 2 hours.

6.3   Filtration and Weighing

6.3.1   Mix a little ashless filter paper pulp with the $BaSO_4$ and filter at room temperature.

6.3.2   Wash the precipitate with small portions of warm distilled water until the washings are free of chloride as indicated by testing with silver nitrate-nitric acid reagent (5.4).

6.3.3   Dry the filter and precipitate.

6.3.4   Ignite at 800°C for 1 hour. DO NOT LET THE FILTER PAPER FLAME.

6.3.5   Cool in a desiccator and weigh.

7.   Calculation

$$mg/1\ SO_4 = \frac{mg\ BaSO_4 \times 411.5}{ml\ sample}$$

8.   Precision and Accuracy

    8.1   A synthetic unknown sample containing 259 mg/1 sulfate, 108 mg/1 Ca, 82 mg/1 Mg, 3.1 mg/1 K, 19.9 mg/1 Na, 241 mg/1 chloride, 250 ug/1 nitrite N, 1.1 mg/1 nitrate N and 42.5 mg/1 alkalinity (contributed by $NaHCO_3$), was analyzed in 32 laboratories by the gravimetric method, with a relative standard deviation of 4.7% and a relative error of 1.9%.

Bibliography

1.   Annual Book of ASTM Standards, Part 31, "Water", Standard D516–68, Method A, p 429 (1976).

2.   Standard Methods for the Examination of Water and Wastewater, 14th Edition, p 493, Method 427A, (1975).

SULFIDE

Method 376.2 (Colorimetric, Methylene Blue)

STORET NO. Total 00745
Dissolved 00746

1. Scope and Application
   1.1 This method is applicable to the measurement of total and dissolved sulfides in drinking, surface and saline waters, domestic and industrial wastes.
   1.2 Acid insoluble sulfides are not measured by this method. Copper sulfide is the only common sulfide in this class.
   1.3 The method is suitable for the measurement of sulfide in concentrations up to 20 mg/l.
2. Summary of Method
   2.1 Sulfide reacts with dimethyl-p-phenylenediamine (p-aminodimethyl aniline) in the presence of ferric chloride to produce methylene blue, a dye which is measured at a wavelength maximum of 625 nm.
3. Comments
   3.1 Samples must be taken with a minimum of aeration. Sulfide may be volatilized by aeration and any oxygen inadvertently added to the sample may convert the sulfide to an unmeasurable form. Dissolved oxygen should not be present in any water used to dilute standards.
   3.2 The analysis must be started immediately.
   3.3 Color and turbidity may interfere with observations of color or with photometric readings.
4. Apparatus
   4.1 Matched test tubes, approximately 125 mm long and 15 mm O.D.
   4.2 Droppers, delivering 20 drops/ml. To obtain uniform drops, hold dropper in vertical position and allow drops to form slowly.
   4.3 Photometer, use either 4.3.1 or 4.3.2.
      4.3.1 Spectrophotometer, for use at 625 nm with cells of 1 cm and 10 cm light path.
      4.3.2 Filter photometer, with filter providing transmittance near 625 nm.
5. Reagents
   5.1 Amino-sulfuric acid stock solution: Dissolve 27 g N,N-dimethyl-p-phenylenediamine oxalate (p-aminodimethylaniline) in a cold mixture of 50 ml conc. H_2SO_4 and 20 ml distilled water in a 100 ml volumetric flask. Cool and dilute to the mark. If dark discard and purchase fresh reagent. Store in dark glass bottle.
   5.2 Amino-sulfuric acid reagent: Dissolve 25 ml amino-sulfuric acid stock solution (5.1) with 975 ml of $1 + 1$ H_2SO_4 (5.4). Store in a dark glass bottle. This solution should be clear.
   5.3 Ferric chloride solution: Dissolve 100 g $FeCl_3 \cdot 6H_2O$ in 40 ml distilled water.

Approved for NPDES
Issued 1978

5.4 Sulfuric acid solution, H_2SO_4, $1+1$

5.5 Diammonium hydrogen phosphate solution: Dissolve 400 g $(NH_4)_2HPO_4$ in 800 ml distilled water.

5.6 Methylene blue solution I: Dissolve 1.0 g of methylene blue in distilled water in a 1 liter volumetric flask and dilute to the mark. Use U.S.P. grade or one certified by the Biological Stain Commission. The dye content reported on the label should be 84% or more. Standardize (5.8) against sulfide solutions of known strength and adjust concentration so that 0.05 ml (1 drop) equals 1.0 mg/1 sulfide.

5.7 Methylene blue solution II: Dilute 10.00 ml of adjusted methylene blue solution I (5.6) to 100 ml with distilled water in a volumetric flask.

5.8 Standardization of methylene blue I solution:

    5.8.1 Place several grams of clean, washed crystals of sodium sulfide $Na_2S \cdot 9H_2O$ in a small beaker.

    5.8.2 Add somewhat less than enough water to cover the crystals.

    5.8.3 Stir occasionally for a few minutes. Pour the solution into another vessel. This reacts slowly with oxygen but the change is insignificnat over a few hours. Make the solution daily.

    5.8.4 To 1 liter of distilled water add 1 drop of solution and mix.

    5.8.5 Immediately determine the sulfide concentration by the methylene blue procedure (6) and by the titrimetric iodide procedure (Method 376.1, this manual).

    5.8.6 Repeat using more than one drop of sulfide solution or less water until at least five tests have been made in the range of 1 to 8 mg/1 sulfide.

    5.8.7 Calculate the average percent error of the methylene blue procedure (6) as compared to the titrimetric iodide procedure (Method 376.1).

    5.8.8 Adjust by dilution or by adding more dye to methylene blue solution I (5.6).

6. Procedure

    6.1 Color development

        6.1.1 Transfer 7.5 ml of sample to each of two matched test tubes using a special wide tipped pipet or filling to a mark on the test tubes.

        6.1.2 To tube A add 0.5 ml amine-sulfuric acid reagent (5.2) and 0.15 ml (3 drops) $FeCl_3$ solution (5.3).

        6.1.3 Mix immediately by inverting the tube only once.

        6.1.4 To tube B add 0.5 ml $1+1$ H_2SO_4 (5.4) and 0.15 ml (3 drops) $FeCl_3$ solution (5.3) and mix.

        6.1.5 Color will develop in tube A in the presence of sulfide. Color development is usually complete in about 1 minute, but a longer time is often required for the fading of the initial pink color.

        6.1.6 Wait 3 to 5 minutes.

        6.1.7 Add 1.6 ml $(NH_4)_2HPO_4$ solution (5.5) to each tube.

        6.1.8 Wait 3 to 5 minutes and make color comparisons. If zinc acetate was used wait at least 10 minutes before making comparison.

6.2 Color comparison

    6.2.1 Visual

        6.2.1.1   Add methylene blue solution I (5.6) and/or II (5.7) (depending on sulfide concentration and accuracy desired) dropwise to tube B (6.1.4) until the color matches that developed in the first tube.

        6.2.1.2   If the concentration exceeds 20 mg/1, repeat 6.2.1.1 using a portion of the sample diluted to one tenth.

    6.2.2 Photometric

        6.2.2.1   Use a 1 cm cell for 0.1 to 2.0 mg/1. Use a 10 cm cell for up to 20 mg/1.

        6.2.2.2   Zero instrument with portion of sample from tube B (6.1.4).

        6.2.2.3   Prepare calibration curve from data obtained in methylene blue standardization (5.8), plotting concentraton obtained from titrimetric iodide procedure (Method 376.1) versus absorbance. A straight line relationship can be assumed from 0 to 1.0 mg/1.

        6.2.2.4   Read the sulfide concentration from the calibration curve.

7.   Calculations

    7.1   Visual comparison: With methylene blue solution I (5.6), adjusted so that 0.05 ml (1 drop) = 1.0 mg/1 sulfide and a 7.5 ml sample

    mg/1 sulfide = number drops methylene blue solution I (5.6) + 0.1 x [number of drops methylene blue solution II (5.7)].

    7.2   Photometric: see 6.2.2.4

8.   Precision and Accuracy:

    8.1   The precision has not been determined. The accuracy is about ±10%.

Bibliography

1.   Standard Methods for he Examination of Water and Wastewater, 14th edition, p. 503, Method 428C (1975).

SULFITE

Method 377.1 (Titrimetric)

STORET NO. 00740

1. Scope and Application
   1.1 This method is applicable to drinking and surface waters, sewage and industrial wastes. The primary application has been to cooling, process and distribution water systems and boiler feedwaters to which sulfide is added in order to reduce dissolved oxygen and eliminate corrosion.
   1.2 The minimum detectable limit is 2–3 mg/1 SO_3.

2. Summary of Method
   2.1 An acidified sample containing a starch indicator is titrated with a standard potassium iodide-iodate titrant to a faint permanent blue end point which appears when the reducing power of the sample has been completely exhausted.

3. Interferences
   3.1 The temperature of the sample must be below 50°C.
   3.2 Care must be taken to allow as little contact with air as possible. For example, do not filter the sample. Keep the buret tip below the surface of the sample.
   3.3 Other oxidizable substances, such as organic compounds, ferrous iron and sulfide are positive interferences. Sulfide may be removed by adding 0.5g of zinc acetate and analyzing the supernatant of the settled sample.
   3.4 Nitrite gives a negative interference by oxidizing sulfite when the sample is acidified; this is corrected by either using a proprietary indicator which eliminates nitrite or by adding sulfamic acid.
   3.5 Copper and possibly other heavy metals catalyze the oxidation of sulfite; EDTA is used to complex metals.
   3.6 A blank must be run to correct for interferences present in the reagents.

4. Apparatus
   4.1 Standard laboratory glassware is used.

5. Reagents
   5.1 Sulfuric acid, H_2SO_4, 1 + 1.
   5.2 Starch indicator: Amylose, Mallinckrodt Chemical Works; Thyodene, Magnus Chemical Co. or equivalent.
   5.3 Dual-Purpose Sulfite Indicator Powder: a proprietary formulation containing sulfamic acid to destroy nitrite.
   5.4 Standard potassium iodide-iodate titrant, 0.0125N: Dissolve 445.8 mg anhydrous potassium iodate, KIO_3 (primary standard grade dried for several hours at 120°C), 4.35g

Approved for NPDES
Issued 1974
Editorial revision 1978

KI and 310 mg $NaHCO_3$ in distilled water and dilute to 1 liter. This titrant is equivalent to 500 ug $SO_3/1.00$ ml.

5.5 Sulfamic Acid: Crystalline

5.6 EDTA Reagent: Dissolve 2.5g EDTA in 100 ml distilled water.

6. Procedure

6.1 Sampling

Contact with air must be minimized. If the sample temperature is greater than 50°C, it must be cooled in a special apparatus described elswhere (see Bibliography). Immediately add 1 ml of EDTA Solution (5.6) per 100 ml of sample.

6.2 Starch Indicator

6.2.1 Place 1 ml H_2SO_4 (5.1) in titration vessel.

6.2.2 Add 0.1 g sulfamic acid crystals (5.5).

6.2.3 Add 50 ml sample.

6.2.4 Add approximately 0.1 g starch indicator (5.2).

6.2.5 Titrate with potassium iodide-iodate titrant (5.4) until a faint permanent blue color develops. Keep the pipet tip below the surface of the sample. View the color change against a white background. Record the ml titrant.

6.2.6 Run a reagent blank using distilled water instead of sample (6.2.3).

6.3 Dual Purpose Sulfite Indicator Powder.

6.3.1 Place 50 ml sample in a titration vessel.

6.3.2 Add 3–4 drops phenolphthalein indicator.

6.3.3 Add sufficient scoops (1g) of indicator (5.3) to discharge the red color.

6.3.4 Titrate with potassium iodide-iodate titrant (5.4) until a faint permanent blue color develops. View the color change against a white background. Record the ml titrant.

6.3.5 Run a reagent blank using distilled water instead of sample (6.3.1).

7. Calculations

7.1 Use the formula:

$$\text{mg/l } SO_3 = \frac{A \times N \times 40,000}{\text{ml sample}}$$

where:

A = ml titrant (6.2.5 or 6.3.4) B = ml
B = ml titrant for the blank (6.2.6 or 6.3.5) and
N = normality of $KI\text{-}KIO_3$ titrant (5.4)

7.2 To calculate as Na_2SO_3
mg/l Na_2SO_3 = mg/l SO_3 x 1.57

8. Precision and Accuracy

8.1 Precision and accuracy data are not available at this time.

Bibliography

1.   Annual Book of ASTM Standards, Part 31, "Water", Standard D1339–72, Method C, p 440 (1976).
2.   Standard Methods for the Examination of Water and Wastewater, 14th Edition, p 508, Method 429, (1975).

METHYLENE BLUE ACTIVE SUBSTANCES (MBAS)

Method 425.1 (Colorimetric)

STORET NO. 38260

1. Scope and Application
   1.1 This method is applicable to the measurement of methylene blue active substances (MBAS) in drinking waters, surface waters, domestic and industrial wastes. It is not applicable to measurement of surfactant-type materials in saline waters.
   1.2 It is not possible to differentiate between linear alkyl sulfonate (LAS) and alkyl benzene sulfonate (ABS) or other isomers of these types of compounds. However, LAS has essentially replaced ABS on the surfactant market so that measurable surfactant materials will probably be LAS type materials.
   1.3 The method is applicable over the range of 0.025 to 100 mg/1 LAS.
2. Summary of Method
   2.1 The dye, methylene blue, in aqueous solution reacts with anionic-type surface active materials to form a blue colored salt. The salt is extractable with chloroform and the intensity of color produced is proportional to the concentration of MBAS.
3. Comments
   3.1 Materials other than man-made surface active agents which react with methylene blue are organically bound sulfates, sulfonates, carboxylates, phosphates, phenols, cyanates, thiocyanates and some inorganic ions such as nitrates and chlorides. However, the occurrence of these materials at interference levels is relatively rare and with the exception of chlorides may generally be disregarded.
   3.2 Chlorides at concentration of about 1000 mg/1 show a positive interference but the degree of interference has not been quantified. For this reason the method is not applicable to brine samples.
   3.3 Naturally occurring organic materials that react with methylene blue are relatively insignificant. Except under highly unusual circumstances, measurements of MBAS in finished waters, surface waters and domestic sewages may be assumed to be accurate measurements of man-made surface active agents.
4. Precision and Accuracy
   4.1 On a sample of filtered river water, spiked with 2.94 mg LAS/liter, 110 analysts obtained a mean of 2.98 mg/1 with a standard deviation of ±0.272.
   4.2 On a sample of tap water spiked with 0.48 mg LAS/liter, 110 analysts obtained a mean of 0.49 mg/1 with a standard deviation of ±0.048.
   4.3 On a sample of distilled water spiked with 0.27 mg LAS/liter, 110 analysts obtained a mean of 0.24 mg/1 with a standard deviation of ±0.036.
   4.4 Analytical Reference Service, Water Surfactant No. 3, Study No. 32, (1968).

Approved for NPDES
Issued 1971

5.   References
  5.1   The procedure to be used for this determination is found in:
         Standard Methods for the Examination of Water and Wastewaters, 14th Edition, p 600, Method No. 512A (1975).
         Annual Book of ASTM Standards, Part 31, "Water", Standard D 2330–68, Method A, p 494 (1976).

Appendix F-3 Biological Pollutants

Parameter	Water/Wastewater Method EPA 600/4-79-020 and EPA 600/8-78-017	Sample Volume Required (mL)	Container Type	Preservative	Holding Time
BOD	405.1	1000	HDPE	Cool 4° C	48 Hours
CBOD	405.1	1000	HDPE	Cool 4° C	48 Hours
COD	410.1	50	HDPE	Cool 4° C, H$_2$SO$_4$ to pH < 2	28 Days
Organic Carbon	415.1	25	HDPE	Cool 4° C, HCl or H$_2$SO$_4$ to pH < 2	28 Days
Coliform, Fecal & Total	Section B, Section C	100	HDPE	0.008% Na$_2$S$_2$O$_3$ and Cool 4° C	6 Hours
Fecal Streptococci	Section D	100	HDPE	0.008% Na$_2$S$_2$O$_3$ and Cool 4° C	6 Hours

BIOCHEMICAL OXYGEN DEMAND

Method 405.1 (5 Days, 20°C)

STORET NO. 00310
Carbonaceous 80082

1.   Scope and Application
   1.1   The biochemical oxygen demand (BOD) test is used for determining the relative oxygen requirements of municipal and industrial wastewaters. Application of the test to organic waste discharges allows calculation of the effect of the discharges on the oxygen resources of the receiving water. Data from BOD tests are used for the development of engineering criteria for the design of wastewater treatment plants.
   1.2   The BOD test is an empirical bioassay-type procedure which measures the dissolved oxygen consumed by microbial life while assimilating and oxidizing the organic matter present. The standard test conditions include dark incubation at 20°C for a specified time period (often 5 days). The actual environmental conditions of temperature, biological population, water movement, sunlight, and oxygen concentration cannot be accurately reproduced in the laboratory. Results obtained must take into account the above factors when relating BOD results to stream oxygen demands.

2.   Summary of Method
   2.1   The sample of waste, or an appropriate dilution, is incubated for 5 days at 20°C in the dark. The reduction in dissolved oxygen concentration during the incubation period yields a measure of the biochemical oxygen demand.

3.   Comments
   3.1   Determination of dissolved oxygen in the BOD test may be made by use of either the Modified Winkler with Full-Bottle Technique or the Probe Method in this manual.
   3.2   Additional information relating to oxygen demanding characteristics of wastewaters can be gained by applying the Total Organic Carbon and Chemical Oxygen Demand tests (also found in this manual).
   3.3   The use of 60 ml incubation bottles in place of the usual 300 ml incubation bottles, in conjunction with the probe, is often convenient.

4.   Precision and Accuracy
   4.1   Eighty-six analysts in fifty-eight laboratories analyzed natural water samples plus an exact increment of biodegradable organic compounds. At a mean value of 2.1 and 175 mg/1 BOD, the standard deviation was ±0.7 and ±26 mg/1, respectively (EPA Method Research Study 3).
   4.2   There is no acceptable procedure for determining the accuracy of the BOD test.

Approved for NPDES CBOD: pending approval for Section 304(h), CWA
Issued 1971
Editorial revision 1974

5.   References

5.1   The procedure to be used for this determination is found in:
Standard Methods for the Examination of Water and Wastewater, 15th
Edition, p. 483, Method 507 (1980).

5.2   Young, J. C., "Chemical Methods for Nitrification Control," J. Water
Poll. Control Fed., 45, p. 637 (1973).

CHEMICAL OXYGEN DEMAND

Method 410.1 (Titrimetric, Mid-Level)

STORET NO. 00340

1. Scope and Application
   1.1 The Chemical Oxygen Demand (COD) method determines the quantity of oxygen required to oxidize the organic matter in a waste sample, under specific conditions of oxidizing agent, temperature, and time.
   1.2 Since the test utilizes a specific chemical oxidation the result has no definite relationship to the Biochemical Oxygen Demand (BOD) of the waste or to the Total Organic Carbon (TOC) level. The test result should be considered as an independent measurement of organic matter in the sample, rather than as a substitute for the BOD or TOC test.
   1.3 The method can be applied to domestic and industrial waste samples having an organic carbon concentration greater than 50 mg/1. For lower concentrations of carbon such as in surface water samples, the Low Level Modification should be used. When the chloride concentration of the sample exceeds 2000 mg/1, the modification for saline waters is required.

2. Summary of Method
   2.1 Organic and oxidizable inorganic substances in the sample are oxidized by potassium dichromate in 50% sulfuric acid solution at reflux temperature. Silver sulfate is used as a catalyst and mercuric sulfate is added to remove chloride interference. The excess dichromate is titrated with standard ferrous ammonium sulfate, using orthophenanthroline ferrous complex as an indicator.

3. Sampling and Preservation
   3.1 Collect the samples in glass bottles, if possible. Use of plastic containers is permissible if it is known that no organic contaminants are present in the containers.
   3.2 Biologically active samples should be tested as soon as possible. Samples containing settleable material should be well mixed, preferably homogenized, to permit removal of representative aliquots.
   3.3 Samples should be preserved with sulfuric acid to a pH < 2 and maintained at 4°C until analysis.

4. Interferences
   4.1 Traces of organic material either from the glassware or atmosphere may cause a gross, positive error.
      4.1.1 Extreme care should be exercised to avoid inclusion of organic materials in the distilled water used for reagent preparation or sample dilution.
      4.1.2 Glassware used in the test should be conditioned by running blank procedures to eliminate traces of organic material.

Approved for NPDES
Issued 1971
Editorial revision 1978

4.2   Volatile materials may be lost when the sample temperature rises during the sulfuric acid addition step. To minimize this loss the flask should be cooled during addition of the sulfuric acid solution.

4.3   Chlorides are quantitatively oxidized by dichromate and represent a positive interference. Mercuric sulfate is added to the digestion flask to complex the chlorides, thereby effectively eliminating the interference on all but brine and estuarine samples.

5.   Apparatus

5.1   Reflux apparatus: Glassware should consist of a 500 ml Erlenmeyer flask or a 300 ml round bottom flask made of heat-resistant glass connected to a 12 inch Allihn condenser by means of a ground glass joint. Any equivalent reflex apparatus may be substituted provided that a ground-glass connection is used between the flask and the condenser.

6.   Reagents

6.1   Distilled water: Special precautions should be taken to insure that distilled water used in this test be low in organic matter.

6.2   Standard potassium dichromate solution (0.250 N): Dissolve 12.259 g $K_2Cr_2O_7$, primary standard grade, previously dried at 103°C for two hours, in distilled water and dilute to 1000 ml.

6.3   Sulfuric acid reagent: Conc. H_2SO_4 containing 23.5g silver sulfate, Ag_2SO_4, per 4.09kg bottle. With continuous stirring, the silver sulfate may be dissolved in about 30 minutes.

6.4   Standard ferrous ammonium sulfate (0.25 N): Dissolve 98.0 g of $Fe(NH_4)_2(SO_4)_2 \cdot 6H_2O$ in distilled water. Add 20 ml of conc. H_2SO_4 (6.8), cool and dilute to 1 liter. This solution must be standardized daily against standard $K_2Cr_2O_7$ solution (6.2).

6.4.1   Standardization: To approximately 200 ml of distilled water add 25.0 ml of 0.25 N $K_2Cr_2O_7$ (6.2) solution. Add 20 ml of H_2SO_4 (6.8) and cool. Titrate with ferrous ammonium sulfate (6.4) using 3 drops of ferroin indicator (6.6). The color change is sharp, going from blue-green to reddish-brown.

$$\text{Normality} = \frac{(\text{ml } K_2Cr_2O_7)(0.25)}{\text{ml Fe (NH}_4)_2 (SO_4)_2}$$

6.5   Mercuric sulfate: Powdered $HgSO_4$.

6.6   Phenanthroline ferrous sulfate (ferroin) indicator solution: Dissolve 1.48 g of 1–10 (ortho) phenanthroline monohydrate, together with 0.70 g of $FeSO_4 \cdot 7H_2O$ in 100 ml of water. This indicator may be purchased already prepared.

6.7   Silver sulfate: Powdered Ag_2SO_4.

6.8   Sulfuric acid (sp. gr. 1.84): Concentrated H_2SO_4.

7.   Procedure

7.1   Place several boiling stones in the reflux flask, followed by 50.0 ml of sample or an aliquot diluted to 50.0 ml and 1 g of $HgSO_4$ (6.5). Add 5.0 ml conc. H_2SO_4 (6.8); swirl until the mercuric sulfate has dissolved. Place reflux flask in an ice bath and slowly add, with swirling, 25.0 ml of 0.25 N $K_2Cr_2O_7$ (6.2). Now add 70 ml of sulfuric acid-silver

ORGANIC CARBON, TOTAL

Method 415.1 (Combustion or Oxidation)

STORET NO. Total 00680
Dissolved 00681

1. Scope and Application
    1.1 This method includes the measurement of organic carbon in drinking, surface and saline waters, domestic and industrial wastes. Exclusions are noted under Definitions and Interferences.
    1.2 The method is most applicable to measurement of organic carbon above 1 mg/l.
2. Summary of Method
    2.1 Organic carbon in a sample is converted to carbon dioxide (CO_2) by catalytic combustion or wet chemical oxidation. The CO_2 formed can be measured directly by an infrared detector or converted to methane (CH_4) and measured by a flame ionization detector. The amount of CO_2 or CH_4 is directly proportional to the concentration of carbonaceous material in the sample.
3. Definitions
    3.1 The carbonaceous analyzer measures all of the carbon in a sample. Because of various properties of carbon-containing compounds in liquid samples, preliminary treatment of the sample prior to analysis dictates the definition of the carbon as it is measured. Forms of carbon that are measured by the method are:
    A) soluble, nonvolatile organic carbon; for instance, natural sugars.
    B) soluble, volatile organic carbon; for instance, mercaptans.
    C) insoluble, partially volatile carbon; for instance, oils.
    D) insoluble, particulate carbonaceous materials, for instance; cellulose fibers.
    E) soluble or insoluble carbonaceous materials adsorbed or entrapped on insoluble inorganic suspended matter; for instance, oily matter adsorbed on silt particles.
    3.2 The final usefulness of the carbon measurement is in assessing the potential oxygen-demanding load of organic material on a receiving stream. This statement applies whether the carbon measurement is made on a sewage plant effluent, industrial waste, or on water taken directly from the stream. In this light, carbonate and bicarbonate carbon are not a part of the oxygen demand in the stream and therefore should be discounted in the final calculation or removed prior to analysis. The manner of preliminary treatment of the sample and instrument settings defines the types of carbon which are measured. Instrument manufacturer's instructions should be followed.

Approved for NPDES
Issued 1971
Editorial revision 1974

4.   Sample Handling and Preservation

4.1   Sampling and storage of samples in glass bottles is preferable. Sampling and storage in plastic bottles such as conventional polyethylene and cubitainers is permissible if it is established that the containers do not contribute contaminating organics to the samples. **NOTE 1:** A brief study performed in the EPA Laboratory indicated that distilled water stored in new, one quart cubitainers did not show any increase in organic carbon after two weeks exposure.

4.2   Because of the possibility of oxidation or bacterial decomposition of some components of aqueous samples, the lapse of time between collection of samples and start of analysis should be kept to a minimum. Also, samples should be kept cool (4°C) and protected from sunlight and atmospheric oxygen.

4.3   In instances where analysis cannot be performed within two hours (2 hours) from time of sampling, the sample is acidified (pH $\leq$ 2) with HCl or H_2SO_4

5.   Interferences

5.1   Carbonate and bicarbonate carbon represent an interference under the terms of this test and must be removed or accounted for in the final calculation.

5.2   This procedure is applicable only to homogeneous samples which can be injected into the apparatus reproducibly by means of a microliter type syringe or pipette. The openings of the syringe or pipette limit the maximum size of particles which may be included in the sample.

6.   Apparatus

6.1   Apparatus for blending or homogenizing samples: Generally, a Waring-type blender is satisfactory.

6.2   Apparatus for total and dissolved organic carbon:

6.2.1 A number of companies manufacture systems for measuring carbonaceous material in liquid samples. Considerations should be made as to the types of samples to be analyzed, the expected concentration range, and forms of carbon to be measured.

6.2.2 No specific analyzer is recommended as superior.

7.   Reagents

7.1   Distilled water used in preparation of standards and for dilution of samples should be ultra pure to reduce the carbon concentration of the blank. Carbon dioxide-free, double distilled water is recommended. Ion exchanged waters are not recommended because of the possibilities of contamination with organic materials from the resins.

7.2   Potassium hydrogen phthalate, stock solution, 1000 mg carbon/liter: Dissolve 0.2128 g of potassium hydrogen phthalate (Primary Standard Grade) in distilled water and dilute to 100.0 ml.
NOTE 2: Sodium oxalate and acetic acid are not recommended as stock solutions.

7.3   Potassium hydrogen phthalate, standard solutions: Prepare standard solutions from the stock solution by dilution with distilled water.

7.4   Carbonate-bicarbonate, stock solution, 1000 mg carbon/liter: Weigh 0.3500 g of sodium bicarbonate and 0.4418 g of sodium carbonate and transfer both to the same 100 ml volumetric flask. Dissolve with distilled water.

7.5 Carbonate-bicarbonate, standard solution: Prepare a series of standards similar to step 7.3.

NOTE 3: This standard is not required by some instruments.

7.6 Blank solution: Use the same distilled water (or similar quality water) used for the preparation of the standard solutions.

8. Procedure

8.1 Follow instrument manufacturer's instructions for calibration, procedure, and calculations.

8.2 For calibration of the instrument, it is recommended that a series of standards encompassing the expected concentration range of the samples be used.

9. Precision and Accuracy

9.1 Twenty-eight analysts in twenty-one laboratories analyzed distilled water solutions containing exact increments of oxidizable organic compounds, with the following results:

Increment as TOC mg/liter	Precision as Standard Deviation TOC, mg/liter	Accuracy as	
		Bias, %	Bias, mg/liter
4.9	3.93	+15.27	+0.75
107	8.32	+ 1.01	+1.08

(FWPCA Method Study 3, Demand Analyses)

Bibliography

1. Annual Book of ASTM Standards, Part 31, "Water", Standard D 2574–79, p 469 (1976).

2. Standard Methods for the Examination of Water and Wastewater, 14th Edition, p 532, Method 505, (1975).

APPENDIX G
GAC ISOTHERM DATA
(US EPA, PUBLICATION NO.
EPA-600/8-80-023)

ALPHABETICAL LIST OF COMPOUNDS

Compound	Formula	Molecular Weight	Calculated Carbon Requirements to Achieve Indicated Change in Concentration (a)					
			Single Stage Powdered Carbon C_f, mg/l				Granular Carbon Column	
			Co, mg/l	0.1	0.01	0.001	Co, mg/l	
Acenaphthene	$C_{12}H_{10}$	154.21	1.0	10	30	60	1.0	5.2
			0.1		2.4	6.1	0.1	1.2
			0.01			0.6	0.01	0.3

Compound	Formula	Molecular Weight	Calculated Carbon Requirements to Achieve Indicated Change in Concentration (a)					
			Single Stage Powdered Carbon C_f, mg/l				Granular Carbon Column	
			Co, mg/l	0.1	0.01	0.001	Co, mg/l	
Acenaphthylene	$C_{12}H_8$	152.21	1.0	18	47	110	1.0	8.7
			0.1		4.3	11	0.1	2.0
			0.01			1.0	0.01	0.5

Compound	Formula	Molecular Weight	Calculated Carbon Requirements to Achieve Indicated Change in Concentration (a)	
			Single Stage Powdered Carbon C_f, mg/l	Granular Carbon Column
Acetone Cyanohdyrin	C_4H_7NO	85.11		

Compound	Formula	Molecular Weight	Calculated Carbon Requirements to Achieve Indicated Change in Concentration (a)					
			Single Stage Powdered Carbon C_f, mg/l				Granular Carbon Column	
			Co, mg/l	0.1	0.01	0.001	Co, mg/l	
Acetophenone	C_8H_8O	120.14	1.0	34	100	230	1.0	14
			0.1		9.2	28	0.1	3.7
			0.01			2.5	0.01	1.0

Compound	Formula	Molecular Weight	Calculated Carbon Requirements to Achieve Indicated Change in Concentration (a)					
			Single Stage Powdered Carbon C_f, mg/l				Granular Carbon Column	
			Co, mg/l	0.1	0.01	0.001	Co, mg/l	
2- Acetylmaniofluorene	$C_{15}H_{13}NO$	222.28	1.0	3.7	5.4	7.2	1.0	3.1
			0.1		0.5	0.7	0.1	0.42
			0.01			0.06	0.01	0.06

Compound	Formula	Molecular Weight	Calculated Carbon Requirements to Achieve Indicated Change in Concentration (a)					
			Single Stage Powdered Carbon C_f, mg/l				Granular Carbon Column	
	$C_{17}H_{19}N_3HC_1$	273.77	Co, mg/l	0.1	0.01	0.001	Co, mg/l	
Acridine Orange			1.0	5.1	7.4	9.9	1.0	4.3
			0.1		0.67	0.98	0.1	0.6
			0.01			0.09	0.01	0.1

Compound	Formula	Molecular Weight	Calculated Carbon Requirements to Achieve Indicated Change in Concentration (a)					
			Single Stage Powdered Carbon C_f, mg/l				Granular Carbon Column	
			Co, mg/l	0.1	0.01	0.001	Co, mg/l	
Acrolein	C_3H_4O	56.06	1.0	3,500	17,000	76,800	1.0	860
			0.1		1,500	7,600	0.1	380
			0.01			690	0.01	170

Compound	Formula	Molecular Weight	Calculated Carbon Requirements to Achieve Indicated Change in Concentration (a)					
			Single Stage Powdered Carbon C_f, mg/l				Granular Carbon Column	
			Co, mg/l	0.1	0.01	0.001	Co, mg/l	
Acrylonitrile	C_3H_3N	53.06	1.0	2,200	7,700	25,000	1.0	710
			0.1		700	2,500	0.1	240
			0.01			230	0.01	80

Compound	Formula	Molecular Weight	Calculated Carbon Requirements to Achieve Indicated Change in Concentration (a)					
			Single Stage Powdered Carbon C_f, mg/l				Granular Carbon Column	
			Co, mg/l	0.1	0.01	0.001	Co, mg/l	
Adenine	$C_5H_5N_5$	135.13	1.0	30	80	190	1.0	14
			0.1		7.3	19	0.1	3.4
			0.01			1.7	0.01	0.8

Compound	Formula	Molecular Weight	Calculated Carbon Requirements to Achieve Indicated Change in Concentration (a)					
			Single Stage Powdered Carbon C_f, mg/l				Granular Carbon Column	
			Co, mg/l	0.1	0.01	0.001	Co, mg/l	
Adipic Acid	$C_6H_{10}O_4$	146.14	1.0	130	430	1300	1.0	50
			0.1		39	130	0.1	15
			0.01			12	0.01	4.2

Compound	Formula	Molecular Weight	Calculated Carbon Requirements to Achieve Indicated Change in Concentration (a)					
			Single Stage Powdered Carbon C_f, mg/l				Granular Carbon Column	
			Co, mg/l	0.1	0.01	0.001	Co, mg/l	
Aldrin	$C_{12}H_8Cl_6$	365.0	1.0	11	110	880	1.0	1.5
			0.1		9.7	88	0.1	1.3
			0.01			8.0	0.01	1.1

Compound	Formula	Molecular Weight	Calculated Carbon Requirements to Achieve Indicated Change in Concentration (a)					
			Single Stage Powdered Carbon C_f, mg/l				Granular Carbon Column	
			Co, mg/l	0.1	0.01	0.001	Co, mg/l	
4-Aminobiphenyl	$C_{12}H_{11}N$	169.12	1.0	8.2	16	30	1.0	5.1
			0.1		1.5	3.0	0.1	1.0
			0.01			0.3	0.01	0.2

Compound	Formula	Molecular Weight	Calculated Carbon Requirements to Achieve Indicated Change in Concentration (a)					
			Single Stage Powdered Carbon C_f, mg/l				Granular Carbon Column	
			Co, mg/l	0.1	0.01	0.001	Co, mg/l	
Anthracene	$C_{14}H_{10}$	178.24	1.0	12	67	340	1.0	2.7
			0.1		6.1	34	0.1	1.3
			0.01			3.1	0.01	0.7

Compound	Formula	Molecular Weight	Calculated Carbon Requirements to Achieve Indicated Change in Concentration (a)					
			Single Stage Powdered Carbon C_f, mg/l				Granular Carbon Column	
			Co, mg/l	0.1	0.01	0.001	Co, mg/l	980
Benzene	C_6H_6	78.12	1.0	35,000	>100,000	>100,000	1.0	4,000
			0.1		>100,000	>100,000	0.1	14,000
			0.01			>100,000	0.01	

Compound	Formula	Molecular Weight	Calculated Carbon Requirements to Achieve Indicated Change in Concentration (a)					
			Single Stage Powdered Carbon C_f, mg/l				Granular Carbon Column	
Benzidine Dihydrochloride	$C_{12}H_{12}N_2 2HCL$	257.16	Co, mg/l	0.1	0.01	0.001	Co, mg/l	
			1.0	9.4	24	58	1.0	4.5
			0.1		2.2	5.7	0.1	1.0
			0.01			0.52	0.01	0.2

Compound	Formula	Molecular Weight	Calculated Carbon Requirements to Achieve Indicated Change in Concentration (a)					
			Single Stage Powdered Carbon C_f, mg/l				Granular Carbon Column	
Benzoic Acid	$C_7H_6O_2$	122.12	Co, mg/l	0.1	0.01	0.001	Co, mg/l	
			1.0	85,000	>100,000	>100,000	1.0	1,300
			0.1		>100,000	>100,000	0.1	9,300
			0.01			>100,000	0.01	67,000

Compound	Formula	Molecular Weight	Calculated Carbon Requirements to Achieve Indicated Change in Concentration (a)					
			Single Stage Powdered Carbon C_f, mg/l				Granular Carbon Column	
3,4-Benzofluoranthene (Benzo(b)fluoranthene	$C_{20}H_{12}$	252.32	Co, mg/l	0.1	0.01	0.001	Co, mg/l	
			1.0	37	95	230	1.0	18
			0.1		8.7	22	0.1	4.2
			0.01			2.0	0.01	1.0

Compound	Formula	Molecular Weight	Calculated Carbon Requirements to Achieve Indicated Change in Concentration (a)					
			Single Stage Powdered Carbon C_f, mg/l				Granular Carbon Column	
Benzo(k)fluoranthene	$C_{20}H_{12}$	252.32	Co, mg/l	0.1	0.01	0.001	Co, mg/l	
			1.0	18	76	280	1.0	5.5
			0.1		6.7	28	0.1	2.1
			0.01			2.6	0.01	0.8

Compound	Formula	Molecular Weight	Calculated Carbon Requirements to Achieve Indicated Change in Concentration (a)					
			Single Stage Powdered Carbon C_f, mg/l				Granular Carbon Column	
Benzo(ghi)perylene	$C_{22}H_{12}$	276.34	Co, mg/l	0.1	0.01	0.001	Co, mg/l	
			1.0	200	510	1200	1.0	93
			0.1		46	120	0.1	22
			0.01			11	0.01	5.5

Compound	Formula	Molecular Weight	Calculated Carbon Requirements to Achieve Indicated Change in Concentration (a)					
			Single Stage Powdered Carbon C_f, mg/l				Granular Carbon Column	
Benzo(a)pyrene	$C_{20}H_{12}$	252.30	Co, mg/l	0.1	0.01	0.001	Co, mg/l	
			1.0	74	220	621	1.0	29
			0.1		20	62	0.1	8.3
			0.01			5.6	0.01	2.2

Compound	Formula	Molecular Weight	Calculated Carbon Requirements to Achieve Indicated Change in Concentration (a)					
			Single Stage Powdered Carbon C_f, mg/l				Granular Carbon Column	
Benzothiazole	C_7H_5NS	135.19	Co, mg/l	0.1	0.01	0.001	Co, mg/l	
			1.0	14	28	52	1.0	8.2
			0.1		2.5	5.2	0.1	1.5
			0.01			0.47	0.01	0.3

Compound	Formula	Molecular Weight	Calculated Carbon Requirements to Achieve Indicated Change in Concentration (a)					
			Single Stage Powdered Carbon C_f, mg/l				Granular Carbon Column	
			Co, mg/l	0.1	0.01	0.001	Co, mg/l	
a-BHC	$C_6H_6CL_6$	290.83	1.0	8.0	24	64	1.0	3.3
			0.1		2.2	6.4	0.1	0.9
			0.01			0.6	0.01	0.2

Compound	Formula	Molecular Weight	Calculated Carbon Requirements to Achieve Indicated Change in Concentration (a)					
			Single Stage Powdered Carbon C_f, mg/l				Granular Carbon Column	
			Co, mg/l	0.1	0.01	0.001	Co, mg/l	
B-BHC	$C_6H_6CL_6$	290.83	1.0	13	43	130	1.0	4.5
			0.1		3.9	13	0.1	1.4
			0.01			1.2	0.01	0.4

Compound	Formula	Molecular Weight	Calculated Carbon Requirements to Achieve Indicated Change in Concentration (a)					
			Single Stage Powdered Carbon C_f, mg/l				Granular Carbon Column	
			Co, mg/l	0.1	0.01	0.001	Co, mg/l	
y-BHC (Lindane)	$C_6H_6CL_6$	290.83	1.0	11	40	115	1.0	3.9
			0.1		3.4	11	0.1	1.2
			0.01			1.0	0.01	0.4

Compound	Formula	Molecular Weight	Calculated Carbon Requirements to Achieve Indicated Change in Concentration (a)					
			Single Stage Powdered Carbon C_f, mg/l				Granular Carbon Column	
			Co, mg/l	0.1	0.01	0.001	Co, mg/l	
Bromoform	$CHBr_3$	252.75	1.0	150	560	1,900	1.0	51
			0.1		51	190	0.1	17
			0.01			17	0.01	5.7

Compound	Formula	Molecular Weight	Calculated Carbon Requirements to Achieve Indicated Change in Concentration (a)					
			Single Stage Powdered Carbon C_f, mg/l				Granular Carbon Column	
			Co, mg/l	0.1	0.01	0.001	Co, mg/l	
4-Bromophenyl phenyl ether	$C_{12}H_9OCL$	249.11	1.0	30	160	770	1.0	7.0
			0.1		14	76	0.1	3.3
			0.01			6.9	0.01	1.6

Compound	Formula	Molecular Weight	Calculated Carbon Requirements to Achieve Indicated Change in Concentration (a)					
			Single Stage Powdered Carbon C_f, mg/l				Granular Carbon Column	
			Co, mg/l	0.1	0.01	0.001	Co, mg/l	
5-Bromouracil	$C_4H_3N_2O_2Br$	190.99	1.0	60	200	580	1.0	23
			0.1		18	58	0.1	6.6
			0.01				0.01	2.0

Compound	Formula	Molecular Weight	Calculated Carbon Requirements to Achieve Indicated Change in Concentration (a)					
			Single Stage Powdered Carbon C_f, mg/l				Granular Carbon Column	
			Co, mg/l	0.1	0.01	0.001	Co, mg/l	
Butylamine	$C_4H_{11}N$	73.14	1.0				1.0	
			0.1				0.1	
			0.01				0.01	

Compound	Formula	Molecular Weight	Calculated Carbon Requirements to Achieve Indicated Change in Concentration (a)					
			Single Stage Powdered Carbon C_f, mg/l				Granular Carbon Column	
			Co, mg/l	0.1	0.01	0.001	Co, mg/l	
Butylbenzyl Phtalate	$C_{19}H_{20}O_4$	312.36	1.0	11	220	4,000	1.0	0.7
			0.1		20	390	0.1	1.2
			0.01			36	0.01	2.2

Compound	Formula	Molecular Weight	Calculated Carbon Requirements to Achieve Indicated Change in Concentration (a)					
			Single Stage Powdered Carbon C_f, mg/l				Granular Carbon Column	
			Co, mg/l	0.1	0.01	0.001	Co, mg/l	
N-Butyl Phthalate	$C_{16}H_{22}O_4$	278.35	1.0	12	37	100	1.0	4.7
			0.1		3.3	10	0.1	1.3
			0.01			0.94	0.01	0.4

Compound	Formula	Molecular Weight	Calculated Carbon Requirements to Achieve Indicated Change in Concentration (a)					
			Single Stage Powdered Carbon C_f, mg/l				Granular Carbon Column	
			Co, mg/l	0.1	0.01	0.001	Co, mg/l	
Carbon Tetrachloride	CCL_4	153.85	1.0	550	4,100	28,000	1.0	90
			0.1		370	2,800	0.1	61
			0.01			250	0.01	42

Compound	Formula	Molecular Weight	Calculated Carbon Requirements to Achieve Indicated Change in Concentration (a)					
			Single Stage Powdered Carbon C_f, mg/l				Granular Carbon Column	
			Co, mg/l	0.1	0.01	0.001	Co, mg/l	
Chlorobenzene	C_6H_5CL	112.56	1.0	92	970	9,400	1.0	11
			0.1		88	930	0.1	11
			0.01			84	0.01	11

Compound	Formula	Molecular Weight	Calculated Carbon Requirements to Achieve Indicated Change in Concentration (a)					
			Single Stage Powdered Carbon C_f, mg/l				Granular Carbon Column	
			Co, mg/l	0.1	0.01	0.001	Co, mg/l	
Chlordane	$C_{10}H_6CL_8$	409.80	1.0	8.8	23	56	1.0	4.1
			0.1		2.1	5.6	0.1	1.0
			0.01			0.5	0.01	0.2

Compound	Formula	Molecular Weight	Calculated Carbon Requirements to Achieve Indicated Change in Concentration (a)					
			Single Stage Powdered Carbon C_f, mg/l				Granular Carbon Column	
			Co, mg/l	0.1	0.01	0.001	Co, mg/l	
Chloroethane	C_2H_5CL	64.52	1.0	14,000	>100,000	>100,000	1.0	1,700
			0.1		12,000	>100,000	0.1	1,400
			0.01			11,000	0.01	1,400

Compound	Formula	Molecular Weight	Calculated Carbon Requirements to Achieve Indicated Change in Concentration (a)					
			Single Stage Powdered Carbon C_f, mg/l				Granular Carbon Column	
			Co, mg/l	0.1	0.01	0.001	Co, mg/l	
Bis(2-chloroethoxy)methane	$C_5H_{10}O_2CL_2$	173.1	1.0	350	1,700	7,800	1.0	88
			0.1		160	770	0.1	38
			0.01			70	0.01	17

Compound	Formula	Molecular Weight	Calculated Carbon Requirements to Achieve Indicated Change in Concentration (a)					
			Single Stage Powdered Carbon C_f, mg/l				Granular Carbon Column	
			Co, mg/l	0.1	0.01	0.001	Co, mg/l	
Bis(2-chloroethyl)ether	$C_4H_8OCL_2$	143.02	1.0	>100,000	>100,000	>100,000	1.0	11,600
			0.1		>100,000	>100,000	0.1	100,000
			0.01			>100,000	0.01	>100,000

Compound	Formula	Molecular Weight	Calculated Carbon Requirements to Achieve Indicated Change in Concentration (a)					
			Single Stage Powdered Carbon C_f, mg/l				Granular Carbon Column	
			Co, mg/l	0.1	0.01	0.001	Co, mg/l	
2-Chloroethyl vinyl ether	C_4H_7OCL	106.55	1.0	1500	10,000	64,000	1.0	260
			0.1		920	6,400	0.1	170
			0.01			580	0.01	100

Compound	Formula	Molecular Weight	Calculated Carbon Requirements to Achieve Indicated Change in Concentration (a)					
			Single Stage Powdered Carbon C_f, mg/l				Granular Carbon Column	
			Co, mg/l	0.1	0.01	0.001	Co, mg/l	
Chloroform	$CHCL_3$	119.38	1.0	1900	11,000	50,000	1.0	4,300
			0.1		1,000	6,000	0.1	210
			0.01			540	0.01	111

Compound	Formula	Molecular Weight	Calculated Carbon Requirements to Achieve Indicated Change in Concentration (a)					
			Single Stage Powdered Carbon C_f, mg/l				Granular Carbon Column	
			Co, mg/l	0.1	0.01	0.001	Co, mg/l	
Bis(2-chloroisopropyl) ether	$C_6H_{12}OCL_2$	171.07	1.0	140	580	2,200	1.0	43
			0.1		55	220	0.1	16
			0.01			20	0.01	5.9

Compound	Formula	Molecular Weight	Calculated Carbon Requirements to Achieve Indicated Change in Concentration (a)					
			Single Stage Powdered Carbon C_f, mg/l				Granular Carbon Column	
			Co, mg/l	0.1	0.01	0.001	Co, mg/l	
Parachlorometa Cresol	$C_7H_7C_{10}$	142.59	1.0	11	17	25	1.0	8.1
			0.1		1.6	2.5	0.1	1.2
			0.01			0.2	0.01	0.2

Compound	Formula	Molecular Weight	Calculated Carbon Requirements to Achieve Indicated Change in Concentration (a)					
			Single Stage Powdered Carbon C_f, mg/l				Granular Carbon Column	
			Co, mg/l	0.1	0.01	0.001	Co, mg/l	
2-Chloronaphthalene	$C_{10}H_7CL$	162.62	1.0	9.3	29	86	1.0	3.6
			0.1		2.7	8.5	0.1	1.0
			0.01			0.8	0.01	0.3

Compound	Formula	Molecular Weight	Calculated Carbon Requirements to Achieve Indicated Change in Concentration (a)					
			Single Stage Powdered Carbon C_f, mg/l				Granular Carbon Column	
			Co, mg/l	0.1	0.01	0.001	Co, mg/l	
1-Chloro-2-Nitrobenzene	$C_6H_4CLNO_2$	157.6	1.0	20	64	180	1.0	7.7
			0.1		5.8	18	0.1	2.2
			0.01			1.7	0.01	0.6

Compound	Formula	Molecular Weight	Calculated Carbon Requirements to Achieve Indicated Change in Concentration (a)					
			Single Stage Powdered Carbon C_f, mg/l				Granular Carbon Column	
			Co, mg/l	0.1	0.01	0.001	Co, mg/l	
2-Chlorophenol	C_6H_5OCL	125.56	1.0	45	130	330	1.0	20
			0.1		12	33	0.1	5.0
			0.01			3.0	0.01	1.3

Compound	Formula	Molecular Weight	Calculated Carbon Requirements to Achieve Indicated Change in Concentration (a)					
			Single Stage Powdered Carbon C_f, mg/l				Granular Carbon Column	
			Co, mg/l	0.1	0.01	0.001	Co, mg/l	
4-Chlorophenyl Phenyl Ether	$C_{12}H_9OCL$	204.66	1.0	15	30	55	1.0	9.0
			0.1		2.7	5.4	0.1	1.6
			0.01			0.5	0.01	0.3

Compound	Formula	Molecular Weight	Calculated Carbon Requirements to Achieve Indicated Change in Concentration (a)					
			Single Stage Powdered Carbon C_f, mg/l				Granular Carbon Column	
			Co, mg/l	0.1	0.01	0.001	Co, mg/l	40
5-Chlorouracil	$C_4H_3N_2O_2CL$	146.54	1.0	140	570	2,200	1.0	15
			0.1		52	220	0.1	5.8
			0.01			20	0.01	

Compound	Formula	Molecular Weight	Calculated Carbon Requirements to Achieve Indicated Change in Concentration (a)					
			Single Stage Powdered Carbon C_f, mg/l				Granular Carbon Column	
			Co, mg/l	0.1	0.01	0.001	Co, mg/l	
Choline Chloride	$C_5H_{14}CLNO$	139.63	1.0				1.0	
			0.1				0.1	
			0.01				0.01	

Compound	Formula	Molecular Weight	Calculated Carbon Requirements to Achieve Indicated Change in Concentration (a)					
			Single Stage Powdered Carbon C_f, mg/l				Granular Carbon Column	
			Co, mg/l	0.1	0.01	0.001	Co, mg/l	
Cyclohexanone	$C_6H_{10}O$	98.14	1.0	820	5,100	29,000	1.0	160
			0.1		470	2,900	0.1	91
			0.01			260	0.01	52

Compound	Formula	Molecular Weight	Calculated Carbon Requirements to Achieve Indicated Change in Concentration (a)					
			Single Stage Powdered Carbon C_f, mg/l				Granular Carbon Column	
			Co, mg/l	0.1	0.01	0.001	Co, mg/l	
Cyclohexylamine	$C_6H_{13}N$	99.2	1.0				1.0	
			0.1				0.1	
			0.01				0.01	

Compound	Formula	Molecular Weight	Calculated Carbon Requirements to Achieve Indicated Change in Concentration (a)					
			Single Stage Powdered Carbon C_f, mg/l				Granular Carbon Column	
			Co, mg/l	0.1	0.01	0.001	Co, mg/l	
Cytosine	$C_4H_5N_3O$	111.10	1.0	>30,000	>100,000	>100,000	1.0	935
			0.1		>100,000	>100,000	0.1	3,300
			0.01			>100,000	0.01	12,500

Compound	Formula	Molecular Weight	Calculated Carbon Requirements to Achieve Indicated Change in Concentration (a)					
			Single Stage Powdered Carbon C_f, mg/l				Granular Carbon Column	
			Co, mg/l	0.1	0.01	0.001	Co, mg/l	
DDE	$C_{14}H_8CL_4$	318.03	1.0	9.0	23	55	1.0	4.3
			0.1		2.1	5.5	0.1	1.0
			0.01			0.5	0.01	0.2

Compound	Formula	Molecular Weight	Calculated Carbon Requirements to Achieve Indicated Change in Concentration (a)					
			Single Stage Powdered Carbon C_f, mg/l				Granular Carbon Column	
			Co, mg/l	0.1	0.01	0.001	Co, mg/l	
DDT	$C_{14}H_9C_{15}$	354.50	1.0	8.8	31	98	1.0	3.1
			0.1		2.8	9.7	0.1	1.0
			0.01			0.9	0.01	0.3

Compound	Formula	Molecular Weight	Calculated Carbon Requirements to Achieve Indicated Change in Concentration (a)					
			Single Stage Powdered Carbon C_f, mg/l				Granular Carbon Column	
			Co, mg/l	0.1	0.01	0.001	Co, mg/l	
Dibenzo(a,h)anthracene	$C_{22}H_{14}$	278.33	1.0	73	450	2,600	1.0	14
			0.1		41	250	0.1	8.3
			0.01			23	0.01	4.8

Compound	Formula	Molecular Weight	Calculated Carbon Requirements to Achieve Indicated Change in Concentration (a)					
			Single Stage Powdered Carbon C_f, mg/l				Granular Carbon Column	
			Co, mg/l	0.1	0.01	0.001	Co, mg/l	
Dibromochloromethane	CHBrCL	208.29	1.0	410	980	2,200	1.0	210
			0.1		89	210	0.1	45
			0.01			19	0.01	9.9

Compound	Formula	Molecular Weight	Calculated Carbon Requirements to Achieve Indicated Change in Concentration (a)					
			Single Stage Powdered Carbon C_f, mg/l				Granular Carbon Column	
			Co, mg/l	0.1	0.01	0.001	Co, mg/l	
1,2-Dibromo-3-chloropropane	$C_3H_4Br_2CL$	235.34	1.0	50	160	490	1.0	19
			0.1		15	48	0.1	5.6
			0.01			4.4	0.01	1.7

Compound	Formula	Molecular Weight	Calculated Carbon Requirements to Achieve Indicated Change in Concentration (a)					
			Single Stage Powdered Carbon C_f, mg/l				Granular Carbon Column	
			Co, mg/l	0.1	0.01	0.001	Co, mg/l	
1,2-Dichlorobenzene	$C_6H_4CL_2$	147.00	1.0	19	57	160	1.0	7.7
			0.1		5.2	15	0.1	2.1
			0.01			1.4	0.01	0.6

Compound	Formula	Molecular Weight	Calculated Carbon Requirements to Achieve Indicated Change in Concentration (a)					
			Single Stage Powdered Carbon C_f, mg/l				Granular Carbon Column	
			Co, mg/l	0.1	0.01	0.001	Co, mg/l	
1,3-Dichlorobenzene	$C_6H_4CL_2$	147.00	1.0	22	68	200	1.0	8.5
			0.1		6.2	19	0.1	2.4
			0.01			1.8	0.01	0.7

Compound	Formula	Molecular Weight	Calculated Carbon Requirements to Achieve Indicated Change in Concentration (a)					
			Single Stage Powdered Carbon C_f, mg/l				Granular Carbon Column	
			Co, mg/l	0.1	0.01	0.001	Co, mg/l	
Dichlorobromothane	CHCL₂BR	163.83	1.0	500	2,100	8,700	1.0	130
			0.1		190	860	0.1	52
			0.01			78	0.01	21

Compound	Formula	Molecular Weight	Calculated Carbon Requirements to Achieve Indicated Change in Concentration (a)					
			Single Stage Powdered Carbon C_f, mg/l				Granular Carbon Column	
			Co, mg/l	0.1	0.01	0.001	Co, mg/l	
1,1-Dichloroethane	C₂H₄CL₂	98.96	1.0	1,000	6,500	22,000	1.0	560
			0.1		600	2,200	0.1	190
			0.01			200	0.01	70

Compound	Formula	Molecular Weight	Calculated Carbon Requirements to Achieve Indicated Change in Concentration (a)					
			Single Stage Powdered Carbon C_f, mg/l				Granular Carbon Column	
			Co, mg/l	0.1	0.01	0.001	Co, mg/l	
1,2-Dichloroethane	C₂H₄CL₂	98.96	1.0	1,700	13,000	86,000	1.0	280
			0.1		1,200	8,600	0.1	190
			0.01			780	0.01	120

Compound	Formula	Molecular Weight	Calculated Carbon Requirements to Achieve Indicated Change in Concentration (a)					
			Single Stage Powdered Carbon C_f, mg/l				Granular Carbon Column	
			Co, mg/l	0.1	0.01	0.001	Co, mg/l	
1,2-trans-Dichloroethene	C₂H₂CL₂	96.94	1.0	950	3,400	11,000	1.0	330
			0.1		310	1,100	0.1	110
			0.01			100	0.01	34

Compound	Formula	Molecular Weight	Calculated Carbon Requirements to Achieve Indicated Change in Concentration (a)					
			Single Stage Powdered Carbon C_f, mg/l				Granular Carbon Column	
			Co, mg/l	0.1	0.01	0.001	Co, mg/l	
1,1-Dichloroethene (1,1-Dichloroethylene)	C₆H₄OCL₂	163.00	1.0	8.0	12	17	1.0	6.4
			0.1		1.1	1.7	0.1	0.9
			0.01			0.2	0.01	0.1

Compound	Formula	Molecular Weight	Calculated Carbon Requirements to Achieve Indicated Change in Concentration (a)					
			Single Stage Powdered Carbon C_f, mg/l				Granular Carbon Column	
			Co, mg/l	0.1	0.01	0.001	Co, mg/l	
1,2-Dichloropropane	C₃H₆CL₂	112.99	1.0	600	2700	11,000	1.0	170
			0.1		240	1,100	0.1	68
			0.01			96	0.01	27

Compound	Formula	Molecular Weight	Calculated Carbon Requirements to Achieve Indicated Change in Concentration (a)					
			Single Stage Powdered Carbon C_f, mg/l				Granular Carbon Column	
			Co, mg/l	0.1	0.01	0.001	Co, mg/l	
1,2-Dichloropropene	C₃H4CL₂	110.98	1.0	320	1,000	3,000	1.0	120
			0.1		93	300	0.1	35
			0.01			27	0.01	10

Compound	Formula	Molecular Weight	Calculated Carbon Requirements to Achieve Indicated Change in Concentration (a)					
			Single Stage Powdered Carbon C_f, mg/l				Granular Carbon Column	
			Co, mg/l	0.1	0.01	0.001	Co, mg/l	
Dieldrin	$C_{12}H_8OCL_6$	380.92	1.0	4.8	17	56	1.0	1.7
			0.1		1.6	5.5	0.1	0.5
			0.01			0.5	0.01	0.2

Compound	Formula	Molecular Weight	Calculated Carbon Requirements to Achieve Indicated Change in Concentration (a)					
			Single Stage Powdered Carbon C_f, mg/l				Granular Carbon Column	
			Co, mg/l	0.1	0.01	0.001	Co, mg/l	
Diethylene Glycol	$C_4H_{10}O_3$	106.12	1.0				1.0	
			0.1				0.1	
			0.01				0.01	

Compound	Formula	Molecular Weight	Calculated Carbon Requirements to Achieve Indicated Change in Concentration (a)					
			Single Stage Powdered Carbon C_f, mg/l				Granular Carbon Column	
			Co, mg/l	0.1	0.01	0.001	Co, mg/l	
4-Dimethylaminoazobenzene	$C_{14}H_{15}N_3$	225.3	1.0	6.3	12	21	1.0	4.0
			0.1		1.1	2.1	0.1	0.71
			0.01			0.2	0.01	0.12

Compound	Formula	Molecular Weight	Calculated Carbon Requirements to Achieve Indicated Change in Concentration (a)					
			Single Stage Powdered Carbon C_f, mg/l				Granular Carbon Column	
			Co, mg/l	0.1	0.01	0.001	Co, mg/l	
N-Dimethylnitrosamine	$(CH_3)_2NNO$	74.08	1.0	>100,000	>100,000	>100,000	1.0	>100,000
			0.1		>100,000	>100,000	0.1	>100,000
			0.01				0.01	>100,000

Compound	Formula	Molecular Weight	Calculated Carbon Requirements to Achieve Indicated Change in Concentration (a)					
			Single Stage Powdered Carbon C_f, mg/l				Granular Carbon Column	
			Co, mg/l	0.1	0.01	0.001	Co, mg/l	
2,4-Dimethylphenol	$C_8H_{10}O$	122.17	1.0	36	110	300	1.0	14
			0.1		98	30	0.1	4.0
			0.01			2.7	0.01	1.0

Compound	Formula	Molecular Weight	Calculated Carbon Requirements to Achieve Indicated Change in Concentration (a)					
			Single Stage Powdered Carbon C_f, mg/l				Granular Carbon Column	
			Co, mg/l	0.1	0.01	0.001	Co, mg/l	
Dimethylphenylcarbinol	$C_9H_{12}O$	136.20	1.0	9.0	21	46	1.0	4.7
			0.1		1.9	4.5	0.1	1.0
			0.01			0.41	0.01	0.2

Compound	Formula	Molecular Weight	Calculated Carbon Requirements to Achieve Indicated Change in Concentration (a)					
			Single Stage Powdered Carbon C_f, mg/l				Granular Carbon Column	
			Co, mg/l	0.1	0.01	0.001	Co, mg/l	
Dimethyl Phthalate	$C_{10}H_{10}O_4$	194.18	1.0	24	67	180	1.0	9.9
			0.1		6.1	17	0.1	2.6
			0.01			1.6	0.01	0.7

Compound	Formula	Molecular Weight	Calculated Carbon Requirements to Achieve Indicated Change in Concentration (a)				
			Single Stage Powdered Carbon C_f, mg/l			Granular Carbon Column	
4,6-Dinitro-o-cresol	$C_7H_6N_2O_5$	198.14	Co, mg/l	0.1	0.01	0.001	Co, mg/l	
			1.0	12	28	63	1.0	6.0
			0.1		2.6	6.2	0.1	1.3
			0.01			0.6	0.01	0.3

Compound	Formula	Molecular Weight	Calculated Carbon Requirements to Achieve Indicated Change in Concentration (a)				
			Single Stage Powdered Carbon C_f, mg/l			Granular Carbon Column	
2,4-Dinitrophenol	$C_6H_4N_2O_5$	184.11	Co, mg/l	0.1	0.01	0.001	Co, mg/l	
			1.0	110	500	2,100	1.0	30
			0.1		45	200	0.1	13
			0.01			18	0.01	5.2

Compound	Formula	Molecular Weight	Calculated Carbon Requirements to Achieve Indicated Change in Concentration (a)				
			Single Stage Powdered Carbon C_f, mg/l			Granular Carbon Column	
2,4-Dinitrotoluene	$C_7H_6N_2O_4$	182.14	Co, mg/l	0.1	0.01	0.001	Co, mg/l	
			1.0	13	29	59	1.0	6.9
			0.1		2.6	5.9	0.1	1.4
			0.01			0.5	0.01	0.3

Compound	Formula	Molecular Weight	Calculated Carbon Requirements to Achieve Indicated Change in Concentration (a)				
			Single Stage Powdered Carbon C_f, mg/l			Granular Carbon Column	
2,6-Dinitrotoluene	$C_7H_6N_2O_4$	182.14	Co, mg/l	0.1	0.01	0.001	Co, mg/l	
			1.0	13	30	62	1.0	6.9
			0.1		2.7	6.2	0.1	1.4
			0.01			0.6	0.01	0.3

Compound	Formula	Molecular Weight	Calculated Carbon Requirements to Achieve Indicated Change in Concentration (a)				
			Single Stage Powdered Carbon C_f, mg/l			Granular Carbon Column	
Diphenylamine	$C_{12}H_{11}N$	169.24	Co, mg/l	0.1	0.01	0.001	Co, mg/l	
			1.0	16	35	72	1.0	8.5
			0.1		3.2	7.2	0.1	1.8
			0.01			0.65	0.01	0.4

Compound	Formula	Molecular Weight	Calculated Carbon Requirements to Achieve Indicated Change in Concentration (a)				
			Single Stage Powdered Carbon C_f, mg/l			Granular Carbon Column	
1,1-Diphenylhydrazine	$C_{12}H_{12}N_2$	184.24	Co, mg/l	0.1	0.01	0.001	Co, mg/l	
			1.0	10	15	22	1.0	7.4
			0.1		1.4	2.2	0.1	1.1
			0.01			0.2	0.01	0.2

Compound	Formula	Molecular Weight	Calculated Carbon Requirements to Achieve Indicated Change in Concentration (a)				
			Single Stage Powdered Carbon C_f, mg/l			Granular Carbon Column	
1,2-Diphenylhydrazine	$C_{12}H_{12}N_2$	184.24	Co, mg/l	0.1	0.01	0.001	Co, mg/l	
			1.0	5.7	630	63,000	1.0	0.06
			0.1		57	6,200	0.1	0.64
			0.01			570	0.01	6.7

Compound	Formula	Molecular Weight	Calculated Carbon Requirements to Achieve Indicated Change in Concentration (a)					
			Single Stage Powdered Carbon C_f, mg/l				Granular Carbon Column	
			Co, mg/l	0.1	0.01	0.001	Co, mg/l	
a-Endosulfan	$C_9H_6SO_3$	406.93	1.0	15	50	160	1.0	160
			0.1		4.6	16	0.1	1.6
			0.01			1.4	0.01	0.5

Compound	Formula	Molecular Weight	Calculated Carbon Requirements to Achieve Indicated Change in Concentration (a)					
			Single Stage Powdered Carbon C_f, mg/l				Granular Carbon Column	
			Co, mg/l	0.1	0.01	0.001	Co, mg/l	
b-Endosulfan	$C_9H_6CL_6\ SO_3$	406.93	1.0	10	74	500	1.0	1.6
			0.1		6.7	50	0.1	1.1
			0.01			4.5	0.01	0.7

Compound	Formula	Molecular Weight	Calculated Carbon Requirements to Achieve Indicated Change in Concentration (a)					
			Single Stage Powdered Carbon C_f, mg/l				Granular Carbon Column	
			Co, mg/l	0.1	0.01	0.001	Co, mg/l	
Endosulfan Sulfate	$C_9H_6CL_6SO_4$	422.93	1.0	8.5	60	390	1.0	1.5
			0.1		5.5	39	0.1	1.0
			0.01			3.5	0.01	0.6

Compound	Formula	Molecular Weight	Calculated Carbon Requirements to Achieve Indicated Change in Concentration (a)					
			Single Stage Powdered Carbon C_f, mg/l				Granular Carbon Column	
			Co, mg/l	0.1	0.01	0.001	Co, mg/l	
Endrin	$C_{12}H_8CL_6O$	381.0	1.0	8.5	60	380	1.0	1.5
			0.1		5.4	37	0.1	0.9
			0.01			3.4	0.01	0.6

Compound	Formula	Molecular Weight	Calculated Carbon Requirements to Achieve Indicated Change in Concentration (a)					
			Single Stage Powdered Carbon C_f, mg/l				Granular Carbon Column	
			Co, mg/l	0.1	0.01	0.001	Co, mg/l	
Ethanol	C_2H_6O	46.07	1.0				1.0	
			0.1				0.1	
			0.01				0.01	

Compound	Formula	Molecular Weight	Calculated Carbon Requirements to Achieve Indicated Change in Concentration (a)					
			Single Stage Powdered Carbon C_f, mg/l				Granular Carbon Column	
			Co, mg/l	0.1	0.01	0.001	Co, mg/l	
Ethylbenzene	C_8H_{10}	106.16	1.0	110	710	4,400	1.0	19
			0.1		65	440	0.1	12
			0.01			40	0.01	7.2

Compound	Formula	Molecular Weight	Calculated Carbon Requirements to Achieve Indicated Change in Concentration (a)					
			Single Stage Powdered Carbon C_f, mg/l				Granular Carbon Column	
			Co, mg/l	0.1	0.01	0.001	Co, mg/l	
Ethylenediamine	$C_2H_8N_2$	50.10	1.0				1.0	
			0.1				0.1	
			0.01				0.01	

Compound	Formula	Molecular Weight	Calculated Carbon Requirements to Achieve Indicated Change in Concentration (a)					
			Single Stage Powdered Carbon C_f, mg/l				Granular Carbon Column	
Ethylenediaminetetraaceticacid (EDTA)	$C_{10}H_{16}N_2O_8$	292.3	Co, mg/l	0.1	0.01	0.001	Co, mg/l	
			1.0	36,000	>100,000	>100,000	1.0	1,160
			0.1		>100,000	>100,000	0.1	3,970
			0.01			>100,000	0.01	13,600

Compound	Formula	Molecular Weight	Single Stage Powdered Carbon C_f, mg/l				Granular Carbon Column	
			Co, mg/l	0.1	0.01	0.001	Co, mg/l	
Bis(2-Ethylhexyl)phthlate	$C_{24}H_{38}O_4$	390.56	1.0	2.5	88	2,800	1.0	0.1
			0.1		8.0	280	0.1	0.3
			0.01			25	0.01	1.0

Compound	Formula	Molecular Weight	Calculated Carbon Requirements to Achieve Indicated Change in Concentration (a)					
			Single Stage Powdered Carbon C_f, mg/l				Granular Carbon Column	
			Co, mg/l	0.1	0.01	0.001	Co, mg/l	
Fluoranthene	$C_{16}H_{10}$	202.26	1.0	6.0	24	100	1.0	1.5
			0.1		2.2	9.9	0.1	0.6
			0.01			0.9	0.01	0.2

Compound	Formula	Molecular Weight	Calculated Carbon Requirements to Achieve Indicated Change in Concentration (a)					
			Single Stage Powdered Carbon C_f, mg/l				Granular Carbon Column	
			Co, mg/l	0.1	0.01	0.001	Co, mg/l	
Fluorene	$C_{13}H_{10}$	166.22	1.0	5.3	11	22	1.0	3.0
			0.1		1.0	2.1	0.1	0.6
			0.01			0.2	0.01	0.1

Compound	Formula	Molecular Weight	Calculated Carbon Requirements to Achieve Indicated Change in Concentration (a)					
			Single Stage Powdered Carbon C_f, mg/l				Granular Carbon Column	
			Co, mg/l	0.1	0.01	0.001	Co, mg/l	
5-Fluorouracil	$C_4H_3N_2O_2F$	130.08	1.0	1,800	21,000	>100,000	1.0	183
			0.1		1,900	23,000	0.1	198
			0.01			2,100	0.01	215

Compound	Formula	Molecular Weight	Calculated Carbon Requirements to Achieve Indicated Change in Concentration (a)					
			Single Stage Powdered Carbon C_f, mg/l				Granular Carbon Column	
			Co, mg/l	0.1	0.01	0.001	Co, mg/l	
Guanine	$C_5H_5N_5$	151.13	1.0	19	51	130	1.0	8.2
			0.1		4.7	13	0.1	2.0
			0.01			1.2	0.01	0.5

Compound	Formula	Molecular Weight	Calculated Carbon Requirements to Achieve Indicated Change in Concentration (a)					
			Single Stage Powdered Carbon C_f, mg/l				Granular Carbon Column	
			Co, mg/l	0.1	0.01	0.001	Co, mg/l	
Heptachlor	$C_{10}H_5CL_7$	373.5	1.0	6.6	64	580	1.0	0.8
			0.1		5.9	57	0.1	0.7
			0.01			5.2	0.01	0.7

Heptachlor Epoxide — Formula: $C_{10}H_5CL_7O$ — Molecular Weight: 389.32

Calculated Carbon Requirements to Achieve Indicated Change in Concentration (a)					
Single Stage Powdered Carbon C_f, mg/l				Granular Carbon Column	
Co, mg/l	0.1	0.01	0.001	Co, mg/l	
1.0	4.3	24	120	1.0	1.0
0.1		2.2	12	0.1	0.5
0.01			1.1	0.01	0.2

Hexachlorobenzene — Formula: C_6CL_6 — Molecular Weight: 284.78

Calculated Carbon Requirements to Achieve Indicated Change in Concentration (a)					
Single Stage Powdered Carbon C_f, mg/l				Granular Carbon Column	
Co, mg/l	0.1	0.01	0.001	Co, mg/l	
1.0	8.0	35	140	1.0	2.2
0.1		3.2	14	0.1	1.0
0.01			1.3	0.01	0.4

Hexachlorobutadiene — Formula: C_4CL_6 — Molecular Weight: 260.76

Calculated Carbon Requirements to Achieve Indicated Change in Concentration (a)					
Single Stage Powdered Carbon C_f, mg/l				Granular Carbon Column	
Co, mg/l	0.1	0.01	0.001	Co, mg/l	
1.0	48	150	430	1.0	3.9
0.1		14	43	0.1	1.1
0.01			3.9	0.01	0.3

Hexachlorocyclopentadiene — Formula: C_5CL_6 — Molecular Weight: 272.77

Compound	Formula	Molecular Weight	Single Stage Powdered Carbon C_f, mg/l				Granular Carbon Column	
			Co, mg/l	0.1	0.01	0.001	Co, mg/l	
			1.0	3.6	5.9	8.9	1.0	2.7
			0.1		0.54	0.88	0.1	0.40
			0.01			0.08	0.01	0.06

Hexachloroethane — Formula: C_2CL_6 — Molecular Weight: 236.74

Calculated Carbon Requirements to Achieve Indicated Change in Concentration (a)					
Single Stage Powdered Carbon C_f, mg/l				Granular Carbon Column	
Co, mg/l	0.1	0.01	0.001	Co, mg/l	
1.0	20	60	140	1.0	10
0.1		5.3	14	0.1	2.5
0.01			1.3	0.01	0.6

Hexamethylenediamine — Formula: $C_6H_{16}N_2$ — Molecular Weight: 116.12

Compound	Formula	Molecular Weight	Single Stage Powdered Carbon C_f, mg/l				Granular Carbon Column	
			Co, mg/l	0.1	0.01	0.001	Co, mg/l	
			1.0				1.0	
			0.1				0.1	
			0.01				0.01	

Hydroquinone — Formula: $C_6H_6O_2$ — Molecular Weight: 110.12

Calculated Carbon Requirements to Achieve Indicated Change in Concentration (a)					
Single Stage Powdered Carbon C_f, mg/l				Granular Carbon Column	
Co, mg/l	0.1	0.01	0.001	Co, mg/l	
1.0	18	35	62	1.0	11
0.1		3.2	6.2	0.1	2.0
0.01			0.56	0.01	0.3

Isophorone — Formula: $C_9H_{14}O$ — Molecular Weight: 138.21

Calculated Carbon Requirements to Achieve Indicated Change in Concentration (a)					
Single Stage Powdered Carbon C_f, mg/l				Granular Carbon Column	
Co, mg/l	0.1	0.01	0.001	Co, mg/l	
1.0	70	190	460	1.0	31
0.1		17	46	0.1	7.6
0.01			4.2	0.01	1.9

Compound	Formula	Molecular Weight	Calculated Carbon Requirements to Achieve Indicated Change in Concentration (a)					
			Single Stage Powdered Carbon C_f, mg/l				Granular Carbon Column	
			Co, mg/l	0.1	0.01	0.001	Co, mg/l	
Methylene Chloride	CH₂CL₂	84.94	1.0	10,000	>100,000	>100,000	1.0	770
			0.1		14,000	>100,000	0.1	1,100
			0.01			21,000	0.01	1,700

Compound	Formula	Molecular Weight	Single Stage Powdered Carbon C_f, mg/l				Granular Carbon Column	
			Co, mg/l	0.1	0.01	0.001	Co, mg/l	
4,4¹-Methylene-Bis (2-Chloroaniline)	C₁₃H₁₂CL₂N₂	264.28	1.0	21	99	440	1.0	5.3
			0.1		9.0	43	0.1	2.3
			0.01			3.9	0.01	1.0

Compound	Formula	Molecular Weight	Calculated Carbon Requirements to Achieve Indicated Change in Concentration (a)					
			Single Stage Powdered Carbon C_f, mg/l				Granular Carbon Column	
			Co, mg/l	0.1	0.01	0.001	Co, mg/l	
Morpholine	C₄H₉NO	87.12	1.0				1.0	
			0.1				0.1	
			0.01				0.01	

Compound	Formula	Molecular Weight	Calculated Carbon Requirements to Achieve Indicated Change in Concentration (a)					
			Single Stage Powdered Carbon C_f, mg/l				Granular Carbon Column	
			Co, mg/l	0.1	0.01	0.001	Co, mg/l	
Naphthalene	C₁₀H₈	128.18	1.0	18	52	140	1.0	7.6
			0.1		4.7	13	0.1	2.0
			0.01			1.2	0.01	0.5

Compound	Formula	Molecular Weight	Calculated Carbon Requirements to Achieve Indicated Change in Concentration (a)					
			Single Stage Powdered Carbon C_f, mg/l				Granular Carbon Column	
			Co, mg/l	0.1	0.01	0.001	Co, mg/l	
a-Naphthol	C₁₀H₈O	144.2	1.0	10	23	48	1.0	5.7
			0.1		2.1	4.7	0.1	1.2
			0.01			0.43	0.01	0.2

Compound	Formula	Molecular Weight	Calculated Carbon Requirements to Achieve Indicated Change in Concentration (a)					
			Single Stage Powdered Carbon C_f, mg/l				Granular Carbon Column	
			Co, mg/l	0.1	0.01	0.001	Co, mg/l	
B-Naphthol	C₁₀H₈O	144.2	1.0	8.4	17	31	1.0	5.1
			0.1		1.5	3.0	0.1	1.0
			0.01			0.28	0.01	0.2

Compound	Formula	Molecular Weight	Calculated Carbon Requirements to Achieve Indicated Change in Concentration (a)					
			Single Stage Powdered Carbon C_f, mg/l				Granular Carbon Column	
			Co, mg/l	0.1	0.01	0.001	Co, mg/l	
a-Naphthylamine	C₁₀H₉N	143.18	1.0	12	29	64	1.0	6.1
			0.1		2.6	6.3	0.1	1.3
			0.01			0.58	0.01	0.3

Compound	Formula	Molecular Weight	Calculated Carbon Requirements to Achieve Indicated Change in Concentration (a)					
			Single Stage Powdered Carbon C_f, mg/l				Granular Carbon Column	
			Co, mg/l	0.1	0.01	0.001	Co, mg/l	
B-Naphthylamine	$C_{10}H_9N$	143.19	1.0	12	26	53	1.0	6.7
			0.1		2.4	5.2	0.1	1.3
			0.01			0.5	0.01	0.3

Compound	Formula	Molecular Weight	Calculated Carbon Requirements to Achieve Indicated Change in Concentration (a)					
			Single Stage Powdered Carbon C_f, mg/l				Granular Carbon Column	
			Co, mg/l	0.1	0.01	0.001	Co, mg/l	
p-Nitroaniline	$C_6H_6N_2O_2$	138.13	1.0	12	25	48	1.0	7.4
			0.1		2.3	4.7	0.1	1.4
			0.01			0.43	0.01	0.2

Compound	Formula	Molecular Weight	Calculated Carbon Requirements to Achieve Indicated Change in Concentration (a)					
			Single Stage Powdered Carbon C_f, mg/l				Granular Carbon Column	
			Co, mg/l	0.1	0.01	0.001	Co, mg/l	
Nitrobenzene	$C_6H_5NO_2$	123.11	1.0	36	110	290	1.0	15
			0.1		9.6	28	0.1	4.0
			0.01			2.6	0.01	1.1

Compound	Formula	Molecular Weight	Calculated Carbon Requirements to Achieve Indicated Change in Concentration (a)					
			Single Stage Powdered Carbon C_f, mg/l				Granular Carbon Column	
			Co, mg/l	0.1	0.01	0.001	Co, mg/l	
4-Nitrobiphenyl	$C_{12}H_9NO_2$	199.21	1.0	4.5	9.3	18	1.0	2.7
			0.1		0.8	1.7	0.1	0.5
			0.01			0.2	0.01	0.1

Compound	Formula	Molecular Weight	Calculated Carbon Requirements to Achieve Indicated Change in Concentration (a)					
			Single Stage Powdered Carbon C_f, mg/l				Granular Carbon Column	
			Co, mg/l	0.1	0.01	0.001	Co, mg/l	
2-Nitrophenol	$C_6H_5NO_3$	139.11	1.0	20	47	100	1.0	10
			0.1		4.3	10	0.1	2.2
			0.01			1.0	0.01	0.5

Compound	Formula	Molecular Weight	Calculated Carbon Requirements to Achieve Indicated Change in Concentration (a)					
			Single Stage Powdered Carbon C_f, mg/l				Granular Carbon Column	
			Co, mg/l	0.1	0.01	0.001	Co, mg/l	
4-Nitrophenol	$C_6H_5NO_3$	139.11	1.0	21	41	74	1.0	13
			0.1		3.7	7.3	0.1	2.3
			0.01			0.7	0.01	0.4

Compound	Formula	Molecular Weight	Calculated Carbon Requirements to Achieve Indicated Change in Concentration (a)					
			Single Stage Powdered Carbon C_f, mg/l				Granular Carbon Column	
			Co, mg/l	0.1	0.01	0.001	Co, mg/l	
N-Nitrosodiphenylamine	$C_{12}H_{10}N_2O$	198.07	1.0	9.8	25	60	1.0	4.6
			0.1		2.3	5.9	0.1	1.1
			0.01			0.54	0.01	0.3

Compound	Formula	Molecular Weight	Single Stage Powdered Carbon C_f, mg/l				Granular Carbon Column	
			Co, mg/l	0.1	0.01	0.001	Co, mg/l	
N-Nitrosodi-n-propylamine	$C_6H_{14}N_2O$	130.19	1.0	67	130	250	1.0	42
			0.1		12	24	0.1	7.7
			0.01			2.2	0.01	1.4

Compound	Formula	Molecular Weight	Calculated Carbon Requirements to Achieve Indicated Change in Concentration (a)					
			Single Stage Powdered Carbon C_f, mg/l				Granular Carbon Column	
			Co, mg/l	0.1	0.01	0.001	Co, mg/l	
P-Nonylphenol	$C_{15}H_{24}O$	220.34	1.0	8.3	21	51	1.0	3.9
			0.1		2.0	5.0	0.1	0.9
			0.01			0.5	0.01	0.2

Compound	Formula	Molecular Weight	Calculated Carbon Requirements to Achieve Indicated Change in Concentration (a)					
			Single Stage Powdered Carbon C_f, mg/l				Granular Carbon Column	
			Co, mg/l	0.1	0.01	0.001	Co, mg/l	
PCB 1221	$C_{12}H_9CL$ — 51% $C_{12}H_8CL_2$ — 32% $C_{12}H_{10}$ — 11%	~ 200.7	1.0	19	100	520	1.0	4.1
			0.1		9.3	52	0.1	2.1
			0.01			5.7	0.01	1.1

Compound	Formula	Molecular Weight	Calculated Carbon Requirements to Achieve Indicated Change in Concentration (a)					
			Single Stage Powdered Carbon C_f, mg/l				Granular Carbon Column	
			Co, mg/l	0.1	0.01	0.001	Co, mg/l	
PCB-1232	$C_{12}H_9CL$ — 31% $C_{12}H_7CL_3$ — 28% $C_{12}H_8CL_2$ — 24%	~ 232.2	1.0	7.7	45	240	1.0	1.6
			0.1		4.1	24	0.1	0.8
			0.01			2.2	0.01	0.5

Compound	Formula	Molecular Weight	Calculated Carbon Requirements to Achieve Indicated Change in Concentration (a)					
			Single Stage Powdered Carbon C_f, mg/l				Granular Carbon Column	
			Co, mg/l	0.1	0.01	0.001	Co, mg/l	
Pentachlorophenol	C_6HOCL_5	266.4	1.0	16	47	130	1.0	6.9
			0.1		4.3	12	0.1	1.8
			0.01			1.1	0.01	0.5

Compound	Formula	Molecular Weight	Calculated Carbon Requirements to Achieve Indicated Change in Concentration (a)					
			Single Stage Powdered Carbon C_f, mg/l				Granular Carbon Column	
			Co, mg/l	0.1	0.01	0.001	Co, mg/l	
Phenanthrene	$C_{14}H_{10}$	178.24	1.0	11	34	95	1.0	4.7
			0.1		3.1	9.4	0.1	1.3
			0.01			0.9	0.01	0.3

Compound	Formula	Molecular Weight	Calculated Carbon Requirements to Achieve Indicated Change in Concentration (a)					
			Single Stage Powdered Carbon C_f, mg/l				Granular Carbon Column	
			Co, mg/l	0.1	0.01	0.001	Co, mg/l	
Phenol	C_6H_6O	94.11	1.0	150	570	2,000	1.0	47
			0.1		52	200	0.1	17
			0.01			18	0.01	5.8

Compound	Formula	Molecular Weight	Calculated Carbon Requirements to Achieve Indicated Change in Concentration (a)					
			Single Stage Powdered Carbon C_f, mg/l				Granular Carbon Column	
			Co, mg/l	0.1	0.01	0.001	Co, mg/l	
Phenylmercuric Acetate	$C_8H_8HgO_2$	336.74	1.0	9.2	28	77	1.0	3.7
			0.1		2.5	7.7	0.1	1.0
			0.01			0.70	0.01	0.3

Compound	Formula	Molecular Weight	Calculated Carbon Requirements to Achieve Indicated Change in Concentration (a)					
			Single Stage Powdered Carbon C_f, mg/l				Granular Carbon Column	
			Co, mg/l	0.1	0.01	0.001	Co, mg/l	
Styrene	C_8H_8	104.14	1.0	27	110	400	1.0	8.3
			0.1		9.8	39	0.1	3.0
			0.01			3.6	0.01	1.1

Compound	Formula	Molecular Weight	Calculated Carbon Requirements to Achieve Indicated Change in Concentration (a)					
			Single Stage Powdered Carbon C_f, mg/l				Granular Carbon Column	
			Co, mg/l	0.1	0.01	0.001	Co, mg/l	
1,1,2,2-Tetrachloroethane	$C_2H_2CL_4$	167.85	1.0	360	940	2200	1.0	95
			0.1		90	220	0.1	22
			0.01			20	0.01	5.3

Compound	Formula	Molecular Weight	Calculated Carbon Requirements to Achieve Indicated Change in Concentration (a)					
			Single Stage Powdered Carbon C_f, mg/l				Granular Carbon Column	
			Co, mg/l	0.1	0.01	0.001	Co, mg/l	
Tetrachloroethene (Tetrachloroethylene)	C_2CL_4	165.83	1.0	64	260	940	1.0	20
			0.1		23	93	0.1	7.1
			0.01			8.5	0.01	2.6

Compound	Formula	Molecular Weight	Calculated Carbon Requirements to Achieve Indicated Change in Concentration (a)					
			Single Stage Powdered Carbon C_f, mg/l				Granular Carbon Column	
			Co, mg/l	0.1	0.01	0.001	Co, mg/l	
1,2,3,4-Tetrahydronaphthalene	$C_{10}H_{12}$	132.21	1.0	78	560	3,600	1.0	14
			0.1		51	360	0.1	8.8
			0.01			33	0.01	5.7

Compound	Formula	Molecular Weight	Calculated Carbon Requirements to Achieve Indicated Change in Concentration (a)					
			Single Stage Powdered Carbon C_f, mg/l				Granular Carbon Column	
			Co, mg/l	0.1	0.01	0.001	Co, mg/l	
Thymine	$C_5H_6N_2O_2$	126.11	1.0	110	380	1,200	1.0	36
			0.1		34	120	0.1	12
			0.01			11	0.01	3.8

Compound	Formula	Molecular Weight	Calculated Carbon Requirements to Achieve Indicated Change in Concentration (a)					
			Single Stage Powdered Carbon C_f, mg/l				Granular Carbon Column	
			Co, mg/l	0.1	0.01	0.001	Co, mg/l	
Toluene	C_7H_8	92.14	1.0	96	290	820	1.0	38
			0.1		27	81	0.1	11
			0.01			7.4	0.01	2.9

Compound	Formula	Molecular Weight	Calculated Carbon Requirements to Achieve Indicated Change in Concentration (a)					
			Single Stage Powdered Carbon C_f, mg/l				Granular Carbon Column	
			Co, mg/l	0.1	0.01	0.001	Co, mg/l	
1,2,4-Trichlorobenzene	$C_6H_3CL_3$	181.45	1.0	12	26	52	1.0	6.4
			0.1		2.3	5.2	0.1	1.3
			0.01			0.5	0.01	0.3

Compound	Formula	Molecular Weight	Calculated Carbon Requirements to Achieve Indicated Change in Concentration (a)					
			Single Stage Powdered Carbon C_f, mg/l				Granular Carbon Column	
			Co, mg/l	0.1	0.01	0.001	Co, mg/l	
1,1,1-Trichloroethane	$C_2H_3CL_3$	133.41	1.0	800	1,900	4,300	1.0	400
			0.1		180	430	0.1	90
			0.01			39	0.01	20

Compound	Formula	Molecular Weight	Calculated Carbon Requirements to Achieve Indicated Change in Concentration (a)					
			Single Stage Powdered Carbon C_f, mg/l				Granular Carbon Column	
			Co, mg/l	0.1	0.01	0.001	Co, mg/l	
1,1,2-Trichloroethane	$C_2H_3CL_3$	133.41	1.0	620	2,700	11,000	1.0	170
			0.1		250	1,100	0.1	69
			0.01			99	0.01	28

Compound	Formula	Molecular Weight	Calculated Carbon Requirements to Achieve Indicated Change in Concentration (a)					
			Single Stage Powdered Carbon C_f, mg/l				Granular Carbon Column	
			Co, mg/l	0.1	0.01	0.001	Co, mg/l	
Trichloroethene (Trichloroethylene)	C_2HCL_3	131.39	1.0	130	620	2,600	1.0	36
			0.1		56	260	0.1	15
			0.01			23	0.01	6.3

Compound	Formula	Molecular Weight	Calculated Carbon Requirements to Achieve Indicated Change in Concentration (a)					
			Single Stage Powdered Carbon C_f, mg/l				Granular Carbon Column	
			Co, mg/l	0.1	0.01	0.001	Co, mg/l	
Trichlorofluoromethane	CCL_3F	137.4	1.0	280	530	930	1.0	180
			0.1		48	92	0.1	31
			0.01			8.4	0.01	5.6

Compound	Formula	Molecular Weight	Calculated Carbon Requirements to Achieve Indicated Change in Concentration (a)					
			Single Stage Powdered Carbon C_f, mg/l				Granular Carbon Column	
			Co, mg/l	0.1	0.01	0.001	Co, mg/l	
2,4,6-Trichlorophenol	$C_6H_9OCL_3$	197.45	1.0	15	41	105	1.0	6.4
			0.1		3.7	10.4	0.1	1.6
			0.01			0.9	0.01	0.4

Compound	Formula	Molecular Weight	Calculated Carbon Requirements to Achieve Indicated Change in Concentration (a)					
			Single Stage Powdered Carbon C_f, mg/l				Granular Carbon Column	
			Co, mg/l	0.1	0.01	0.001	Co, mg/l	
Triethanolamine	$C_6H_{15}NO_3$	149.19	1.0				1.0	
			0.1				0.1	
			0.01				0.01	

Compound	Formula	Molecular Weight	Calculated Carbon Requirements to Achieve Indicated Change in Concentration (a)					
			Single Stage Powdered Carbon C_f, mg/l				Granular Carbon Column	
			Co, mg/l	0.1	0.01	0.001	Co, mg/l	
Uracil	$C_4H_4N_2O_2$	112.06	1.0	350	1,700	7,100	1.0	91
			0.1		150	710	0.1	38
			0.01			64	0.01	17

Compound	Formula	Molecular Weight	Calculated Carbon Requirements to Achieve Indicated Change in Concentration (a)					
			Single Stage Powdered Carbon C_f, mg/l				Granular Carbon Column	
			Co, mg/l	0.1	0.01	0.001	Co, mg/l	
P-Xylene	C_8H_{10}	106.2	1.0		15	24	1.0	12
			0.1			2.2	0.1	1.9
			0.01				0.01	0.3

Notes:
1. The second Granular Carbon Column (GAC) column of figures indicates the mg/l or GAC required to reach breakthrough (when Cf = Co)
2. The following table is a summary of carbon adsorption capacities at neutral pH arranged in descending order when the equilibrium concentration of the compound is 1.0 mg/l.
3. The following compounds are not adsorbed on GAC:

NOT ADSORBED

Acetone Cyanohydrin	Adipic Acid	Butylamine	Choline Chloride
Cyclohexylamine	Diethylene Glycol	Ethanol	Hexamethylenediamine
Hydroquinone	Morpholine	Triethanolamine	

SUMMARY OF CARBON ADSORPTION CAPACITIES

Compound	Adsorption[a] Capacity, mg/g	Compound	Adsorption[a] Capacity, mg/g
Bis(2-Ethylhexyl) phthalate	11,300	Phenanthrene	215
Butylbenzyl phthalate	1,520	Dimethylphenylcarbinol*	210
Heptachlor	1,220	4-Aminobiphenyl	200
Heptachlor epoxide	1,038	Beta-Naphthol*	200
Endosulfan sulfate	686	Alpha-Endosulfan	194
Endrin	666	Acenaphthene	190
Fluoranthene	664	4,4' Methylene-bis-(2-chloroaniline)	190
Aldrin	651	Benzo (k) fluoranthene	181
PCB-1232	630	Acridine orange*	180
Beta-Endosulfan	615	Alpha-Naphthol	180
Dieldrin	606	4,6-Dinitro-o-cresol	169
Hexachlorobenzene	450	Alpha-Naphthylamine	160
Anthracene	376	2,4-Dichlorophenol	157
4-Nitrobiphenyl	370	1,2,4-Trichlorobenzene	157
Fluorene	330	2,4,6-Trichlorophenol	155
DDT	322	Beta-Naphthylamine	150
2-Acetylaminofluorene	318	Pentachlorophenol	150
Alph-BHC	303	2,4-Dinitrotoluene	146
Anethole*	300	4-Bromophenyl phenyl ether	144
3,3-Dichlorobenzidine	300	p-Nitroaniline*	140
2-Chloronaphthalene	280	1,1-Diphenylhydrazine	135
Phenylmercuric Acetate	270	Naphthalene	132
Hexachlorobutadiene	258	1-Chloro-2-nitrobenzene	130
Gamma-BHC (lindane)	256	1,2-Dichlorobenzene	129
p-Nonylphenol	250	p-Chlorometacresol	124
4-Dimethylaminoazobenzene	249	1,4-Dichlorobenzene	121
Chlordane	245	Benzothiazole*	120
PCB-1221	242	Diphenylamine	120
DDE	232	Guanine*	120
Acridine yellow*	230	Styrene	120
Benzidine dihydrochloride	220	1,3-Dichlorobenzene	118
Beta-BHC	220	Acenaphthylene	115
N-Butylphthalate	220	4-Chlorophenyl phenyl ether	111
N-Nitrosodiphenylamine	220	Diethyl phthalate	110
2-Nitrophenol	99	Bromoform	20
Dimethyl phthalate	97	Carbon tetrachloride	11
Hexachloroethane	97	Bis(2-Chloroethoxy) methane	11
Chlorobenzene	91	Uracil*	11
p-Xylene	85	Benzo(ghi)perylene	11
2,4-Dimethylphenol	78	1,1,2,2-Tetrachloroethane	11
4-Nitrophenol	76	1,2-Dichloropropene	8.2
Acetophenone	74	Dichlorobromomethane	7.9
1,2,3,4-Tetrahydronaphthalene	74	Cyclohexanone*	6.2
Adenine*	71	1,2-Dichloropropane	5.9
Dibenzo(a,h)anthracene	69	1,1,2-Trichloroethane	5.8
Nitrobenzene	68	Trichlorofluoromethane	5.6
3,4-Benzofluoranthene	57	5-Fluorouracil*	5.5
1,2-Dibromo-3-chloropropane	53	1,1-Dichloroethylene	4.9
Ethylbenzene	53	Dibromochloromethane	4.8
2-Chlorophenol	51	2-Chloroethyl vinyl ether	3.9
Tetrachloroethene	51	1,2-Dichloroethane	3.6
o-Anisidine*	50	1,2-trans-Dichloroethene	3.1
5 Bromouracil	44	Chloroform	2.6
Benzo(a)pyrene	34	1,1,1-Trichloroethane	2.5
2,4-Dinitrophenol	33	1,1-Dichloroethane	1.8
Isophorone	32	Acrylonitrile	1.4
Trichloroethene	28	Methylene chloride	1.3
Thymine*	27	Acrolein	1.2
Toluene	26	Cytosine*	1.1
5-Chlorouracil*	25	Benzene	1.0
N-Nitrosodi-n-propylamine	24	Ethylenediaminetetra-acetic acid	0.86
Bis(2-Chloroisopropyl) ether	24	Benzoic acid	0.76
Phenol	21	Chloroethane	0.59
		N-Dimethylnitrosamine	6.8×10^{-5}

INDEX